Nanomaterials and Plant Potential

Azamal Husen • Muhammad Iqbal
Editors

Nanomaterials and Plant Potential

Springer

Editors
Azamal Husen
Department of Biology
College of Natural and Computational Sciences
University of Gondar
Gondar, Ethiopia

Muhammad Iqbal
Department of Botany
Faculty of Science
Jamia Hamdard (Deemed University)
New Delhi, India

ISBN 978-3-030-05568-4 ISBN 978-3-030-05569-1 (eBook)
https://doi.org/10.1007/978-3-030-05569-1

Library of Congress Control Number: 2019930586

This Springer imprint is published by the registered company Springer Nature Switzerland AG
The registered company address is: Gewerbestrasse 11, 6330 Cham, Switzerland

Preface

Various plants and plant parts are used for green synthesis of the metal and metal-oxide nanoparticles (NPs), as they contain metabolites such as alkaloids, flavonoids, phenols, terpenoids, alcohols, sugars, and proteins, which act as reducing agents to produce NPs and, in some cases, also as capping agents and stabilizers for NPs. Plants with either aroma or color in their leaves, flowers, or roots drew special attention because they all contain such chemicals that reduce the metal ions to metal NPs. The size and morphology of the metal and metal-oxide NPs are dependent on the biogenic-synthetic route, incubation time, temperature, concentration, and pH of the solution. Since the application of the metal and metal-oxide NPs on plant systems has begun only recently, little is known about their possible effects on plant growth, development, and productivity. Accumulation and translocation of NPs in plants and the consequent growth response and stress modulation in plant systems are yet to be understood fully. Plants responses to NPs may be positive or negative and vary with species as well as with the type of the NP. Cytotoxicity of NPs often depends on their concentration, size, and shape. Their impact on vegetative growth and fruit/seed production is often positive at lower concentrations and negative at higher ones. Moreover, certain NPs enhance the antioxidant enzymatic activity and the free radical scavenging potential of plants and alter the micro RNAs expression, which regulates different morphophysiological traits and metabolic processes in plants, ultimately affecting their growth and yield. NPs also help in genetic reforms by effecting an efficient transfer of DNA or complete plastid genome into the respective plant genome due to their miniscule size and improved capacity of site-specific penetration. Moreover, a controlled application of nanomaterials in the form of nanofertilizer offers a more synchronized nutrient fluidity for the uptake by the plants exposed, thus ensuring an increased nutrient availability and utilization. All these issues need to be addressed in details. The book in hand is an attempt to meet this requirement. We hope that the graduate, postgraduate, and research students of nanoscience and plant science will equally find it interesting, informative, and worth reading.

This book, comprising of 23 chapters in total, highlights the latest developments in phyto-mediated fabrication of the metal and metal-oxide NPs and their characterization by a variety of modern techniques. Application of NPs in various fields including drug delivery, therapeutic treatment, catalysis, photoluminescence, paints and coatings, environmental sensing and renewable energy, plant production and protection, development of lubricants and fuel additives, and antimicrobial and antioxidant formulations, inter alia, has been discussed in various chapters. Special emphasis has been laid on the coverage of the impact of NP application on plant growth and productivity, toxicity caused, if any, and plant responses to the stress imposed. The book contents have been divided broadly in two sections, viz., (a) plant-mediated synthesis and applications of nanomaterials and (b) interaction of plants and nanomaterials. The first part includes 13 chapters, while the second one contains 10. The vast coverage of diverse aspects of the subject by these chapters can be realized well from the table of contents.

We feel obliged not only to those colleagues who so kindly acceded to our requests to contribute chapters for this volume but also to those who helped us in reviewing these contributions. Professor Imran Ali (Chemistry) and Dr. M Irfan Qureshi (Biotechnology) of Jamia Millia Islamia (Central University), New Delhi, and Dr. Rabea Parveen (Pharmaceutics), Dr. Abida Parveen (Phytomedicine), and Dr. Bushra Parveen (Pharmacology) of Jamia Hamdard (Deemed University), New Delhi, extended specific assistance in vetting and finalizing different chapter manuscripts. We are indebted to Mr. Eric Stannard, Senior Editor (Botany) at Springer, and all his associates for their sustained cooperation during the preparation of this book. We shall be happy receiving comments and criticism, if any, from subject experts and general readers of this book.

Gondar, Ethiopia
New Delhi, India
September, 2018

Azamal Husen
Muhammad Iqbal

Contents

Part I
Plant-Mediated Synthesis and Applications of Nanomaterials

Chapter 1
Nanomaterials and Plant Potential: An Overview

Azamal Husen and Muhammad Iqbal

1.1 Introduction

During the last few decades, development of nanotechnology and use of nanomaterials (NMs) have been extensively on the rise. NMs are characterized by having at least one dimension in nanometre (nm) range. A nanometre is one billionth of a metre, i.e. 10^{-9} m, which denotes a length equivalent to 10 hydrogen or 5 silicon atoms aligned in a line. Several terms, including nanoparticles (NPs), nanoscale particles, nanosized particles, nanoscale materials, nanosized materials, nano-objects and/or nanostructured materials, are used to describe the NMs. They possess some exclusive physico-chemical properties, viz. high reactivity and surface area, tunable pore size and particle morphology. These materials are in great demand for their fascinating properties and diverse technological applications. They find potential applications in areas of disease management, biomedical sciences, electronics, nanosensors, biomarkers, display devices, pollution trace detection, environmental remediation, agriculture (quality improvement, growth and nutritional value enhancement, gene preservation, etc.), automotive, information and communication technology, energy, textile, construction and so on; the list of uses is expanding fast (Husen and Siddiqi 2014a, b; Siddiqi and Husen 2016a, b, c, d; Siddiqi et al. 2018a, b, c).

NMs are formed naturally or from anthropogenic sources and can therefore be divided into (i) natural and (ii) engineered NMs (Lidén 2011). Nanosized particles are present in the nature where they are automatically synthesized due to the natural processes like biomineralization and biodegradation (Uddin et al. 2013). NPs in

A. Husen (✉)
Department of Biology, College of Natural and Computational Sciences, University of Gondar, Gondar, Ethiopia

M. Iqbal
Department of Botany, Faculty of Science, Jamia Hamdard (Deemed University), New Delhi, India

A. Husen, M. Iqbal (eds.), *Nanomaterials and Plant Potential*,
https://doi.org/10.1007/978-3-030-05569-1_1

vivo are produced by several natural processes such as volcanic eruption, erosion, friction, forest fires, marine wave strokes, chemical and physical weathering of rocks and so on. However, in our daily life, incidental NPs are produced by many anthropogenic processes/sources such as fossils fuel combustion, industrial effluents, automobile exhaust and welding fumes.

During the ancient times, NPs were used in some form by the Damascans to create swords with very sharp edges and by the Romans to craft iridescent glassware. For instance, Romans in the fourth century AD used the Lycurgus Cups (Fig. 1.1), made of a glass that changes colour when light strikes upon them. These contain silver-gold alloy NPs distributed in such a way that they look green in reflected light and brilliant red in transmitted light (Leonhardt 2007; Freestone et al. 2007). Richard Feynman, the Nobel Prize winner in Physics, presented an idea of atom-by-atom construction of molecules in a scientific manner (Feynman 1960). In his lecture 'There is Plenty of Room at the Bottom' delivered in 1959, he predicted that the entire encyclopaedia would one day fit on the head of a pin, a library with all the world books would be adjusted in 3 square yards, and microcomputers and nanomedicines would float in veins. However, the scientific community did not take much notice of these comments at that point of time. The term 'nanotechnology' was first coined and defined in the year 1974 by Prof. Norio Taniguchi of the Tokyo Science University, Japan, who asserted that 'nanotechnology mainly consists of processing, separation, consolidation and deformation of materials by one atom or one molecule'. Ekimov and Onuschchenko (1981) reported three-dimensional quantum confinements of NPs in 1981. However, publication of the book 'Engines of Creation: The Coming Era of Nanotechnology' by Eric Drexler (1987) is considered to be the beginning of the present-day nanotechnology revolution. Eric Drexler proposed that the molecular nanotechnology is a branch of science and technology

Fig. 1.1 The Lycurgus Cup of glass that appears red (right) in transmitted light and green (left) in reflected light. © Trustees of the British Museum

that would allow manufacturers to fabricate products from the bottom-up with enhanced molecular control. This technology would allow every molecule to be inserted into its specific place so that the manufacturing systems using this process would be clean, efficient and highly productive. Around the same time, development of scanning probe microscopy by IBM scientists made it possible to fulfil the Feynman's vision of atom-by-atom construction of molecules. Furthermore, the discovery of two allotropes of carbon in nanosize form, i.e. fullerene by Smalley in 1986 (O'Brien et al. 1988) and carbon nanotubes by Iijima (1991), gave a new boost to nanotechnology research.

In recent years, nanosized particles, mostly of metals, metal oxides, carbon or fullerene, have been engineered to meet the specific objectives. Numerous methods including chemical methods (Sotiriou and Pratsinis 2010; Sotiriou et al. 2011; Zhang et al. 2011; Roldán et al. 2013), physical methods (Tien et al. 2008; Abou El-Nour et al. 2010; Asanithi et al. 2012) and biological methods (Husen and Siddiqi 2014b, c; Husen 2017; Siddiqi et al. 2018a) are now available for NP synthesis. NPs are produced in two ways, viz. 'bottom-up' (buildup of material from the bottom: atom by atom, molecule by molecule or cluster by cluster) and 'top-down' (slicing or successive cutting of a bulk material to get the nanosized particle) (Fig. 1.2). The 'bottom-up' approach is usually a superior choice for NPs preparation, as it involves a homogeneous system wherein catalysts (reducing agents and enzymes) synthesize the nanostructures that are controlled by the catalyst itself (Iravani 2011; Husen and Siddiqi 2014b). The 'top-down' approach generally works with the material in its bulk form, and the size reduction to nanoscale is achieved by specialized ablations, for instance, thermal decomposition, mechanical grinding, etching, cutting and sputtering. The main demerit of this approach is the surface structural defects, which have a significant impact on physical features and surface chemistry of the NPs produced.

The physical methods for NP synthesis normally include laser ablation and evaporation-condensation techniques (Jung et al. 2006), while the chemical methods employ chemical reductants ($NaBH_4$, ethanol, ethylene glycol, etc.), aerosol technique, electrochemical or sonochemical deposition, photochemical reduction and laser irradiation technique (Sotiriou and Pratsinis 2010; Liu et al. 2011; Sotiriou

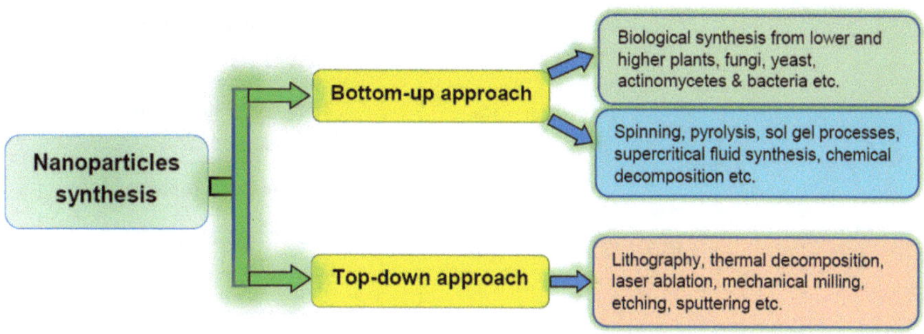

Fig. 1.2 The bottom-up and top-down approaches for NPs synthesis

et al. 2011; Zhang et al. 2011; Roldán et al. 2013). Physical methodologies are known to be costly due to high and continuous energy consumption to maintain the high pressure and temperature employed in NPs synthesis and require a highly sophisticated instrumentation. On the other side, chemical approach for NPs synthesis releases hazardous by-products and leads to environmental contamination. Moreover, certain chemicals are too costly and some of these may give rise to noxious chemical species tangled on the surface of NPs (Husen and Siddiqi 2014b). Biological methods involving the use of plants and/or microorganisms have several advantages over the chemical and physical methods. These are simple, cost-effective and eco-friendly and can be easily scaled up for high yields/production (Husen and Siddiqi 2014b; Husen 2017) (Fig. 1.3). Given this, biosynthesis of NPs, using the biological agents, viz. bacteria, fungi, yeast, plant and algal extracts, has rapidly gained the ground (Siddiqi and Husen 2016a, b, d, 2017a; Siddiqi et al. 2018a).

In recent years, the term 'green synthesis' has been used for the plant-based synthesis of NPs. Owing to rich biodiversity, several higher and lower plants and their parts (leaves, stems, roots, shoots, flowers, barks and seeds) and a big range of metabolic products or compounds (alkaloids, flavonoids, saponins, steroids, tannins, nutritional compounds) have been identified to possess a tremendous potential for this purpose and are being used successfully for an efficient and rapid green

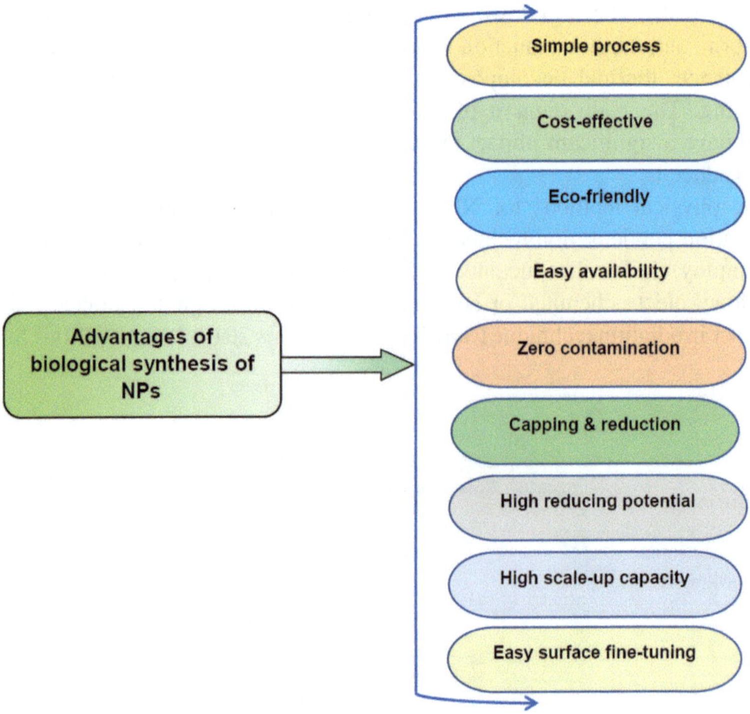

Fig. 1.3 Advantages of biological synthesis of nanoparticles

synthesis of NPs in nonhazardous ways (Fig. 1.4). Several factors determine the quality of synthesized NPs for their potential applications. The shape and size of NPs achieved in the plant-mediated synthesis depend on the protocol used, plant-extract concentration, nature of reducing agents, pH of reaction mixture, incubation time, temperature and light intensity, etc. (Fig. 1.5). For instance, Tippayawat et al. (2016) have fabricated silver NPs from *Aloe vera* plant extract. The reaction time as well as the temperature of reaction mixture markedly influenced the process. Their size (70.70–192.02 nm) changed with reaction time and temperature of reaction mixture used during the fabrication. Chapters 3 and 5 of this book discuss these aspects in detail. Green synthesis has been used mainly for producing silver and gold NPs, possibly because of their wider applications, but the technique is also being applied for obtaining NPs of many other metal and metal oxides (e.g. Fe, Pd, Pt, Se, Ru, PbS, CdS, CuO, CeO_2, TiO_2, ZnO, Fe_2O_3, Fe_3O_4, Al_2O_3, NiO, MoS_3 and SiO_2). Fabrication of NMs is followed by their characterization so as to determine the size distribution, aggregation, surface area and porosity, solubility, hydrated surface analysis, zeta potential, wettability, adsorption potential, shape and size of

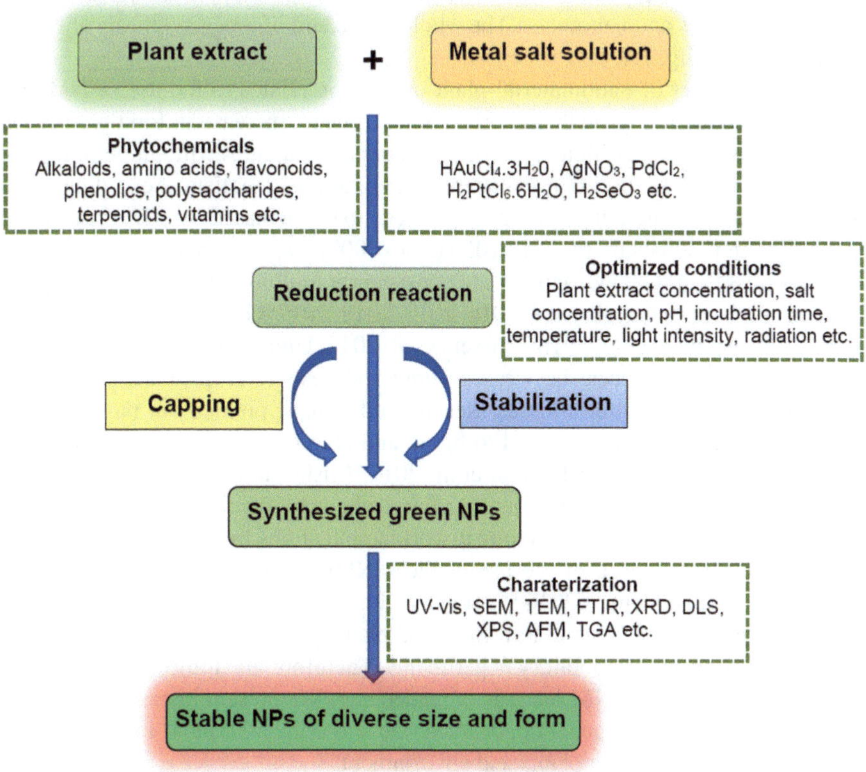

Fig. 1.4 Detailed scheme of plant-mediated (green) synthesis of NPs, optimization, stabilization and characterization techniques

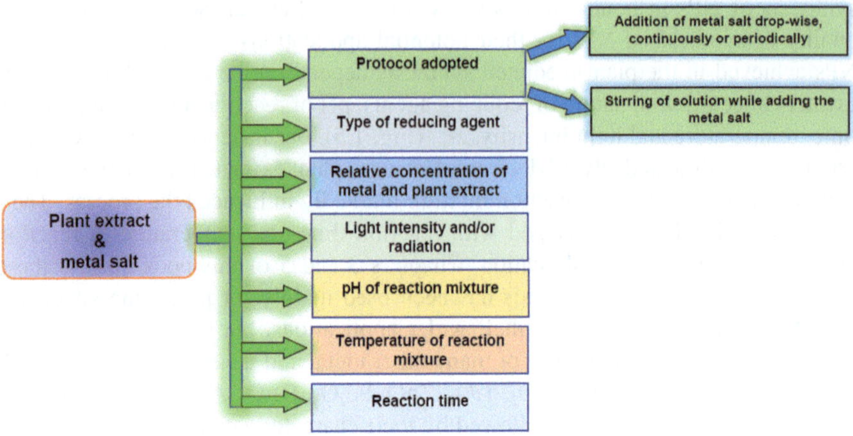

Fig. 1.5 A schematic illustration of optimization and the growth factors that affect plant-mediated synthesis of NPs

interactive surface, crystallinity, fractal dimensions, orientation, intercalation and dispersion of NPs and nanotubes. The whole procedure of green synthesis of NPs is summarized in Fig. 1.4.

Another aspect of nanomaterial-plant interaction relates to the use of NMs for enhanced production, growth and protection of plants. Increase in food production with the minimum and efficient use of fertilizer and pesticides without polluting the environment is now the main cynosure of agriculture scientists. According to an FAO estimate, productivity of food crops needs to be increased by 70% for an extra 2.3 billion individuals by the year 2050 (FAO 2009). On the other hand, increasing abiotic stresses in the environment are consistently inhibiting the growth and yield of food crops (Husen et al. 2014, 2016, 2017, 2018, 2019; Getnet et al. 2015; Yousuf et al. 2015; Embiale et al. 2016; Hussein et al. 2017; Umar et al. 2018). In addition, infections caused by bacteria, viruses, fungi and pests to crop plants are another major cause for agricultural losses. Recent studies have pointed out the beneficial effects of NMs on plants exposed to biotic and abiotic stresses (de Oliveira et al. 2014; Hossain et al. 2015; Iavicoli et al. 2017; Mishra et al. 2017; Rastogi et al. 2017; Siddiqi and Husen 2017b). Only few techniques involving the use of NMs could be successfully translated to real-world applications by now, and many questions concerning the risk of the use of these NMs in plant system remain unanswered (Siddiqi and Husen 2016c, 2017b). Although collation of information on the toxicity of engineered NMs to terrestrial and plant system is in progress, most of it is based on short-term laboratory experiments. Therefore, the information available on the inherent role of NM hazards is inadequate. We present here an account of efforts hitherto made to generate, characterize and utilize the plant-mediated NMs and assess their impact on plants and environment.

1.2 Type of Engineered NMs

Based on their morphology, size (physical) and chemical properties, the engineered NMs are categorized as carbon NPs, metal NPs, ceramics NPs, semiconductor NPs, polymeric NPs and lipid-based NPs. Carbon-based NPs represent two major classes, i.e. carbon nanotubes (CNTs) and fullerenes. CNTs are elongated, tubular structure and 1–2 nm in diameter (Ibrahim 2013).

According to Aqel et al. (2012), CNTs are reliant on their diameter telicity and look like a graphite sheet rolling upon itself. They are called single-walled, double-walled or multi-walled CNTs, depending on the number of walls they have. Fullerenes contain NMs made of globular hollow cage, i.e. allotropic form of carbon. They are commercially important due to their high strength, structure, electrical conductivity, electron affinity and versatility (Astefanei et al. 2015).

Metal NPs are made of metals [gold (Au), silver (Ag), palladium (Pd), platinum (Pt), selenium (Se), copper (Cu), iron (Fe), etc.] and/or metal oxides [titanium dioxide (TiO_2), zinc oxide (ZnO), copper oxide (CuO), cuprous oxide (CuO), indium oxide (In_2O_3), etc.] showing specific opto-electrical properties due to their localized surface plasmon resonance (SPR) features (SPR is a phenomenon occurring at metal surfaces, typically gold or silver, when an incident light beam strikes the surface at a particular angle. Depending on the thickness of a molecular layer at the metal surface, SPR makes a graded reduction in intensity of the reflected light. Many standard tools for measuring adsorption of material onto planar metal surface or the surface of metal NPs are based on it).

Ceramics materials are the inorganic non-metallic solid materials synthesized through successive heating and cooling and are found in amorphous, polycrystalline, dense, porous or hollow forms (Sigmund et al. 2006). These particles are used in catalysis, photocatalysis, photodegradation of dyes and imaging applications (Thomas et al. 2015). Semiconductor NPs carry properties intermediate of metals and non-metals and are therefore useful in photocatalysis, photo-optics, electronic devices, etc.

Polymeric NPs are solid colloidal particles containing macromolecular materials that attach, adsorb, dissolve and encapsulate drugs or therapeutic compounds. These particles are identified as nanospheres or nano-capsules. The former are matrix particles whose overall mass is generally solid and the other molecules are adsorbed at the outer boundary of their spherical surface. However, in nano-capsules, the solid mass is encapsulated within the particle completely (Rao and Geckeler 2011). They protect active compounds and facilitate easy delivery and permeability of drugs into the target cells with higher efficacy and efficiency.

Lipid-based NPs are in general spherical with their diameter ranging from 10 to 1000 nm. Like polymeric NPs, lipid NPs contain a solid core made of lipid and a matrix that comprises of soluble lipophilic molecules. Surfactants or emulsifiers stabilize the external core of these NPs (Rawat et al. 2011). These days, lipid nanotechnology is recognized as a special area of science, which mainly focuses on designing and synthesis of lipid NPs that are used as drug carriers for delivery and

RNA release, for instance, in cancer treatment (Puri et al. 2009; Gujrati et al. 2014). These particles have the advantage of better reproducibility using multiple strategies and larger scale-up feasibility.

1.3 Characterization Techniques

Currently, several techniques such as UV-visible spectroscopy (UV-vis), transmission electron microscopy (TEM), high-resolution transmission electron microscopy (HRTEM), scanning electron microscopy (SEM), X-ray diffraction (XRD), Fourier-transform infrared spectroscopy (FTIR), atomic force microscopy (AFM), X-ray photoelectron spectroscopy (XPS), thermogravimetric analysis (TGA), energy-dispersive X-ray spectroscopy (EDS), dynamic light scattering (DLS), zeta potential, surface-enhanced Raman spectroscopy (SERS), nuclear magnetic resonance spectroscopy (NMR), matrix-assisted laser desorption/ionization time-of-flight mass spectrometry (MALDI-TOF), dual polarization interferometry and many others are available for NPs characterization (Fig. 1.6).

Out of these, some techniques (spectroscopy and microscopy) that include UV-vis, DLS, AFM, TEM, SEM, XRD and FTIR, are used more frequently. The AFM, SEM and TEM are regarded as the direct techniques to obtain data from

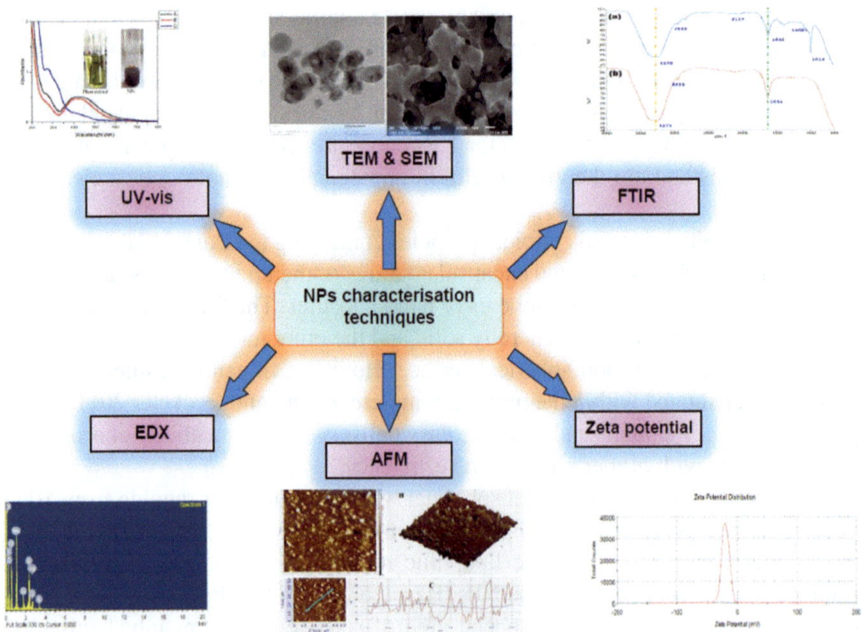

Fig. 1.6 Frequently used techniques for NPs characterization

images at various resolutions. Moreover, SEM and TEM have been widely used to determine the shape and size of NPs, as shown in Fig. 1.7.

With the help of TEM, small particles can be observed and the crystallographic structure of a sample can be imaged at an atomic scale. The arrangement of atoms and their local microstructures such as lattice fringe, glide plane, lattice vacancies and defects, screw axes and the surface atomic arrangement of crystalline NPs can be analysed using HRTEM (Brice-Profeta et al. 2005). With the help of SEM, normally data for selected areas of the surface of NM are collected, and a two-dimensional image with spatial variations is displayed. AFM can be used in either liquid or gas medium. This technique facilitates the study of morphology of NPs and biomolecules. Unlike TEM and SEM, it produces three-dimensional images so that the particle volume and height can be assessed (Vesenka et al. 1993; Mucalo et al. 2002). This technique is proficient of ultra-high resolution for particle size measurement and is based on the physical scanning of samples at the submicron level using a probe tip (Zur Mühlen et al. 1996). Using AFM, quantitative ideas regarding individual NPs and groups of particles, viz. size (length, width and height), morphology and surface texture, can be evaluated with the help of software-based image processing.

The UV-vis, DLS, XRD, EDS, FTIR and Raman spectroscopy techniques are the indirect procedures for observing the details related to structure, composition,

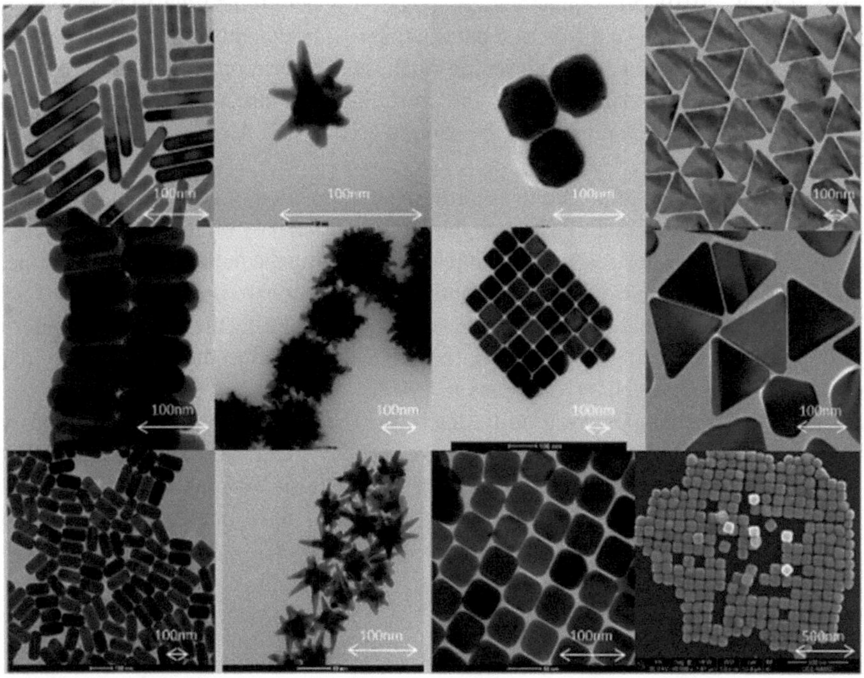

Fig. 1.7 TEM and SEM micrographs depicting various morphologically different types of NPs. (Adopted from: www.ucd.ie/cbni/newsevents/cbni-in-the-news/name,196987,en.html)

crystal phase and properties of NPs. It is known that UV-vis spectroscopy covers the UV range between 190 and 380 nm, while the visible range extends between 380 and 800 nm. Both the radiations interact with matter and promote electronic transitions from the ground state to higher energy states. Poinern (2014) has mentioned that the wavelengths between 300 and 800 nm are commonly used for characterization of metallic NPs ranging in size from 2 to about 100 nm. Absorption measurements are detected by the presence of peaks usually between 500 and 550 nm for gold NPs (Siddiqi and Husen 2017a; Husen 2017) and between 400 and 450 nm for silver NPs (Siddiqi and Husen 2016b; Siddiqi et al. 2018a).

DLS spectroscopy can be used to determine the pattern of size distribution and surface charge quantification of NPs suspended in a liquid (Jiang et al. 2009; Siddiqi and Husen 2017a). This technique is extensively used to record the size of Brownian NPs in colloidal suspensions (De Jaeger et al. 1991). Once a monochromatic beam of light (laser) is directed onto a solution of spherical particles in Brownian motion, a Doppler shift occurs when the light hits the moving particles, thereby changing the wavelength of the incoming beam of light by a value related to the particle size. Consequently, DLS allows computation of size distribution and NP motion in the medium and can be computed by recording the diffusion coefficient of the particle.

The XPS technique is used to investigate the mechanism of reaction that occurs on the surface of magnetic NPs, to measure the bonding features of different elements involved and to approve the structure of elements present in the magnetic NPs (Faraji et al. 2010). NPs elemental composition is determined via EDS mapping. XRD analysis exhibits a diffraction pattern that is subsequently compared with data contained in a standard crystallographic database to determine the structural details. XRD analysis of the data identifies the crystallite size, structure, preferred crystal orientation and phases present in the samples (Klug and Alexander 1974; Barrett et al. 1986).

TGA technique is practiced to confirm the composition of coatings, such as surfactants or polymers, and to estimate the binding efficiency on the surface of magnetic NPs. Raman spectroscopy is suitable for detecting the molecules in vibrational modes. This technique is used to identify the vibrational signals of a variety of chemical species that are attached to the surface of NPs during the synthesis (Poinern 2014). Moreover, using SERS it was possible to determine the single molecular attachments on silver NPs (Kneipp et al. 1997).

The FTIR techniques are used to determine the surface chemistry and identify surface residues, viz. functional groups like carbonyls and hydroxyls moieties, that adhere to the surface of the product during NP synthesis. Availability of functional groups in the NMs can be easily examined by the size of peaks of the spectrum (Faraji et al. 2010; Chauhan et al. 2012). Further details of NPs characterization techniques are given in Table 1.1.

Table 1.1 Characterization techniques used during NPs synthesis

Type of studies	Characterization techniques and their uses
NPs formation	*UV-vis*: Offers information on the size, structure, stabilization and aggregation of NPs
NPs shape and size determination	*TEM*: Examines the shape, size (10^{-10} m), morphology and allographic structure of NPs *HTEM*: Examines the arrangement of the atoms and their local microstructures, viz. lattice fringe, glide plane, lattice vacancies and defects, screw axes and surface atomic arrangement of crystalline NPs *SEM*: Examines the morphology by direct visualization *AFM*: Examines the size and form, i.e. length, width and height, and other physical properties, viz. morphology and surface texture *DLS*: Displays the particle size distribution
Surface charge	*FTIR*: Describes NPs to understand their functional groups and determine the emission, absorption, photoconductivity or Raman scattering of a solid, liquid or gas *XPS*: Examines the mechanism of reaction that occurs on the surface of magnetic NPs and the characteristics of bonding of different elements involved, in addition to confirming the structure and speciation of elements present in the chemical composition of the magnetic NPs *TGA*: Approves the formation of coatings, viz. surfactants or polymers, to estimate the binding efficiency on the surface of magnetic NPs *Zeta potential*: Determines the stability and surface charge of colloidal NPs as well as the nature of materials encapsulated inside the NPs or coated on their surface
Crystallinity	*XRD*: Identifies and quantifies the various crystalline forms or elemental compositions of NPs
Magnetic properties	*Vibrating sample magnetometry*: Estimates the magnetization of magnetic NPs *Superconducting quantum interference device magnetometry*: Examines the magnetic properties of magnetic NPs
Other techniques	*Chromatography and related techniques*: Separate NPs on the basis of their affinity towards the mobile phase *EDX*: Identifies the elemental composition of NPs *Field flow flotation*: Separates different NPs based on their magnetic susceptibility *Filtration and centrifugation technique*: Fractionates the preparative size of NPs *Laser-induced breakdown detection*: Examines the concentration and size of colloids *Mass spectrometry*: Examines the fluorescent labelled NPs *Small angle X-ray scattering*: Performs the structural characterization of solid and fluid materials in the nanometre range *X-ray fluorescence spectroscopy*: Recognizes and examines the concentrations of elements present in solid, powdered or liquid samples *Hyperspectral imaging*: Identifies the type of NPs, studies the fate and transformation of these particles in water samples and characterizes the unique surface chemistry and functional groups added to the NM

1.4 Physical and Chemical Characters of NMs

The large surface area, shape, composition, mechanical strength, optically dynamic and chemically reactive physico-chemical characters of NMs make them suitable for use in different disciplines of science and technology. The surface area of NMs plays a critical role in their toxic manifestations (Holgate 2010). With a decrease in the particle size, the surface area increases, leading to a concentration-dependent enhancement in oxidation and DNA-damaging capabilities (Risom et al. 2005). Additionally, other characteristics, like surface roughness, hydrophobicity and charge of NPs, affect their cellular uptake (Nel et al. 2009). Plant responses (absorption, translocation, accumulation and overall interaction) to NMs also depend on particle size (Husen and Siddiqi 2014a, b; Siddiqi and Husen 2017a). Lin and Xing (2007) found no sign of toxicity in radish, rape, ryegrass, lettuce and cucumber after the application of Al_2O_3 NPs of 60 nm size, although root elongation in maize was reduced by 35%. Doshi et al. (2008) reported that the treatment of 100 nm Al_2O_3 NPs had no adverse effect on plant growth in *Phaseolus vulgaris* and *Lolium perenne*, while *Arabidopsis thaliana* did not show any phytotoxicity with 150 nm NPs (Lee et al. 2010). Sadiq et al. (2011) reported the negative effect of below 50 nm Al_2O_3 NPs on the development of some microalgae. Chen et al. (2006) used zebrafish to examine the in vivo toxicity of different gold and silver NPs in the size range of 3, 10, 50 and 100 nm. Silver NPs caused a size-dependent mortality, while the impact of gold NPs was independent of size. On the whole, the size and surface area are important factors in determining the impact of NPs on plants as well as animals. The chemical nature of the constituents also has a role to play. Both the electronic and optical characters of NPs are interdependent to a large extent. For example, size-dependent optical characters are observed in metal NPs. They exhibit a strong UV-vis excitation band, which is not seen in the spectrum of the bulk metal. This band appears when the incident photon frequency is persistent through the collective excitation of the conduction electrons and is termed as the localized SPR. The wavelength peak of the localized SPR spectrum depends on the shape, size, dielectric features and interparticle spacing of NPs along with the conditions of solvents, substrate and adsorbates (Eustis and El-Sayed 2006).

The magnetic property of NPs, a result of uneven electronic distribution pattern, which in turn is dependent on the types of fabrication process such as the solvothermal, co-precipitation, microemulsion and thermal decomposition techniques, decides their suitability for several applications (Wu et al. 2015; Qi et al. 2016). Various mechanical features such as hardness, stress and strain, elastic modulus, adhesion and friction are examined to understand the exact mechanical nature of NMs. These characters facilitate their application in tribology, surface engineering, nanofabrication and nanomanufacturing. Additional features like surface coating, coagulation and lubrication also help in defining their mechanical characters (Guo et al. 2014). Solid metallic NPs exhibit higher thermal conductivities in comparison to fluid ones. For instance, thermal conductivity of copper at room temperature is ~700 times higher in comparison to water and ~3000 times higher in comparison to the engine oil. Hence, fluids containing the suspended solid materials or particles are supposed to have a higher thermal conductivity than the conventional heat transfer fluids. For instance, Cao (2002) reported that the nanofluids containing CuO or Al_2O_3 NPs in water or ethylene show a higher thermal conductivity.

1.5 Application and Impact of NMs

The statement 'There's plenty of room at the bottom' made by Richard Feynman in 1959 regarding the nonotechnology (Feynman 1960) still holds good. This technology has started influencing our lifestyle with a lot of promise and is expected to boom further in the near future through its tremendous impact on material science, electronics, medicine, pharmacology and agriculture sectors. By now, the NMs of carbon, fullerene, Au, Ag, Cu, Zn, Pt, Pd, Se, Fe, ZnO, TiO_2, Fe_2O_3, Fe_3O_4, Al_2O_3, NiO, CeO_2, MoS_3, SiO_2, CdSe, TiO_2 and nanoclay have found significant applications in the areas of agri-food, agrochemicals, biosensors, cosmetics, catalysts, lubricants, fuel additives, paints and coatings, food packaging, nanomedicine and nanocarriers, among others (Fig. 1.8 and Table 1.2). Many research institutions and commercial organizations have taken up the nanotechnology-based R&D programmes all over the globe. The number of nanotechnology-based products available in the market now runs into thousands. The global market for nanocomposites totalled US \$ 2.0 billion in 2017 and is estimated to reach US \$ 7.3 billion by 2022, growing at a compound annual growth rate (CAGR) of 29.5% for the period of 2017–2022 (McWilliams 2018).

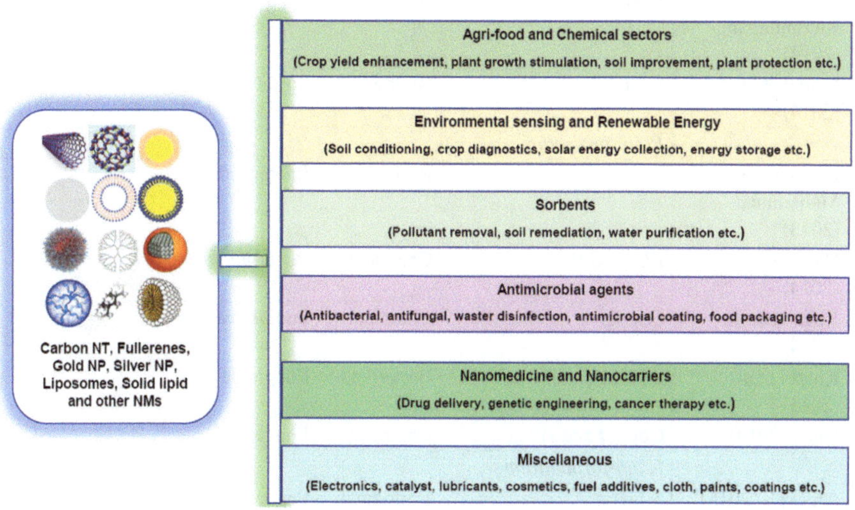

Fig. 1.8 Various applications of nanomaterials

Table 1.2 Role of NMs in plant system

Key references	NMs	Fabrication process	Specific role
	Modulation of plant growth		
Arora et al. (2012)	Gold	Chemical	Increased germination rate
Barrena et al. (2009)	Gold	Chemical	Increased germination rate
Jasim et al. (2016)	Silver	Chemical	Increased plant growth
Khodakovskaya et al. (2009)	MWCT	Chemical	Increased seed germination and growth rate
Laware and Raskar (2014)	Zinc oxide	Chemical	Increased flowering and seed productivity
Ngo et al. (2014)	Iron, cobalt, copper	Chemical	Increased germination rate
Nhan et al. (2015)	Cerium oxide	Chemical	Destruction of vascular bundles
Qian et al. (2013)	Silver	Chemical	Root growth inhibition
Soares et al. (2016)	Nickle oxide	Chemical	Decrease in leaf surface area, chlorophyll and carotenoids
Song et al. (2016)	Copper oxide	Chemical	Inhibition of plant growth
Syu et al. (2014)	Silver	Chemical	ROS generation, root growth promotion, activation of gene expression of indoleacetic acid protein 8, 9-cis-epoxycarotenoid dioxygenase and dehydration-responsive RD22
Vinković et al. (2017)	Silver	Chemical	Dose-dependent decrease in plant growth
	Seed treatment		
Anand and Kulothungan (2014)	Silver	Biological	Seed dressing
Morsy et al. (2014)	Silver	Biological	Surface sterilizer
	Plant protection		
El-Rahman and Mohammad (2013)		Biological	Bactericidal
Hafez et al. (2014)	Zinc oxide	Chemical	Bactericidal
Kanhed et al. (2014)	Copper	Chemical	Fungicidal
Khadri et al. (2013)	Silver	Biological	Fungicidal
Ouda (2014)	Copper and silver	Chemical	Fungicidal
	Nano-formulation		
Ali et al. (2015)	Silver	Biological	Pesticidal

(continued)

Table 1.2 (continued)

Key references	NMs	Fabrication process	Specific role
Grillo et al. (2014)	Chitosan	Chemical	Herbicidal
Pereira et al. (2014)	Polyepsiloncaprolactone	Chemical	Herbicidal
Yu et al. (2015)	Carboxymethyl chitosan	Chemical	Herbicidal
	Biosensing		
Boro et al. (2011)	Gold	Chemical	Herbicide detection
Dubas and Pimpan (2008)	Silver	Chemical	Herbicide detection
Kang et al. (2010)	Gold	Chemical	Organophosphates detection
Luo et al. (2014)	Carbon	Chemical	Herbicide detection
Zhao et al. (2011)	Graphene	Chemical	Herbicide detection

By tradition, silver and gold acquire elite properties and have numerous applications. In general, silver NMs have gained more attraction due to their intrinsic antimicrobial activity and are already used in a wide range of commercially available medical and other consumer stuffs (Salata 2004; Sondi and Salopek-Sondi 2004; Asha-Rani et al. 2009; Husen and Siddiqi 2014b; Siddiqi et al. 2018a). These are also effective in crop protection and disease management (Khot et al. 2012; Ocsoy et al. 2013; Siddiqi and Husen 2016a) and can be used to control many types of plant pathogen in a benign way, compared to the conventional fungicides (Park et al. 2006). They have also shown activity against various cancer cell lines and plasmodial pathogens (Sukirtha et al. 2012; Husen and Siddiqi 2014b; Murugan et al. 2015). Over 320 tons of silver NMs were being manufactured annually during the last decade and used in nanomedical imaging, biosensing and food products (Chen and Schluesener 2008; Ahamed et al. 2010). Silver-based NMs are also used in wound dressings, catheters and various home care products due to their antimicrobial activity (Asha-Rani et al. 2009). Their applications are discussed in detail in Chap. 5 of this book.

Likewise, advancement in synthesis and surface functionalization of NMs, i.e. their effective manipulation, has increased the scope of gold NMs application. Facile tuneable size and functionality make them a valuable scaffold for efficient recognition and delivery of biomolecules, and they can also deliver large molecule drugs with ease. Because of their size and exclusive optical and chemical properties, they have the capacity to gather inactively around tumours and can therefore be used in thermal treatment procedures (Hirsch et al. 2003; Zheng and Sache 2009). They also find use in diagnostics, biological imaging, biosensors, therapeutic agent delivery, photodynamic therapy, electronics, textile, energy harvesting, water disinfection and environmental monitoring/cleanup and possess catalytic, antioxidant and antimicrobial activities (Salata 2004; Chen et al. 2008; Holzinger et al. 2014; Shaalan et al. 2016; Yu et al. 2016; Husen 2017; Siddiqi and Husen 2017a). Applications of gold NMs are discussed in details in Chaps. 3 and 4 of this book.

Silver and gold NMs exhibit both beneficial and adverse effects on growth and yield of plants (Husen and Siddiqi 2014b; Jasim et al. 2017; Rui et al. 2018; Siddiqi and Husen 2016a, 2017a; Kim et al. 2018). It needs to be identified which trace elements are useful for the given plant species so that the same NPs may be used.

Platinum NMs have proved to be excellent therapeutic agents for use in chemotherapy, especially in the treatment of cancer cells (Hou et al. 2013; Wang et al. 2013; Min et al. 2014; Pandey et al. 2014). In combination with ion irradiation, they enhance the efficiency of cancer therapy (Periasamy and Alshatwi 2013). Further, these are used in water electrolysis (Soundarrajan et al. 2012).

Palladium NMs synthesized from *Hippophae rhamnoides* leaf extract have been examined for their heterogeneous catalytic activity in Suzuki-Miyaura coupling reaction (Nasrollahzadeh et al. 2015). Selenium NMs have shown growth inhibition of *Staphylococcus aureus* and can be used as a medicine against the *S. aureus* infection (Tran and Webster 2011). They also act as photovoltaic and semiconductor, antioxidant and chemoprotective agents (Chen et al. 2008).

Alumina-, titania- and carbon-based NMs are frequently used to get the desirable mechanical properties in coatings industries (Shao et al. 2012; Mallakpour and Sirous 2015; Kot et al. 2016). Coating enhances the mechanical strength, as it increases the toughness and wear resistance. Copper-based NMs have shown strong antimicrobial, antioxidant, anticancer as well as catalytic activity (Stoimenov et al. 2002; Akhavan and Ghaderi 2012; Hassan et al. 2012; Duman et al. 2016; Nagajyothi et al. 2017; Ojha et al. 2017). Titanium dioxide-based NMs are used in degrading the organic contamination and wastewater disinfection and in antiseptic and antibacterial compositions for their photocatalytic properties (Zhang and Chen 2009; Mahmoud et al. 2017). Moreover, these particles are used in UV-resistant material, printing ink, paper industry, self-cleaning ceramics and glass, coating and cosmetic products such as sunscreen creams, whitening creams, morning and night creams and skin milks (Wolf et al. 2003; Kaida et al. 2004; Wang et al. 2007; Weir et al. 2012; Husen and Siddiqi 2014b).

Iron oxide NMs (e.g. Fe_3O_4 or Fe_2O_3) have also gained ground in medical and some other sectors such as pesticide detection and removal of dyes (Ali et al. 2016; Siddiqi et al. 2016). Arokiyaraj et al. (2013) have reported the antibacterial activity of *Argemone mexicana*-mediated iron oxide NPs against *Proteus mirabilis* and *Escherichia coli*. Iron NPs synthesized from the *Lawsonia inermis* and *Gardenia jasminoides* leaves extract proved toxic to various bacterial strains (Naseem and Farrukh 2015). Iron oxide NMs are also used as a nano-fertilizer. For instance, a study on the effectiveness of Fe_2O_3 NPs as fertilizer for *Arachis hypogaea* has revealed that the Fe_2O_3 NPs and Fe_2O_3-EDTA effectively increased the root length and plant height and biomass by regulating the phytohormones and antioxidant enzymes' activity (Rui et al. 2016). The Fe_2O_3 NPs were adsorbed onto the soil, increasing easy availability of iron to peanut plants. Likewise, growth parameters of *Solanum lycopersicum* were improved under the influence of Fe_2O_3 NPs (Shankramma et al. 2016). In this study, NPs were dumped on root hairs and root tips, followed by the nodal and middle zones of the plant. The authors proposed that biomineralization of NPs occurred due to the presence of rich phytochemicals in

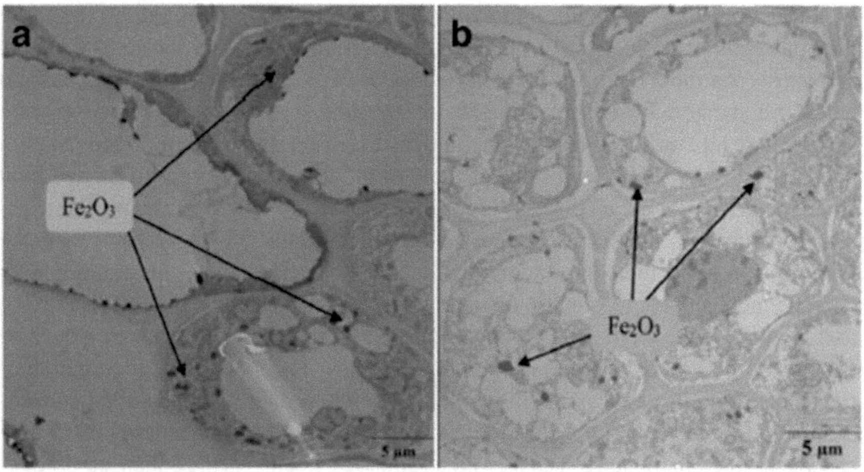

Fig. 1.9 TEM images of root sections of non-transgenic cotton (**a**) and Bt transgenic cotton (**b**) plants obtained after 10 days of treatment with Fe_2O_3 NPs. (Adopted from: Nhan et al. 2016)

plants. Further, exposure of Bt transgenic and non-transgenic cotton to Fe_2O_3 NPs (1000 mg L^{-1}) showed the presence of dark dots (particles) primarily localized in the endodermis and vascular cylinder (Fig. 1.9) (Nhan et al. 2016). Absorption of Fe_2O_3 NPs and their aggregation in roots were apparent; iron content in shoots and roots increased in a dose-dependent manner. It was speculated that the bioaccumulation of Fe_2O_3 NPs in these cotton might cause potential risk for environmental and human health.

Zinc oxide NMs also exhibit antimicrobial activity, photodegradation or photocatalytic property and show potential in drug delivery and anticancer therapy (Vimala et al. 2014; Bhuyan et al. 2015; Ali et al. 2016; Karnan and Selvakumar 2016; Thatoi et al. 2016; Nava et al. 2017; Siddiqi et al. 2018b). These particles carrying phycomolecule ligands acted as a novel growth promotor in cotton plants (Venkatachalam et al. 2016) and helped in increasing the plant productivity. Other studies have revealed the potential of zinc oxide NPs in stimulating seed germination and plant growth (Siddiqi and Husen 2017b).

Carbon nanotubes are able to penetrate the seed coat and plant cell wall, depending on their size, concentration and solubility. This penetration can bring changes in the metabolic functions of plants, leading to increase in plant biomass and the fruit/grain yield. However, carbon NMs have shown toxicity in some plant species (Husen and Siddiqi 2014a). Details of plant response to carbon NMs are discussed in Chap. 22 of this book.

In general, plants may show both positive and negative responses to metal oxide NPs with respect to the developmental, physiological and biochemical processes (Siddiqi and Husen 2017b). These responses are summarized in Table 1.1 of Chap. 22 of this book. NPs are capable to infiltrate in living plant tissues and migrate to different regions of the plant system (Corredor et al. 2009). In the aqueous medium or soil matrix, they move through the symplastic or apoplastic region to penetrate the epidermis of roots, pass through the cortex and finally translocate to the stem and leaves via the xylem and phloem (Corredor et al. 2009; Wang et al. 2012) and hence, often cause a widespread impact on the plant system. Their toxicity on plant system depends on their dose, shape and size (Husen and Siddiqi 2014b; Siddiqi and Husen 2016a, 2017b). Besides, the fate of these NPs in the ecosystem and food chain integrity is of great concern (Fig. 1.10). Application of NPs to some plant species has shown remarkable changes in the antioxidant enzyme activity and the upregulation of heat-shock proteins. Various plant species have evolved antioxidant defence mechanisms to prevent oxidative damage and enhance plant resistance to toxicity caused by the metal or metal oxide NMs; these mechanisms are yet to be fully explored and understood (Anjum et al. 2015). The NMs absorption and their translocation in different parts of the plant depend on their bioavailability, concentration, solubility and exposure time. Despite their useful applications, NMs may cause toxicities in the terrestrial as well as aquatic system. As the enhanced production, application and disposal of NMs will inexorably increase their release into the ecosystem, biodiversity is likely to be affected. Therefore, consequences of their

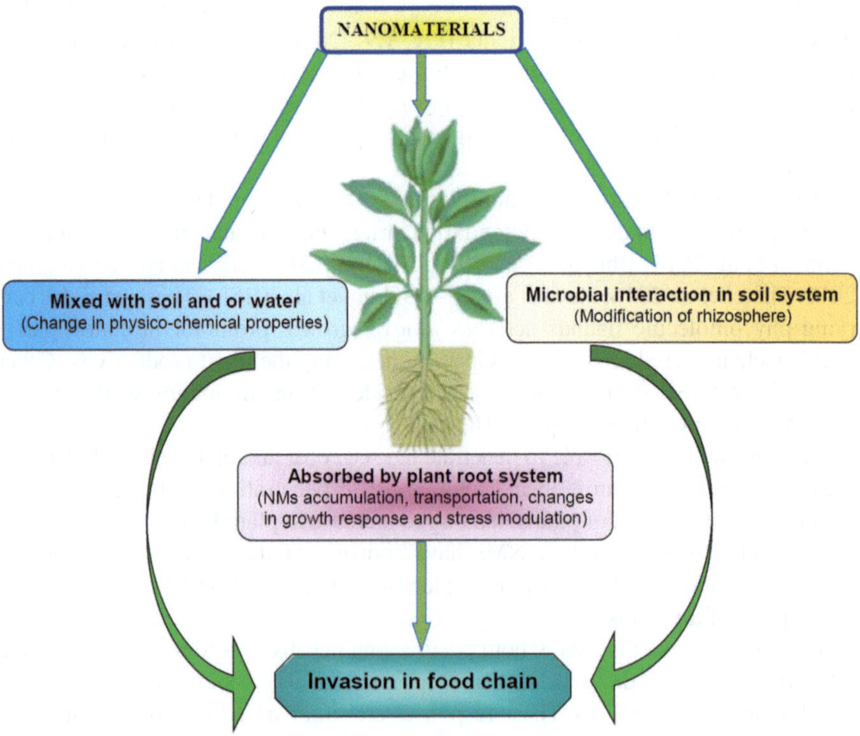

Fig. 1.10 Interaction of NMs with soil-plant system and their invasion in food chain or ecosystem

interaction with living beings (plants, animals and humans), direct or through the food chain, need to be fully investigated and understood.

1.6 Conclusion

Plant-mediated production of NMs is a green synthetic route, which is clean, safe, cost-effective and eco-friendly, giving conveniently utilizable NMs for application in various disciplines of science and technology. Plant-based green synthesis of NMs has a great promise for applications in cutting-edge technologies such as the fluorescent labelling, drug delivery, tumour destruction via heating and development of effective antimicrobial bandages. The amount of accumulation of NPs varies with the reduction potential of ions, and the reducing capacity of the plant depends on the presence of various types of metabolites in its tissues. Concentrations of plant extract and the substrate, temperature and pH of reaction mixture and exposure time largely determine the morphology, size and stability of NPs. Many issues concerning the required morphology, size and stability of NMs, reproducibility of synthesis process and the exact mechanisms involved still remain unresolved or partially resolved. Physico-chemical characters of NMs such as surface area, shape, composition, mechanical strength, optical dynamicity and chemical reactivity have a role in determining their toxicity to plant and animal systems. These materials can enable the antimicrobial compounds' delivery for use as pesticides in plant protection practices. Some of these may replace toxic chemicals and fertilizers or at least minimize their use in agriculture/horticulture in the near future. A variety of NMs have caused both beneficial and adverse effects in plants. The absorption, transportation and accumulation of NMs in plant system depend on the exposure time, shape and size of the NM and on the nature of plants species. In lower concentrations, NMs often have promotive role. At higher doses, they may cause toxicity, ultimately enhancing the generation of reactive oxygen species, which leads to disruption of the cellular metabolism and the consequent modulation of antioxidant defence system. By the way, almost all the NM vs plant studies have been carried out on plants in early developmental stages, maintained in laboratories. Long-term investigations based on field trials in natural environmental conditions are still required to develop a comprehensive and clear concept. Instead of making an arbitrary use of NMs, it is important to know which trace elements, and in what quantity, are useful for a given plant species so that a precise and effective treatment is applied. The overall cost involved should also be worked out to evaluate the economical feasibilities. The future research should also focus on the impact of NMs (1) on food chain and ultimately on the health of consumers (animals and humans) and (2) on the habitat environment as a whole.

References

Abou El-Nour KMM, Eftaiha A, Al-Warthan A, Ammar RAA (2010) Synthesis and applications of silver nanoparticles. Arab J Chem 3:135–140

Ahamed M, AlSaalhi MS, Siddiqui MKJ (2010) Silver nanoparticle applications and human health. Clin Chim Acta 411:1841–1848

Akhavan O, Ghaderi E (2012) Cu and CuO nanoparticles immobilized by silica thin films as antibacterial materials and photocatalysts. Surf Coat Technol 205:219–223

Ali SM, Yousef NMH, Nafady NA (2015) Application of biosynthesized silver nanoparticles for the control of land snail *Eobania vermiculata* and some plant pathogenic fungi. J Nanomater 2015:218904

Ali A, Zafar H, Zia M, Ul Haq I, Phull AR, Ali JS, Hussain A (2016) Synthesis, characterization, applications, and challenges of iron oxide nanoparticles. Nanotechnol Sci Appl 9:49–67

Anand R, Kulothungan S (2014) Silver mediated bacterial nanoparticles as seed dressing against crown rot pathogen of groundnut. Arch Appl Sci Res 6:109–113

Anjum NA, Adam V, Kizek R, Duarte AC, Pereira E, Iqbal M, Lukatkin AS, Ahmad I (2015) Nanoscale copper in the soil-plant system: toxicity and underlying potential mechanisms. Environ Res 138:306–325

Aqel A, El-Nour KMMA, Ammar RAA, Al-Warthan A (2012) Carbon nanotubes, science and technology part (I) structure, synthesis and characterization. Arab J Chem 5:1–23

Arokiyaraj S, Saravanan M, Udaya Prakash NK (2013) Enhanced antibacterial activity of iron oxide magnetic nanoparticles treated with *Argemone mexicana* L. leaf extract: an in vitro study. Mat Res Bull 48:3323–3327

Asanithi P, Chaiyakun S, Limsuwan P (2012) Growth of silver nanoparticles by DC magnetron sputtering. J Nanomater 2012:963609

Asha-Rani PV, Mun GLK, Hande MP, Valiyaveettil S (2009) Cytotoxicity and genotoxicity of silver nanoparticles in human cells. ACS Nano 3:279–290

Astefanei A, Núñez O, Galceran MT (2015) Characterisation and determination of fullerenes: a critical review. Anal Chim Acta 882:1–21

Arora S, Sharma P, Kumar S, Nayan R, Khanna PK, Zaidi MGH (2012) Gold nanoparticle induced enhancement in growth and seed yield of Brassica juncea. Plant Growth Regul 66:303–310

Barrena R, Casals E, Colon J, Font X, Sanchez A, Puntes V (2009) Evaluation of the ecotoxicity of model nanoparticles. Chemosphere 75:850–857

Barrett CS, Cohen JB, Faber J, Jenkins R, Leyden DE, Russ JC, Predecki PK (1986) Advances in X-ray analysis, vol 29. Plenum Press, New York

Bhuyan T, Mishra K, Khanuja M, Prasad R, Varma A (2015) Biosynthesis of zinc oxide nanoparticles from *Azadirachta indica* for antibacterial and photocatalytic applications. Mater Sci Semicond Process 32:55–61

Boro RC, Kaushal J, Nangia Y, Wangoo N, Bhasinc A, Suri CR (2011) Gold nanoparticles catalyzed chemiluminescence immunoassay for detection of herbicide 2,4-dichlorophenoxyacetic acid. Analyst 136:2125–2130

Brice-Profeta S, Arrio MA, Troncetal E (2005) Magneticorderin ūFE;-Fe2O3 nanoparticles: a XMCD study. J Magne Magne Mat 288:354–365

Cao YC (2002) Nanoparticles with Raman spectroscopic fingerprints for DNA and RNA detection. Science 80:1536–1540

Chauhan RPS, Gupta C, Prakash D (2012) Methodological advancements in green nanotechnology and their applications in biological synthesis of herbal nanoparticles. Int J Bioassays 7:6–10

Chen X, Schluesener HJ (2008) Nanosilver: a nanoproduct in medical application. Toxicol Lett 176:1–12

Chen Z, Meng HA, Xing GM, Chen C, Zhao Y, Jia G, Wang T, Yuan H, Ye C, Zhao F, Chai Z, Zhu C, Fang X, Ma B, Wan L (2006) Acute toxicological effects of copper nanoparticles in vivo. Toxicol Lett 163:109–120

Chen T, Wong YS, Zheng W, Bai Y, Huang L (2008) Selenium nanoparticles fabricated in *Undaria pinnatifida* polysaccharide solutions induce mitochondria-mediated apoptosis in A375 human melanoma cells. Coll Surf B 67:26–31

Corredor E, Testillano PS, Coronado MJ, González-Melendi P, Fernández-Pacheco R, Marquina C, Ibarra MR, de la Fuente JM, Rubiales D, Pérez-de-Luque A, Risueño MC (2009) Nanoparticle penetration and transport in living pumpkin plants: in situ subcellular identification. BMC Plant Biol 9:45

de Oliveira JL, Campos EVR, Bakshi M, Abhilash PC, Fraceto LF (2014) Application of nanotechnology for the encapsulation of botanical insecticides for sustainable agriculture: prospects and promises. Biotechnol Adv 32:1550–1561

De Jaeger N, Demeyere H, Finsy R, Sneyers R, Vanderdeelen J, van der Meeren P, van Laethem M (1991) Particle sizing by photon correlation spectroscopy. Part I. Monodisperse latices. Influence of scattering angle and concentration of dispersed material. Part Part Sys Charact 8:179–186

Doshi R, Braida W, Christodoulatos C, Wazne M, O'Connor G (2008) Nanoaluminum: transport through sand columns and environmental effects on plant and soil communication. Environ Res 106:296–303

Dubas ST, Pimpan V (2008) Humic acid assisted synthesis of silver nanoparticles and its application to herbicide detection. Mater Lett 62:2661–2663

Duman F, Ismail Ocsoy I, Kup FO (2016) Chamomile flower extract-directed CuO nanoparticle formation for its antioxidant and DNA cleavage properties. Mater Sci Engg C 60:333–338

Ekimov AI, Onuschchenko AA (1981) Quantum size effect in three-dimensional microscopic semiconductor crystals. JETP Lett 34:345–348

El-Rahman AFA, Mohammad TGM (2013) Green synthesis of silver nanoparticle using *Eucalyptus globulus* leaf extract and its antibacterial activity. J Appl Sci Res 9:6437–6440

Embiale A, Hussein M, Husen A, Sahile S, Mohammed K (2016) Differential sensitivity of *Pisum sativum* L. cultivars to water-deficit stress: changes in growth, water status, chlorophyll fluorescence and gas exchange attributes. J Agron 15:45–57

Eric D (1987) Engines of creation: the coming area of nanotechnology. Anchor Library of Science

Eustis S, El-Sayed MA (2006) Why gold nanoparticles are more precious than pretty gold: noble metal surface plasmon resonance and its enhancement of the radiative and nonradiative properties of nanocrystals of different shapes. Chem Soc Rev 35:209–217

FAO (2009) How to feed the world in 2050. FAO Report Oct. 2009, pp 1–35

Faraji M, Yamini Y, Rezaee M (2010) Magnetic nanoparticles: synthesis, stabilization, functionalization, characterization, and applications. J Iran Chem Soc 7:1–37

Feynman RP (1960) There's plenty of room at the bottom. Eng Sci 23:22–36

Freestone I, Meeks N, Sax M, Higgitt C (2007) The Lycurgus cup – a Roman nanotechnology. Gold Bull 40:270–277

Getnet Z, Husen A, Fetene M, Yemata G (2015) Growth, water status, physiological, biochemical and yield response of stay green sorghum {*Sorghum bicolor* (L.) Moench} varieties-a field trial under drought-prone area in Amhara regional state, Ethiopia. J Agron 14:188–202

Grillo R, Pereira AES, Nishisaka CS, Lima RD, Oehlke K, Greiner R, Leonardo F, Fraceto LF (2014) Chitosan/tripolyphosphate nanoparticles loaded with paraquat herbicide. An environmentally safer alternative for weed control. J Hazard Mater 278:163–171

Gujrati M, Malamas A, Shin T, Jin E, Sun Y, Lu ZR (2014) Multifunctional cationic lipid-based nanoparticles facilitate endosomal escape and reduction-triggered cytosolic siRNA release. Mol Pharm 11:2734–2744

Guo D, Xie G, Luo J (2014) Mechanical properties of nanoparticles: basics and applications. J Phys D Appl Phys 47:13001

Hafez EE, Hassan HS, Elkady M, Salama E (2014) Assessment of antibacterial activity for synthesized zinc oxide nanorods against plant pathogenic strains. Int J Sci Technol Res 3:318–324

Hassan MS, Amna T, Yang OB, El-Newehy MH, Al-Deyab SS, Khil MS (2012) Smart copper oxide nanocrystals: synthesis, characterization, electrochemical and potent antibacterial activity. Colloids Surf B Biointerfaces 97:201–206

Hirsch LR, Stafford RJ, Bankson JA, Sershen SR, Rivera B, Price RE, Hazle JD, Halas NJ, West JL (2003) Nanoshell-mediated near-infrared thermal therapy of tumours under magnetic resonance guidance. Proc Natl Acad Sci U S A 100:13549–13554

Holgate ST (2010) Exposure, uptake, distribution and toxicity of nanomaterials in humans. J Biomed Nanotechnol 6:1–19

Holzinger M, Le Goff A, Cosnier S (2014) Nanomaterials for biosensing applications: a review. Front Chem 2:63

Hossain Z, Mustafa G, Komatsu S (2015) Plant responses to nanoparticle stress. Int J Mol Sci 16:26644–26653

Hou J, Shang J, Jiao C, Jiang P, Xiao H, Luo L, Liu T (2013) A core crosslinked polymeric micellar platium(IV) prodrug with enhanced anticancer efficiency. Macromol Biosci 13:954–965

Husen A (2017) Gold nanoparticles from plant system: synthesis, characterization and their application. In: Ghorbanpour M, Manika K, Varma A (eds) Nanoscience and plant–soil systems, vol 48. Springer, Cham, pp 455–479

Husen A, Siddiqi KS (2014a) Carbon and fullerene nanomaterials in plant system. J Nanobiotechnol 12:16

Husen A, Siddiqi KS (2014b) Phytosynthesis of nanoparticles: concept, controversy and application. Nanoscale Res Lett 9:229

Husen A, Siddiqi KS (2014c) Plants and microbes assisted selenium nanoparticles: characterization and application. J Nanobiotechnol 12:28

Husen A, Iqbal M, Aref MI (2014) Growth, water status and leaf characteristics of *Brassica carinata* under drought stress and rehydration conditions. Braz J Bot 37:217–227

Husen A, Iqbal M, Aref IM (2016) IAA-induced alteration in growth and photosynthesis of pea (*Pisum sativum* L.) plants grown under salt stress. J Environ Biol 37:421–429

Husen A, Iqbal M, Aref IM (2017) Plant growth and foliar characteristics of faba bean (*Vicia faba* L.) as affected by indole-acetic acid under water-sufficient and water-deficient conditions. J Environ Biol 38:179–186

Husen A, Iqbal M, Sohrab SS, Ansari MKA (2018) Salicylic acid alleviates salinity-caused damage to foliar functions, plant growth and antioxidant system in Ethiopian mustard (*Brassica carinata* A. Br.). Agri Food Sec 7:44

Husen A, Iqbal M, Khanum N, Aref IM, Sohrab SS, Meshresa G (2019) Modulation of salt-stress tolerance of niger (*Guizotia abyssinica*), an oilseed plant, by application of salicylic acid. J Environ Biol 40:40:94–104

Hussein M, Embiale A, Husen A, Aref IM, Iqbal M (2017) Salinity-induced modulation of plant growth and photosynthetic parameters in faba bean (*Vicia faba*) cultivars. Pak J Bot 49:867–877

Iavicoli I, Leso V, Beezhold DH, Shvedova AA (2017) Nanotechnology in agriculture: opportunities, toxicological implications, and occupational risks. Toxicol Appl Pharmacol 329:96–111

Ibrahim KS (2013) Carbon nanotubes-properties and applications: a review. Carbon Lett 14:131–144

Iijima S (1991) Helical microtubules of graphitic carbon. Nature 354:56–58

Iravani S (2011) Green synthesis of metal nanoparticles using plants. Green Chem 13:2638–2650

Jasim B, Thomas R, Mathew J, Radhakrishnan EK (2016) Plant growth and diosgenin enhancement effect of silver nanoparticles in Fenugreek (*Trigonella foenum-graecum* L.). Saudi Pharm J 25:443–447

Jasim B, Thomas R, Mathew J, Radhakrishnan EK (2017) Plant growth and diosgenin enhancement effect of silver nanoparticles in Fenugreek (*Trigonella foenum-graecum* L.). Saudi Pharma J 25:443–447

Jiang J, Oberdörster G, Biswas P (2009) Characterization of size, surface charge, and agglomeration state of nanoparticle dispersions for toxicological studies. J Nanopart Res 11:77–89

Jung JH, Cheol OH, Soo NH, Ji JH, Soo KS (2006) Metal nanoparticle generation using a small ceramic heater with a local heating area. J Aerosol Sci 37:1662–1670

Kaida T, Kobayashi K, Adachi M, Suzuki F (2004) Optical characteristics of titanium oxide interference film and the film laminated with oxides and their applications for cosmetics. J Cosmet Sci 55:219–220

Kang T, Wang F, Lu L, Zhang Y, Liu T (2010) Methyl parathion sensors based on gold nanoparticles and Nafion film modified glassy carbon electrodes. Sensors Actuators B Chem 145:104–109

Kanhed P, Birla S, Gaikwad S, Gade A, Seabra AB, Rubilar O, Duran N, Rai M (2014) In vitro antifungal efficacy of copper nanoparticles against selected crop pathogenic fungi. Mater Lett 115:13–17

Karnan T, Selvakumar SAS (2016) Biosynthesis of ZnO nanoparticles using rambutan (*Nephelium lappaceum* L.) peel extract and their photocatalytic activity on methyl orange dye. J Mol Struct 1125:358–365

Khadri H, Alzohairy M, Janardhan A, Kumar AP, Narasimha G (2013) Green synthesis of silver nanoparticles with high fungicidal activity from olive seed extract. Adv Nanopart 2:241–246

Khodakovskaya M, Dervishi E, Mahmood M, Xu Y, Li Z, Watanabe F, Biris AS (2009) Carbon nanotubes are able to penetrate plant seed coat and dramatically affect seed germination and plant growth. ACS Nano 3:3221–3227

Khot LR, Sankaran S, Maja JM, Ehsani R, Schuster EW (2012) Applications of nanomaterials in agricultural production and crop protection: a review. Crop Prot 35:64–70

Kim DY, Saratale RG, Shinde S, Syed A, Ameen F, Ghodake G (2018) Green synthesis of silver nanoparticles using *Laminaria japonica* extract: characterization and seedling growth assessment. J Clean Prod 172:2910–2918

Klug HP, Alexander LE (1974) X-ray diffraction procedures for poly-crystallite and amorphous materials, 2nd edn. Wiley, New York

Kneipp K, Wang Y, Kneipp H, Perelman LT, Itzkan I, Dasari RR, Fela MS (1997) Single molecule detection using surface-enhanced Raman scattering (SERS). Phys Rev Lett 78:1667–1670

Kot M, Major Ł, Lackner JM, Chronowska-Przywara K, Janusz M, Rakowski W (2016) Mechanical and tribological properties of carbon-based graded coatings. J Nanomater 2016:1–14

Laware S, Raskar S (2014) Influence of zinc oxide nanoparticles on growth, flowering and seed productivity in onion. Int J Curr Microbiol App Sci 3:874–881

Lee CW, Mahendra S, Zodrow K, Li D, Tsai YC, Braam J, Alvarez PJ (2010) Developmental phytotoxicity of metal oxide nanoparticles to *Arabidopsis thaliana*. Environ Toxicol Chem 29:669–675

Leonhardt U (2007) Optical metamaterials: invisibility cup. Nat Photon 1:207–208

Lidén G (2011) The European Commission tries to define nanomaterials. Ann Occup Hyg 55:1–5

Lin D, Xing B (2007) Phytotoxicity of nanoparticles: inhibition of seed germination and root growth. Environ Pollut 150:243–250

Liu J, Qiao SZ, Hu QH (2011) Magnetic nanocomposites with mesoporous structures: synthesis and applications. Small 7:425–443

Luo M, Liu D, Zhao L, Han J, Liang Y, Wang P, Zhou Z (2014) A novel magnetic ionic liquid modified carbon nanotube for the simultaneous determination of aryloxyphenoxy-propionate herbicides and their metabolites in water. Anal Chim Acta 852:88–96

Mahmoud WMM, Rastogi T, Kümmerer K (2017) Application of titanium dioxide nanoparticles as a photocatalyst for the removal of micropollutants such as pharmaceuticals from water. Curr Opin Green Sust Chem 6:1–10

Mallakpour S, Sirous F (2015) Surface coating of α-Al_2O_3 nanoparticles with poly(vinyl alcohol) as biocompatible coupling agent for improving properties of bio-active poly(amide-imide) based nanocomposites having l-phenylalanine linkages. Prog Org Coat 85:138–145

McWilliams A (2018) Nanocomposites, nanoparticles, nanoclays and nanotubes: global markets to 2022. BCC Research Report Overview (June 2018), pp 1–7

Min Y, Li J, Liu F, Yeow EK, Xing B (2014) NIR light mediated photoactivation pt based antitumor prodrug and simultaneous cellular apoptosis imaging via upconversion nanoparticles. Angew Chem Int Ed Engl 53:1012–1016

Mishra S, Keswani C, Abhilash PC, Fraceto LF, Singh HB (2017) Integrated approach of Agri-nanotechnology: challenges and future trends. Front Plant Sci 8:471

Morsy MK, Khalaf HH, Sharoba AM, El-Tanahi HH, Cutter CN (2014) Incorporation of essential oils and nanoparticles in pullulan films to control foodborne pathogens on meat and poultry products. J Food Sci 79:M675–M682

Mucalo MR, Bullen CR, Manley-Harris M, McIntire TM (2002) Arabinogalactan from the Western larch tree: a new, purified and highly water-soluble polysaccharide-based protecting agent for maintaining precious metal nanoparticles in colloidal suspension. J Mat Sci 37:493–504

Murugan K, Labeeba MA, Panneerselvam C, Dinesh D, Suresh U, Subramaniam J, Madhiyazhagan P, Hwang J, Wang L, Nicoletti M, Benelli G (2015) *Aristolochia indica* green synthesized silver nanoparticles: a sustainable control tool against the malaria vector *Anopheles stephensi*. Res Vet Sci 102:127–135

Nagajyothi PC, Muthuraman P, Sreeknath TVM, Kim DH, Shim J (2017) Anticancer activity of copper oxide nanoparticles against human cervical carcinoma cells. Arab J Chem 10:215–225

Naseem T, Farrukh MA (2015) Antibacterial activity of green synthesis of iron nanoparticles using *Lawsonia inermis* and *Gardenia jasminoides* leaves extract. J Chem 2015:912342

Nasrollahzadeh M, Sajadi SM, Maham M (2015) Green synthesis of palladium nanoparticles using *Hippophae rhamnoides* Linn leaf extract and their catalytic activity for the Suzuki–Miyaura coupling in water. J Mol Catal A Chem 396:297–303

Nava OJ, Soto-Robles CA, Góomez-Gutiérreza CM, Vilchis-Nestor AR, Castro-Beltrán A, Olivas A, Luque PA (2017) Fruit peel extract mediated green synthesis of zinc oxide nanoparticles. J Mol Struct 1147:1–6

Nel AE, Mädler L, Velegol D, Xia T, Hoek EM, Somasundaran P, Klaessig F, Castranova V, Thompson M (2009) Understanding biophysicochemical interactions at the nano-bio interface. Nat Mater 8:543–557

Ngo QB, Dao TH, Nguyen HC, Tran XT, Nguten TV, Khuu TD, Huynh TH (2014) Effects of nanocrystalline powders (Fe, Co and Cu) on the germination, growth, crop yield and product quality of soybean (Vietnamese species DT-51). Adv Nat Sci Nanosci Nanotechnol 5:1–7

Nhan LV, Ma C, Rui Y, Liu S, Li X, Xing B, Liu L (2015) Phytotoxic mechanism of nanoparticles: destruction of chloroplasts and vascular bundles and alteration of nutrient absorption. Sci Rep 5:11618

Nhan LV, Ma C, Rui Y, Cao W, Deng Y, Liu L, Xing B (2016) The effects of Fe_2O_3 nanoparticles on physiology and insecticide activity in non-transgenic and Bt-transgenic cotton. Front Plant Sci 6:1263

O'Brien SC, Heath JR, Curl RF, Smalley RE (1988) Photophysics of buckminsterfullerene and other carbon cluster ions. J Chem Phys 88:220

Ocsoy I, Paret ML, Ocsoy MA, Kunwar S, Chen T, You M, Tan W (2013) Nanotechnology in plant disease management: DNA-directed silver nanoparticles on graphene oxide as an antibacterial against *Xanthomonas perforans*. ACS Nano 7:8972–8980

Ojha NK, Zyryanov GV, Majee A, Charushin VN, Chupakhin ON, Santra S (2017) Copper nanoparticles as inexpensive and efficient catalyst: a valuable contribution in organic synthesis. Coord Chem Rev 353:1–57

Ouda SM (2014) Antifungal activity of silver and copper nanoparticles on two plant pathogens, *Alternaria alternata* and *Botrytis cinerea*. Res J Microbiol 9:34–42

Pandey A, Kulkarni A, Roy B, Goldman A, Sarangi S, Sengupta P, Phipps C, Kopparam J, Oh M, Basu S, Kohandel M, Sengupta S (2014) Sequential application of a cytotoxic nanoparticle and a PI3 K inhibitor enhances antitumor efficacy. Cancer Res 74:675–685

Park HJ, Kim SH, Kim HJ, Choi SH (2006) A new composition of nanosized silica-silver for control of various plant diseases. Plant Pathol J 22:295–302

Pereira AES, Grillo R, Mello NFS, Rosa AH, Fraceto LF (2014) Application of poly(epsilon-caprolactone) nanoparticles containing atrazine herbicide as an alternative technique to control weeds and reduce damage to the environment. J Hazard Mater 268:207–215

Periasamy VS, Alshatwi AA (2013) Tea polyphenols modulate antioxidant redox system on cisplatin-induced reactive oxygen species generation in a human breast cancer cell. Basic Clin Pharmacol Toxicol 112:374–384

Poinern GEJ (2014) A laboratory course in nanoscience and nanotechnology, 1st edn. CRC Press Taylor & Francis, Boca Raton

Puri A, Loomis K, Smith B, Lee JH, Yavlovich A, Heldman E, Blumenthal R (2009) Lipid-based nanoparticles as pharmaceutical drug carriers: from concepts to clinic. Crit Rev Ther Drug Carrier Syst 26:523–580

Qi M, Zhang K, Li S, Wu J, Pham-Huy C, Diao X, Xiao D, He H (2016) Superparamagnetic Fe_3O_4 nanoparticles: synthesis by a solvothermal process and functionalization for a magnetic targeted curcumin delivery system. New J Chem 4480:4480–4491

Qian H, Peng X, Han X, Ren J, Sun L, Fu Z (2013) Comparison of the toxicity of silver nanoparticles and silver ions on the growth of terrestrial plant model *Arabidopsis thaliana*. J Environ Sci 25:1947–1955

Rao JP, Geckeler KE (2011) Polymer nanoparticles: preparation techniques and size-control parameters. Prog Polym Sci 36:887–913

Rastogi A, Zivcak M, Sytar O, Kalaji HM, He X, Mbarki S, Brestic M (2017) Impact of metal and metal oxide nanoparticles on plant: a critical review. Front Chem 5:78

Rawat MK, Jain A, Singh S, Mehnert W, Thunemann AF, Souto EB, Mehta A, Vyas SP (2011) Studies on binary lipid matrix based solid lipid nanoparticles of repaglinide: in vitro and in vivo evaluation. J Pharm Sci 100:2366–2378

Risom L, Møller P, Loft S (2005) Oxidative stress-induced DNA damage by particulate air pollution. Mutat Res 592:119–137

Roldán MV, Pellegri N, de Sanctis O (2013) Electrochemical method for Ag-PEG nanoparticles synthesis. J Nanopart 2013:524150

Rui M, Ma C, Hao Y, Guo J, Rui Y, Tang X, Zhao Q, Fan X, Zhang Z, Hou T, Zhu S (2016) Iron oxide nanoparticles as a potential iron fertilizer for peanut (*Arachis hypogaea*). Front Plant Sci 7:815

Rui M, Ma C, Tang X, Yang J, Jiang F, Pan Y, Xiang Z, Hao Y, Rui Y, Cao W, Xing B (2018) Phytotoxicity of silver nanoparticles to peanut (*Arachis hypogaea* L.): physiological responses and food safety. ACS Sustain Chem Eng 5:6557–6567

Sadiq IM, Pakrashi S, Chandrasekaran N, Mukherjee A (2011) Studies on toxicity of aluminum oxide (Al2O3) nanoparticles to microalgae species: *Scenedesmus* sp. and *Chlorella* sp. J Nanopart Res 13:3287–3299

Salata OV (2004) Applications of nanoparticles in biology and medicine. J Nanobiotechnol 2:3

Shaalan M, Saleh M, El-Mahdy M, El-Matbouli M (2016) Recent progress in applications of nanoparticles in fish medicine: a review. Nanomed Nanotechnol Biol Med 12:701–710

Shankramma K, Yallappa S, Shivanna MB, Manjanna J (2016) Fe_2O_3 magnetic nanoparticles to enhance *Solanum lycopersicum* (tomato) plant growth and their biomineralization. Appl Nanosci 6:983–990

Shao W, Nabb D, Renevier N, Sherrington I, Luo JK (2012) Mechanical and corrosion resistance properties of TiO_2 nanoparticles reinforced Ni coating by electrodeposition. IOP Conf Ser Mater Sci Eng 40:12043

Siddiqi KS, Husen A (2016a) Fabrication of metal nanoparticles from fungi and metal salts: scope and application. Nanoscale Res Lett 11:98

Siddiqi KS, Husen A (2016b) Fabrication of metal and metal oxide nanoparticles by algae and their toxic effects. Nanoscale Res Lett 11:363

Siddiqi KS, Husen A (2016c) Engineered gold nanoparticles and plant adaptation potential. Nanoscale Res Lett 11:400

Siddiqi KS, Husen A (2016d) Green synthesis, characterization and uses of palladium/platinum nanoparticles. Nanoscale Res Lett 11:482

Siddiqi KS, Husen A (2017a) Recent advances in plant-mediated engineered gold nanoparticles and their application in biological system. J Trace Elem Med Biol 40:10–23

Siddiqi KS, Husen A (2017b) Plant response to engineered metal oxide nanoparticles. Nanoscale Res Lett 12:92

Siddiqi KS, Rahman A, Tajuddin, Husen A (2016) Biogenic fabrication of iron/iron oxide nanoparticles and their application. Nanoscale Res Lett 11:498

Siddiqi KS, Husen A, Rao RAK (2018a) A review on biosynthesis of silver nanoparticles and their biocidal properties. J Nanobiotechnol 16:14
Siddiqi KS, Rahman A, Tajuddin, Husen A (2018b) Properties of zinc oxide nanoparticles and their activity against microbes. Nano Res Lett 13:141
Siddiqi KS, Husen A, Sohrab SS, Osman M (2018c) Recent status of nanomaterials fabrication and their potential applications in neurological disease management. Nano Res Lett 13:231
Sigmund W, Yuh J, Park H, Maneeratana V, Pyrgiotakis G, Daga A, Taylor J, Nino JC (2006) Processing and structure relationships in electrospinning of ceramic fiber systems. J Am Ceram Soc 89:395–407
Soares C, Branco-Neves S, de-Sousa A, Pereira R, Fidalgo F (2016) Ecotoxicological relevance of nano-NiO and acetaminophen to *Hordeum vulgare* L.: combining standardized procedures and physiological endpoints. Chemosphere 165:442–452
Sondi I, Salopek-Sondi B (2004) Silver nanoparticles as antimicrobial agent: a case study on *E. coli* as a model for Gram-negative bacteria. J Colloid Interface Sci 275:177–182
Song G, Hou W, Gao Y, Wang Y, Lin L, Zhang Z, Niu Q, Ma R, Mu L, Wang H (2016) Effects of CuO nanoparticles on *Lemna minor*. Bot Stud 57:3
Sotiriou GA, Pratsinis SE (2010) Antibacterial activity of nanosilver ions and particles. Environ Sci Technol 44:5649–5654
Sotiriou GA, Teleki A, Camenzind A, Krumeich F, Meyer A, Panke S, Pratsinis SE (2011) Nanosilver on nanostructured silica: antibacterial activity and Ag surface area. Chem Eng J 170:547–554
Soundarrajan C, Sankari A, Dhandapani P, Maruthamuthu S, Ravichandran S, Sozhan G, Palaniswamy N (2012) Rapid biological synthesis of platinum nanoparticles using *Ocimum sanctum* for water electrolysis applications. Bioprocess Biosyst Eng 35:827–833
Stoimenov PK, Klinger RL, Marchin RL, Klabunde KJ (2002) Metal oxide nanoparticles as bactericidal agents. Langmuir 18:6679–6686
Sukirtha R, Priyanka KM, Antony JJ, Kamalakkannan S, Thangam R, Gunasekaran P, Krishnan M, Achiraman S (2012) Cytotoxic effect of green synthesized silver nanoparticles using *Melia azedarach* against in vitro HeLa cell lines and lymphoma mice model. Process Biochem 47:273–279
Syu Y, Hung J, Chen JC, Chuang H (2014) Impacts of size and shape of silver nanoparticles on *Arabidopsis* plant growth and gene expression. Plant Physiol Biochem 83:57–64
Thatoi P, Kerry RG, Gouda S, Das G, Pramanik K, Thatoi H, Patra JK (2016) Photo-mediated green synthesis of silver and zinc oxide nanoparticles using aqueous extracts of two mangrove plant species, *Heritiera fomes* and *Sonneratia apetala* and investigation of their biomedical applications. J Photochem Photobiol B Biol 163:311–318
Thomas S, Harshita BSP, Mishra P, Talegaonkar S (2015) Ceramic nanoparticles: fabrication methods and applications in drug delivery. Curr Pharm Des 21:6165–6188
Tien DC, Tseng KH, Liao CY, Huang JC, Tsung TT (2008) Discovery of ionic silver in silver nanoparticle suspension fabricated by arc discharge method. J Alloys Compd 463:408–411
Tippayawat P, Phromviyo N, Boueroy P, Chompoosor A (2016) Green synthesis of silver nanoparticles in *Aloe vera* plant extract prepared by a hydrothermal method and their synergistic antibacterial activity. Peer J 4:e2589
Tran PA, Webster TJ (2011) Selenium nanoparticles inhibit *Staphylococcus aureus* growth. Int J Nanomedicine 6:1553–1558
Uddin I, Poddar P, Kumar U, Phogat N (2013) A novel microbial bio-milling technique for the size reduction of micron sized Gd_2O_3 particles into nanosized particles. J Green Sci Tech 1:48–53
Umar S, Anjum NA, Ahmad P, Iqbal M (2018) Drought-induced changes in growth, photosynthesis, and yield traits in mungbean: Role of potassium and sulfur nutrition. In: Ozturk M, Hakeem KR, Ashraf M (eds) Crop production technologies for sustainable use and conservation: physiological and molecular advances. Apple Academic Press, Waretown, NJ, pp 79–89
Venkatachalam P, Priyanka N, Manikandan K, Ganeshbabu I, Indiraarulselvi P, Geetha N, Muralikrishna K, Bhattacharya RC, Tiwari M, Sharma N, Sahi SV (2016) Enhanced plant

growth promoting role of phycomolecules coated zinc oxide nanoparticles with P supplementation in cotton (*Gossypium hirsutum* L.). Plant Physiol Biochem 110:118–127

Vesenka J, Manne S, Giberson R, Marsh T, Henderson E (1993) Colloidal gold particles as an incompressible atomic force microscope imaging standard for assessing the compressibility of biomolecules. Biophys J 65:992–997

Vimala K, Sundarraj S, Paulpandi M, Vengatesan S, Kannan S (2014) Green synthesized doxorubicin loaded zinc oxide nanoparticles regulates the Bax and Bcl-2 expression in breast and colon carcinoma. Process Biochem 49:160–172

Vinković T, Novák O, Strnad M, Goessler W, Jurašin DD, Paradiković N, Vrček IV (2017) Cytokinin response in pepper plants (*Capsicum annuum* L.) exposed to silver nanoparticles. Environ Res 156:10–18

Wang JJ, Sanderson BJ, Wang H (2007) Cyto-and genotoxicity of ultrafine TiO_2 particles in cultured human lymphoblastoid cells. Mutat Res 628:99–106

Wang Z, Xie X, Zhao J, Liu X, Feng W, White JC, Xing B (2012) Xylem- and phloem-based transport of CuO nanoparticles in maize (*Zea mays* L.). Environ Sci Technol 46:4434–4441

Wang J, Wang X, Song Y, Zhu C, Wang K, Guo Z (2013) Detecting and delivering platinum anticancer drugs using fluorescent maghemite nanoparticles. Chem Commun (Camb) 49:2786–2788

Weir A, Westerhoff P, Fabricius L, Hristovski K, von Goetz N (2012) Titanium dioxide nanoparticles in food and personal care products. Environ Sci Technol 46:2242–2250

Wolf R, Matz H, Orion E, Lipozencic J (2003) Sunscreens–the ultimate cosmetic. Acta Dermatovenerol Croat 11:158–162

Wu W, Wu Z, Yu T, Jiang C, Kim WS (2015) Recent progress on magnetic iron oxide nanoparticles: synthesis, surface functional strategies and biomedical applications. Sci Technol Adv Mater 16:023501

Yousuf PY, Ahmad A, Hemant, Ganie AH, Aref IM, Iqbal M (2015) Potassium and calcium application ameliorates growth and oxidative homeostasis in salt-stressed Indian mustard (*Brassica juncea*) plants. Pak J Bot 47(5):1629–1639

Yu Z, Sun X, Song H, Wang W, Ye Z, Shi L, Ding K (2015) Glutathione-responsive carboxymethyl chitosan nanoparticles for controlled release of herbicides. Mater Sci Appl 6:591–604

Yu J, Xu D, Guan HN, Wang C, Huang LK, Chi DF (2016) Facile one-step green synthesis of gold nanoparticles using *Citrus maxima* aqueous extracts and its catalytic activity. Mater Lett 166:110–112

Zhang H, Chen G (2009) Potent antibacterial activities of Ag/TiO_2 nanocomposite powders synthesized by a one-pot sol-gel method. Environ Sci Technol 43:2905–2910

Zhang Q, Li N, Goebl J, Lu Z, Yin Y (2011) A systematic study of the synthesis of silver nanoplates: is citrate a "magic" reagent? J Am Chem Soc 133:18931–18939

Zhao G, Song S, Wang C, Wu Q, Wang Z (2011) Determination of triazine herbicides in environmental water samples by high-performance liquid chromatography using graphene-coated magnetic nanoparticles as adsorbent. Anal Chim Acta 708:155–159

Zheng Y, Sache L (2009) Gold nanoparticles enhance DNA damage induced by anti-cancer drugs and radiation. Radiat Res 172:114–119

Zur Mühlen A, Zur Mühlen E, Niehus H, Mehnert W (1996) Atomic force microscopy studies of solid lipid nanoparticles. Pharm Res 13:1411–1416

Chapter 2
Basic Chemistry and Biomedical Significance of Nanomaterials

Mahmoud Nasrollahzadeh, S. Mohammad Sajadi, and Muhammad Iqbal

2.1 Introduction

Nanotechnology that has emerged in the recent past has progressed dramatically during the last two decades and is expected to have an indelible impact on different agricultural and industrial products of the present-day world and has to play a special role in the fields of medicine and biology. With their unique chemical, physical, and mechanical properties, nanomaterials (NMs) have a wide range of their industrial and biologicalapplications. They work as highly organized, self-repairing, self-replicating, and information-rich molecular tools and can enable easy and efficient transfer of biochemical materials at the cellular and subcellular levels. Given this, NMs can be used for developing a better understanding about the molecular motors, enzyme activities, protein dynamics, DNA transcription, cell signaling, and molecular, genomic, proteomic, and metabolic reactions.

Nanomaterials are the entities of extremely small size, often measuring 1–100 nm, occasionally up to 1000 nm (Table 2.1). Research in this area is progressing leaps and bounds; development of quantum-confined nanocrystalline and doped nanocrystalline materials and evolution of smart drug delivery system through nanostructures are the major cynosures of experts today (Albrecht et al. 2006; Bennet and Kim 2014; Rajput 2015).

Nanostructures have unique physicochemical properties and an array of potential applications due to their minute size and large surface area (Fig. 2.1). Based on the

M. Nasrollahzadeh (✉)
Department of Chemistry, Faculty of Science, University of Qom, Qom, Iran

S. M. Sajadi
Scientific Research Center, Soran University, Kurdistan Regional Government, Soran, Iraq

M. Iqbal
Department of Botany, Faculty of Science, Jamia Hamdard (Deemed University), New Delhi, India

A. Husen, M. Iqbal (eds.), *Nanomaterials and Plant Potential*,
https://doi.org/10.1007/978-3-030-05569-1_2

Table 2.1 The nanostructures and their sizes

Type of nanostructure	Size (diameter)	Materials
Nanocrystals and clusters	1–10 nm	Metals, semiconductors, magnetic materials
Other nanoparticles	1–100 nm	Ceramic oxides
Nanowires	1–100 nm	Metals, semiconductors, oxides, sulfides, nitrides
Nanotubes	1–100 nm	Carbon, layered metal chalcogenides
Nanoporous solids	0.5–10 nm	Zeolites, phosphates, etc.
2-D arrays (of nanoparticles)	Several nm^2–μm^2	Metals, semiconductors, magnetic materials
Surfaces and thin films	1–1000 nm	A variety of materials
3-D structures (superlattices)	Several nm in 3-D	Metals, semiconductors, magnetic materials

specific nanomaterial characteristics, improved efficiencies or new functionalities can be achieved for a wide range of nanoproducts (Husen and Siddiqi 2014; Siddiqi and Husen 2016a, b, 2017a, b; Siddiqi et al. 2016, 2018a, b, c; Husen 2017). However, these efforts can cause increased loads and the consequent harms to the environment and living beings (humans in particular) when nanomaterials are released from the NM-based products and applications.

In fact, nanosized materials are common natural constituents of proteins, enzymes, nucleic acids, viruses, magnetite, ferritin, atmospheric tiny particles and fires, and volcanic eruptions. The nanoparticle properties are distinct from those of the analogous bulk materials mainly in their chemical reactivity, molecular and electronic structure, and mechanical behavior (Ehrman et al. 1999; Fendler 2001). The nanosized structures, typically those less than 10 nm in size, are markedly different from the bulk materials, depending on their surface area, bond, shape, and energy in nanometer dimensions, which cause a profound effect on the structure, phase transformations, strain, and reactivity of materials. Certain phases may exist

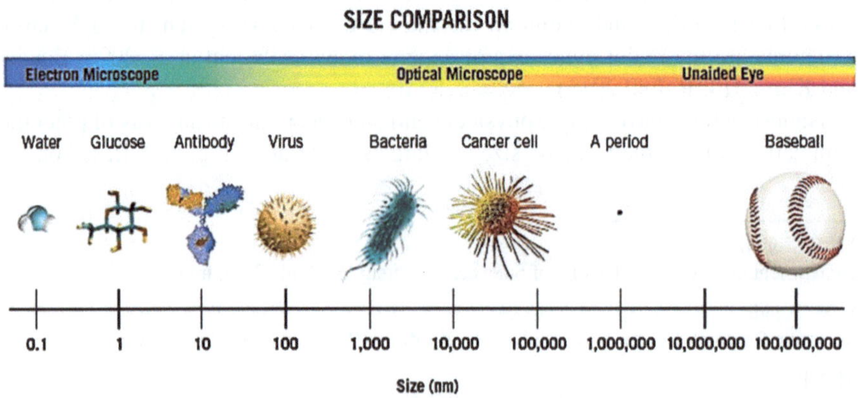

Fig. 2.1 Schematic wide spectrum of small structures of nano-dimensions in comparison to objects visible with the naked eye. (Mitchnick et al. 1991)

only as nanoparticles and require transformations in chemistry, stoichiometry, and structure with their evolution to larger sizes. It is because in the nanometer dimension, materials exhibit novel properties, different from isolated atoms and bulk materials, and thus the properties of materials depend largely on the size of particles they are composed of (Adachi 2000). This chapter aims to elucidate some important chemical/biochemical characteristics of nanomaterials with emphasis on their use in bio-systems and, hence, their influence on the quality of life.

2.2 Importance of Nanoscale

The wave properties of electrons are affected by changes in the nanoscale. By patterning the material over the nanometer, it is possible to change the basic characteristics of materials such as melting point, magnetic effect, and electrical charge without changing their chemical composition. Furthermore, nano promises to let us introduce new artificial components and assemblies inside the cell to make new systems. The large surface areas of nanostructures make them ideal for use in composite materials, reacting systems, drug delivery, and energy storage. The finite size of material entities, as compared to the molecular scale, determines an increase of the relative importance of surface tension and local electromagnetic effects, making the nanostructured materials harder and less brittle (Jain et al. 2009; Jiang et al. 2007). In short, nanostructures now have potential application almost in every sphere of science and technology such as agriculture (nanofertilizers, nanopesticides), chemicals and cosmetics (paints, coatings), electronics (nanoparticles, carbon nanotubes, biopolymers), energy and environment (water- and air-purifying filters, fuel cells, solar cells, photovoltaics), food science (nanocapsules, nutraceuticals, food processing), materials (semiconductor chips, memory storage, optoelectronics, photonics), microscopy (atomic force microscope, scanning tunneling microscope), military equipment (biosensors, nanoweapons, sensory enhancement technology), nanomedicine (nano drugs, medical devices, tissue engineering), etc.

2.3 Nanochemistry

Nanochemistry is a branch of solid-state chemistry, which synthesizes nanoscale materials in one, two, or three dimensions. Synthesis and organization of a nanosized structure under controlled situations should provide a reproducible method of developing materials that are perfect in size and shape down to atoms. Structures and properties of nanosized systems are designed to obtain materials with new chemical, physical, and pharmaceutical behavior. Chemical properties of nanomaterials (NMs) considerably change at the nanoscale. As the percentage of surface atoms in nanoparticles (NPs) is large, compared to bulk objects, the reactivity of NPs is greater than that of the bulk materials. The nanoscale materials owe for their main chemical properties to the increased population of surface atoms in nanoscale dimensions, the higher average energy of atoms than in the bulk material, the structure and nature of chemical bonding at the surface, and the mutual interaction among NPs. Although the composition, structure, and molecular weight may be important for some NMs, yet the properties like particle shape, size, and distribution, electronic surface characteristics, state of dispersion/agglomeration, and conductivity have a pivotal role in the majority of these materials. The quantum confinement, phase transition, and surface plasmon resonance are some other important characteristics of nanosized structures (Brune et al. 1998; Jolivet et al. 2004).

2.3.1 Quantum Confinement

Quantum confinement refers to the change of electronic and optical properties when the material attains a small size of 10 nm or less. It is indicative of a restriction on the motion of randomly moving electrons present in a material to specific discrete energy levels rather than the quasi-continuum of energy bands. When dimensions of a material are small enough to be comparable to the de Broglie wavelength of the electrons involved, the electrons present in the material behave like those in atoms. Plainly speaking, electrons occupy discrete energy levels rather than a quasi-continuum of energy in a band.

Quantum confinement effect is one of the most popular terms in the nano world. As explained above, this effect sets in due to change in the atomic structure brought about by the impact of ultrasmall length scale on the energy band structure. As a result of "geometrical" constraints, electrons "feel" the presence of the particle boundaries and respond to changes in the particle size by adjusting their energy, and this is what we call the quantum confinement effect. When the particle dimension of a semiconductor nears the Bohr exciton radius, properties of the material become size-dependent. Quantum confinement results in a collapse of the continuous energy bands of a bulk material into discrete energy levels as in atoms. This creates a discrete absorption spectrum, in contrast to the continuous absorption spectrum of a bulk semiconductor. Thus, in a quantum-confined structure, the motion of the charge carriers (electrons and holes) is confined by potential barriers. Based on the nature of confinement, a quantum-confined structure is classified as a quantum dot (or nanocrystal), quantum wire, or quantum well. In quantum dots, the charge carriers are confined in all three dimensions, and the electrons exhibit a discrete energy spectrum as in atoms. Quantum wires are formed when two dimensions of the system are confined. In quantum well, only one dimension is confined, and the charge carriers are free to move in two dimensions (Arivazhagan 2013; Parker 2017).

The confinement effect usually varies with different classes of material, each of which normally has a characteristic length scale, e.g., (a) exciton Bohr radius (i.e., electron-hole pair radius) for semiconductors, (b) typical size of a domain for ferromagnetic materials, and (c) coherence length of Cooper pairs for superconductors. So, when the size of a given material is comparable to these characteristic length scales, the electrons present in the material are said to be confined to discrete energy levels. The spacing between the energy levels increases with a decrease in the size of the material. Moreover, nanocrystals have a large surface area and a large population of surface atoms depending on the size of the particle (Chang and Waclawik 2014).

Thus, significant changes in the electrical and optical properties of materials are observed for the descending size. In small nanocrystals, the electronic energy levels are discrete – and not continuous as in the bulk material – due to confinement of electronic wave function to the physical dimensions of the particles (Fig. 2.2).

2.3.2 *Surface Plasmon Resonance (SPR)*

Surface plasmon resonance (SPR) is an optical effect that can be utilized to measure the binding of molecules in real time without using labels. It is used to measure the binding kinetics and affinity of molecular interactions; for instance, it can measure the binding between two proteins, a protein and an antibody, a protein and DNA, and so on. Unlike the traditional techniques such as ELISA, SPR allows determination of binding kinetics and not just binding affinity, because it provides real-time binding data of both the association and dissociation phases of the interaction and hence offers deep insight into the binding strength and stability of the interaction (Schasfoort 2017). This physical process can occur when plane-polarized light hits

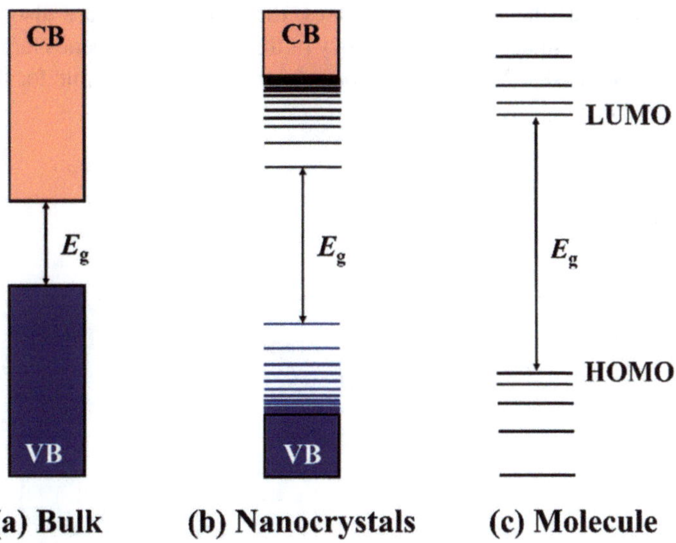

Fig. 2.2 A comparison of the electronic energy states of different types of semiconductor materials: (**a**) bulk inorganic semiconductors, (**b**) inorganic semiconductor nanocrystals, and (**c**) molecular semiconductors (Chang and Waclawik 2014). CB conduction band, Eg band gap energy, HOMO highest occupied molecular orbital, LUMO lowest unoccupied molecular orbital, VB valence band

a thin metal film under total internal reflection (TIR) conditions. In fact, the photon and electron behavior can be described only when they have both wave and particle properties. As per the quantum theory, a plasmon is the particle name of the electron density wave. When the photons convert to plasmons, a "gap" occurs in the reflected light intensity.

SPR is size-dependent; the density of states and the spatial length scale of the electronic motion are reduced, when dimensions of the matter decrease. In fact, when a nanoparticle is smaller than the wavelength of light, coherent oscillation of the conduction band electrons is induced by interaction with an electromagnetic field which causes surface plasmon resonance to occur (Levchenko et al. 2006; Louchet et al. 2006; Zielińska-Jurek 2014). When noble metal nanoparticles (MNPs) are influenced by electromagnetic radiations, their conduction electrons show collective oscillations, i.e., SPR (Fig. 2.3). Beside the size of NPs, the other parameters affecting the SPR are shape and dielectric properties of NPs, which cause selective photon absorption, scattering, and local electromagnetic field enhancement in the SPR phenomenon (Johnson and Johal 2018).

SPR signals of nano-dimensions are desirable in several technological applications such as coupling in linear chains of metallic NPs, light transportation, and the direction of the chain. For example, when Au NPs absorb light, the oscillating electromagnetic field of the light triggers polarization of the conduction band electrons on the surface of NPs, and thus the polarized electrons pass through the collective coherent oscillations with respect to the positive ions in the metallic lattice; these oscillations are called surface plasmon oscillations (Fig. 2.4). Because of having the same frequency as the incident light does, these oscillations are also known as surface plasmon resonance. The frequency of this parameter depends largely on the size and shape of NPs. A single plasmonic frequency is responsible for the intense

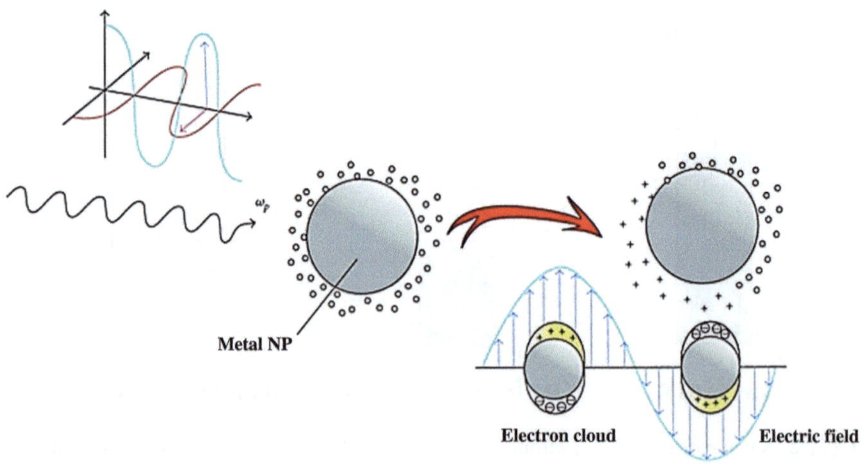

Fig. 2.3 Localized surface plasmon resonance of noble metal (Ag, Au) nanoparticles, a collective electron density oscillation caused by the electric field component of incoming light. (Zielińska-Jurek 2014)

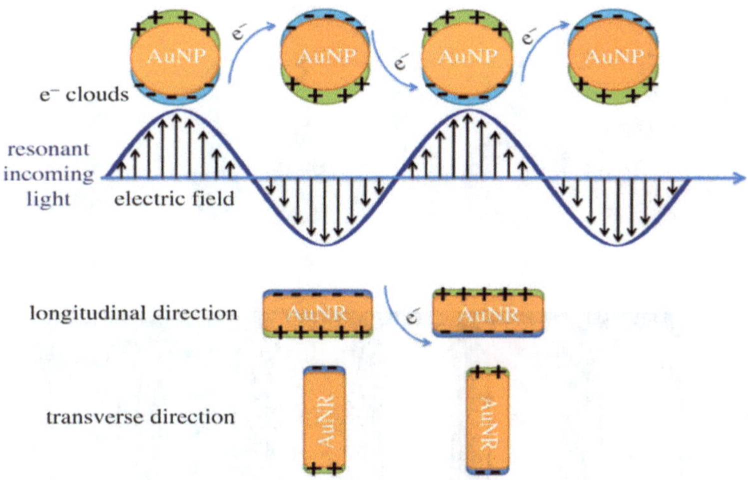

Fig. 2.4 Surface plasmon oscillations in spherical gold nanoparticles (Au NPs) and gold nanorods (Au NRs). (Reproduced with permission from Yasun et al. (2013))

red color of the spherical gold NPs. For gold nanorods (Au NRs), however, there are two plasmonic frequencies, known as the longitudinal and transverse bands. The transverse band depends on the electron oscillations that occur along the transverse direction as a weak absorption band in the visible region similar to the SPR frequency of the spherical Au NPs. The longitudinal band, related to the electron oscillations along the longitudinal direction, is a strong absorption band in the vis-NIR region (Yasun et al. 2013).

The dataobtained from SPR have proved to be of great help in (a) screening and developing new pharmaceuticals and new bio-therapeutics, (b) controlling quality in bioprocess monitoring, (c) developing new diagnostic assays, and (d) discovering and characterizing protein function and disease mechanism, etc.

After the light absorption, the plasmonic electrons cause plasmonic scattering or convert the absorbed energy to thermal energy via its transfer to the metal lattice. In this case, the high temperature of metal network is decreased by photon interaction and transferred to the surrounding environment (Park et al. 2004). This forms the basis of all plasmonic NP-based photothermal therapy applications. Because the plasmon resonance band of NPs contains both scattering and absorption components, tuning of the shape and size of the NPs can dramatically change their scattering and absorptionproperties (Fig. 2.5).

2.3.3 Nanoparticle Size Effects

Nanoparticles benefit from their small size and dimensions. In fact, if the surface energies of polymorphs differ significantly, at small sizes, the order of phase stability can be changed (Sanders 2018). Also, decreasing the size of particles changes

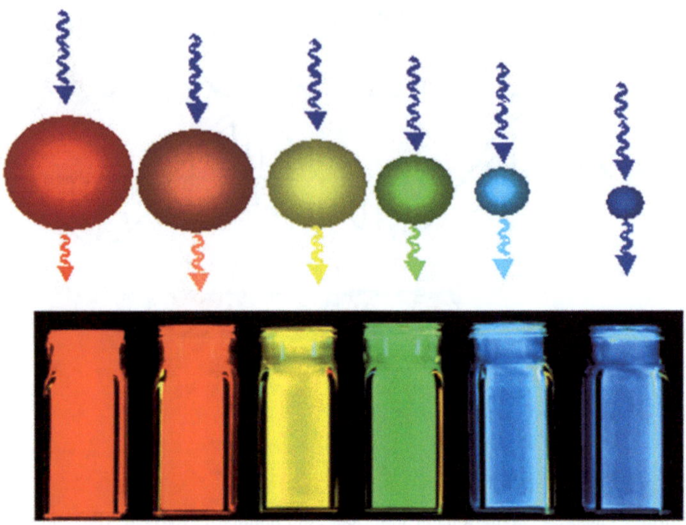

Fig. 2.5 Effect of the size of nanocrystals on the emitted color of the absorbed light. (Lundquist et al. 2017)

Table 2.2 Correlation among the size, abundance, and surface area of particles

Particle diameter (nm)	Abundance (N/cm^3)	Surface Area (μm^2 cm^{-3})
5	153,000,000	12,000
20	2,400,000	3016
250	1200	240
5000	0.15	12

the crystalline habit such as morphology, crystallinity, and miller indices. Furthermore, evolution of structural, thermodynamic, electronic, spectroscopic, electromagnetic, and chemical features of these finite systems is related to change in particle size. Therefore, properties of a material depend on its electron movements (Lv et al. 2009; Kang et al. 2012; Sanders 2018). If the physical size of the material is reduced to the nanoscale, its properties change dramatically and become sensitive to size and shape. Size effects thus constitute a peculiar and fascinating aspect of nanomaterials (Table 2.2).

2.3.4 *Size Distribution of Nanostructures*

The particles in nanoscale have a high proportion of atoms near their surfaces, a feature responsible for several important deviations from the bulk structure and chemistry at different size scales (Fig. 2.6). Other aspects influencing variations between the bulk material and nanomaterial properties include restriction on wave function radius, separation of defects and interacting strain fields, relative dominance of bulk or surface energy, and changes in vibrational properties (Johnson and Johal 2018). Assessment of nanoscale characteristics requires information on solid-state properties and their dimensions as well as on nucleation and initial growth of precipitates. Thus, according to the classical nucleation theory (CNT), all substances pass through a nanosized regime, either on or just after nucleation. As the nucleation is generally a crucial rate-controlling step in precipitation, understanding of NP characteristics is a prelude to evaluating the kinetics of precipitation and other phase transformations (Li et al. 2010; Nasir 2010).

Among the various methods for calculating the NP size distribution, small-angle X-ray scattering (SAXS) technique, which takes into consideration the mean particle radius, the width of the size distribution, and the particle concentration, is held to be sufficiently accurate and reliable. This uncomplicated bulk nanostructural quantification technique is particularly sensitive to the smaller end of the nanoscale. Results from SAXS have repeatedly been demonstrated to agree well with findings from electron microscopy and have also shown inter-instrument reproducibility. It is a suitable laboratory-independent reference method for in situ nanoparticle analysis, at least for monomodally distributed particles in suspension (Pauw et al. 2017).

2.3.5 *Shape of Nanoparticles*

Movement of electrons often determines the properties of materials. When the electrons move on a nanoscale dimension, unusual effects appear. For example, gold NPs in size less than 100 nm have red color in transparent media, but, in size less than 3 nm, they act as a catalyst for chemical reactions (Fig. 2.5). Furthermore, the optical properties of nanostructures change on changing their size, shape, aggregation state, and local environment. The nanostructures have various shapes such as particles, rods, and prisms (Fig. 2.7), which determine their functions and properties. There are many synthetic methods, such as top-down, bottom-up, bulk material applications; physical, chemical, and mechanical processes; application of high

Fig. 2.6 Percentage of surface atoms in a particle is strongly dependent on the particle size. (Nasir 2010)

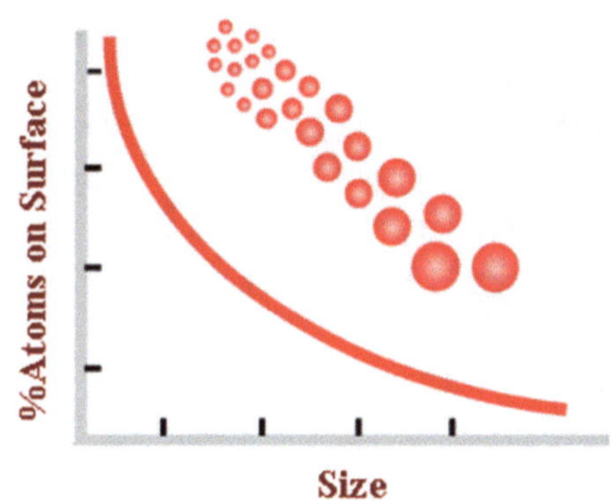

temperature; and assembly from building blocks and solution-based methods, to produce the shape-controlled NPs. These processes of synthesizing nanostructures are classified into metallic (monometallic, bimetallic, and magnetite or metal oxides) and organic (mainly lipids or polymers) types (Sajanlal and Pradeep 2009; Ragaei and Sabry 2014).

Despite a particular seed shape of nanomaterials, their shape and size essentially relate to such factors during synthesis as the type and concentration of reducing agents, stabilizing agents and temperature of the reaction solution. Among the NP-shape-controlling methods, reduction of metal cation using the reducing agents (such as sodium borohydride), which also act as the stabilizing agents and affect the growth of the particles, is very important (Maham et al. 2017; Maryami et al. 2017; Momeni et al. 2017; Nasrollahzadeh et al. 2016a, b, 2017, 2018a, b, c, d, 2019; Sajjadi et al. 2017). Any change in the stabilizing agent and the molar ratio of stabilizer to metal source and also the variation of temperature alter the shape (of seeds) and size of NPs by influencing their growth in a particular direction (Murphy et al. 2005). For instance, to achieve the spherical NPs, the whole surface of NPs should be covered by the stabilizing agent during anisotropic growth process (Fig. 2.8). Likewise, increasing the reaction temperature tends to increase the average adsorp-

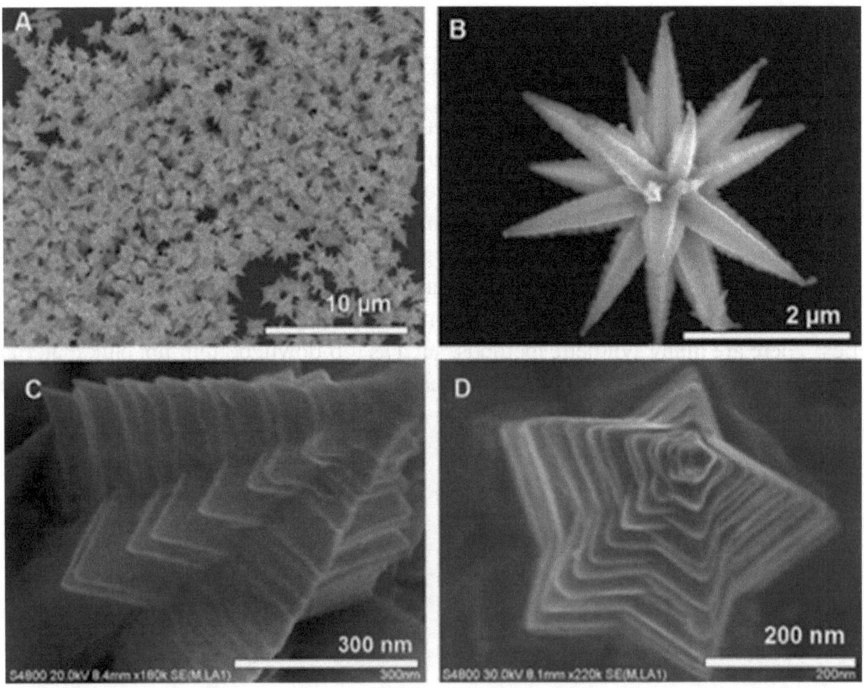

Fig. 2.7 Large area (**a**) and corresponding single particles. (**b**) Field-emission scanning electron microscopy (FESEM) images of gold mesoflowers (Au MFs). (**c**) An enlarged FESEM image of a single stem of the MF showing ridges along the edges. (**d**) Top view of a single stem of the MF showing the pentagonal structure. (Sajanlal and Pradeep 2009)

tion of capping molecules on NPs. In fact, the particle morphology depends largely on supersaturation of the solution, which is regulated by its temperature. Reduction in the surface is responsible for the grain boundary enlargement, and therefore the particle size increases as a function of the temperature. With increase in the NP size, its sensing response decreases. It is also found that operable temperature of sensors fabricated with small particles is lower than those with large particles (Cao and Wang 2011).

Lee et al. (2014) obtained shape transformation of Ag NPs (from spherical to others) under irradiation of green light-emitting diodes (LEDs) at different temperatures. The spherical NPs got converted to nanoplates at 60 °C and decahedral NPs at 0 °C. Moreover, the tips and edges of decahedral NPs gradually became blunt at ambient temperature, while the nanoplates could retain their morphology for long. Both the nanoplate colloids and the decahedral NP colloid synthesized at 60 °C and 0 °C, respectively, exhibited good surface-enhanced Raman spectroscopy (SERS) activities for the probe molecule R6G in the absence of polyvinylpyrrolidone (PVP).

In most of the cases, NP size grows through the mechanism of "Ostwald ripening" (for more soluble materials) or "oriented attachment" (for less soluble crystals). The former facilitates the growth of large particles due to dissolution of smaller ones, whereas the latter involves merging of smaller particles. Both processes are influenced by the temperature (at least through the diffusion via kT). Ostwald ripening occurs because the larger particles are energetically more stable than the smaller ones. Temperature affects the process by influencing the interfacial energy, growth rate coefficients, and solubility. On the other hand, oriented attachment takes place because aggregation reduces the interphase boundary and the total (surface) energy of the system. On the whole, heating or cooling of the reaction system heavily affects the reaction capability of components in reduction, surfactant adsorption/desorption, formation and growth rate, and hence the shape, size, and size distributions.

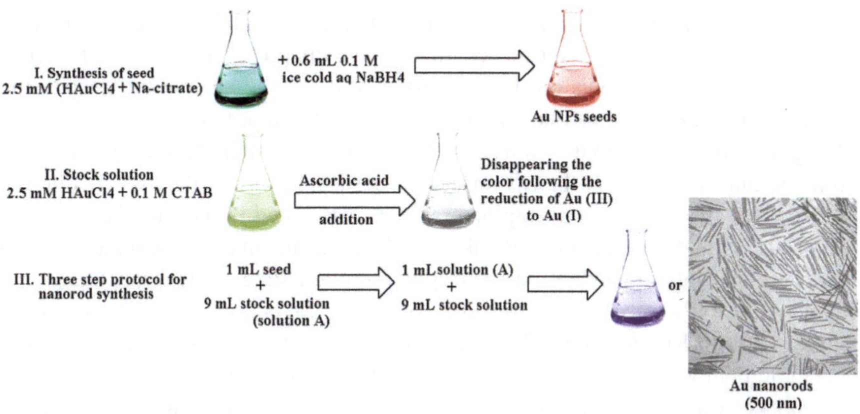

Fig. 2.8 Seed-mediated growth approach to making gold nanorods of controlled aspect ratio. Bottom right is the transmission electron micrograph of gold nanorods of 500 nm long in average. (Murphy et al. 2005)

2.3.6 Agglomeration of Nanoparticles

Agglomeration refers to a loose particle assemblage in a suspension that can be broken simply by mechanical force. It is different from aggregation which is a definite pattern of molecule clubbing. Agglomeration represents a mechanism that leads to destabilization of colloidal systems. During this process, particles dispersed in the liquid phase stick to each other and spontaneously form irregular clusters. The high surface area of nanoparticles and the strong attraction among them result in aggregation/agglomeration.

The agglomeration of NPs reduces the potential enhancement of mechanical properties in nanocomposites due to the restriction of interfacial area. Therefore, the main challenge in production of nanocomposites includes not only the achievement of small NPs but also their good dispersion (Ashraf et al. 2018). In some applications, agglomeration is a beneficial process causing enlargement of particles to improve powder properties and is widely used in chemical processes to overcome problems such as segregation, difficult flow, low bulk density, and particle-size distribution monitoring. The way in which NPs may agglomerate stems from the type of forces between NPs in the solution. Under certain conditions NPs can assemble in a crystallographically oriented mode, leading to a type of crystal growth with different kinetics than one characterizing the classical layer-by-layer growth, and consequently differ in growth forms and habits (Joshi et al. 2012; Kuntworbe et al. 2012).

Upon exposure to biological systems, NPs may interact with the outer surface of the cellular membrane and subsequently enter the cells by different endocytic routes. Targeting of specific cellular structures, the release of NPs by the cells, and, on the contrary, their degradation in lysosomes are vital features that can markedly influence the NP toxicity/safety and also the efficacy of novel nanomedicines. Various endocytic pathways may be involved in the NP uptake depending on the features of the cell lines and the NPs used (Halamoda-Kenzaoui et al. 2017).

It is well explored that the mechanism of the cellular uptake and endocytosis is affected by the size, shape, surface chemistry, and charge of NPs, but the effect of agglomeration state of NPs on this process is still poorly understood, despite the fact that agglomeration is one of the predominant features of NP suspensions. Changes to the pH and ionic strength or the presence of biomolecules, particularly proteins, can easily modify the NP surface properties, leading to the loss of colloidal stability and the formation of agglomerates. It was noted by Halamoda-Kenzaoui et al. (2017) that the level of cell uptake and the mechanism of endocytosis of silica NPs were strongly dependent on their agglomeration state. Well-dispersed 80 nm Rubipy-SiO_2 NPs were internalized mainly by the caveola-mediated endocytosis, whereas 30 nm Rubipy-SiO_2 NPs entered the cells via a combination of different endocytic pathways. Interestingly, with the increase of NP agglomeration, the cellular uptake was highly enhanced, and the mechanism of endocytosis was slightly modified with a predominant role of macropinocytosis. This indicates that a modified environment can easily induce NP agglomeration and consequently influence a biological response.

Agglomeration, a process involving mass conservation and a reduction in surface area and number of particles, shifts the particle distribution toward larger sizes, covering the aerosols and colloids that tend to settle more rapidly under gravity but diffuse more slowly. In other words, agglomeration principally occurs because of the high surface energy of NMs, due to which they tend to agglomerate to diminish this energy. Agglomeration of NPs is influenced by environmental factors. For instance, adding bad solvent into the NP solution leads to agglomeration of NPs due to minimization of interface. Likewise, when the temperature is lowered, the enthalpy change is negative; to compensate this change, the system tries to maximize the entropy by separating the media (gas/liquid molecules) from huge NPs (Peddieson and Chamkha 2016). Several factors such as pH, temperature, ionic strength, and mixing rate may affect agglomeration, or the breakup of agglomerates (Fig. 2.9).

2.3.7 Effect of pH, Ionic Strength, and Temperature on Agglomeration

In the absence of agglomeration, a colloidal dispersion is stable as the potential barrier in this state is sufficiently high to prevent particles from joining one another. The stability of dispersions relates to the surface electrostatic potential (which depends on the pH of the solution) and the ion concentration of the solution. The net interaction potential between particles can be used to predict the pH and salt concentration regimes expected to promote agglomeration. For instance, if the repulsive barrier to agglomeration is less than or equal to the thermal energy in the system (K_BT), the stability map demonstrates that agglomeration appears near the

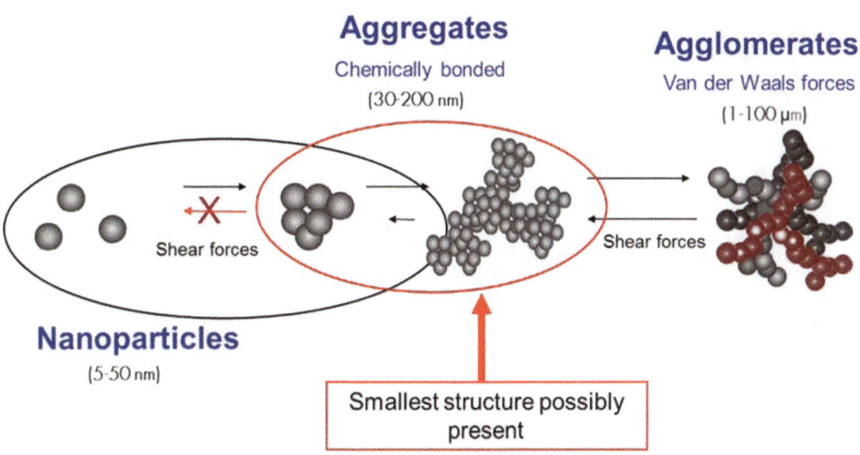

Fig. 2.9 Agglomeration and aggregation of nanoparticles. (Peddieson and Chamkha 2016)

isoelectric point around the neutral pH (Tan et al. 2001). As the salt concentration increases, instability regime widens. The schematic form of a stable and non-stable colloidal system is shown in Fig. 2.10.

Study of size distribution of NPs using spectrophotometric techniques is a valid way to investigate the surface charge in agglomeration process. At a low pH with all particles positively charged, the particles are dispersed, and the agglomerate size is almost identical to the primary particle size. At the isoelectric point, the particles have little surface charge, and the primary particles stick to each other to form large agglomerates. At an alkaline pH, with all the particles negatively charged, one might expect to see primary particles. However, the particles are somewhat agglomerated due to high salt content of the solution.

Higher temperature of the system during the process of agglomeration affects the process gradually and favors the formation of more regular agglomerates with mechanically stronger and denser clusters. Maghsoodi and Yari (2014) obtained spherical, dense, and strong agglomerates with optimized temperature. Esmailpour et al. (2015) observed that the size of agglomerates decreased by increasing the gas velocity at a constant temperature. Moreover, the minimum fluidization velocity and the agglomerate size increased with rise in temperature due to increasing van der Waals cohesive force. Fluidization, a process similar to liquefaction, converts the granular material consisting of micron-sized particles from a static solid-like state to a dynamic fluid-like state, when a fluid (liquid or gas) is passed over the granular material. NPs are not fluidized individually but as agglomerates (very dilute clusters of around 200 μm consisting of ~10^{10} primary particles). The NP fluidization is identified as agglomerate particulate fluidization (APF) and agglomerate bubbling fluidization (ABF). The APF is characterized by smooth fluidization, high bed expansion, and uniform distribution of agglomerates throughout the bed, while the ABF exhibits large bubbles and the low bed expansion ratio by increasing

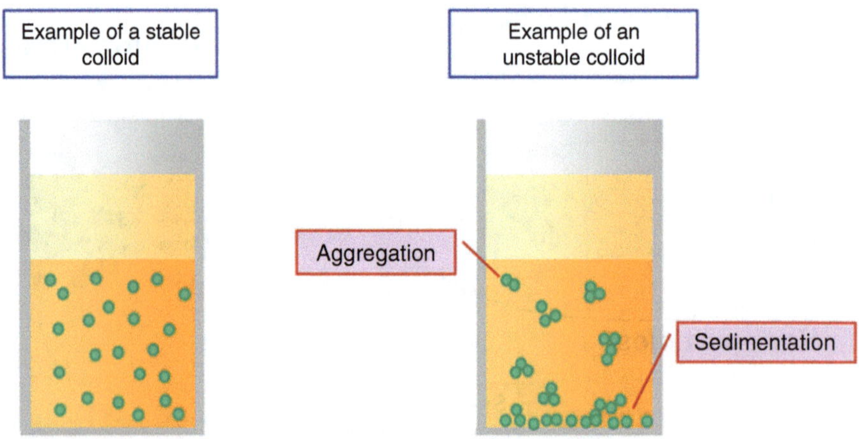

Fig. 2.10 Schematic presentation of the stable and non-stable colloidal systems. (Elimelech et al. 1998)

the gas velocity. Smooth fluidization gives rise to highly porous agglomerates in the range of several hundred microns. They keep breaking and reforming during fluidization because of the contrasting cohesive forces among NPs and the separation forces originating from the fluid (Esmailpour et al. 2015).

In a recent study on hydrophobic silica (R972) and hydrophilic titania (P25) nanoparticles, Esmailpour et al. (2018) observed that increasing the bed temperature can convert the fluidization regime from APF to ABF for hydrophobic silica nanopowder. Large agglomerates and bubbles were formed in the bed at higher temperatures. In contrast, hydrophilic titania NPs fluidized in the ABF way at lower temperatures but in the APF way at elevated temperatures.

2.3.8 Solubility and Phase Transition of Nanoparticles

In the bulk crystalline materials, properties of the material are independent on its particle size and chemical composition. In the nano regime, the minute size of the particle increases the surface area as well as the contribution of the surface energy to the overall energy of the crystalline system. The significant shift toward surface atoms of nanoparticulate matter is illustrated in Fig. 2.6 where the percentage of surface atoms in a particle is plotted against the particle size. This figure shows that in nanometer regime the large population of particle surface atoms markedly determines the properties of the material because the atoms located on the surface of a solid have less adjacent coordination state and therefore they are chemically more active, compared to the bulk atoms, and a large amount of energy is associated with this surface. The surface of NPs often plays a crucial role in determining the NP properties, including the stability, reactivity, solubility, and phase transition (Rector and Bunker 1995; LaFemina 1995a, b).

As mentioned earlier, almost all characteristic features of nanostructures stem from their size and the number of surface atoms; the solubility and reactivity of materials can, therefore, be optimized by altering the particle size. According to the Gibbs-Thomson theory, the NP solubility increases with the decreasing particle size. A solution is a homogeneous mixture of two or more substances, of which the solute dispersed among the molecules or ions of the solvent has the particles of less than 10 nm, which pass through a filter paper easily. A suspension is a heterogeneous mixture in which particle size of one or more components is greater than a few microns and particles are big enough to scatter light. After some time, the particles in aqueous suspension settle under water due to the influence of gravity. A colloidal solution is a mixture in which particles of a substance are of the size intermediate between those of a solution and suspension (10 nm–1 μm). These are too small to be filtered and/or seen with the naked eye. The particles in a colloid system are larger than in a solution but small enough to be dispersed evenly and maintain a homogeneous appearance; however they are large enough to scatter light. At a certain temperature, when a solid (solute) is mixed with a liquid (solvent) to form a solution, the molecules or ions on the solid disperse uniformly into the liquid and move in the solvent. When they touch the solid surface, they are adsorbed back on it. Suspensions of NPs exhibit phase behavior similar to that of the molecular solution systems in the equilibrium state. The phases in the system can be manipulated through interactions between the solute particles and between the solute and the solvent particles (Batra et al. 2016). The effect of different variables such as density, temperature, concentration, and strength of colloidal interactions and pressure is taken into account while describing the phase transitions. In fact, the strength of interaction distinguishes colloidal suspensions most strongly from molecular systems. As the range and strength of attractions and repulsions can vary over an enormous range with colloidal particles, suspensions of NPs offer a unique test bed for treatments of the molecular phase behavior (Table 2.3).

2.4 Nanomaterials in Bio-systems

Among the nanostructures, nanocrystals find significant place in the modern medical technology, like biomolecular detection and diagnostics, and antimicrobial therapeutics (Fig. 2.11). In fact, the large surface area-to-mass ratio of nanomaterials heavily affects their reactivity. The most common nanostructures used in medicinalapplications, mainly as delivery systems, are ceramic-based NPs, polymeric NPs, metal NPs, micelles, liposomes, and dendrimers (Selmer-Olsen et al. 1996).

Table 2.3 Differences between true solution, colloidal solution, and suspension

Property	True solution	Colloidal solution	Suspension
Size of particles	<1 nm	1 nm–1000 nm (1 μm)	>1 μm
Nature of solution Appearance	Homogeneous Transparent	Heterogeneous Translucent	Heterogeneous Opaque
Visibility	Solute particles not visible with the naked eye or through microscope	Solute particles not visible with the naked eye; can be seen with ultramicroscope	Suspension particles can be seen with the naked eye
Filterability	Solute particles pass through a filter paper or a parchment membrane	Solute particles pass through a filter paper but not through a parchment membrane	Solute particles cannot pass through a filter paper or a parchment membrane
Settling ability	Solute particles do not settle	Solute particles do not settle but can be made to settle by centrifugation	Solute particles settle down due to the force of gravitation
Light scattering	Solution does not scatter light (no Tyndall effect)	Solution shows Tyndall effect	Suspension may or may not show Tyndall effect
Particle movement	Solution does not show Brownian movement	Brownian movement of particles is visible	Brownian movement may or may not be visible

Several other NP systems such as solid lipid NPs, inorganic NPs, and microemulsions have also been used in the formulation, encapsulation, and release of active compounds extracted or derived from natural resources. The main objective of nanomedicine is to ensure drug transport to action sites, to maximize the desired pharmacological influence of drugs, and to overcome the factors that may hinder effectiveness of the treatment (El-Say and El-Sawy 2017). The controlled drug delivery system comprises of four major modes of delivery, viz., (a) rate-programmed drug delivery, where drug diffusion from the system has to follow a specific release rate profile (in this type of delivery, the therapeutic formulation is totally or partially loaded in the reservoir space, which is covered by the pre-programmed polymeric membrane, the function of which can be optimized with block copolymers through multifunctionalization); (b) activation-modulated drug delivery, where the drug release is induced by various physical, chemical/biochemical, or environmental stimuli (e.g., various pressures, magnetics, electricity, salt concentration, pH, light, temperature, hypoxia) and facilitated by external supply of energy; (c) feedback-regulated drug delivery, where the rate of release is determined by the concentration of biochemical substance (triggering agent) via some feedback mechanism; and (d) site-targeting drug delivery, where diffusion rate and partitioning of drug release are regulated by the specific targeting moiety, solubilizer, and drug moiety (Bennet and Kim 2014).

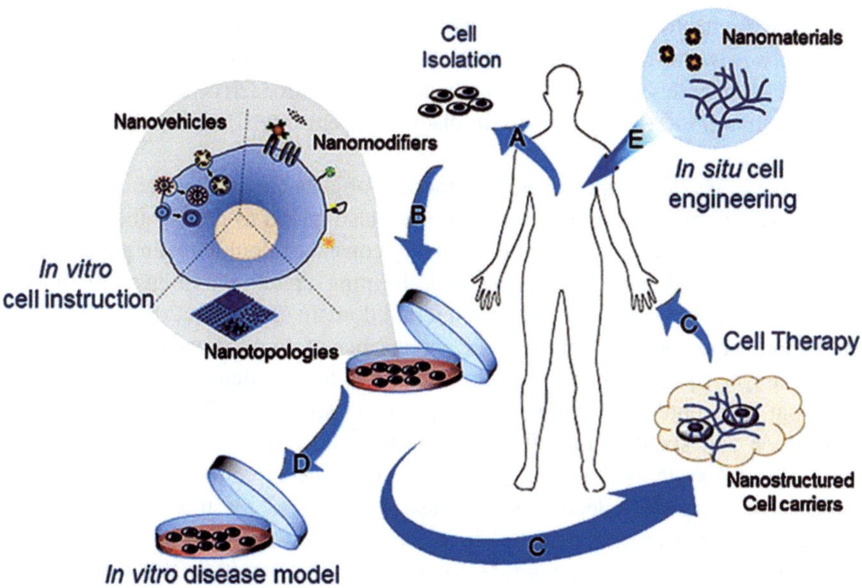

Fig. 2.11 Multifacetedapplications of nanomaterials in the cell engineering and therapy. (Wang et al. 2016b)

2.4.1 Micelles and Liposomes

Certain nanostructures with walls composed of phospholipid moieties form a "core-shell structure," which provides a suitable platform for drug delivery to the central nervous system (CNS). They carry diverse amounts and types of therapeutic materials and, in some cases, accomplish target delivery to specific cell types within the CNS. The ability of some of these structures to traverse the blood-brain barrier (BBB), avoiding the checks of the immune surveillance system, enables them to enter the CNS without any neurosurgical procedure. Micelles and liposomes are most prominent among these structures. Nanomicelles and nanoliposomes typically range in size from ten to several hundred nanometers. Micelles are typically spherical and have an outer surface of charged or hydrophilic moieties and an inner lipophilic region. Phospholipid micelles can be utilized to carry other amphiphilic molecules or hydrophobic substances within their inner lipophilic region. The use of amphiphilic block copolymers to produce clinical nanomicelles for drug delivery is quite common these days (Fiandaca and Bankiewicz 2013).

Thus, micelles, which are lipid molecules with amphipathic nature of fatty acids due to the presence of both hydrophilic and hydrophobic regions, arrange themselves in spherical form in the aqueous solutions (Fig. 2.12). Their hydrophilic head faces to water, but hydrophobic tails are inside and away from water. Fatty acids from micelles usually have a single hydrocarbon chain as opposed to two hydrocarbon tails. This allows them to conform to spherical shape for lesser steric hindrance within a fatty acid. Fatty acids from glycolipids and phospholipids have two hydrophobic chains that are too bulky to fit into the spherical shape as micelles do and, therefore, prefer to form glycolipids and phospholipids. Micelles are formed spontaneously in water due to the amphipathic nature of the molecule. In fact, when lipids form micelles, the hydrophobic tails interact with each other, releasing water from them and increasing the system disorder or entropy (Du et al. 2003, Li et al. 2009).

On the other hand, liposomes are spherical objects comprising mainly of lipids (Fig. 2.12). Sometimes other constituents are also added to modify their chemical and physical properties. Liposomes typically consist of double-chain phospholipid amphiphiles combined with cholesterol, forming spheroidal bilayer membrane structures that encompass an aqueous internal domain (Torchilin 2005). The length of the fatty acid chains and the presence or absence of double bonds within the liposome bilayer lipids influence the membrane fluidity, as does the combination of different phospholipids within the membrane structure. Cholesterol moieties strengthen and stabilize the bilayer membrane and reduce the cation leakage in physiological systems. Increasing the molar cholesterol content of the liposomal drug carriers typically decreases the release kinetics of the therapeutic from the nanocarrier (Panwar et al. 2010). Therefore, specific liposomal properties can be tailored by regulating the membrane components (Panwar et al. 2010). Liposomes are typically formed by adding energy to amphiphilic phospholipids in aqueous solution. Depending on lipid monomer concentration and environmental factors,

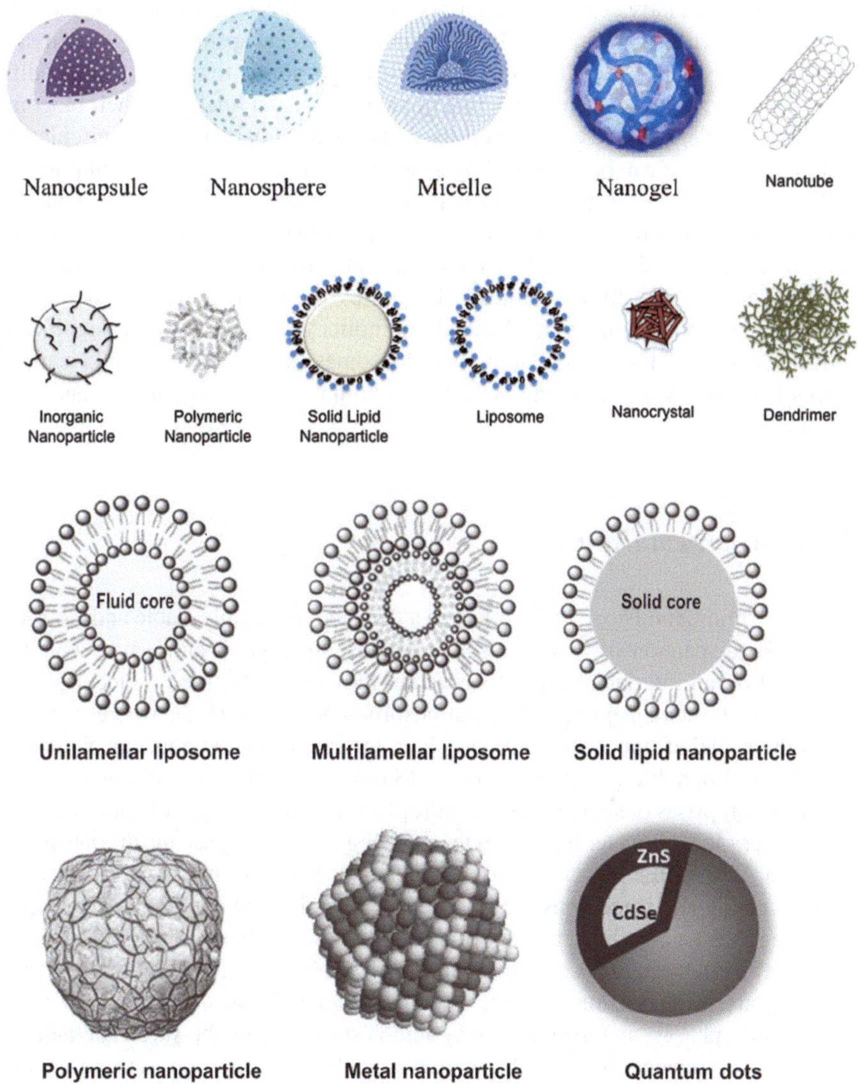

Fig. 2.12 Schematic morphological representation of different types of nanostructures. (Veszelka et al. 2015)

liposomal structures can range from long tubules to spheres, with dimensions ranging from several hundred angstroms to several hundred micron meters. A prototypical liposomal vesicle has a single, closed lipid unilamellar bilayer confining a single internal aqueous volume (Fiandaca and Bankiewicz 2013).

Liposomes are different from micelles as they are composed of a lipid bilayer, separating an aqueous internal compartment from the bulk aqueous phase. Micelles, by contrast, are closed lipid monolayers with a fatty acid core and polar surface or a polar core with fatty acids on the surface (inverted micelle). Thus, liposomes are an important drug-carrier system due to their ability to encapsulate drugs, their stability and long shelf life, controllable size and charge, ability to function and modify the surface due to presence of many functional groups, and finally for their biocompatibility and degradability. However, they have a short half-life in the circulation system, although it can be enhanced by a better control of the size and composition of liposome vesicles. Although liposomes are suitable to encapsulate nonpolar drugs in the hydrophobic bilayer of the vesicle, sometimes such drugs affect the integrity of these vesicles rendering them unsuitable for nonpolar drugs (Endo et al. 2010). Wang et al. (2017) synthesized a series of novel derivatives of EEDQ (*N*-ethoxycarbonyl-2-ethoxy-1, 2-dihydroquinoline), which showed significant anticancer activity. Using the Tf-modified liposomes as carriers, they could achieve successful delivery of therapeutic, which could improve tumor cell uptake and antitumor effect.

2.4.2 Microemulsions

The term "microemulsion" (ME) refers to a thermodynamically stable and isotropically clear dispersion of two immiscible liquids, such as oil and water, which is stabilized by an interfacial film of surfactant molecules. As these molecules have both polar and apolar groups, they get adsorbed at the interface, where they can fulfill their dual affinity with the hydrophilic groups located in aqueous phase and with the hydrophobic groups in oil or air (Saini et al. 2014). The dispersed phase typically comprises of small particles or droplets, with a size range of 5 nm–200 nm, and has very low oil/water interfacial tension (Fig. 2.12). Because the droplet size is less than 25% of the wavelength of visible light, MEs are transparent. They are formed readily and sometimes spontaneously, generally without high-energy input. In many cases a co-surfactant or cosolvent is used in addition to the surfactant, the oil phase and the water phase (Saini et al. 2014). Their formation is facilitated by the ultralow interfacial tension of the component systems. Thus, they consist of aqueous and oily phases, and the disperse system is stabilized by the surfactant and co-surfactant components. MEs are thermodynamically stable (in contrast to emulsions) and exhibit characteristics of macroemulsion (particles can be measured with diffraction of laser light) but sometimes behave as a real solution (active substances have a saturation solubility and do not show active substance distribution as in macroemulsions). When MEs are added into water, their lipid phase precipitates to form small particles.

As one of their several important applications, MEs offer an efficient drug delivery system and enable the selective release of active pharmaceutical ingredients along specified lengths in the gastrointestinal tract by inverting a water-in-oil emulsion to an oil-in-water emulsion. MEs provide promising drug delivery systems due to their easy formulation, thermodynamic stability, and ability to facilitate delivery of lipophilic and hydrophilic drugs into the skin. The composition and quantities of the ME components, as also the included active substances, affect the droplet size, distribution, and viscosity of the ME. Their particle size (5–200 nm) enhances their penetration through cellular membranes, making them suitable as drug carriers. Due to the presence of amphipathic components in MEs, they are very good dissolving agents (Mocan 2013). The surfactant and co-surfactant may enhance drug penetration, disrupting the lipids of the stratum corneum (Juškait et al. 2015). The concept of ME was first introduced by Hoar and Schulman (1943) who developed the formulation by dispersing oil in an aqueous surfactant solution and adding an alcohol as a co-surfactant.

2.4.3 Other Relevant Materials

Solid lipid nanoparticles (SLNs), introduced in 1991, represent an alternative carrier system to traditional colloidal carriers such as emulsions, liposomes, and polymeric micro- as well as nanostructures. This system is composed of nanosized spherical solid lipid particles, which are dispersed in water or some aqueous surfactant solution. It resembles an oil-in-water emulsion for parenteral nutrition with the difference that the liquid lipid (oil) of the emulsion has been replaced here by a solid lipid (Bagul et al. 2018). SLNs are primarily made of a solid lipid core with a monolayer phospholipid shell. The solid state of the nanoparticulate matrix provides protection to chemically labile drugs and facilitates a prolonged drug release (Lin et al. 2017). The solid core contains the drug dissolved or dispersed in the solid high-melting fat matrix. The hydrophobic chains of phospholipids are embedded in the fat matrix. They have the potential to carry lipophilic or hydrophilic drugs or diagnostics (Ramteke et al. 2012).

The SLNs (50–1000 nm) thus consist of physiologically tolerated lipid components that remain in solid state at room temperature and are dispersed in water or in aqueous surfactant solution. They possess the characteristic positive traits including small size, large surface area, high drug loading, and the interaction of phases at the interface. Their hydrophobic core provides a suitable environment for entrapment of hydrophobic drugs to improve their bioavailability (Bagul et al. 2018; Lingayat et al. 2017). They resemble the other types of nanocarriers in being suitable to encapsulate nonpolar insoluble drugs in their polymeric core and shielding them from the outside environment, to increase drug stability and reduce its toxicity to the body (Fig. 2.12). Besides, they are easier to prepare and cheaper for the scale-up productions in comparison to other drug delivery systems. They ensure a sustained and slow release of the drug in the targeted site.

SLNs have certain advantage over other nano-delivery systems, e.g., they have lower chronic or acute toxicity, enhanced bioavailability and productivity, higher reproducibility, limited use of organic solvents in preparation, ability to protect liable drugs, possibility of incorporating both hydrophilic and hydrophobic compounds, and ability to bypass the spleen or liver filtration for 120–200 nm particle size. Given this, the SLNs are most appropriate for oral delivery of phyto-bioactive compounds, such as curcumin, resveratrol, quercetin, and other polyphenols. However, the bulk release of such compounds in the stomach at a lower pH of about 1–3 renders the SLN delivery system inadequate. To meet this challenge, SLNs are subjected to surface modification, and the surface-modified SLN (SMSLN) could improve the delivery output, preventing the high release of phyto-bioactive compounds in the stomach (Ganesan et al. 2018).

Polymer nanoparticles, derived from biodegradable polymers, constitute a special type of nontoxic drug delivery system. The special features of this system, besides nontoxicity, are biocompatibility, biodegradability, prolonged circulation, controlled release, and a broad payload spectrum of a therapeutic agent (El-Say and El-Sawy 2017). Polymer nanoparticles may carry sugars, proteins, and many other naturally occurring macromolecules. Of late, encapsulation of anticancer agents within the polyhydroxyalkanoates, poly-(lactic-co-glycolic acid), and cyclodextrin-based nanoparticles has been tried to target exactly the specific cancer sites (Masood 2016).

Dendritic polymers and dendrimers belong to a special class of macromolecules composed of many monomer units that are chemically linked together (Fig. 2.12). They are good encapsulating agents for hydrophobic drugs due to their nonpolar core and are known for their structural perfection, water solubility, and monodispersity (Núñeza et al. 2014). Dendrimers are nanosized, radially symmetric molecules having a well-defined homogeneous and monodisperse structure comprising of a typically symmetrical core, an inner shell, and an outer shell. All varieties of dendrimers have the properties of polyvalency, solubility, self-assembling, electrostatic interactions, chemical stability, and low cytotoxicity (Abbasi et al. 2014). Dendrimers and dendritic polymers with unique inherent supramolecular features and multivalent properties are most suitable carriers in the fields of gene and drug delivery and biomimicry. Whereas dendritic polymers do not have the perfect dendrimer branched structure, they exhibit high surface functionality and are easy to produce. Dendrimer-based technologies provide a platform for mimicking the naturally occurring biological assemblies to design synthetic alternatives in the field of nanomedicine (Kretzmann et al. 2017).

Inorganic nanoparticles are classified into three main classes, viz., transition metal NPs, ceramic NPs, and carbon NPs. Transition metal nanoparticles have many applications as drug carriers (e.g., application of gold NPs as shuttles for site-specific delivery of toxic drugs) and as drugs themselves when excited by light radiation to damage the DNA and/or modify proteins, promote lipid peroxidation, and destroy the cell microenvironment, causing cell death in cancer therapy (Fig. 2.12). Also, they are used for imaging in diagnosis as well as therapy monitoring. Ceramic NPs are developed mostly from oxides, nitrides, and carbides with silica (SiO_2) and used as hollow shells or cores coated with biodegradable and biocompatible polymers (Fig. 2.12). These surface modifications enable them to be used as targeted delivery systems (Veszelka et al. 2015).

2.5 Preparation of Nanostructures for Use in Medicine

The mode of preparation has a role in determining the physicochemical characteristics of the polymer and the drug to be loaded. The primary manufacturing methods of nanoparticles include:

2.5.1 *Emulsion-Solvent Evaporation Method*

This maximally used method of NP preparation comprises of (a) emulsification of the polymer solution into an aqueous phase and (b) evaporation of polymer solvent containing the polymer precipitation as nanospheres (Wang et al. 2016c). Finally, the NPs obtained are centrifuged and washed with distilled water to remove the possible contaminants and then lyophilized for storage (Fig. 2.13).

2.5.2 *Double Emulsion and Evaporation Method*

This method is suitable for encapsulating the hydrophilic drugs. It involves addition of aqueous drug solutions to organic polymer solution under vigorous stirring to form water/oil (w/o) emulsion, which is then added into a second aqueous phase with continuous stirring to form the w/o/w emulsion. Finally, the solvent is removed from this emulsion by evaporation, and NPs are isolated by centrifugation at high speed (Noviendri 2014) and washed thoroughly before lyophilization. The amount of hydrophilic drug to be incorporated, the concentration of stabilizer used, the polymer concentration, and the volume of aqueous phase are some variables that affect the characterization of NPs in this process (Fig. 2.14).

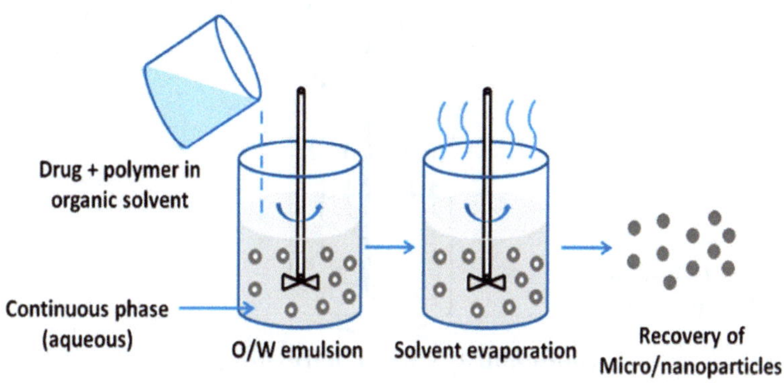

Fig. 2.13 Emulsion-solvent evaporation method (Wang et al. 2016c)

2.5.3 *Salting-Out Method*

This method is based on separating the water-miscible solvent from aqueous solution. The polymer and the drug are initially dissolved in an emulsified media of an aqueous gel, containing an electrolyte or nonelectrolyte agent as the salting-out agent and a colloidal stabilizer (Wang et al. 2016c). This water/oil emulsion is diluted with a sufficient volume of aqueous solution to enhance the diffusion of solvent into the aqueous phase to form the nanospheres (Fig. 2.15).

2.5.4 *Emulsion-Diffusion Method*

In this method, the encapsulating polymer is dissolved in a partially water-miscible solvent and saturated with water to achieve a thermodynamic equilibrium of both liquids (Fig. 2.16). Subsequently, the polymer-water saturated solvent phase is emulsified in an aqueous solution containing stabilizer based on the oil-to-polymer ratio, which leads to solvent diffusion to the external phase and the formation of nanospheres or nanocapsules (Esmaeili et al. 2013).

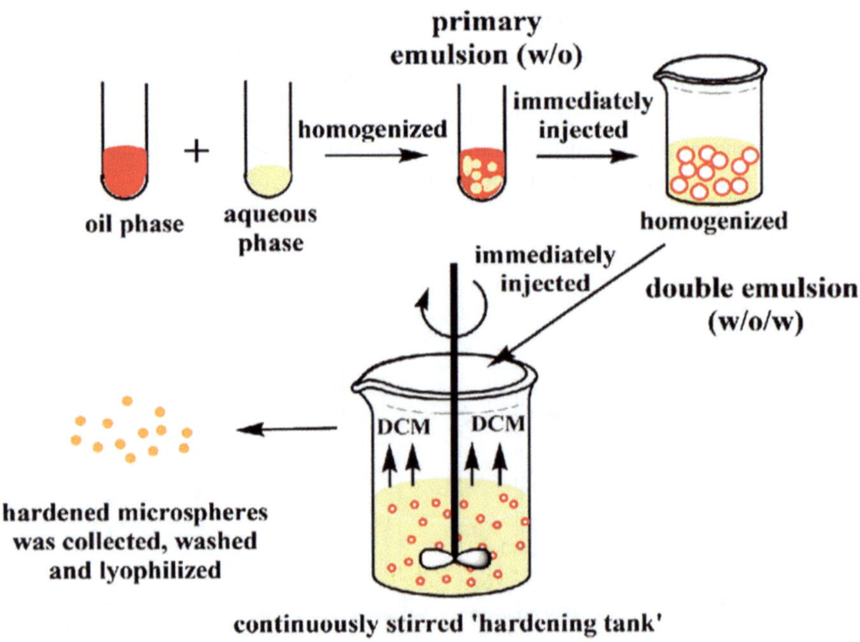

Fig. 2.14 Double emulsion and evaporation method. (Noviendri 2014)

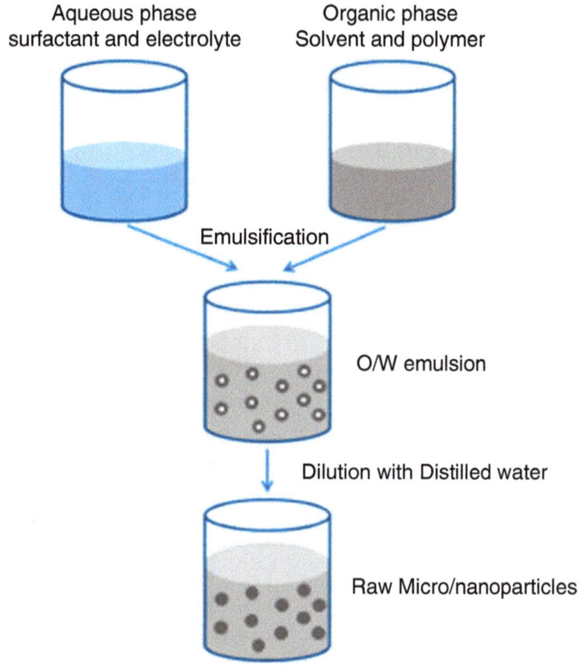

Fig. 2.15 The salting-out method. (Reproduced with permission from Wang et al. 2016c)

2.5.5 Solvent Displacement/Precipitation Method

Solvent displacement involves the precipitation of a polymeric organic solution and diffusion of an organic solvent in the aqueous medium, with or without a surfactant (Fig. 2.17). The polymers, drug, and lipophilic surfactant are dissolved in a polar water-miscible solvent such as acetone or ethanol. The solution is then transferred to an aqueous solution containing stabilizer under magnetic stirring to enable immediate formation of NPs (Ezhilarasi et al. 2012).

2.6 Nanoencapsulation and Nanoencapsulated Materials

Nanoencapsulation is a technology used to encapsulate substances in miniature form and pack bioactive materials at the nanoscale range. Delivery of bioactive material to different sites within a body is affected by the particle size, and, therefore, nanoencapsulation has the potential to improve bioavailability and controlled release of bioactive compounds and ensure their precise targeting. The nanocarriers (NCs) thus produced (10–1000 nm) are expressed as nanocapsules and nanospheres. Nanocapsules are vesicular systems in which the bioactive compound is confined to a cavity surrounded by a unique polymer membrane, while nanospheres are matrix systems where the bioactive compound is uniformly dispersed (Suganya and

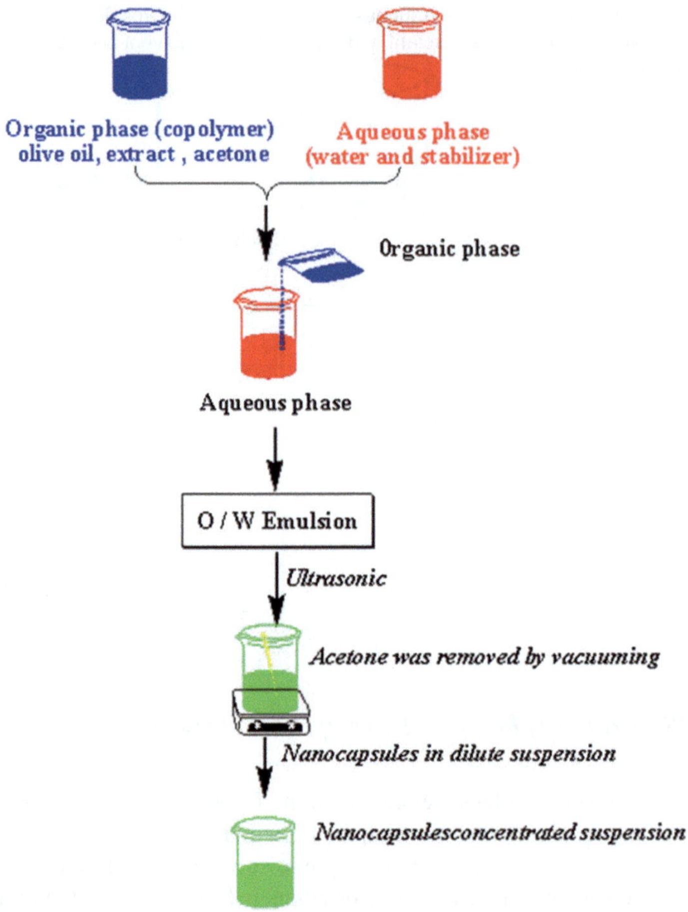

Fig. 2.16 Emulsion-diffusion method. (Esmaeili et al. 2013)

Anuradha 2017). NCs protect their payload from premature degradation in the biological environment, enhance bioavailability, and prolong presence in blood and cellular uptake.

Recent nanoencapsulation methodology encompasses nanoemulsification, electrospinning, electrospraying, formation of nanostructures via cyclodextrins, and synthesis of nanoliposomes, solid lipid nanoparticles, nanostructured lipid carriers, etc. (Jafari 2017). The selection of technique for NC synthesis is made according to the chemical structure of therapeutic agent, type of application, and time of retention inside the body. NCs of different dimensions can be synthesized by using different matrices. Size and size distribution of NCs affect their cellular uptake and penetration across the biological barriers. Size and surface chemistry of NCs determine their in vivo performance. Drug release mechanisms can also be modulated depending upon the nature of therapeutic agent and type of NCs (Kumari et al.

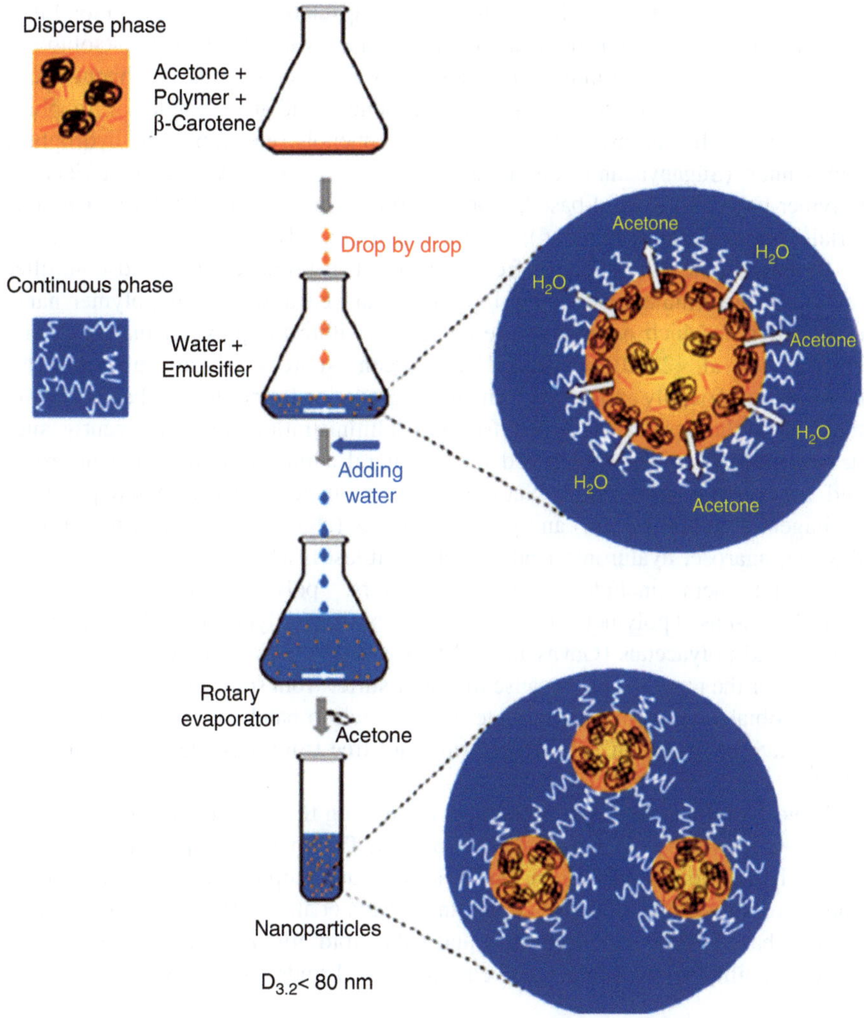

Fig. 2.17 The solvent displacement/precipitation method (Ezhilarasi et al. 2012)

2014). The mononuclear phagocytic system (MPS) in the body recognizes NCs as foreign particles and rapidly removes them from the systemic circulation. Given this, if a prolonged presence of NCs in blood is desired, surface of NCs has to be modified to avoid phagocytosis. Surface modification of NCs is conducted either by tagging ligand or hydrophilic polymers on their surface. Further, surface charge is another important factor that affects the cellular response of NCs. In general, NCs with cationic charge are taken up readily by negatively charged cell membranes, compared to neutral or positively charged ones (Kumari et al. 2014).

The encapsulated material is concerned with the internal phase, core material, and filler. Particle size is also an important factor for the formulations of encapsulated materials. Core materials used for nanoencapsulation are lipophilic (soluble in lipids and organic solvents but insoluble in water) as well as hydrophilic (soluble in water but insoluble in lipids and organic solvents) compounds. Lycopene, beta-carotene, lutein, phytosterols, and docosahexaenoic acid are examples of lipophilic compounds, whereas ascorbic acid and polyphenols, etc. represent hydrophilic compounds (Suganya and Anuradha 2017). The commonly used materials are polymer-based, solid lipid-based, inorganic porous-based, and clay-based nanomaterials (Nuruzzaman et al. 2016), as depicted in Fig. 2.18.

Biodegradable polymers are frequently used to produce nanosized controlled release drug formulations. Active ingredients are encapsulated as polymer nanocomposites in which nanofillers are dispersed within the polymer matrix (Mora-Huertas et al. 2010). Produced by green sources, these structures are environment-friendly and do not form any degradation by-products. The majority of these formulations are designed for oral administration, though recently such devices have also been introduced for parenteral administration, ocular insertion, and transdermal application. Natural polymers include protein-based polymers (collagen, albumin, gelatin) and polysaccharides (alginate, cyclodextrin, chitosan, dextran, agarose, hyaluronic acid, starch, cellulose), whereas biodegradable synthetic polymers include several polyesters, polyanhydrides, polyamides, phosphorus-based polymers, and many others such as polycyanoacrylates, polyurethanes, and polyacetals (Gavasane and Pawar 2014). Synthetic polymers are preferred over the natural ones because the latter suffer from some disadvantages, such as microbial contamination, climate-based batch-to-batch variation, and uncontrolled rate of hydration, while the former are free from these defects (Shah et al. 2011).

Lipid-based nanovectors (liposomes) are among the best delivery systems with better encapsulating efficiency and low toxicity. They have great potential to encapsulate the ingredients having various polarities and simplify the in vivo dispersion and absorption of the bioactive compounds (Aina et al. 2007). Of the various types of lipid-based NMs, nanoliposomes and solid lipid NPs have already established their suitability to encapsulate active ingredients. Lipids, especially charged lipids,

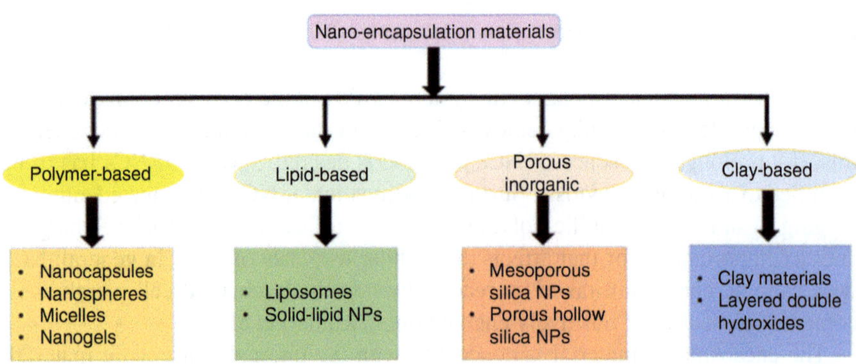

Fig. 2.18 Various types of nanoencapsulation materials and the structures they produce

are used to design NPs characterized by a core-shell structure, wherein a lipid shell interacts with a core having different biomaterials. Drugs carrying a net charge can be condensed in the core, which is then covered by the lipid shell. This approach works well with the delivery of macromolecular drug (e.g., nucleic acids) and small molecules (e.g., bisphosphonates) (Campani et al. 2018).

Inorganic porous nanomaterials have also attracted the interest of researchers as being highly capable of encapsulating the bioactive compounds. In fact, the polymeric nanoencapsulated materials suffer from various limitations such as poor thermal and chemical stability, rapid elimination of the plant enzyme system, degradation of some polymers obtained during the formation of acidic monomers, and decreased pH value within the polymer matrix. In comparison to polymeric nanoencapsulated materials, these inorganic materials offer a nontoxic, biocompatible, and stable alternative and have been used for controlled release applications (Hillyer and Albrecht 2001, Lao et al. 2010).

Both natural and synthetic nanoporous materials consist of pores with diameter roughly in the range 0.4–100 nm and are classified on the basis of their pore size as microporous (≤2 nm), mesoporous (2–50 nm), and macroporous (>50 nm) materials. Utilization of the pores in impregnation of nanoparticles/proteins/ions or in transporting the latter across the pores of membranes is of practical and scientific interest. Their surface area design and pore size determine their applications in diverse areas such as fuel cells, solar cells, Li-ion batteries, super capacitors, hydrogen storage, catalysis, gas purification, separation technologies, drug delivery, and cell imaging. Typical examples of nanoporous solids are zeolites, activated carbon, metal-organic frameworks, ceramics, silicates, aerogels, pillared materials, various polymers, and inorganic porous hybrid materials. Zeolites and mesoporous silica have come up as important materials for applications in drug delivery and imaging. Properties of biocompatibility, low toxicity, large surface areas, and their ability to control the physicochemical properties make them especially apt for biomedical applications (Datt et al. 2012; Bhaumik 2017).

These porous nanomaterials can be synthesized by using inorganic, organic, or organic-inorganic hybrid framework building units/metal ions with or without using template molecules as the structure-directing agents. Historically, the field has evolved from synthetic Al-rich zeolites, followed by high-silica zeolites and then metallosilicates, followed by aluminophosphates, mesoporous silica, and a range of allied mesoporous materials. The development of periodic mesoporous organosilicas (PMOs), mesoporous carbons, metal-organic frameworks (MOFs), zeolitic imidazolate frameworks (ZIFs), porous organic polymers (POPs), covalent organic frameworks (COFs), and porous metal phosphates/phosphonates has added further dimensions to the family of porous nanomaterials. Their pore topologies, sizes, and surfaces can be tuned as per the requirement (Bhaumik 2017).

The use of various inorganic nanoparticles as drug carriers is becoming increasingly common. The functionality of nanocarriers in real-life environments explains the enthusiasm for their use. However, the foremost consideration regarding the use of NMs in medicine pertains to their safety. Several functionalities are typically added onto nanocarriers, but the most crucial feature to be administered is that they should possess a long residence time in blood circulation. This sufficiently relates to their coatings because it is the outmost layer which dictates their first interactions with the surroundings and often determines their biofate (Tamarov et al. 2018). The use of porous inorganic nanoparticles as drug carriers for cancer therapy has the potential to improve the life expectancy of the patients affected by this disease. However, much work is needed to overcome their drawbacks, which are aggravated by their hard nature (Baeza et al. 2017).

Nanoclays provide opportunities for developing economical and multifunctional nanocarriers. Functionalization of clay nanoparticles with different polymers and surfactants is essential to manipulate the electrostatic interactions between chemical loading and clay particles (Zhang et al. 2013). Nanoclays are naturally occurring clay minerals with at least one dimension in the range of 1–100 nm. These are found in the form of anionic as well as cationic clays, depending on the surface-layer charge and the types of interlayer ions. Montmorillonite, kaolinite, laponite, halloysite, bentonite, hectorite, laponite, sepiolite, saponite, and vermiculite are among the major nanoclays (Peña-Parás et al. 2018). Nanoclays are used widely as reinforcements for polymer matrix composites to improve the mechanical, thermal, and anticorrosion properties. Because of being nontoxic, nanoclays and their composites have been studied for biomedical applications such as bone cement, tissue engineering, drug delivery, wound healing, and enzyme immobilization, among others (Dasan 2015; Peña-Parás et al. 2018).

2.7 Medical Significance of Nanostructures

Nanotechnology-enabled drug delivery has created a great impact on every mode of drug administration, from oral to injectable. The nanotechnology-based drugs are able to permeate through cell walls. In various bone disorders, such as bone fractures, osteoarthritis, osteoporosis, or bone cancers, the traditional implant materials have a short lifetime and resistance inside the body and are affected by the loosening, inflammation, infection, osteolysis, and wear debris side effects. In fact, the bone is a nanocomposite consisting of a nano-dimension protein-based soft hydrogel template, water, and hard inorganic components such as hydroxyapatite, HA (20–80 nm long and 2–5 nm thick). This self-assembled nanostructure closely surrounds and affects the mesenchymal stem cell; osteoblast (bone-forming cell), osteoclast, and fibroblast adhesion; proliferation; and differentiation (Wang et al. 2016b). Another bone structure composed of nanosystems is the cartilage, which is a low regenerative tissue, containing a small percentage of chondrocytes but dense nanostructure rich in collagen fibers, proteoglycans, and elastin fibers. The limited regenerative capacity of cartilage relates to the lack of chondrocyte mobility in this dense nanostructure and to the absence of progenitor cells and vascular networks necessary for an efficient tissue repair (Wang et al. 2016b). The excellent mechanical and biomimetic properties of nanostructures have rendered them suitable for improving the bone cell and chondrocyte functions.

Due to their biocidal, anti-adhesive, and delivery characteristics, nanomaterials prevent the formation of biofilms within the oral cavity. Metal NPs in the size range of 1–10 nm have particularly shown significant biocidal activity against bacteria. The most common NPs used for orthodonticsapplications are those of Ag, Cu, Au, Zn, TiO_2, magnetite chitosan, quaternary ammonia NPs, fluorapatite, fluorohydroxyapatite, or hydroxyapatite (Khurshid et al. 2015). The success of various forms of NPs (including liposomes, polymer particles, micelles, dendrimers, quantum dots, gold NPs, and carbon nanotubes) that have been synthesized and tested for therapeutic applications depends primarily on their ability to avoid/minimize accumulation at undesired sites and reach the therapeutic site at necessary doses. In fact, their ultimate biodistribution is dependent on a variety of biological barriers they come across in the human body. For intravascular delivery of NPs, for instance, the barriers arise in the form of (a) immune clearance in the reticuloendothelial system comprising of the liver and spleen, (b) permeation across the endothelium in the target tissues, (c) penetration through the dense interstitial space and extracellular matrix of the target tissue, (d) endocytosis and intracellular localization in the target cells, and (e) diffusion in the cell vesicles and cytoplasm/nucleus (Barua and Mitragotri 2014). The alternative modes of NP penetration into the human body are mainly through the skin, lungs, and gastrointestinal (GI) tract (Yah et al. 2011; Wang and Wang 2014; Riasat et al. 2016).

The skin, the outermost and the largest organ surface, functions as the first-line barrier between the external environment and the internal organs of the human body. With increasing exposure of the human skin to NPs, the issue of the capacity of NPs to penetrate through the skin has become a crucial subject of research (Liang et al. 2013). NP penetration through the skin barrier depends on their size; particles of 500–1000 nm can penetrate to lower levels of the human skin in which the smaller particles penetrate deeper (Ryman-Rasmussen et al. 2007; Nasir 2010). Given the potential of solid NPs to penetrate the stratum corneum and diffuse into the underlying structures, their topical use raises a genuine health and safety issue (Wang and Wang 2014). These ultrafine particles can enter the body through skin pores, debilitated tissues, injection, and the olfactory, respiratory, and intestinal tracts and may lead to various adverse biological effects. Some efforts have been made to determine the portal routes of nanoscale materials on experimental animals (Yah et al. 2011; Lin et al. 2016).

The skin is known to protect the body by resisting the penetration of molecules and microorganisms present in the external environment and preventing excessive loss of water to maintain homeostasis. The main resistance comes from the stratum corneum, which is made of layers of flattened corneocytes surrounded by lipid bilayers composed of ceramides in particular. However, this also resists penetration of most of the topically applied compounds, although the transcellular route through corneocytes may help to some extent. Hair follicles and associated sebaceous glands as well as sweat glands, despite forming a minute portion (nearly 0.1%) of the total area of skin surface, provide potential routes of access into the skin and hence have importance for nanosystems (Nastiti et al. 2017).

Microemulsion (ME) and nanoemulsion (NE) have also been used as the potential delivery vehicles for transferring drug molecules through the stratum corneum. Terminology suggests that NE should have a smaller particle size than ME, as nano and micro refer to 10^{-9} and 10^{-6}, respectively, but in fact the size range of the two systems is hardly different. Further, both have a low polydispersity (up to nearly 10%) and are similar in physical appearance and texture. However, while ME shows thermodynamic stability, NE is thermodynamically unstable but kinetically stable (McClements 2012; Gupta et al. 2016). Moreover, ME is broken by changes in temperature and/or dilution, whereas NE remains stable. So, the basic difference between NE and ME pertains only to their thermodynamic stability, the same being responsible for the higher energy input required for NE preparation (Nastiti et al. 2017).

Currently, lung ailments such as cystic fibrosis, COPD, and asthma are treated with inhaled drugs such as corticosteroids that adhere to the walls of air passages. Various polymeric nanoparticles, such as liposomes and dendrimers, among others, are now used as carriers, sometimes in combination with small molecules, cytokines, growth factors, and/or pluripotent stem cells. Thick mucus often built up on the route passages lessens the effectiveness of the delivery system of these muco-adhesive particles (MAPs). The hydrophobic and electrostatic forces within this mucus layer facilitate trapping of particles and preventing their access to the airway epithelium. Proper size and charge of NPs help overcome this challenge. Coating the NPs with a high density of low molecular weight polyethylene glycol (PEG) renders their surface neutrally charged, thus improving their transport across the mucus layer (Bonner 2016). Recently, researchers turned to NPs that are small enough to make their way through mucus membranes direct to the lining of the lungs, providing medication to affected areas. These mucus-penetrating particles (MPP) remain in the lungs for long, releasing medication for an extended period of time (Iyer et al. 2015; Schneider et al. 2017). Aerosolization or inhalation of colloidal systems has shown huge potential for targeted drug delivery. Moreover, the surfactant-associated proteins present at the interface strengthen the impact of these formulations by reducing surface tension (Paranjpe and Müller-Goymann 2014; Yeagle 2017).

On the other hand, certain NPs such as ceria (cerium dioxide NPs) could cause mitochondrial damage leading to a decrease in cell viability and a progress of apoptosis and induce autophagy in human peripheral blood monocytes (Hussain et al. 2012). Likewise, TiO_2nanoparticles could induce oxidative stress-mediated acute lung inflammation. Similarly, pulmonary exposure to ZnO NPs might cause transient increases in acute lung inflammation (Xia et al. 2016). Surface area of NPs and their dissolution property are important in deciding their toxic potential. Thus, toxicity caused by the ultrafine NP penetration to the lungs, if any, merits special attention. Exposure to NPs in occupational or environmental settings often causes pulmonary diseases or exacerbates the preexisting ones on one hand, while the same NPs prove highly useful for therapeutic applications in nanomedicine, on the other.

The gastrointestinal (GI) tract offers extensive surface area (300–400 m^2) for drug absorption by absorptive epithelial cells (enterocytes). It contains many other types of cells including mucin-secreting goblet cells, specialized M cells associated with Peyer's patches responsible for antigen transportation through dendritic cells, endocrine cells, and Paneth cells, which may facilitate the process of drug absorption. However, many hydrophobic and hydrophilic drugs have poor bioavailability, when administered orally, due to their inadequate physicochemical (solubility, stability) and/or biopharmaceutical (permeability, metabolic stability) properties. Oral delivery is even more challenging for biologics (e.g., peptides, proteins, and nucleic acids) due to their hydrophilicity (leading to low permeability), high molecular weight, and poor chemical/enzymatic stability in the GI tract (Date et al. 2016). Penetration of NPs into the human body through intestinal barrier depends strongly on their size; the smaller the particle diameter, the faster they could penetrate the mucus to reach the colonic eutrocytes; 14 nm diameter permeated within 2 min, 415 nm particles took 30 min, while 1000 nm particles were unable to cross this barrier (Hillyer and Albrecht 2001).

Due to the presence of marked physiological and biochemical barriers to peptide absorption in the GI tract, oral delivery of peptide drugs remains a challenge. Nano formulations can improve drug stability in the harsh GI tract environment, increasing drug solubility and bioavailability, enabling the targeting of specific sites, and providing sustained release in the GI tract. However, the unique and diverse physiology along the GI tract, including the widely varying pH, mucus that varies in thickness and structure, and numerous cell types, forms a significant barrier to effective delivery (Date et al. 2016; Riasat et al. 2016). Especially designed NPs, which may be able to (i) protect their cargo against enzymatic breakdown in the gut lumen and by intestinal cells, (ii) take the peptide safe across the mucus barrier, and also (iii) pass the intestinal epithelium that lines the intestinal lumen, are expected to deliver goods. To overcome the problem, absorption enhancers have been incorporated in many oral peptide delivery systems. Efforts have been made in pharmacological trials to adopt the paracellular route via tight junctions and the transcellular transcytosis route to pass the epithelial layer. Tight junctions can be modulated by MLC phosphorylation via MLCP inhibition, resulting in significant absorption of peptides. The transcellular pathways via relatively well-explored transcytotic pathways (as those for vitamin B12 and IgG) have proved promising for the oral delivery of peptides encapsulated in NPs (Lundquist and Artursson 2016). However, it is to be seen how far the promising results observed with tight junction regulation and transcytosis in small animals can be applicable to humans. It is believed that muco-adhesive surface properties on particles delivered to the GI tract improve oral absorption or local targeting of various difficult-to-deliver drug classes. Delivering drugs in non-muco-adhesive MPP might provide enhanced particle (drug) distribution in the GI tract (Maisel et al. 2015).

2.8 Conclusion

Besides the important role of size in nanosystems, the chemical properties of NMs, such as composition, structure, and molecular weight, determine their effectively significant roles vis-à-vis human health. Further, their applications as drug delivery systems and their potential for diagnosis and therapy are the major tools in nanomedicine. The nanosized systems are vital factors in developing the intracellular systems, architecting the biomimetic polymers, and controlling the delivery and action of advanced polymers in bio-systems for therapeutic purposes. In conclusion, it can be genuinely expected that a variety of nanostructures, with their unique physical and chemical properties, are likely to bring a great revolution in the field of medicine and healthcare, provided serious investigations are undertaken to understand the various apparent and hidden aspects of their toxicity and its adverse effects on the human biological systems.

References

Abbasi E, Aval SF, Akbarzadeh A, Milani M, Nasrabadi HT, Joo SW, Hanifehpour Y, Nejati-Koshki K, Pashaei-Asl R (2014) Dendrimers: synthesis, applications, and properties. Nanoscale Res Lett 9:247

Adachi E (2000) Three-dimensional self-assembly of gold nanocolloids in spheroids due to dialysis in the presence of sodium mercaptoacetate. Langmuir 16:6460–6469

Aina V, Perardi A, Bergandi L, Malavasi G, Menabue L, Morterra C, Ghigo D (2007) Cytotoxicity of zinc-containing bioactive glasses in contact with human osteoblasts. Chem Biol Interact 167:207–218

Albrecht MA, Evans CW, Raston CL (2006) Green chemistry and the health implications of nanoparticles. Green Chem 8:417–432

Arivazhagan V (2013) Investigation of quantum confinement effect in pbse/znse multiple quantum well structures prepared by thermal evaporation technique. PhD thesis, Department of Physics, Karunya University, Coimbatore, India

Ashraf MA, Peng W, Zare Y, Rhee KY (2018) Effects of size and aggregation/agglomeration of nanoparticles on the interfacial/interphase properties and tensile strength of polymer nanocomposites. Nanoscale Res Lett 13:214

Baeza A, Ruiz-Molina D, Vallet-Regi M (2017) Recent advances in porous nanoparticles for drug delivery in antitumoral applications: inorganic nanoparticles and nanoscale metal-organic frameworks. Expert Opin Drug Deliv 14:783–796

Bagul US, Pisal VV, Solanki NV, Karnavat A (2018) Current status of solid lipid nanoparticles: a review. Mod Appl Bioequiv Bioavail 3(MS.ID.555617):001–009

Barua S, Mitragotri S (2014) Challenges associated with penetration of nanoparticles across cell and tissue barriers: a review of current status and future prospects. Nano Today 9:223–243

Batra P, Mushtaq A, Mazumder J, Rizvi MS, Miglani R (2016) Nanoparticles and their applications in orthodontics. Adv Dent Oral Health 2:555584–555597

Bennet D, Kim S (2014) Polymer nanoparticles for smart drug delivery. In: Sezer AD (ed) Application of nanotechnology in drug delivery, Chapter 8, InTech, London, pp 257–310. https://doi.org/10.5772/58422.

Bhaumik A (2017) Porous nanomaterials for energy, environment and biomedical applications. J Mater Sci Nanomater 1:e109

Bonner JC (2016) Nanotechnology in pulmonary disease. In: Bhushan B (ed) Encyclopedia of nanotechnology. Springer Science + Business Media, Dordrecht, pp 2880–2885

Brune H, Giovannini M, Bromann K, Kern K (1998) Self-organized growth of nanostructure arrays on strain-relief patterns. Nature 394:451–453

Campani V, Giarra S, De Rosa G (2018) Lipid-based core-shell nanoparticles: evolution and potentialities in drug delivery. Open Nano 3:5–17

Cao G, Wang Y (2011) Nanostructures and nanomaterials: synthesis, properties and applications. Imperial College Press, London

Chang J, Waclawik ER (2014) Colloidal semiconductor nanocrystals: controlled synthesis and surface chemistry in organic media. RSC Adv 4:23505–23511

Dasan KP (2015) Nanoclay/polymer composites: recent developments and future prospects. In: Thakur V, Thakur M (eds) Eco-friendly polymer nanocomposites. Advanced structured materials, vol 75. Springer, New Delhi

Date AA, Hanes J, Ensign LM (2016) Nanoparticles for oral delivery: design, evaluation and state-of-the-art. J Control Release 240:504–526

Datt A, Ndiege N, Larsen SC (2012) Development of porous nanomaterials for applications in drug delivery and imaging. In: Nanomaterials for biomedicine, ACS Symposium Series, vol 1119. American Chemical Society, Washington, D.C, pp 239–258

Du J, Chen Y, Zhang Y, Han CC, Fischer F, Schmidt M (2003) Organic/inorganic hybrid vesicles based on a reactive block copolymer. J Am Chem Soc 125:14710–14711

Ehrman SH, Friedlander SK, Zachariah MR (1999) Phase segregation in binary SiO_2/TiO_2 and SiO_2/Fe_2O_3 nanoparticle aerosols formed in a premixed flame. J Mater Res 14:4551–4561

Elimelech M, Jia X, Gregory J, Williams R (1998) Particle deposition and aggregation: measurement, modelling and simulation, Colloid and Surface Engineering Series. Butterworth-Heinemann, Oxford, p 124

El-Say KM, El-Sawy HS (2017) Polymeric nanoparticles: promising platform for drug delivery. Int J Pharm 528:675–691

Endo Y, Sato K, Anzai J-I (2010) Preparation of avidin-containing polyelectrolyte microcapsules and their uptake and release properties. Polym Bull 66:711–720

Esmaeili A, Rahnamoun S, Sharifnia F (2013) Effect of O/W process parameters on *Crataegus azarolus* L. nanocapsule properties. J Nanobiotechnol 11:16–21

Esmailpour AA, Zarghami R, Mostoufi N (2015) Effect of temperature on the nanoparticles agglomerates fluidization. In: Proc. Int. Conf. modelling, simulation and applied mathematics (MSAM 2015). Atlantis Press, Tehran, pp 242–245

Esmailpour AA, Mostoufi N, Zarghami R (2018) Effect of temperature on fluidization of hydrophilic and hydrophobic nanoparticle agglomerates. Exp Thermal Fluid Sci 96:63–74

Ezhilarasi PN, Karthik P, Chhanwal N, Anandharamakrishnan C (2012) Nanoencapsulation techniques for food bioactive components: a review. Food Bioprocess Technol 6:628–647

Fendler JH (2001) Colloid chemical approach to nanotechnology. Korean J Chem Eng 18:1–13

Fiandaca MS, Bankiewicz KS (2013) Micelles and liposomes: lipid nanovehicles for intracerebral drug delivery. In: Kateb B, Heiss JD (eds) The textbook of nanoneuroscience and nanoneurosurgery. CRC Press, Taylor & Francis Group, Boca Raton, pp 51–64

Ganesan P, Ramalingam P, Karthivashan G, Ko YT, Choi D-K (2018) Recent developments in solid lipid nanoparticle and surface-modified solid lipid nanoparticle delivery systems for oral delivery of phyto-bioactive compounds in various chronic diseases. Int J Nanomedicine 13:1569–1583

Gavasane AJ, Pawar HA (2014) Synthetic biodegradable polymers used in controlled drug delivery system: an overview. Clin Pharmacol Biopharm 3:121

Gupta A, Eral HB, Hatton TA, Doyle PS (2016) Nanoemulsions: formation, properties and applications. Soft Matter 12:2826–2841

Halamoda-Kenzaoui B, Ceridono M, Urbán P, Bogni A, Ponti J, Gioria S, Kinsner-Ovaskainen A (2017) The agglomeration state of nanoparticles can influence the mechanism of their cellular internalisation. J Nanobiotechnol 15:48

Hillyer JF, Albrecht RM (2001) Gastrointestinal presorption and tissue distribution of differently sized colloidal gold nanoparticles. J Pharmacol Sci 90:1927–1936

Hoar TP, Schulman JH (1943) Transparent water in oil dispersions: the oleopathic hydromicelle. Nature 152:102–107

Husen A (2017) Gold nanoparticles from plant system: synthesis, characterization and application. In: Ghorbanpourn M, Manika K, Varma A (eds) Nanoscience and plant–soil systems, vol 48. Springer International Publication, Cham, pp 455–479

Husen A, Siddiqi KS (2014) Phytosynthesis of nanoparticles: concept, controversy and application. Nano Res Lett 9:229

Hussain S, Al-Nsour F, Rice AB, Marshburn J, Yinling B, Ji Z, Zink JI, Walker NJ, Garantziotis S (2012) Cerium dioxide nanoparticles induce apoptosis and autophagy in human peripheral blood monocytes. ACS Nano 6:5820–5829

Iyer R, Hsia CCW, Nguyen KT (2015) Nano-therapeutics for the lung: state-of-the-art and future perspectives. Curr Pharm Des 21:5233–5244

Jafari SM (2017) An overview of nanoencapsulation techniques and their classification. In: Jafari SM (ed) Nanoencapsulation technologies for the food and nutraceutical industries. Academic Press, London, pp 1–34

Jain D, Daima HK, Kachhwala S, Kothari SL (2009) Synthesis of plant-mediated silver nanoparticles using papaya fruit extract and evaluation of their antimicrobial activities. Digest J Nanomater Biostruct 4:557–563

Jiang J, Chen D-R, Biswas P (2007) Synthesis of nanoparticles in a flame aerosol reactor with independent and strict control of their size, crystal phase and morphology. Nanotechnology 18:285603–285611

Johnson LE, Johal MS (2018) Understanding nanomaterials, 2nd edn. CRC Press, Boca Raton

Jolivet JP, Froidefond C, Pottier A, Chanéac C, Cassaignon S, Tronc E, Euzen P (2004) Size tailoring of oxide nanoparticles by precipitation in aqueous medium. A semi-quant model. J Mater Chem 14:3281–3288

Joshi MD, Unger WJ, Storm G, van Kooyk Y, Mastrobattista E (2012) Targeting tumor antigens to dendritic cells using particulate carriers. J Control Release 161:25–37

Juškait V, Ramanauskien K, Briedis V (2015) Design and formulation of optimized microemulsions for dermal delivery of resveratrol. Evid Based Complement Alternat Med 540916:10

Kang G, Son H, Lim JM, Kweon H-S, Lee IS, Kang D, Jung JH (2012) Functionalized Fe_3O_4 nanoparticles for detecting zinc ions in living cells and their cytotoxicity. Chem Eur J 18:5843–5847

Khurshid Z, Zafar M, Qasim S, Shahab S, Naseem M, AbuReqaiba A (2015) Advances in nanotechnology for restorative dentistry. Mater 8:717–731

Kretzmann JA, Evans CW, Norret M, Iyer KS (2017) Supramolecular assemblies of dendrimers and dendritic polymers in nanomedicine. In: Atwood J (ed) Comprehensive supramolecular chemistry II. Academic Press (Elsevier Inc.), USA, pp 237–256

Kumari A, Singla R, Guliani A, Yadav SK (2014) Nanoencapsulation for drug delivery. EXCLI J 13:265–286

Kuntworbe N, Martini N, Shaw J, Al-Kassas R (2012) Malaria intervention policies and pharmaceutical nanotechnology as a potential tool for malaria management. Drug Dev Res 73:167–184

LaFemina JP (1995a) Tank waste treatment. Science Task Quarterly Report for January–March 1995. PNL10763

LaFemina JP (1995b) Tank waste treatment. Science Task Quarterly Report for April–June 1995. PNL1076x

Lao S-B, Zhang Z-X, Xu H-H, Jiang G-B (2010) Novel amphiphilic chitosan derivatives: synthesis, characterization and micellar solubilization of rotenone. Carbohydr Polym 82:1136–1142

Lee S-W, Chang S-H, Lai Y-S, Lin C-C, Tsai C-M, Lee Y-C, Chen J-C, Huang C-L (2014) Effect of temperature on the growth of silver nanoparticles using plasmon-mediated method under the irradiation of green LEDs. Materials 7:7781–7798

Levchenko AA, Li G, Boerio-Goates J, Woodfield BF, Navrotsky A (2006) TiO_2 stability landscape: polymorphism, surface energy and bound water energetics. Chem Mater 18:6324–6332

Li X, Lu T, Zhang J, Xu J, Hu Q, Zhao S, Shen J (2009) A study of properties of "micelle-enhanced" polyelectrolyte capsules: structure, encapsulation and in vitro release. Acta Biomater 5:2122–2131

Li X, Si Z, Lei Y, Tang J, Wang S, Su S, Song S, Zhao L, Zhang H (2010) Direct hydrothermal synthesis of single crystalline triangular Fe_3O_4 nanoprisms. Cryst Eng Comm 12:2060–2063

Liang XW, Xu ZP, Grice J, Zvyagin AV, Roberts MS, Liu X (2013) Penetration of nanoparticles into human skin. Curr Pharm Des 19:6353–6366

Lin LL, Yamada M, Prow TW (2016) Imaging nanoparticle skin penetration in humans. In: Hamblin MR, Avci P, Eds PTW (eds) Nanoscience in dermatology. Academic Press, London, pp 351–364

Lin CH, Chen CH, Lin ZC, Fang JY (2017) Recent advances in oral delivery of drugs and bioactive natural products using solid lipid nanoparticles as the carriers. J Food Drug Anal 25:219–234

Lingayat VJ, Zarekar NS, Shendge RS (2017) Solid lipid nanoparticles: a review. Nanosci Nanotechnol Res 4:67–72

Louchet F, Weiss J, Richeton T (2006) Hall-Petch Law revisited in terms of collective dislocation dynamics. Phys Rev Lett 97:075504–075509

Lundquist P, Artursson P (2016) Oral absorption of peptides and nanoparticles across the human intestine: opportunities, limitations and studies in human tissues. Adv Drug Deliv Rev (B) 106:256–276

Lundquist B, Rawstern R, Varga B, Liu L, Bergeson L (2017) The next big thing is really small: how nanotechnology will change the future of your business, Available at: http://www.nanotech-now.com/current-uses.htm. Accessed 5 Feb 2015

Lv Y, Wang H, Wang X, Bai J (2009) Synthesis, characterization and growing mechanism of monodisperse Fe_3O_4 microspheres. J Cryst Growth 311:3445–3450

Maghsoodi M, Yari Z (2014) Effect of temperature on wet agglomeration of crystals. Iran J Basic Med Sci 17:344–350

Maham M, Nasrollahzadeh M, Sajadi SM, Nekoei M (2017) Biosynthesis of Ag/reduced graphene oxide/Fe_3O_4 using Lotus garcinii leaf extract and its application as a recyclable nanocatalyst for the reduction of 4-nitrophenol and organic dyes. J Colloid Interf Sci 497:33–42

Maisel K, Ensign L, Reddy M, Cone R, Hanes J (2015) Effect of surface chemistry on nanoparticle interaction with gastrointestinal mucus and distribution in the gastrointestinal tract following oral and rectal administration in the mouse. J Control Release 197:48–57

Maryami M, Nasrollahzadeh M, Mehdipour E, Sajadi SM (2017) Green synthesis of the Pd/perlite nanocomposite as a heterogeneous catalyst for reduction of nitroarenes and organic dyes in water. Sep Purif Technol 184:298–307

Masood F (2016) Polymeric nanoparticles for targeted drug delivery system for cancer therapy. Mat Sci Engg C 60:569–578

McClements DJ (2012) Nanoemulsions versus microemulsions: clarification of critical differences. Soft Matter 8:1719–1729

Mitchnick M, Lee R, Cohen J, Becker B, Frank B, Gwozdz G, Zubris K, Okoh J, Goldman L (1991) Particle sciences drug development services, Available at: www.particlesciences.com. Accessed 15 May 2017

Mocan L (2013) Drug delivery applications of gold nanoparticles. Biotechnol Mol Bio Nanomed 1:1–7

Momeni SS, Nasrollahzadeh M, Rustaiyan A (2017) Biosynthesis and application of Ag/bone nanocomposite for the hydration of cyanamides in Myrica gale L. extract as a green solvent. J Colloid Interf Sci 499:93–101

Mora-Huertas CE, Fessi H, Elaissari A (2010) Polymer-based nanocapsules for drug delivery. Int J Pharm 385:113–142

Murphy CJ, Sau TK, Gole AM, Orendorff CJ, Gao J, Gou L, Hunyadi SE, Li T (2005) Anisotropic metal nanoparticles: synthesis, assembly, and optical applications. J Phys Chem B 109:13857–13870

Nasir A (2010) Nanodermatology: a Glimpse of Caution Just Beyond the Horizon – Part II, Available at: http://www.skintherapyletter.com/2010/15.9/2.html. Accessed 20 May 2017

Nasrollahzadeh M, Atarod M, Jaleh B, Gandomi M (2016a) In situ green synthesis of Ag nanoparticles on graphene oxide/TiO_2 nanocomposite and their catalytic activity for the reduction of 4-nitrophenol, Congo red and Methylene blue. Ceram Int 42:8587–8596

Nasrollahzadeh M, Sajadi SM, Hatamifard A (2016b) Waste chicken eggshell as a natural valuable resource and environmentally benign support for biosynthesis of catalytically active Cu/eggshell, Fe_3O_4/eggshell and Cu/Fe_3O_4/eggshell nanocomposites. Appl Catal B Environ 191:209–227

Nasrollahzadeh M, Momeni SS, Sajadi SM (2017) Green synthesis of copper nanoparticles using *Plantago asiatica* leaf extract and their application for the cyanation of aldehydes using $K_4Fe(CN)_6$. J Colloid Interf Sci 506:471–477

Nasrollahzadeh M, Issaabadi Z, Sajadi SM (2018a) Green synthesis of a Cu/MgO nanocomposite by Cassytha filiformis L. extract and investigation of its catalytic activity in the reduction of methylene blue, congo red and nitro compounds in aqueous media. RSC Adv 8:3723–3735

Nasrollahzadeh M, Issaabadi Z, Sajadi SM (2018b) Green synthesis of Pd/Fe_3O_4 nanocomposite using *Hibiscus tiliaceus* L. extract and its application for reductive catalysis of Cr(VI) and nitro compounds. Sep Purif Technol 197:253–260

Nasrollahzadeh M, Sajadi SM, Maham M, Kohsari I (2018c) Biosynthesis, characterization and catalytic activity of the Pd/bentonite nanocomposite for base- and ligand-free oxidative hydroxylation of phenylboronic acid and reduction of Chromium (VI) and nitro compounds. Micropor Mesopor Mater 271:128–137

Nasrollahzadeh M, Sajjadi M, Dasmeh HR, Sajadi SM (2018d) Green synthesis of the Cu/sodium borosilicate nanocomposite and investigation of its catalytic activity. J Alloy Compd 763:1024–1034

Nasrollahzadeh M, Issaabadi Z, Sajadi SM (2019) Green synthesis of Cu/Al_2O_3 NPs as an efficient and recyclable catalyst for reduction of 2,4-dinitrophenylhydrazine, Methylene blue and Congo red. Compos B Eng 166:112–119

Nastiti CMRR, Ponto T, Abd E, Grice JE, Benson HAE, Roberts MS (2017) Topical nano and microemulsions for skin delivery. Pharmaceutics 9:37

Noviendri D (2014) Microencapsulation of fucoxanthin by water-in-oil-in water (w/o/w) double emulsion solvent evaporation method: a review. Squalen Bull Mar Fish Postharvest Biotechnol 9:137–150

Núñeza JD, Benito AM, González R, Aragón J, Arenal R, Maser WK (2014) Integration and bioactivity of hydroxyapatite grown on carbon nanotubes and graphene oxide. Carbon 79:590–604

Nuruzzaman M, Rahman MM, Liu Y, Naidu R (2016) Nanoencapsulation, nano-guard for pesticides: a new window for safe application. J Agric Food Chem 64:1447–1483

Panwar P, Pandey B, Lakhera PC, Singh KP (2010) Preparation, characterization, and in vitro release study of albendazole-encapsulated nanosize liposomes. Int J Nanomedicine 5:101–108

Paranjpe M, Müller-Goymann CC (2014) Nanoparticle-mediated pulmonary drug delivery: a review. Int J Mol Sci 15:5852–5873

Park J, An K, Hwang Y, Park J-G, Noh H-J, Kim J-Y, Park J-H, Hwang N-M, Hyeon T (2004) Ultra-large-scale syntheses of monodisperse nanocrystals. Nat Mater 3:891–895

Parker R (2017) Quantum confinement: effects, observations and insights. Nova Science Publishers, New York

Pauw BR, Kastner C, Thunemann AF (2017) Nanoparticle size distribution quantification: results of a small-angle X-ray scattering inter-laboratory comparison. J Appl Cryst 50(5):1280–1288

Peddieson J, Chamkha AJ (2016) Modeling of nanofluid aggregation. Curr Nanomater 1(2):117–123

Peña-Parás L, Sánchez-Fernández JA, Vidaltamayo R (2018) Nanoclays for biomedical applications. In: Martínez L, Kharissova O, Kharisov B (eds) Handbook of ecomaterials. Springer, Cham, pp 1–19

Ragaei M, Sabry AH (2014) Nanotechnology for insect pest control. Int J Sci Environ Technol 3:528–545

Rajput N (2015) Methods of preparation of nanoparticles-a review. Int J Adv Res Technol 7:1806–1811

Ramteke KH, Joshi SA, Dhole SN (2012) Solid lipid nanoparticle: A review. IOSR J Pharm 2(6):34–44

Rector DR, Bunker BC (1995) Effect of colloidal aggregation on the sedimentation and rheological properties of tank waste. United States: N. p., 1995. Web. Pacific Northwest Lab., Richland, WA, USA. https://doi.org/10.2172/113874.PNL-10761

Riasat R, Guangjun N, Riasat Z, Aslam I (2016) Effects of nanoparticles on gastrointestinal disorders and therapy. J Clin Toxicol 6:313

Ryman-Rasmussen JP, Riviere JE, Monteiro-Riviere NA (2007) Surface coatings determine cytotoxicity and irritation potential of quantum dot nanoparticles in epidermal keratinocytes. J Invest Dermatol 127:143–153

Saini JK, Nautiyal U, Kumar MS, Singh D, Anwar F (2014) Microemulsions: a potential novel drug delivery system. Int J Pharm Med Res 2:15–20

Sajanlal PR, Pradeep T (2009) Mesoflowers: a new class of highly efficient surface-enhanced Raman active and infrared-absorbing materials. Nano Res 2:306–320

Sajjadi M, Nasrollahzadeh M, Sajadi SM (2017) Green synthesis of Ag/Fe_3O_4 nanocomposite using Euphorbia peplus L. leaf extract and evaluation of its catalytic activity. J Colloid Interf Sci 497:1–13

Sanders WC (2018) Basic principles of nanotechnology. CRC Press, Boca Raton

Schasfoort RBM (2017) Introduction to surface plasmon resonance. In: Schasfoorst RBM (ed) Handbook of surface plasmon resonance, 2nd edn. Royal Soc Chem, London, pp 1–26

Schneider CS, Craig S, Xu Q, Boylan NJ, Chisholm J, Tang BC (2017) Nanoparticles that do not adhere to mucus provide uniform and long-lasting drug delivery to airways following inhalation. Sci Adv 3(4):e1601556. https://doi.org/10.1126/sciadv.1601556

Selmer-Olsen E, Ratnaweera HC, Pehrson R (1996) A novel treatment process for dairy wastewater with chitosan produced from shrimp-shell waste. Wat Sci Tech 34:33–40

Shah N, Mewada RK, Shah T (2011) Application of biodegradable polymers in controlled drug delivery. Proc Int Conf on current trends in technology. Nirma University, Ahmedabad, pp 1–6

Siddiqi KS, Husen A (2016a) Fabrication of metal nanoparticles from fungi and metal salts: scope and application. Nanoscale Res Lett 11:98

Siddiqi KS, Husen A (2016b) Green synthesis, characterization and uses of palladium/platinum nanoparticles. Nanoscale Res Lett 11:482

Siddiqi KS, Husen A (2017a) Recent advances in plant-mediated engineered gold nanoparticles and their application in biological system. J Trace Elem Med Biol 40:10–23

Siddiqi KS, Husen A (2017b) Plant response to engineered metal oxide nanoparticles. Nanoscale Res Lett 12:92

Siddiqi KS, Rahman A, Tajuddin, Husen A (2016) Biogenic fabrication of iron/iron oxide nanoparticles and their application. Nanoscale Res Lett 11:498

Siddiqi KS, Husen A, Rao RAK (2018a) A review on biosynthesis of silver nanoparticles and their biocidal properties. J Nanobiotechnol 16:14

Siddiqi KS, Rahman A, Tajuddin HA (2018b) Properties of zinc oxide nanoparticles and their activity against microbes. Nanoscale Res Lett 13:141

Siddiqi KS, Husen A, Sohrab SS, Osman M (2018c) Recent status of nanomaterials fabrication and their potential applications in neurological disease management. Nanoscale Res Lett 13:231

Suganya V, Anuradha V (2017) Microencapsulation and Nanoencapsulation: a review. Int J Pharm Clin Res 9(3):233–239

Tamarov K, Näkki S, Xu W, Lehto V-P (2018) Approaches to improve the biocompatibility and systemic circulation of inorganic porous nanoparticles. J Mater Chem B 6:3632–3649

Tan C, Fung BM, Newman JK, Vu C (2001) Organic aerogels with very high impact strength. Adv Mater 13:644–651

Torchilin VP (2005) Recent advances with liposomes as pharmaceutical carriers. Nat Rev Drug Discov 4:145–160

Veszelka S, Bocsik A, Walter FR, Hantosi D, Deli MA (2015) Blood-brain barrier co-culture models to study nanoparticle penetration: focus on co-culture systems. Acta Biol Szeged 59:157–168

Wang L-P, Wang J-Y (2014) Skin penetration of inorganic and metallic nanoparticles. J Shanghai Jiaotong Univ (Sci) 19:691–697

Wang Q, Yan J, Yang J, Li B (2016b) Nanomaterials promise better bone repair. Mater Today 19:451–463

Wang Y, Li P, Tran TT-D, Zhang J, Kong L (2016c) Manufacturing techniques and surface engineering of polymer based nanoparticles for targeted drug delivery to cancer. Nano 6:26–32

Wang M, Lee RJ, Bi Y, Li L, Yan G, Lu J, Meng Q, Teng L, Xie J (2017) Transferrin-conjugated liposomes loaded with novel dihydroquinoline derivatives as potential anticancer agents. PLoS One 12:e0186821

Xia T, Zhu Y, Mu L, Zhang Z-F, Liu S (2016) Pulmonary diseases induced by ambient ultrafine and engineered nanoparticles in twenty-first century. Nat Sci Rev 3:416–429

Yah CS, Iyuke SE, Simate GS (2011) A review of nanoparticles toxicity and their routes of exposures. Iranian J Pharm Sci 8:299–314

Yasun E, Kang H, Erdal H, Cansiz S, Ocsoy I, Huang Y-F, Tan W (2013) Cancer cell sensing and therapy using affinity tag-conjugated gold nanorods. Interface Focus 3:1–9

Yeagle P (2017) Nanoparticles for drug delivery in lungs. Science 356:37–38

Zhang Y, Nypelö T, Salas C, Rojas OJ (2013) Cellulose nanofibrils: from strong materials to bioactive surfaces. J Renew Mater 1:195–206

Zielińska-Jurek A (2014) Progress, challenge, and perspective of bimetallic TiO_2-based photocatalysts. J Nanomater 4:1–17

Chapter 3
Plant-Mediated Fabrication of Gold Nanoparticles and Their Applications

Azamal Husen, Qazi Inamur Rahman, Muhammad Iqbal, Mansur Osman Yassin, and Rakesh Kumar Bachheti

3.1 Introduction

Nanotechnology, dealing with tiny particles of 1–100 nm, has gained increasing attention over the last three decades. These particles are commonly used in the household, industrial, and healthcare products and also have enormous potential in the nanotechnology-driven smart agriculture (Boxi et al. 2016; Fraceto et al. 2016; Siddiqi et al. 2016, 2018a, b, c, d; Siddiqi and Husen 2016, 2017a, b; Ovais et al. 2017). At present, above 1000 commercial products containing nanoparticles (NPs) are available in the market (Vance et al. 2015). The main challenges encountered during the fabrication of NPs relate to creating the desired shape, size, and monodispersity; and hence a refinement in the fabrication process is consistently required.

In general, NPs are fabricated by using two modes of preparation, i.e., the "bottom-up" (buildup of a material from the bottom: atom by atom, molecule by molecule, or cluster by cluster) and "top-down" (slicing or successive cutting of a bulk

A. Husen (✉)
Department of Biology, College of Natural and Computational Sciences, University of Gondar, Gondar, Ethiopia

Q. I. Rahman
Department of Chemistry, College of Natural and Computational Sciences, University of Gondar, Gondar, Ethiopia

M. Iqbal
Department of Botany, Faculty of Science, Jamia Hamdard (Deemed University), New Delhi, India

M. O. Yassin
Department of Surgery, College of Medicine and Health Sciences, University of Gondar, Gondar, Ethiopia

R. K. Bachheti
Department of Industrial Chemistry, Addis Ababa Science and Technology University, Addis Ababa, Ethiopia

A. Husen, M. Iqbal (eds.), *Nanomaterials and Plant Potential*,
https://doi.org/10.1007/978-3-030-05569-1_3

material to get nano-sized particle) procedures (Husen and Siddiqi 2014). The "bottom-up" procedure is usually preferred in both the chemical and biological syntheses of NPs (Vijayaraghavan and Nalini 2010; Narayanan and Sakthivel 2010a). On the other hand, the "top-down" procedure usually works with the material in bulk form, and the size reduction to the nanoscale is then achieved by specialized ablations, for instance, thermal decomposition, mechanical grinding, etching, cutting, lithography, laser ablation, and sputtering. The main demerit of this procedure is the surface structural defects, which have a significant impact on the physical features and surface chemistry of the metallic NPs.

Synthesis of metallic NPs through chemical reduction of metal salts in solution phase is most common (Lin et al. 2010), while physical approaches to synthesize metallic NPs include ultraviolet irradiation (Kundu et al. 2007), laser ablation (Tsuji et al. 2003), radiolysis (Meyre et al. 2008), sonochemistry (Okitsu et al. 2007), and so forth. Both the chemical and physical methods have been successful in producing well-defined NPs, but the use of plants or herbal extracts in NP fabrication has emerged as an alternative approach during the last few decades. This methodology is simple, cost-effective, and eco-friendly and can be easily scaled up for high yields (Husen and Siddiqi 2014, 2017a; Husen 2017). The extraction is done by soaking of plant samples in a green solvent; the extract so obtained contains flavonoids, terpenoids, proteins, reducing sugars, alkaloids, and other metabolites that act as the reducing and capping agents for reducing the metallic ions, and their concentrations are critical in governing the particle shape. The dried plants and their parts can be stored for longer time at room temperature, while fresh samples should be preserved at –20 °C to avoid deterioration. Moreover, since the seasonal and ontogenetic variations in phytochemical constituents are very common (Iqbal et al. 2011, 2018), dried plant samples collected at a proper time can be stored and used when needed. Thus, biogenic fabrication of NPs can occur with living as well as inactivated plant biomasses. Gardea-Torresdey et al. (2002, 2003) reported the possibility of using live alfalfa plants for the bioreduction of Au(III) to Au(0), which produced gold NPs ranging in size from 6 to 10 nm. *Brassica juncea* and *Medicago sativa* were also used to produce gold NPs at room temperature (Bali and Harris 2010). Aqueous extracts obtained from several plant leaves, roots, bark, seeds, fruits, galls, and petals have been used for this purpose (Table 3.1). Certain parameters such as pH, concentration, and temperature of reaction mixtures have to be adjusted to obtain certain size range, shape, and stability of the particles (Husen 2017; Siddiqi and Husen 2017a).

Gold NPs have drawn greater attention in the recent years due to their widespread uses. They have a larger surface area, higher dispersion owing to their very small size, and are highly stable and biocompatible. Some recent studies have elucidated the plausible positive and negative effects of gold NPs on plant growth and development (Siddiqi and Husen 2016). This chapter presents an overview of (a) the recent techniques of plant-mediated fabrication of gold NPs; (b) their characterization by UV-vis spectroscopy, thermogravimetric analysis, X-ray diffractometry, and SEM/TEM, among others; and (c) their application in some cutting-edge areas.

Table 3.1 Plant-mediated synthesis of gold nanoparticles, characterization techniques employed, and their application

Botanical name	Plant part used	Solvent used	Synthesis condition	Characterization techniques	Shape and Size	Phytoconstituents responsible for reduction of gold ion	Application	References
Abelmoschus esculentus	Pulp	Distilled water	Extract mixed with ingredient; reaction at room temperature with continuous stirring for 6 h	XRD, UV-vis, FTIR, RTEM, EDX, DLS	Spherical, triangle, and hexagonal; 4–32 nm	Phytochemicals, viz., vitamins and proteins	Anticancer and antimicrobial	Mollick et al. (2014)
Acacia nilotica	Leaf	Distilled water	Room temperature and mild condition	HRTEM, EDX, FTIR, XRD	Spherical; 6–12 nm	Flavonoids, tannins, triterpenoids, saponin, and polyphenolic compounds	Photocatalysis	Majumdar et al. (2013)
Aegle marmelos	Leaf	Distilled water	Solution based; mixing of ingredients; color change	UV-vis, XRD, FTIR, TEM, EDX, SAED, zeta potential	Mostly spherical; few irregular; 10.5–38 nm	Extract rich in polyphenolic tannin molecules	Detection of vitamin B or thiamine	Rao and Paria (2014)
Aloe vera	Leaf	Distilled water	Solution based; mixing of ingredients; reaction set for completion	UV-vis, XRD, FTIR, FESEM, EDX, HRTEM, SAED	Spherical and triangular; ~ 350 nm	Carbonyl, alcohols, phenols, and carboxylic acid derivative	–	Chandran et al. (2006)
Amomum subulatum	Black cardamom extract	Distilled water	Solution based; mixing of ingredients; reaction at appropriate pH; change in the solution color	UV-vis, XRD, FTIR, TEM, EDX	Different shapes and sizes; varying ratio of $AuCl_4$ ions/plant extract	1,8-cineole, β-pinene, and α-terpineol	–	Singh and Srivastava (2015)
Andrographis paniculata	Leaf	Distilled water	Solution based; sonication	UV-vis, XRD, FESEM, FTIR, EDX, HRTEM, SAED	Spherical; 5–75 nm	Phenolic acids, antioxidants, and flavonoids	Anticancner	Babu et al. (2012)
Antigonon leptopus	Leaf	Distilled water	Solution based; mixing of ingredients; reaction set at particular pH	UV-vis, XRD, FTIR, HRTEM, EDX, SAED, DLS	Spherical; 13–28 nm	Carbonyl, amide, and carboxylic groups	Antioxidant and anticancer	Balasubramani et al. (2015)
Artemisia capillaris	*A. capillaris* water extract	Distilled water	Solution based and used sonication	UV-vis, FTIR, XRD, TEM	Spherical; 17–29 nm	Flavonoids, amino acid, and phenolic compounds	Catalyst activity	Lim et al. (2016)
Beta vulgaris	Pulp	Distilled water	Solution based; mixing of ingredients; fixed at appropriate pH	UV-vis, FTIR, EDX, HRTEM, SAED	Nanowires: shape and size vary with extract concentration and pH	Protein and polysaccharides	–	Castro et al. (2011)

(continued)

Table 3.1 (continued)

Botanical name	Plant part used	Solvent used	Synthesis condition	Characterization techniques	Shape and Size	Phytoconstituents responsible for reduction of gold ion	Application	References
Cacumen platycladi	Leaf	Distilled water	Solution based; mixing of ingredients; reaction set at water bath at particular pH	UV-vis, XRD, FTIR, HRTEM, EDX, SAED, TG	Spheres, triangles, and hexahedrons; changes with respect to pH	Flavonoid and reducing sugar	–	Zhan et al. (2011)
Camellia sinensis	Leaf	Distilled water	Solution based; mixing of ingredients; the change in the color	UV-vis, FTIR, HRTEM, CV	Nanorods and nanoprisms; mean diameter ~ 20 nm	Polyphenolic compounds especially polyphenols and flavonoids	–	Begum et al. (2009)
Cassia tora	Leaf	Distilled water	Solution based mixing of ingredients	UV-vis, FTIR, HRTEM, DLS, zeta potential	Spherical; mean diam. ~ 5 nm	Amides, alcohols, and aromatic compounds	Anticancer and antioxidant	Abel et al. (2016)
Chenopodium album	Leaf	Distilled water	Solution method: change in color of solution	UV-vis, XRD, FTIR, EDX, TEM	Quasi-spherical; 10–30 nm	Oxalic acid	–	Dwivedi et al. (2010)
Cinnamomum zeylanicum	Leaf	Distilled water	Solution based; mixing of ingredients; change in the color of solution	UV-vis, XRD, FTIR, HRTEM, SAED, PL	Spherical; an average size 25 nm	Terpenoids like eugenol, cinnamaldehyde, tannin, and sucrose	Photoluminescence (PL)	Smitha et al. (2009)
Commelina nudiflora	Whole plants	Distilled water	Solution based; mixing of ingredients; change in color	UV-vis, XRD, FESEM, FTIR, EDX	Spherical; 50–150 nm	Polysaccharides (sugar) and proteins	Antibacterial and antioxidant	Kuppusamy et al. (2015)
Coriandrum sativum	Leaf	Distilled water	Solution based; mixing of ingredients; change of color of solution	UV-vis, XRD, FTIR, EDX, TEM, SAED	Spherical, triangles, and decahedral; 8–58 nm	Active biomolecules, viz., amine, amide, and carboxylic group in protein	–	Narayanan and Sakthivel (2008)
Cymbopogon flexuosus	Leaf	Distilled water; chloroform	Column chromatography	UV-vis, FTIR, TEM, AFM, SAED, NMR	Triangular; mean diameter ~80 nm	Ketones, aldehydes, and carboxylic acids	–	Shankar et al. (2004)
Cystoseira baccata	Leaf	Distilled water	Mixing of ingredients with leaf extract; solution kept under stirring for 24 h	UV-vis, XRD, FTIR, HRTEM, EDS, STEM, zeta potential	Spherical; ~2.2 ± 8.4 nm	Polysaccharides, phenolic compounds, proteins, vitamins, and terpenoids	Anticancer	González-Ballesteros et al. (2017)

Dillenia indica	Fruit	Distilled water	Solution based; mixing of ingredients; change in the color of solution	UV-vis, XRD, FTIR, HRTEM, TGA, DSC	Spherical, triangular, tetragonal, pentagonal; irregular contours; 5–50 nm	High phenolic content	Cytotoxicity study	Sett et al. (2016)
Diospyros kaki and *Magnolia kobus*	Leaf	Distilled water	Mixing of reactant and then reaction set for reflux at water bath, 25–95 °C	FTIR, ESEM, AFM, RTEM, XPS, EDX	A mixture of plate (triangles, pentagons, hexagons) and spheres; 5–300 nm	Some proteins, terpenoids having functional groups of amines, alcohols, ketones, aldehydes, and carboxylic acids	–	Song et al. (2009)
Euphorbia hirta	Leaf	Distilled water	Solution based; mixing of ingredients; reaction set for a week	UV-vis, XRD, FTIR, EDX, HRTEM, AFM, SAED, Raman	Spherical; 10–50 nm	Various biomolecules	Antibacterial	Annamalai et al. (2013)
Guggulutiktam kashayam	Herb	Distilled water	Solution method: change in color of solution	UV-vis, XRD, FTIR, EDX, TEM	Spherical; 20–35 nm	Several biomolecules	Catalytic activity	Suvith and Philip (2014)
Madhuca longifolia	Leaf	Distilled water	Mixing of ingredients; reaction at fixed pH 2, under continuous stirring	UV-vis, FTIR, TEM, fluorescence emission	Triangular nanoplates; 7 nm to 3 μm	Plant proteins, viz., tyrosine	IR blocker	Fayaz et al. (2011)
Momordica charantia	Fruit	Distilled water	Solution based; mixing of ingredients; set at 100 °C	UV-vis, XRD, FTIR, HRTEM, EDX, SAED	Spherical; 30–100 nm	Triterpenes, proteins, and steroids	Depletion of nitrate reductase activity	Panday et al. (2012)
Moringa oleifera	Flower	Distilled water	Solution based; mild condition	TEM, UV-vis, SEM, EDX, Zeta potential	Spherical; 3–6 nm	Trace aromatic but abundant aliphatic compounds, i.e., proteins and lipids	Catalytic reduction and anticancer	Anand et al. (2015)
Morinda citrifolia	Root	Distilled water	Solution based; mixing of ingredients; changes in color of solution	UV-vis, XRD, FTIR, FESEM, EDX, HRTEM, AFM	Spherical and triangular; 12–38 nm	Protein associated with amide, carbonyl, aldehyde groups	–	Suman et al. (2014)
Musa paradisiaca	Peel	Distilled water	Solution based; mixing of ingredients; solution at 353 K for 20 min, change in solution color	UV-vis, XRD, FTIR, SEM, EDX, TEM, zeta potential	Spherical and triangular; ~ 50 nm	Phenols and carboxylic acids and amide compounds	Antibiotic resistance and anticancer	Vijayakumar et al. (2017)

(continued)

Table 3.1 (continued)

Botanical name	Plant part used	Solvent used	Synthesis condition	Characterization techniques	Shape and Size	Phytoconstituents responsible for reduction of gold ion	Application	References
Naregamia alata	Leaf	Distilled water	Mixing of ingredients with leaf extract; solution put in the microwave reactor for 1mint	UV-vis, XRD, FTIR, FESEM, EDX, HRTEM, AFM	Polyshaped nanoparticles; ~9.19 ± 27.92 nm	Secondary flavonoids	Catalysis	Francis et al. (2017b)
Nigella sativa	Seeds	Distilled water	Seed extract in water by soaking and boiling and mixed with ingredient change in color of solution	TEM, UV-vis, FTIR, EDX	Spherical, triangular, and hexagonal; 15–29 nm	Vapors rich in amide, alcohol, phenolic group compounds, and their derivatives	Antibacterial and anticancer	Manju et al. (2016)
Olea europaea	Leaf	Distilled water	Solution method: mixing of reactant at appropriate pH; change in color of solution	UV-vis, FTIR, XRD, TEM, TGA	Triangular, hexagonal, and spherical; 50–100 nm	Potential biomolecules, viz., oleuropein, apigenin-7-glucoside, and/or luteolin-7-glucoside	–	Khalil et al. (2012)
Pistacia integerrima	Gall extract of *Pistacia integerrima*	Distilled water	Solution based; mixing of ingredients; color change at particular pH	UV-vis, XRD, FTIR, SEM, EDX, TEM	Spherical; shapes changing with pH change; 20–200 nm	Terpenoids having functional groups of amines, alcohol, phenol, aldehyde, ketones, and carboxylic acid	Enzyme inhibition, antibacterial, antifungal, antinociceptive, muscle relaxant, and sedative	Islam et al. (2015a, b)
Plumbago zeylanica	Bark	Distilled water	Solution based; mixing of ingredients with bark extract; change in color of solution	UV-vis, XRD, FTIR, SEM, TEM, EDX	Spherical; ~ 28 nm	Plumbagin and hydroplumbagin glucoside containing many hydroxyl groups	Antioxidant, antimicrobial, and cytotoxic	Velammal et al. (2016)
Punica granatum	Juice	Distilled water	Solution based; after mixing ingredients set the reaction for 24 h; change in color	UV-vis, XRD, FTIR, HRTEM, SAED	Spherical, triangular, and hexagonal; 23–35 nm	Terpenoids, alkaloids, sugar, amino acid, polyphenols, fatty acids, aromatic compounds	Catalytic reduction	Dash and Bag (2014)
Rosa hybrid	Petal	Distilled water	Solution based; mixing of ingredients; reaction set on magnetic heater stirrer at 80 °C for 1 h	UV-vis, XRD, FTIR, EDX, XPS, HRTEM, SAED, DLS	Polydispersed spherical, triangular, and hexagonal; ~10 nm	Sugar and proteins	–	Noruzi et al. (2011)

Salicornia brachiata	Extract	Distilled water	Solution based; mixing of ingredients; solution at 60 °C with trace amount of $NaBH_4$, change in color of solution	UV-vis, XRD, FTIR, FESEM, EDX, HRTEM, SAED	Polydispersed spherical; 22–35 nm	Polyphenols, glycosides, flavonoids, carbohydrates, and protein	Catalytic and antibacterial	Ahmad et al. (2014)
Salix alba	Leaf	Distilled water	Solution based; reactant-mixed; color change recorded	UV-vis, FTIR, SEM, AFM	Spherical; 50–80 nm	Amines, amide, and aromatic groups	Enzyme inhibition, antibacterial, antifungal, antinociceptive, muscle relaxant and sedative	Islam et al. (2015)
Sapindus mukorossi	Pericarp (soap nut shells)	Distilled water	Solution method: change in color of solution	UV-vis, FTIR, XRD, TEM, EDX, SAED	Quasi-sphere; 6–15 nm	Carboxylic groups in the saponins and the carbonyl groups in the flavonoids	Catalytic activity	Reddy et al. (2012)
Sphaeranthus amaranthoides	Leaf	Distilled water	Solution based: mixing of ingredients; change in the color	UV-vis, FTIR, EDX, HRTEM	Spherical; 39–47 nm	Carbohydrate, tannins, saponins, steroids, glycosides, terpenoids, and alkaloids	–	Nellore et al. (2012)
Stevia rebaudiana	Leaf	Distilled water	Solution based; mixing of ingredients; set at a rotation rate of 150 rpm at 30 °C	UV-vis, XRD, FTIR, HRTEM, EDX, SAED, TG	Octahedral and some nanohexagons; 8–20 nm	Phytochemical present in extract	–	Mishra et al. (2010)
Stevia rebaudiana	Leaf	Distilled water	Solution based; mixing of ingredients; set for reaction with continuous stirring	UV-vis, XRD, FTIR, SEM, EDX, TEM, Zeta Potential	Spherical, 5–20 nm	Terpenoids and proteins	–	Sadeghi et al. (2015)
Terminalia catappa	Leaf	Distilled water	Solution method, mixing of reactant; changes in the color of solution	UV-vis, XRD, FTIR, TEM	Spherical; 10–35 nm	Hydrolysable tannin, polyphenols and carboxylic compounds	–	Ankamwar (2010)
Terminalia arjuna	Fruit extract	Distilled water	Solution based: mixing of ingredients; change in the color of solution	UV-vis, XRD, FTIR, TEM, EDX, SAED, AFM, DLS, zeta potential	Spherical; 5–50 nm	Tannin, terpenoid, saponins, flavonoids, glycosides, and polyphenolic compounds	Seed germination enhancer	Gopinath et al. (2014)

(continued)

Table 3.1 (continued)

Botanical name	Plant part used	Solvent used	Synthesis condition	Characterization techniques	Shape and Size	Phytoconstituents responsible for reduction of gold ion	Application	References
Terminalia arjuna	Leaf	Distilled water	Solution based; mixing of ingredients; change the color of solution	UV-vis, XRD, FTIR, SEM, TEM, EDX, AFM, SAED	Spherical; 20–50 nm	Arjunetin, leucoanthocyanidins, and hydrolyzable tannins	Induces mitotic cell division and pollen germination	Gopinath et al. (2013)
Terminalia arjuna	Bark	Distilled water	Aliquot of bark extract mixed with reactant; solution fixed at 80 °C for 15 min; change in the solution color	UV-vis, XRD, FTIR, FESEM, HRTEM, EDX, DLS, zeta potential	Spherical and triangular; 20–50 nm	Polyphenols	Neuroprotective potential via antioxidant, anticholinesterase, and antiamyloidogenic effects	Suganthy et al. (2018)
Withania somnifera	Leaf	Distilled water; ethanol	Solution based; mixing of reactant; the reaction set at particular pH	UV-vis, FTIR, TEM, EDX	Spherical and hexagonal	The phenolic groups (–OH) residue	–	Bindhani and Panigrahi (2014)

3.2 Fabrication and Characterization of Gold Nanoparticles

The principal biomolecules such as amines, amino acids, aldehydes, ketones, carboxylic acids, phenols, proteins, flavonoids, saponins, steroids, alkaloids, and tannins and different nutritional compounds present in plant parts and their extracts reduce the metal ion to NP (Fig. 3.1 and Table 3.1). Gold NPs show a distinct optical response usually ascribed to the localized surface plasmon resonance (SPR), i.e., the collective oscillation of electrons in the conduction band of gold NPs in resonance with a specific wavelength of incident light. The SPR of gold NPs results in a strong absorbance band in the visible region of 500–600 nm, which can be measured by UV-Vis spectroscopy, the first technique used to characterize gold NPs. In addition, several other techniques such as transmission electron microscopy, scanning electron microscopy, X-ray diffraction, Fourier transform infrared spectroscopy, atomic force microscopy, energy-dispersive X-ray spectroscopy, dynamic light scattering, zeta potential, surface-enhanced Raman spectroscopy, nuclear magnetic resonance spectroscopy, and others are also used.

3.2.1 Fabrication of Gold Nanoparticles

As the seasonal changes and phyto-developmental stages considerably affect the chemical constituents in plant tissues (Iqbal et al. 2011, 2018), it is advised to collect the relevant plant material at a proper time and optimal stage of plant development and store it in dried form. The dried plant materials can be stored for long durations at room temperature. However, to avoid any deterioration, the material may be preserved at –20 °C. The extracts of the whole plant or plant parts (leaves, stems, roots, bark, seeds, flowers, or floral parts) in appropriate solvents contain the capping and reducing agents that are required to reduce the metallic ions. These biomolecules actively participate in the bioreduction process. Shankar et al. (2003) obtained gold NPs from geranium (*Pelargonium graveolens*) leaf extract. This fabrication process, resulting in gold NPs of spherical, triangular, decahedral, and icosahedral shapes, was accomplished within 48 h. The presence of terpenoids in the extract was held responsible for the reduction of gold ions and the formation of gold NPs. In another study, Chandran et al. (2006) produced gold NPs from *Aloe vera* leaf extract and controlled their shape and size; these were triangular in shape and 50–350 nm in size. Both the shape and size were dependent on the leaf extract quantity. Low concentration of the leaf extract added to chloroauric acid ($HAuCl_4$) solution increased the production of triangular NPs. With a high concentration of leaf extract, the ratio of nanogold triangles to sphericals was reduced. It was proposed that the carbonyl functional groups found in the leaf extract were responsible for the reduction of gold ions and NP production.

With the advancement in the plant-mediated NP fabrication techniques, some researchers used sun-dried leaf powder dissolved in water at ambient temperature as

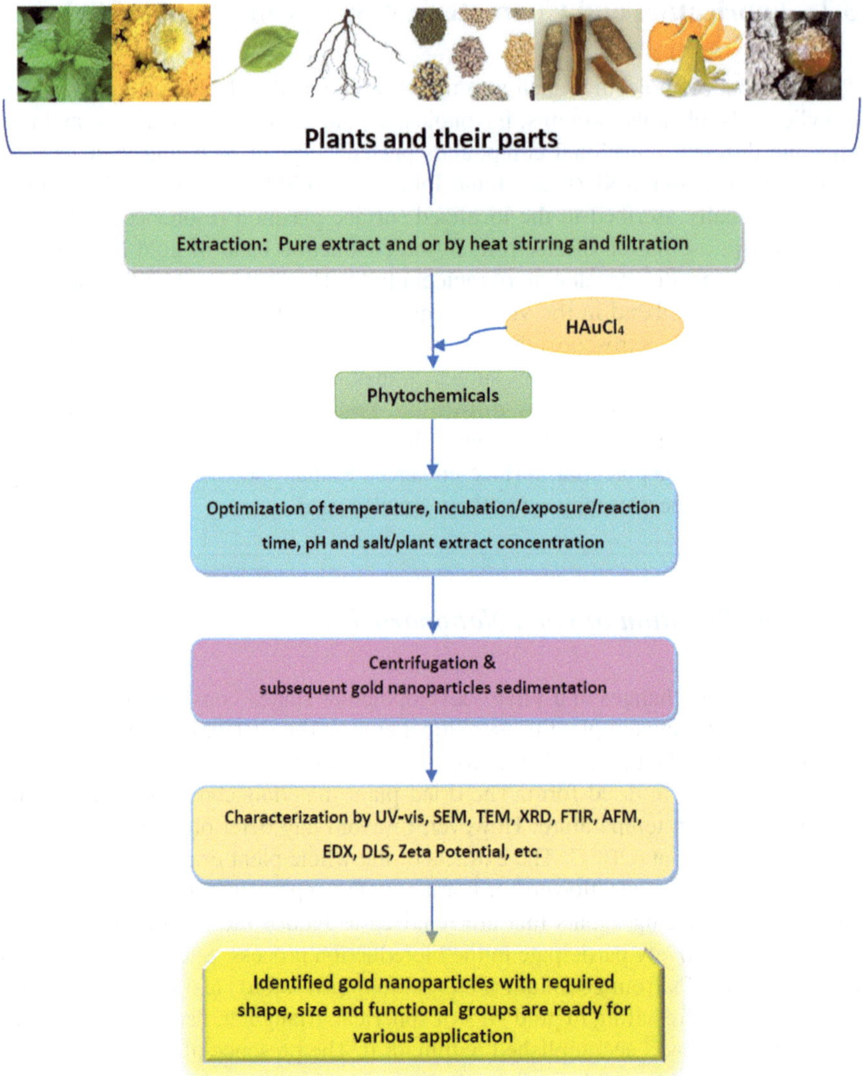

Fig. 3.1 Fabrication and characterization of gold nanoparticles using various plant parts

an alternative of the plant extract obtained by boiling. For this procedure, no accelerator/moderator, viz., ammonia, was required; nonetheless, the concentration of various plant leaf extracts was the rate-determining step. For instance, Huang et al. (2007a) used a sun-dried powder of *Cinnamomum camphora* leaves for the fabrication of gold and silver NPs at ambient temperature. They executed a series of experiments by adding 0.1 g and 0.5 g dried leaf extract to 50 mL of 1 mM aqueous $HAuCl_4$ and synthesized gold NPs of 100 nm and 200 nm size, respectively. Fourier transform infrared (FTIR) spectroscopy confirmed that the biomass of dried leaves was rich in polyols, which were responsible for the bioreduction of Au^{3+} to Au^0. In another study, Phillip (2009) used the dried leaf powder extract of *Mangifera indica* for the synthesis of spherical gold NPs at room temperature. These particles were smaller, more uniform in size, and stable for more than 5 months. Further, FTIR spectroscopy has demonstrated the role of water-soluble compounds, such as flavonoids, terpenoids, and thiamine, as stabilizing agents in the gold NP fabrication.

Narayanan and Sakthivel (2010b) obtained gold NPs by using the aqueous extract of *Coleus amboinicus* leaves. The extract was rich in aromatic amines, amide (II) groups, and secondary alcohols, which acted as capping agent during the bioreduction of Au(III) to Au(0). Bioreduction of $HAuCl_4$ was successfully observed by monitoring the change in color of the solution by gradual addition of leaf extract; a distinct UV-vis absorption peak was seen at 536 nm, which corresponded to the SPR of gold NPs. The particles of different shapes (spherical, triangular, truncated triangular, hexagonal, and decahedral) measured 4.6–55.1 nm. Noruzi et al. (2011) described an inexpensive and easy method for facile synthesis of gold NPs from rose petals extract and $HAuCl_4$. With a 10% extract solution, 2 mM Au^{3+} solution failed to form gold NPs; on increasing the extract concentration, however, anisotropic gold NPs were successfully formed, which was confirmed by the peak at 750 nm in SPR spectrum. As the desired concentration of the extract is mixed with the Au^{3+} solution, the solution color changed from yellow (extract color) to violet within 5 min at room temperature, which indicated the formation of gold NPs. The SPR band at ≈525 nm confirmed the formation of gold NPs, which were polydispersed with different shapes (viz., spherical, triangular, and hexagonal) and an average particle size of 10 nm, as determined by the dynamic light-scattering (DLS) method. Philip et al. (2011) proposed an easy and cost-effective protocol based on the leaf extract of *Murraya koenigii* and $HAuCl_4$ solution. The leaf extract was reportedly rich in polyphenols, alkaloids, carbazole, and flavonoids, which were responsible for bioreduction of Au(III) into Au(0). The NPs produced were spherical, with an average size of 20 nm. These were quite stable, showing no aggregation for more than 2 months.

Nellore et al. (2012) reported a simple procedure to synthesize gold NPs by gradual addition of the leaf extract of *Sphaeranthus amaranthoides* into the $HAuCl_4$ solution, which changed the solution color from pale yellow to purple-red within 5 min, indicating the formation of NPs that exhibited a well-resolved SPR band at 525 nm. The SPR spectrum remained unchanged even after 30 days, indicating that the NPs produced were quite stable in aqueous solution without aggregation. FTIR spectroscopy of the leaf extract before and after the addition of gold solution revealed an abundance of carbohydrate, tannins, steroids, glycosides, terpenoids, and alkaloids in the leaf extract. High-resolution transmission electron microscopy (*HRTEM*) images confirmed that the synthesized NPs were polydispersed and predominantly spherical in shape with a size range of 39–47 nm. Ghosh et al. (2012) used flower extract (*Gnidia glauca*) for gold NP fabrication, which was evident by the change in color from yellow to dark-red in the visible range of the spectrum (450–600 nm). The reaction started 2 min after the interaction of flower extract with $HAuCl_4$ solution and was accomplished in 20 min, thus showing a greater efficiency in comparison to the earlier work of Vankar and Bajpai (2010), wherein the reaction completed in ~2 h. Gopinath et al. (2013) synthesized spherical gold NPs of 20–50 nm by using the leaf extract of *Terminalia arjuna* with $HAuCl_4$ solution, whereas Annamalai et al. (2013) used *Euphorbia hirta* leaf extract and $HAuCl_4$ solution to obtain the monodispersed, almost spherical NPs of 10–50 nm. Majumdar et al. (2013) used $HAuCl_4$ salt solution with *Acacia nilotica* leaf extract rich in flavonoids, tannins, triterpenoids, and saponins, which facilitated the reduction of auric salt to neutral metal ions. The reaction resulted in highly crystalline monodispersed spherical gold NPs of 6–12 nm size. The particle size significantly decreased as the leaf extract concentrations were increased. Dash and Bag (2014) used *Punica granatum*juice, rich in phytochemicals (viz., terpenoids, sugar, polyphenols, alkaloids, fatty acids, and aromatic compounds) for fabrication of gold NPs, which were triangular, pentagonal, hexagonal, and spherical in shape. With 1440–2400 mg L^{-1} concentration of the juice extract, the average particle size exhibited a gradual decrease from 35.8 to 23.1 nm.

With progress in the plant-mediated fabrication of gold NPs, some researchers started using microwave radiation for rapid and easy NP synthesis. For instance, Yasmin et al. (2014) could fabricate the stable spherical gold NPs of 16–30 nm from the fresh chopped leaves of *Hibiscus rosa-sinensis* by using microwave heating for 3 min. The authors suggested that alkaloids and flavonoids present in the leaf tissue played a key role in the fabrication process. Likewise, Joseph and Mathew (2015a) used microwave radiation for this purpose using the extract of chopped fresh leaves of *Aerva lanata* boiled in distilled water. The extract was rich in various alkaloids, flavonoids, and other phytochemicals which brought about bioreduction of Au^{3+} to Au^0. The synthesized NPs were polydispersed and mostly spherical in general, with some being triangular, hexagonal, and plate-like in shape. The average diameter of spherical particles was 17.97 nm, exhibiting the characteristic SPR band at ~535 nm.

Stable gold NPs (triangular, hexagonal, and nearly spherical) were also fabricated by using 1 M $HAuCl_4$ and flower extract of *Moringa oleifera* (Anand et al. 2015). Free from impurity, these NPs of 5 nm average diameter were homogeneously distributed throughout and displayed a well-resolved SPR band at 540 nm. The ^{1}H-NMR spectroscopy and FTIR studies have confirmed that the secondary metabolites, viz., trace, aromatic but abundant aliphatic compounds (proteins and lipids), were involved in the bioreduction of $Au^{3+} \rightarrow Au^0$. Manju et al. (2016) made use of the *Nigella sativa* seed oil for this purpose. The NPs (spherical, triangle, and hexagonal) were in the range of 15–28.4 nm and showed a distinct SPR band at 540 nm.

In a recent study, Vijayakumar et al. (2017) have fabricated gold NPs using the extract of banana (*Musa paradisiaca*) peels and $HAuCl_4$ solution. Banana peel contains antioxidant compounds, viz., gallocatechin and dopamine, that caused bioreduction of Au(III) to Au(0). The triangular to spherical NPs of about 50 nm exhibited well-resolved SPR band at 541 nm which was characteristic of gold NPs. Suganthy et al. (2018) synthesized gold NPs by using bark extract of *Terminalia arjuna*. The phytochemicals present in the bark extract, including polyphenols such as (+)-catechin, ellagic acid, gallic acid, and their derivatives, were responsible to facilitate the reduction of Au^{3+} ions. The authors also succeeded in obtaining Au and Pd bimetallic NPs of different shapes. The gold NPs were anisotropic ranging in size from 3 to 70 nm with average diameter of 30 nm and showing a broad SPR band near 536 nm. In another study, Raouf et al. (2017) used the dried powder as well as ethanolic extract of a red alga (*Galaxaura elongata*) to synthesize gold NPs. A strong SPR band was seen at ~535 nm and ~536 nm for the NPs formed by the algal ethanolic extract and the algal powder, respectively. The FTIR study showed a high percentage of andrographolide and alloaromadendrene oxides and suggested for the reduction of $HAuCl^-$, which acted as the stabilizing agent during synthesis. The particles were spherical in shape along with a few rods, triangular, truncated triangular, and hexagonal ones, and exhibited a wide range of size from 3.85 to 77.13 nm, which was inconsistent with zeta potential results.

3.2.2 Characterization of Gold Nanoparticles

The fabricated gold NPs exhibited distinct optical and physical properties, depending on their size (diameter), shape, surface structure, and agglomeration state. The various characterization techniques are summarized in Table 3.1.

3.2.2.1 Ultraviolet-Visible (UV-Vis) Spectroscopy

UV-vis spectroscopy is the most important analytical technique to characterize the formation and stability of gold NPs. The formation is determined by monitoring the change in color during the synthesis. When the reactant is mixed with appropriate plant extract (e.g., leaf, seed, flower, fruit, etc.), the color of the mixture changes (from yellow to violet) immediately. This is attributed to electromagnetic radiation with free electron present in the conduction band of gold NPs and exhibits strong absorbance band in the visible region (500–600 nm) known as SPR. Figure 3.2

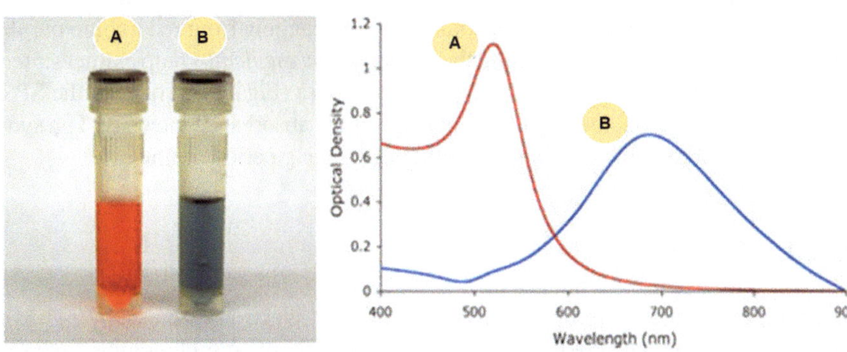

Fig. 3.2 Shape-dependent localized SPR as indicated by the visual appearance and UV-vis spectra of (**a**) the spherical and (**b**) urchin-shaped gold nanoparticles. (Adopted from: www.cytodiagnostics.com)

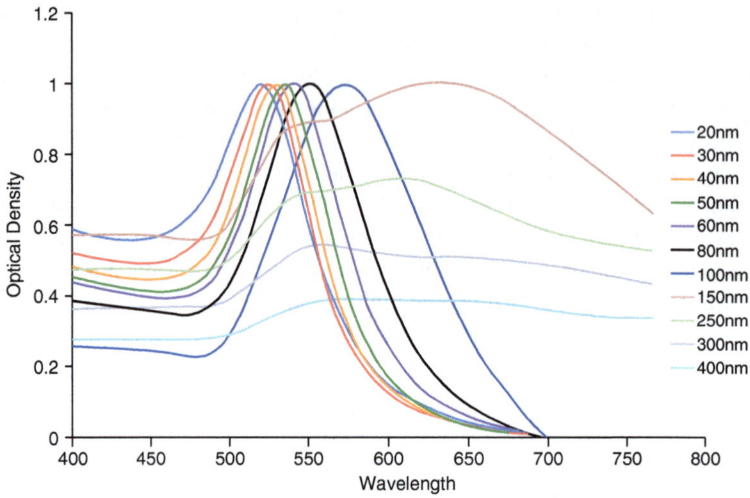

Fig. 3.3 Size-dependent SPR spectrum of gold nanoparticles, showing the presence of red shift of the absorption maximum, as the size of nanoparticles increases. (Adopted from: www.cytodiagnostics.com)

shows the shape-dependent SPR spectrum of gold NPs. As the particle diameter increases, absorption peak shifts toward higher wavelength. For uneven shaped particles such as gold nano-urchins, the absorption peak position shifts into the far red region of the spectrum as compared to the spherical particle of the same diameter. Figure 3.3 also reveals UV-vis spectra of gold NPs as a function of reaction time; it has been observed that there is no significant difference between the intensity of SPR bands in minutes 0, 5, 10, 15, 20, and 25, indicating that the reaction has been completed during the first minute.

The stability of synthesized gold NPs is strictly dependent on the degree of aggregation, i.e., irreversible interparticle coupling, which is accompanied by shifting of the characteristic absorption peak toward the red region of the electromagnetic radiation. Mollick et al. (2014) found the solution color changing from red to blue/purple while synthesizing gold NPs by using the *Abelmoschus esculentus* pulp extract stored at room temperature for 5 months. The UV-vis spectra (Fig. 3.4) produced the SPR band at 538 nm without any significant change in the absorbance intensity. The synthesized NPs were stable at room temperature for a long period of time.

3.2.2.2 Microscopy

Scanning electron microscopy (SEM), transmission electron microscopy (TEM), and atomic force microscopy (AFM) are used to study the morphological features of synthesized gold NPs effectively. SEM is used to examine the morphology of as-synthesized nanomaterials to determine the shapes of particles and their surface features at nanoscale level. It measures electron scattering from the surface of the sample, which requires highly accelerated short wavelength electrons for

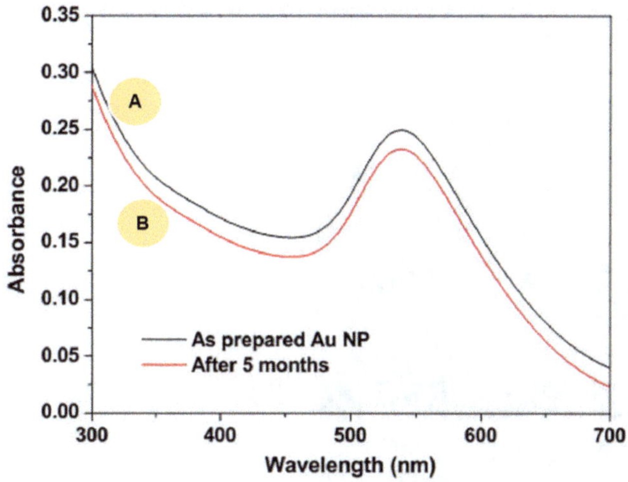

Fig. 3.4 UV-vis absorption spectra of gold nanoparticles synthesized by using pulp extract of *Abelmoschus esculentus*; (**a**) after reaction accomplished; (**b**) after 5 months at room temperature. (Adopted from: Mollick et al. 2014)

high-resolution images, which could be magnified up to 200,000 times. Chandran et al. (2014) successfully synthesized gold NPs by using leaf extracts of two medicinally important plants *Cucurbita pepo* and *Malva crispa* as shown in Fig. 3.5. The elemental and chemical compositions of as-synthesized gold nanomaterials are determined by energy-dispersive spectroscopy (EDS), which is usually attached with either FESEM or TEM. When the highly energetic electron beam is focused over the nanomaterial, each of its constituent elements emits characteristic energy X-rays by electron beam irradiation.

In the case of TEM, high-energy electrons accelerated with 200 KV are transmitted through an ultrathin specimen. After successful interaction with the sample, electrons are transmitted, and an image appears on the screen which provides information of the bulk material from very low to high magnification. TEM is typically used to determine the physical size of as-synthesized materials and their nature (crystalline or amorphous). Figure 3.6 includes TEM micrographs revealing different features of gold NPs synthesized by using the leaf extract of *Acacia nilotica* at 60 (Fig. 3.6a–e), 100 (Fig. 3.6f–g), and 200 mg L^{-1} (Fig. 3.6i–k) concentrations. The synthesized NPs were predominantly spherical in shape. As the concentration of the extract increased from 60 to 200 mg L^{-1}, the average particle size varied from 12.24 to 5.99 nm. Figure 3.6c shows HRTEM image of gold NPs with d-spacing value of 0.24 nm, which matched fully with the expected d-spacing of the [111] plane of face-centered cubic crystalline Au (JCPDS, no. 04–0784). Selected area electron diffraction (SAED) is a crystallographic experimental technique that can be linked to TEM in order to obtain valuable insight regarding the crystalline nature of the material synthesized and its analogy to X-ray powder diffraction. The SAED patterns can be used to identify the crystal structures and measure the lattice parameters. Figure 3.6h shows the SAED pattern obtained from a gold NP; the diffraction

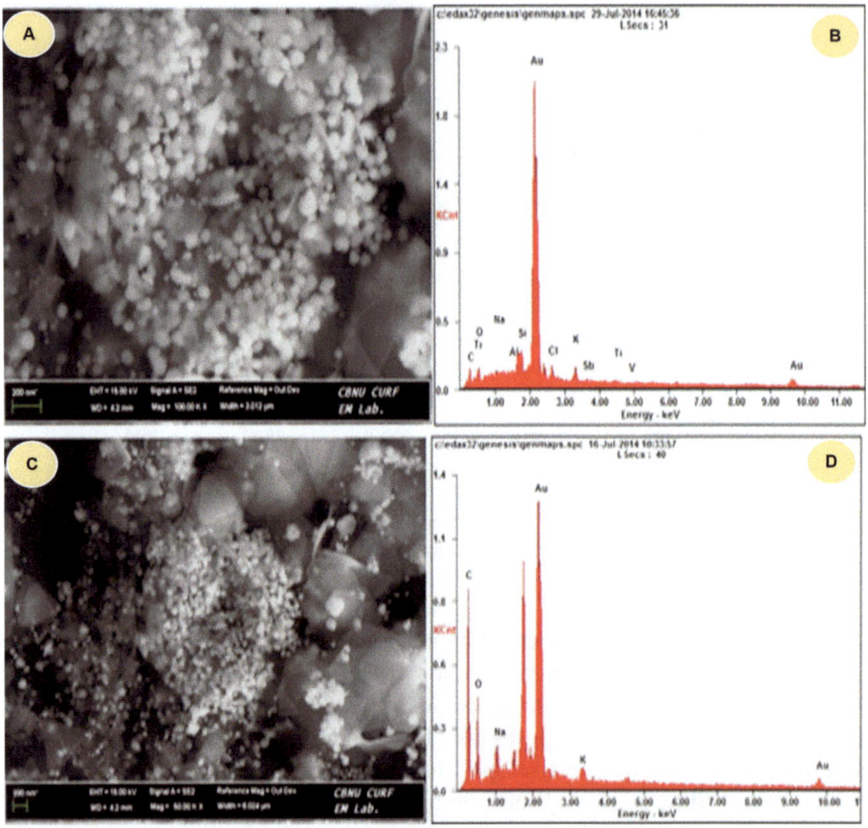

Fig. 3.5 FE-SEM image of gold nanoparticles synthesized from pumpkin leaf extract (**a**) and their EDS analysis (**b**); the FE-SEM image of gold NPs synthesized from curled mallow leaves (**c**) and their EDS analysis (**d**). (Adopted from: Chandran et al. 2014)

rings from inner to outer associated with [111], [200], [220], and [311] atomic planes of Au indicate the formation of crystalline gold NPs.

AFM analytical technique is effectively used to study the topology of a sample in three dimensions (x, y, and z). It can measure the height of sample which is not possible with FESEM. Gopinath et al. (2014) worked out the surface morphology of the as-synthesized gold NPs by AFM analysis as shown in Fig. 3.7. The micrograph reveals that the as-synthesized gold NPs possess spherical shape and are 20–50 nm in size.

3.2.2.3 X-Ray Diffraction (XRD)

This technique is used to analyze the crystal structure, phase, and other structural parameters such as average grain size, crystallinity, strain, and crystal defects. When X-ray interacts with crystalline phase of materials, it produces diffraction peaks.

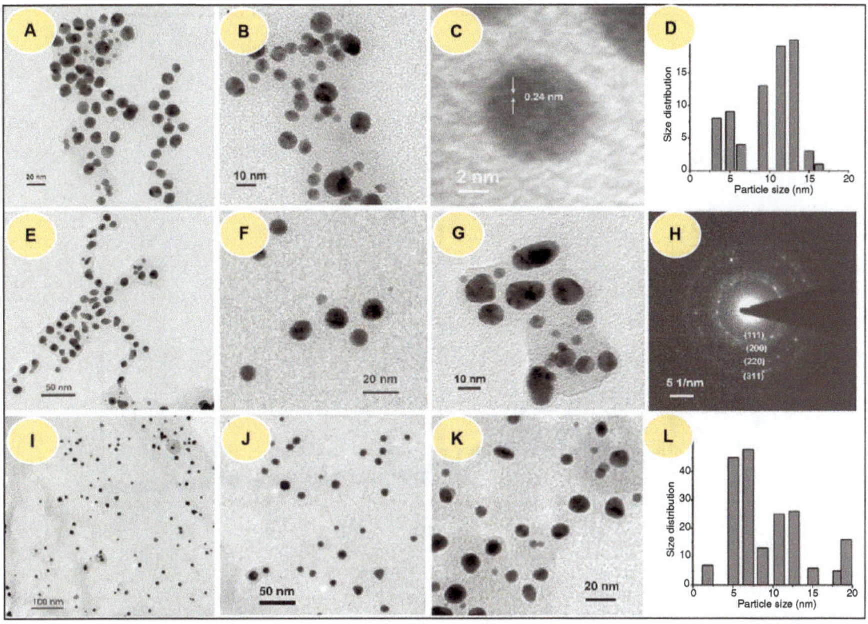

Fig. 3.6 TEM images, SAED, and histograms showing the effect of the concentration of leaf extract of *Acacia nilotica* on the size of gold nanoparticles: (**a–e**) of gold nanoparticles at 60 mgL^{-1}, TEM images (**f–g**) of gold nanoparticles at 100 mgL^{-1}, TEM images (**i–k**) of nanoparticles at 200 mgL^{-1}, (**h**) SAED of gold nanoparticle, and (**d, l**) histograms of gold nanoparticles at 60 and 200 mgL^{-1}, respectively. (Adopted from: Majumdar et al. 2013)

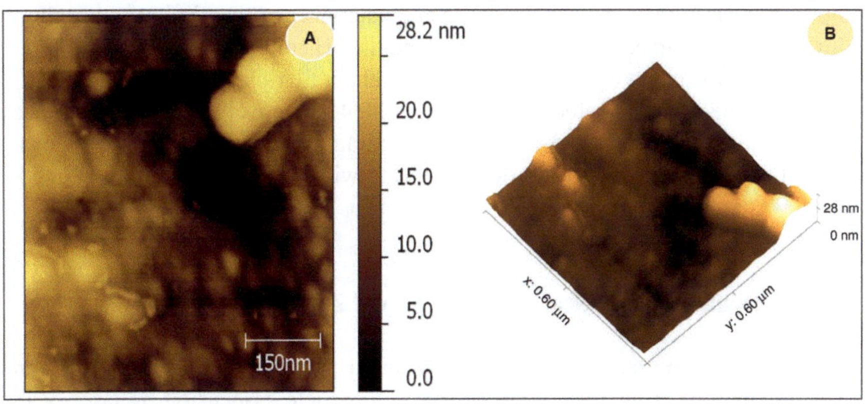

Fig. 3.7 AFM image of gold nanoparticles using aqueous fruit extract of *Terminalia arjuna*: (**a**) 2D image and (**b**) 3D image. (Adopted from: Gopinath et al. 2014)

The intensity of peaks reflects the distribution of atoms within the lattice, and these peaks can be used as "fingerprints" for identification of solid phases. Figure 3.8a shows a diffractogram of gold NPs synthesized by using *Terminalia arjuna* fruit; major reflections appear at 38.24°, 44.45°, and 66.30° corresponding to (111),

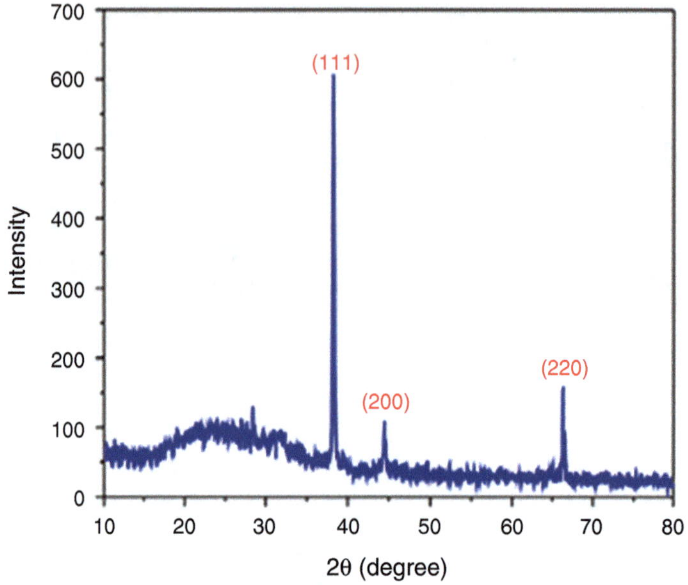

Fig. 3.8 XRD pattern of gold nanoparticles synthesized by using fruit extract of *T. arjuna* with $HAuCl_4$ aqueous solution. (Adopted from: Gopinath et al. 2014)

(200), and (220) planes, which confirm that these NPs belong to face-centered cubic phase of gold, as per the JCPDS card no. 04–0784 and the Au° nature of NPs (Gopinath et al. 2014). Singh and Srivastava (2015) reported that pH plays an important role in the synthesis of gold nanomaterials and that different morphologies of Au NMs can be observed by changing the pH conditions. Figure 3.8b shows that there is no change in d-spacing value, as all similar planes in XRD peaks correspond to the same 2θ value. Suman et al. (2014) effectively calculated the mean size of gold NPs (synthesized from the root extract of *Morinda citrifolia*) by using the Debye-Scherrer equation to determine the width of (111) Bragg's reflection, which was around 15 nm.

3.2.2.4 Dynamic Light Scattering (DLS) and Zeta Potential Analysis

DLS provides important information regarding the size of as-synthesized NPs and their distribution. It involves illumination of a particle suspension by laser beam, which results in temporal fluctuation of the elastic scattering intensity of light, i.e., Rayleigh scattering4125 induced from the Brownian motion of the particles of a size much smaller than the incident light wavelength, at a fixed scattering angle. Gopinath et al. (2014) reported the size of gold NPs in the range 5–60 nm with an average of 25 nm; however, the size of some particles increased due to agglomeration as revealed in Fig. 3.9a showing the DLS result. Zeta potential is a physical property, which is used to analyze the stability of synthesized NPs; it measures the

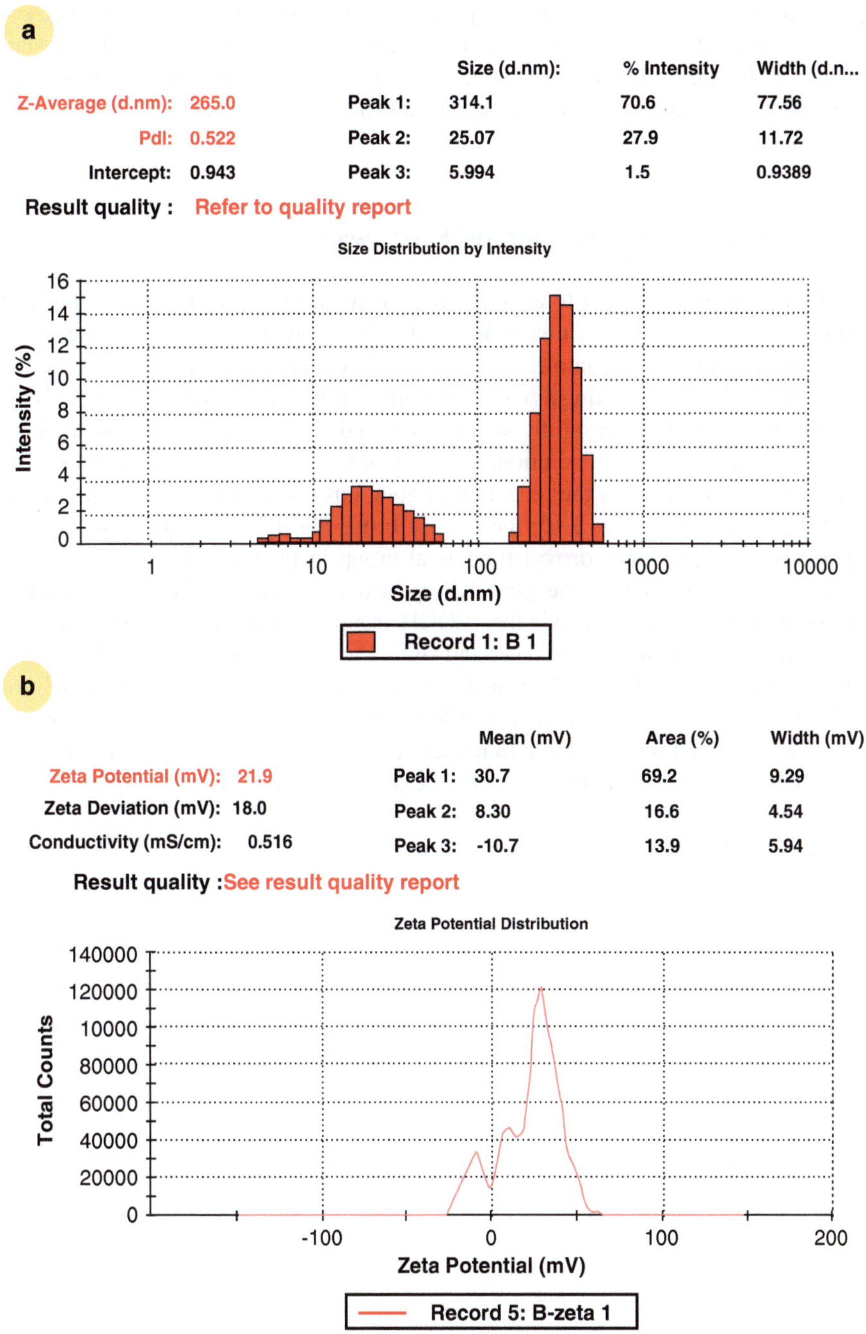

Fig. 3.9 (**a**) Particle size distribution of gold nanoparticles synthesized with the assistance of *Terminalia arjuna* fruit extract through the DLS method; (**b**) measurement of zeta potential analysis to determine the stability of these particles. (Adopted from: Gopinath et al. 2014)

potential difference between two suspended particles present in the colloidal suspension. Its value may be positive or negative ranging between –30 mV and + 30 mV. Figure 3.9b shows the zeta potential measurement of synthesized gold NPs having a value of 21.9 mV, which means that the surface of the NPs has a net positive charge.

3.2.2.5 Fourier Transform Infrared Spectroscopy (FTIR)

FTIR spectroscopy is used to examine the purity and composition of synthesized NPs. It measures the interaction of infrared radiation with molecules. In the green route of NP synthesis, biomolecules of secondary metabolites are responsible for reducing gold ions and acting as capping agents for the facile synthesis of gold NPs. Geetha et al. (2013) successfully synthesized gold NPs by using *Couroupita guianensis* extract and $HAuCl_4$ solution. Interpretation of FTIR spectra reveals the biomolecules responsible for the reduction of gold ions and capping of the bioreduced AuNPs, as shown in Fig. 3.10; the intense broad absorption at 3414 cm^{-1} is the characteristic peak for hydroxyl functional group in phenols and alcoholic compounds. A comparison of the gold NP spectrum, a and b, indicates correlation between the peak and the –OH in the –COOH group; the peak appeared at 3414 cm^{-1} in raw material, but it was narrower and shifted to longer wave number 3615 cm^{-1} after encapsulation of NPs. The peaks at 1073, 1288, 1383, and 1648 cm^{-1} are reduced, and a new peak appeared at 1744 cm^{-1} which indicates that the alcoholic group is converted into aldehyde to reduce Au^{3+} to Au^0. Similarly, Jayaseelan et al. (2013) reported the FTIR pattern of the gold NPs obtained with the aqueous seed

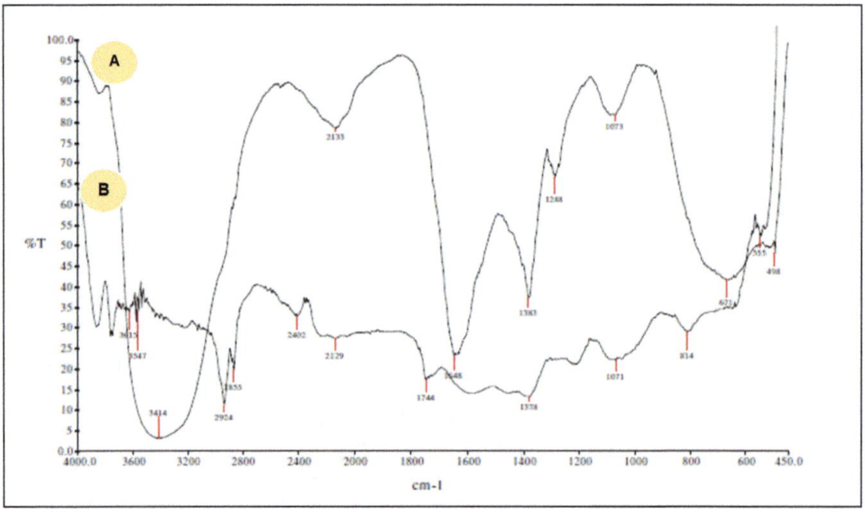

Fig. 3.10 FTIR spectra: (**a**) flower extract of *Couroupita guianensis*, (**b**) gold nanoparticles synthesized by *C. guianensis* flower extract. (Adopted from: Geetha et al. 2013)

extract of *Abelmoschus esculentus* exhibiting bands at (1047 and 1064 cm^{-1}), (1321 and 1388 cm^{-1}), (1643 and 1616 cm^{-1}), and (3425 and 3438 cm^{-1}). In this study, the intense band pattern at 1047 and 1064 cm^{-1} was characteristic of C – OH stretching of secondary alcohols. The peaks 1321 and 1388 cm^{-1} correspond to C – N stretching vibrations of aromatic amines. The absorption peaks located at 1643 and 1616 cm^{-1} were identified as the amide I and arose due to the carbonyl stretch vibrations in the amide linkages of the proteins. The broad peaks of 3425 and 3438 cm^{-1} corresponding to NH stretching in amide (II) were observed in gold NPs and in the aqueous seed extract of *A. esculentus*, respectively. In this study, the FTIR spectrum results have shown that the extracts containing OH as a functional group act in capping the NPs synthesis.

3.3 Factors Affecting the Fabrication

During the fabrication of gold NPs, the challenges frequently encountered are (a) to control the particles' shape and size and (b) to achieve monodispersity in solution phase. The size, shape, and stability of gold NPs are influenced by various factors such as incubation time, temperature, pH, pressure, and concentration of plant extract/biomass and the gold solution. The desired morphology of NPs can be achieved by altering these growth factors in the synthesis medium. The major works related to the control of growth factors have been summarized in Table 3.2, and a brief account of these studies is given below.

Table 3.2 Effect of the relevant factors (temperature, pH, and incubation time of the solution) on the particle shape during the plant-mediated fabrication of gold nanoparticles

Level	Shapes	References
25 °C	Triangular	Song et al. (2009)
60 °C	Pentagonal	Song et al. (2009)
90 °C	Hexagonal	Song et al. (2009)
30 °C	Spherical, triangle, truncated triangles, and decahedral	Narayanan and Sakthivel (2010b)
Room temperature	Polydisperse with spherical, triangular, and hexagonal	Noruzi et al. (2011)
80 °C	Spherical	Das et al. (2011)
50 °C	Spherical, nanoprisms, nanotriangles, hexagons, trapezoids	Ghosh et al. (2012)
40 °C	Spherical	Gonnelli et al. (2015)
100 °C	Spherical	Mishra et al. (2016)
pH 4.8	Spherical	Sharma et al. (2007)
pH 9	Spherical	Castro et al. (2011)
pH 10	Rod shaped	Castro et al. (2011)
pH 11	Nanowires	Castro et al. (2011)
pH 2	Rod shaped, large	Armendariz et al. (2004)

Table 3.2 (continued)

Level	Shapes	References
pH 3–4	Rod shaped, small	Armendariz et al. (2004)
pH 8	Spherical, oval, polyhedral	Pasca et al. (2014)
pH 4–6	Spherical	Mishra et al. (2016)
5 h	Spherical	Chandran et al. (2006)
25 h	Triangular	Chandran et al. (2006)
0–6 days	Spherical	Sharma et al. (2007)
12 h	Spherical, triangular, truncated, decahedral	Narayanan and Sakthivel (2008)
2 min	Smaller, spherical	Philip (2010)
<5 min	Spherical, triangular, and hexagonal	Noruzi et al. (2011)
1 h	Spherical, triangle, truncated triangular, and decahedral	Narayanan and Sakthivel (2010b)
30 min	Spherical	Das et al. (2011)
20 min	Spherical	Kumar et al. (2011)
20 min	Spherical, nanoprisms, nanotriangles, hexagons, trapezoids	Ghosh et al. (2012)
20 s	Spherical	Kumar et al. (2012)
3 min	Spherical	Yasmin et al. (2014)
30 min	Spherical	Gonnelli et al. (2015)
6 min	Spherical	Mishra et al. (2016)

3.3.1 Temperature

Control of temperature is an important factor with respect to the shape and size of NPs in plant-mediated synthesis. During the gold NP fabrication from *Nyctanthes arbor-tristis* ethanolic flower extract, predominantly spherical NPs were obtained at high temperatures, while they assumed different shapes (triangular, pentagonal, rod-like, as well as spherical) at low temperatures (Das et al. 2011). This shape variation was ascribed to exposure of the nucleation procedure of metallic NPs to the reaction temperature. The high reaction rate at a high temperature more effectively facilitates the ingesting of gold ions in the establishment of nuclei, which could inhibit the secondary reduction process on the surface of the preformed nuclei, thus resulting in spherical particles. The secondary nucleation is favored at low temperature, which could result in particles of different shapes (Das et al. 2011). Similarly, Song et al. (2009) demonstrated a decrease in size of gold NPs synthesized by *Magnolia kobus* and *Diospyros kaki* leaf broth, when temperature was increased. They suggested that most of the gold ions were consumed in the formation of nuclei, and the secondary reduction phenomenon on the surface of nuclei was inhibited on increasing the temperature (Song et al. 2009). Thus, the suppression of secondary nucleation phenomenon at high temperatures might be a reason for the formation of particles with more uniform morphology. Gericke and Pinches (2006) obtained nanorod or platelet-shaped gold NPs at higher temperatures, while spherical NPs at lower temperatures. Dubey et al. (2010b) noted that an increase in temperature from 25 to 150 °C leads to increase in the sharpness of absorption peaks for both gold and silver NPs in *Tanacetum vulgare* fruit extract. High temperature also increases the rate of reaction, which enhances NP fabrication (Phillip 2009; Dwivedi and Gopal 2010). As the sharpness in absorbance peak depends on the size of NPs, the small particle size at high temperature results in the sharpening of the plasmon resonance band of gold and silver NPs (Fayaz et al. 2009; Shaligram et al. 2009). Ghodake et al. (2010) used pear fruit extract for fabrication of gold nanoplates at room temperature. Ability to produce the metallic NPs at room temperature makes the processing easy and has the advantage of energy-saving. However, further investigation is still required regarding the control of shape and size of gold NPs by adjusting the temperature; availability of temperature-sensitive metabolites in the plant is of great importance in this context.

3.3.2 pH

Several studies have laid emphasis on the role of pH of reaction mixtures on the shape and size of particles during the NP fabrication from plant extracts. The reaction pH has the ability to change the electrical charges of biomolecules, which possibly affect their stabilizing and capping capabilities and then the NPs growth. This may help in forming the NPs of some particular shapes at a certain pH range so that a greater stability could be accomplished. However, extracts of various plants or different parts of the same plant species may have varying values of pH; hence optimization of fabrication process is desirable for an effective NP synthesis. Gardea-Torresdey et al. (1999) confirmed the significant role of pH in the fabrication of colloidal gold from alfalfa biomass and suggested that the NP size varies with the change in pH. This was endorsed by Mock et al. (2002). Armendariz et al. (2004) then examined the gold NPs biosynthesized with *Avena sativa* biomass and found their size to be highly dependent on the pH value. These authors obtained smaller gold NPs at pH values of 3 and 4 and the larger ones at pH 2. The process of gold NP aggregation to form the larger ones at pH 2 was ideal over the nucleation to form new NPs. However, at pH 3 and 4, perhaps more functional groups are accessible for Au(III) complexes to bind with the biomass at the same time, which facilitates the successive formation of larger amounts of NPs of smaller diameters. Nevertheless, at pH 5 the biomass carries an overall negative charge due to the functional group (carboxyl) available in the biomass. Thus, the negatively charged $[AuCl4]^-$ does not approach the binding sites easily, and this prevents the binding of Au(III) and its reduction to Au(0), leading to the formation of fewer NPs. Moreover, electrostatic repulsion may also prevent aggregation and growth of small NPs and, therefore, result in the formation of small gold NPs of irregular shape (Armendariz et al. 2004). Ghodake et al. (2010) used pear fruit extract for fabricating gold nanoplates and found that alkaline condition was more effective for triangular and hexagonal nanoplates, while the structures of these nanoplates were barely detected at acidic pH.

3.3.3 Incubation Time

Incubation time, i.e., the time duration needed for a successful achievement of all reaction steps, influences the fabrication process of NPs. For instance, in the case of *Tanacetum vulgare* fruit extract-mediated fabrication of silver and gold NPs, the reaction started within 10 min, and an increase in the incubation time led to the sharpening of peaks for these NPs (Dubey et al. 2010a, b). Dwivedi and Gopal (2010) also found that by increasing the gold salt concentration, the size of gold NPs could be increased. In addition, increase in the reaction temperature increased the reaction rate also. Using the extract of *Rosa* hybrid petals, Noruzi et al. (2011) recorded a higher rate of reaction in comparison to previous studies, as the gold NP fabrication process was finished within 5 min (Noruzi et al. 2011). However, using the *Terminalia chebula* plant extracts, Kumar et al. (2012) produced gold NPs in 20 s only.

3.3.4 Plant Biomass Concentration

The plant biomass and/or extract concentrations used during the formation of NPs is another significant factor, which regulates the level of reduction and stabilization by the biomolecules and affects the NP morphology, as noticed in the case of gold NPs obtained with the help of *Cymbopogon flexuosus* (Shankar et al. 2005). Anisotropic gold NPs were biosynthesized by the reaction of *C. flexuosus* extract with Au(III) ions. The size of nanotriangles could be changed by altering the extract concentration. As the extract amounts were increased, the size of the triangular and hexagonal particles decreased, whereas the ratio of the number of spherical NPs to triangular/hexagonal particles increased. Song et al. (2009) used *Magnolia kobus* and *Diospyros kaki* leaf extracts for the biosynthesis of gold NPs. At higher extract concentrations, smaller and mainly spherical NPs were formed, while a variety of other morphologies in larger sizes were produced at lower extract concentrations. Kasthuri et al. (2009) also reported a change in particle morphology with the phyllanthin-assisted and clove extract-assisted biosynthesis of gold NPs. At lower concentrations of plant extract, particle size increased, and hexagonal or triangular gold NPs were formed. Dubey et al. (2010a) used the aqueous extract of *Sorbus aucuparia* leaf for the synthesis of silver and gold NPs. They recorded an accelerated particle formation and a reduced size of both the silver and gold nanocolloids when the amount of extract was increased. Similarly, an increase in the amount of the *Chenopodium album* leaf extract caused a decrease in the particle size (Dwivedi and Gopal 2010). During the biosynthesis of gold NPs with the help of *Sapindus mukorossi* extract, few aggregations and heterogeneous structures were observed at a lower (15%) concentration, while spherical NPs and more dispersion occurred at a higher (45%) concentration (Reddy et al. 2012).

3.4 Applications of Gold Nanoparticles

The use of gold NPs for staining of glass and enamels and for treating certain diseases dates back to the sixteenth century. Now their applications are common in biomedicine, bio-sensing, catalysis, agriculture, electronic and magnetic devices, etc. (Husen and Siddiqi 2014; Husen 2017; Siddiqi and Husen 2017a; Ovais et al. 2017). As they are nontoxic and readily adsorb on DNA, Au NPs are used for bombardment during delivery of genetic materials into plants (Moaveni et al. 2011) and in the quantitative determination of heavy metals (Liu and Lu 2003), blood glucose (Luo et al. 2004), and pesticides (Lisha and Pradeep 2009). Their applications in various areas are discussed below and shown in Fig. 3.11.

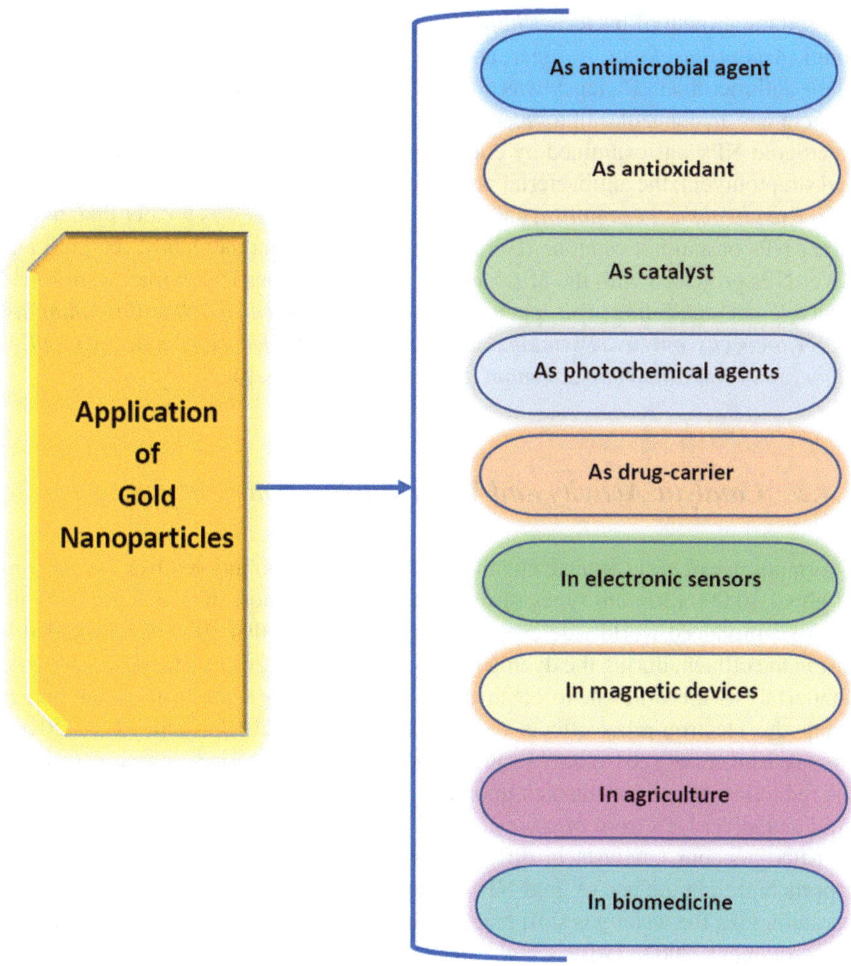

Fig. 3.11 Various applications of gold nanoparticles

3.4.1 Antimicrobial Agents

Microbial resistance to antibiotics has an undesirable consequence on human health by causing side effects due to consumption of higher drug doses and/or a prolonged treatment period. The biosynthesized NPs offer an auspicious solution in the drug-resistance management. Gold NPs fabricated with plant material have shown strong antimicrobial activity (Sreelakshmi et al. 2011; Husen and Siddiqi 2014; Piruthiviraj et al. 2016; Balasubramani et al. 2017). The growth of gram-negative and gram-positive bacteria was effectively inhibited with functionalized gold NPs (Li et al. 2014). The gold and silver NPs produced from *Mentha piperita* have shown a strong antibacterial effect against *Staphylococcus aureus* and *Escherichia coli* (Ali et al. 2011). Similar impact of honey-mediated gold and silver NPs was noticed by Sreelakshmi et al. (2011). Gold NPs obtained from *Brassica oleracea* also suppressed the growth of bacteria (*Staphylococcus aureus*, *Klebsiella pneumonia*) and fungi (*Aspergillus flavus*, *A. niger*, and *Candida albicans*) (Piruthiviraj et al. 2016). Their antimicrobial efficiency was comparable with that of the standard drugs gentamicin and fluconazole. In a recent study, bioactivity of *Nitzschia (*diatom)-fabricated gold NPs was examined by coupling them with antibiotics such as penicillin and streptomycin; the antibacterial potential of the combination against *Escherichia coli*, *Pseudomonas aeruginosa*, and *Staphylococcus aureus* was greater than that of either NPs or antibiotics alone (Borase et al. 2017). In another study, the gold and silver NPs produced with the *Mussaenda glabrata* aqueous leaf extract were found effective in inhibiting the pathogenic microbes *such as Bacillus pumilus*, *Staphylococcus aureus*, *Pseudomonas aeruginosa*, *Escherichia coli*, *Aspergillus niger*, and *Penicillium chrysogenum* (Francis et al. 2017a, b).

3.4.2 Catalytic Activity and Water Purification

Enormous use of toxic organic chemicals leads to water pollution. Over 7×10^5 tons of about 10,000 different types of dyes and pigments, most of which are carcinogenic, is produced worldwide every year. It is estimated that 10–15% of the dye is lost in the effluent during the dyeing process (Gupta et al. 2011). The silver and gold nanocatalysts are efficient to clean these cancer-causing dyes from water bodies using the electron-relay effects (Joseph and Mathew 2015b; Lim et al. 2016). Ramakrishna et al. (2016) have produced gold NPs using marine algae as catalyst in the reduction of various nitro compounds. Francis et al. (2017a, b) have fabricated gold and silver NPs from *Mussaenda glabrata* aqueous leaf extract and used them as heterogeneous catalysts in different dye degradation processes. Several other reports highlight the use of gold NPs as catalyst for selective reactions at low temperature, viz., the water gas shift reaction and selective oxidation of carbon monoxide (Andreeva 2002; Grisel et al. 2002; Hutchings and Haruta 2005), glycerol (Carrettin et al. 2002), and methanol (Hernández et al. 2006), hydrogenation of unsaturated materials (Claus et al. 2000), and reduction of aromatic nitro compounds (Kundu et al. 2009) and a toxic pollutant 4-nitrophenol to 4-aminophenol (Ghosh et al. 2012; Yu et al. 2016). Gold NPs of different sizes fabricated with the *Tribulus terrestris* fruit extract exhibited catalytic reduction of p-nitroaniline to p-phenylenediamine (Gopinath et al. 2019). Recently, Manjari et al. (2017) achieved an ultrarapid homogeneous and heterogeneous complete degradation of methylene blue and Congo red within few seconds by using as catalyst the silver and gold NPs fabricated with the *Aglaia elaeagnoidea* flower extract. They found above 90% conversion of 4-nitrophenol to 4-aminophenol within few minutes for homogenous and within few seconds for heterogeneous degradation. Pradeep and Anshup (2009) also found gold NPs suitable for the removal of heavy metals through alloy formation by varying the composition such as Au_3Hg, AuHg, and $AuHg_3$ phases; thus these NPs can be used for removing Hg ions from contaminated water. Gold NPs are also used for the detection and removal of organic compounds such as pesticides like malathion, endosulfan, and chlorpyrifos (Han et al. 2003; Nair et al. 2003). Burns et al. (2006) noted that salt-induced aggregation of gold NPs is useful for detecting pesticides at low concentration and removing them from drinking water. Furthermore, removal of diesel oil droplets floating on water through swelling and absorption of the gold NP composite has also been observed (Gupta and Kulkarni 2011). All these findings have a lot of promise for the wastewater purification by degrading its toxic contaminants.

3.4.3 *Antioxidant Potential*

Oxidation is a process in which free radicals are generated due to chemical reaction; free radicals cause damage to cells and disturb their functioning. Antioxidants are known to prevent cell damage, DNA damage, malignant transformation, heart diseases, cancer, and oxidative stress (Watters et al. 2007). Both natural and synthetic antioxidants are used to cure these problems. Anagnostopoulou et al. (2006) and Carocho and Ferreira (2013) have suggested that the synthetic antioxidants are suspected to cause adverse health effects. In a study, Au NP-embedded flavones with other dietary nutrients have exhibited significant enhancement in the antioxidant activity (Medhe et al. 2014). Leu et al. (2012) have reported that the antioxidative ability of known antioxidants was much increased when used in combination with gold NPs for the purpose of wound healing. Abel et al. (2016) produced gold NPs from the *Cassia tora* secondary metabolite conjugate and noted their higher bioavailability and the antioxidant and anticancer effect against the colon cancer cell line (Col320). Manjari et al. (2017) found enhanced antioxidant activity on increasing the concentration of gold and silver NPs obtained from the *Aglaia elaeagnoidea* flower extract. Sathish et al. (2016) have reported extraordinary antioxidant potential of gold NPs produced by using the aqueous fruit extract of *Couroupita guianensis*. Balasubramani et al. (2015) found the *Antigonon leptopus*-mediated gold NPs as a free radical scavenger and an anticancer agent.

3.4.4 Photochemical Agents

Gold NPs are used as inert carriers to modify the excited state photochemistry and prevent the unwanted phototoxicity effects of the adsorbed drug molecules (Wang et al. 2016). A study of the uptake of gold NPs by *Arabidopsis thaliana* roots and their subsequent translocation to leaves has shown the generation of nanobubbles and acoustic signals in plant tissues (Koo et al. 2016). In addition, the leaves containing gold NPs exhibited a higher temperature across the leaf surface and induced expression of heat shock-regulated genes, when exposed to laser beam. These results demonstrate that gold NPs in the leaves can act as the photochemical agents and raise the leaf temperature by absorbing the light waves to induce photochemical activity even when the temperature is very low.

3.4.5 Plant Response to Gold Nanoparticles

Being stationary and having a large leaf area, plants are prone to exposure to a wide range of NPs available in their surrounding environment (Dietz and Herth 2011). They significantly control metal ions or NPs by accumulating them into their biomass (Husen and Siddiqi 2014; Iqbal et al. 2015; Siddiqi and Husen 2016). Upon entering into plants, gold NPs may activate or prevent plant growth and biomass accumulation. In general, higher doses of gold solutions may cause toxicity and thus affect the plant growth negatively, producing changes at the physiological, biochemical, and molecular levels (Siddiqi and Husen 2016). Although treatment of *Arabidopsis thaliana* root with a lower dose (10 ppm $KAuCl_4$) markedly increased the root length, a significant decrease was noted at higher doses (25, 50 and100 ppm). Further, at higher concentrations of $KAuCl_4$, iron content was depleted, while Zn and P contents declined (Jain et al. 2014). Another study of *A. thaliana* revealed a significant role of gold NPs in seed germination, modulation of antioxidant system, and regulation of microRNA expression that controls different morphological, physiological, and metabolic processes (Kumar et al. 2013). Judy et al. (2012) recorded accumulation of 2–54 μg gold g^{-1} in the dried tobacco plants after their exposure for 7 days to 30 μg gold mL^{-1} in the nutrient solution.

Earlier, Barrena et al. (2009) found a low or zero toxicity of the gold, silver, and magnetite NPs at concentrations of 62, 100, and 116 μg mL^{-1}, respectively, in cucumber and lettuce plants. Many other studies have also shown that gold can accumulate to varying degrees in different plant species such as *Brassica juncea*, *B. campestris*, *Trifolium repens*, *Sorghum helense*, *Raphanus sativus*, *Kalanchoe serrata*, and *Helianthus annuus* (Wilson-Corral et al. 2012; Siddiqi and Husen 2016). Gold NP application improved the seed germination in cucumber and lettuce (Barrena et al. 2009), *Brassica juncea* (Arora et al. 2012) and *Gloriosa superba* (Gopinath et al. 2014). *B. juncea* seedlings exhibited changes both in growth and seed yield; foliar spray of gold NPs enhanced the oil production with a concomitant decline in reducing sugar and total sugar contents (Arora et al. 2012). Redox status of the treated *B. juncea* was also improved. Seed germination percentage increased with spray and/or inoculation with 25 ppm gold NPs, but the higher concentrations had a negative impact. Gunjan et al. (2014) reported that gold NPs (in 100–400 ppm doses) reduced the *B. juncea* growth possibly due to rise in free radical stress, as indicated by the general rise in activities of antioxidative enzymes, viz., catalase, ascorbate peroxidase, guaiacol peroxidase, and glutathione reductase. Furthermore, proline and hydrogen peroxide contents were also enhanced due to the formation of reactive oxygen species (ROS). These observations suggest that the ROS production, which makes for the physiological and biochemical stress in plants, is dependent on the dose of gold NPs. Of late, Balalakshmi et al. (2017) produced spherical, 25 nm gold NPs, using *Sphaeranthus indicus*, and examined them at 0, 1, 3, 5, 7, and 10% concentrations of the plant *extract* for the mitotic cell division assays, pollen germination experiments, and in vivo toxicity trials against an aquatic crustacean. In this experiment, gold NPs did not show any toxic effects on plant cells and aquatic crustacean. Most of the studies on plant responses to gold NPs have so far been focused on early stages of plant growth (Siddiqi and Husen 2016). Random application of metal particles may not produce the desired results with reference to plant growth or yield. It is necessary to identify which trace elements are useful for a given plant species so that the same may be used in the form of NPs to increase the quality and quantity of plant products.

3.4.6 Biomedical Application

Gold NPs are successfully used for transporting drugs, genes, and other biomolecules due to their low toxicity, high surface area, and tunable stability. For instance, gold NPs with 2 nm core diameter have been used as carrier in the preparation of the chemotherapeutic drug agent, paclitaxel (Gibson et al. 2007). Moreover, their tunable size and functionality also allow them to work as a useful scaffold for the delivery of peptides, proteins, or nucleic acids like DNA or RNA (Bhumkar et al. 2007; Ghosh et al. 2008; Rana et al. 2012; Ding et al. 2014). They were also adhered to the vascular endothelial growth factor antibodies, which are used in treating the B-chronic lymphocytic leukemia (Mukherjee et al. 2007). Gold NPs are conjugated oligonucleotide/DNA and employed in clinical diagnosis and colorimetric detection of targeted DNA, virus, etc. (Upadhyay et al. 2006; Baptista et al. 2008; Costa et al. 2010; Nun et al. 2013). Gold NP-bifunctional oligonucleotide probe conjugate for the detection of dsDNA even in very minute quantity has also been reported (Dharanivasan et al. 2016). Thanh and Rosenzweig (2002) developed a unique, sensitive, and highly specific immunoassay system for antibodies by using gold NPs. The colloidal gold NPs transmit drug to the site of the use regardless of their shape, if both of them (NPs and drug) are biocompatible, stable, and easily form bonds with each other. They are finally deposited in cells and on irradiation with visible light; the heat mediated by gold NPs destroys the cancerous tissues locally (Huang et al. 2006; Huff et al. 2007; Lowery et al. 2006). They are also used in the Carter-Wallace home pregnancy's "First Response" kit. During its use, when the latex micro- and gold NPs derivatized with a hormone released by pregnant women were mixed, the micro- and nanoparticles coagulated, forming clumps of pink color (Bangs 1996). Bio-detection sensitivity obtained from spherical NPs was not strong enough to trace the interaction of biomolecules (Orendorff et al. 2005). In its place, irregular-shaped NPs could improve the biological detection sensitivity greatly. NP fabrication from plants may give irregular-shaped nanostructures in a one-pot process due to the complexity of the bio-reducing and bio-capping agents (Lukman et al. 2011).

Thus, NPs fabricated with the help of plants have a potential and could be utilized in various applications that need irregular shape of nanostructures. Raman spectroscopy is a vibrational method for collecting information about the chemical structure (González-Solís et al. 2013). Gold NPs also increase the intensity of Raman scattering of adjacent molecules and hence are extensively used in surface-enhanced Raman scattering (SERS) for the detection and quantitative analysis of Raman active materials (Qian and Nie 2008; Zamarion et al. 2008; Dasary et al. 2009; Ding et al. 2013). This technique is used to increase the sensitivity and reproducibility of the analyses (Kumar and Yadav 2009). SERS technique is used to distinguish cancer cells from normal cells, when gold nanorods conjugate to anti-epidermal growth factor receptor antibodies (Huang et al. 2007b). SERS with gold NPs has been employed in cancer research to detect tumors (Cai et al. 2008; Huang and El-Sayed 2010) and in immunoassays as well as the examination of living cells (Kneipp et al. 2002; Grubisha et al. 2003; Neng et al. 2010). Gold NPs have also been used in the cell and phantom imaging (Huang et al. 2007b; Cai et al. 2008), mostly during cell culture studies. Peleg et al. (1999) have reported the benefit of the enhanced second harmonic signal by antibody-conjugated gold NPs for functional cellular imaging around single molecules. Krishnaraj et al. (2014) have reported the in vitro cytotoxic effect of silver and gold NPs (obtained by an *Acalypha indica-mediated* synthesis) against the MDA-MB-231, human breast cancer cells, and could cause significant cytotoxic effects and apoptotic features. Subsequently, many other research groups have reported anticancer properties of plant-mediated gold NPs such as those involving *Moringa oleifera* (Anand et al. 2015; Tiloke et al. 2016), *Portulaca grandiflora (*Ashokkumar et al. 2016), *Musa paradisiaca* (Vijayakumar et al. 2017), *Spinacia oleracea* (Ramachandran et al. 2017), and so on.

3.5 Conclusion

The principal biomolecules in plant extracts, which are responsible for the reduction of metal ions to form NPs, include amines, amino acids, aldehydes, ketones, carboxylic acids, phenols, proteins, flavonoids, saponins, steroids, alkaloids, tannins, vitamins, etc. The complex nature of plant extracts and their involvement in the gold

NP fabrication require a comprehensive investigation. Many factors such as incubation time, temperature, plant extract concentration, pH values, etc. affect the growth, morphology, and stability of particles. Optimization techniques for a homogenous production of gold NPs are needed for better processing and application in various sectors. The random and excessive use of gold NPs may not give the desired output in terms of plant growth and yield in all the species and may rather lead to ecological risks. It is essential to explore which specific metal/trace element works well with which species, so that NPs of the same material may be used to increase the quality and quantity of plant products in a cost-effective manner.

References

Abel EE, Poonga PRJ, Panicker SG (2016) Characterization and in vitro studies on anticancer, antioxidant activity against colon cancer cell line of gold nanoparticles capped with *Cassia tora* SM leaf extract. Appl Nanosci 6:121–129

Ali MD, Thajuddin N, Jeganathan K, Gunasekaran M (2011) Plant extract mediated synthesis of silver and gold nanoparticles and its antibacterial activity against clinically isolated pathogens. Colloids Surf B Biointerfaces 85:360–365

Anagnostopoulou MA, Kefalas P, Papageorgiou VP, Assimopoulou AN, Boskou D (2006) Radical scavenging activity of various extracts and fractions of sweet orange peel (Citrus sinensis). J Food Chem 94:19–25

Anand K, Gengan RM, Phulukdaree A, Chuturgoon A (2015) Agroforestry waste *Moringa oleifera* petals mediated green synthesis of gold nanoparticles and their anti-cancer and catalytic activity. J Ind Eng Chem 21:1105–1111

Andreeva D (2002) Low temperature water gas shift over gold catalysts. Gold Bull 35:82–88

Ankamwar B (2010) Biosynthesis of gold nanoparticles (green-gold) using leaf extract of *Terminalia Catappa*. E-J Chem 7:1334–1339

Annamalai A, Christina VLP, Sudha D, Kalpana M, Lakshmi PTV (2013) Green synthesis, characterization and antimicrobial activity of Au NPs using *Euphorbia hirta* L. leaf extract. Colloids Surf B: Biointerfaces 108:60–65

Armendariz V, Herrera I, Peralta-Videa JR, Jose-Yacaman M, Troiani H, Santiago P, Gardea-Torresdey JL (2004) Size controlled gold nanoparticle formation by Avena sativa biomass: use of plants in nanobiotechnology. J Nano Res 6:377–382

Arora S, Sharma P, Kumar S, Nayan R, Khanna PK, Zaidi MGH (2012) Gold-nanoparticle induced enhancement in growth and seed yield of *Brassica juncea*. Plant Growth Regul 66:303–310

Ashokkumar T, Arockiaraj J, Vijayaraghavan K (2016) Biosynthesis of gold nanoparticles using green roof species *Portulaca grandiflora* and their cytotoxic effects against C6 glioma human cancer cells. Environ Prog Sustain Energy 35:1732–1740

Babu PJ, Saranya S, Sharma P, Tamuli R, Bora U (2012) Gold nanoparticles: sonocatalytic synthesis using ethanolic extract of *Andrographis paniculata* and functionalization with polycaprolactone–gelatin composites. Front Mater Sci 6:236–249

Balalakshmi C, Gopinath K, Govindarajan M, Lokesh R, Arumugam A, Alharbi NS, Kadaikunnan S, Khaled JM, Benelli G (2017) Green synthesis of gold nanoparticles using a cheap *Sphaeranthus indicus* extract: impact on plant cells and the aquatic crustacean *Artemia nauplii*. J Photochem Photobiol B 173:598–605

Balasubramani G, Ramkumara R, Krishnaveni N, Pazhanimuthu A, Natarajan T, Sowmiya R, Perumal P (2015) Structural characterization, antioxidant and anticancer properties of gold nanoparticles synthesized from leaf extract (decoction) of *Antigonon leptopus* Hook. & Arn. J Trace Elem Med Biol 30:83–89

Balasubramani G, Ramkumar R, Raja RK, Aiswarya D, Rajthilak C, Perumal P (2017) *Albizia amara* Roxb. Mediated gold nanoparticles and evaluation of their antioxidant, antibacterial and cytotoxic properties. J Clust Sci 28:259–275

Bali R, Harris AT (2010) Biogenic synthesis of Au nanoparticles using vascular plants. Ind Eng Chem Res 49:12762–12772

Bangs LB (1996) New developments in particle-based immunoassays: introduction. Pure Appl Chem 68:1873–1879

Baptista P, Pereira E, Eaton P, Doria G, Miranda A, Gomes I, Quaresma P, Franco R (2008) Gold nanoparticles for the development of clinical diagnosis methods. Anal Bioanal Chem 391:943–950

Barrena R, Casals E, Colón J, Font X, Sánchez A, Puntes V (2009) Evaluation of the ecotoxicity of model nanoparticles. Chemosphere 75:850–857

Begum NA, Mondal S, Basu S, Laskar RA, Mandal D (2009) Biogenic synthesis of Au and Ag nanoparticles using aqueous solutions of *Black Tea* leaf extracts. Colloids Surf B Biointerfaces 71:113–118

Bhumkar DR, Joshi HM, Sastry M, Pokharkar VB (2007) Chitosan reduced gold nanoparticles as novel carriers for transmucosal delivery of insulin. Pharm Res 24:1415–1426

Bindhani BK, Panigrahi AK (2014) Green synthesis and characterization of gold nanoparticles using leaf extracts of *Withania somnifera* (Linn.) (Ashwagandha). Int J Mat Sci Appl 3:279–284

Borase HP, Patil CD, Suryawanshi RK, Koli SH, Mohite BV, Benelli G, Patil SV (2017) Mechanistic approach for fabrication of gold nanoparticles by *Nitzschia* diatom and their antibacterial activity. Bioprocess Biosyst Eng 40:1437–1446

Boxi SS, Mukherjee K, Paria S (2016) Ag doped hollow TiO_2 nanoparticles as an effective green fungicide against *Fusarium solani* and *Venturia inaequalis* phytopathogens. Nanotechnology 27:085103–085116

Burns C, Spendel WU, Puckett S, Pacey GE (2006) Solution ionic strength effect on gold nanoparticle solution color transition. Talanta 69:873–876

Cai W, Gao T, Hong H, Sun J (2008) Applications of gold nanoparticles in cancer nanotechnology. Nanotechnol Sci Appl 1:17–32

Carocho M, Ferreira ICFR (2013) A review on antioxidants, prooxidants and related controversy: natural and synthetic compounds, screening and analysis methodologies and future perspectives. Food Chem Toxicol 51:15–25

Carrettin S, McMorn P, Johnston P, Griffin K, Hutchings GJ (2002) Selective oxidation of glycerol to glyceric acid using a gold catalyst in aqueous sodium hydroxide. Chem Commun 7:696–697

Castro L, Blazquez ML, Munoz JA, Gonzalez F, Garcia-Balboa C, Ballester A (2011) Biosynthesis of gold nanowires using *sugar beet* pulp. Process Biochem 46:1076–1082

Chandran SP, Chaudhary M, Pasricha R, Ahmad A, Sastry M (2006) Synthesis of gold nanotriangles and silver nanoparticles using *Aloe vera* plant extract. Biotechnol Prog 22:577–583

Chandran K, Song S, Yun SII (2014) Effect of size and shape controlled biogenic synthesis of gold nanoparticles and their mode of interactions against food borne bacterial pathogens. Arab J Chem. https://doi.org/10.1016/j.arabjc.2014.11.041

Claus P, Brückner A, Mohr C, Hofmeister H (2000) Supported gold nanoparticles from quantum dot to mesoscopic size scale: effect of electronic and structural properties on catalytic hydrogenation of conjugated functional groups. J Am Chem Soc 122:11430–11439

Costa P, Amaro A, Botelho V, Inacio J, Baptista PV (2010) Gold nanoprobe assay for the identification of mycobacteria of the Mycobacterium tuberculosis complex. Clin Microbiol Infect 16:1464–1469

Daizy P (2009) Honey mediated green synthesis of gold nanoparticles. Spectrochim Acta A Mol Biomol Spectrosc 73:650–653

Das RK, Gogoi N, Bora U (2011) Green synthesis of gold nanoparticles using *Nyctanthes arbortristis* flower extract. Bioprocess Biosyst Eng 34:615–619

Dasary SS, Singh AK, Senapati D, Yu H, Ray PC (2009) Gold nanoparticle based label-free SERS probe for ultrasensitive and selective detection of trinitrotoluene. J Am Chem Soc 131:13806–13812

Dash SS, Bag BG (2014) Synthesis of gold nanoparticles using renewable *Punica granatum* juice and study of its catalytic activity. Appl Nanosci 4:55–59

Dharanivasan G, Riyaz SUM, Jesse DMJ, Muthuramalingam TR, Rajendran G, Kathiravan K (2016) DNA templated self-assembly of gold nanoparticle clusters in the colorimetric detection of plant viral DNA using a gold nanoparticle conjugated bifunctional oligonucleotide probe. RSC Adv 6:11773–11785

Dietz KJ, Herth S (2011) Plant nanotoxicology. Trends Plant Sci 16:582–589

Ding X, Kong L, Wang J, Fang F, Li D, Liu J (2013) Highly sensitive SERS detection of Hg2+ ions in aqueous media using gold nanoparticles/graphene heterojunctions. ACS Appl Mater Interfaces 5:7072–7078

Ding Y, Jiang Z, Saha K, Kim CS, Kim ST, Landis RF, Rotello VM (2014) Gold nanoparticles for nucleic acid delivery. Mol Ther 22:1075–1083

Dubey SP, Lahtinen M, Särkkä H, Sillanpää M (2010a) Bioprospective of *Sorbus aucuparia* leaf extract in development of silver and gold nanocolloids. Colloids Surf B Biointerfaces 80:26–33

Dubey SP, Lahtinen M, Sillanpaa M (2010b) Tansy fruit mediated greener synthesis of silver and gold nanoparticles. Process Biochem 45:1065–1071

Dwivedi AD, Gopal K (2010) Biosynthesis of silver and gold nanoparticles using *Chenopodium album* leaf extract. Colloids Surf A Physicochem Eng Asp 369:27–33

Fayaz AM, Balaji K, Kalaichelvan PT, Venkatesan R (2009) Fungal based synthesis of silver nanoparticles-an effect of temperature on the size of particles. Colloids Surf B Biointerfaces 74:123–126

Fayaz AM, Girilal M, Venkatesan R, Kalaichelvan PT (2011) Biosynthesis of anisotropic gold nanoparticles using Maduca longifolia extract and their potential in infrared absorption. Colloids Surf B Biointerfaces 88:287–291

Fraceto LF, Grillo R, de Medeiros DGA, Scognamiglio V, Rea G, Bartolucci C (2016) Nanotechnology in agriculture: which innovation potential does it have? Front Environ Sci 4:20–25

Francis S, Joseph S, Koshy EP, Mathew B (2017a) Green synthesis and characterization of gold and silver nanoparticles using *Mussaenda glabrata* leaf extract and their environmental applications to dye degradation. Environ Sci Pollut Res 24:17347–17357

Francis S, Joseph S, Koshy EP, Mathew B (2017b) Synthesis and characterization of multifunctional gold and silver nanoparticles using leaf extract of *Naregamia alata* and their applications to catalysis and control of mastitis. New J Chem 41:14288–14298

Gardea-Torresdey JL, Tiemann KJ, Gamez G, Dokken K, Tehuacamanero S, Jose-Yacaman M (1999) Gold nanoparticles obtained by bio-precipitation from gold (III) solutions. J Nanopart Res 1:397–404

Gardea-Torresdey JL, Parsons JG, Gomez E, Peralta-Videa J, Troiani HE, Santiago P, Yacaman MJ (2002) Formation and growth of au nanoparticles inside live alfalfa plants. Nano Lett 2:397–401

Gardea-Torresdey JL, Gomez E, Peralta-Videa JR, Parsons JG, Troiani H, Jose-Yacaman M (2003) Alfalfa sprouts: a natural source for the synthesis of silver nanoparticles. Langmuir 19:1357–1361

Geetha R, Kumar TA, Tamilselvan S, Govindaraju K, Sadiq M, Singaravelu G (2013) Green synthesis of gold nanoparticles and their anticancer activityCancer. Nano 4:91–98

Gericke M, Pinches A (2006) Biological synthesis of metal nanoparticles. Hydrometallurgy 83:132–140

Ghodake GS, Deshpande NG, Lee YP, Jin ES (2010) Pear fruit extract-assisted room-temperature biosynthesis of gold nanoplates. Colloids Surf B Biointerfaces 75:584–589

Ghosh P, Han G, De M, Kim CK, Rotello VM (2008) Gold nanoparticles in delivery applications. Adv Drug Deliv Rev 60:1307–1315

Ghosh S, Patil S, Ahire M, Kitture R, Gurav DD, Jabgunde AM, Kale S, Pardesi K, Shinde V, Bellare J, Dhavale DD, Chopade BA (2012) *Gnidia glauca* flower extract mediated synthesis of gold nanoparticles and evaluation of its chemocatalytic potential. J Nanobiotechnol 10:17

Gibson JD, Khanal BP, Zubarev ER (2007) Paclitaxel-functionalized gold nanoparticles. J Am Chem Soc 129:11653–11661

Gonnelli C, Cacioppo F, Cristiana G, Capozzoli L, Salvatici C, Salvatici MC, Colzi I, Bubba MD, Ancillotti C, Ristori S (2015) *Cucurbita pepo* L. extracts as a versatile hydrotropic source for the synthesis of gold nanoparticles with different shapes. Green Chem Lett Rev 8:39–47

González-Ballesteros N, Prado-López S, Rodríguez-González JB, Lastra M, Rodríguez-Argüelles MC (2017) Green synthesis of gold nanoparticles using brown algae *Cystoseirabaccata*: its activity in colon cancer cells. Colloids Surf B Biointerfaces 153:190–198

González-Solís JL, Luévano-Colmenero GH, Vargas-Mancilla J (2013) Surface enhanced Raman spectroscopy in breast cancer cells. Laser Ther 22:37–42

Gopinath K, Venkatesha KS, Ilangovana R, Sankaranarayanan K, Arumugama A (2013) Green synthesis of gold nanoparticles from leaf extract of *Terminalia arjuna* for the enhanced mitotic cell division and pollen pollen germination activity. Ind Crop Prod 50:737–742

Gopinath K, Gowri S, Karthika V, Arumugam A (2014) Green synthesis of gold nanoparticles from fruit extract of *Terminalia arjuna* for the enhanced seed germination activity of Gloriosa superb. J Nanostruct Chem 4:115–125

Gopinath V, Priyadarshini S, Ali MD, Loke MF, Thajuddin N, Alharbi NS, Yadavalli T, Alagiri M, Vadivelu J (2019) Anti-helicobacter pylori, cytotoxicity and catalytic activity of biosynthesized gold nanoparticles: multifaceted application. Arab J Chem 12:33–40

Grisel R, Weststrate KJ, Gluhoi A, Nieuwenhuys BE (2002) Catalysis by gold nanoparticles. Gold Bull 35:39–45

Grubisha DS, Lipert RJ, Park H-Y, Driskell J, Porter MD (2003) Femtomolar detection of prostate-specific antigen: an immunoassay based on surface-enhanced Raman scattering and immunogold labels. Anal Chem 75:5936–5943

Gunjan B, Zaidi MGH, Sandeep A (2014) Impact of gold nanoparticles on physiological and biochemical characteristics of *Brassica juncea*. J Plant Biochem Physiol 2:133–138

Gupta R, Kulkarni GU (2011) Removal of organic compounds from water by using a gold nanoparticle-poly(dimethylsiloxane) nanocomposite foam. ChemSusChem 4:737–743

Gupta VK, Jain R, Saleh TA, Nayak A, Malathi S, Agarwal S (2011) Equilibirum and thermodynamic studies on the removal and recovery of Safranine-T Dye from industrial effluents. Sep Sci Technol 46:839–846

Han A, Dufva M, Belleville E, Christensen CB (2003) Detection of analyte binding to microarrays using gold nanoparticle labels and a desktop scanner. Lab Chip 3:329–332

Hernández J, Solla-Gullón J, Herrero E, Aldaz A, Feliu JM (2006) Methanol oxidation on gold nanoparticles in alkaline media: unusual electrocatalytic activity. Electrochim Acta 52:1662–1669. http://www.cytodiagnostics.com/store/pc/Introduction-to-Gold-Nanoparticle-Characterization-d3.htm (as Assessed on 3 December, 2017)

Huang X, El-Sayed MA (2010) Gold nanoparticles: optical properties and implementations in cancer diagnosis and photothermal therapy. J Adv Res 1:13–28

Huang XH, Jain PK, El-Sayed IH, El-Sayed MA (2006) Determination of the minimum temperature required for selective photothermal destruction of cancer cells with the use of immune targeted gold nanoparticles. Photochem Photobiol 82:412–417

Huang J, Li Q, Sun D, Lu Y, Su Y, Yang X, Wang H, Wang Y, Shao W, He N, Hong CC (2007a) Biosynthesis of silver and gold nanoparticles by novel sundried *Cinnamomum camphora* leaf. Nanotechnology 18:105104–105115

Huang X, El-Sayed IH, Qian W, El-Sayed MA (2007b) Cancer cells assemble and align gold nanorods conjugated to antibodies to produce highly enhanced, sharp, and polarized surface Raman spectra: a potential cancer diagnostic marker. Nano Lett 7:1591–1597

Huff TB, Tong L, Zhao Y, Hansen MN, Cheng JX, Wei A (2007) Hyper thermic effects of gold nanorods on tumor cells. Nanomedicine (Lond) 2:125–132

Husen A (2017) Gold nanoparticles from plant system: synthesis, characterization and their application. In: Ghorbanpour M, Manika K, Varma A (eds) Nanoscience and plant–soil systems, vol 48. Springer, Cham, pp 455–479

Husen A, Siddiqi KS (2014) Phytosynthesis of nanoparticles: concept, controversy and application. Nanoscale Res Lett 9:229

Hutchings GJ, Haruta M (2005) A golden age of catalysis: a perspective. Appl Catal A 291:2–5

Iqbal M, Ahmad A, Siddiqi TO (2011) Characterization of controversial plant drugs and effect of changing environment on active ingredients. In: Ahmad A, Siddiqi TO, Iqbal M (eds) Medicinal plants in changing environment. Capital Publishing Company, New Delhi, pp 1–10

Iqbal M, Ahmad A, Ansari MKA, Qureshi MI, Aref IM, Khan PR, Hegazy SS, El-Atta H, Husen A, Hakeem KR (2015) Improving the phytoextraction capacity of plants to scavenge metal(loid)-contaminated sites. Environ Rev 23:1–22

Iqbal M, Parveen R, Parveen A, Parveen B, Aref IM (2018) Establishing the botanical identity of plant drugs based on their active ingredients under diverse growth conditions. J Environ Biol 39(1):123–136

Islam NU, Jalil K, Shahid M, Muhammad N, Rauf A (2015a) *Pistacia integerrima* gall extract mediated green synthesis of gold nanoparticles and their biological activities. Arab J Chem. https://doi.org/10.1016/j.arabjc.2015.02.014

Islam NU, Jalil K, Shahid M, Rauf A, Muhammad N, Khan A, Raza Shah MR, Khan MA (2015b) Green synthesis and biological activities of gold nanoparticles functionalized with *Salix alba*. Arab J Chem. https://doi.org/10.1016/j.arabjc.2015.06.025

Jain A, Sinilal B, Starnes DL, Sanagala R, Krishnamurthy S, Sahi SV (2014) Role of Fe-responsive genes in bioreduction and transport of ionic gold to roots of Arabidopsis thaliana during synthesis of gold nanoparticles. Plant Physiol Biochem 84:189–196

Jayaseelan C, Ramkumar R, Rahuman AA, Perumal P (2013) Green synthesis of gold nanoparticles using seed aqueous extract of *Abelmoschus esculentus* and its antifungal activity. Ind Crop Prod 45:423–429

Joseph S, Mathew B (2015a) Microwave assisted facile green synthesis of silver and gold nanocatalysts using the leaf extract of *Aerva lanata*. Spectrochim Acta A Mol Biomol Spectrosc 136:1371–1379

Joseph S, Mathew B (2015b) Microwave-assisted green synthesis of silver nanoparticles and the study on catalytic activity in the degradation of dyes. J Mol Liq 204:184–191

Judy JD, Unrine JM, Rao W, Wirick S, Bertsch PM (2012) Bioavailability of gold nanomaterials to plants: importance of particle size and surface coating. Environ Sci Technol 46:8467–8474

Kasthuri J, Kathiravan K, Rajendiran N (2009) Phyllanthin assisted biosynthesis of silver and gold nanoparticles: a novel biological approach. J Nanopart Res 11:1075–1085

Khalil MMH, Ismail EH, Magdoub FE (2012) Biosynthesis of Au nanoparticles using *olive leaf* extract. Arab J Chem 5:431–437

Kneipp K, Haka AS, Kneipp H, Badizadegan K, Yoshizawa N, Boone C, Shafer-Peltier KE, Motz JT, Dasari RR, Feld MS (2002) Surface-enhanced Raman spectroscopy in single living cells using gold nanoparticles. Appl Spectrosc 56:150–154

Koo Y, Lukianova-Hleb EY, Pan J, Thompson SM, Lapotko DO, Braam J (2016) In planta response of *Arabidopsis* to photothermal impact mediated by gold nanoparticles. Small 12:623–630

Krishnaraj C, Muthukumaran P, Ramachandran R, Balakumaran MD, Kalaichelvan PT (2014) *Acalypha indica* Linn: biogenic synthesis of silver and gold nanoparticles and their cytotoxic effects against MDA-MB-231, human breast cancer cells. Biotechnol Rep 4:42–49

Kumar V, Yadav SK (2009) Plant-mediated synthesis of silver and gold nanoparticles and their applications. J Chem Technol Biotechnol 84:151–157

Kumar KP, Paul W, Sharma CP (2011) Green synthesis of gold nanoparticles with Zingiber officinale extract: characterization and blood compatibility. Process Biochem 46:2007–2013

Kumar KM, Mandal BK, Sinha M, Krishnakumar V (2012) *Terminalia chebula* mediated green and rapid synthesis of gold nanoparticles. Spectrochim Acta A Mol Biomol Spectrosc 86:490–494

Kumar V, Guleria P, Kumar V, Yadav SK (2013) Gold nanoparticle exposure induces growth and yield enhancement in Arabidopsis thaliana. Sci Total Environ 461–462:462–468

Kundu S, Panigrahi S, Praharaj S, Basu S, Ghosh SK, Pal A, Pal T (2007) Anisotropic growth of gold clusters to gold nanocubes under UV irradiation. Nanotechnology 18:075712

Kundu S, Lau S, Liang H (2009) Shape-controlled catalysis by cetyltrimethylammonium bromide terminated gold nanospheres, nanorods, and nanoprisms. J Phys Chem C 113:5150–5156

Kuppusamy P, Yusoff MM, Ichwan SJA, Parine NR, Maniam GP, Govindan N (2015) *Commelina nudiflora* L. edible weed as a novel source for gold nanoparticles synthesis and studies on different physical–chemical and biological properties. J Ind Eng Chem 27:59–67

Leu JG, Chen SA, Chen HM, Wu WM, Hung CF, Yao YD, Tu CS, Liang YJ (2012) The effects of gold nanoparticles in wound healing with antioxidant epigallocatechin gallate and α-lipoic acid. Nanomedicine 8:767–775

Li X, Robinson SM, Gupta A, Saha K, Jiang Z, Moyano DF, Sahar A, Riley MA, Rotello VM (2014) Functional gold nanoparticles as potent antimicrobial agents against multi-drug-resistant bacteria. ACS Nano 8:10682–10686

Lim SH, Ahn EY, Park Y (2016) Green synthesis and catalytic activity of gold nanoparticles synthesized by *Artemisia capillaris* water extract. Nanoscale Res Lett 11:474–484

Lin L, Wang W, Huang J, Li Q, Sun D, Yang X, Wang H, He N, Wang Y (2010) Nature factory of silver nanowires: plant mediated synthesis using broth of *Cassia fistula* leaf. Chem Eng J 162:852–858

Lisha KP, Pradeep T (2009) Enhanced visual detection of pesticides using gold nanoparticles. J Environ Sci Health B 44:697–705

Liu J, Lu Y (2003) A colorimetric lead biosensor using DNAzyme-directed assembly of gold nanoparticles. J Am Chem Soc 125:6642–6643

Lowery AR, Gobin AM, Day ES, Halas NJ, West JL (2006) Immuno nano shells for targeted photothermal ablation of tumor cells. Int J Nanomedicine 1:149–154

Lukman AI, Gong B, Marjo CE, Roessner U, Harris AT (2011) Facile synthesis, stabilization, and anti-bacterial performance of discrete ag nanoparticles using Medicago sativa seed exudates. J Colloid Interface Sci 353:433–444

Luo XL, Xu JJ, Du Y, Chen HY (2004) A glucose biosensor based on chitosan–glucose oxidase–gold nanoparticles biocomposite formed by one-step electrodeposition. Anal Biochem 334:284–289

Majumdar R, Bag BG, Maity N (2013) *Acacia nilotica* (Babool) leaf extract mediated size-controlled rapid synthesis of gold nanoparticles and study of its catalytic activity. Int Nano Lett 3:53–58

Manjari G, Saran S, Arun T, Devipriya SP, Rao AVB (2017) Facile *Aglaia elaeagnoidea* mediated synthesis of silver and gold nanoparticles: antioxidant and catalysis properties. J Clust Sci 28:2041–2056

Manju S, Malaikozhundan B, Vijayakumar S, Shanthi S, Jaishabanu A, Ekambaram P, Vaseeharan B (2016) Antibacterial, antibiofilm and cytotoxic effects of *Nigella sativa* essential oil coated gold nanoparticles. Microb Pathog 91:129–135

Medhe S, Bansal P, Srivastava MM (2014) Enhanced antioxidant activity of gold nanoparticle embedded 3,6-dihydroxyflavone: a combinational study. Appl Nanosci 4:153–161

Meyre ME, Tréguer-Delapierre M, Faure C (2008) Radiation induced synthesis of gold nanoparticles within lamellar phases. Formation of aligned colloidal gold by radiolysis. Langmuir 24:4421–4425

Mishra AN, Bhadauria S, Gaur MS, Pasricha R, Kushwah BS (2010) Synthesis of gold nanoparticles by leaves of zero-calorie sweetener herb (*Stevia rebaudiana*) and their nanoscopic characterization by spectroscopy and microscopy. Int J Green Nanotechnol Phys Chem 1:118–124

Mishra P, Ray S, Sinha S, Das B, Khan MI, Behera SK, Il Yun S, Tripathy SK, Mishra A (2016) Facile bio-synthesis of gold nanoparticles by using extract of *Hibiscus sabdariffa* and evaluation of its cytotoxicity against U87 glioblastoma cells under hyperglycemic condition. Biochem Eng J 105:264–272

Moaveni P, Karimi K, Valojerdi MZ (2011) The nanoparticles in plants. J Nano Struct Chem 2:59–78

Mock JJ, Barbic M, Smith DR, Schultz DA, Schultz S (2002) Shape effects in plasmon resonance of individual colloidal silver nanoparticles. J Chem Phys 116:6755–6759

Mollick MMR, Bhowmick B, Mondal D, Maity D, Rana D, Dash SK, Chattopadhyay S, Roy S, Sarkar J, Acharya K, Chakrabortye M, Chattopadhyay D (2014) Anticancer (in vitro) and antimicrobial effect of gold nanoparticles synthesized using *Abelmoschus esculentus* (L.) pulp extract via a green route. RSC Adv 4:37838–37849

Mukherjee P, Bhattacharya R, Bone N, Lee YK, Patra C, Wang S, Lu L, Secreto C, Banerjee PC, Yaszemski MJ, Kay NE, Mukhopadhyay D (2007) Potential therapeutic application of gold nanoparticles in B-chronic lymphocytic leukemia (BCLL): enhancing apoptosis. J Nanobiotechnol 5:4

Nair AS, Tom RT, Pradeep T (2003) Detection and extraction of endosulfan by metal nanoparticles. J Environ Monit 5:363–365

Narayanan KB, Sakthivel N (2008) *Coriander leaf* mediated biosynthesis of gold nanoparticles. Mater Lett 62:4588–4590

Narayanan KB, Sakthivel N (2010a) Biological synthesis of metal nanoparticles by microbes. Adv Colloid Interf Sci 156:1–13

Narayanan KD, Sakthivel N (2010b) Phytosynthesis of gold nanoparticles using leaf extract of *Coleus amboinicus Lour.* Mater Charact 61:1232–1238

Nellore J, Paulineb PC, Amarnathc K (2012) Biogenic synthesis by *Sphearanthus Amaranthoids;* towards the efficient production of the biocompatible gold nanoparticles. Dig J Nanomater Biostruct 7:123–133

Neng J, Harpster MH, Zhang H, Mecham JO, Wilson WC, Johnson PA (2010) A versatile SERS-based immunoassay for immunoglobulin detection using antigen-coated gold nanoparticles and malachite green-conjugated protein A/G. Biosens Bioelectron 26:1009–1015

Noruzi M, Zare D, Khoshnevisan K, Davoodi D (2011) Rapid green synthesis of gold nanoparticles using *Rosa hybrida* petal extract at room temperature. Spectrochim Acta A Mol Biomol Spectrosc 79:1461–1465

Nun YS, Jaroenrama W, Sriurairatana S, Suebsing R, Kiatpathomchai W (2013) Visual detection of white spot syndrome virus using DNA-functionalized gold nanoparticles as probes combined with loop-mediated isothermal amplification. Mol Cell Probes 27:71–79

Okitsu K, Mizukoshi Y, Yamamoto TA, Maeda Y, Nagata Y (2007) Sonochemical synthesis of gold nanoparticles on chitosan. Mater Lett 61:3429–3431

Orendorff CJ, Gole A, Sau TK, Murphy CJ (2005) Surface enhanced Raman spectroscopy of self-assembled monolayers: Sandwich architecture and nanoparticle shape dependence. Anal Chem 77:3261–3266

Ovais M, Raza A, Naz S, Islam NU, Khalil AT, Ali S, Khan MA, Shinwari ZA (2017) Current state and prospects of the phytosynthesized colloidal gold nanoparticles and their applications in cancer theranostics. Appl Microbiol Biotechnol 101:3551–3565

Pandey S, Oza G, Mewada A, Madhuri S (2012) Green synthesis of highly stable gold nanoparticles using *Momordica charantia* as Nano fabricator. Arch Appl Sci Res 4:1135–1141

Pasca RD, Mocanu A, Cobzac SC, Petean I, Horovitz O, Tomoaia-Cotisel M (2014) Biogenic syntheses of gold nanoparticles using plant extracts. Part Sci Technol 32:131–137

Peleg G, Lewis A, Linial M, Loew LM (1999) Nonlinear optical measurement of membrane potential around single molecules at selected cellular sites. Proc Natl Acad Sci U S A 96:6700–6704

Phillip D (2009) Biosynthesis of Ag, Au and Au-Ag nanoparticles using edible mushrooms extracts. Spectrochim Acta A Mol Biomol Spectrosc 73:374–381

Philip D (2010) Rapid green synthesis of spherical gold nanoparticles using *Mangifera indica* leaf. Spectro Acta A Mol Biomol Spectrosc 77:807–810

Philip D, Unnib C, Aromala SA, Vidhua VK (2011) Murraya Koenigii leaf-assisted rapid green synthesis of silver and gold nanoparticles. Spectrochim Acta A Mol Biomol Spectrosc 78:899–904

Piruthiviraj P, Margret A, Krishnamurthy PP (2016) Gold nanoparticles synthesized by *Brassica oleracea* (Broccoli) acting as antimicrobial agents against human pathogenic bacteria and fungi. Appl Nanosci 6:467–473

Pradeep T, Anshup A (2009) Noble metal nanoparticles for water purification: a critical review. Thin Solid Films 517:6441–6478

Qian XM, Nie S (2008) Single-molecule and single-nanoparticle SERS: from fundamental mechanisms to biomedical applications. Chem Soc Rev 37:912–920

Ramachandran R, Krishnaraj C, Sivakumar AS, Prasannakumar P, Kumar VKA, Shim KS, Song CG, Yun SII (2017) Anticancer activity of biologically synthesized silver and gold nanoparticles on mouse myoblast cancer cells and their toxicity against embryonic zebrafish. Mater Sci Eng C 73:674–683

Ramakrishna M, Babu DR, Gengan RM, Chandra S, Rao GN (2016) Green synthesis of gold nanoparticles using marine algae and evaluation of their catalytic activity. J Nanostructure Chem 6:1–13

Rana S, Bajaj A, Mout R, Rotello VM (2012) Monolayer coated gold nanoparticles for delivery applications. Adv Drug Deliv Rev 64:200–216

Rao KJ, Paria S (2014) Green synthesis of gold nanoparticles using aqueous *Aegle marmelos* leaf extract and their application for thiamine detection. RSC Adv 4:28645–28652

Raouf NA, Al-Enazib NM, Ibraheem IBM (2017) Green biosynthesis of gold nanoparticles using *Galaxaura elongata* and characterization of their antibacterial activity. Arab J Chem 10:S3029–S3039

Reddy V, Torati RS, Oh S, Kim C (2012) Biosynthesis of gold nanoparticles assisted by *Sapindus mukorossi* gaertn. Fruit pericarp and their catalytic application for the reduction of *p*-nitroaniline. Ind Eng Chem Res 52:556–564

Sadeghi B, Mohammadzadeh M, Babakhani B (2015) Green synthesis of gold nanoparticles using *Stevia rebaudiana* leaf extracts: characterization and their stability. J Photochem Photobiol B 148:101–106

Sathish KG, Jha PK, Vignesh V, Rajkuberan C, Jeyaraj M, Selvakumar M, Jha R, Sivaramakrishnan S (2016) Cannonball fruit (*Couroupita guianensis*, Aubl.) extract mediated synthesis of gold nanoparticles and evaluation of its antioxidant activity. J Mol Liq 215:229–236

Sett A, Gadewar M, Sharma P, Deka M, Bora U (2016) Green synthesis of gold nanoparticles using aqueous extract of *Dillenia indica*. Adv Nat Sci Nanosci Nanotechnol 7:025005–025013

Shaligram NS, Bule M, Bhambure R, Singhal RS, Singh SK, Szakacs G, Pandey A (2009) Biosynthesis of silver nanoparticles using aqueous extract from the compactin producing fungal strain. Process Biochem 44:939–943

Shankar SS, Ahmad A, Pasricha R, Sastry M (2003) Bioreduction of chloroaurate ions by geranium leaves and its endophytic fungus yields gold nanoparticles of different shapes. J Mater Chem 13:1822–1826

Shankar SS, Rai A, Ankamwar B, Singh A, Ahmad A, Sastry M (2004) Biological synthesis of triangular gold Nanoprisms. Nat Mater 3:482–488

Shankar SS, Rai A, Ahmad A, Sastry M (2005) Controlling the optical properties of lemongrass extract synthesized gold nanotriangles and potential application in infrared-absorbing optical coatings. Chem Mater 17:566–572

Sharma NC, Sahi SV, Nath S, Parsons JG, Gardea-Torresdey JL, Tarasankar P (2007) Synthesis of plant-mediated gold nanoparticles and catalytic role of biomatrix-embedded nanomaterials. Environ Sci Technol 41:5137–5142

Siddiqi KS, Husen A (2016) Engineered gold nanoparticles and plant adaptation potential. Nanoscale Res Lett 11:400

Siddiqi KS, Husen A (2017a) Recent advances in plant-mediated engineered gold nanoparticles and their application in biological system. J Trace Elem Med Biol 40:10–23

Siddiqi KS, Husen A (2017b) Plant response to engineered metal oxide nanoparticles. Nanoscale Res Lett 12:92

Siddiqi KS, Rahman A, Tajuddin HA (2016) Biogenic fabrication of iron/iron oxide nanoparticles and their application. Nanoscale Res Lett 11:498

Siddiqi KS, Husen A, Rao RAK (2018a) A review on biosynthesis of silver nanoparticles and their biocidal properties. J Nanobiotechnol 16:14

Siddiqi KS, Rahman A, Tajuddin, Husen A (2018b) Properties of zinc oxide nanoparticles and their activity against microbes. Nanoscale Res Lett 13:141

Siddiqi KS, Husen A, Sohrab SS, Osman M (2018c) Recent status of nanomaterials fabrication and their potential applications in neurological disease management. Nanoscale Res Lett 13:231
Siddiqi KS, Rashid M, Rahman A, Tajuddin, Husen A, Rehman S (2018d) Biogenic fabrication and characterization of silver nanoparticles using aqueous-ethanolic extract of lichen (*Usnea longissima*) and their antimicrobial activity. Biomat Res 22:23
Singh AK, Srivastava ON (2015) One-step green synthesis of gold nanoparticles using black cardamom and effect of pH on its synthesis. Nanoscale Res Lett 10:353–364
Smitha SL, Philip D, Gopchandran KG (2009) Green synthesis of gold nanoparticles using *Cinnamomum zeylanicum* leaf broth. Spectrochim Acta A Mol Biomol Spectrosc 74:735–739
Song JY, Jang H-K, Kim BS (2009) Biological synthesis of gold nanoparticles using *Magnolia kobus* and *Diospyros kaki* leaf extracts. Process Biochem 44:1133–1138
Sreelakshmi C, Datta K, Yadav J, Reddy B (2011) Honey derivatized Au and Ag nanoparticles and evaluation of its antimicrobial activity. J Nanosci Nanotechnol 11:6995–7000
Suganthy N, Ramkumar VS, Pugazhendhi A, Benelli G, Archunan G (2018) Biogenic synthesis of gold nanoparticles from *Terminalia arjuna* bark extract: assessment of safety aspects and neuroprotective potential via antioxidant, anticholinesterase, and antiamyloidogenic effects. Environ Sci Pollut Res 25:10418–10433
Suman TY, Rajasree SRR, Ramkumar R, Rajthilak C, Perumal P (2014) The Green synthesis of gold nanoparticles using an aqueous root extract of *Morinda citrifolia* L. Spectrochim Acta A Mol Biomol Spectrosc 118:11–16
Suvith VS, Philip D (2014) Catalytic degradation of methylene blue using biosynthesized gold and silver nanoparticles. Spectrochim Acta A Mol Biomol Spectrosc 118:526–532
Thanh NTK, Rosenzweig Z (2002) Development of an aggregation-based immunoassay for anti-protein a using gold nanoparticles. Anal Chem 74:1624–1628
Tiloke C, Phulukdaree A, Anand K, Gengan RM, Chuturgoon AA (2016) *Moringa oleifera* gold nanoparticles modulate oncogenes, tumor suppressor genes, and caspase-9 splice variants in A549 cells. J Cell Biochem 117:2302–2314
Tsuji T, Kakita T, Tsuji M (2003) Preparation of nano-size particles of silver with femtosecond laser ablation in water. Appl Surf Sci 206:314–320
Upadhyay P, Hanif M, Bhaskar S (2006) Visual detection of IS6110 of Mycobacterium tuberculosis in sputum samples using a test based on colloidal gold and latex beads. Clin Microbiol Infect 12:1118–1122
Vankar PS, Bajpai D (2010) Preparation of gold nanoparticles from Mirabilis jalapa flowers. Ind J Biochem Biophys 47:157–160
Vance ME, Kuiken T, Vejerano EP, McGinnis SP, Hochella MF Jr, Rejeski D, Hull MS (2015) Nanotechnology in the real world: redeveloping the nanomaterial consumer products inventory. Beilstein J Nanotechnol 6: 1769–1780
Velammal SP, Devi TA, Amaladhas TP (2016) Antioxidant, antimicrobial and cytotoxic activities of silver and gold nanoparticles synthesized using *Plumbago zeylanica* bark. J Nanostruct Chem 6:247–260
Vijayakumar S, Vaseeharan B, Malaikozhundan B, Gopi N, Ekambaram P, Pachaiappan R, Velusamy P, Murugan K, Benelli G, Kumar RS, Suriyanarayanamoorthy M (2017) Therapeutic effects of gold nanoparticles synthesized using *Musa paradisiaca* peel extract against multiple antibiotic resistant Enterococcus faecalis biofilms and human lung cancer cells (A549). Microb Pathog 102:173–183
Vijayaraghavan K, Nalini SPK (2010) Biotemplates in the green synthesis of silver nanoparticles. Biotechnol J 5:1098–1110
Wang R, Yue L, Yu Y, Zou X, Song D, Liu K, Liu Y, Su H (2016) Gold nanoparticles modify the photophysical and photochemical properties of 6-thioguanine: preventing DNA oxidative damage. J Phys Chem C 120:14410–14415
Watters JL, Satia JA, Kupper LL, Swenberg JA, Schroeder JC, Switzer BR (2007) Associations of antioxidant nutrients and oxidative DNA damage in healthy African-American and white adults. Cancer Epidemiol Biomark Prev 16:1428–1436

Wilson-Corral V, Anderson CWN, Rodriguez-Lopez M (2012) Gold phytomining. A review of the relevance of this technology to mineral extraction in the 21st century. J Environ Manag 111:249–257

Yasmin A, Ramesh K, Rajeshkumar S (2014) Optimization and stabilization of gold nanoparticles by using herbal plant extract with microwave heating. Nano Converg 1:12

Yu J, Xu D, Guan HN, Wang C, Huang LK, Chi DF (2016) Facile one-step green synthesis of gold nanoparticles using *Citrus maxima* aqueous extracts and its catalytic activity. Mater Lett 166:110–112

Zamarion VM, Timm RA, Araki K, Toma HE (2008) Ultrasensitive SERS nanoprobes for hazardous metal ions based on trimercaptotriazine-modified gold nanoparticles. Inorg Chem 47:2934–2936

Zhan G, Huang J, Lin L, Lin W, Emmanuel K, Li Q (2011) Synthesis of gold nanoparticles by *Cacumen Platycladi* leaf extract and its simulated solution: toward the plant-mediated biosynthetic mechanism. J Nanopart Res 13:4957–4968

Chapter 4
Green Synthesis of Gold Nanoparticles by Using Natural Gums

Alle Madhusudhan, Ganapuram Bhagavanth Reddy, and Indana Murali Krishana

4.1 Introduction

Nanomaterials are cornerstones of nanoscience and nanotechnology, which constitute an interdisciplinary field that uses principles of chemistry, biology, physicsand engineering to design and fabricate nanoscale materials (Farokhzad and Langer 2009; Husen and Siddiqi 2014a, b, c; Siddiqi and Husen 2016a, b, c, d, 2017a, b; Siddiqi et al. 2016, 2018a, b). It has already caused a significant commercial impact, which will surely increase in the near future. Nanoscale materials are defined as a set of substances where at least one of their dimensions is less than 100 nm. Metallic nanoparticles (NPs) are definitely among the most widely studied systems in the modern nanoscience (Mirkin 2005) due to the fact that metals often have totally different properties when brought down to the nanometer dimensions. Metal NPs are important due to their extreme small size and large surface-to-volume ratio, which make them differ in properties from their native metals (Daniel and Astruc 2004; Bogunia-Kubik and Sugisaka 2002; Zharov et al. 2005). The design and manufacture of inorganic nanomaterials with novel applications can be achieved by controlling the size and shape at nanoscale. Nanoparticles (NPs) exhibit shape- and size-dependent properties, which are suitable for various applications such as

A. Madhusudhan (✉)
Department of Chemistry, Samskruti College of Engineering & Technology, Kondapur, Hyderabad, Telangana State, India

Department of Chemistry, University College of Science, Osmania University, Hyderabad, Telangana State, India

G. B. Reddy
Department of Chemistry, Palamuru University, Mahbubnagar, Telangana State, India

I. M. Krishana
Department of Chemistry, Samskruti College of Engineering & Technology, Kondapur, Hyderabad, Telangana State, India

A. Husen, M. Iqbal (eds.), *Nanomaterials and Plant Potential*,
https://doi.org/10.1007/978-3-030-05569-1_4

antimicrobial activity, chemical sensors, catalysis, drug delivery systems, filters and medical radiographic images. Because of their biocompatibility, stability, nontoxicity and oxidation resistance, gold nanoparticles have a specific significance in biomedical applications and in the emerging interdisciplinary fields of science (Husen 2017; Siddiqi and Husen 2017a). They show distinguished surface plasmon resonance (SPR) absorption properties, which are strongly related to their size, shape and interparticle distance. Their large surface-to-volume ratio can become a basis for a large variety of catalytic applications (Corma and Serna 2006; Dimitratos et al. 2012). For example, they exhibit (a) catalytic reduction in the gas-phase selective hydrogenation of 1,3-butadiene to butane, (b) catalytic reduction of 4-nitrophenol and hexacyanoferrate (III) by sodium borohydride, and (c) the Suzuki–Miyaura cross-coupling reaction (Bond and Sermon 1973; Sermon et al. 1979; Atnafu et al. 2016; Jie et al. 2009). Nanomedicine involves extensive use of gold NPs and nanorods for diagnosis and therapeutic purposes (Siddiqi and Husen 2017a; Llevot and Astruc 2012; Sánchez et al. 2012) as carriers for delivering drugs, antigens and genetic materials to targeting sites without any side effects (Madhusudhan et al. 2014; Dykman and Bogatyrev 2007; Pissuwan et al. 2011) and as drugs or diagnostic tools in the therapy of tumors and rheumatoid arthritis (Bhattacharya and Mukherjee 2008; Brown et al. 2008). They also find application in electronics, photonics, sensing and imaging (El-Sayed 2001; Husen and Siddiqi 2014b).

Among the various ways to synthesize AuNPs, the most common are the reduction of gold ions with reducing agents such as sodium borohydride, hydrazine, N,N-dimethylformamide, citrate, or other organic compounds (Zhang and Wu 2010; Rivas et al. 2001) followed by surface modification with a suitable capping agent to prevent aggregation of particles by electrostatic or physical repulsion (Erik et al. 2012). However, the use of such reducing agents may show environmental toxicity or pose biological risks. Therefore, utilization of nontoxic chemicals, environmentally benign solvent (water) and renewable materials (plant products), agriculture waste (banana peel) and/or living organisms (viruses, bacteria, fungi, algae), which would decrease the carcinogenic chemical waste and prevent pollution, are some of the key issues in the green chemistry research (Dahl et al. 2007; Siddiqi et al. 2016). The use of plant-based biomaterials or biomolecules to synthesize AuNPs is expected to cause a minimum quantity of hazardous waste and consume less energy, compared to the chemical synthesis methods.

Biosynthesis of AuNPs and gold nanorods (AuNRs) has several advantages such as simplicity and biocompatibility. Biomolecules act as the reducing and capping agents in the large-scale commercial production of AuNPs. Biosynthesis of AuNPs using naturally occurring gums has proved to be cost-effective, nontoxic and eco-friendly. Plant extracts, such as natural gums, may act as both the reducing and capping agents in AuNP synthesis.

Natural gums are considered to be the pathological products that ooze out following injury to the plant. Gums are hydrocolloids, which are hydrophilic in nature. These are effective water adsorbents and may be solubilized by water, producing viscous aqueous systems. On hydrolysis, gums yield a mixture of sugars and uronic acids (Choudhary and Pawar 2014). Most of them are heterogeneous polysaccharides with complicated structures, which would vary depending on the source and their age and possess extremely high molecular masses. Owing to the presence of a large number of hydroxyl groups, water is bonded inside the molecular structure by hydrogen bonding and additionally within the voids created by the complex molecular configuration. Therefore, it is not possible to provide defined structural formulae of these biopolymers (Rana et al. 2011). Linear polysaccharides occupy extra space than the highly branched compounds of a similar mass.

Generally, gum exudates contain galacturonic acid, arabinose, uronic acids, galactose, rhamnose, protein, Mg and Ca as the major structural constituents and also mannose, glucose, protein, xylose and fat as the minor constituents. Natural gums are biodegradable, biocompatible, nontoxic, environment-friendly, having a low-cost and processing edible sources. Gums of different sources and their derivatives, represent a distinct group of polymers widely used in food industry, pharmacy and medicine and in the manufacture of cosmetics, textiles, adhesives, lithography, paints and paper (Bhardwaj et al. 2000; Rana et al. 2011; Mirhosseini and Amid 2012; Ibrahim et al. 2010; Prajapati et al. 2013a, b).

The plant-based naturally occurring gums, such as gum kondagogu, gum karaya, gum tragacanth, gum arabic, gum salmalia, guar gum, locust bean gum, gellan gum, xanthan gum, olibanum gum, gum katira, gum ghatti, bael gum and Prunus domestica gum, are used as the reducing and capping agents for fabrication of platinum, silver, gold and palladium NPs through green synthesis (Reddy et al. 2015a, b, c, 2017; Dhar et al. 2011; Atnafu et al. 2016; Saikat et al. 2012; Alam et al. 2017; Wu and Chen 2010; Pandey et al. 2013; Pooja et al. 2014; Rao et al. 2017; Huang et al. 2007a, b). Natural gums contain specific functional groups, such as -OH, $-NH_2$, -CHO, $-CONH_2$,-COOH, etc., which help in surface stabilization of NPs better than in other biological syntheses. Biosynthesis of AuNPs of various shapes and sizes by using various natural gums is shown in Table 4.1. The long polymer chain restricts the process of agglomeration of NPs. Therefore, NPs biosynthesized with natural gums were more stable through several months. The NPs stabilized with naturally occurring gums are harmless to cells and suitable for safe delivery of drugs, cancer detection and bio-imaging (Wu et al. 2006; Pooja et al. 2015). The attributes such as biocompatibility and antibacterial activity render the gold and silver NPs suitable as coating material in food packing and biomedical engineering. The present chapter is focused on green synthesis of AuNPs using natural gums and on their diverse applications.

4.2 Description of Various Gums

Gum kondagogu is a naturally available polysaccharide component extracted from the bark of *Cochlospermum gossypium* DC (family Bixaceae). It is made up of sugars such as galactose, arabinose, mannose, glucose, gluconic acid, rhamnose and galacturonic acid with sugar linkage of (1 → 6) β-D-Gal p, (1 → 2) β-D-Gal p, (1 → 2) α-L-Rha, 4-O-Me-α-D-Glc p A, (1 → 4) β-D-Glc p and (1 → 4) α-D-Gal p

Table 4.1 Gold nanoparticles of various size and shape synthesized by using the naturally occurring gums

Gum	Size	Shape	Reference
Bael gum	2 µm to 90 nm	Triangular	Subramanian et al. (2016)
Carboxymethyl Gum karaya	14 ± 2 nm	Spherical	Reddy et al. (2017)
Gellan gum	13 ± 1 nm	Spherical	Dhar et al. (2008)
Gellan gum	14 nm	Spherical	Dhar et al. (2011)
Gellan gum	47 ± 10 nm	Nanorods	Vieira et al. (2015)
Gum acacia	4–29 nm	Spherical	Reddy et al. (2015b)
Gum arabic	10 ± 2 nm	Spherical	Kattumuri et al. (2007)
Gum arabic	21.1 ± 4.6 nm	Spherical	Wu and Chen (2010)
Gum arabic	1.6 ± 0.7 nm	Spherical	Liu et al. (2013)
Gum arabic	4.045 ± 0.99 nm	Spherical	Joshita et al. (2014)
Gum arabic	Average size 5.5 nm	Spherical	Thanaa et al. (2015)
Gum ghatti	30 ± 1.2 nm	Spherical	Alam et al. (2017)
Guar gum	~6.5 nm	Spherical	Pandey et al. (2013)
Gum karaya	5.0 ± 1.2 nm	Spherical	Padil and Černík (2015)
Gum karaya	12 ± 2 nm	Spherical	Pooja et al. (2015)
Gum kondagogu	7.8 ± 2.3 nm	Spherical	Vinod et al. (2011)
Gum kondagogu	12 ± 2 nm	Spherical	Reddy et al. (2015a)
Gum kondagogu	4.08–12.73 ± 0.75 nm	Spherical	Selvi et al. (2017)
Katira gum	~6.9 nm	Spherical	Saikat et al. (2012)
Locust bean gum	16 ± 3 nm	Spherical	Tagad et al. (2014)
Olibanum gum	3 ± 2 nm	Spherical	Atnafu et al. (2016)
Prunus domestica gum	7–30 nm	Spherical	Islam et al. (2017)
Salmalia malabarica gum	12 ± 2 nm	Spherical	Reddy et al. (2015c)
Tamarind gum	50–150 nm	Spherical	Biswala et al. (2013)
Xanthan gum	15–20 nm	Spherical	Pooja et al. (2014)

A. It is an acidic gum and has carboxylic acid, acetyl, hydroxyl, and carbonyl groups as the major functional groups (Kumar and Ahuja 2012).

Gum karaya obtained from *Sterculia urens* Roxb. (family Sterculiaceae) is a highly branched acidic polysaccharide. The carbohydrate structure has the main chain of rhamnogalacturonan consisting of α-(1 → 4)-linked D-galacturonic acid and α-(1 → 2)-linked L-rhamnosyl residues. The side chain is composed of (1 → 3)-linked β-D-glucuronic acid, or (1 → 2)-linked β-D-galactose on the galacturonic acid unit, where one-half of the rhamnose is substituted by (1 → 4)-linked β-D-galactose (Galla and Dubasi 2010).

Gum tragacanth (GT) is a natural gum obtained from dried sap of several species of the genus *Astragalus* (e.g., *A. tragacantha*, *A. adscendens*, *A. gummifer* and *A. brachycalyx*). The gum contains pectinaceous arabinogalactans and fucose-substituted xylogalacturonans. It is a heterogeneous and acidic polysaccharide, which contains a mixture of two polysaccharides, of which the water-soluble component is called tragacanthin and the acidic water-swellable component is called bassorin. The water-soluble tragacanthin is a neutral, highly branched arabinogalactan comprising (1 → 6)- and (1 → 3)-linked core chain with galactose and arabinose units and side groups of (1 → 2)-, (1 → 3)- and (1 → 5)-linked arabinose units occurring as monosaccharides or oligosaccharides. Bassorin, a pectic component, has a chain of (1 → 4)-linked α-D-galacturonic acid units, some of which are substituted at O-3 with β-D-xylopyranosyl units and some are terminated with D-Gal or L-Fuc (Kora and Arunachalam 2012).

Gum arabic, obtained from *Acacia nilotica* (L.) Willd. ex Delile (family Mimosaceae), is a slightly acidic or neutral polysaccharide complex obtained as a mixed magnesium, potassium and calcium salt. The backbone consists of 1 → 3-linked β-D-galactopyranosyl units, and the side chains contain units of β-D-glucuronopyranosyl, α-L-rhamnopyranosyl, α-L-arabinofuranosyl and 4-O-methyl-β-D-glucuronopyranosyl (Dror et al. 2006).

Gum salmalia extracted from the plant *Bombax ceiba* L. (former *Salmalia malabaricum*) (family Bombacaceae). On hydrolysis, it shows the presence of various sugars such as L-arabinose, D-galactose, D-galacturonic acid, rhamnose, thiamine, riboflavin, 2,3,5-tri- and 2,5-di-O-methyl-L-arabinose and 2,3,4,6-tetra-,2,6-di- and 2,4-di-O-methyl-D-galactose (Das et al. 1990).

Guar gum (GG) is a non-ionic edible polysaccharide derived from the seeds of *Cyamopsis tetragonolobus* (L.) Taub. (family Fabaceae). It consists of linear chains of (1, 4)-β-D-mannopyranosyl units with α-D-galactopyranosyl units attached via (1 → 6) linkages (Dodi et al. 2011).

Locust bean gum (LBG) is extracted from the endosperms of hard seeds of the locust bean tree, *Ceratonia siliqua* L. (family Fabaceae). It is a non-starch polysaccharide and polyhydroxylated biopolymer consisting of galactomannan units with mannose backbone and galactose units as single side chain (Dey et al. 2011).

Xanthan gum (XG) is a natural extracellular polysaccharide derived from a cabbage plant bacterium, known as *Xanthomonas campestris*, which is an anaerobic, gram-negative rod made up of penta saccharide repeat units, comprising glucose, mannose and glucuronic acid in the molar ratio 2.0:2.0:1.0 (Maity and Sa 2014).

Olibanum gum (OG) is obtained from *Boswellia serrata* Roxb. (family Burseraceae). Its chemical composition depends on its three principal areas of origin, Eritrea, Aden/Somalia, and India, which contains approximately 20–30% polysaccharides, 13–17% acid resin, 3–8% volatile oil and 40–60% boswellic acid (Kora et al. 2012).

Gellan gum (GG) is an anionic polysaccharide produced by a microorganism *Sphingomonas elodea*. It has a tetrasaccharide repeating unit composed of (1–4)-α-L-rhamnose, (1–4)-β-D-glucose and (1–3)-β-D-glucose and (1–4)-β-D-glucuronic acid as the backbone (Ahuja et al. 2013).

Katira gum (KG) is exuded from the *Cochlospermum religiosum* L. (Alston) (family Bixaceae) and consists of L-rhamnose, D-galactose and D-galacturonic acid in a molar ratio 3:2:1, together with traces of a ketohexose (Ojha et al. 2008).

Bael gum (BG) is obtained from fruits of *Aegle marmelos* L. Correa (family Rutaceae). Purified bael gum polysaccharides contain D-galacturonic acid (7%), L-arabinose (12.5%), D-galactose (71%) and L-rhamnose (6.5%) (Arati et al. 1976).

Prunus domestica gum (PDG) is isolated from the plum tree (*Prunus insititia* L. of the family Rosaceae) and contains L-arabinose, D-galactose and D-mannose, in a molar ratio of 3:2:1 and 3% of D-xylose (Islam et al. 2017).

4.3 Synthesis of AuNPs

In general, bioreduction of Au^{3+} with gum solution to form a ruby red color solution indicates the formation of AuNPs (Punuri et al. 2012). The size, shape, and morphology of AuNPs depend on the (i) concentration and volume of $HAuCl_4$ and gum, (ii) reaction time and (iii) pH medium of the solution. Brief procedure for fabrication of AuNPs involves mixing of various concentrations of gum solution with aliquot amount of precursor ($HAuCl_4$). The mixture is subjected to autoclaving or microwave irradiation or direct heating or ultra-sonication. The color of the resultant changes from yellow to ruby red indicating the formation of AuNPs, as depicted in Fig. 4.1. The major advantage of gum-capped AuNPs is that these NPs are more stable for several months without aggregation, owing to the presence of multifunctional groups and possess negative surface which acts as a stabilizing agent.

Dhar and his group demonstrated synthesis of AuNPs with gold ion solution heating with gellan gum; pH of the solution was adjusted to 11–12 with NaOH to

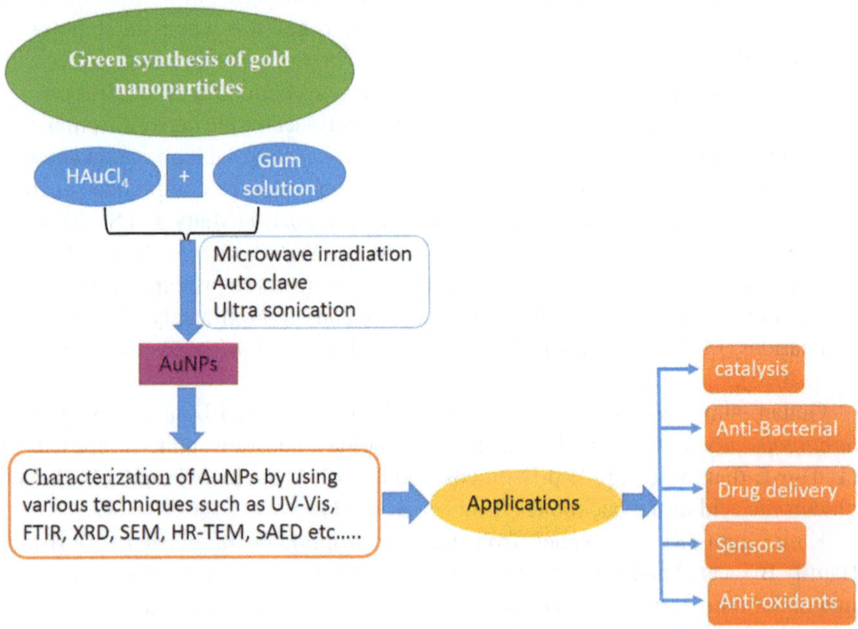

Fig. 4.1 Schematic representation of fabrication of AuNPs by using natural gums, their characterization and applications

yield ruby red AuNPs. Formation of AuNPs was confirmed by UV-Vis absorbance maximum at 520 nm. The TEM image reveals that NPs are well-dispersed with narrow size distribution and an average size of 13 ± 1 nm (Dhar et al. 2011). The SAED pattern of the AuNPs is indexed to the 111, 200, 220 and 311 Bragg reflections which represents a face-centered cubic (fcc) structure (Dhar et al. 2011). These gum-capped AuNPs have greater stability to pH and electrolyte changes relative to the citrate- and borohydride-reduced NPs (Dhar et al. 2011). Sílvia Vieira and coworkers developed layer-by-layer-coated gold nanorods (AuNRs) by using poly(acrylic acid), poly(allylamine hydrochloride), and gellan gum, which allowed formation of a GG hydrogel-like shell with 7 nm thickness around individual AuNPs. This method produces nanorods with an average length of 47 ± 10 nm (Vieira et al. 2015).

Gum kondagogu-stabilized AuNPs were prepared by mixing of gum solution with chloroauric acid in a boiling tube and kept in an autoclave at 120 °C and 15 psi pressure for 10 min (Reddy et al. 2015a), but the prepared NPs are larger in size (12 ± 2 nm) and of unequal shape. The surface plasma resonance (SPR) of prepared NPs exhibits at around 515–560 nm. The formation of NPs increases with increase in the concentration of gum, $HAuCl_4$ and autoclave time. FTIR spectra reveal that hydroxyl and carboxyl group peaks shifted from 3418 to 3445 cm^{-1} and 1606 to 1598 cm^{-1}, thus indicating the involvement of both the functional groups in the synthesis process. Vinod et al. (2011) showed that much smaller (7.8 ± 2.3 nm diameter) AuNPs could be produced from the reduction of $HAuCl_4$ with GK solution by using an orbital shaker at 75 °C, 250 rpm for 1 h. FTIR spectra indicate that hydroxyl functional group in the gum was responsible for the reduction of metal cation to NPs. UV-Vis analysis showed SPR at 525 nm due to the formation of AuNPs (Vinod et al. 2011).

Gold NPs of various sizes and shapes are fabricated by using heteropolysaccharide extracted from the gum of *Cochlospermum religiosum* (katira gum). In this method, small (6.9 nm) and mostly spherical, with a few having a rod-like or decahedral shape, NPs are produced from the aqueous polysaccharide solution mixed with $HAuCl_4$. This mixture was then heated at 70 °C with constant stirring with the help of a magnetic stirrer. A typical SPR band was obtained at 540 nm initiallyand after 6 h of the reaction time, SPR band shifted to 520 nm, beyond which no change in SPR band was observed. The fcc crystalline nature of these AuNPs was identified by XRD analysis and SAED pattern (Saikat et al. 2012).

GG-capped AuNPs are spherical in shape, 6.5 nm in size with d-spacing (HR-TEM) of 0.23, 0.20 and 0.14 nm, which reveals fcc crystal lattice with (111), (200) and (220) planes, and the same was further confirmed by SAED pattern in HR-TEM (Pandey et al. 2013). These NPs were prepared by mixing Au^{3+} ions with the guar gum solution of predefined concentration, added with few drops of NaOH solution. The resultant solution was then stirred gently at the desired temperature (80 °C) for 160 min to yield gold NPs. The formed AuNP SPR band at 539 nm and peak intensity of the resultant solution gradually increase with increase in time, indicating the formation of AuNPs.

LBG-stabilized AuNPs of various sizes and shapes are synthesized by employing the autoclave method, as described by Tagad et al. (2014). The AuNPs produced are spherical in shape and have the fcc crystal structure and SPR band at 537 nm. The authors of this procedure (Tagad et al. 2014) conclude that the hydroxyl groups and the hemiacetal reducing ends of LBG act as the active reaction centers to facilitate the reduction of Au^{3+} to Au^0 as shown in Fig. 4.2. This is further confirmed by FTIR spectra, which show a peak at 1728 cm^{-1} which arises from the carbonyl stretching vibrations at the reducing end of the gum after addition of Au^{3+} ions in the reaction mixture. The carbonyl functional groups at the ends of the gum chain get oxidized to the carboxyl group; increase in the peak intensity was observed at 1642 and 1554 cm^{-1} and the peak disappeared at 1728 cm^{-1}. It indicates that the reduction of Au^{3+} is coupled with the oxidation of hemiacetal/aldehyde groups in the formation of LBG-stabilized AuNPs.

Formation of the xanthan gum-capped AuNPs is monitored through various reaction parameters such as (i) reaction time, (ii) temperature, (iii) concentration and volume of gum solution and (iv) concentration and volume of gold solution. In this method, XG-capped NPs are prepared by heating the aqueous solution of $HAuCl_4$ at 80 °C in the presence of xanthan gum solution (1.5 mg mL^{-1}) with a reaction time of 3 h. The NPs produced show a mean particle size of 15–20 nm and a zeta potential of −47.2 ± 2.59 mV. These NPs are stable at a pH range of 5–9 and the NaCl concentration up to 0.5 M (Pooja et al. 2014). Muddineti et al. (2016) reported the synthesis of AuNPs by using ascorbic acid as the reducing agent and xanthan gum as the stabilizing agent. Synthesis of AuNPs was carried out by mixing xanthan gum with hydrogen tetrachloroaurate (III) hydrate, followed by addition of ascorbic acid to a mixture of xanthan gum Au^{3+} solution to form AuNPs. The effect of concentration of $HAuCl_4$, ascorbic acid and methoxy polyethylene glycol thiol (mPEG800-SH)

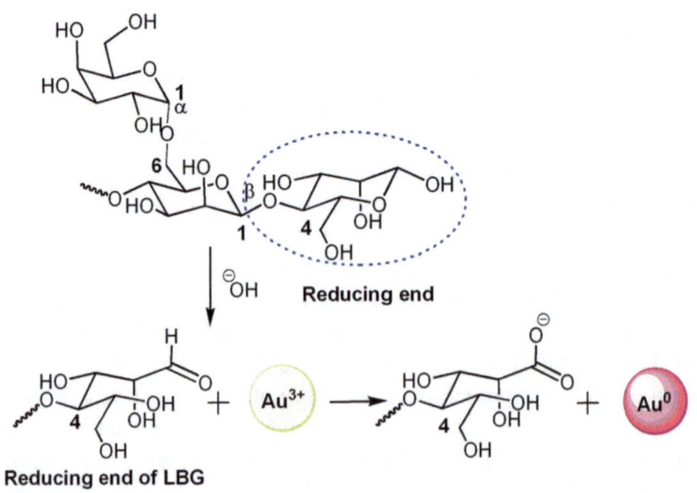

Fig. 4.2 LBG structure and reaction mechanism of AuNP synthesis at the reducing end of LBG biopolymer (Tagad et al. 2014)

was optimized. Stable AuNPs were formed at a concentration of 0.25 mM, 50 μM, and 1 mM for $HAuCl_4$, ascorbic acid and PEG-800SH, respectively. The so-produced AuNPs showed an absorption maximum at 540 nm and a hydrodynamic diameter of 80 ± 3 nm. During the reaction, the Au^{3+} ions were reduced to Au^0 (metallic gold) by using ascorbic acid. Au^0 atoms were further stabilized by the XG-coating and PEGylation, resulting in the formation of PX-AuNPs with improved stability. This is explained by the following chemical equation:

$$2HAuCl_4 + 3C_6H_8O_6 \rightarrow 2Au^O + 3C_6H_6O_6 + 8HCl$$

Gum acacia-stabilized AuNPs are prepared by mixing a homogeneous GA solution with chloroauric acid in a boiling tube and autoclaving at 121 °C and 15 psi pressure for 15 min. The NPs formed exhibit SPR within 520–550 nm. As the gum and chloroauric acid concentrations increase, formation of NPs also increases, as confirmed by the UV-Vis spectra. The NPs show fcc crystal structure and have the particle size around 4–29 nm. FTIR analysis reveals that –OH groups available in the gum matrix are responsible for reducing the $HAuCl_4$ to AuNPs (Reddy et al. 2015b). The authors also developed SMG-stabilized stable AuNPs with a particle size of 12 ± 2 nm, which is adjusted by varying the amounts of gum and chloroauric acid used in the NP synthesis. These NPs have SPR band around 520–530 nm and are highly crystalline in structure with fcc geometry. FTIR analysis indicates that –OH groups available in the gum matrix might be responsible for the reduction of Au^{3+}into AuNPs (Reddy et al. 2015c).

Likewise, one-step microwave-assisted green synthesis of well-stabilized AuNPs by reduction of $HAuCl_4$ with water-soluble olibanum gum was reported by Atnafu et al. (2016). Formation of AuNPs was confirmed with the help of UV-Vis, FTIR, XRD and TEM spectra. The fabricated NPs were well-dispersed and spherical in shape and their average diameter was around 3 ± 2 nm. The rate of NP formation increased and the NP size decreased when gum concentration and irradiation time were increased. The authors concluded that the –OH functional group of gum was responsible for the formation and stabilization of AuNPs.

Pooja et al. (2015) reported the synthesis and characterization of gum karaya (GK)-stabilized AuNPs. In this method GK solution is mixed with $HAuCl_4$ and stirred magnetically for 1 h at 90 °C. The solution color change from colorless to wine red at the absorbance of 536 nm indicates the formation of AuNPs. The observed particle size of optimized AuNPs is 91.7 nm with 0.389 PDI and −21.8 mV zeta potentialand particle size ranges between 20 and 25 nm, as determined by TEM. These authors also found that a 90 °C reaction temperature, 60 min reaction time, 2.5 mL of 15 mg mL^{-1} gum solution and 100μL of gold solution are the optimal conditions for the synthesis of NPs (Pooja et al. 2015). Carboxyl-methylated GK-capped AuNPs were prepared by Reddy et al. (2017). Carboxymethyl gum-capped AuNPs are smaller in size (14 ± 2 nm), compared to normal gum-capped AuNPs because carboxymethyl gum is a more effective capping agent than the normal gum. As the gum is converted to carboxymethyl gum, its solubility increases and NP formation accelerates (Reddy et al. 2017).

A facile and complete green synthesis of AuNPs by gum arabic was reported by Wu and Chen (2010). Various concentrations of hydrogen tetrachloroaurate (0.1, 0.3, 0.5, 0.7, and 0.9 mM) were mixed with concentrations (1, 5, 10, 15, and 20 mg mL^{-1}) of arabic gum solution and stirred gently at the desired temperatures (25, 40, 55, and 70 °C) to yield AuNPs. The synthesis reaction was finished in 2–4 h. Increase in the reaction temperature increased the NP formation rate but had no significant effect on the optical property and the size of AuNPs. The particle size distribution was broader at a higher Au^{3+} ion concentration or a lower gum arabic concentration due to insufficient protection. A too high gum arabic concentration was not suitable for stabilization of AuNPs because the increased intermolecular force of the gum arabic might hinder the dispersion of AuNPs.

The use of gum ghatti for reduction, stabilization and surface functionalization during the synthesis of stable AuNPs was reported by Alam et al. (2017). In this method, preheated gum solution was added dropwise to a required volume of $HAuCl_4$ (0.1 M) under constant stirring until the color changed from yellow to ruby red color, which indicated the formation of AuNPs. The synthesized and optimized AuNPs were spherical, having SPR band at 530 nm, DLS for its hydrodynamic size (112.5 nm), PDI (0.222), and zeta potential (−21.3 mV). The authors explained that a 4% of gum solution was just sufficient for nucleation, moderate size and growth; temperature above 60 °C ensured the best performance.

Gum tragacanth-stabilized AuNPs were prepared by mixing gum solution with gold solution and magnetically stirred (150 rpm) for 4 h at 65 °C. The resultant solution had SPR at 548 nmand synthesis of GT-AuNPs was optimized by changing various parameters like concentration and volume of GT and gold solutions. The maximum absorbance of SPR was observed for 5 mg mL^{-1} GT and 5 mM gold solution concentrations. GT-capped AuNPs were stable against salt concentration, pH and plasma contact time (Rao et al. 2017).

An interesting study by Subramanian et al. (2016) illustrated an eco-friendly and selective synthesis of triangular AuNPs by the reduction of $HAuCl_4$ with bael gum (*Aegle marmelos* gum); the NPs were in the range of 92 ± 20 nm. The size and shape (such as triangular with a tail, triangular, triangular truncated at the apex, and hexagonal) depend on the concentration of bael gum and reaction temperature. The concentration ratio of bael gum and $HAuCl_4$ for the reduction of $HAuCl_4$ 1:1 at 30 °C was very slow and resulted in the formation of plates largely triangular, but the ratio increment 1:2.5 at 30 °C formed the polydispersed AuNPs with tail. Further increment in ratio (1:4) at room temperature gave perfect nanoplates (triangular as well as hexagonal) of ≈ 2 μm. The increase in weight percentage of bael gum from 1:7 to 1:12.5 led to a reduction in size from ≈ 1 μm to ≈ 500 nm. In this stage, predominantly 67% of NPs were hexagonal in shape, the rest being triangular and truncated triangular. Further increase in the concentration of bael gum gave rise to spherical AuNPs with average size 20 nm. At fixed bael gum and $HAuCl_4$ concentrations, the NP size was 260 ± 66 nm at 50 °C and 92 ± 20 nm at 70 °C. This suggests that 1:10 ratio of $HAuCl_4$ and bael gum forms pure triangular NPs at 70 °C and spherical AuNPs at more than 90 °C.

4.3.1 Reaction Mechanism

The naturally occurring gums act as a reducing agent for gold ions and a capping agent after the formation of AuNPs. The generally accepted mechanism for fabrication of AuNPs involves a two-step process, i.e., atom formation and then polymerization of the atom. In the first step, gum solution is heated by various means such as autoclaving, microwave irradiation, normal heating, or sonochemical treatment. The biopolymer obtained may expand and become more accessible for gold ions to interact with the available functional groups of gums. Gums contain carbohydrate polymers such as rhamnose, galactose and uranic acid and the functional groups such as hydroxyl, carbonyl and carboxyl groups. Compared to other biopolymers, a gum is an anionic functional group with a greater charge density. The presence of negatively charged groups is also confirmed from the negative zeta potential value for the gum. This negative charge facilitates the attraction of positively charged gold ion to the polymeric chain. Subsequently, these gold ions oxidize the hydroxyl and carbonyl groups to carboxyl group and conversely get themselves reduced to gold atoms. In addition to this inherent oxidation, the dissolved air also causes oxidation of the existing hydroxyl groups to carbonyl (-CHO and -COOH) groups. In turn, these powerful reducing aldehyde groups, along with the other abundantly present carbonyl groups, reduce more and more of the gold ions to gold atoms. In the second step, these NPs are capped and stabilized by the polysaccharides present in the gum. The stabilization of AuNPs occurs through a strong association between AuNP surface and "O" atoms of the hydroxyl and carbonyl functional groups of the natural gums. The resulting stabilized AuNPs avoid coalescing with the neighboring one due to the electrostatic repulsion and steric effect (Tagad et al. 2014; Vinod et al. 2011; Subramanian et al. 2016; Pandey et al. 2013; Selvi et al. 2017).

4.4 Characterization of AuNPs

Synthesized NPs are characterized by using various techniques such as UV-Vis spectroscopy, FTIR spectroscopy, X-ray diffraction (XRD), scanning electron microscopy (SEM), TEM, selected area electron diffraction (SAED) and dynamic light scattering (DLS) zeta potential. These techniques are useful to determine the crystal structure, size, shape, charge, disparity, surface area and surface modification of NPs.

UV-Vis spectroscopy is a simple and sensitive technique for the characterization of AuNPs due to its extension to SPR. In general trend, SPR band shows a red shift with increasing particle size, while aggregation of colloidal gold causes a decrease in the intensity of the main peak and also results in a long tail on the long-wavelength side of the peak. The shape-dependent radioactive properties of metallic particles can be treated by the Gans modification of Mie theory, which predicts that a shift in the SPR occurs when particles deviate from spherical geometry. In this condition, the longitudinal and transverse dipole polarizability no longer produces equivalent resonances. Therefore, there appear two plasma resonances: a broadened and red-shifted longitudinal plasmon resonance and a transverse plasmon resonance. For example, gold nanorods (AuNRs) exhibit two plasmon bands: strong longitudinal band in the near-IR region and a weak transverse band, similar to that of spherical AuNPs (520 nm), in the visible region. The resonance of the longitudinal mode, a red shift and strong, depends on the aspect ratio (defined as the ratio of the length to width of the rod) of the nanorods. An absorption spectrum of colloidal AuNRs (with accepted ratio 4:1), showing the presence of two absorption maxima, is presented in Fig. 4.3a. The spherical AuNPs show only one SPR around 500–600 nm (Fig. 4.3b), and this is attributed to the oscillation of the conduction band electrons induced by the interacting electromagnetic field with the concerned metallic NPs (Saha et al. 2012; Huang and El-sayed 2010; Daniel and Astruc 2004; Mayer and Hafner 2011).

In order to monitor the effect of different parameters to the formation of AuNPs, the UV-Vis absorption spectra of synthesized AuNPs were recorded against various parameters such as gum concentration, $HAuCl_4$ and reaction time. The SPR absorption band appears approximately at 500–600 nm, which is associated with the formation of AuNPs. The efficacy of the formation of AuNPs increases with increase in the concentration of gum at fixed concentration of $HAuCl_4$. Further increase in the concentration of gum results in a decrease in the peak intensity and the peaks shifted toward longer wavelength are shown in Fig. 4.4a. The production of AuNPs with different concentrations of $HAuCl_4$ and reaction times with fixed concentration of gum is shown in Fig. 4.4b. It can be observed that the strong absorption band increases with increase in concentration and reaction time. The stability of synthesized NPs was monitored at different pH values, electrolytes, and reaction times. The red shift, normally observed in UV-Vis spectra, is an indication of agglomeration of NPs or increase in the size of the particle or both. The synthesized NPs did not show any shift in the peak, which indicates that NPs are stable and aggregation of gold NPs is protected by the gum.

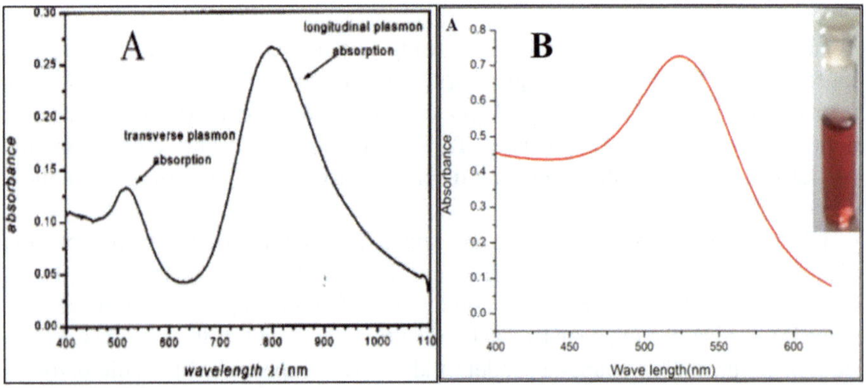

Fig. 4.3 UV-Vis absorption spectra of (**a**) gold nanorods (Ghosh and Pal 2007), (**b**) gold nanoparticles (Madhusudhan et al. 2014)

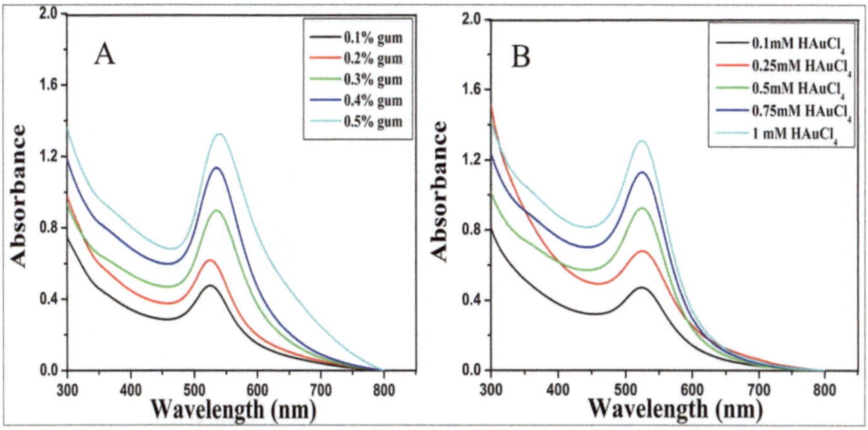

Fig. 4.4 UV-Vis absorption spectra of the AuNPs synthesized: (**a**) different concentrations of olibanum gum solution containing 1 mM $HAuCl_4$ by MWI 9 min and (**b**) different concentrations of $HAuCl_4$ solution containing 0.5% olibanum gum by MWI 9 min (Atnafu et al. 2016)

The active sites on natural gums involved in the reduction of gold ions and AuNP formation were investigated by using the Fourier-transform infrared (FTIR) spectroscopy. The obtained aqueous AuNP dispersion was first centrifuged or lyophilized, and then the sample was used for FTIR analysis, prepared in the form of a thin transparent pellet with potassium bromide (KBr). A pure KBr was used as a background and the same was subtracted from the FTIR spectra of the gum-capped AuNPs. The FTIR spectra were recorded with a scanning range from 400 to 4000 cm^{-1}. The interaction between gum and AuNPs through hydroxyl and carbonyl groups showed the stretching vibration of hydroxyl group at 3445 cm^{-1}, carbonyl stretching vibration at 1722 cm^{-1}, and asymmetric stretch of carboxylate group at 1606 cm^{-1} (Reddy et al. 2015a). After the formation of AuNPs, the peaks for hydroxyl group and asymmetric stretch of carboxyl group were shifted from 3418 to 3445 cm^{-1} and from 1606 to 1598 cm^{-1}, respectively, in comparison with those in the native gum kondagogu, suggesting that both hydroxyl and carbonyl functional groups were involved in the synthesis and stabilization of AuNPs (Fig. 4.5). In another study, Atnafu et al. (2016) demonstrated that the FTIR spectra of olibanum gum show major peaks located at 3421, 2932, 1726, 1427, 1244 and 1033 cm^{-1}. The peaks 3421 and 2932 cm^{-1} can be assigned to the stretching vibration of the hydroxyl and methylene group. The bands observed at 1726, 1617 and 1427 cm^{-1} could be attributed to the characteristic asymmetrical and symmetrical stretches of the CO_2^- (carboxylate ion) group connected to the olibanum gum. The absorption bands of the olibanum gum-capped AuNPs were detected at 3356, 2932, 1719, 1595, 1448, 1259 and 1065 cm^{-1}. These peaks were shifted from 3421 to 3356, 1726 to 1719, 1617 to 1595, and 1427 to 1438 cm^{-1} when compared to those in the native olibanum gum, and other peaks were found to remain unchanged. Rao et al. (2017) reported that –OH, –C=O and –C–O functional groups of gum tragacanth are actively involved in the reduction and stabilization of AuNPs. These

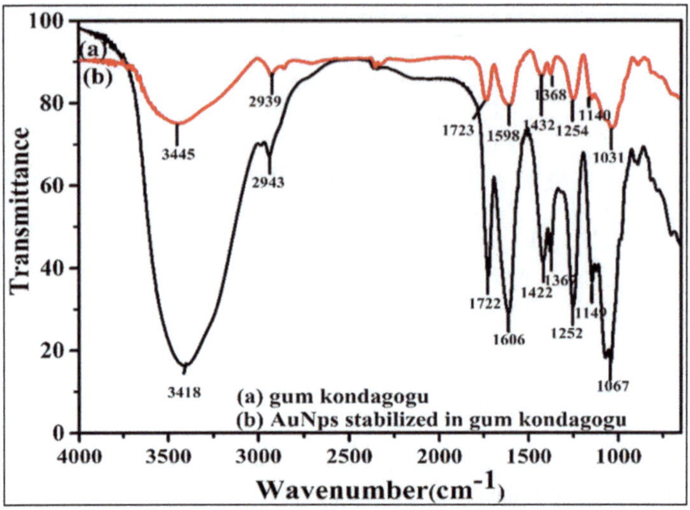

Fig. 4.5 FTIR spectra of (a) gum kondagogu and (b) AuNPs stabilized in gum kondagogu (Reddy et al. 2015a)

results indicated that hydroxyl groups and the carboxyl groups were involved in the synthesis and stabilization of AuNPs. In general, formation of AuNPs is a two-step process. In the first step, gum containing hydroxyl functional group is oxidized to form carbonyl functional group, and gold ion is reduced to gold atom. In the second step, the oxygen atom which is present in the functional carbonyl and hydroxyl groups helps in forming the lid to protect aggregation of NPs as it has a strong association with the AuNPs.

The shape, size and surface morphology of NPs are studied microscopically using the SEM, FE-SEM, TEM and AFM techniques. Among all these, TEM has the maximum magnification and resolution. To understand better the effects of synthesis conditions on the size and shape of AuNPs, TEM was used to evaluate the morphology and size of some representative AuNPs obtained. In order to estimate the effect of capping nature of gum extract over the growth and formation of AuNPs, experiments were conducted with different concentrations (0.1 and 0.5%) of olibanum gum with fixed concentration of precursor (1 mM $HAuCl_4$) and 9 min microwave irradiation (MWI) time (Atnafu et al. 2016), as shown in Fig. 4.6a, c. The prepared NPs were found to be spherical in shape and well-dispersed in the gum matrix. The average particle diameter obtained from these micrographs was about 8 ± 2 (0.1% olibanum gum) to 3 ± 4 nm (0.5% olibanum gum) as shown in Fig. 4.6b, d. These findings clearly indicate that as the concentration of gum increases from 0.1% to 0.5%, the average size of the AuNPs formed decreases.

The selected area electron diffraction (SAED) pattern, which exhibited diffused ring pattern, indicates that these AuNPs are highly polycrystalline in nature. The four rings associated with the pattern can be ascribed to the diffraction from (111), (200), (220) and (311) lattice planes of the face-centered cubic (fcc) gold (Fig. 4.7a).

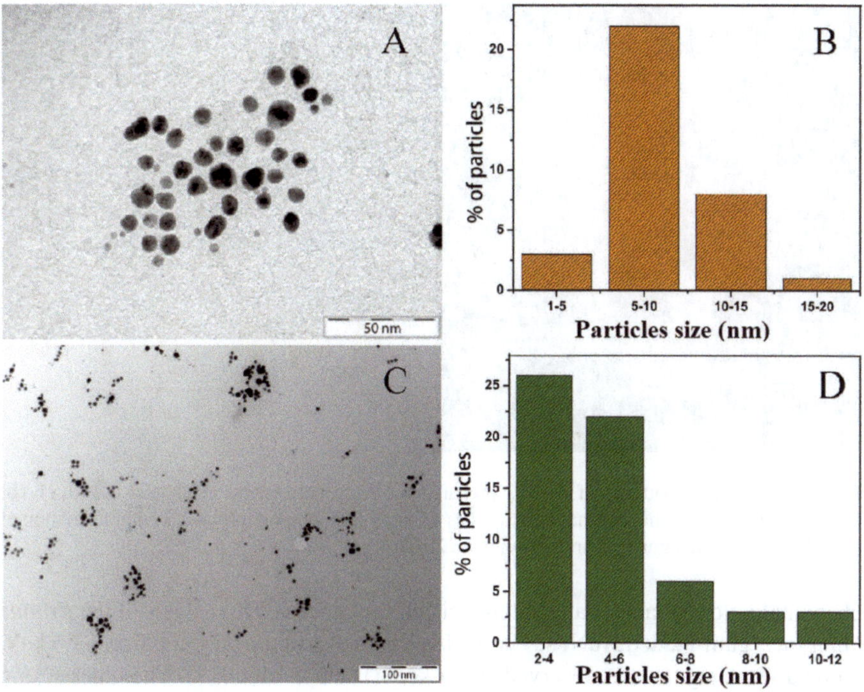

Fig. 4.6 TEM photomicrographs showing (**a**) 0.1% olibanum gum, 1 mM $HAuCl_4$, and 9 min MWI time and (**c**) 0.5% olibanum gum, 1 mM of $HAuCl_4$, and 9 min MWI time. The (**b**) and (**d**) are histograms of the particle size distribution of 0.1% and 0.5% of gum (Atnafu et al. 2016)

Subramanian et al. (2016) demonstrated that the size and shape of NPs depend critically on the concentration of the bael gum. The bael gum-capped NPs with triangular, triangular with truncation at the apex, and hexagonal plates are the predominant shapes over a wide concentration range, at room temperature. At higher concentration of bael gum and at room temperature, the NPs are spherical. However, fine-tuning of temperature, at a particular concentration of bael gum, leads to the formation of perfect triangular NPs, as shown in Fig. 4.7b. Rao et al. (2017) determined the shape and morphology of NPs through AFM microscopy and established that the prepared NPs were spherical in shape with nano-range size and stabilized by the multifunctional groups of gum tragacanth.

4.4.1 EDX Analysis

The presence of Au atoms in the gum-capped AuNPs was confirmed by Reddy et al. (2015c) and Atnafu et al. (2016), using the energy-dispersive X-ray analysis (EDXA), which gives additional evidence for the reduction of $HAuCl_4$ into

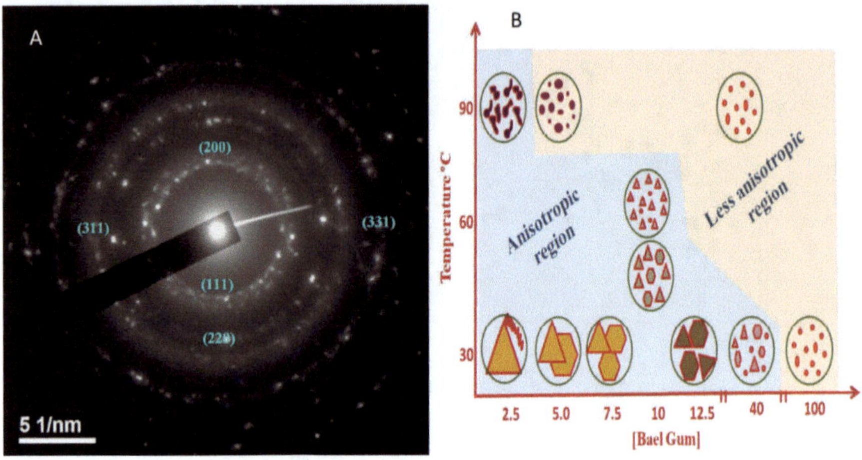

Fig. 4.7 (**a**) SAED patterns of AuNPs prepared by using katira gum (Saikat et al. 2012). (**b**) Representation of the shape of nanoparticles as a function of temperature and weight formation of AuNPs by using bael gum (Subramanian et al. 2016)

elemental gold. Of late, Islam and associates explained EDX analysis of the *Prunus domestica* gum-loaded AuNPs. These NPs showed strong peaks at 0.6 and 2.6 keV, while a weak signal was observed at 9.7 keV (Islam et al. 2017). The other strong signals for C and O were also there due to the presence of bioorganic molecules capping the gold NPs.

4.4.2 Dynamic Light Scattering (DLS)

Dynamic light scattering (DLS) is the most versatile and useful set of techniques for measuring the average particle size and zeta potential. It was observed that the obtained DLS and TEM differ considerably for small AuNPs. While TEM gives the diameter of the individual NPs, DLS also takes the capping shell into account, thus providing information on the size of the whole conjugate. In suspensions of small AuNPs, the number of saccharide molecules per particle is high, and hence the hydrodynamic size obtained from DLS is greater than that obtained by TEM. The stability and surface charge of synthesized AuNPs were also studied by zeta potential. Particles with highly negative or positive surface charge are considered to be stable. An absolute zeta potential value of ±30 mV is a general indication that the colloidal solution is highly stable (Venkatpurwar et al. 2011; Sankar et al. 2017). Reddy et al. (2015b) reported that the gum acacia-stabilized AuNPs have an average particle size of about 24.29 nm and zeta potential value of −29.3 mV, as shown in Fig. 4.8. These results indicate that AuNPs capped with gum acacia carried negatively charged groups and were dispersed in the medium, proving that they were stable. Pooja et al. (2014) demonstrated that the xanthan gum-stabilized AuNPs had a mean hydrodynamic diameter of 41 ± 3.78 nm. The NPs showed a high poly-disparity (0.35–0.6), which can be attributed to the presence of a wide range of particles. The high negative surface potential (−47.2 ± 2.59) indicated a high stability of NPs in the presence of gum molecules.

4.4.3 XRD Analysis

The XRD technique is used to determine and confirm the crystalline structure of fabricated AuNPs. XRD pattern of the gum-capped AuNPs is shown in Fig. 4.9. In a study of Reddy et al. (2015a), AuNPs exhibited four well-defined peaks at $2\theta = 38.1$, 44.3, 64.5 and 77.7. All the four peaks correspond to standard Bragg reflections for (111), (200), (220) and (311) planes of the face center cubic (fcc) crystal structure of metallic gold. The existence of diffraction peaks was matched with the standard data files (the JCPDS card No. 04–0784) for all reflections. Furthermore, no extra peaks were found in XRD spectrum, indicating the high purity of the resultant AuNPs. The diffraction peak at 38.1 was a highly intense peak among the peaks observed. Saikat et al. (2012) (katira gum-capped AuNPs), Tagad et al. (2014) (locust bean gum-capped AuNPs), Islam et al. (2017) (*Prunus domestica* gum-stabilized AuNPs) and Reddy et al. (2015c) (*Salmalia malabarica* gum-capped AuNPs) reported similar findings on the formation of AuNPs. The mean particle diameter of AuNPs was calculated by using XRD data. AuNP diameter was derived by using Debye Scherer's formula ($D = k\,\lambda / \beta^{1/2} \cos\theta$), which exploits the reference peak width at an angle θ, where λ is the X-ray wavelength (1.5418), $\beta^{1/2}$ is

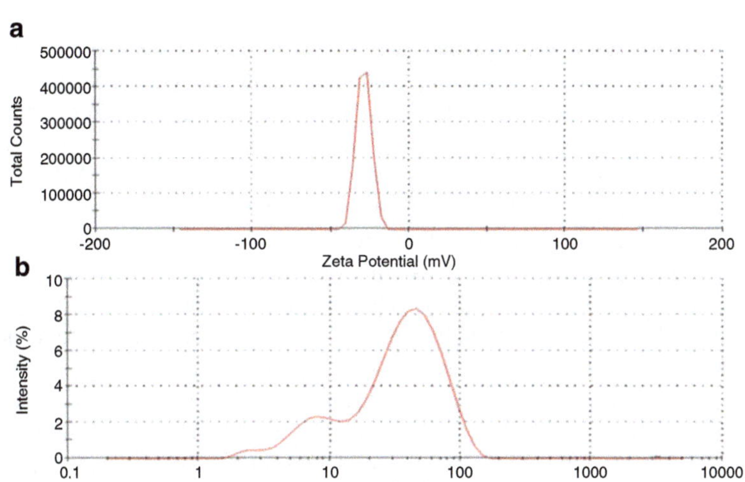

Fig. 4.8 (**a**) Zeta potential and (**b**) Zeta sizer distribution of the AuNPs with gum acacia (Reddy 2015b)

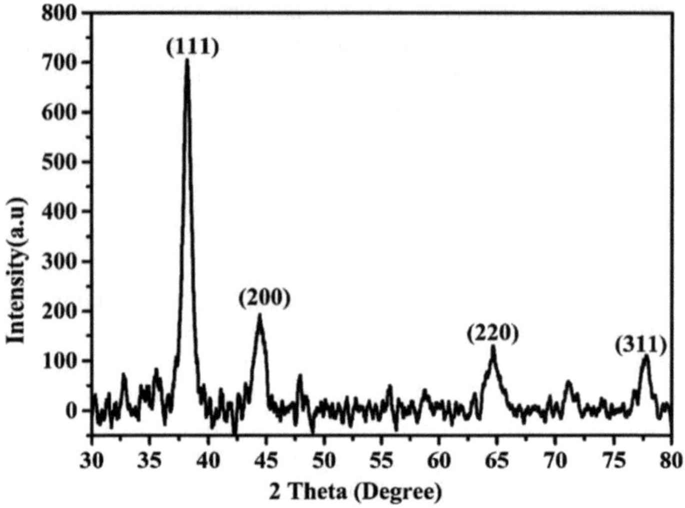

Fig. 4.9 XRD pattern of AuNPs stabilized by gum kondagogu (Reddy et al. 2015a)

the width of the XRD peak at half height and k is the shape factor. Average diameter of particle was 20 nm for the *Prunus domestica* gum-stabilized AuNPs, 13.2 nm for the *Salmalia malabarica* gum-capped AuNPs, and 13.1 nm for the gum kondagogu-capped AuNPs.

4.5 Applications

AuNPs have attracted a widespread interest due to their distinctive electrical and optical properties, such as unique and tunable SPR, surface-enhanced emission and surface-enhanced Raman scattering. Convenient and easy approach for particle size modification throughout the synthesis process and surface functionalization with different kinds of materials adds to the researcher's interest. The AuNPs remained largely overlooked till the 1850's, as they were used merely to decorate ceramics and glass. Until the Middle Ages, the soluble gold was used to disclose fabulous curative powers for various diseases, such as venereal problems, dysentery and tumors, and for diagnosis of syphilis (Dykman and Khlebtsov 2011, 2012). Thus, the remarkable features exhibited by colloidal gold have been utilized for centuries. Now, the advancement in synthesis and surface functionalization of AuNPs has led to various promising applications. Recent reports suggest that AuNPs capped with natural gums are being used in the fields of drug delivery, catalysis, antimicrobial activity, colorimetric metal ion detection and antiproliferative and vapor sensing.

Gold NPs are one of the most commonly explored and used carriers for the delivery of anticancer drug, such as doxorubicin hydrochloride (DOX), due to their controlled size, shape, surface functionality, improved efficacy and target specificity.

Dhar et al. (2008) demonstrated that gellan gum (GG)-capped most stable AuNPs have been used as carriers for the delivery of the cationic anthracycline drug, DOX, with a loading efficiency of about 75%. The reduction in zeta potential value from −38.25 to −30.00 mV indicates the electrostatic stability due to interaction between cationic DOX and anionic GG components in the DOX-loaded GG-AuNPs. DOX-loaded AuNPs are stable at different pH levels ranging from 4.0 to 8.0. Pooja et al. (2014) reported that the DOX-loaded AuNPs capped with xanthan gum were stable under different pH, electrolyte and serum conditions. Drug loading takes place on the surface of NPs due to interaction between the positive-charged amine group of DOX and the negative acidic group of gum, which was confirmed by the decrease in the surface charge to −29.1 ± 2.78 mV.

The study by Rao et al. (2017) suggests that gum tragacanth-capped AuNPs cargos increase the therapeutic efficiency of naringin by enhancing its bactericidal activity through their destabilizing effect on the bacterial surface morphology. Gum arabic-capped small-sized AuNPs can be used as a photothermal agent, which exhibits a strong photothermal effect for killing cancer cells, as reported by Liu et al. (2013).

The distinctive properties of AuNPs allow their use in the detection of copper metal, which was an identified potable and surface water contaminant. Carboxymethyl gum karaya (CMGK)-functionalized AuNPs were reported for the sensitive and selective detection of Cu^{2+} in aqueous solution, in the presence of 13 other metal ions. The Cu^{2+} sensing relies on the aggregation of CMGK-capped AuNPs owing to the addition of Cu^{2+} resulting in an increase in absorbance of NPs at 647 nm and a change in their color from red to blue (Reddy et al. 2017).

The guar gum-capped AuNPs were exploited for optical sensor applications for detecting the aqueous ammonia based on SPR. The developed method is found to be simple, with low cost, low detection limit of 1 ppb, high sensitivity and great reproducibility (Pandey et al. 2013). Locust bean gum (LBG)-stabilized AuNPs were doped in SnO_2 matrix and their utilization for ethanol vapor sensing was investigated. LBG-capped AuNPs doped in SnO_2 showed a fast response (~5 s) and good ethanol sensing behavior in the range of 10–120 ppm at room temperature (Tagad et al. 2014).

Gold NPs exhibit antibacterial activity by directly interacting with the bacteria cell walls and cause lysis. The antibacterial effect of AuNPs mostly depends on the size of particles; NPs smaller than 10 nm have a direct interaction with the bacteria and produce electronic effects, which enhance the reactivity of NPs. Reddy et al. (2015a) suggested that gram-negative strains of bacteria (*E. coli*) with thin cell walls are more susceptible to cell wall damage, compared to gram-positive strain bacteria (*B. subtilis*) with a thick cell wall. Islam et al. (2017) demonstrated that *Prunus domestica* gum-stabilized gold and silver NPs show antibacterial activity against gram-positive strains of *S. aureus* and gram-negative strains of *E. coli* and *P. aeruginosa*, but AuNPs are less effective compared to AgNPs.

Gold NPs possess a large surface area and high surface energy, making them the best catalysts in many reactions such as reduction of methylene blue (MB), Congo red (CR) and reduction of p-nitrophenol (4-NP) to p-aminophenol (4-AP). The catalytic activity of synthesized AuNPs using gum kondagogu, gum acacia, katira gum locust bean gum, and olibanum gum was established in the reduction of 4-NP to 4-AP in the presence of $NaBH_4$. The rate of reduction was influenced by parameters such as concentration of AuNPs and reaction temperature (Reddy et al. 2015a, b; Saikat et al. 2012; Tagad et al. 2014; Atnafu et al. 2016).

The SMG-capped AuNPs operate through the mechanism of electron relay effect and accomplish the reduction of MB dye in 9 min and CR dye in 10 min in the presence of $NaBH_4$ (Reddy et al. 2015c). The catalytic electron transfer reaction between $K_3[Fe(CN)_6]$ and $NaBH_4$, resulting in the formation of $K_4[Fe(CN)_6]$ and dihydrogen borate ion, is effectively carried out in the presence of acacia gum-capped AuNPs and this reaction depends upon the concentration of AuNPs and temperature. Similar results were obtained for the olibanum gum-capped AuNPs (Reddy et al. 2015b; Atnafu et al. 2016).

4.6 Conclusion

Fabrication of AuNPs accomplished by using natural gums is nonhazardous, cost-effective, renewableand eco-friendly, compared to conventional methods. These gums serve the role of self-reducing and capping agents and thus the hazards associated with the use of additional capping agents are avoided. Gold NPs with controlled size and shape were obtained simply by varying the ratio of $HAuCl_4$ to gum concentrations. Gum-capped AuNPs are relatively more stable and do not show any sign of agglomeration even after storage for several months at room temperature, compared to those prepared through other bioreduction methods. Gums are capable to reduce metal ions faster than other bioreduction materials. Furthermore, in the scale-up industrial production of well-dispersed AuNPs, naturally occurring gums are certainly better than other bioreduction materials. Gum-capped gold NPs are more effective in a range of applications including catalysis, diagnosis, photonics and therapeutics due to their novel properties, biocompatibility and low toxicity.

References

Ahuja M, Singh S, Kumar A (2013) Evaluation of carboxymethyl gellan gum as a mucoadhesive polymer. Int J Biol Macromol 53:114–121

Alam MS, Garg A, Pottoo FH, Saifullah MK, Tareq AI, Manzoor O, Mohsin M, Javed MN (2017) Gum ghatti mediated, one pot green synthesis of optimized gold nanoparticles: investigation of process-variables impact using *Box-Behnken* based statistical design. Int J Biol Macromol 104:758–767

Arati R, Bhattacharya SB, Mukherjee AK, Rao CVN (1976) The structure of degraded bael (*Aegle marmelos*) gum. Carbohydr Res 50:87–96

Atnafu GA, Ayal AM, Akele ML, Addis KA, Reddy GB, Veerabhadram G, Madhusudhan A (2016) Microwave-assisted green synthesis of gold nanoparticles using olibanum gum (*Boswellia ser-*

rata) and its catalytic reduction of 4-Nitrophenol and Hexacyanoferrate (III) by sodium borohydride. J Clust Sci 28:917–935

Reddy GB, Madhusudhan A, Ramakrishna D, Ayodhya D, Veerabhadram G (2015a) Green chemistry approach for the synthesis of gold nanoparticles with gum kondagogu: characterization, catalytic and antibacterial activity. J Nanostruct Chem 5:185–193

Reddy GB, Madhusudhan A, Ramakrishna D, Ayodhya D, Veerabhadram G (2015b) Catalytic reduction of p-Nitrophenol and Hexacyanoferrate (III) by borohydride using green synthesized gold nanoparticles. J Chin Chem Soc 62:420–428

Reddy GB, Madhusudhan A, Ramakrishna D, Ayodhya D, Veerabhadram G (2015c) Catalytic reduction of methylene blue and Congo red dyes using green synthesized gold nanoparticles capped by salmalia malabarica gum. Int Nano Lett 5:215–222

Reddy GB, Rajkumar B, Ramakrishna D, Girija MK, Veerabhadram G (2017) Facile green synthesis of gold nanoparticles with carboxymethyl gum karaya, selective and sensitive colorimetric detection of copper (II) ions. J Clust Sci 28:2873–2890

Bhardwaj TR, Kanwar M, Lal R, Gupta A (2000) Natural gums and modified natural gums as sustained-release carriers. Drug Dev Ind Pharm 26:1025–1038

Bhattacharya R, Mukherjee P (2008) Biological properties of "naked" metal nanoparticles. Adv Drug Deliv Rev 60:1289–1306

Biswala SK, Parida UK, Bindhani BK (2013) Gold nanoparticles capped with tamarind seed polysaccharide blended with chitosan composite for the growth of phosphate mineral. Int J Cur Eng Tech 3:1104–1108

Bogunia-Kubik K, Sugisaka M (2002) From molecular biology to nanotechnology and nanomedicine. Biosystems 65:123–138

Bond GC, Sermon PA (1973) Gold catalysts for olefin hydrogenation. Gold Bull 6:102–105

Brown CL, Whitehouse MW, Tiekink ERT, Bushell GR (2008) Colloidal metallic gold is not bio-inert. Inflammopharmacology 16:133–137

Choudhary PD, Pawar HA (2014) Recently investigated natural gums and mucilages as pharmaceutical excipients: an overview. J Pharm 2014:204849

Corma A, Serna P (2006) Chemo selective hydrogenation of nitro compounds with supported gold catalysts. Science 313:332–334

Dahl JA, Maddux BLS, Hutchison JE (2007) Toward greener nano synthesis. Chem Rev 107:2228–2269

Daniel MC, Astruc D (2004) Gold nanoparticles: assembly, supramolecular chemistry, quantum-size-related properties, and applications toward biology, catalysis, and nanotechnology. Chem Rev 104:293–346

Das S, Ghosal PK, Ray B (1990) Structural studies of a polysaccharide from the seeds of *salmalia malabarica*. Carbohydr Res 207:336–339

Dey P, Sa B, Maiti S (2011) Carboxymethyl ethers of locust bean gum a review. Int J Pharm Pharm Sci 2:4–7

Dhar S, Maheswara Reddy E, Shiras A, Pokharkar V, Prasad BLV (2008) Natural gum reduced/stabilized gold nanoparticles for drug delivery formulations. Chem Eur J 14:10244–10250

Dhar S, Mali V, Bodhankar S, Shiras A, Prasad BL, Pokharkar V (2011) Biocompatible gellan gum-reduced gold nanoparticles: cellular uptake and subacute oral toxicity studies. J Appl Toxicol 31(5):411–420

Dimitratos N, Lopez-Sanchez JA, Hutchings JG (2012) Selective liquid phase oxidation with supported metal nanoparticles. Chem Sci 3:20–44

Dodi G, Hritcu D, Popa MI (2011) Carboxyl methylation of guar gum: synthesis and characterization. Cellulose Chem Technol 45:171–176

Dror Y, Cohen Y, Yerushalmi-rozen R (2006) Structure of gum arabic in aqueous solution. J Polym Sci Part B Polym Phys 44:3265–3271

Dykman LA, Bogatyrev VA (2007) Gold nanoparticles: preparation, functionalization and applications in biochemistry and immunochemistry. Russ Chem Rev 76:181–191

Dykman LA, Khlebtsov NG (2011) Gold nanoparticles in biology and medicine: recent advances and prospects. Acta Nat 3:34–55

Dykman L, Khlebtsov N (2012) Gold nanoparticles in biomedical applications: recent advances and perspectives. Chem Soc Rev 41:2256–2282

El-Sayed MA (2001) Some interesting properties of metals confined in time and nanometer space of different shapes. Acc Chem Res 34:257–264

Erik CD, Alaaldin MA, Xiaohua H, Catherine JM, Mostafa AE (2012) The golden age: gold nanoparticles for biomedicine. Chem Soc Rev 41:2740–2779

Farokhzad OC, Langer R (2009) Impact of nanotechnology on drug delivery. ACS Nano 3:16–20

Galla NR, Dubasi GR (2010) Chemical and functional characterization of gum karaya (*Sterculia urens L*) seed meal. Food Hydrocoll 24:479–485

Ghosh SK, Pal T (2007) Interparticle coupling effect on the surface plasmon resonance of gold nanoparticles: from theory to applications. Chem Rev 107:4797–4862

Huang X, El-sayed MA (2010) Gold nanoparticles: optical properties and implementations in cancer diagnosis and photothermal therapy. J Adv Res 1:13–28

Huang I, Li Q, Sun D, Lu Y, Su Y, Yang X, Wang H, Wang Y, Shao W, He N, Hong J, Chen C (2007a) Biosynthesis of silver and gold nanoparticles by novel sundried *Cinnamomum camphora* leaf. Nanotechnology 18:105104–105114

Huang CC, Yang Z, Lee KH, Chang HT (2007b) Synthesis of highly fluorescent gold nanoparticles for sensing mercury(II). Angew Chem Int Ed 46:6824–6828

Husen A (2017) Gold nanoparticles from plant system: synthesis, characterization and their application. In: Ghorbanpourn M, Manika K, Varma A (eds) Nanoscience and plant–soil systems, vol 48. Springer, Cham, pp 455–479

Husen A, Siddiqi KS (2014a) Carbon and fullerene nanomaterials in plant system. J Nanobiotechnol 12:16

Husen A, Siddiqi KS (2014b) Phytosynthesis of nanoparticles: concept, controversy and application. Nano Res Lett 9:229

Husen A, Siddiqi KS (2014c) Plants and microbes assisted selenium nanoparticles: characterization and application. J Nanobiotechnol 12:28

Ibrahim NA, Abo-Shosha MH, Allam EA, El-Zairy EM (2010) New thickening agents based on tamarind seed gum and karaya gum polysaccharides. Carbohydr Polym 81:402–408

Islam NU, Amin R, Shahid M, Amin M, Zaib S, Iqbal J (2017) A multi-target therapeutic potential of *Prunus domestica* gum stabilized nanoparticles exhibited prospective anti-inflammatory and analgesic properties. BMC Complement Altern Med 17:276–293

Jie H, Yan L, Rong G (2009) Facile synthesis of highly stable gold nanoparticles and their unexpected excellent catalytic activity for suzuki−miyaura cross-coupling reaction in water. J Am Chem Soc 131:2060–2061

Joshita D, Sutriyo PP, Anung P (2014) Antioxidant activity of gold nanoparticles using gum arabic as a stabilizing agent. Int J Pharm Pharm Sci 6:462–465

Kattumuri V, Katti K, Bhaskaran S, Boote EJ, Casteel SW, Fent GM, Robertson DJ, Chandrasekhar M, Kannan R, Katti KV (2007) Gum arabic as a phytochemical construct for the stabilization of gold nanoparticles: *in vivo* pharmacokinetics and X-ray-contrast-imaging studies. Small 3:333–341

Kora AJ, Arunachalam J (2012) Green fabrication of silver nanoparticles by gum tragacanth (*Astragalus gummifer*): a dual functional reductant and stabilizer. J Nanomater 2012:869765

Kora AJ, Sashidhar RB, Arunachalam J (2012) Aqueous extract of gum olibanum (*Boswellia serrata*): a reductant and stabilizer for the biosynthesis of antibacterial silver nanoparticles. Process Biochem 47:1516–1520

Kumar A, Ahuja M (2012) Carboxymethyl gum kondagogu: synthesis, characterization and evaluation as mucoadhesive polymer. Carbohydr Polym 90:637–643

Liu CP, Lin FS, Chien CT, Tseng SY, Luo CW, Chen CH, Chen JK, Tseng FG, Hwu Y, Lo LW, Yang CS, Lin SY (2013) In-situ formation and assembly of gold nanoparticles by gum arabic as efficient photothermal agent for killing cancer cells. Macromol Biosci 13:1314–1320

Llevot A, Astruc D (2012) Applications of vectorized gold nanoparticles to the diagnosis and therapy of cancer. Chem Soc Rev 41:242–257

Madhusudhan A, Reddy GB, Venkatesham M, Veerabhadram G, Kumar DA, Sumathi N, Ming YY, Anren H, Surya SS (2014) Efficient pH dependent drug delivery to target cancer cells by gold nanoparticles capped with carboxymethyl chitosan. Int J Mol Sci 15:8216–8234

Maity S, Sa B (2014) Ca-carboxymethyl xanthan gum mini-matrices: swelling, erosion and their impact on drug release mechanism. Int J Biol Macromol 68:78–85

Mayer KM, Hafner JH (2011) Localized surface plasmon resonance sensors. Chem Rev 111:3828–3857

Mirhosseini H, Amid BT (2012) A review study on chemical composition and molecular structure of newly plant gum exudates and seed gums. Food Res Int 46:387–398

Mirkin CA (2005) The beginning of a *small* revolution. Small 1:14–16

Muddineti OS, Kumari P, Ajjarapu S, Lakhani PM, Bahl R, Ghosh B, Biswas S (2016) Xanthan gum stabilized PEGylated gold nanoparticles for improved delivery of curcumin in cancer. Nanotechnology 27(32):325101–325112

Ojha AK, Maiti D, Chandra K, Mondal S, Das Sadhan K, Roy D, Ghosh K, Islam SS (2008) Structural assignment of a heteropolysaccharide isolated from the gum of *Cochlospermum religiosum* (Katira gum). Carbohydr Res 19:1222–1231

Padil VV, Černík M (2015) Poly (vinyl alcohol)/gum karaya electrospun plasma treated membrane for the removal of nanoparticles (Au, Ag, Pt, CuO and Fe_3O_4) from aqueous solutions. J Hazard Mater 287:102–110

Pandey S, Goswami GK, Nanda KK (2013) Green synthesis of polysaccharide/gold nanoparticle nanocomposite: an efficient ammonia sensor. Carbohydr Polym 94:229–234

Pissuwan D, Niidome T, Cortie MB (2011) The forthcoming applications of gold nanoparticles in drug and gene delivery systems. J Control Release 5:65–71

Pooja D, Panyaram S, Kulhari H, Rachamalla SS, Sistla R (2014) Xanthan gum stabilized gold nanoparticles: characterization, biocompatibility, stability and cytotoxicity. Carbohydr Polym 110:1–9

Pooja D, Panyaram S, Kulhari H, Reddy B, Rachamalla SS, Sistla R (2015) Natural polysaccharide functionalized gold nanoparticles as biocompatible drug delivery carrier. Int J Biol Macromol 80:48–56

Prajapati VD, Jani GK, Moradiya NG, Randeria NP (2013a) Pharmaceutical applications of various natural gums, mucilages and their modified forms. Carbohydr Polym 92:1685–1699

Prajapati VD, Jani GK, Moradiya NG, Randeria NP, Nagar BJ (2013b) Locust bean gum: a versatile biopolymer. Carbohydr Polym 94:814–821

Punuri JB, Sharma P, Sibyala S, Tamuli R, Bora U (2012) *Piper betle*-mediated green synthesis of biocompatible gold nanoparticles. Int Nano Lett 2:18–27

Rana V, Rai P, Tiwary AK, Singh RS, Kennedy JF, Knill CJ (2011) Modified gums: approaches and applications in drug delivery. Carbohydr Polym 83:1031–1047

Rao K, Imran M, Jabri T, Ali I, Perveen S, Shafiullah AS, Shah MR (2017) Gum tragacanth stabilized green gold nanoparticles as cargos for naringin loading: a morphological investigation through AFM. Carbohydr Polym 15:243–252

Rivas L, Sanchez-Cortes S, Garcia-Ramos JV, Morcillo G (2001) Growth of silver colloidal particles obtained by citrate reduction to increase the raman enhancement factor. Langmuir 17:574–577

Saha K, Agasti SS, Kim C, Li X, Rotello VM (2012) Gold nanoparticles in chemical and biological sensing. Chem Rev 112:2739–2779

Saikat M, Ipsita KS, Syed SI (2012) Green synthesis of gold nanoparticles using gum polysaccharide of *Cochlospermum religiosum* (katira gum) and study of catalytic activity. Phys E 45:130–134

Sánchez MP, Boulaiz H, Ortega-Vinuesa JL, Peula-García JM, Aránega A (2012) Novel drug delivery system based on docetaxel-loaded nanocapsules as a therapeutic strategy against breast cancer cells. Int J Mol Sci 13:4906–4919

Sankar R, Rahman PKSM, Varunkumar K, Anusha C, Kalaiarasia A, Subramanian K, Shivashangaric KS, Ravikumara V (2017) Facile synthesis of *Curcuma longa* tuber powder engineered metal nanoparticles for bioimaging applications. J Mol Struct 1129:8–16

Selvi SK, Mahesh J, Sashidhar RB (2017) Anti-proliferative activity of gum kondagogu (*Cochlospermum gossypium*)-gold nanoparticle constructs on B16F10 melanoma cells: an *in vitro* model. Bioact Carbohydrates Diet Fibre 11:38–47

Sermon PA, Bond GC, Wells PB (1979) Hydrogenation of alkenes over supported gold. J Chem Soc Farady Trans 1(75):385–394

Siddiqi KS, Husen A (2016a) Fabrication of metal nanoparticles from fungi and metal salts: scope and application. Nano Res Lett 11:98

Siddiqi KS, Husen A (2016b) Fabrication of metal and metal oxide nanoparticles by algae and their toxic effects. Nano Res Lett 11:363

Siddiqi KS, Husen A (2016c) Engineered gold nanoparticles and plant adaptation potential. Nano Res Lett 11:400

Siddiqi KS, Husen A (2016d) Green synthesis, characterization and uses of palladium/platinum nanoparticles. Nano Res Lett 11:482

Siddiqi KS, Husen A (2017a) Recent advances in plant-mediated engineered gold nanoparticles and their application in biological system. J Trace Elements Med Biol 40:10–23

Siddiqi KS, Husen A (2017b) Plant response to engineered metal oxide nanoparticles. Nano Res Lett 12:92

Siddiqi KS, Rahman A, Tajuddin HA (2016) Biogenic fabrication of iron/iron oxide nanoparticles and their application. Nano Res Lett 11:498

Siddiqi KS, Husen A, Rao RAK (2018a) A review on biosynthesis of silver nanoparticles and their biocidal properties. J Nanobiotechnol 16:14

Siddiqi KS, Rahman A, Tajuddin HA (2018b) Properties of zinc oxide nanoparticles and their activity against microbes. Nano Res Lett 13:141

Siddiqi KS, Husen A, Sohrab SS, Osman M (2018c) Recent status of nanomaterials fabrication and their potential applications in neurological disease management. Nano Res Lett 13:231

Subramanian SB, Bezawada SR, Dhamodharan R (2016) Green, selective, seedless and one-pot synthesis of triangular Au nanoplates of controlled size using bael gum and mechanistic study. ACS Sustain Chem Eng 4:3830–3839

Tagad CK, Rajdeo KS, Kulkarni A, More P, Aiyer RC, Sabharwal S (2014) Green synthesis of polysaccharide stabilized gold nanoparticles: chemo catalytic and room temperature operable vapor sensing application. RSC Adv 4:24014–24019

Thanaa IS, Rasha SSE, Suzan AAE (2015) Green synthesis of gold nanoparticles using cumin seeds and gum arabic: studying their photothermal efficiency. Nanosci Nanotechnol 5:89–96

Venkatpurwar V, Shiras A, Pokharkar V (2011) Porphyran capped gold nanoparticles as a novel carrier for delivery of anticancer drug: *in vitro* cytotoxicity study. Int J Pharm 409:314–320

Vieira S, Vial S, Maia FR, Carvahlo M, Reis RL, Granja PL, Oliveira M (2015) Gellan gum-coated gold nanorods: an intracellular nanosystem for bone tissue engineering. RSC Adv 5:77996–78005

Vinod VT, Saravanan P, Sreedhar B, Devi DK, Sashidhar RB (2011) A facile synthesis and characterization of Ag, Au and Pt nanoparticles using a natural hydrocolloid gum kondagogu (*Cochlospermum gossypium*). Colloids Surf B Biointerfaces 3:291–298

Wu CC, Chen DH (2010) Facile green synthesis of gold nanoparticles with gum arabic as a stabilizing agent and reducing agent. Gold Bull 43:234–240

Wu YL, Li YN, Liu P, Gardner S, Ong BS (2006) Studies of gold nanoparticles as precursors to printed conductive features for thin film transistors. Chem Mater 18:4627–4632

Zhang Z, Wu Y (2010) Investigation of the $NaBH_4$-induced aggregation of Au nanoparticles. Langmuir 26:9214–9223

Zharov VP, Kim JW, Curiel DT, Everts M (2005) Self-assembling nanoclusters in living systems: application for integrated photothermal nanodiagnostics and nanotherapy. Nanomed Nanotechnol Biol Med 1:326–345

Chapter 5
Plant-Based Fabrication of Silver Nanoparticles and Their Application

Vinod Kumar Mishra, Azamal Husen, Qazi Inamur Rahman, Muhammad Iqbal, Sayed Sartaj Sohrab, and Mansur Osman Yassin

5.1 Introduction

Nanoparticles (NPs) are now being used in many research areas, such as food industry, material science, medical science, and plant science. Of the various NPs, silver nanoparticles (Ag NPs) are widely used in biosensing, cryogenic superconducting, environmental care, food technology, catalyst, and antistatic and antimicrobial activities (Lara et al. 2011; Lokina et al. 2014; Husen and Siddiqi 2014b; Wei et al. 2015; He et al. 2016; Siddiqi et al. 2018a, b, c). Several procedures such as chemical reduction, electrochemical reduction, irradiation reduction, and photocatalytic reduction have been used for fabrication of Ag NPs (Zhang et al. 2007). Size, morphology, stability, and properties of NPs are strongly affected by experimental

V. K. Mishra
Department of Biotechnology, Doon P.G. Paramedical College,
Dehra Dun, Uttarakhand, India

A. Husen (✉)
Department of Biology, College of Natural and Computational Sciences,
University of Gondar, Gondar, Ethiopia

Q. I. Rahman
Department of Chemistry, College of Natural and Computational Sciences,
University of Gondar, Gondar, Ethiopia

M. Iqbal
Department of Botany, Faculty of Science, Jamia Hamdard (Deemed University),
New Delhi, India

S. S. Sohrab
Special Infectious Agents Unit, King Fahd Medical Research Center (KFMRC), King
Abdulaziz University, Jeddah, Kingdom of Saudi Arabia

M. O. Yassin
Department of Surgery, College of Medicine and Health Sciences, University of Gondar,
Gondar, Ethiopia

A. Husen, M. Iqbal (eds.), *Nanomaterials and Plant Potential*,
https://doi.org/10.1007/978-3-030-05569-1_5

conditions, kinetics of interaction of metal ions with reducing agents, and the mode of adsorption of stabilizing agent on metal NPs (Knoll and Keilmann 1999; Sengupta et al. 2005). Thus, design of a fabrication method in which the size, morphology, stability, and properties of NPs are controlled has the pivotal significance (Husen and Siddiqi 2014b, c; Siddiqi and Husen 2016a, b, c, 2017a, b; Siddiqi et al. 2016; Husen 2017). Tao et al. (2006) have reported that chemical reduction is frequently used for fabrication of stable, colloidal, and well-dispersed Ag NPs in water as well as organic solvents. In general, sodium borohydride, pyridine, and ethylene glycol are used for this purpose, but these compounds are often expensive, lethal, and highly reactive. On the other hand, plant-mediated protocols for Ag NP fabrication are safe and comparatively less complex. In addition, the chemicals used are environment-friendly and less expensive and can be conveniently used for a large-scale production (Iravani 2011; Husen and Siddiqi 2014b).

Development of plant-mediated (biological) experimental processes for fabrication of Ag NPs is currently evolving as an important branch of nanotechnology. In the biological process, extracts from living systems serve as the reducing and capping agents. Many routes have been developed for biological fabrication of Ag NPs from the corresponding salts. For instance, aqueous extracts of *Ferocactus echidne* have been used to reduce silver nitrate ($AgNO_3$) to form Ag NPs within 6 h (Cinelli et al. 2015). The polyphenols and ascorbic acid present in the aqueous extracts of *F. echidne* easily reduced silver ions and also acted as capping agents to prevent their agglomeration. Similarly, extracts of several other plants like *Areca catechu* (Rajan et al. 2015), *Erigeron bonariensis* (Kumar et al. 2015), *Momordica charantia* (Ajitha et al. 2015), *Euphorbia amygdaloides* (Cicek et al. 2015), *Impatiens balsamina* (Nalavothula et al. 2015), *Aloe vera* (Logaranjan et al. 2016), *Artemisia absinthium* (Ali et al. 2016), *Chelidonium majus* (Barbinta-Patrascu et al. 2016), *Terminalia chebula* (Edison et al. 2016a), *Cerasus serrulata* (Karthik et al. 2016), *Solanum indicum* (Sengottaiyan et al. 2016), and *Fraxinus excelsior* (Parveen et al. 2016), among others, have been examined and found capable for a rapid intra- or extracellular synthesis of Ag NPs. The present chapter aims at summarizing the efforts concerning the plant-mediated synthesis of Ag NPs and focusing on the phytochemicals and factors affecting this biogenesis. It also discusses the numerous applications of Ag NPs in different disciplines of biological sciences.

5.2 Fabrication and Characterization of Ag NPs

Ag NPs are being extensively synthesized using plant extracts, but the exact mechanism for this synthesis still needs to be fully investigated. The biochemical synthesis of Ag NPs is a complex phenomenon involving an array of biomolecules such as enzymes/proteins, vitamins, organic acids such as citrates, amino acids, and polysaccharides for reduction and capping (Fig. 5.1). Recent studies have shown that phytochemicals such as proteins, flavonoids, polyphenols, alkaloids, saponins, phenols, essential oils, and polyols play a major role in the bioreduction of silver ions

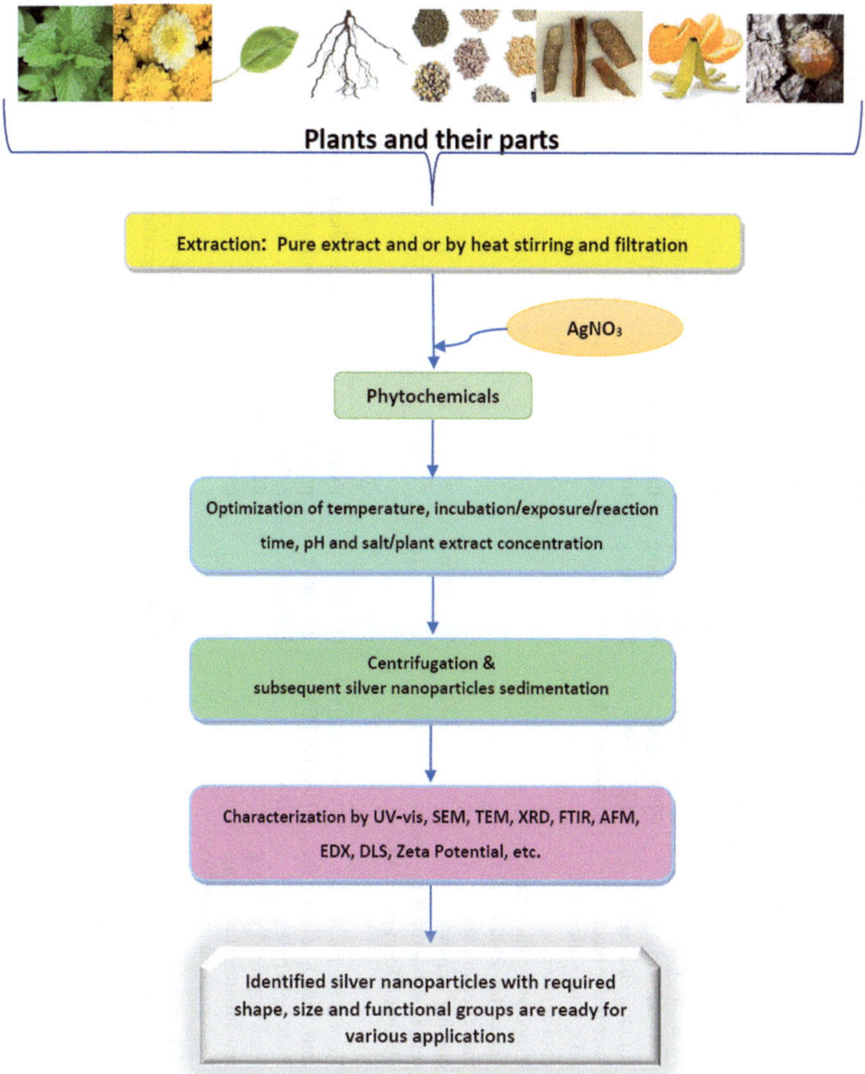

Fig. 5.1 Synthesis and characterization of Ag NPs using various plant parts

and capping of the synthesized NPs (Table 5.1). Most interestingly, there are reports of biological synthesis of Ag NPs using isolated pure compounds, such as apiin (apigenin-7-apiosyl-glucoside), phyllanthin, and flavonoids (such as pinocembrin and galangin) (Kasthuri et al. 2009a, b; Roy et al. 2010).

The most commonly used techniques for characterization of NPs are ultraviolet-visible spectroscopy, scanning electron microscopy (SEM), transmission electron microscopy (TEM), atomic force microscopy (AFM), energy-dispersive spectroscopy (EDS), X-ray diffraction (XRD), dynamic light scattering (DLS), infrared

Table 5.1 Major studies (during 2015–2018) on plant-mediated synthesis of Ag NPs; their morphology, characterization, and uses; and the techniques employed

Plant	Part of plant used	Solvent used	Synthesis conditions	Characterization techniques	Shape and size	Phytoconstituents responsible for reduction of silver ion	Uses	Key references
Aloe vera	Whole plant	Distilled water	Aqueous plant extract was added to $AgNO_3$ solution and kept in microwave reactor for 60 s; it was allowed to cool at room temperature and again heated for another 60 s; these heating and cooling steps were repeated 3–4 times to obtain uniform-sized Ag NPs	UV-Vis, XRD, FTIR, SEM, TEM, EDX, Raman, AFM	Spherical and octahedron; 5–50 nm	Extract rich in flavanones and terpenoids	Antimicrobial activity	Logaranjan et al. (2016)
Alpinia calcarata	Root	Distilled water	Root extract was mixed into $AgNO_3$ solution for 24 h; light pale pink color appeared within a few minutes, indicating the formation of Ag NPs	UV-Vis, XRD, FTIR, HRTEM, EDX	Quasi-spherical; 5–15 nm	Extract rich in protein, polyphenolic and flavonoidal compounds	Bactericidal and nonlinear optics	Pugazhendhi et al. (2015)
Anacardium occidentale	Testa	Distilled water	Aqueous extract of *A. occidentale* was added to $AgNO_3$. Formation of Ag NPs was observed by change in color of the solution	UV-Vis, XRD, FTIR, HRTEM, EDX	Spherical; 25 nm	Phytoconstituents viz., catechin, epicatechin, epigallocatechin, syringic acid, gallic acid, and p-coumaric acid	Dyes degradation	Edison et al. (2016c)

Andean blackberry	Fruit extract	Distilled water	Fruit extract was heated (62–65 °C) in 50 mL of deionized water for 60 min and allowed to cool; gradually added into silver salt solution at room temperature Change in color of solution confirmed the formation of Ag NPs	UV-Vis, XRD, TEM, FTIR, DLS, zeta potential	Spherical; 12–50 nm	Extract rich in flavonoids/phenolic compounds	Antioxidant efficacy	Kumar et al. (2017a)
Andrographis echioides	Leaf	Distilled water	Aqueous leaf extract was mixed with $AgNO_3$; color of solution turned brown indicating formation of Ag NPs; confirmed by UV-Vis absorption	UV-Vis, XRD, FTIR, HRSEM, TEM, EDX, AFM, DLS	Cubic, pentagonal, and hexagonal; 68–91nm	Extract rich in phenols and their derivatives	Anticancer and antibacterial activities	Elangovan et al. (2015)
Anisomeles indica	Leaf	Distilled water	Aqueous leaf extract solution gradually added into $AgNO_3$ solution which was incubated at room temperature; observed change in color of the solution	UV-Vis, XRD, FTIR, SEM, TEM, EDX	Spherical; 50–100 nm	Extract rich in flavonoids and terpenoids	Mosquitocidal potential against malaria, dengue, and Japanese encephalitis vectors	Govindarajan et al. (2016a)
Anoectochilus elatus	Leaf	Distilled water	Aqueous leaf extract mixed into $AgNO_3$ solution for 60 min, colorless suspension turned brown; attributed to Ag NP formation	UV-Vis, XRD, FTIR, SEM, TEM	Spherical; 20 nm	Flavonoids and reducing sugars	Antimicrobial activity	Gopinath et al. (2017)

(continued)

Table 5.1 (continued)

Plant	Part of plant used	Solvent used	Synthesis conditions	Characterization techniques	Shape and size	Phytoconstituents responsible for reduction of silver ion	Uses	Key references
Averrhoa carambola	Fruit	Distilled water	Silver salt solution is mixed with fruit extract and pH of resultant solution fixed; the reaction to settle down is observed	UV-Vis, XRD, FTIR, TEM, DLS, zeta potential	Spherical; size varies with respect to pH	Fruit extract rich in polyols, aldehydes, amines, and organic acids	-	Chowdhury et al. (2015)
Azadirachta indica	Leaf	Distilled water	Aqueous leaf extract was mixed with $AgNO_3$ in the dark; incubation of mixture solution at room temperature. Change in color from colorless to brown indicated reduction of Ag+ into Ag^0	UV-Vis, XRD, FTIR, SEM, TEM, EDX, DLS, PL	Spherical; 34 nm	Extract rich in flavanoids and terpenoids	Antimicrobial activity	Ahmed et al. (2016)
Calliandra haematocephala	Leaf	Distilled water	Aqueous leaf extract was mixed into $AgNO_3$ solution and heated at 80 °C for 10 min; change in color of the solution was observed	UV-Vis, XRD, FTIR, SEM, EDX, DLS, zeta potential	Spherical; 13–91 nm	Extract rich in phenolic groups present in gallic acid	Antibacterial activity and hydrogen peroxide sensing capability	Raja et al. (2017)
Cardiospermum halicacabum	Leaf	Distilled water	Aqueous leaf extract was mixed into $AgNO_3$ solution; observed change in color of the solution under incubation of 16h	UV-Vis, XRD, FTIR, SEM, TEM, EDX, DLS, zeta potential	Spherical; ~74 nm	Extract rich in polyphenols	Antimicrobial, antioxidant, and super oxide scavenging activity	Sundararajan et al. (2016)

Carica papaya	Peel	Distilled water	Peel extract solution was mixed with silver salt solution and the reduction reaction observed by change in color solution within 30 min	UV-Vis, XRD, FTIR, SEM, TEM, EDX, DLS, AFM, zeta potential	Spherical; 10–30 nm	Extract rich in water soluble antioxidant constituents	Antioxidant and antimicrobial activities	Kokila et al. (2016)
Cinnamomum tsoi	Leaf	Distilled water	Aqueous silver salt solution mixed with extract of leaf and set on specific temperature to obtain colloids; change of color indicated formation of Ag NPs	UV-Vis, XRD, HRTEM, SEAD, FTIR, DLS, zeta potential	Spherical, triangular, and hexagonal; 10– 20 nm	Polyphenolic functionalities present in the extract	Antidiabetic activity	Maddinedi et al. (2017)
Chelidonium majus	Root	Distilled water	Aqueous root extract mixed into silver ion solution and after few min the formation of Ag NPs was noticed	UV-Vis, XRD, TEM	Spherical; 15–35 nm	Extract rich in phytochemicals	Antibacterial activity	Alishah et al. (2016)
Cydonia oblonga	Seed	Distilled water	Seed extract was added into $AgNO_3$ solution, incubated for 4 h. Change in color of solution indicated the formation of Ag NPs	UV-Vis, XRD, FTIR, SEM	~38 nm	Extract rich in flavanones and terpenoids	-	Zia et al. (2016)
Euphorbia amygdaloides	Whole plant	Distilled water	Purified peroxidase enzyme from *E. amygdaloides* was added into $AgNO_3$ solution and incubated for 4 h; appearance of brown color indicated Ag NP formation	UV-Vis, XRD, SEM	Spherical; 7–20 nm	Extract rich in phytochemicals	Antibacterial activity	Cicek et al. (2015)

(continued)

Table 5.1 (continued)

Plant	Part of plant used	Solvent used	Synthesis conditions	Characterization techniques	Shape and size	Phytoconstituents responsible for reduction of silver ion	Uses	Key references
Eucalyptus	Wood	Distilled water	Pre-hydrolyzed liquor of *Eucalyptus* wood was mixed into $AgNO_3$; observed change in color of the solution	UV-Vis, XRD, FTIR, FESEM, TEM, EDX, Raman, AFM	Spherical; 25–30 nm	Extract rich in polyphenols, hemicelluloses, and their derivates	Antimicrobial activity	Shivakumar et al. (2017)
Erigeron bonariensis	Leaf	Distilled water	Aqueous leaf extract mixed with $AgNO_3$ solution under continuous stirring and exposed to direct sunlight for instant biosynthesis of Ag NPs	UV-Vis, XRD, FTIR, SEM, EDX, TEM, XPS, AFM	Spherical; ~ 13 nm	Extract rich in phytochemicals, viz., flavonoids, triterpenes, steroids, caffeoyl derivatives	Catalytic activity	Kumar et al. (2015)
Isatis tinctoria	Whole plant	Distilled water	Aqueous extract was mixed with $AgNO_3$ solution, and NP synthesis was detected by color change from light yellow to deep brown	UV-Vis, XRD, FTIR, SEM, TEM, EDX, DLS, zeta potential	Spherical; size range varies with respect to extract concentration	The plant extract rich in the saponins and flavonoids	Antileishmanial activity	Ahmad et al. (2016)
Fraxinus excelsior	Leaf	Distilled water and ethanol	Aqueous leaf extract solution was mixed into $AgNO_3$. Microwave radiation caused change in color of solution from yellow to brown, indicating Ag NP fabrication	UV-Vis, FTIR, SEM, TEM, EDX	Spherical; 25–40 nm	Extract rich in flavonoids, alkaloids, glycosides, terpenoids, phenolic compounds, amino acid residues and peptides of protein	Antioxidant assay	Parveen et al. (2016)

Helicteres isora	Root	Distilled water	$AgNO_3$ solution and aqueous root extracts were mixed in the ratio of 1:1 and incubated at room temperature for 6 h	UV-Vis, XRD, FTIR, TEM, EDX, DLS, zeta potential	Spherical; 16–95 nm	Root extract rich in aromatic and carbonyl derivative of the protein	Antioxidant and antibacterial activity	Bhakya et al. (2016)
Melia azedarach	Leaf	Distilled water	Aqueous leaf extract solution was mixed into $AgNO_3$ solution at room temperature and change in color of the solution observed	UV-Vis, FTIR, SEM, EDX	Spherical; 34–48 nm	Extract rich in phytochemicals	Antibacterial activity	Mehmood et al. (2017)
Momordica charantia	Leaf	Distilled water	Leaf extract was added into aqueous $AgNO_3$ solution, and the colorless reaction mixture changed to brownish yellow after few minutes	UV-Vis, XRD, FTIR, FESEM, TEM, XPS, EDS, DLS, zeta potential	Spherical; ~13 nm	Flavonoids, alkaloids, tannins, phytosterols	Antimicrobial and catalytic activity	Ajitha et al. (2015)
Momordica cymbalaria	Fruit	Distilled water	Aqueous $AgNO_3$ solution mixed into fruit extract and incubated for 8 h at room temperature; observed change in color of the solution due to the reduction reaction	UV-Vis, XRD, FTIR, SEM, AFM	Spherical; 15.4 nm	Phenolics, flavonoids, terpenoids, and vitamins	Antimicrobial, antioxidant and cytotoxicity activities	Swamy et al. (2015)

(continued)

Table 5.1 (continued)

Plant	Part of plant used	Solvent used	Synthesis conditions	Characterization techniques	Shape and size	Phytoconstituents responsible for reduction of silver ion	Uses	Key references
Musa	Peel	Distilled water	Aqueous $AgNO_3$ solution mixed with banana peel extract, incubated for 72 h; extract reduces the silver solution within 5 min after heating the mixture at 40–100 °C; the color changed to reddish brown	UV-Vis, XRD, FTIR, SEM, EDX, TEM, DLS	Spherical; 23.7 nm	Peel extract rich in phytochemicals	Antimicrobial activity	Ibrahim (2016)
Nigella sativa	Leaf	Distilled water and methanol	Aqueous silver salt solution gradually mixed with sodium borohydrate solution under continuous stirring; added leaf extract of *N. sativa* drop-wise to the resultant solution with continuous stirring at 50–60 °C	UV-Vis, SEM, FTIR	Spherical; size varies with respect to reaction condition	Flavonoids and terpenoids present in the extract	Cytotoxicity and phytotoxicity	Amooaghaie et al. (2015)
Parkia roxburghii	Leaf	Distilled water	One g of dry leaf powder was mixed into $AgNO_3$ solution at room temperature for 12 h. The initial clear solution turned to reddish brown indicating the formation of colloidal Ag NPs	UV-Vis, XRD, FTIR, SEM, TEM	Spherical; 5–25 nm	Extract rich in polypeptides and carboxylate groups in the amino acid residues of the protein molecules	Photocatalytic and antibacterial activity	Paul et al. (2016)

Physalis angulata	Leaf	Distilled water	Aqueous extract of *P. angulata* was mixed with silver ion solution and put in bright sunlight and in the dark; formation of Ag NPs occurred under bright sunlight	UV-Vis, XRD, FTIR, SEM, TEM, EDX, SAED, AFM	Spherical; 11–96 nm	Extract rich in flavonoids, phenol, saponin, tannins, phytosterols, and glycosides	Antibacterial and antioxidant activity	Kumar et al. (2017b)
Picrasma quassioides	Bark	Distilled water	Prepared colloidal solution of graphene oxide with water poured drop-wise into silver ion solution. Suitable amount of bark extract was mixed to this solution at room temperature. It led to the formation of GO-Ag NP nanocomposites	UV-Vis, SEM, EDX, FTIR, TEM, SAED	Spherical; size varies with respect to reaction condition	Bark extracts rich in phytochemicals	Catalytic activity	Sreekanth et al. (2015)
Pimpinella anisum	Seed	Distilled water	Aqueous seed extract was added into $AgNO_3$ solution at room temperature; after 172 h, yellow color solution turned into dark brown, indicating the formation of Ag NPs	UV-Vis, FTIR, TEM, EDX	Spherical; 80–85 nm	Extract rich in phytochemicals	Anticancer activity	Devanesan et al. (2017)

(continued)

Table 5.1 (continued)

Plant	Part of plant used	Solvent used	Synthesis conditions	Characterization techniques	Shape and size	Phytoconstituents responsible for reduction of silver ion	Uses	Key references
Piper nigrum	Seed	Distilled water	Appropriate amount of reactant mixed with seed extract and resultant solution exposed under sunlight for different time durations; change in color of the solution was observed	UV-Vis, XRD, AFM	Spherical; 10–60 nm	Extract contains biomolecules, viz., vitamins, polysaccharides, amino acids, piperine, and alkaloids	-	Mohapatra et al. (2015)
Rosa andeli	Petals	Distilled water	Aqueous $AgNO_3$ solution was reduced by using petal extract at 0 °C for 15 min and then kept at room temperature. Appearance of yellow color indicated the formation of Ag NPs	UV-Vis, SEM, EDX, FTIR, TEM, SAED	Spherical; 1.5–3.5 nm	Flavonoids and polyphenols present in the Rosa "Andeli" petal extract	Photocatalytic activity	Suarez-Cerda et al. (2015)
Salvinia molesta	Leaf	Distilled water	Aqueous leaf extract was mixed into $AgNO_3$; rapid change in color of reaction mixture from yellowish green to reddish brown within 20 s under sunlight exposure, a primary visual indication for Ag NP fabrication	UV-Vis, XRD, FTIR, FESEM, EDX, HRTEM, AFM	Spherical; 12.46 nm	Extract rich in hydroxyl and amino groups	Antibacterial activity	Verma et al. (2016)

Sargassum polycystum	Whole plant	Distilled water	Aqueous extract mixed with $AgNO_3$ solution and the solution kept into incubator at 20 °C. Change in color of the solution indicated the formation of Ag NPs	UV-Vis, XRD, FTIR, SEM, TEM, EDX	Spherical; ≈ 28 nm	Aqueous extract rich in alkaloids, steroids, flavanoids, protein, terpenoids, amino acids, carbohydrate, quinones, phenols, and tannins	Antioxidant and anticancer activity	Palanisamy et al. (2017)
Sida cordifolia	Leaf	Distilled water	The leaf extract mixed with ingredients under continuous stirring at room temperature and the resultant solution exposed to sunlight for 10 min to produce Ag NPs	UV-Vis, FTIR, SEM, TEM, EDX, zeta potential	Spherical and prism; 10–30 nm	Extract rich in bio-functional organic molecules that contain O-H and N-H functional groups	Antioxidant and anticancer activity	Srinithya et al. (2016)
Skimmia laureola	Leaf	Distilled water	Aqueous leaf extract of *S. laureola* and $AgNO_3$ were mixed in the ratio of 3:7 at room temperature and change in color of the solution was observed	UV-Vis, XRD, SEM, FTIR	Spherical and hexagonal; 38 ± 0.27 nm	Extract rich in biomolecules, skimmidiol	Antibacterial activity	Ahmed et al. (2015)
Sterculia acuminata	Fruit	Distilled water	Fruit extract solution was mixed with $AgNO_3$; change of color of solution from light yellow to reddish brown confirmed the formation of Ag NPs	UV-Vis, XRD, FTIR, TEM, HPLC, zeta potential	Spherical; 10 nm	High concentration of secondary metabolites, viz., gallic acid	Dyes degradation	Bogireddy et al. (2016)

spectroscopy, and so on. Microscopic techniques such as SEM, TEM, and AFM are mainly used for morphological studies of NPs. At the very outset, synthesis of NPs has to be standardized using plants or their extracts. On addition of a reducing agent to the reaction medium, a change in the color of the medium serves as indicator of NP synthesis. Ag NPs generally exhibit yellowish-brown color in the aqueous solution as a result of surface plasmon resonance (SPR) (Song and Kim 2009). A progressive increase in the characteristic peak with increase in reaction time and concentration of plant extracts is a clear indicator of NP formation. Metallic NPs absorb light in the wavelength range between 300 and 800 nm. UV-Vis spectroscopy is generally used for initial NP characterization. Ag NPs are characterized by their SPR peaks from 400 to 450 nm (Mittal et al. 2013). SEM and TEM are incredibly versatile in characterizing the size, morphology, and composition of NPs, the TEM having a higher resolution compared to SEM (Sattler 2010). For three-dimensional structure of NPs, AFM has an advantage over SEM and TEM. Powder XRD has become a cornerstone technique for deriving crystallite size in nanoscience. X-rays have photon energies in the range of 100 eV–100 keV. The X-rays can penetrate deep into the materials and provide information about the bulk structure. For diffraction studies, only short-wavelength X-rays in the range of a few angstroms to 0.1 Å (1–120 keV) are used. The wavelength of X-rays is comparable to the size of atoms; therefore, they are very much useful for studying the 3D structure in a wide range of materials including NPs. XRD is used to determine the crystalline nature of NPs using Debye Scherrer's equation (Prathna et al. 2011). DLS measures light scattered from the laser that passes through a colloid. Modulation of scattered light intensity as a function of time is analyzed, and hydrodynamic size of particles can be determined (Berne and Pecora 2000). EDS is employed to determine the elemental composition of NPs. IR radiation is absorbed by a molecule possessing dipole moment, and its oscillating frequency is the same as the frequency of incident IR light (Johal, 2011). Surface modification of NPs by organic functional groups may be characterized by infrared spectroscopy. As the infrared light interacts with a molecule, chemical bonds will stretch, contract, and bend. The functional group in the molecule tends to adsorb infrared radiation in a specific wave number. The correlation of the band wavenumber position with the chemical structure is used to identify a functional group in a NP-associated molecule in a sample. This technique has been used for characterizing silver and gold NPs and their associated molecules from plant extracts. Raman spectroscopy offers a unique and noninvasive tool for exploring the behavior of the components within a given biomaterial and its surrounding microenvironment. The use of surface-enhanced Raman spectroscopy (SERS) in material characterization, in concept development, and in identifying the NP applications has been thoroughly reviewed (Dieringer et al. 2006).

Paul et al. (2016) approached green route for synthesis of Ag NPs by using dried leaf biomass of *Parkia roxburghii.* When the dried leaf powder was added to silver nitrate solution under continuous stirring, the resultant clear solution turned to a brown color solution within few minutes due to SPR and shows a relatively broad peak at around 440 nm, which was characteristic of Ag NPs, while the steady increase in the intensity of SPR suggests an increase in the yield of NPs with time,

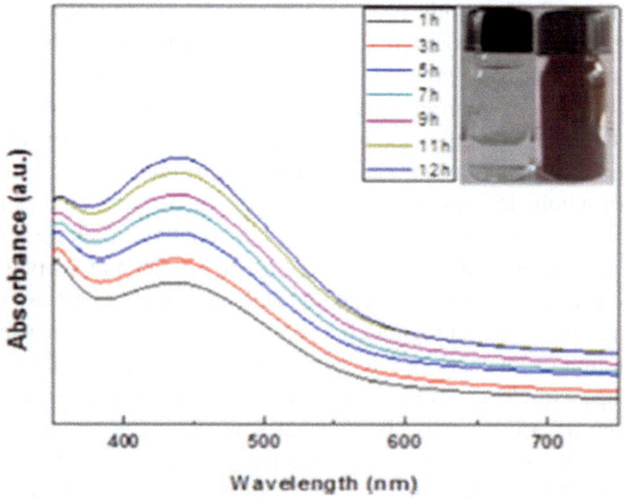

Fig. 5.2 Time-dependent UV-visible absorption spectra of Ag NPs synthesized by using dried leaf biomass of *Parkia roxburghii* (Paul et al. 2016)

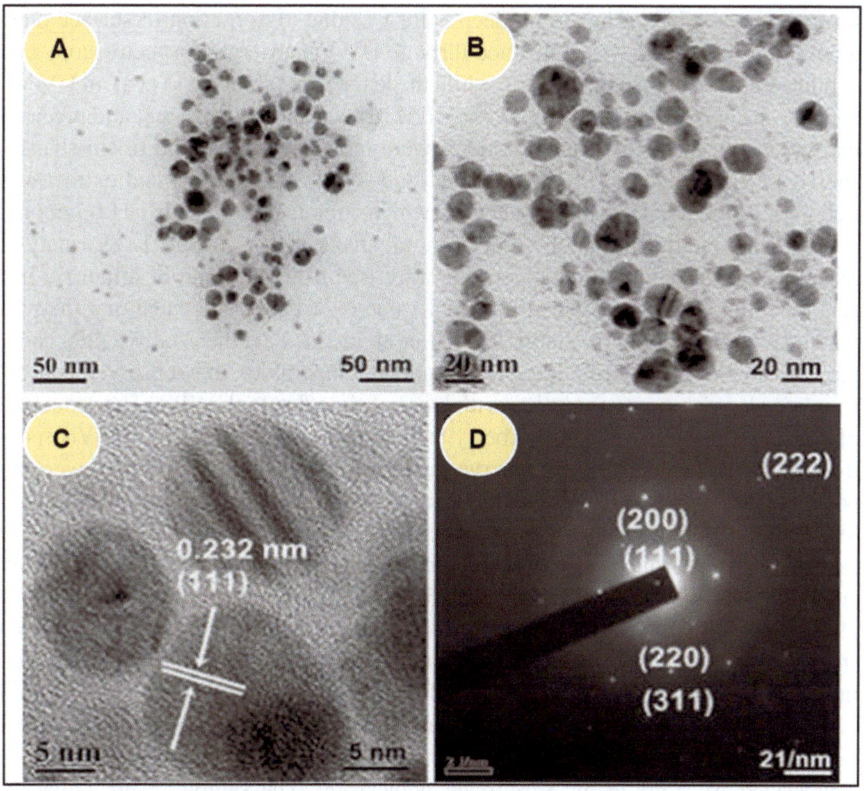

Fig. 5.3 TEM micrograph images of Ag NPs synthesized by dried leaf biomass of *Parkia roxburghii*: (**a**, **b**) TEM images, (**c**) HRTEM image showed distance between two lattice planes, and (**d**) SAED pattern of Ag NPs (Paul et al. 2016)

as shown in Fig. 5.2. The synthesized Ag NPs were highly crystalline in nature. The transmission electron micrograph (Fig. 5.3) showed that these NPs were quasi-spherical in shape and 5–25 nm in size. The FTIR analysis showed that polypeptides and carboxylate groups in protein play a pivotal role in the reduction of silver ions.

Ahmed et al. (2016) used *Azadirachta indica* leaf extract for Ag NP fabrication. On incubation in the dark, at ambient temperature, the mixture changed from colorless to brown within 15 min due to reduction of Ag^+ to Ag^0. The distinct UV-Vis absorption peak was observed in the range of 436–448 nm, which was characteristic to Ag NPs. On changing the concentration of leaf extract and keeping that of the aqueous silver nitrate solution constant, the authors observed increase in the intensity of absorption peaks after regular time intervals, and the color intensity increased with the duration of incubation. The synthesized Ag NPs were spherical in shape, and the average particle size was ~34 nm. Further, the FTIR analysis confirmed that the leaf extract was rich in active biomolecules including flavanoids and terpenoids and their derivatives and acted as the reducing agent during Ag NP synthesis. Srinithya et al. (2016) synthesized Ag NPs by using dried leaf extract of *Sida cordifolia*. Filtrate was mixed with silver nitrate solution under continuous stirring, and a stable solution was exposed to sunlight for 10 m at room temperature. This could change the color of solution from light yellow to dark brown, which indicated the formation of Ag NPs. The strong deep color was due to excitation of surface plasmon vibrations which was confirmed through UV-Vis absorption spectroscopy and exhibited a distinct peak at 454 nm, while the leaf extract showed no peak in UV-Vis absorption spectroscopy. The TEM and SEM results were inconsistent and revealed that the synthesized silver nanoparticles were mostly spherical and prism-shaped, showing a size range of 10–20 nm. FTIR study confirmed that the leaf extract was rich in bio-functional organic molecules, which contained O-H and N-H functional groups, and facilitated the bioreduction of silver nitrate, while EDX analysis revealed that the synthesized silver NPs were free from any kind of impurity. The zeta potential and high negative potential value (−30.5 eV) indicated that the prepared Ag NPs were highly stable. Kumar et al. (2017a) synthesized Ag NPs using the Andean blackberry fruit extract. At ambient temperature, they observed appearance of yellowish-orange solution with lapse of time due to the SPR. The reduction of aqueous Ag+ ions by the blackberry fruit extract was analyzed by UV-Vis spectroscopy. It was interesting to observe through UV-Vis spectra that there were no peaks in the range of 380–480 nm, but after 3.5 h, a new peak appeared around 380–480 nm. This indicates that synthesis of Ag NPs started after 3.5 h with the assistance of the fruit extract. The synthesized Ag NPs were spherical and highly crystalline, showing a size range of 12–50 nm. The presence of peaks near 3270, 2933, and 1642 cm^{-1} (Fig. 5.4a) could be due to the O–H, aliphatic C–H, and C=O stretching vibration of flavonoids/phenolic groups in the fruit extract, which played important role in the synthesis of NPs. Figure 5.4b shows deviation of 3275 and 1634 cm^{-1} for Ag NPs, which suggests that the O–H and C=O groups were adsorbed on the surface of NPs and involved in the reduction process. Palanisamy et al. (2017) synthesized Ag NPs, using *Sargassum polycystum*. The samples were collected from sea, dried, powdered, mixed with silver nitrate solution, and incubated at 20 °C. Color change due to SPR vibration in the particles, showing distinct UV-Vis

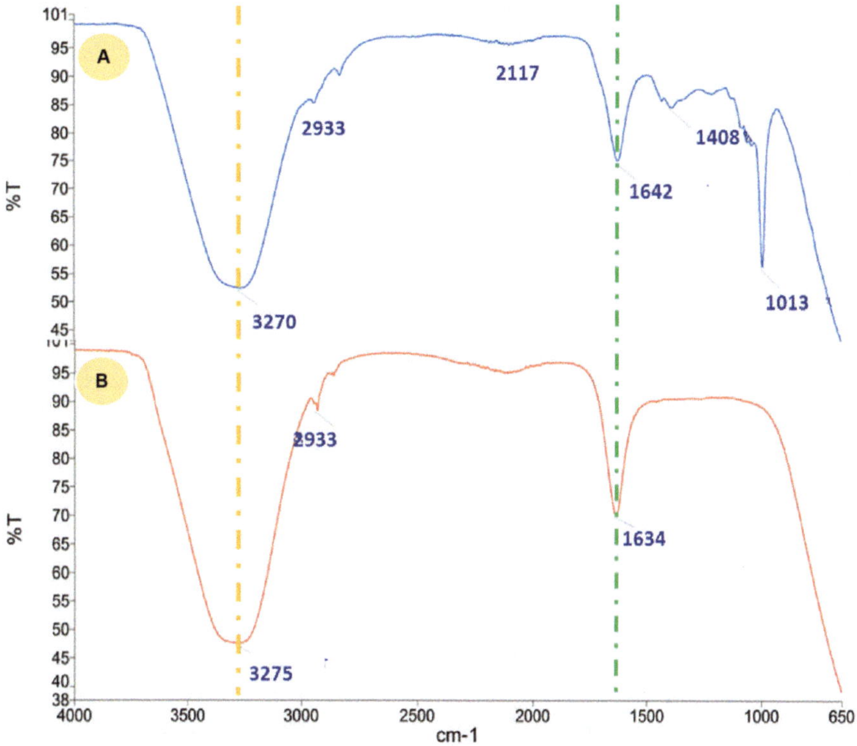

Fig. 5.4 FTIR spectra of (**a**) Andean blackberry fruit extract (ABFE) and (**b**) Ag NPs synthesized by using ABFE extract (Kumar et al. 2017a)

absorption peak at 405 nm, confirmed the formation of Ag NPs. The XRD revealed that the NPs synthesized were highly crystalline and free from any impurity. The SEM and TEM studies revealed that the NPs were spherical with average size around ~28 nm. FTIR analysis indicated that the extract was rich in alkaloids, flavonoids, steroids, terpenoids, protein, and C-O stretching ether group which played important role in the reduction of silver ion. Rao and Tang (2017) used *Eriobotrya japonica* leaf extract for synthesizing the Ag NPs. The aqueous leaf extract was mixed with silver nitrate solution and incubated at 80 °C. It turned from colorless to yellowish brown due to reduction of silver ion and displayed distinct UV-Vis absorption band at around 435 nm, which indicated the formation of Ag NPs. The synthesized NPs were spherical, with average size 19.75 nm, as determined by TEM analysis. They were highly stable and showed excellent zeta potential value about −20.4 mV, as shown in Fig. 5.5, which indicates the significant stability of Ag NPs in the suspension. The NPs were face-centered cubic in nature and exhibited diffraction peaks at the 2θ values of 38.11°, 44.33°, 64.35°, and 77.62° corresponding to 111, 200, 220, and 311 crystallographic planes, respectively, as shown in Fig. 5.6. The FTIR analysis confirmed that the leaf extract was rich in triterpenic acids, flavonoids, sesquiterpene glycosides, polysaccharides, and proteins, which were responsible for the bioreduction of silver ions.

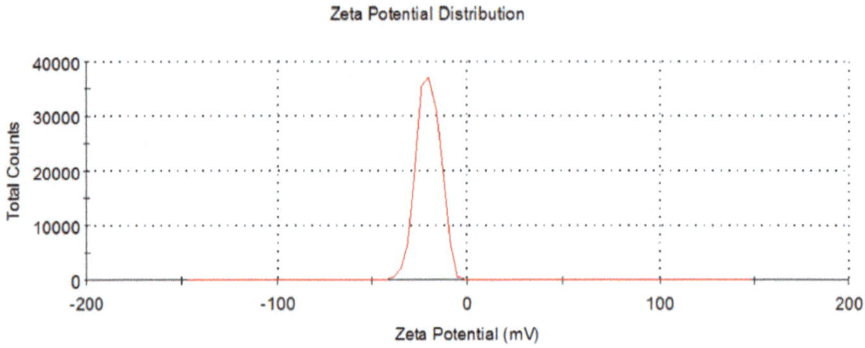

Fig. 5.5 Zeta potential curve of Ag NP synthesized by using *Eriobotrya japonica* leaf extract (Rao and Tang 2017)

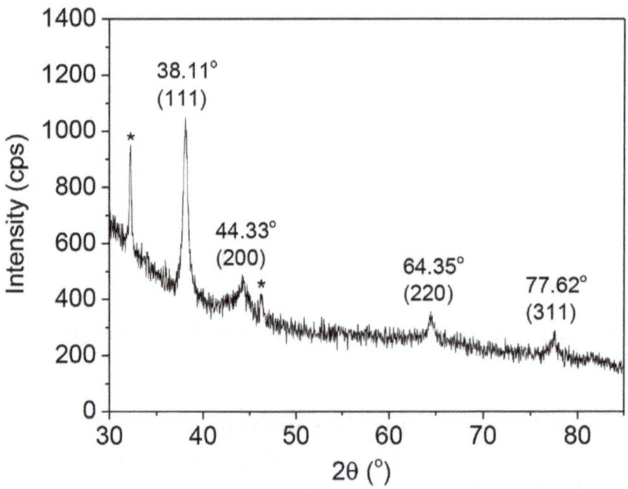

Fig. 5.6 XRD pattern of Ag NPs synthesized by using *Eriobotrya japonica* leaf extract (Rao and Tang 2017)

Kumar et al. (2017b) used the extract of *Physalis angulata* leaves for Ag NP fabrication. They have suggested that the SPR raised through collective oscillations of free conduction electrons was responsible for Ag NP fabrication which revealed change in color of the reaction mixture from watery to reddish brown. The reaction mixture exposed to bright sunlight exhibited instant color change and SPR spectra within few minutes of the inoculation of aqueous extract of *P. angulata* into $AgNO_3$ solution, whereas the reaction mixture kept in dark condition did not attain the same color change and SPR spectra even in 12 h, as shown in Fig. 5.7a, b. FTIR analysis confirmed that the leaf extract was rich in phytochemicals such as flavonoids, phenol, saponin, tannins, phytosterols, and glycosides. The authors have suggested that when the resultant mixture is exposed to sunlight, these phytochemicals get photosensitized by absorbing photons from the sunlight and released excited electrons,

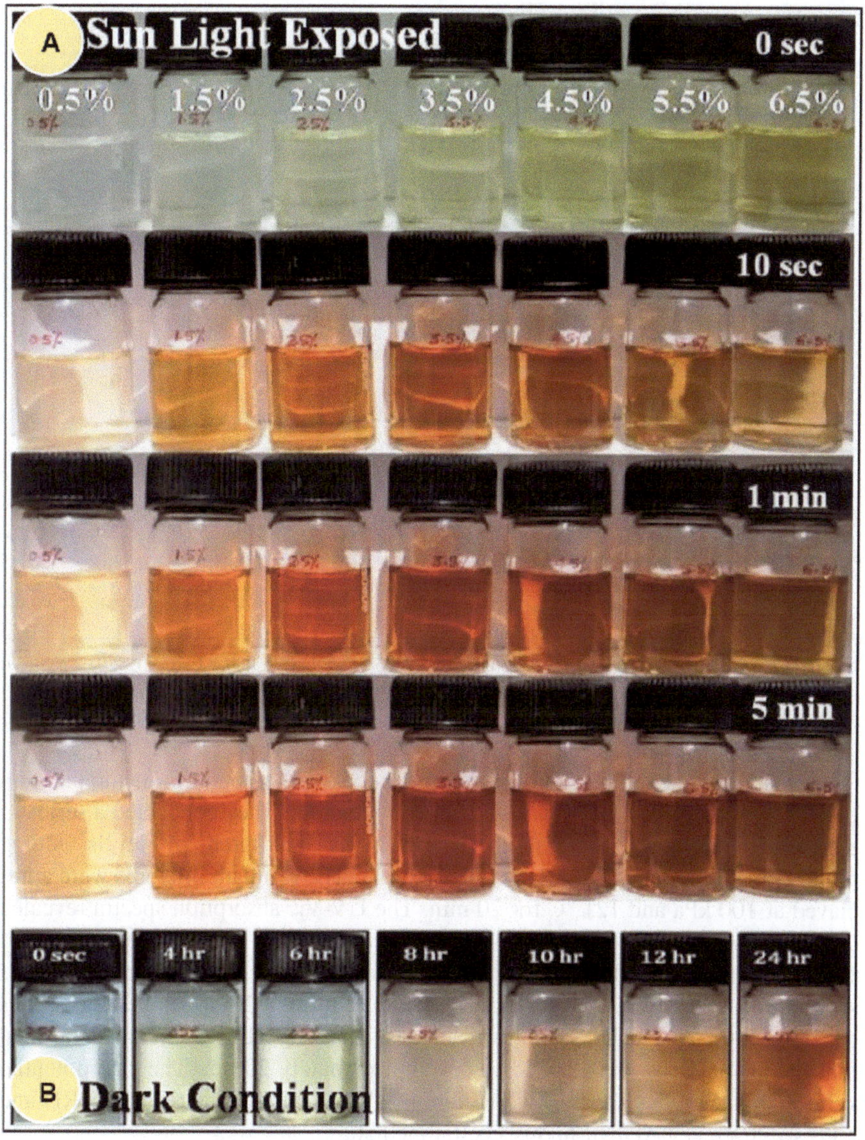

Fig. 5.7 Reaction extent of Ag NP biosynthesis showing a distinct pattern of change in color of the reaction mixture (**a**) in bright sunlight condition and (**b**) in dark condition (Kumar et al. 2017b)

which induced reduction of Ag^+ to Ag^0. The NPs were spherical in shape with average size of 36 nm. The XRD and SAED patterns confirmed the crystalline nature of synthesized NPs, while AFM studies provided the 2D and 3D topographical view and roughness profile of the NPs, respectively, as shown in Fig. 5.8. The maximum profile peak height and valley depth were 17.5 nm and 14.3 nm, respectively.

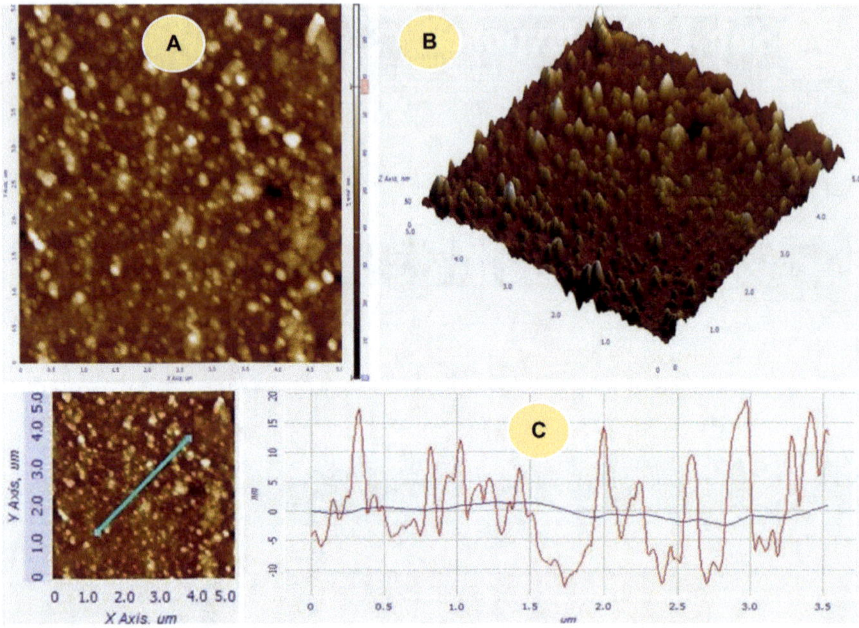

Fig. 5.8 AFM images of Ag NPs synthesized by using aqueous extract of *Physalis angulata* leaves assisted by sunlight: (**a**) lateral and (**b**) three-dimensional view and (**c**) roughness profile of surface (Kumar et al. 2017b)

Recently, Kim et al. (2018) have demonstrated a green route for the preparation of Ag NPs, using the *Laminaria japonica* algal extract. They conducted a series of experiments to optimize the synthesis conditions at different pH levels, adding molar concentrations of NaOH and algal extract. The resultant mixtures were autoclaved at 100 kPa and 121 °C for 20 min. The UV-Vis absorption spectra revealed the intensity of peak at 405 nm. This increased remarkably as the volume of the leaf extract was increased. The synthesized Ag NPs were spherical in shape, while their size appreciably varied with respect to the molar ratio of NaOH to leaf extract, as shown in Fig. 5.9. The DLS result was inconsistent with TEM result and confirmed that the average size of Ag NPs was of approximately 20 nm. The zeta potential values confirmed that the NPs were highly stable in nature, whereas the XRD and XPS spectra revealed their highly crystalline nature.

5.3 Factors Affecting Fabrication of Ag NPs

Several factors such as the concentration, pH, incubation time, and temperature of plant extract or biomass affect the conversion of metal ions to NPs and control the shape and size of NPs during fabrication.

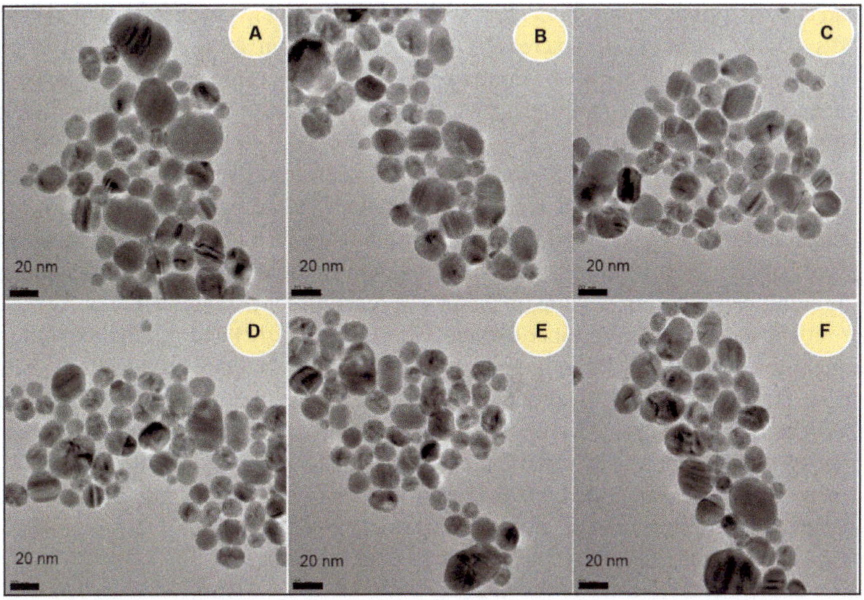

Fig. 5.9 High-resolution TEM (**a–f**) of Ag NPs synthesized by using initial concentrations of 2 mM $AgNO_3$ and 2 mL (3% w/v) *Laminaria japonica* extract and various molar ratios of NaOH (5, 7.5, 15, 20, 30, and 40 mM, respectively) (Kim et al. 2018)

5.3.1 Temperature

Silver nanotriangles were fabricated with *Medicago sativa* seed exudates at high temperatures (above 30 °C) due to suppression of shape-directing agents at lower temperatures (Lukman et al. 2011). Lin et al. (2010) used *Cassia fistula* leaf broth for fabrication of silver NPs. Silver nanowires were formed at room temperature, while spherical NPs and short nanorods appeared at 60 °C. The authors suggested that interaction between the biomolecules and the faces of silver perhaps changed at elevated temperatures which hindered the NP coalescence in solution. Cruz et al. (2010) used *Lippia citriodora* to fabricate Ag NPs and showed that the reduction rate increased as the reaction temperature was raised, showing that the NP nucleation was facilitated at a high temperature. The authors obtained highly polycrystalline Ag NPs at high temperature (95 °C) and less crystalline ones at the room temperature. Andreescu et al. (2007) and Sathishkumar et al. (2010) demonstrated that fabrication of Ag NPs has a positive correlation with the increase in temperature. This is possibly because of the crystal growth of faces through the silver atom deposition on cubic faces rather than the nucleation of new silver crystals at elevated temperatures (Lengke et al. 2007). Furthermore, increase in SPR with an increase in temperature speaks of a positive correlation between NP production and temperature (Sathishkumar et al. 2010). So, all these investigations suggest that temperature is one of the important growth factors for fabrication of NPs of the desired shape and size. In a study, using *Tanacetum vulgare* fruit extract to produce silver and gold NPs (Dubey et al. 2010), increase in temperature from 25 to 150 °C led to an increase in the sharpness of absorption peaks, possibly because a rise in temperature enhanced the reaction rate and hence the NP fabrication process (Phillip 2009; Dwivedi and Gopal 2010).

Upadhyay and Verma (2014) have shown that reaction time is temperature dependent for fabrication of Ag NPs from potato extracts. A change from room temperature to 60 °C reduced the reaction time from 16 h to 15 min and produced spherical NPs with an average size of 61 nm. Ghosh et al. (2012) also investigated the significance of temperature on the rate of reaction for the fabrication of Ag NPs from *Dioscorea bulbifera* tuber extract (Ghosh et al. 2012) and got the maximum rate of reaction at 50 °C. Similar conclusions were drawn while producing Ag NPs using *Pulicaria glutinosa* extract (Khan et al. 2013) and *Solanum xanthocarpum* berry extract (Amin et al. 2012). Sun et al. (2014) used tea leaf extract for developing Ag NPs and produced spherical NPs of 91 and 175 nm at 25, 40, and 55 °C, respectively. They suggested that the reaction rates of Ag NP synthesis increased as the temperature increased, and hence the particle sizes increased.

5.3.2 *pH*

Sathishkumar et al. (2009) used *Cinnamon zeylanicum* bark powder and extract over a wider (1–11) pH range and suggested that the pH of the solution dropped in most of the cases after the fabrication of Ag NPs. At lower or acidic pH, large-sized ellipsoidal Ag NPs were obtained, whereas highly dispersed small-sized and spherical NPs were formed at higher or alkaline pH. The authors suggested that at higher pH, availability of a large number of functional groups facilitates a higher number of Ag(I) to bind and then form a large number of NPs with smaller diameter. In another study, Sathishkumar et al. (2010) used *Curcuma longa* for Ag NP fabrication and obtained large NPs at lower pH while small and highly dispersed NPs at higher pH. With the mangosteen leaf extract, aggregation of Ag NPs facilitated overnucleation at low pH, whereas higher pH favored nucleation and subsequent development of a large number of NPs with smaller diameter (Veerasamy et al. 2011). Tripathy et al. (2010) used an aqueous extract of *Azadirachta indica* leaves for fabricating Ag NPs and found improved stability of cluster distribution at alkaline pH range. They have proposed that complete charging of the clusters was obtained at alkaline pH possibly due to the availability of large amounts of OH^-. The repulsive electrostatic/electrosteric relations were enhanced, which led to increased stability and reduced aggregation (Tripathy et al. 2010). This study demonstrates a reduced Ag NP aggregation under alkaline conditions, which corresponds to the lesser particle production at higher pH. Maria et al. (2015) used *Ziziphus xylopyrus* bark extracts for Ag NP fabrication and reported that the optimum condition is silver nitrate solution at 10 mM; and the alkaline initial pH of 11 facilitates SPR peak of high intensity, showing increased fabrication of Ag NPs.

Thus, the acidic medium suppressed the production of Ag NPs, whereas the basic medium enhanced it.

5.3.3 *Incubation Time*

The contact or incubation time, i.e., the time duration needed for accomplishment of all reaction steps, also affects the process of NP formation. For instance, Li et al. (2007) fabricated Ag NPs by treating silver ions with *Capsicum annuum* extract and found that the crystalline phase of NPs was altered from polycrystalline to single crystalline and their size was enhanced by increasing the reaction time. In this experiment, 5 h reaction time produced spherical and polycrystalline-shaped NPs (10 ± 2 nm), whereas the particle size was increased from 25 ± 3 nm to 40 ± 5 nm by increasing the reaction time from 9 to 13 h, respectively. In another study, sharpness of UV absorption spectra peaks was increased with an increase in the contact time during the fabrication of silver and gold NPs using the leaf extract of *Chenopodium album* (Dwivedi and Gopal 2010). The authors observed that NPs appeared within 15 min of reaction and increased up to 2 h and thereafter only a slight variation took place. Fabrication of silver and gold NPs started with the help of *Tanacetum vulgare* fruit extract within 10 min of the reaction (Dubey et al. 2010). The authors found that an increase in contact time was favorable for sharpening of peaks in both silver and gold NPs. They reported a shorter reaction contact time in comparison to the previous reports of Fayaz et al. (2009) and Shaligram et al. (2009). Veerasamy et al. (2011) used mangosteen leaf extract for Ag NP fabrication and suggested that the optimum time needed for completion of reaction was 60 min. They pointed out that due to the instability of NPs formed, an optimum duration is needed for complete nucleation and subsequent stability of NPs.

5.3.4 *Plant Biomass Concentration*

This is another crucial growth factor for optimum Ag NP production. The process starts when suitable precursor concentration is available in a suitable range for nucleation. The availability of reducing and capping agents determines whether the metal precursors could be reduced and ultimately participate in NP fabrication. Thus, plant biomass concentration at the time of NP fabrication cannot be ignored, as it facilitates the reduction and stabilization set in by the active biomolecules that usually influence the final outcome. Prathna et al. (2011) have used *Citrus limon* aqueous extract for fabrication of Ag NPs and found that by increasing the mixing ratio of metal solutions and plant biomass, smaller size of NPs was obtained. This could be due to the increased amount of electron-rich bioreducing agents when a higher concentration of plant biomass was used in the reaction medium (Prathna et al. 2011).

5.4 Applications of Ag NPs

In recent years, one of the significant opportunities for NP use lies in biological sectors. From the ancient time, silver-based compounds have been used as nontoxic, inorganic, and antibacterial agents in numerous applications due to their biocidal effects, for example, in wood preservatives or for water purification in hospitals. In general, nanomaterials revealed unique and modified physical and chemical properties, compared to their macroscaled counterparts. Ag NPs are important for their antibacterial properties, even against the multidrug-resistant human pathogens (Gopinath et al. 2012; Das et al. 2013; Prakash et al. 2013; Sankar et al. 2013; Gogoi et al. 2015; Chung et al. 2016; Karatoprak et al. 2017). Singh et al. (2013) have revealed that Ag NPs are effective against yeast and fungal pathogens. Antioxidant activity of plant-mediated Ag NPs is also observed (Raghunandan et al. 2011; Dipankar and Murugan 2012; Chung et al. 2016). Some reports on their use in destruction of cancer cells are also available (Subramanian and Suja 2012; Vivek et al. 2012; Sankar et al. 2013; Chung et al. 2016). Ag NPs are already integrated into applications such as wound treatment, sterilization, food sanitation, antibacterial textiles, X-ray, and gene/drug delivery (Husen and Siddiqi 2014b; Chung et al. 2016; Siddiqi and Husen 2016a, b; Abasi et al. 2016; Mattea et al. 2017). They are also used for enhancing plant growth and development (Yin et al. 2012; Husen and Siddiqi 2014b; Baskar et al. 2015; Vinković et al. 2017). In fact, Ag NPs are one of the most widely used and studied engineered NPs.

5.4.1 Antimicrobial

A clear understanding on the interactions of Ag NPs with the cell surface is vital for the assessment of bactericidal activity and for advanced biomedical and environmental applications. One important feature of Ag NPs is that they have an antibacterial effect (Siddiqi et al. 2018a, d). However, different bacteria have different cell wall structure and are therefore different in their susceptibility. The study of antibacterial activity of Ag NPs by different methods has shown that they are effective toward different bacterial species in different media. The difference in the sensitivity of bacteria toward Ag NPs is due to their ability to form capsules and to the variation in thickness of peptidoglycan layer in their cell wall (de Aragao et al. 2016; Cicek et al. 2015). A higher antibacterial activity of Ag NPs against gram-positive bacteria than gram-negative bacteria has been reported by many researchers (Yakout and Mostafa 2015; Jadhav et al. 2015; Paul et al. 2016). However, the exact mechanism for this specific response remains unknown. On the other hand, some workers have reported the use of Ag NPs for inhibiting the growth of both gram-positive and gram-negative bacteria (Swamy et al. 2015; Pugazhendhi et al. 2015; Kokila et al. 2016; El-Sherbiny et al. 2016). Ag NP toxicity or the growth rate of bacterial population is also dependent on the NP and pH of the medium and exposure time to pathogens (Siddiqi et al.

2018a). In general, the minimum inhibitory concentration (MIC) values are used to investigate the antibacterial potential of NPs (Kamala-Kannan et al. 2016; Sundararajan et al. 2016; Jadhav et al. 2016). Lok et al. (2006) suggested that the response of Ag NPs is connected with bacterial cell wall structure and the variation of antibacterial activity may derive from differences in bacterial cell wall structure and the thickness of peptidoglycan layer of this wall. It has also been proposed that cationic silver denatures proteins by getting attached to the thiol group and phosphorus contained in DNA present in the bacterial cell wall. This inhibits replication of DNA and causes deterioration of plasma membrane leading to cell death. It was hypothesized that Ag NPs may bind to mesosomes (cell organelle) which further inhibit the production of energy in the bacterial cell (Balamanikandan et al. 2015; Emmanuel et al. 2015; Awad et al. 2015; Ahmed et al. 2015). It has also been suggested that Ag NPs produce reactive oxygen species (ROS) and free radicals which cause apoptosis leading to cell death, thus checking their replication (Siddiqi et al. 2018a) (Fig. 5.10). High concentrations of Ag NPs exhibit strong anti-bactericidal performance (Lu et al. 2015; Ali et al. 2015; Mohan et al. 2016). It was reported that Ag NPs synthesized from Azadirachta indica aqueous leaf extract or *Emblica officinalis* fruit extract show higher antibacterial activity, compared to those prepared from conventional methods (Ahmed et al. 2016; Ramesh et al. 2015). Recently, Tareq et al. (2017) have fabricated Ag NPs from *Bryophyllum pinnatum* leaf extract and found them strongly active against food pathogen (*Escherichia coli*-MTCC-443) and agriculture pathogen (*Bacillus megaterium*-MTCC-2412).

The antifungal property of Ag NPs against different strains of fungi including *Aspergillus*, *Candida*, and *Trichophyton* has been studied by several researchers. The growth of fungal strains *Aspergillus flavus*, *A. niger*, *A. ochraceus*, and *A. terres* was inhibited by the Ag NPs synthesized with *Allium cepa* extract (Balamanikandan et al. 2015). Compared to the standard antifungal agent fluconazole, green synthe-

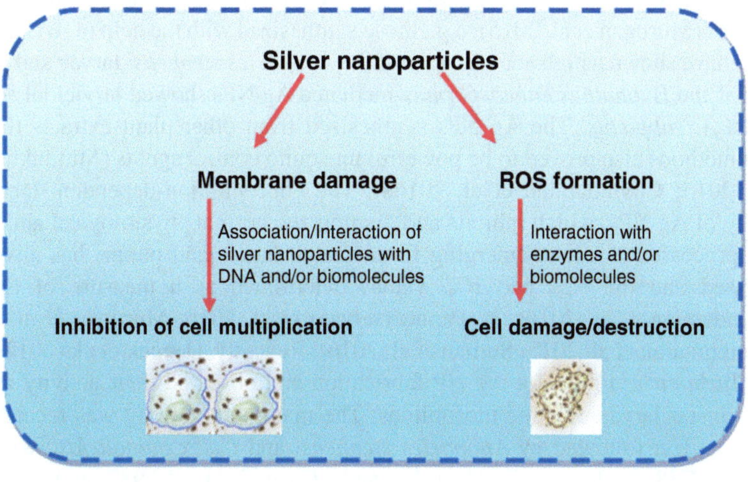

Fig. 5.10 Mechanism of action of Ag NPs against bacterial cells (Siddiqi et al. 2018a)

sized Ag NPs were more effective, showing a higher activity against *Candida albicans* than against some other fungal species such as *Candida tropicalis*, *Trichophyton mentagrophytes*, and *Aspergillus flavus*. Fungal strains *C. albicans* and *C. tropicalis* were significantly more susceptible to Ag NPs synthesized from ribose sugar than to the fungicidal drug Amphotericin B (Balamanikandan et al. 2015; Sathyaseelan et al. 2015; Mallmann et al. 2015). The Ag NPs synthesized from fresh stem and freeze-dried roots of *Amaranthus dubius* exhibited a high antifungal activity, whereas leaf extracts (fresh and freeze-dried) showed insignificant response (Sigamioney et al. 2016). *Penicillium* spp. were found to be more sensitive toward Ag NPs synthesized from *Momordica charantia* than from *Aspergillus flavus* and *A. niger*. It was proposed that cell lysis occurs because of deactivation of sulfhydryl contained in the fungal cell wall and this leads to the formation of insoluble compounds followed by a disruption of membrane-bound enzymes and lipids (Ajitha et al. 2015). Balashanmugam et al. (2016) have used *Cassia roxburghii* aqueous leaf extract for Ag NP fabrication. The particles produced showed a higher antifungal activity in comparison to the conventional fungicide amphotericin B against human pathogenic (*Aspergillus niger*, *A. fumigatus*, *A. flavus*, *Penicillium* sp., *Candida albicans*) as well as plant pathogenic (*Rhizoctonia solani*, *Fusarium oxysporum*, and *Curvularia* sp.) fungi. In a recent study, Bahrami-Teimoori et al. (2017) used the leaf extract of *Amaranthus retroflexus* to produce Ag NPs; worked out their MIC_{50} against *Alternaria alternata*, *Fusarium oxysporum*, and *Macrophomina phaseolina*; and suggested that these particles could be used as a potent antifungal agent against some plant pathogenic fungi.

5.4.2 Biomedical Application

The biosynthesized Ag NPs are found to possess mosquitocidal and antiplasmodial activities (Murugan et al. 2015), e.g., those synthesized with the help of *Aristolochia indica* have shown a high activity against the *Anopheles stephensi* larvae and pupae, whereas the *Hybanthus enneaspermus*-mediated Ag NPs showed larvicidal activity against *A. subpictus*. The Ag NPs synthesized from other plant extracts through green methods also proved to be powerful mosquito control agents (Muthukumaran et al. 2015; Govindarajan et al. 2016a). The concentration-dependent larvicidal activity of Ag NPs, which inhibits and disrupts the normal physiological and metabolic processes of adults emerging from treated larvae and pupae, has also been discussed and the IC_{50} and IC_{90} values determined as a measure of toxicity (Govindarajan et al. 2016a, b; Panneerselvam et al. 2016; Murugan et al. 2015; Muthukumaran et al. 2015; Suman et al. 2016). Recently, Deepak et al. (2017) have used *Turbinaria ornata* for Ag NP fabrication and observed their activity against fourth instar larvae of three mosquitoes. The maximum activity was recorded on *Aedes aegypti* followed by *Anopheles stephensi* and *Culex quinquefasciatus* with LC_{50} of 0.738, 1.134, and 1.494 μg L^{-1} and LC_{90} of 3.342, 17.982, and 22.475 μg L^{-1} lethal concentration values, respectively.

In addition, Ag NPs interact with HIV-1 depending on their size (Elechiguerra et al. 2005). Particle size from 1 to 10 nm easily interacts with HIV-1 virus via preferential binding to gp120 glycoprotein knobs. This interaction inhibits the virus from binding to host cells, as shown in an in vitro investigation (Elechiguerra et al. 2005). So, these particles hopefully find an application in preventing and controlling HIV infection. They also show antiviral activity against herpes simplex virus type 1 (HSV-1), influenza virus, and hepatitis B virus (Baram-Pinto et al. 2009; Gaikwad et al. 2013; Lu et al. 2008; Xiang et al. 2011) and have been used in wound dressings due to their antimicrobial activity (Liu et al. 2010; Kwan et al. 2011). Becker (1999) found them useful in topical ointments and creams for preventing infection of burns and open wounds. Thus, Ag NPs are believed to promote wound healing in a dose-dependent manner (Tian et al. 2007).

The biosynthesized Ag NPs have shown anticancer properties due to structurally diverse chemical constituents present in them. They have exhibited potent cytotoxic effects against various cancer cell lines. With increase in the concentration of NPs, cell viability decreased. This is explained on the basis of the active physiochemical interaction of silver ions with the functional groups of intracellular proteins, along with nitrogenous bases and phosphate groups as well (Varghese et al. 2015; Mata et al. 2015; Nalavothula et al. 2015; Sreekanth et al. 2016). Manikandan et al. (2015) have demonstrated that Ag NPs fabricated from *Rosa indica* have anticancer and anti-inflammatory properties. Those obtained from *Piper longum* and *Melia azedarch* were also shown to have a cytotoxic effect on HEp-2 (Jacob et al. 2012) and HeLa cell lines (Sukirtha et al. 2012). Silver ions can produce reactive oxygen species (ROS) which may cause apoptosis or necrosis in human cancer cells through nuclear condensation and fragmentation.

Valodkar et al. (2011) tested the toxicity of plant-latex-capped Ag NPs toward human lung carcinoma cells in vitro and suggested that these particles are toxic to A549 cells in a concentration-dependent manner. They also suggested that plant latex can solubilize Ag NPs in water and act as a potential biocompatible vehicle for transport of Ag NPs to target tumor cells. Nevertheless, a detailed study of biocompatibility of these plant-latex-mediated NPs is still awaited. Moreover, Ag NPs have been examined to cure other diseases also including leukemia (Guo et al. 2013, 2014), hepatocellular carcinoma (Kim et al. 2009; Sahu et al. 2014; Faedmaleki et al. 2014), breast cancer (Franco-Molina et al. 2010; Gurunathan et al. 2013a, b), lung cancer (Foldbjerg et al. 2011), and skin and/or oral carcinoma (Austin et al. 2011). In an experiment, Jeyaraj et al. (2013) have claimed that Ag NPs fabricated from *Podophyllum hexandrum* have a significantly enhanced anticancer activity in comparison to Cisplatin, the standard anticancer drug (Jeyaraj et al. 2013). *Dimocarpus longan* peel's aqueous extract was used to produce Ag NPs (He et al. 2016), which showed cytotoxicity in a dose-dependent manner against prostate cancer (PC-3) cells through a decrease of stat 3, bcl-2, and survivin, as well as an increase in caspase-3. The authors have suggested that these particles could be used for prostate cancer treatments, although a comprehensive investigation to understand the molecular mechanism and in vivo effects of Ag NPs on prostate cancer is still required. Ag NPs obtained with the help of many other plants such as *Allium sativum* (Ahamed et al. 2011), *Annona squamosa* (Vivek et al. 2012), *Citrullus colocynthis* (Satyavani et al. 2011a, b), *Piper longum* (Reddy et al. 2014), *Melia dubia* (Kathiravan et al. 2014), *Mentha arvensis* (Banerjee et al. 2017), and *Pimpinella anisum* (Devanesan et al. 2017), among others, have been examined against different human cancer cell lines.

5.4.3 Antioxidant Potential and Catalytic Activity

The antioxidant property of synthesized Ag NPs is determined by their free radical-scavenging capacity. The better scavenging activity of green synthesized Ag NPs is due to capping agents, the phytochemicals present in the plant extract. One of the tools used for this purpose is the DPPH (1,1-diphenyl-l-2-picrylhydrazyl) radical-scavenging assay. The hydrogen-donating activity of antioxidants that converts the color of DPPH from violet to yellow by making it a stable diamagnetic molecule is the measure of antioxidant property (Priya et al. 2015; Rajan et al. 2015; Arunachalam et al. 2015; Parveen et al. 2016; Srinithya et al. 2016; Sriranjani et al. 2016; Lateef et al. 2016; Kumar et al. 2016). The biosynthesized Ag NPs have been very effective in preventing oxidative stress and therefore act as a powerful antioxidant agent. *Artocarpus altilis* aqueous leaf extract was used for Ag NP fabrication, and the product showed a moderate antimicrobial and antioxidant property (Ravichandran et al. 2016). Parveen et al. (2016) have reported a microwave-assisted Ag NP fabrication from *Fraxinus excelsior* leaf extract. The NPs obtained have shown a promising ability to diffuse the toxic free radicals and hence can be used as a food additive or in the nutraceutical and biopharmaceutical industries. Furthermore, silver NPs have been used for catalytic activity in a number of partial oxidation reactions, for instance, the oxidation of methanol to formaldehyde and ethylene to ethylene oxide. Ag NPs obtained from *Terminalia chebula* fruit extract (Edison and Sethuraman 2012) and from *Saraca indica* flower extract (Vidhu and Philip 2014) also revealed an effective catalytic activity in the reduction and removal of methylene blue, a cationic dye. The aqueous extract of *Dalbergia spinosa leaves was used for the production of Ag NPs (*Muniyappan and Nagarajan 2014), which *were* effective in the catalytic reduction of 4-nitrophenol into 4-aminophenol. *Salvadora persica* aqueous stem extract was also used for fabrication of Ag NPs, which accomplished 96% photodegradation activity against methylene blue in 80 min (Tahir et al. 2015). Kumar et al. (2016) have claimed a zero-cost approach for producing stable and spherical Ag NPs from the aqueous extract of *Erigeron bonariensis* which acted as both reducing and stabilizing agent in the presence of direct sunlight and exhibited catalytic activity toward the degradation of acridine orange without involving any hazardous reducing agent. In addition, Edison et al. (2016b) have used a marine green alga (*Caulerpa racemosa*) for eco-friendly synthesis of Ag NPs, which showed excellent catalytic activity toward the degradation of methylene blue.

5.4.4 Agricultural Application

To get the maximum advantage of nanotechnological applications in agriculture, the first important point is to examine the biological influence of NPs on plant growth and developmental processes. In general, most of the investigations have been focused on the NPs uptake and accumulation, biodistribution, and phenotypic responses such as root/shoot length, biomass, and seed germination and DNA damage (Rico et al. 2011; Yin et al. 2011; Atha et al. 2012; Husen and Siddiqi 2014a, b; Baskar et al. 2015). Dimpka et al. (2013) have reported that Ag NPs damage root cells, impair plant growth and seed germination, and affect leaf transpiration, root elongation, and plant biomass of *Triticum aestivum*. The dual toxicity of Ag NPs is associated with the Ag^+ ions released from Ag NPs and with the size of Ag NPs. For example, the ROS-mediated oxidative burst and inhibition of respiratory enzyme activity caused by the dissolved ionic Ag^+ released from Ag NPs have been reported by Kim et al. (2009). The size of NPs also affects toxicity; the smaller Ag NPs (<5 nm) caused an increased toxicity compared to the dissolved silver or the large-sized Ag NPs at the same concentration (Choi and Hu 2008). However, many other factors, viz., plant species, age, mode of treatment, concentration of NPs, and duration of treatment, are also important in determining the impact of Ag NPs in plants (Husen and Siddiqi 2014b; Siddiqi and Husen 2017b; Baskar et al. 2015; Vinković et al. 2017).

Seed germination percentage and shoot length declined by 10 mg L^{-1} Ag NPs in *Hordeum vulgare* and *Linum usitatissimum* (El-Temsah and Jone 2010), whereas seed germination was not affected in *Cucumis sativus* and *Lactuca sativa* even at 100 mg L^{-1} (Barrena et al. 2009). The positive effects of Ag NPs on growth activities have been observed in various species like *Brassica juncea*, *Panicum virgatum*, *Phytolacca americana*, *Phaseolus vulgaris*, and *Zea mays* (Sharma et al. 2012; Yin et al. 2012; Salama 2012). Baskar et al. (2015) found the biologically synthesized Ag NPs to act as a growth promoter at a lower dose (100 mg L^{-1}) and as a growth inhibitor at a higher dose (500 mg L^{-1}) in seedlings of *Brassica rapa* ssp. *pekinensis*. Bioaccumulation of Ag NPs in seedlings occurred in a dose-dependent manner. Similarly, the increased intracellular ROS generation and MDA accumulation were more pronounced at higher doses (250 and 500 mg L^{-1}) of Ag NP treatments. DNA damage, anthocyanin production, and antioxidant gene expression were also elevated at higher doses. In addition, Baskar et al. (2015) claimed that most of the glucosinolate biosynthesis and regulatory-related genes were induced at higher doses of Ag NPs. Thus, in general the higher dose of Ag NP treatments leads to phytotoxicity by suppressing the growth-related parameters, as these were also associated with the molecular and physiological deviations (Thuesombat et al. 2014; Baskar et al. 2015; Cvjetko et al. 2017; Rui et al. 2017). However, Qian et al. (2013) observed that 2-week exposure of Ag NPs was toxic to *Arabidopsis thaliana* seedlings; the seedling growth was significantly inhibited even at very low (0.5 and 3 mg L^{-1}) concentrations. Moreover, Ag NPs resulted in varying degrees of toxicity on algal growth, photosynthesis, antioxidant systems, and carbohydrate metabolism

(Huang et al. 2016; Qian et al. 2016). Comprehensive research needs to be taken up on various issues such as NP interaction with biological systems and their toxicity in the natural environment and the potential risks to ecosystems.

5.5 Conclusion

The use of environment-friendly materials, such as plants, for the fabrication of Ag NPs has shown several benefits. Plant-mediated fabrication of Ag NPs is cost-effective and can be utilized as an appropriate alternative route for the large-scale NP production. The principal and active biomolecules present in plant extracts are responsible for the reduction of metal ions to produce NPs. However, due to huge diversity and complex nature of plants, it is important to identify the specific compound that is involved in NP fabrication. Biomolecules, namely, polysaccharides, proteins, terpenoids, flavonoids, and alkaloids, are often considered as the potential bioreducing and stabilizing agents that facilitate NP fabrication. It is also reported that two active compounds belonging to different groups were responsible for the conversion of silver salts into nano-sized silver particles. In this situation, perhaps one group of compounds reduces the metal ions, while the other group acts as capping agent for the NPs synthesized. The bioreduction process of metal ions may possibly stem from interchange of numerous active components available in the plant extract or biomass. Further, these fabrication processes are influenced by several variables, like incubation time, temperature of the plant extract or biomass, its concentration, and pH value, which are difficult to control. In the recent past, due to their unique properties, Ag NPs have been incorporated into antimicrobial applications, biosensor materials, composite fibers, cryogenic superconducting materials, cosmetic products, electronic components, medical imaging appliances, drug delivery and hyperthermia of tumors, biolabeling, sensors, filters, food storage, textile coatings, agriculture, and a number of environmental issues. At present, a better understanding of the mechanism of plant-mediated fabrication of Ag NPs and their application to the cutting-edge areas of technology constitutes a very promising field for future research.

References

Abasi E, Milani M, Aval SF, Kouhi M, Akbarzadeh A, Nasrabadi HT, Nikasa P, Joo SW, Hanifehpour Y, Nejati-Koshki K, Samiei M (2016) Silver nanoparticles: synthesis methods, bio-applications and properties. Cri Rev Microbiol 42:173–180

Ahamed M, Majeed Khan MA, Siddiqui MKJ, AlSalhi MS, Alrokayan SA (2011) Green synthesis: characterization and evaluation of biocompatibility silver nanoparticles. Physica E 43:1266–1271

Ahmad A, Wei Y, Syed F, Khan S, Khan GM, Tahir K, Khan AU, Raza M, Khan FU, Yuan Q (2016) *Isatis tinctoria* mediated synthesis of amphotericin B-bound silver nanoparticles with enhanced photoinduced antileishmanial activity: a novel green approach. J Photoch Photobio B 161:17–24

Ahmed MJ, Murtaza G, Mehmood A, Bhatti TM (2015) Green synthesis of silver nanoparticles using leaves extract of *Skimmia laureola*: characterization and antibacterial activity. Mater Lett 153:10–13

Ahmed S, Saifullah AM, Swami BL, Ikram S (2016) Green synthesis of silver nanoparticles using *Azadirachta indica* aqueous leaf extract. J Radiat Res Appl Sci 9:1–7

Ajitha B, Reddy YAK, Reddy PS (2015) Biosynthesis of silver nanoparticles using *Momordica charantia* leaf broth: evaluation of their innate antimicrobial and catalytic activities. J Photoch Photobio B 146:1–9

Ali M, Kim B, Belfield KD, Norman D, Brennan M, Ali GS (2016) Green synthesis and characterization of silver nanoparticles using *Artemisia absinthium* aqueous extract- a comprehensive study. Mat Sci Eng C 58:359–365

Ali SG, Khan HM, Jalal M, Ansari MA, Mahdi AA, Ahmad MK (2015) Green synthesis of silver nanoparticles using the leaf extract of *Putranjiva roxburghii wall.* and their antimicrobial activity. Asian J Pharm Clinical Res 8:335–338

Alishah H, Seyedi SP, Ebrahimipour SY, Esmaeili-Mahani S (2016) A green approach for silver nanoparticles using root extract of *Chelidonium majus*: characterization and antibacterial evaluation. J Cluster Sci 27:421–429

Amin M, Anwar F, Janjua MRSA, Iqbal MA, Rashid U (2012) Green synthesis of silver nanoparticles through reduction with *Solanum xanthocarpum* L. berry extract: characterization, antimicrobial and urease inhibitory activities against *Helicobacter pylori*. Int J Mol Sci 13:9923–9941

Amooaghaie R, Saeri MR, Azizi M (2015) Synthesis, characterization and biocompatibility of silver nanoparticles synthesized from *Nigella sativa* leaf extract in comparison with chemical silver nanoparticles. Ecotoxicol Environ Saf 120:400–408

Andreescu D, Eastman C, Balantrapu K, Goia DVA (2007) Simple route for manufacturing highly dispersed silver nanoparticles. J Mater Res 22:2488–2496

Arunachalam K, Shanmuganathan B, Sreeja PS, Parimelazhagan T (2015) Phytosynthesis of silver nanoparticles using the leaves extract of *Ficus talboti* king and evaluation of antioxidant and antibacterial activities. Environ Sci Pollut Res 22:18066–18075

Atha DH, Wang H, Petersen EJ, Cleveland D, Holbrook RD, Jaruga P, Dizdaroglu M, Xing B, Nelson BC (2012) Copper oxide nanoparticle mediated DNA damage in terrestrial plant models. Environ Sci Technol 46:1819–1827

Austin LA, Kang B, Yen CW, El-Sayed MA (2011) Nuclear targeted silver nanospheres perturb the cancer cell cycle differently than those of nanogold. Bioconjugate Chem 22:2324–2331

Awad MA, Mekhamer WK, Merghani NM, Hendi AA, Ortashi KMO, Al-Abbas F, Eisa NE (2015) Green synthesis, characterization and anti-bacterial activity of silver/polystyrene nanocomposite. J Nanomater 943821:6

Bahrami-Teimoori B, Nikparast Y, Hojatianfar M, Akhlaghi M, Ghorbani R, Pourianfar HR (2017) Characterisation and antifungal activity of silver nanoparticles biologically synthesised by *Amaranthus retroflexus* leaf extract. J Exp Nanosci 12:129–139

Balamanikandan T, Balaji S, Pandirajan J (2015) Biological Synthesis of silver nanoparticles by using onion (*Allium cepa)* extract and their antibacterial and antifungal activity. World App Sci J 33:939–943

Balashanmugam P, Balakumaran MD, Murugan R, Dhanapal K, Kalaichelvan PT (2016) Phytogenic synthesis of silver nanoparticles, optimization and evaluation of *in vitro* antifungal activity against human and plant pathogens. Microbiol Res 192:52–64

Banerjee PP, Bandyopadhyay A, Harsha SN, Policegoudra RS, Bhattacharya S, Karak N, Chattopadhyay A (2017) *Mentha arvensis* (Linn.)-mediated green silver nanoparticles trigger caspase 9-dependent cell death in MCF7 and MDA-MB-231 cells. Breast Cancer 9:265–278

Baram-Pinto D, Shukla S, Perkas N, Gedanken A, Sarid R (2009) Inhibition of herpes simplex virus type 1 infection by silver nanoparticles capped with mercaptoethane sulfonate. Bioconjugate Chem 20:1497–1502

Barbinta-Patrascu ME, Badea N, Ungureanu C, Constantin M, Pirvu C, Rau I (2016) Silver-based biohybrids "green" synthesized from *Chelidonium majus* L. Opt Mater 56:94–99
Barrena R, Casals E, Colón J, Font X, Sánchez A, Puntes V (2009) Evaluation of the ecotoxicity of model nanoparticles. Chemosphere 75:850–857
Baskar V, Venkatesh J, Park SW (2015) Impact of biologically synthesized silver nanoparticles on the growth and physiological responses in *Brassica rapa* ssp. *pekinensis*. Environ Sci Pollut Res 22:17672–11768
Becker RO (1999) Silver ions in the treatment of local infections. Met Based Drugs 6:297–300
Berne BJ, Pecora R (2000) Dynamic light scattering: with applications to chemistry, biology, and physics. Dover, New York
Bhakya S, Muthukrishnan S, Sukumaran M, Muthukumar M (2016) Biogenic synthesis of silver nanoparticles and their antioxidant and antibacterial activity. Appl Nanosci 6:755–766
Bogireddy NKR, Kumar HAK, Badal Kumar Mandal BM (2016) Biofabricated silver nanoparticles as green catalyst in the degradation of different textile dyes. J Environ Chem Eng 4:56–64
Choi O, Hu ZQ (2008) Size dependent and reactive oxygen species related nanosilver toxicity to nitrifying bacteria. Environ Sci Technol 42:4583–4588
Chowdhury IH, Ghosh S, Roy M, Naskar MK (2015) Green synthesis of water-dispersible silver nanoparticles at room temperature using green carambola (star fruit) extract. J Sol-Gel Sci Technol 73:199–207
Chung MIII, Park I, Seung-Hyun K, Thiruvengadam M, Rajakumar G (2016) Plant-mediated synthesis of silver nanoparticles: their characteristic properties and therapeutic applications. Nano Res Lett 11:40
Cicek S, Gungor AA, Adiguzel A, Nadaroglu H (2015) Biochemical evaluation and green synthesis of nano silver using peroxidase from Euphorbia (*Euphorbia amygdaloides*) and its antibacterial activity. J Chem 486948:7 https://doi.org/10.1155/2015/486948
Cinelli M, Coles SR, Nadagouda MN, Btaszczyński J, Słowiński R, Varma RS, Kirman K (2015) A green chemistry-based classification model for the synthesis of silver nanoparticles. Green Chem 17:2825–2839
Cruz D, Falé PL, Mourato A, Vaz PD, Luisa Serralheiro M, Lino ARL (2010) Preparation and physicochemical characterization of Ag nanoparticles biosynthesized by *Lippia citriodora* (*Lemon verbena*). Colloids Surf B 81:67–73
Cvjetko P, Milošić A, Domijan AM, Vrček IV, Tolić S, Štefanić PP, Letofsky-Papst I, Tkalec M, Balen B (2017) Toxicity of silver ions and differently coated silver nanoparticles in *Allium cepa* roots. Ecotoxicol Environ Saf 137:18–28
de Aragao AP, de Oliveira TM, Quelemes PV, Perfeito MLG, Araujo MC, de Araujo Sousa Santiago J, Cardoso VS, Quaresma P, de Souza de Almeida Leite JR, da Silva DA (2016) Green synthesis of silver nanoparticles using the seaweed *Gracilaria birdiae* and their antibacterial activity. Arabian J Chem. https://doi.org/10.1016/j.arabjc.2016.04.014.
Das J, Paul Das M, Velusamy P (2013) *Sesbania grandiflora* leaf extract mediated green synthesis of antibacterial silver nanoparticles against selected human pathogens. Spectrochim. Acta A Mol Biomol Spectrosc 104:265–270
Deepak P, Sowmiya R, Ramkumar R, Balasubramani G, Aiswarya D, Perumal P (2017) Structural characterization and evaluation of mosquito-larvicidal property of silver nanoparticles synthesized from the seaweed, *Turbinaria ornata* (Turner) J. Agardh 1848. Artif Cells Nanomed Biotechnol 45:990–998
Devanesan S, AlSalhi MS, Vishnubalaji R, Akram A, Alfuraydi AA, Alajez NM, Alfayez M, Murugan K, Sayed SRM, Nicoletti M, Benelli G (2017) Rapid biological synthesis of silver nanoparticles using plant seed extracts and their cytotoxicity on colorectal cancer cell lines. J Clust Sci 28:595–605
Dieringer JA, McFarland AD, Shah NC, Stuart DA, Whitney AV, Yonzon CR, Young MA, Zhang X, Van Duyne RP (2006) Surface enhanced Raman spectroscopy: new materials, concepts, characterization tools, and applications. Faraday Discuss 132:9–26

Dimpka CO, McLean JE, Martineau N, Britt DW, Haverkamp R, Anderson AJ (2013) Silver nanoparticles disrupt wheat (*Triticum aestivum* L.) growth in a sand matrix. Environ Sci Technol 47:1082–1090

Dipankar C, Murugan S (2012) The green synthesis, characterization, and evaluation of the biological activities of silver nanoparticles synthesized from *Iresine herbstii* leaf aqueous extracts. Colloids Surf B Biointerfaces 98:112–119

Dubey SP, Lahtinen M, Sillanpaa M (2010) Tansy fruit mediated greener synthesis of silver and gold nanoparticles. Process Biochem 45:1065–1071

Dwivedi AD, Gopal K (2010) Biosynthesis of silver and gold nanoparticles using *Chenopodium album* leaf extract. Colloids Surf A: Physiol Eng Aspect 369:27–33

Edison TNJI, Apchudan R, Lee YR (2016a) Optical sensor for dissolved ammonia through the green synthesis of silver nanoparticles by fruit extract of *Terminala chebula*. J Cluster Sci 27:683–690

Edison TNJI, Atchudan R, Kamal C, Lee YR (2016b) *Caulerpa racemosa*: a marine green alga for eco-friendly synthesis of silver nanoparticles and its catalytic degradation of methylene blue. Bioproc Biosys Eng 39:1401–1408

Edison TNJI, Atchudan R, Sethuraman MG, Lee YR (2016c) Reductive-degradation of carcinogenic azo dyes using *Anacardium occidentale* testa derived silver nanoparticles. J Photochem Photobiol B 162:604–610

Edison TJI, Sethuraman MG (2012) Instant green synthesis of silver nanoparticles using *Terminalia chebula* fruit extract and evaluation of their catalytic activity on reduction of methylene blue. Proc Biochem 47:1351–1357

Elangovan K, Elumalai D, Anupriya S, Shenbhagaraman R, Kaleena PK, Murugesan K (2015) Phyto mediated biogenic synthesis of silver nanoparticles using leaf extract of *Andrographis echioides* and its bio-efficacy on anticancer and antibacterial activities. J Photochem Photobiol B 151:118–124

Elechiguerra JL, Burt JL, Morones JR, Camacho-Bragado A, Gao X, Lara HH, Yacaman MJ (2005) Interaction of silver nanoparticles with HIV-1. J Nanobiotechnol 3:6

El-Sherbiny IM, El-Shibiny A, Salih E (2016) Photo-induced green synthesized and anti-microbial efficacy of poly (Ɛ- caprolactone)/ curcumin/ grape leaf extract- silver hybrid nanoparticles. J Photochem Photobiol B 160:355–363

El-Temsah YS, Jone EJ (2010) Impact of Fe and Ag nanoparticles on seed germination and differences in bioavailability during exposure in aqueous suspension and soil. Environ Toxicol 27:42–49

Emmanuel R, Palanisamy S, Chen SM, Chelladurai K, Padmavathy S, Saravanan M, Prakash P, Ali MA, Al-Hemaid FMA (2015) Antimicrobial efficacy of green synthesized drug blended silver nanoparticles against dental caries and periodontal disease causing microorganisms. Mater Sci Eng C 56:374–379

Faedmaleki F, H Shirazi F, Salarian AA, Ahmadi Ashtiani H, Rastegar H (2014) Toxicity effect of silver nanoparticles on mice liver primary cell culture and HepG2 cell line. Iran J Pharm Res 13:235–242

Fayaz AM, Balaji K, Kalaichelvan PT, Venkatesan R (2009) Fungal based synthesis of silver nanoparticles-an effect of temperature on the size of particles. Colloids Surf B 74:123–126

Foldbjerg R, Dang DA, Autrup H (2011) Cytotoxicity and genotoxicity of silver nanoparticles in the human lung cancer cell line, A549. Arch Toxicol 85:743–750

Franco-Molina MA, Mendoza-Gamboa E, Sierra-Rivera CA, Gómez-Flores RA, Zapata-Benavides P, Castillo-Tello P, Alcocer-González JM, Miranda-Hernández DF, Tamez-Guerra RS, Rodríguez-Padilla C (2010) Antitumor activity of colloidal silver on MCF-7 human breast cancer cells. J Exp Clin Cancer Res 29:148–154

Gaikwad S, Ingle A, Gade A, Rai M, Falanga A, Incoronato N, Russo L, Galdiero S, Galdiero M (2013) Antiviral activity of mycosynthesized silver nanoparticles against herpes simplex virus and human parainfluenza virus type 3. Int J Nanomed 8:4303–4314

Ghosh S, Patil S, Ahire M, Kitture R, Kale S, Pardesi K, Cameotra SS, Bellare J, Dhavale DD, Jabgunde A, Chopade BA (2012) Synthesis of silver nanoparticles using *Dioscorea bulbifera* tuber extract and evaluation of its synergistic potential in combination with antimicrobial agents. Int J Nanomed 7:483

Gogoi N, Babu PJ, Mahanta C, Bora U (2015) Green synthesis and characterization of silver nanoparticles using alcoholic flower extract of *Nyctanthes arbor-tristis* and in vitro investigation of their antibacterial and cytotoxic activities. Mater Sci Eng C Mater Biol Appl 46:463–469

Gopinath K, Devi NP, Govindarajan M, Bhakyaraj K, Kumaraguru S, Arumugama A, Alharbi NS, Kadaikunnan S, Benelli G (2017) One-Pot green synthesis of silver nanoparticles using the orchid leaf extracts of *Anoectochilus elatus*: growth inhibition activity on seven microbial pathogens. J Clust Sci 28:1541–1550

Gopinath V, Mubarak Ali D, Priyadarshini S, Meera Priyadharsshini N, Thajuddin N, Velusamy P (2012) Biosynthesis of silver nanoparticles from *Tribulus terrestris* and its antimicrobial activity: a novel biological approach. Coll Surf B Biointerf 96:69–74

Govindarajan M, Rajeswary M, Veerakumar K, Muthukumaran U, Hoti SL, Benelli G (2016a) Green synthesis and characterization of silver nanoparticles fabricated using *Anisomeles indica*: mosquitocidal potential against malaria, dengue and Japanese encephalitis vectors. Exp Parasitol 161:40–47

Govindarajan M, Rajeswary M, Veerakumar K, Muthukumaran U, Hoti SL, Mehlhorn H, Barnard DR, Benell G (2016b) Novel synthesis of silver nanoparticles using *Bauhinia variegata*: a recent eco-friendly approach for mosquito control. Parasitol Res 115:723–733

Guo D, Zhao Y, Zhang Y, Wang Q, Huang Z, Ding Q, Guo Z, Zhou X, Zhu L, Gu N (2014) The cellular uptake and cytotoxic effect of silver nanoparticles on chronic myeloid leukemia cells. J Biomed Nanotechnol 10:669–678

Guo D, Zhu L, Huang Z, Zhou H, Ge Y, Ma W, Wu J, Zhang X, Zhou X, Zhang Y, Zhao Y, Gu N (2013) Anti-leukemia activity of PVP-coated silver nanoparticles via generation of reactive oxygen species and release of silver ions. Biomaterials 34:7884–7894

Gurunathan S, Han JW, Eppakayala V, Jeyaraj M, Kim JH (2013a) Cytotoxicity of biologically synthesized silver nanoparticles in MDA-MB-231 human breast cancer cells. Biomed Res Int 2013:535796–535805

Gurunathan S, Raman J, Abd Malek SN, John PA, Vikineswary S (2013b) Green synthesis of silver nanoparticles using *Ganoderma neo-japonicum Imazeki*: a potential cytotoxic agent against breast cancer cells. Int J Nanomed 8:4399–4413

He Y, Du Z, Ma S, Cheng S, Jiang S, Liu Y, Li D, Huang H, Zhang K, Zheng X (2016) Biosynthesis, antibacterial activity and anticancer effects against prostate cancer (PC-3) cells of silver nanoparticles using *Dimocarpus longan* Lour. peel extract. Nano Res Lett 11:300

Huang T, Sui M, Yan X, Zhang X, Yuan Z (2016) Anti-algae efficacy of silver nanoparticles to *Microcystis aeruginosa*: influence of NOM, divalent cations, and pH. Coll Surf A: Physicochem Eng Aspect 509:492–503

Husen A (2017) Gold nanoparticles from plant system: synthesis, characterization and their application. In: Ghorbanpourn M, Manika K, Varma A (eds) Nanoscience and plant–soil systems, vol 48. Springer, Cham, pp 455–479

Husen A, Siddiqi KS (2014a) Carbon and fullerene nanomaterials in plant system. J Nanobiotechnol 12:16

Husen A, Siddiqi KS (2014b) Phytosynthesis of nanoparticles: concept, controversy and application. Nano Res Lett 9:229

Husen A, Siddiqi KS (2014c) Plants and microbes assisted selenium nanoparticles: characterization and application. J Nanobiotechnol 12:28

Ibrahim HMM (2016) Green synthesis and characterization of silver nanoparticles using banana peel extract and their antimicrobial activity against representative microorganisms. J Radiat Res Appl Sci 8:265–275

Iravani S (2011) Green synthesis of metal nanoparticles using plants. Green Chem 13:2638–2650

Jacob SJP, Finub JS, Narayanan A (2012) Synthesis of silver nanoparticles using Piper longum leaf extracts and its cytotoxic activity against Hep-2 cell line. Coll Surf B: Biointerf 91:212–214

Jadhav K, Dhamecha D, Bhattacharya D, Patil M (2016) Green and ecofriendly synthesis of silver nanoparticles: characterization, biocompatibility studies and gel formulation for treatment of infections in burns. J Photochem Photobiol B 155:109–115

Jadhav K, Dhamecha D, Dalvi B, Patil M (2015) Green synthesis of silver nanoparticles using *Salacia chinensis*: characterization and its antibacterial activity. Particul Sci Technol 33:445–455

Jeyaraj M, Rajesh M, Arun R, Mubarak Ali D, Sathishkumar G, Sivanandhan G, Kapildev G, Manickavasagam M, Premkumar K, Thajuddin N, Ganapathi A (2013) An investigation on the cytotoxicity and caspase-mediated apoptotic effect of biologically synthesized silver nanoparticles using *Podophyllum hexandrum* on human cervical carcinoma cells. Coll Surf B: Biointerf 102:708–717

Johal MS (2011) Understanding nanomaterials. CRC Press, Boca Raton

Kamala-Kannan S, Manoharan K, Thiyagarajan P, Govarthanan M, Kim J (2016) Green synthesis of silver nanoparticles using *Solanum indicum* L. and their antibacterial, splenocyte cytotoxic potentials. Res Chem Intermediat 42:3095–3103

Kasthuri J, Kathiravan K, Rajendiran N (2009a) Phyllanthin assisted biosynthesis of silver and gold nanoparticles: a novel biological approach. J Nanopart Res 11:1075–1085

Kasthuri J, Veerapandian S, Rajendiran N (2009b) Biological and synthesis of silver and gold nanoparticles using apiin as reducing agent. Colloids Surf B Biointerfaces 68:55–60

Karatoprak GS, Aydin G, Altinsoy B, Altinkaynak C, Kos M, Ocsoy I (2017) The effect of *Pelargonium endlicherianum* fenzl. root extracts on formation of nanoparticles and their antimicrobial activities. Enzy Micro Technolo 97:21–26

Karthik R, Hou Y, Chen S, Elangovan A, Ganesan M (2016) Eco-friendly synthesis of Ag-NPs using *Cerasus serrulata* plant extract – Its catalytic, electrochemical reduction of 4-NPh and antibacterial activity. J Ind Eng Chem 37:330–339

Kathiravan V, Ravi S, Kumar SA (2014) Synthesis of silver nanoparticles from *Melia dubia* leaf extract and their in vitro anticancer activity. Spectro Acta Part A: Mol Biomole Spectro 130:116–121

Khan M, Khan M, Adil SF, Tahir MN, Tremel W, Alkhathlan HZ, Al-Warthan A, Siddiqui MRH (2013) Green synthesis of silver nanoparticles mediated by *Pulicaria glutinosa* extract. Int J Nanomed 8:1507–1516

Kim DY, Saratale RG, Shinde S, Syed A, Ameen F, Ghodakea G (2018) Green synthesis of silver nanoparticles using *Laminaria japonica* extract: characterization and seedling growth assessment. J Clean Produc 172:2910–2918

Kim S, Choi JE, Choi J, Chung KH, Park K, Yi J, Ryu DY (2009) Oxidative stress-dependent toxicity of silver nanoparticles in human hepatoma cells. Toxicol In Vitro 23:1076–1084

Knoll B, Keilmann F (1999) Near-field probing of vibrational absorption for chemical microscopy. Nature 399:134–137

Kokila T, Ramesh PS, Geetha D (2016) Biosynthesis of AgNPs using *Carica Papaya* peel extract and evaluation of its antioxidant and antimicrobial activities. Ecotoxicol Environ Saf 134:467–473

Kumar B, Smita K, Cumbal L, Debut A (2017a) Green synthesis of silver nanoparticles using Andean blackberry fruit extract. Saudi J Bio Sci 24:45–50

Kumar V, Singh DK, Mohan S, Gundampati RK, Hasan SH (2017b) Photoinduced green synthesis of silver nanoparticles using aqueous extract of *Physalis angulata* and its antibacterial and antioxidant activity. J Environ Chem Eng 5:744–756

Kumar V, Singh DK, Mohan S, Hasan SH (2015) Photo induced biosynthesis of silver nanoparticles using aqueous extract of *Erigeron bonariensis* and its catalytic activity against Acridine orange. J Photoch Photobio B 55:39–50

Kumar B, Smita K, Cumbal L, Debut A (2016) *Ficus carica* (Fig) Fruit mediated green synthesis of silver nanoparticles and its antioxidant activity: a comparison of thermal and ultrasonication approach. Bio Nano Sci 6:15–21

Kwan KH, Liu X, To MK, Yeung KW, Ho C-m, Wong KK (2011) Modulation of collagen alignment by silver nanoparticles results in better mechanical properties in wound healing. Nanomed Nanotechnol Biol Med 7:497–504

Lara HH, Garza-Trevino EN, Ixtepan-Turrent L, Singh DK (2011) Silver nanoparticles are broad-spectrum bactericidal and virucidal compounds. J Nanobiotechnol 9:30

Lateef A, Azeez MA, Asafa TB, Yakeen TA, Akinboro A, Oladipo LC, Azeez L, Ojo SA, Gueguim-Kana EB, Beukes LS (2016) Cocoa pod husk extract-mediated biosynthesis of silver nanoparticles: its antimicrobial.; antioxidant and larvicidal activities. J Nanostructure Chem 6:159–169

Lengke MF, Fleet ME, Southam G (2007) Biosynthesis of silver nanoparticles by filamentous cyanobacteria from a silver (I) nitrate complex. Langmuir 23:2691–2699

Li S, Shen Y, Xie A, Yu X, Qiu L, Zhang L, Zhang Q (2007) Green synthesis of silver nanoparticles using *Capsicum annuum* L. extract. Green Chem 9:852–885

Lin L, Wang W, Huang J, Li Q, Sun D, Yang X, Wang H, He N, Wang Y (2010) Nature factory of silver nanowires: plant mediated synthesis using broth of *Cassia fistula* leaf. Chem Eng J 162:852–858

Liu X, Lee PY, Ho CM, Lui VC, Chen Y, Che CM, Tam PK, Wong KK (2010) Silver nanoparticles mediate differential responses in keratinocytes and fibroblasts during skin wound healing. Chem Med Chem 5:468–475

Logaranjan K, Raiza AJ, Gopinath SCB, Chen Y, Pandian K (2016) Shape- and size-controlled synthesis of silver nanoparticles using *Aloe vera* plant extract and their antimicrobial activity. Nano Res Lett 11:520–528

Lok CN, Ho CM, Chen R, He QY, Yu WY, Sun H, Tam PK, Chiu JF, Che CM (2006) Proteomic analysis of the mode of antibacterial action of silver nanoparticles. J Proteome Res 5:916–924

Lokina S, Stephen A, Kaviyarasan V, Arulvasu C, Narayanan V (2014) Cytotoxicity and antimicrobial activities of green synthesized silver nanoparticles. Euro J Med Chem 76:256–263

Lu L, Sun RW, Chen R, Hui CK, Ho CM, Luk JM, Lau GK, Che CM (2008) Silver nanoparticles inhibit hepatitis B virus replication. Antivir Ther 13:253–262

Lu Z, Xiao J, Wang Y, Meng M (2015) In situ synthesis of silver nanoparticles uniformly distributed on polydopamine-coated silk fibres for antibacterial application. J Colloid Interf Sci 452:8–14

Lukman AI, Gong B, Marjo CE, Roessner U, Harris AT (2011) Facile synthesis, stabilization, and anti-bacterial performance of discrete Ag nanoparticles using *Medicago sativa* seed exudates. J Colloid Interface Sci 353:433–444

Maddinedi SB, Mandal BK, Maddili SK (2017) Biofabrication of size controllable silver nanoparticles – A green approach. J Photoch Photobio B 167:236–241

Mallmann EJJ, Cunha FA, Castro BNMF, Maciel AM, Menezes EA, Fechine PBA (2015) Antifungal Activity of silver nanoparticles obtained by green synthesis. Rev Inst Med Trop Sao Paulo 57:165–167

Manikandan R, Manikandan B, Raman T, Arunagirinathan K, Prabhu NM, Basu MJ, Perumal M, Palanisamy S, Munusamy A (2015) Biosynthesis of silver nanoparticles using ethanolic petals extract of *Rosa indica* and characterization of its antibacterial, anticancer and anti-inflammatory activities. Spectro Acta Part A Mol Biomol Spectro 138:120–129

Maria BS, Devadiga A, Kodialbail VS, Saidutta MB (2015) Synthesis of silver nanoparticles using medicinal *Zizyphus xylopyrus* bark extract. Appl Nanosci 5:755–762

Mata R, Nakkala JR, Sadras SR (2015) Catalytic and biological activities of green silver nanoparticles synthesized from *Plumeria alba* (frangipani) flower extract. Mater Sci Eng C 51:216–225

Mattea F, Vedelago J, Malano F, Gomez C, Strumia MC, Valent M (2017) Silver nanoparticles in X-ray biomedical applications. Rad Phy Chem 130:442–450

Mehmood A, Murtaza G, Bhatti TM, Kausar R (2017) Phyto-mediated synthesis of silver nanoparticles from *Melia azedarach* L. leaf extract: characterization and antibacterial activity. Arabian J Chem 10:3048–3053

Mittal AK, Chisti Y, Banerjee UC (2013) Synthesis of metallic nanoparticles using plant extracts. Biotechnol Adv 31:346–356

Mohan S, Oluwafemi OS, Songca SP, Jayachandran VP, Rouxel D, Joubert O, Kalarikkal N, Thomas S (2016) Synthesis, antibacterial, cytotoxicity and sensing properties of starch-capped silver nanoparticles. J Mol Liq 213:75–81

Mohapatra B, Kuriakose S, Mohapatra S (2015) Rapid green synthesis of silver nanoparticles and nanorods using *Piper nigrum* extract. J Alloy Compd 637:119–126

Muniyappan N, Nagarajan NS (2014) Green synthesis of silver nanoparticles with *Dalbergia spinosa* leaves and their applications in biological and catalytic activities. Proc Biochem 49:1054–1061

Murugan K, Labeeba MA, Panneerselvam C, Dinesh D, Suresh U, Subramaniam J, Madhiyazhagan P, Hwang J, Wang L, Nicoletti M, Benelli G (2015) *Aristolochia indica* green synthesized silver nanoparticles: a sustainable control tool against the malaria vector *Anopheles stephensi*. Res Vet Sci 102:127–135

Muthukumaran U, Govindarajan M, Rajeswary M (2015) Green synthesis of silver nanoparticles from *Cassia roxburghii* – a most potent power for mosquito control. Parasitol Res 114:4385–4395

Nalavothula R, Alwala J, Nagati VB, Manthurpadigya PR (2015) Biosynthesis of silver nanoparticles using *Impatiens balsamina* leaf extracts and its characterization and cytotoxic studies using human cell lines. Inter J Chem Tech Res 7:2460–2468

Palanisamy S, Rajasekar P, Vijayaprasath G, Ravi R, Manikandan R, Prabhu MN (2017) A green route to synthesis silver nanoparticles using *Sargassum polycystum* and its antioxidant and cytotoxic effects: an in vitro analysis. Mater Lett 189:196–200

Panneerselvam C, Murugan K, Roni M, Aziz AT, Suresh U, Rajaganesh R, Madhiyazhagan P, Subramaniam J, Dinesh D, Nicoletti M, Higuchi A, Alarfaj AA, Munusamy MA, Kumar S, Desneux N, Benelli G (2016) Fern-synthesized nanoparticles in the fight against malaria: LC/MS analysis of *Pteridium aquilinum* leaf extract and biosynthesis of silver nanoparticles with high mosquitocidal and antiplasmodial activity. Parasitol Res 115:997–1013

Parveen M, Ahmad F, Malla AM, Azaz S (2016) Microwave-assisted green synthesis of silver nanoparticles from *Fraxinus excelsior* leaf extract and its antioxidant assay. Appl Nanosci 6:267–276

Paul B, Bhuyan B, Purkayastha DD, Dhar SS (2016) Photoctalytic and antibacterial activities of gold and silver nanoparticles synthesized using biomass of *Parkia roxburghii* leaf. J Photochem Photobiol B 154:1–7

Phillip D (2009) Biosynthesis of Ag, Au and Au-Ag nanoparticles using edible mushrooms extracts. Spectrochim Acta Part A 73:374–381

Prakash P, Gnanaprakasam P, Emmanuel R, Arokiyaraj S, Saravanan M (2013) Green synthesis of silver nanoparticles from leaf extract of *Mimusops elengi* Linn. for enhanced antibacterial activity against multi drug resistant clinical isolates. Coll Surf B: Biointerf 108:255–259

Prathna TC, Chandrasekaran N, Raichur AM, Mukherjee A (2011) Biomimetic synthesis of silver nanoparticles by *Citrus limon* (lemon) aqueous extract and theoretical prediction of particle size. Coll Surf B: Biointerf 82:152–159

Priya RS, Geetha D, Ramesh PS (2015) Antioxidant activity of chemically synthesized AgNPs and biosynthesized *Pongamia pinnata* leaf extract mediated AgNPs- a comparative study. Ecotoxicol Environ Saf 134:308–318

Pugazhendhi S, Kirubha E, Palanisamy PK, Gopalakrishnan R (2015) Synthesis and characterization of silver nanoparticles from *Alpinia calcarata* by green approach and its applications in bactericidal and nonlinear optics. Appl Surf Sci 357:1801–1808

Qian H, Peng X, Han X, Ren J, Sun L, Fu Z (2013) Comparison of the toxicity of silver nanoparticles and silver ions on the growth of terrestrial plant model *Arabidopsis thaliana*. J Environ Sci 25:1947–1956

Qian H, Zhu K, Lu H, Lavoie M, Chen S, Zhou Z, Deng Z, Chen J, Fu Z (2016) Contrasting silver nanoparticle toxicity and detoxification strategies in *Microcystis aeruginosa* and *Chlorella vulgaris*: new insights from proteomic and physiological analyses. Sci Tot Environ 572:1213–1221

Raghunandan D, Borgaonkar PA, Bendegumble B, Bedre MD, Bhagawanraju M, Yalagatti MS, Huh DS, Abbaraju V (2011) Microwave-assisted rapid extracellular biosynthesis of silver nanoparticles using Carom seed (*Trachyspermum copticum*) extract and in vitro studies. Am J Anal Chem 2:475–483

Raja S, Ramesh V, Thivaharan V (2017) Green biosynthesis of silver nanoparticles using *Calliandra haematocephala* leaf extract, their antibacterial activity and hydrogen peroxide sensing capability. Arabian J Chem 10:253–261

Rajan A, Vilas V, Philip D (2015) Catalytic and antioxidant properties of biogenic silver nanoparticles synthesized using *Areca catechu* nut. J Mol Liq 207:231–236

Ramesh PS, Kokila T, Geetha D (2015) Plant mediated green synthesis and antibacterial activity of silver nanoparticles using *Emblica officinalis* fruit extract. Spectrochim Acta Mol Biomol Spectrosc 142:339–343

Rao B, Tang RC (2017) Green synthesis of silver nanoparticles with antibacterial activities using aqueous *Eriobotrya japonica* leaf extract. Adv Nat Sci Nanosci Nanotechnol 8:015014

Ravichandran V, Vasanthi S, Shalini S, Shah SAA, Harish R (2016) Green synthesis of silver nanoparticles using *Atrocarpus altilis* leaf extract and the study of their antimicrobial and antioxidant activity. Mat Lett 180:264–267

Reddy NJ, Vali DN, Rani M, Rani SS (2014) Evaluation of antioxidant, antibacterial and cytotoxic effects of green synthesized silver nanoparticles by *Piper longum* fruit. Mat Sci Eng: C 34:115–122

Rico CM, Majumdar S, Duarte-Gardea M, Peralta-Videa JR, Gardea-Torresdey JL (2011) Interaction of nanoparticles with edible plants and their possible implications in the food chain. J Agric Food Chem 59:3485–3498

Roy N, Mondal S, Laskar RA, Basu S, Mandal D, Begum NA (2010) Biogenic synthesis of Au and Ag nanoparticles by Indian propolis and its constituents. Colloids Surf B Biointerfaces 76:317–325

Rui M, Ma C, Tang X, Yang J, Jiang F, Pan Y, Xiang Z, Hao Y, Rui Y, Cao W, Xing B (2017) Phytotoxicity of silver nanoparticles to peanut (*Arachis hypogaea* L.): physiological responses and food safety. ACS Sus Chem Eng 5:6557–6567

Sahu SC, Zheng J, Graham L, Chen L, Ihrie J, Yourick JJ, Sprando RL (2014) Comparative cytotoxicity of nanosilver in human liver HepG2 and colon Caco2 cells in culture. J Appl Toxicol 34:1155–1166

Salama HMH (2012) Effects of silver nanoparticles in some crop plants, common bean (*Phaseolus vulgaris* L.) and corn (*Zea mays* L.). Int Res J Biotechnol 3:190–197

Sankar R, Karthik A, Prabu A, Karthik S, Shivashangari KS, Ravikumar V (2013) *Origanum vulgare* mediated biosynthesis of silver nanoparticles for its antibacterial and anticancer activity. Coll Surf B Biointerf 108:80–84

Sathishkumar M, Sneha K, Yun YS (2010) Immobilization of silver nanoparticles synthesized using *Curcuma longa* tuber powder and extract on cotton cloth for bactericidal activity. Bioresour Technol 101:7958–7965

Sathishkumar M, Sneha K, Won SW, Cho CW, Kim S, Yun YS (2009) Cinnamon zeylanicum bark extract and powder mediated green synthesis of nano-crystalline silver particles and its bactericidal activity. Colloids Surf B Biointerfaces 73:332–338

Sathyaseelan T, Subbiah M, Sivamugugan V (2015) Green synthesis of silver nano particles using marine brown *Alga Lobophora variegata* and its efficiency in antifungal activity. World J Pharma Res 4:2137–2145

Sattler KD (2010) Handbook of nanophysics: Nanoparticles and quantum dots. CRC Press (Taylor and Francis Group), London

Satyavani K, Gurudeeban S, Ramanathan T, Balasubramanian T (2011a) Biomedical potential of silver nanoparticles synthesized from calli cells of *Citrullus colocynthis* (L.) Schrad. J Nanobiotechnol 9:43–50

Satyavani K, Ramanathan T, Gurudeekan S (2011b) Green synthesis of silver nanoparticles using stem dried callus extract of bitter apple (*Citrullus colocynthis*). Dig J Nanomater Biostruct 6:1019–1024

Sengottaiyan A, Mythili R, Selvankumar T, Aravinthan A, Kamala-Kannan S, Manoharan K, Thiyagarajan P, Govarthanan M, Kim J (2016) Green synthesis of silver nanoparticles using *Solanum indicum* L. and their antibacterial, splenocyte cytotoxic potentials. Res Chem Intermed 42:3095–3103

Sengupta S, Eavarone D, Capila I, Zhao G, Watson N, Kiziltepe T, Sasisekharan R (2005) Temporal targeting of tumour cells and neovasculature with a nanoscale delivery system. Nature 436:568–572

Shaligram NS, Bule M, Bhambure R, Singhal RS, Singh SK, Szakacs G, Pandey A (2009) Biosynthesis of silver nanoparticles using aqueous extract from the compactin producing fungal strain. Process Biochem 44:939–943

Sharma P, Bhatt D, Zaidi MG, Saradhi PP, Khanna PK, Arora S (2012) Silver nanoparticle-mediated enhancement in growth and antioxidant status of *Brassica juncea*. Appl Biochem Biotechnol 167:2225–2233

Shivakumar M, Nagashree KL, Yallappa S, Manjappa S, Manjunath KS, Dharmaprakash MS (2017) Biosynthesis of silver nanoparticles using pre-hydrolysis liquor of Eucalyptus wood and its effective antimicrobial activity. Enzy Micro Techol 97:55–62

Siddiqi KS, Husen A (2016a) Fabrication of metal nanoparticles from fungi and metal salts: scope and application. Nano Res Lett 11:98

Siddiqi KS, Husen A (2016b) Fabrication of metal and metal oxide nanoparticles by algae and their toxic effects. Nano Res Lett 11:363

Siddiqi KS, Husen A (2016c) Green synthesis, characterization and uses of palladium/platinum nanoparticles. Nano Res Lett 11:482

Siddiqi KS, Husen A (2017a) Recent advances in plant-mediated engineered gold nanoparticles and their application in biological system. J Trace Elements Med Biol 40:10–23

Siddiqi KS, Husen A (2017b) Plant response to engineered metal oxide nanoparticles. Nano Res Lett 12:92

Siddiqi KS, Husen A, Rao RAK (2018a) A review on biosynthesis of silver nanoparticles and their biocidal properties. J Nanobiotechnol 16:14

Siddiqi KS, Rahman A, Tajuddin HA (2018b) Properties of zinc oxide nanoparticles and their activity against microbes. Nano Res Lett 13:141

Siddiqi KS, Husen A, Sohrab SS, Osman M (2018c) Recent status of nanomaterials fabrication and their potential applications in neurological disease management. Nano Res Lett 13:231

Siddiqi KS, Rashid M, Rahman A, Tajuddin, Husen A, Rehman S (2018d) Biogenic fabrication and characterization of silver nanoparticles using aqueous-ethanolic extract of lichen (*Usnea longissima*) and their antimicrobial activity. Biomat Res 22:23

Siddiqi KS, Rahman A, Tajuddin HA (2016) Biogenic fabrication of iron/iron oxide nanoparticles and their application. Nano Res Lett 11:498

Sigamioney M, Shaik S, Govender P, Krishna SBN, Sershen (2016) African leafy vegetables as bio-factories for silver nanoparticles: a case study on *Amaranthus dubius* C Mart. Ex Thell S Afr J Bot 103:230–240

Singh M, Kumar M, Kalaivani R, Manikandan S, Kumaraguru AK (2013) Metallic silver nanoparticle: a therapeutic agent in combination with antifungal drug against human fungal pathogen. Bioprocess Biosyst Eng 36:407–415

Song JY, Kim BS (2009) Biological synthesis of bimetallic Au/Ag nanoparticles using Persimmon (Diopyros kaki) leaf extract. Korean J Chem Eng 25:808–811

Sreekanth TVM, Jung M, Eom I (2015) Green synthesis of silver nanoparticles, decorated on graphene oxide nanosheets and their catalytic activity. Appl Surf Sci 361:102–106

Sreekanth TVM, Pandurangan M, Jung M, Lee YR, Eom I (2016) Eco-friendly decoration of graphene oxide with green synthesized silver nanoparticles: cytotoxic activity. Res chem Intermediat 42:5665–5676

Srinithya B, Kumar VV, Vadivel V, Pemaiah B, Anthony SP, Muthuraman MS (2016) Synthesis of biofunctionalized AgNPs using medicinally important *Sida cordifolia* leaf extract for enhanced antioxidant and anticancer activities. Mater Lett 170:101–104

Sriranjani R, Srinthiya B, Vadivel V, Pemaiah B, Anthony SP, Sivasubramanian A, Muthuraman MS (2016) Silver nanoparticles synthesis using *Clerodendrum phlomidis* leaf extract and preliminary investigation of its antioxidant and anticancer activity. J Mol Liq 220:926–930

Suarez-Cerda J, Alonso-Nunez G, Espinoza-Gomez H, Flores-Lopez LZ (2015) Synthesis, kinetics and photocatalytic study of "ultra-small" Ag-NPs obtained by a green chemistry method using an extract of Rosa 'Andeli' double delight petals. J Coll Interf Sci 458:169–177

Subramanian V, Suja S (2012) Green synthesis of silver nanoparticles using *Coleus amboinicus* Lour, antioxidant activity and in vitro cytotoxicity against Ehrlich's Ascite carcinoma. J Pharm Res 5:1268–1272

Sukirtha R, Priyanka KM, Antony JJ, Kamalakkannan S, Thangam R, Gunasekaran P, Krishnan M, Achiraman S (2012) Cytotoxic effect of green synthesized silver nanoparticles using *Melia azedarach* against in vitro HeLa cell lines and lymphoma mice model. Process Biochem 47:273–279

Suman TY, Rajasree SR, Jayaseelan C, Mary R, Gayathri S, Aranganathan L, Remya RR (2016) GC-MS analysis of bioactive components and biosynthesis of silver nanoparticles using *Hybanthus enneaspermus* at room temperature evaluation of their stability and its larvicidal activity. Environ Sci Pollut Res 23:2705–2714

Sun Q, Cai X, Li J, Zheng M, Chen Z, Yu CP (2014) Green synthesis of silver nanoparticles using tea leaf extract and evaluation of their stability and antibacterial activity. Coll Surf A Physicochem Eng Aspect 444:226–231

Sundararajan B, Mahendran G, Thamaraiselvi R, Kumari BDR (2016) Biological activities of synthesized silver nanoparticles from *Cardiospermum halicacabum* L. Bull Mater Sci 39:423–431

Swamy MK, Akhtar MS, Mohanty SK, Sinniah UR (2015) Synthesis and characterization of silver nano particles using fruit extract of *Momordica cymbalaria* and assessment of their in vitro antimicrobial, antioxidant and cytotoxicity activities. Spectrochim Acta Part A Mol Biomol Spectrosc 151:939–944

Tao A, Sinsermsuksaku P, Yang P (2006) Polyhedral silver nanocrystals with distinct scattering segnetures. Angew Chem Int Ed 45:4597–2601

Tahir K, Nazir S, Li B, Khan AU, Khan ZUH, Ahmad A, Khan FU (2015) An efficient photo catalytic activity of green synthesized silver nanoparticles using *Salvadora persica* stem extract. Separ Purifi Technol 150:316–324

Tareq FK, Fayzunnesa M, Kabir MS (2017) Antimicrobial activity of plant-median synthesized silver nanoparticles against food and agricultural pathogens. Microb Pathogen 109:228–232

Thuesombat P, Hannongbua S, Akasit S, Chadchawan S (2014) Effect of silver nanoparticles on rice (*Oryza sativa* L. cv. KDML 105) seed germination and seedling growth. Ecotoxicol Environ Saf 104:302–309

Tian J, Wong KK, Ho CM, Lok CN, Yu WY, Che CM, Chiu JF, Tam PK (2007) Topical delivery of silver nanoparticles promotes wound healing. Chem Med Chem 2:129–136

Tripathy A, Raichur AM, Chandrasekaran N, Prathna TC, Mukherjee A (2010) Process variables in biomimetic synthesis of silver nanoparticles by aqueous extract of *Azadirachta indica* (Neem) leaves. J Nanopart Res 12:237–246

Upadhyay LSB, Verma M (2014) Synthesis and characterization of cysteine functionalized silver nanoparticles for biomolecule immobilization. Bioprocess Biosyst Eng 37:2139–2148

Valodkar M, Jadeja RN, Thounaojam MC, Devkar RV, Thakore S (2011) In vitro toxicity study of plant latex capped silver nanoparticles in human lung carcinoma cells. Mater Sci Eng C 31:1723–1728

Varghese A, Anandhi P, Arunadevi R, Boovisha A, Sounthari P, Saranya J, Parameswari K, Chitra S (2015) Satin Leaf (*Chrysophyllum oliviforme*) extract mediated green synthesis of silver nanoparticles: antioxidant and anticancer activities. JPSR 7:266–273

Veerasamy R, Xin TZ, Gunasagaran S, Xiang TFW, Yang EFC, Jeyakumar N, Dhanraj SA (2011) Biosynthesis of silver nanoparticles using mangosteen leaf extract and evaluation of their antimicrobial activities. J Saudi Chem Soc 15:113–120

Verma DK, Hasan SH, Banik RM (2016) Photo-catalyzed and phyto-mediated rapid green synthesis of silver nanoparticles using herbal extract of *Salvinia molesta* and its antimicrobial efficacy. J Photochem Photobiol B 155:51–59

Vidhu VK, Philip D (2014) Spectroscopic, microscopic and catalytic properties of silver nanoparticles synthesized using *Saraca indica* flower. Spectro Acta Part A: Mol Biomol Spect 117:102–108

Vinković T, Novák O, Strnad M, Goessler W, Jurašin DD, Parađiković N, Vrček IV (2017) Cytokinin response in pepper plants (*Capsicum annuum* L.) exposed to silver nanoparticles. Environ Res 156:10–18

Vivek R, Thangam R, Muthuchelian K, Gunasekaran P, Kaveri K, Kannan S (2012) Green biosynthesis of silver nanoparticles from *Annona squamosa* leaf extract and its in vitro cytotoxic effect on MCF-7 cells. Process Biochem 47:2405–2410

Wei L, Lu J, Xu H, Patel A, Chen ZS, Chen G (2015) Silver nanoparticles: synthesis, properties, and therapeutic applications. Drug Discov Today 20:595–601

Xiang DX, Chen Q, Pang L, Zheng CL (2011) Inhibitory effects of silver nanoparticles on H1N1 influenza A virus in vitro. J Virol Methods 178:137–142

Yakout SM, Mostafa AA (2015) A novel green synthesis of silver nanoparticles using soluble starch and its antibacterial activity. Int J Clin Exp Med 8:3538–3544

Yin L, Colman BP, McGill BM, Wright JP, Bernhardt ES (2012) Effects of silver nanoparticle exposure on germination and early growth of eleven wetland plants. PLoS ONE 7:e47674

Yin L, Cheng Y, Espinasse B, Colman BP, Auffan M, Wiesner M, Bernhardt ES (2011) More than the ions: the effects of silver nanoparticles on *Lolium multiflorum*. Environ Sci Technol 45:2360–2367

Zhang W, Qiao X, Chen J (2007) Synthesis of silver nanoparticles – effects of concerned parameters in water/oil microemulsion. Mat Sci Eng B 142:1–15

Zia F, Ghafoor N, Iqbal M, Mehboob S (2016) Green synthesis and characterization of silver nanoparticles using *Cydonia oblong* seed extract. Appl Nanosci 6:1023–1029

[illegible]

Chapter 6
Plant Protein-Based Nanoparticles and Their Biomedical Applications

Siavash Iravani and Ashutosh Kumar Shukla

6.1 Introduction

The plant protein-based nanoparticles (NPs), such as soy proteins, zein, and wheat gliadins, have several benefits and are widely available, compared to the animal-derived proteins. They can be used as drug carriers for lipophilic or anticancer drugs and as delivery systems for bioactive ingredients (Wan et al. 2015; Malekzad et al. 2017). These proteins can be used to encapsulate drugs in order to protect them from a rapid degradation by environmental stress. These carriers are biodegradable and metabolizable and can be prepared under soft conditions without the use of toxic and hazardous organic solvents or materials. They can incorporate a wide variety of drugs in a relatively non-specific fashion. Moreover, these NPs may offer various possibilities for surface modification and covalent attachment of drugs and ligands. Corn, wheat, and soybeans contain proteins, which are readily available, biodegradable, and considerably less allergic in contrast to animal proteins such as bovine collagen (Malekzad et al. 2017). Plant protein-based NPs can be used in drug delivery and gene delivery systems, in bioactive compound delivery, and in improvement of oral bioavailability of drugs. Moreover, they have applications in food industry and tissue engineering. Some important advantages of plant protein-based NPs have been summarized in Table 6.1.

Plant proteins are generated as coproducts when cereal grains are processed for food or fuel and have limited non-food applications. Zein, soy proteins, and wheat proteins (gluten, gliadin, and glutenin) are the major plant proteins. Peanuts, sorghum, millets, and other cereal grains also contain some lesser amounts of proteins.

S. Iravani
Faculty of Pharmacy and Pharmaceutical Sciences, Isfahan University of Medical Sciences, Isfahan, Iran

A. K. Shukla (✉)
Physics Department, Ewing Christian College, Prayagraj, UP, India

A. Husen, M. Iqbal (eds.), *Nanomaterials and Plant Potential*,
https://doi.org/10.1007/978-3-030-05569-1_6

Table 6.1 Important advantages of plant protein-based NPs

Important advantages of plant protein-based nanoparticles	No risk of infectious pathogens due to contamination with animal tissues
	Can incorporate a wide variety of drugs
	Less expensive than animal proteins
	Can be prepared under soft and nontoxic conditions
	Biodegradable and metabolizable
	Possess functional groups which can be easily employed either to adsorb or to covalently couple the molecules that can modify the targeting properties of NPs

Plant proteins are more widely available but have lower molecular weights than collagen and silk. They carry higher net negative charges than collagen and silk and would, therefore, be more suitable for the delivery of positively charged drugs. Plant proteins also have highly polar amino acids that make them hydrophilic and more capable to attract cells. The wide range of isoelectric points for plant proteins allows researchers to choose appropriate proteins for delivering specific drugs into the body (Orecchioni et al. 2006; Reddy and Yang 2011; Nahete et al. 2013; Gadad et al. 2014; Mohammadinejad et al. 2016). Proteins are based on a series of L-α-amino acids with various characteristics and consist of four distinct structures. The primary structure is the amino acid sequence. The secondary structure consists of regularly repeating local structures stabilized by hydrogen bonds, and the most common examples are the alpha-helix, beta-sheet, and turns. The tertiary structure includes the overall form of a single protein molecule: the spatial relationship of the secondary structures to one another. The tertiary structure not only is generally stabilized by nonlocal interactions, most commonly the formation of a hydrophobic core, but also through salt bridges, hydrogen bonds, disulfide bonds, and posttranslational modifications. The quaternary structure, usually called protein subunit in this context, is the structure formed by several protein molecules (polypeptide chains) which act as a single protein complex (Whitford 2013).

In this chapter, we present an overview of issues related to the plant protein-based NPs and highlight their potential biomedical applications.

6.2 Plant Protein-Based NPs and Their Biomedical Applications

6.2.1 The Zein NPs

Zein is a water-insoluble plant storage prolamine protein from maize seeds (*Zea mays* L.), which has been used extensively in many industrial and food applications and for coatings of paper cups, clothing fabrics, adhesives, and binders (Podaralla and Perumal 2012). The hydrophobic nature of zein provides a barrier to the encapsulated actives improving their storage stability in product conditions as well as effectively reducing their degradation in the gastrointestinal tract. Zein has three quarters of lipophilic and one quarter of hydrophilic amino acid residues and consists of three fractions that vary in molecular weight and solubility. These include α-zein (19–24 kDa; 75–80% of total protein), β-zein (17–18 kDa, 10–15%), and γ-zein (27 kDa, 5–10%). The large proportion (>50%) of nonpolar amino acids (leucine, proline, alanine, and phenylalanine) in zein makes it water insoluble. Zein is deficient in essential amino acids, such as lysine and tryptophan, and, hence, is poor in nutritional quality. Commercial zein is available in two, yellow and white grades. Zein particulate systems have been prepared using phase separation based on the differential solubility of zein in ethanol and aqueous solution. Zein is approved by the Food and Drug Administration (FDA) as a safe excipient for pharmaceutical film coatings (Patel and Velikov 2014). Xu et al. (2011) investigated an almost biodegradable hollow zein to remove reactive dyes from simulated post-dyeing wastewater with a remarkably high efficiency. Hollow zein NPs have a higher adsorption for Reactive Blue 19 than solid structures, and the adsorption amount has been shown to increase as the temperature decreases, the pH decreases, or the initial dye concentration increases. The adsorption capacity of hollow zein is also much higher than that of the various biodegradable adsorbents developed to remove reactive dyes.

Zein can be used in controlled and targeted drug delivery and tissue engineering. Furthermore, this biodegradable and biocompatible protein can be used for several industrial applications including agriculture, cosmetics, packaging, and pharmaceuticals. In this regard, there are several advantages including biodegradability, ease of availability, high drug-binding capacity, and availability of large surface area for the drug and cells to be entrapped (summarized in Table 6.2) (Paliwal and Palakurthi 2014). Zein resource and method of extraction have effects on its pharmaceutical, physiochemical, and therapeutic properties. Genetic materials such as DNA, siRNA, and oligonucleotides within intracellular compartments may be tried to develop zein-based NPs for gene therapy (Paliwal and Palakurthi 2014). Actually, the amphiphilic property of zein and the size advantage of nanofibers have been brought

Table 6.2 Some important properties of zein

Properties	Water insoluble but efficiently soluble in aqueous alcohol solution
	Gelling character
	Less costly, biodegradable, and biocompatible
	Adhesive nature, high drug-binding capacity
	Antioxidant and antimicrobial characters
	High molecular weight; brick-like structure
	More sustained drug release than hydrophilic proteins
	Strong EPR effect can be achieved
	Protection of encapsulated sensitive drug from hydrolysis Protection of loaded drug from microbial degradation
	High payload with sustained release of loaded drug

together in developing an ideal delivery system for siRNA. The morphological analysis of the GAPDH-siRNA-loaded zein nanofibers revealed the proper encapsulation of the siRNA in the polymeric matrix. The loading efficiency of this delivery system was found to be about 58.57% (w/w). The agarose gel analysis revealed that the zein nanofibers preserved the integrity of siRNA for a longer period even at the room temperature. The in vitro release studies not only depicted the sustaining potential of the zein nanofibers but also ensured the release of sufficient quantity of siRNA required to induce the gene-silencing effect. The amphiphilic property of zein supported the cell attachment and thereby facilitated the transfection of siRNA into the cells. The potential of the developed system in inducing the desired gene-silencing effect was confirmed by the quantitative real-time PCR analysis (Karthikeyan et al. 2015). In another study, controlled delivery of hollow NPs from zein to different organs of mice was achieved via cross-linking, using citric acid, a nontoxic polycarboxylic acid derived from starch (Karthikeyan et al. 2015). Hollow zein NPs were chemically cross-linked with citric acid in order to achieve a controlled delivery and prolonged accumulation of NPs in the kidney. NPs showed improved stability in aqueous environment at pH 7.4 without affecting the adsorption of 5-fluorouracil (5-FU), a common anticancer chemotherapy drug. Therefore, citric acid cross-linked hollow zein NPs can be the potential vehicles for a controllable delivery of anticancer therapeutics (Xu et al. 2015).

Zou and Gu (2013) designed TPGS 1000 (TPGS) emulsified zein NPs (i.e., TZN) in order to improve the oral bioavailability of daidzin (an isoflavone glycoside with estrogenic activities). They fabricated the zein NPs and TZN using an anti-solvent method and determined that the NPs must be spherical in shape, with a mean size of approximately 200 nm and low polydispersity. The NP zeta potentials were approximately +25 mV at pH 5.5 and −23 mV at pH 7.4. The addition of TPGS as an emulsifier increased the encapsulation efficiency of daidzin in the zein NPs from 53% to 63%. Daidzin-loaded TZN exhibited a slower daidzin release when compared with the daidzin-loaded zein NPs, in both simulated digestive fluids and a pH 7.4 buffer. Confocal laser scanning microscopy suggested that the cellular uptake of the coumarin-6-labeled TZN in human intestinal epithelial Caco-2 cells is significantly higher than that of the fluorescent zein NPs. The cellular uptake and transport studies revealed that daidzin in TZN was taken up more efficiently into Caco-2 cells and transported more quickly through a Caco-2 monolayer than through a daidzin solution (Zou and Gu 2013). A pharmacokinetic study demonstrated that the C_{max} of daidzin in mice, after oral administration of daidzin-loaded TZN, was 5.66 ± 0.16 μM, which was improved by 2.64-fold, compared with that of the daidzin solution (2.14 ± 0.04 μM). Moreover, the areas under the curve (AUC) for daidzin-loaded TZN were enhanced by 2.4-fold, compared with that of the daidzin solution (Zou and Gu 2013).

Xu et al. (2011) developed hollow zein NPs for potential drug delivery applications, with average diameters as small as 65 nm, which were capable of loading a large amount of drug and penetrating into the cell cytoplasm. Hollow zein NPs were capable of loading as high as 369 mg g^{-1} of the drug metformin at an equilibrium concentration of 3 g L^{-1}. Metformin in hollow zein NPs showed a more sustained and controlled release profile than that in solid zein NPs, and the hollow zein NPs

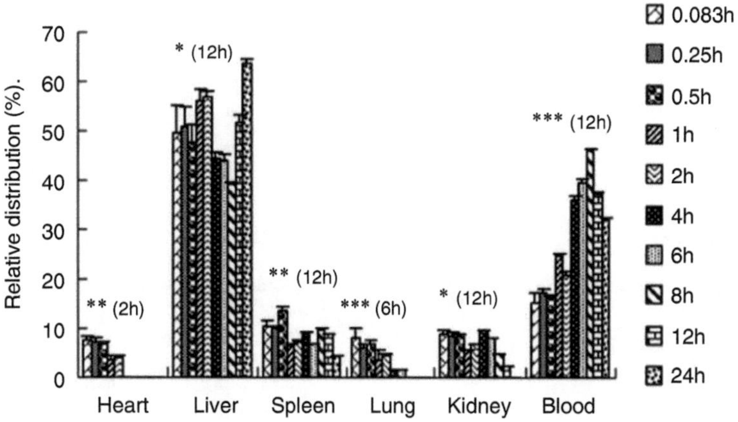

Fig. 6.1 Zein NPs could be distributed significantly in the liver after intravenous administration to mice and adequately remained in blood for at least 24 h. Data were given as mean ± SD ($n = 3$). Statistical significance compared between two groups: *$p < 0.001$, **$p < 0.05$, and ***$p > 0.05$. (Reproduced with permission of Lai and Guo (2011))

were able to enter the fibroblast cells 1 h after incubation. Moreover, Lai and Guo (2011) proposed a new zein NP-encapsulated 5-FU that targets the liver through intravenous delivery. The highest drug loading was obtained using the following conditions: ratio of zein, 5-FU, 3:1 (v/v); zein concentration, 12.5 mg mL^{-1}; pH 9.18; mixing time, 3 h; and ethanol concentration, 40%. The encapsulation efficiency and the drug loading were 60.7 ± 1.74 and 9.17 ± 0.11, respectively. The size of the zein NPs and the zeta potential were 114.9 ± 59.4 nm and − 45 ± 0.3 mV, respectively. In vivo, zein NPs mostly accumulated in the liver following intravenous injection, and the targeting efficiency increased by 31.33%. The relative uptake rate for the liver was 2.79, and the nano-sized ZPs were beneficial for prolonged blood residence (7.2-fold increase) (Fig. 6.1).

Zein NPs (approximately 800 nm) were synthesized and conjugated with quantum dots (ZnS/Mn) (Lai and Guo 2011). They were encapsulated with 5-FU and employed successfully for cellular imaging. Evaluation of the NP biocompatibility showed that NPs at higher concentrations were compatible for cells and were, therefore, expected to be the promising agents for targeted drug delivery (Aswathy et al. 2012). Zhang et al. (2014) prepared and characterized thymol-loaded zein NPs stabilized with sodium caseinate and chitosan hydrochloride. Based on their evaluation, they concluded that the encapsulated thymol was more effective in suppressing the gram-positive bacterium than unencapsulated thymol for a longer period. In another study, Lee et al. (2013) introduced a novel drug delivery system composed of zein and demonstrated that zein NPs protected the therapeutic proteins, catalase and SOD, from the harsh conditions found in the gastrointestinal (GI) tract. Folate-conjugated catalase (CAT) or superoxide dismutase (SOD) in zein NPs was able to target the activated macrophages and scavenge the reactive oxygen species (ROS) generated by macrophages in vitro (Fig. 6.2).

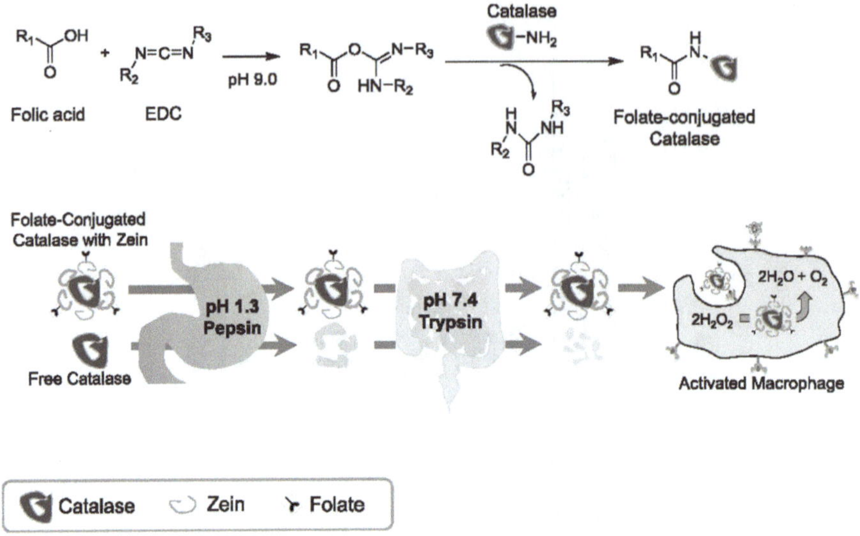

Fig. 6.2 Folate-conjugated catalase or SOD in zein NPs can target the activated macrophages and scavenge the ROS generated by macrophages in vitro. (Reproduced with permission of Lee et al. (2013))

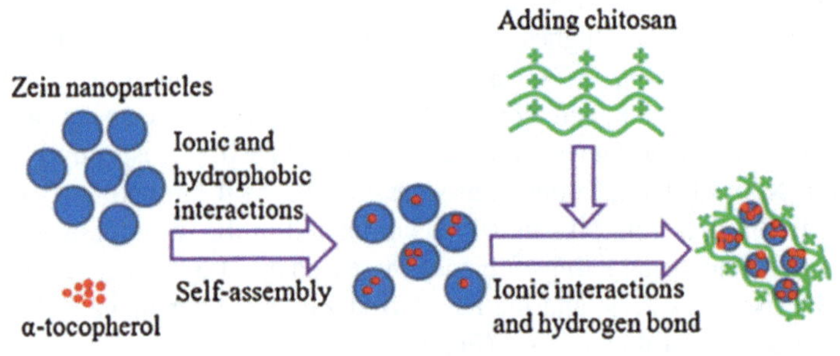

Fig. 6.3 Schematic representation of the complex formation due to electrostatic interactions and hydrogen bonds. (Reproduced with permission of Luo et al. (2011))

Chen et al. (2014) fabricated the tangeretin-loaded protein NPs by mixing an organic phase containing zein and tangeretin, with an aqueous phase containing β-lactoglobulin. Extensive particle aggregation occurred at a high ionic strength (>100 mM) and intermediate pH (4.5–5.5) due to a reduced electrostatic repulsion. Extensive aggregation also occurred at temperatures exceeding 60 °C, which was presumably due to the increased hydrophobic attraction. Overall, they showed that the protein-based NPs could be used to encapsulate the bioactive tangeretin; therefore, it could be readily dispersed in compatible food products. In another study, Luo et al. (2011) encapsulated a hydrophobic nutrient, α-tocopherol, into a zein/chitosan complex. Physicochemical and structural analysis showed that electrostatic interactions and hydrogen bonds were the major forces responsible for the complex formation (Fig. 6.3). Scanning electron microscopy (SEM) revealed the spherical nature and smooth surface of the complex. The particle size and zeta potential of the complex varied from 200 to 800 nm and from +22.8 to +40.9 mV, respectively. The kinetic release profile of the α-tocopherol showed a burst effect, followed by a slow release. Compared with zein NPs, the zein/chitosan complex provided better protection of α-tocopherol release against gastrointestinal conditions, due to the chitosan coatings (Luo et al. 2011).

Luo et al. (2012) prepared zein NPs coated with carboxymethyl chitosan to encapsulate vitamin D_3 and found that this vitamin was first encapsulated into zein NPs using a low-energy phase separation method and was simultaneously coated with carboxymethyl chitosan. Then, calcium was added to cross-link the carboxymethyl chitosan to obtain a thicker and denser coating. The NPs with carboxymethyl chitosan coatings had a spherical structure, with particle size ranging from 86 to 200 nm. The encapsulation efficiency was greatly improved to 87.9% after carboxymethyl chitosan coating, compared with 52.2% for those using zein as a single encapsulant. The NPs with coatings provided a better controlled release of vitamin D_3 in both the phosphate-buffered saline medium and the simulated gastrointestinal tract conditions. Photostability against UV light was significantly improved after encapsulation. Indole-3-carbinol and 3,3′-diindolylmethane are two bioactive compounds that are obtained from cruciferous vegetables. However, the stability of these compounds has been a major limitation for their pharmaceutical applications. Luo et al. (2013) prepared zein and zein/carboxymethyl chitosan NPs to encapsulate indole-3-carbinol and 3,3′-diindolylmethane with a combined liquid-liquid phase separation and ionic gelation method. After the zein NPs were coated with carboxymethyl chitosan, the zeta potential decreased from approximately −10 to −20 mV, while the encapsulation efficiency was greatly improved. Both the NP formulations provided for a controlled release of indole-3-carbinol and 3,3′-diindolylmethane in the buffered saline. Zein and zein/carboxymethyl chitosan NPs demonstrated similar protection for both indole-3-carbinol and 3,3′-diindolylmethane against UV light, which was attributed mainly to the zein protein. Compared with zein NPs, the zein/carboxymethyl chitosan NPs exhibited a better protection of indole-3-carbinol against degradation and a better inhibition against its oligomerization to 3,3′diindolylmethane under thermal conditions (37 °C).

Wu et al. (2012) encapsulated two essential oils, thymol and carvacrol, in zein NPs using a liquid-liquid dispersion method and analyzed their antioxidant properties by the 2,2-diphenyl-1-picrylhydrazyl radical (DPPH radical dot) and ferric ion spectrophotometric assay. The DPPH radical dot was reduced in the range of 24.8–66.8%, depending on the formulation, and more than 65% hydroxyl free radicals were quenched by the samples. Reduction of *Escherichia coli* from 0.8 to 1.8 log CFU/mL was achieved in the presence of NPs encapsulating the essential oils.

Regier et al. (2012) fabricated zein nanospheres encapsulating DNA using a coacervation technique, without the use of harsh solvents or temperatures, resulting in the preservation of DNA integrity and particles with diameters that ranged from 157.8 ± 3.9 nm to 396.8 ± 16.1 nm, depending on the zein to DNA ratio. In this study, DNA encapsulation efficiencies were maximized to 65.3%, with a maximum loading of 6.1 ± 0.2 mg DNA g^{-1} zein. The spheres protected encapsulated DNA from DNase I degradation and exhibited sustained plasmid release for at least 7 days, with minimal burst during the initial release phase. The zein/DNA nanospheres demonstrated a robust biocompatibility, cellular association, and internalization.

Gomez-Estaca et al. (2012) prepared zein NPs with a compact spherical structure and a narrow size distribution by electro-hydrodynamic atomization and showed that zein NPs could be obtained from zein at concentrations ranging from 2.5% to 15% (w/w). The size of these particles, ranging from 175 to 900 nm, increased with increasing polymer concentrations. Compact nanostructures were obtained for 2.5% and 5% zein solutions, whereas 10% and 15% solutions yielded collapsed and shrunken particles. The flow rate was also shown to have an effect, where the lower the flow rate, the smaller the formed NPs. The morphology of the NPs did not change after incorporating curcumin (hydrophobic bioactive molecule) in proportions ranging from 1:500 to 1:10 (curcumin/zein), and the encapsulation efficiency was approximately 85–90%. Fluorescence microscopy images showed that the ensuing nanostructures were similar in form to the matrix systems, with the curcumin homogeneously distributed in the zein matrix. The curcumin remained in an amorphous state in the NPs, as revealed by X-ray diffractometry, which showed a close contact with the polymer. After 3 months of storage at 23 °C and 43% relative humidity in the dark, neither the size nor the morphology of the NPs underwent significant changes, and the curcumin content was not affected. Due to the encapsulation, the curcumin was well dispersed when evaluated in an aqueous food matrix of semi-skim milk.

Hu et al. (2015) prepared biopolymer core-shell NPs using zein (hydrophobic protein as the core) and pectin (hydrophilic polysaccharide as the shell). Particles were prepared by coating cationic zein NPs with anionic pectin molecules using the electrostatic deposition (pH = 4). The core-shell NPs were fortified with curcumin at a high loading efficiency (>86%). The prepared spherical NPs (~250 nm) had a narrow size distribution, and the polydispersity index was about 0.24. The prepared biopolymer NPs could be useful for incorporating curcumin into functional foods and beverages and into dietary supplements and pharmaceutical products.

Jiang et al. (2012) investigated the core-sheath nanofibers prepared using the coaxial electrospinning for providing the biphasic drug release profiles. Using ketoprofen as the model drug and polyvinylpyrrolidone and zein as the sheath polymer and core matrix, respectively, the coaxial process could be carried out smoothly and continuously without any clogging of the spinneret. In this study, SEM and transmission electron microscopy (TEM) demonstrated that the nanofibers were linear with a homogeneous structure and had a clear core-sheath structure with an average diameter of 730 ± 190 nm, in which the sheath had a thickness of approximately 90 nm. Differential scanning calorimetric and X-ray diffraction analyses verified that all the components in the core-sheath nanofibers were present in an amorphous state. The attenuated total reflectance of the Fourier transform infrared (FTIR) spectra demonstrated that both the sheath and core matrix had a good compatibility with ketoprofen due to hydrogen bonding. In vitro dissolution tests showed that the nanofibers provided an immediate release of 42.3% of the contained ketoprofen, followed by a sustained release of the remaining drug over 10 h.

Huang et al. (2013) investigated the preparation of drug-loaded fibers using a modified coaxial electrospinning process, in which only unspinnable solvent was used as the sheath fluid. With a zein/ibuprofen co-dissolving solution and N,N-dimethylformamide as the core and sheath fluids, respectively, the drug-loaded zein fibers were continuously and smoothly generated without any clogging of the spinneret. Field-emission SEM and TEM observations demonstrated that the fibers had a ribbon-like morphology with a smooth surface. The average fiber diameters were 0.94 ± 0.34 and 0.67 ± 0.21 μm, when the sheath-to-core flow rate ratios were 0.11 and 0.25, respectively. X-ray diffraction and differential scanning calorimetry analyses verified that the ibuprofen was amorphous in all of the fiber composites. FTIR spectra showed that the zein exhibited good compatibility with ibuprofen due to hydrogen bonding. In vitro dissolution tests showed that all the fibers provided sustained drug release files via a typical Fickian diffusion mechanism. This modified coaxial electrospinning process could expand the capability for electrospinning to generate fibers and provide a new method for developing novel drug delivery systems.

Sun et al. (2011) tested supercritical CO_2 anti-solvent technology for preparing zein NPs loaded with resveratrol. Results demonstrated that the resveratrol yield was lower when CO_2 pressure increased, while the loading yield was higher with increased temperature and ratio. The structure of resveratrol-loaded zein NPs was a matrix with a well-distributed spherical shape. Furthermore, in vitro drug release studies showed that the products exhibited a slower release than the single resveratrol. In another study, Hu et al. (2012) applied solution-enhanced dispersion by supercritical fluids for the production of lutein/zein NPs and found that NPs with high drug loading and high entrapment efficiency could be prepared by this process. Temperature, pressure, ratio of lutein/zein, and solution flow rate influenced the morphology, drug loading, entrapment efficiency, and mean particle size of the lutein/zein NPs. Lower temperature and solution flow rate, coupled with high pressure, favored smaller and more regular-shaped spheres. The initial burst release was hardly observed in NPs processed at 45 °C/10 MPa. Furthermore, the lutein release profile displayed a near zero-order release, which implied that the NPs played a role in controlled lutein release.

Zou et al. (2012) fabricated cranberry procyanidins (CPs)-zein NPs, using a modified liquid-liquid dispersion method. They found that the particle size of the CPs-zein NPs increased from 392 nm to 447 nm, with increasing the CPs-to-zein mass ratios from 1:8 to 1:2. As observed by SEM, the CPs-zein NPs exhibited a spherical morphology, and the loading efficiency of CPs in the CPs-zein NPs decreased with the increase in the CPs-to-zein mass ratios from 1:8 to 1:2 and ranged from 10% to 86%. The oligomers with higher degrees of polymerization (DP) showed a higher loading efficiency than the oligomers with lower DPs, suggesting a greater binding affinity on zein proteins. The loading capacity of the CPs-zein NPs fabricated using a high CPs-to-zein mass ratio (1:2) was significantly higher than of those fabricated using a low mass ratio (1:8). FTIR spectroscopy suggested that the primary interactions between the CPs and zein were hydrogen bonds and hydrophobic interactions. Cell culture studies using human promyelocytic leukemia HL-60 cells showed that the CPs encapsulated in the NPs decreased the cytotoxicity better in comparison to the CPs.

Zhong and Jin (2009) used spray-drying to encapsulate a model antimicrobial of lysozyme in the corn zein. The effects of the zein/lysozyme (20:1 to 4:1) and zein/thymol (1:0 to 4:1) ratios on the microstructures of the microcapsules, and the in vitro release profiles of the encapsulated lysozyme, were investigated. In all cases, less lysozyme was released at higher pH, resulting from stronger molecular attraction between zein and lysozyme. Nanoscale microcapsule matrix structures were correlated with the release characteristics of the encapsulated lysozyme. At the intermediate zein/lysozyme (10:1) and zein/thymol (50:1) ratios, microcapsules had a continuous matrix structure and showed a sustained release (11.1–65.3%) of lysozyme at pH 6 over 49 days. The biocompatibility, nanoscale diameters, potential for loading a large amount of drugs, and ability to penetrate into cells make the zein NPs ideal candidates for carrying various molecules for intracellular drug delivery and tissue engineering for biomedical applications (Xu et al. 2011).

6.2.2 The Gliadin NPs

Ezpeleta et al. (1996a, b) have studied the feasibility of preparing small-sized carriers from vegetal macromolecules. They selected the gliadin (a vegetal protein fraction from wheat gluten) NPs as drug carriers for all-trans retinoic acid (RA). Their systems were prepared by a desolvation method for macromolecules, which produced gliadin NPs of about 500 nm, with a yield close to 90% of the initial protein in the environmentally acceptable solvents such as ethanol and water. Moreover, due to the low solubility of this protein in water and its high hydrophobicity, the gliadin NPs did not require any further chemical or physical treatment for hardening. Gliadin NPs were quite stable over 4 days in the buffered saline but rapidly degraded over 3 h when incubated in the buffered saline solution containing trypsin. However, chemical cross-linkage of NPs with glutaraldehyde significantly increased their stability. Under conditions used by Ezpeleta et al. 1996a, b), the payload limit was 76.4 μg RA mg^{-1} NPs (for an RA/initial protein ratio of 90 μg mg^{-1}), which corresponded to a RA entrapment efficiency of approximately 75% of the added drug. Nevertheless, the entrapment efficiency was high (between 97 and 85%) for RA/initial protein ratios up to 90 μg mg^{-1}. Finally, the in vitro release profiles of RA-loaded gliadin NPs showed a biphasic pattern, where an initial burst effect (in which approximately 20% RA was released), followed by zero-order diffusion (release rate 0.065 mg RA h^{-1}), was observed.

Arangoa et al. (2001) reported that gliadin NPs dramatically increased the carbazole oral bioavailability up to 49% and provided sustained release properties related to a decline in the carbazole plasma elimination rate. The carbazole release rates from NPs (NP and CL-NP), calculated by deconvolution, were found to be of the same order as the elimination rates of the adhered NP fractions in the stomach mucosa. In addition, a good correlation was found between the carbazole plasma levels and the amount of adhered carriers to the stomach mucosa, during the period when absorption takes place.

α-Tocopherol or vitamin E is widely used as a strong antioxidant in many medical and cosmetic applications. However, it is rapidly degraded because of its light, heat, and oxygen sensitivity. Thus, all vitamin E formulations must avoid contact with light, heat, and air. The drug-loaded vitamin E carriers are an attractive option, particularly if they are made of bio-acceptable macromolecules, such as vegetal proteins. For instance, gliadins generate NPs by a desolvation method and may interact with the epidermal keratin for therapeutic or cosmetic formulations. The vitamin E-loaded gliadin NPs have been characterized based on their size, zeta potential, vitamin E payload, and entrapment efficiency. It has come out that the gliadin particle size is ~900 nm after vitamin E loading, and their charge is close to zero. These particles are suitable vitamin E drug carriers, with an optimum encapsulation rate of ~100 vitamin E μg per gliadin mg, with an efficiency of more than 77%. The release behavior of vitamin E-loaded NPs has been interpreted as a "burst effect," followed by a diffusion process through a homogeneous sphere (Duclairoir et al. 2002).

Kajal and Misra (2011) prepared NPs incorporating the tetanus toxoid and a model antigen ovalbumin and investigated them as the delivery vehicles for oral immunization. Gliadin was used as the carrier because of its biocompatibility, oral bioavailability, and muco-adhesive properties. The size of NPs with 50% w/w of antigen remained stable over 3 weeks of testing. Gulfam et al. (2012) used an electrospray deposition system to synthesize gliadin and the gliadin-gelatin composite NPs for delivery and controlled release of an anticancer drug (e.g., cyclophosphamide). They found that cyclophosphamide was gradually released from the gliadin NPs for 48 h. In contrast, the gliadin-gelatin composite NPs rapidly released the cyclophosphamide. Furthermore, they demonstrated that breast cancer cells cultured with the cyclophosphamide-loaded 7% gliadin NPs for 24 h became apoptotic, which was confirmed by the western blot analysis.

6.2.3 The Legumin and Vicilin NPs

Legumes constitute the third largest family (Fabaceae or Leguminosae) of flowering plants and the second most important plant family in agriculture. They are particularly interesting because of having the capacity to fix the atmospheric nitrogen through mutualistic interactions with the rhizobial soil bacteria, a trait that is important from ecological as well as agricultural viewpoints. The most widely cultivated legumes are pigeon peas, common beans, mung beans, cowpeas, alfalfa, chickpeas, clovers, lentils, garden peas, lupins, and peanuts (Young and Bharti 2012). Legumin is a storage protein from *Pisum sativum*. Mirshahi et al. (1996) prepared legumin NPs of approximately 250 nm via a pH-coacervation method and chemical cross-linking with glutaraldehyde. However, this organic solvent-free preparation method yielded only about 27% of protein as NPs. In addition, no significant differences in size, percentage yield, or surface charge were observed between the legumin NPs cross-linked with different glutaraldehyde concentrations. The legumin NPs were quite stable in the buffered saline and followed a zero-order degradation manner, whereby a longer half-life (t50) was obtained with increasing glutaraldehyde concentrations. The amount of methylene blue (MB) used as a model of hydrophilic drug was approximately 6.2% of the initial dye. Its release from the NPs consisted of a rapid initial phase, followed by a slower second period, in which the rates in the second phase were inversely related to the degree of cross-linking.

Vicilin NPs were used as drug carriers in a complex system in which these particles were conjugated with *Ulex europaeus* lectin molecules due to their sugar-binding property via a two-stage carbodiimide method (NPs had an average diameter of approximately 660 nm). The lectin was fixed by coupling its amino groups to carbodiimide-activated carboxylic groups on the vicilin NPs. The reported delivery system was used in the in vitro studies which revealed a high potential of the prepared conjugate toward the bovine submaxillary gland mucin (BSM) glycoprotein as a biological model (Ezpeleta et al. 1996a, b).

6.2.4 The Glycinin and β-Conglycinin NPs

Soybean (*Glycine max* L.) is currently one of the most abundant sources of plant proteins. The enriched form of soy protein, known as soy protein isolate, has been reported to exhibit high nutritional values and desirable functionalities, and its wide application as a food ingredient has been well documented. The soy protein isolate also possesses a balanced composition of the nonpolar, polar, and charged amino acids, and, therefore, drugs can be incorporated with its various functional groups. The major components of soy protein isolate are glycinin and β-conglycinin. Glycinin is composed of acidic and basic polypeptides linked by disulfide bonds, and β-conglycinin is a glycoprotein (Wang et al. 2014). In one study, Teng et al. (2012) successfully encapsulated curcumin, as a model drug, into NPs. The average size of the curcumin-loaded NPs ranged from 220.1 to 286.7 nm, and their zeta potential was approximately −36 mV. The highest encapsulation efficiency and the loading efficiency achieved in their study were 97.2% and 2.7%, respectively.

6.3 Conclusion

Nowadays, plant protein-based NPs have been used in the pharmaceutical and innovative biomedical applications (e.g., drug encapsulation in plant protein NPs). The use of these proteins as a natural, biodegradable, and food-grade source for the generation of functional colloidal structures has received a widespread attention. The simplicity of generating them (e.g., zein colloidal particles) coupled with the ease of loading them with functional ingredients makes them attractive systems for industrial applications. The inherent slow digestion of some of the plant protein-based NPs such as zein colloidal particles in the intestine lends an interesting aspect to these particles for controlling the release of encapsulated actives. Moreover, plant protein-based NPs have been used extensively for encapsulation of functional actives ranging from the essential oils to natural antioxidants. Moreover, encapsulation can be extended to controlling the optical properties of pigments and colorants and the subsequent development of responsive or structural colors. Future developments are expected in the field of design of colloidal structures in order to improve the control on particle surface activity in their ability to stabilize the fluid-in-fluid dispersions (e.g., foams and emulsions).

References

Arangoa MA, Campanero MA, Renedo MJ, Ponchel G, Irache JM (2001) Gliadin nanoparticles as carriers for the oral administration of lipophilic drugs. Relationships between bioadhesion and pharmacokinetics. Pharm Res 18:1521–1527

Aswathy RG, Sivakumar B, Brahatheeswaran D, Fukuda T, Yoshida Y, Maekawa T, Kumar DS (2012) Biocompatible fluorescent zein nanoparticles for simultaneous bioimaging and drug delivery application. Adv Nat Sci Nanosci Nanotechnol 3:025006

Chen J, Zheng J, McClements DJ, Xiao H (2014) Tangeretin-loaded protein nanoparticles fabricated from zein/β-lactoglobulin: preparation, characterization, and functional performance. Food Chem 158:466–472

Duclairoir C, Orecchioni AM, Depraetere P, Nakache E (2002) α-Tocopherol encapsulation and in vitro release from wheat gliadin nanoparticles. J Microencapsul 19:53–60

Ezpeleta I, Irache JM, Stainmesse S, Chabenat C, Gueguen J, Orecchioni AM (1996a) Preparation of lectin-vicilin nanoparticle conjugates using the carbodiimide coupling technique. Int J Pharm 142:227–233

Ezpeleta I, Irache JM, Stainmesse S, Chabenat C, Gueguen J, Popineau Y, Orecchioni A-M (1996b) Gliadin nanoparticles for the controlled release of all-trans-retinoic acid. Int J Pharm 131:191–200

Gadad AP, Vijay Kuma SV, Dandagi PM, Bolm UB, Pallavi PN (2014) Nanoparticles and their therapeutic applications in pharmacy. Int J Pharm Sci Nanotech 7:2509–2519

Gomez-Estaca J, Balaguer MP, Gavara R, Hernandez-Munoz P (2012) Formation of zein nanoparticles by electrohydrodynamic atomization: effect of the main processing variables and suitability for encapsulating the food coloring and active ingredient curcumin. Food Hydrocoll 28:82–91

Gulfam M, J-eun K, Lee JM, Ku B, Chung BH, Chung BG (2012) Anticancer drug-loaded gliadin nanoparticles induce apoptosis in breast cancer cells. Langmuir 28:8216–8223

Hu D, Lin C, Liu L, Li S, Zhao Y (2012) Preparation, characterization, and in vitro release investigation of lutein/zein nanoparticles via solution enhanced dispersion by supercritical fluids. J Food Eng 109(3):545–552

Hu K, Huang X, Gao Y, Huang X, Xiao H, McClements DJ (2015) Core-shell biopolymer nanoparticle delivery systems: synthesis and characterization of curcumin fortified zein–pectin nanoparticles. Food Chem 182:275–281

Huang W, Zou T, Li S, Jing J, Xia X, Liu X (2013) Drug-loaded zein nanofibers prepared using a modified coaxial electrospinning process. AAPS Pharm Sci Tech 14:675–681

Jiang Y-N, Mo H-Y, Yu D-G (2012) Electrospun drug-loaded core–sheath PVP/zein nanofibers for biphasic drug release. Int J Pharm 438:232–239

Kajal H, Misra A (2011) Preparation of tetanus toxoid and ovalbumin loaded gliadin nanoparticles for oral immunization. J Biomed Nanotechnol 7:211–212

Karthikeyan K, Krishnaswamy VR, Lakra R, Kiran MS, Korrapati PS (2015) Fabrication of electrospun zein nanofibers for the sustained delivery of siRNA. J Mater Sci Mater Med 26:101

Lai LF, Guo HX (2011) Preparation of new 5-fluorouracil-loaded zein nanoparticles for liver targeting. Int J Pharm 404:317–323

Lee S, Ali Alwahab NS, Zainab Moazzam M (2013) Zein-based oral drug delivery system targeting activated macrophages. Int J Pharm 454:388–393

Luo Y, Zhang B, Whent M, Yu LL, Wang Q (2011) Preparation and characterization of zein/chitosan complex for encapsulation of α-tocopherol, and its in vitro controlled release study. Colloids Surf B Biointerfaces 85:145–152

Luo Y, Teng Z, Wang Q (2012) Development of zein nanoparticles coated with carboxymethyl chitosan for encapsulation and controlled release of vitamin D_3. J Agric Food Chem 60:836–843

Luo Y, Wang TTY, Teng Z, Chen P, Sun J, Wang Q (2013) Encapsulation of indole-3-carbinol and 3, 3′-diindolylmethane in zein/carboxymethyl chitosan nanoparticles with controlled release property and improved stability. Food Chem 139:224–230

Malekzad H, Mirshekari H, Sahandi Zangabad P, Moosavi Basri SM, Baniasadi F, Sharifi Aghdam M, Karimi M, Hamblin MR (2017) Plant protein-based hydrophobic fine and ultrafine carrier particles in drug delivery systems. Crit Rev Biotechnol 24:1–21

Mirshahi T, Irache JM, Gueguen J, Orecchioni AM (1996) Development of drug delivery systems from vegetal proteins: legumin nanoparticles. Drug Dev Ind Pharm 22:841–846

Mohammadinejad R, Karimi S, Iravani S, Varma RS (2016) Plant-derived nanostructures: types and applications. Green Chem 18:20–52

Nehete JY, Bhambar RS, Narkhede MR, Gawali SR (2013) Natural proteins: sources, isolation, characterization and applications. Pharmacogn Rev 7:107–116

Orecchioni AM, Duclairoir C, Renard D, Nakache E (2006) Gliadin characterization by sans and gliadin nanoparticle growth modelization. J Nanosci Nanotechnol 6:3171–3178

Paliwal R, Palakurthi S (2014) Zein in controlled drug delivery and tissue engineering. J Control Release 189:108–122

Patel AR, Velikov KP (2014) Zein as a source of functional colloidal nano- and microstructures. Curr Opin Colloid Interface Sci 19:450–458

Podaralla S, Perumal O (2012) Influence of formulation factors on the preparation of zein nanoparticles. AAPS PharmSciTech 13:919–927

Reddy N, Yang Y (2011) Potential of plant proteins for medical applications. Trends Biotechnol 29:490–498

Regier MC, Taylor JD, Borcyk T, Yang Y, Pannier AK (2012) Fabrication and characterization of DNA-loaded zein nanospheres. J Nanobiotechnol 10:44

Sun L-j, Shen P-q, Zhao Y-p (2011) Preparation of resveratrol-loaded zein nanoparticles by method of supercritical CO_2 anti-solvent technology [J]. Fine Chemicals 3:014

Teng Z, Luo Y, Wang Q (2012) Nanoparticles synthesized from soy protein: preparation, characterization, and application for nutraceutical encapsulation. J Agric Food Chem 60:2712–2720

Wan ZL, Guo J, Yang XQ (2015) Plant protein-based delivery systems for bioactive ingredients in foods. Food Funct 6:2876–2889

Wang T, Qin GX, Sun ZW, Zhao Y (2014) Advances of research on glycinin and β-conglycinin: a review of two major soybean allergenic proteins. Crit Rev Food Sci Nutr 54:850–862

Whitford D (2013) Proteins: structure and function. Wiley, Hoboken

Wu Y, Luo Y, Wang Q (2012) Antioxidant and antimicrobial properties of essential oils encapsulated in zein nanoparticles prepared by liquid–liquid dispersion method. LWT-Food Sci Technol 48:283–290

Xu H, Jiang Q, Reddy N, Yang Y (2011) Hollow nanoparticles from zein for potential medical applications. J Mater Chem 21:18227–18235

Xu H, Shen L, Xu L, Yang Y (2015) Controlled delivery of hollow corn protein nanoparticles via non-toxic crosslinking: in vivo and drug loading study. Biomed Microdevices 17:8

Young ND, Bharti AK (2012) Genome-enabled insights into legume biology. Annu Rev Plant Biol 63:283–305

Zhang Y, Niu Y, Luo Y, Ge M, Yang T, Yu LL, Wang Q (2014) Fabrication, characterization and antimicrobial activities of thymol-loaded zein nanoparticles stabilized by sodium caseinate–chitosan hydrochloride double layers. Food Chem 142:269–275

Zhong Q, Jin M (2009) Nanoscalar structures of spray-dried zein microcapsules and in vitro release kinetics of the encapsulated lysozyme as affected by formulations. J Agric Food Chem 57:3886–3894

Zou T, Gu L (2013) TPGS emulsified zein nanoparticles enhanced Oral bioavailability of daidzin: in vitro characteristics and in vivo performance. Mol Pharm 10:2062–2070

Zou T, Li Z, Percival SS, Bonard S, Gu L (2012) Fabrication, characterization, and cytotoxicity evaluation of cranberry procyanidins-zein nanoparticles. Food Hydrocoll 27:293–300

Chapter 7
Natural Product-Based Fabrication of Zinc-Oxide Nanoparticles and Their Applications

Azamal Husen

7.1 Introduction

In nanotechnology, materials with at least one dimension less than 100 nanometres (nm) are created for various applications. The market value of this technology will likely cross US$ 75.8 billion by 2020 due to its significant expansion worldwide (Research and Markets 2015). A range of nanoparticles (NPs), including carbon, metal, metal oxide, dendrimers and composites, have been fabricated over the last few decades and utilized in different disciplines of science and technology.

Zinc (Zn) is an essential trace element for human system and occurs in all the six enzyme classes, namely, oxidoreductases, lyases, isomerases, transferases, hydrolases and ligases (Auld 2001). It modulates many physiological functions in eukaryotes (Jansen et al. 2009; Maremanda et al. 2014). Since time immemorial, zinc oxide (ZnO) is recognized as an antibacterial agent (Frederickson et al. 2005). It was used in various ointments for cure of injuries and boils even in 2000 BC (Halioua and Ziskind 2005). Currently, ZnO is used in sunscreen lotion as a supplement and in photoconductive material, LED, transparent transistors, solar cells, memory devices (Ozgur et al. 2005; Klingshirn 2007), cosmetics (De Graaf et al. 1999; Brahms et al. 2005) and catalysis (Speight 2002). Even though significant amount of ZnO is produced each year, a very small portion of it is used as medicine (Brown 1976), and Zn compounds are considered to be ecotoxic for mammals and plants in traces (Patnaik 2003; Araujo-Lima et al. 2017). However, it is suggested that ZnO NPs and Zn NPs coated with soluble polymeric material may be used for treating wounds, ulcers and many microbial infections besides being used as a drug carrier in cancer therapy (Siddiqi et al. 2018). Due to their specific optical and electrical features (Vayssieres et al. 2001), these materials are also recognized as a possible agent in optoelectronic applications to work in the visible and near to UV

A. Husen (✉)
Department of Biology, College of Natural and Computational Sciences, University of Gondar, Gondar, Ethiopia

A. Husen, M. Iqbal (eds.), *Nanomaterials and Plant Potential*,
https://doi.org/10.1007/978-3-030-05569-1_7

regions. They are used mostly in UV light-emitting devices (Rajalakshmi et al. 2012; Izu et al. 2014), as photo-catalysts (Yang and Park 2008; Deng et al. 2008), in ethanol gas sensors (Guo and Peng 2015; Xie et al. 2016) and in several pharmaceutical and cosmetic industries (Lu et al. 2015). Kim et al. (2014) have suggested that these materials have been shown to decrease the expression mRNA of inflammatory cytokines by inhibiting the activation of NF-kB (nuclear factor kappa-B cells). As mentioned above, due to their optical and electrical features (Vayssieres et al. 2001) and antimicrobial properties (Siddiqi et al. 2018), these materials exhibit a great potential to boost agricultural productivity (Siddiqi and Husen 2017). ZnO NPs show a large surface-area-to-volume ratio, which causes a considerable enhancement in their efficiency in blocking UV radiations, compared to the bulk material. ZnO NPs with luminescent or magnetic properties are used as detection probes and drug carriers both in vitro and in vivo (Xiong 2013). They are, therefore, frequently used in various biomedical applications (Mirzaeia and Darroudi 2017).

The shape of ZnO NPs depends on fabrication process. They are in the form of nanorods, nano plates (Mahmud et al. 2006; Wang and Lou 2008; Jang et al. 2008), nano spheres (Kakiuchi et al. 2006), nano boxes (Mahmud et al. 2006), hexagonals, tripods (Mahmud and Abdullah 2006), tetrapods (Shen et al. 2009), nanowires, nano tubes, nano rings (Ding and Wang 2009; Wang 2004a, b), nano cages and nano flowers (Xie et al. 2009; Moezzi et al. 2011). ZnO NPs are prepared by such techniques as the sol-gel method (Lee et al. 2003; Hubbard et al. 2006), precipitation (Wang and Muhammed 1999), hydrothermal synthesis (Xu et al. 2004) and spray pyrolysis (Tani et al. 2002). As reported in Chaps. 1, 3, and 5 in this book, several protocols are used for the fabrication of NPs, for instance, laser ablation, chemical reduction, milling and so on. These conventional protocols often involve substances hazardous to human health and environment. NPs' fabrication by using biological materials provides a dependable and safe alternative.

In the recent past, several plants (or their parts and products) such as *Parthenium hysterophorus* (Rajiv et al. 2013), *Nephelium lappaceum* (Yuvakkumar et al. (2014), *Ocimum basilicum* L. var. *purpurascens* (Abdul et al. 2014), *Pongamia pinnata* (Sundrarajan et al. 2015), *Vitex trifolia* (Elumalai et al. 2015a), *Ficus benghalensis* (Shekhawat et al. 2015), *Punica granatum* (Mishra and Sharma 2015), *Hibiscus subdariffa* (Bala et al. 2015), *Trifolium pratense* (Dobrucka and Długaszewska 2016), *Couroupita guianensis* (Manokari and Shekhawat 2016), *Calotropis gigantea* (Chaudhuri and Malodia 2017), *Ferulago angulata* (Mehr et al. 2018), *Tabernaemontana divaricata* (Raja et al. 2018) and *Andrographis paniculata* (Kavitha et al. 2017; Rajakumar et al. 2018), among others, have been effectively utilized for efficient and rapid fabrication of ZnO NPs (Fig. 7.1). The objective of this chapter is to present the current information on plant-based products and their role in ZnO NPs' fabrication. It also highlights application of these NPs in various disciplines of science.

Fig. 7.1 Some recently used plants for fabrication of ZnO NPs

7.2 Fabrication and Characterization of ZnO NPs

Plant components (roots, leaves, stems, seeds, flowers and fruits) and their products have been utilized for ZnO NPs' fabrication, as plant extracts are rich in phytochemicals that act as the reducing and/or stabilizing agents. Like other NPs, fabrication process of ZnO NPs depends on various growth factors, including concentration of plant extract or biomass, concentration of salt, incubation or reaction time, temperature and pH of the solution. Hence, standardization of these growth factors is important in obtaining the desirable size and shape of NPs for their maximum exploitation and application. The most commonly used techniques for characterization of ZnO NPs are UV-vis spectroscopy, transmission electron microscopy (TEM), scanning electron microscopy (SEM), X-ray diffraction (XRD) and Fourier transform infrared spectroscopy (FTIR). These techniques facilitate the determination of size, shape, structure, crystal orientation and phases, composition, surface modification and surface chemistry of NPs.

Dried sap (gum tragacanth) of *Astragalus gummifer* was used to fabricate ZnO NPs of less than 50 nm (Darrudi et al. 2013). *Citrus aurantifolia* extract was used to produce ZnO NPs of 50–200 nm (Samat and Nor 2013). *Calotropis gigantea* leaf extract was used in the presence of $ZnNO_3$ salt to fabricate ZnO NPs of 30–35 nm (Vidya et al. 2013). *Acalypha indica* leaf extract was used in the presence of zinc acetate salt to produce NPs of 100–200 nm by Gnanasangeetha and Thambwani (2013). *Parthenium hysterophorus* leaf extract was used in the presence of $ZnNO_3$ salt to fabricate ZnO NPs of 16–108 nm by Sindhura et al. (2014). Oudhia et al. (2015) used *Azadirachta indica* leaf extract in the presence of zinc acetate salt to produce NPs of 25 nm, whereas Divyapriya et al. (2014) used leaf extract of *Murraya koenigii* with $ZnNO_3$ salt to fabricate ZnO NPs of 50 nm size. Senthilkumar and Sivakumar (2014) used *Camellia sinensis* leaf extract with zinc acetate salt to fabricate ZnO NPs of 16 nm. Awwad et al. (2014) used leaf extract of *Olea europaea* with $ZnSO_4$salt to fabricate ZnO NPs of 20 nm diameter. Ramesh et al. (2015) used the leaf extract of *Solanum nigrum* as a capping agent for ZnO NPs' fabrication. Morphologically, they were hexagonal wurtzite and quasi-spherical in shape with their size ranging from 20 to 30 nm. Further details are presented in Table 7.1. Ali et al. (2016) have synthesized ZnO NPs from leaf extract of *Aloe barbadensis*. Particle size ranged from 8 to 18 nm with diverse shapes. Karnan and Selvakumar (2016) fabricated ZnO NPs of 25–40 nm size and spherical in shape from the peel extract of *Nephelium lappaceum.* Milk latex of *Carica papaya* was also used to produce varied shaped nano flowers of ZnO NPs (Sharma 2016). In another study, Azizi et al. (2016) used flower extract of *Anchusa italica* to produce ZnO NPs. They suggested that different plant products such as tannins, alkaloids, triterpenes, saponins and flavonoids are available in the flower extract and facilitated the reduction of zinc ions into hexagonal shaped ZnO NPs. A microwave-assisted synthesis of ZnO NPs has also been done using an aqueous extract of fallen flowers of *Jacaranda mimosifolia* (Sharma et al. 2016). Morphologically, they were spherical and their size ranged between 2 and 4 nm. GC-MS and FTIR studies have shown the role of oleic acid as a stabilizing and capping agent during the fabrication process. Chaudhuri and Malodia (2017) used leaf extract of *Calotropis gigantea* with zinc acetate salt in the presence of 2 M NaOH to fabricate ZnO NPs. The combination of reaction mixture, 200 mM zinc acetate salt with 15 ml of *C. gigantea* leaf extract, was highly favourable for producing highly monodisperse crystalline NPs of less than 20 nm size. The NPs obtained were characterized using UV-vis, SEM, XRD, FTIR, DLS, EDX and AFM. UV-vis spectroscopy revealed a peak near 350 nm, which was characteristic of ZnO NPs (Fig. 7.2). DLS investigation exhibited a

Table 7.1 Major studies (during 2013–2018) on plant-mediated fabrication of ZnO NPs, their morphology, characterization and product or phytoconstituents

Key references	Plant (common name)	Family	Part used	Size in nm (characterization techniques) and shape	Phytoconstituents responsible for ZnO NP fabrication
Mehr et al. (2018)	*Ferulago angulata* (Chavir)	Apiaceae	Plant extract	32–36 (FESEM); spherical	C-O and C-N stretching bands of the aliphatic amines or alcohols/phenols due to the presence of polyphenols
Raja et al. (2018)	*Tabernaemontana divaricata* (pinwheel flower)	Apocynaceae	Leaf	20–50 (TEM); spherical	O-H, C-H, C-O stretching bands; presence of steroids, terpenoids, flavonoids, phenyl propanoids, phenolic acids and enzymes in leaf extract
Rajakumar et al. (2018)	*Andrographis paniculata* (green chirayta)	Acanthaceae	Leaf	96–115 (SEM), 57 ± 0.3 (TEM); spherical and hexagonal	O-H stretching of alcohols, C-H bands for amide I and amide II, C-N stretching bands of aliphatic and aromatic amines, C-O stretching bands of polysaccharides, alcohols and phenolic groups; proteins and metabolites such as polyphenols, alkaloids carboxylic acid and flavonoids also present
Sorbiun et al. (2018)	*Quercus* spp. (oak trees)	Proteaceae	Jaft extract	34 (FESEM); spherical	Alcohols, phenolic groups and C-N stretching bands of aliphatic and aromatic amines; amide I bands of proteins or enzymes; and O-H stretching bands of alcohol or phenol
Suresh et al. (2018)	*Costus pictus* (painted spiral ginger)	Costaceae	Leaf	20–80 (TEM); hexagonal, rod-shaped or spherical	O-H stretching band with a C≡O peak, which indicates that the compound is an aliphatic carboxylic acid; C-F stretching band for mono- and poly-fluorinated compounds; C-O stretching band for saturated primary alcohol; C-H stretching band of an aromatic aldehyde; and carbonyl group of flavonoids
Chaudhuri and Malodia (2017)	*Calotropis gigantea* (crown flower)	Apocynaceae	Leaf	10 (XRD); spherical	-OH and -CH stretching bands

(continued)

Table 7.1 (continued)

Key references	Plant (common name)	Family	Part used	Size in nm (characterization techniques) and shape	Phytoconstituents responsible for ZnO NP fabrication
Kavitha et al. (2017)	*Andrographis paniculata* (green chirayta)	Acanthaceae	Leaf	20.23 (SEM); hexagonal nanorod particle	Terpenoid fractions
Rajendran and Sengodan (2017)	*Sesbania grandiflora* (hummingbird tree)	Fabaceae	Leaf	15–35(SEM); spherical	N-H stretching band, symmetric and asymmetric vibration of C=O, O-H stretching band
Ali et al. (2016)	*Aloe vera* (aloe vera)	Liliaceae	Leaf extract	8–20 (XRD); spherical, oval, hexagonal	O-H of phenol, amines, O-H of alcohol and C-H of alkanes, the amide of protein and enzymes
Dobrucka and Długaszewska (2016)	*Trifolium pratense* (red clover)	Fabaceae	Flower	60–70 (XRD); spherical	Hydroxyl, -C-O, -C-O-C, C = C stretching bands
Kavithaa et al. (2016)	*Santalum album* (Indian sandalwood)	Santalaceae	Leaves	100 (DLS and SEM), 70–140 (TEM); nanorods	N-H stretching band of amide II, carboxylate group, carbonyl stretching band, O-H of alcohol
Krupa and Vimala (2016)	*Cocos nucifera* (coconut)	Arecaceae	Coconut water	20–80 (TEM), 21.2 (XRD); spherical and mainly hexagonal without any agglomeration	O-H of alcohol and carboxylic acid, C=O of ketones, C-N of aromatic and aliphatic amines
Madan et al. (2016)	*Azadirachta indica* (neem)	Meliaceae	Fresh leaves	10–30 (TEM), 9–40 (XRD); hexagonal disk, nano buds	O-H between H_2O and CO_2, carbonate moieties
Vanathi et al. (2016)	*Eichhornia crassipes* (water hyacinth)	Pontederiaceae	Leaf extract	32–36 (SEM, TEM), 32 (XRD); spherical, no aggregation	–
Ambika and Sundrarajan (2015)	*Vitex negundo* (nochi)	Lamiaceae	Leaf	75–80 (SEM & EDX), 38.17 (XRD); spherical	O-H, C-H, C=C stretching bands
Anbuvannan et al. (2015a)	*Phyllanthus niruri* (bhumi amla)	Phyllanthaceae	Leaf extract	25.61 (FESEM, XRD); hexagonal wurtzite, quasi-spherical	O-H, C-H, C-O stretching bands, aromatic aldehyde

Anbuvannan et al. (2015b)	*Anisochilus carnosus* (kapurli)	Lamiaceae	Leaf extract	56.14 (30 mL of extract), 49.55 (40 mL), 38.59 (50 mL) (XRD), 20–40 (FESEM), 30–40 (TEM); hexagonal wurtzite, quasi-spherical	O-H of water, alcohol, phenol C-H of alkane, O-H of carboxylic acid, C=O of the nitro group
Bhuyan et al. (2015)	*Azadirachta indica* (neem)	Meliaceae	Leaf	9.6–25.5 (TEM); spherical	Amide II stretching band, C-N stretching band of aliphatic, aromatic amide, an aliphatic amine, alcohol, phenol, secondary amine, C-H of alkane and aromatics, C = C-H of alkynes, C=O, C-C of an alkane
Elumalai and Velmurugan (2015)	*Azadirachta indica* (neem)	Meliaceae	Fresh leaves	18 (XRD); spherical	Amine, alcohol, ketone, carboxylic acid
Qian et al. (2015)	*Aloe vera* (aloe vera)	Liliaceae	Freeze dried leaf peel	25–65 (SEM, TEM); spherical, hexagonal	–
Ramesh et al. (2015)	*Solanum nigrum* (black nightshade)	Solanaceae	Leaf extract	20–30 (XRD, FESEM) 29.79 (TEM); hexagonal wurtzite, quasi-spherical	O-H, aldehydic C-H, amide III stretching bands of protein, carboxyl side group, C-N of amine, carbonyl group
Sundrarajan et al. (2015)	*Pongamia pinnata* (Indian beech)	Fabaceae	Fresh leaves	26 (XRD), agglomeration of 100 (DLS, SEM, TEM); spherical, hexagonal, nano rod	O-H stretching band, C=O stretching band of carboxylic acid or their ester, C-O-H stretching band
Thema et al. (2015)	*Agathosma betulina* (buchu)	Rutaceae	Dry leaves	15.8 (TEM) 12–26 (HRTEM); quasi-spherical agglomerates	O-H of hydroxyl group, Zn-O stretching band
Aladpoosh and Montazer (2015)	*Gossypium* spp. (cotton)	Malvaceae	Cellulosic fibre	13 (XRD); Wurtzite, spherical, nano rod	O-H, [C=O, C-O, C-O-C] (due to Zn precursor)
Elumalai et al. (2015b)	*Moringa oleifera* (drumstick tree)	Moringaceae	Leaf	24 (XRD), 16–20 (FESEM) spherical and granular nano sized shape with a group of aggregates	O-H, C-H of alkane, C=O of alcohol, carboxylic acid

(continued)

Table 7.1 (continued)

Key references	Plant (common name)	Family	Part used	Size in nm (characterization techniques) and shape	Phytoconstituents responsible for ZnO NP fabrication
Fu and Fu (2015)	*Plectranthus amboinicus* (Mexican mint)	Lamiaceae	Leaf extract	50–180 (SEM); rod shape NPs with agglomerates	Zn-O, C-O of C-O-SO3, phosphorus compound
Sundrarajan et al. (2015)	*Vitex negundo* (nochi)	Lamiaceae	Flowers	38.17 (XRD), 10–130 (DLS) hexagonal	–
Abdul et al. (2014)	*Ocimum basilicum* L. var. *purpurascens* (red rubin basil)	Lamiaceae	Leaf extract	50 (TEM, EDS), 14.28 (XRD); hexagonal wurtzite	–
Nagajyothi et al. (2014)	*Coptidis rhizoma* (coptis)	Ranunculaceae	Dried rhizome	2.9–25.2(TEM); spherical, rod shaped	Primary and secondary amines, aromatic and aliphatic amines, alcohol, carboxylic acid, alkyl halide, alkynes
Rajiv et al. (2013)	*Parthenium hysterophorus* (congress grass)	Asteraceae	Leaf extract	22–35 (50% plant extract), 75–90 (25% plant extract) (XRD, TEM); spherical, hexagonal	N-H and N-H stretching bands, phosphorus compound, secondary sulphonamide, monosubstituted alkyne, amine salt, vinyl cis-trisubstituted
Yuvakkumar et al. (2014)	*Nephelium lappaceum* (rambutan)	Sapindaceae	Fruit peels	50.95 (XRD); needle-shaped forming agglomerate	O-H, H-O-H stretching bands

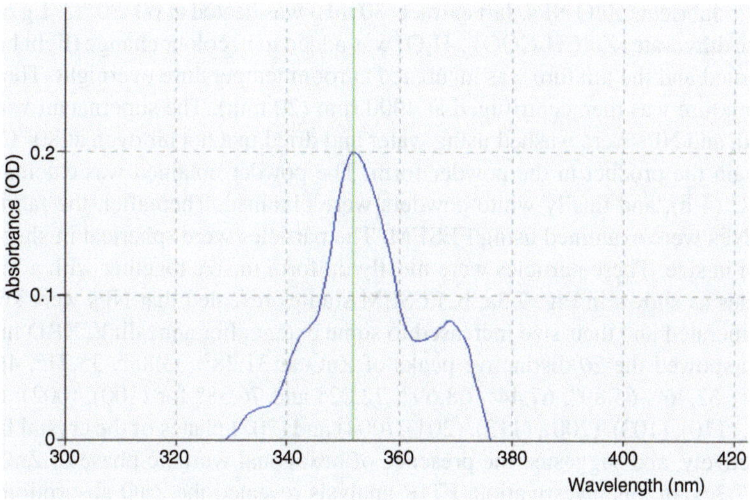

Fig. 7.2 UV-vis spectroscopy of zinc acetate solution, *Calotropis gigantea*leaf extracts and fabricated ZnO NPs after 24 h. (Adapted from Chaudhuri and Malodia (2017))

single peak at 11 nm (100%), and polydispersity index was 0.245. XRD analysis showed that these are highly crystalline ZnO NPs having an average size of 10 nm. FTIR study displayed bands at 4307, 3390, 2825, 871, 439 and 420 cm^{-1} which were the features of –OH stretching vibration and –CH stretching vibration. However, the band at 871 cm^{-1} was due to asymmetrical and symmetrical stretching of zinc carboxylates, resulting in the involvement of carboxylic groups in proteins of *C. gigantea* leaf extract. SEM images confirmed the spherical shape of NPs. EDX have shown the presence of zinc and oxygen; and the atomic percentage of zinc and oxygen was 33.31 and 68.69, respectively. AFM study showed 2D and 3D images of ZnO NPs, indicating that they were monodisperse having a size range of 1.5–8.5 nm.

ZnO NPs were fabricated from *Andrographis paniculata* leaf extract (Rajakumar et al. 2018) and characterized using UV-vis, TEM, SEM, FTIR and XRD techniques. The NPs were spherical and hexagonal, and their size varied from 96 to 115 and 57 ± 0.3 nm, respectively. FTIR analysis suggested that the phenolic compounds, terpenoids and proteins present in the leaf extract were responsible for the NPs' nucleation and stability, while XRD studies confirmed nanocrystal formation. Raja et al. (2018) have fabricated ZnO NPs from aqueous leaf extract of *Tabernaemontana divaricata,* and characterized them using the UV-vis, TEM, SEM, FTIR and XRD techniques. They were spherical in shape and 20–50 nm in size. The steroids, terpenoids, flavonoids, phenyl propanoids, phenolic acids and enzymes present in the leaf extract acted as the capping and reducing agents. In this study, XRD analysis confirmed the presence of pure hexagonal wurtzite crystalline structure of ZnO, and the average crystallite size was 36.82 nm. In an experiment, Sorbiun et al. (2018) used oak dried fruit hull (Jaft) as the reducing and stabilizing agent to fabricate ZnO NPs. Jaft extract (30 mL) was heated at 60–80 °C, 1 g of zinc acetate dihydrate [$Zn(CH_3COO)_2.{}_2H_2O$] was added to it, colour change (light brown) was noted and the mixture was incubated at room temperature overnight. The reaction mixture was then centrifuged at 4000 rpm (20 min). The supernatant was discarded, and NPs were washed using water and dried in a hot air oven at 80 °C (6 h) to obtain the product in the powder form. The powder obtained was calcinated at 500 °C (4 h), and finally white powders were obtained. Thereafter, the fabricated ZnO NPs were examined using FESEM. The particles were spherical in shape and 34 nm in size. These particles were mostly uniform in size together with a few big particles as shown in Fig. 7.3a, b. FESEM studies revealed that NPs were slightly agglomerated and their size increased to some extent after annealing. XRD investigation showed the 2θ distinctive peaks of ZnO at 31.28°, 33.88°, 35.71°, 46.98°, 56.08°, 62.36°, 65.89°, 67.44°, 68.69°, 72.02° and 76.38° for (100), (002), (101), (102), (110), (103), (200), (112), (201), (004) and (202) planes of the crystal lattice, respectively, and suggested the presence of hexagonal wurtzite phase of ZnO NPs (Fig. 7.3c). In this investigation, FTIR analysis revealed the ZnO absorption band near 571 cm^{-1} (Fig. 7.3d). In addition, other peaks noticed at 1032, 1125 and 1382 cm^{-1} were ascribed to alcohols, phenolic groups and C-N stretching vibrations of aliphatic and aromatic amines, respectively. The authors suggested the peak around 1492 cm^{-1} was due to amide I bonds of proteins or enzymes, band at 3386 cm^{-1} represented the O-H stretching alcohol or phenol group and the peak around 860 cm^{-1} was probably due to O-H functional group (Sorbiun et al. 2018). A brief summary of greenfabrication of ZnO NPs from different plant parts and their products is presented in Table 7.1.

7.3 Applications of ZnO NPs

ZnO NPs are used as an antibacterial, antifungal and anticancer drug delivery agent; as a biofertilizer in plant system; as catalysts, biosensors, gas sensors, solar cells, photo detectors, UV absorbers and optical materials; as antivirus agent in coating; and as filter in manufacturing rubber, plastic and cigarettes and in bioimaging, ceramics, cosmetics and electrical industries. Some of the important and widely used applications of ZnO NPs are described below.

7.3.1 Antimicrobial Activity

ZnO NPs obtained from the aqueous leaf extract of *Azadirachta indica* have shown antimicrobial activity against yeast and both gram-positive and gram-negative bacteria (Elumalai and Velmurugan 2015). Further, ZnO NPs fabricated by using the same plant leaf extract were found to show antibacterial activity against *Staphylococcus aureus*, *Streptococcus pyogenes* and *Escherichia coli* at varying

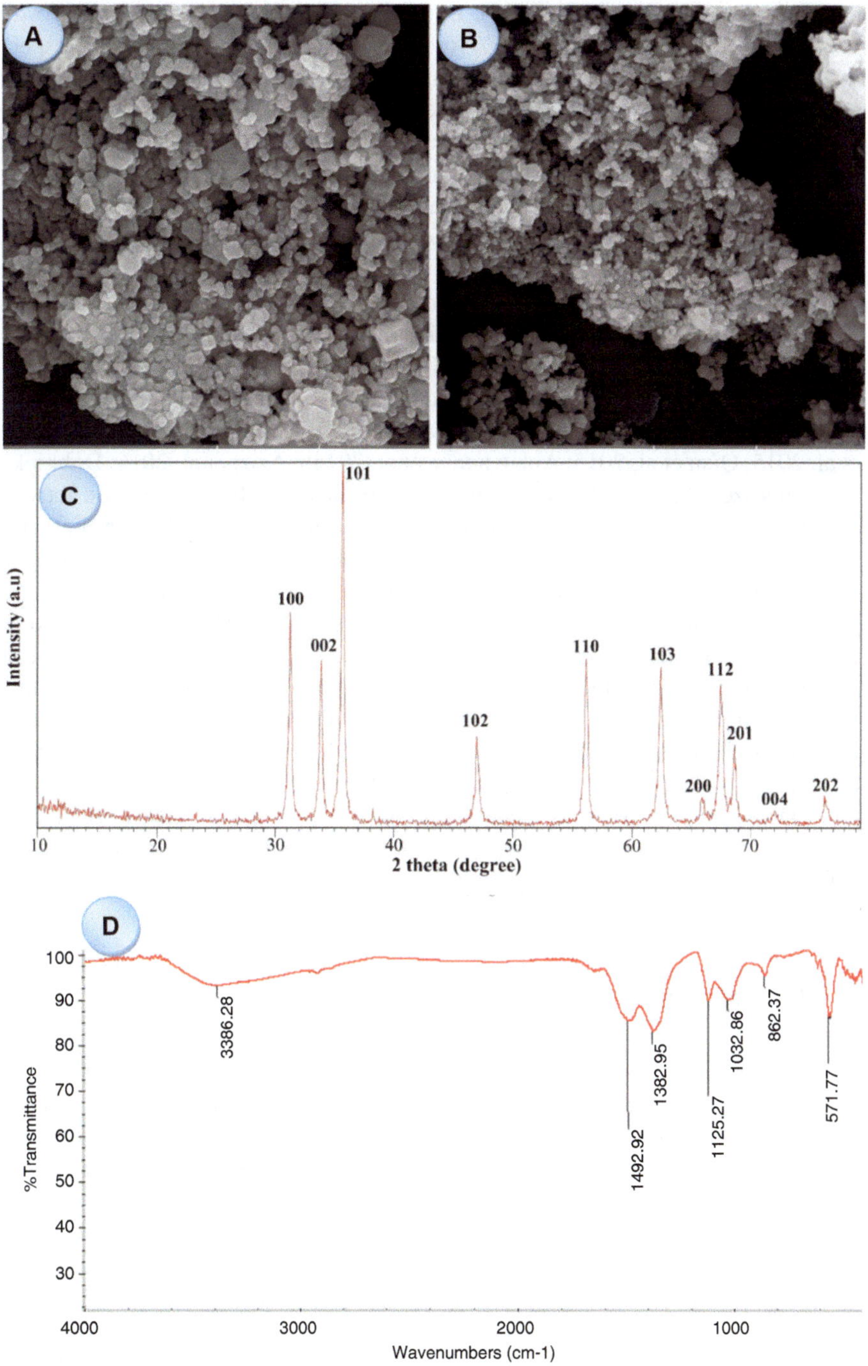

Fig. 7.3 FESEM images of (**a, b**), XRD patterns (**c**) and FTIR spectrum (**d**) of ZnO NPs calcined at 500 °C. (Adapted from Sorbiun et al. (2018))

concentrations (Bhuyan et al. 2015). In terms of percentage, the bacterial growth was reduced by 69.2%, 57% and 52% for *E. coli*, *S. aureus* and *S. pyogenes*, respectively. Ali et al. (2016) reported a remarkable antibacterial activity of ZnO NPs against *E. coli* and *S. aureus*. They opined that the *A. barbadensis*-mediated ZnO NPs induced cellular and tissue damage in the bacterial stains and, therefore, have a higher efficiency than the conventional antibiotics. ZnO NPs obtained from milky latex of *C. papaya* have also shown remarkable photocatalytic activity and antibacterial activity against *Pseudomonas aeruginosa* and *S. aureus* (Sharma 2016). Recently, ZnO NPs fabricated from aqueous leaf extract of *T. divaricata* have displayed a high antibacterial activity against *S. aureus* and *E. coli* and a little less against *Salmonella paratyphi* in comparison to standard pharmaceutical formulations (Raja et al. (2018). Earlier, many other studies have demonstrated the antibacterial activity of ZnO NPs fabricated from various plants and their parts against different microbes (Senthilkumar and Sivakumar 2014; Abdul et al. 2014; Ramesh et al. 2015; Qian et al. 2015; Anbuvannan et al. 2015b; Azizi et al. 2016; Dobrucka and Długaszewska 2016; Vanathi et al. 2016; Madan et al. 2016). Cytotoxic activity possibly involves both the production of ROS and deposition of NPs in the cytoplasm and/or on the outer cell membrane. The mechanism of NPs' action against microbes possibly involves activation of ZnO NPs by light and penetration into the bacterial cell wall via diffusion. It has been established from SEM and TEM studies of the bacterial cells that ZnO NPs disintegrate the cell membrane and accumulate in the cytoplasm where they interact with biomolecules causing cell apoptosis that leads to cell death. It has also been reported that NPs induce production of reactive oxygen species (hydroxyl radicals, superoxides and hydrogen peroxide) and put the cells under oxidative stress, causing damage to cellular components (lipids, proteins, DNA, etc.) (Lovric et al. 2005; Xia et al. 2006; Long et al. 2006), which ultimately triggers apoptosis. However, if ZnO NPs produce ROS, they can damage the skin and cannot be used in sunscreen lotion.

7.3.2 Antifungal Activity

Some reports are also available on the role of ZnO NPs as an antifungal agent. Narendhran and Sivaraj (2016) used *Lantana aculeata* to produce ZnO NPs (spherical, 12 ± 3 nm in size) and tested their antifungal activity. They inhibited *Aspergillus flavus* maximally *Fusarium oxysporum* to a lesser extent. *Parthenium hysterophorus*, a weed plant, was used to fabricate ZnO NPs (spherical and hexagonal; 27 ± 5 nm and 84 ± 2 nm in size, respectively) which were tested against fungal pathogens (Rajiv et al. 2013). The highest zone of inhibition was recorded in 25 μg ml^{-1} of 27 ± 5 nm NPs against *Aspergillus flavus* and *A. niger*. Similarly, some other studies have shown the antifungal efficacy of ZnO NPs fabricated from plants and their parts, against fungal pathogens (Elumalai et al. 2015a,b; Kathiravan et al. 2015; Jamdagni et al. 2018). The antifungal effect of NPs is dependent on their size and concentration; possibly, an increase in NPs' concentration may increase the inhibition of fungal growth, due to burst of fungal cell membrane, and cause a decline in the fungal enzymatic activity (Kim et al. 2012a; Siddiqi and Husen 2016).

7.3.3 Photocatalytic Activity

It is an established fact that the toxic effluents including hazardous organic dyes discharged from various sources (mainly from fabric, printing, manufacturing and other industries) adversely influence the aquatic ecosystem. Thus, degradation of hazardous organic dyes and other toxic wastes in water system has always been a key issue. ZnO NPs fabricated from various plant sources have shown notable photocatalytic degradation potential against various dyes such as methyl orange, methylene blue, Congo red and so on. *Anisochilus carnosus* leaf extract was used to obtain ZnO NPs, which showed photocatalytic degradation of methylene blue dye (Anbuvannan et al. 2015b). Fruit peel extract of *Nephelium lappaceum* was used to produce ZnO NPs, which caused 83.99% photocatalytic degradation of methyl orange azo dye in 120 min under UV light of the wavelength 364 nm (Karnan and Selvakumar 2016). Seaweed-capped ZnO NPs were produced from four zinc source precursors, namely, zinc acetate, zinc nitrate and zinc sulphate, using the brown seaweed (*Padina tetrastromatica*). All these NPs exhibited positive photocatalytic activity against anionic and cationic dyes drimarene turquoise blue S-G and methylene blue in sunlight (Pandimurugan and Thambidurai 2016). Similarly, Momeni et al. (2016) produced ZnO NPs from leaf extract of *Euphorbia prolifera* and found them having a good catalytic potential to degrade methylene blue and Congo red in the presence of $NaBH_4$ in water at room temperature. According to Raja et al. (2018), *T. divaricata*-mediated ZnO NPs brought out a complete degradation of methylene blue dye (in 90 min), suggesting that the prepared catalyst is applicable for bacterial decontamination as well as for industrial wastewater treatment.

7.3.4 Antioxidant, Antidiabetic and Anti-inflammatory Properties

Antioxidant compounds prevent oxidation of other molecules. It is known that the oxidation is a chemical reaction that can produce free radicals, causing chain reactions that may harm or damage cells. ZnO NPs produced from *Cassia fistula* have shown remarkable antioxidant activity through scavenging of 1,1-diphenyl-2-picrylhydrazyl free radicals (Suresh et al. 2015). Those fabricated with the help of *Polygala tenuifolia* root extract (Nagajyothi et al. 2015) exhibited a moderate antioxidant activity by scavenging 45.47% DPPH at 1 mg mL^{-1} and an excellent anti-inflammatory activity in a dose-dependent manner and suppressed the mRNA and protein expressions of intracellular nitric oxide synthase activity, COX-2, interleukin-1β, interleukin-6 and TNF-α (pro-inflammatory cytokines). Aqueous stem extract of *Ruta graveolens* was used as the reducing agent for fabrication of ZnO NPs, which could inhibit the antioxidant activity of 1,1-diphenyl-2-picrylhydrazyl free radicals effectively (Lingaraju et al. 2016). In a recent study, Rajakumar et al. (2018) have suggested that the *A. paniculata* leaf extract-mediated synthesized ZnO NPs have antioxidant, antidiabetic and anti-inflammatory abilities and can therefore be utilized in various biological applications as in cosmetic, food and biomedical industries.

7.3.5 Drug and Gene Delivery

Nanomaterials, for instance, carbon nanotubes, Fe_3O_4 NPs, etc., loaded with drug, are capable to penetrate inside the cells and deliver drugs at target sites efficiently. In vivo studies have shown that ZnO NP can be highly toxic to cancer cells, bacteria and leukemic T cell (Hanley et al. 2008; Wang et al. 2009). ZnO quantum dots with intrinsic blue fluorescence, coated with folate-conjugated chitosan via electrostatic interaction and loaded with doxorubicin (DOX – used in chemotherapy), can entrap the drug, e.g. doxorubicin, and release at the normal physiological pH of 7.4 (Yuan et al. 2010). ZnO NPs coated with aminopolysiloxane have been evaluated in leukaemia K562 and adriamycin-resistant K562/A02 cells (Guo et al. 2008), which were found to be sensitive to ZnO NPs. Cell proliferation is suppressed by UV light after incubation with NPs. The toxicity was dependent on the size of ZnO NPs. ZnO NPs can induce ROS in aqueous solution upon absorption of UV light which can induce necrosis or death of the cell (Wilson and Patterson 2008). ZnO nanorods of 20x50nm exhibited minimum cellular toxicity to normal cells and therefore were used to deliver drugs in cancer patients (Zhang et al. 2011). Kundu et al. (2014) used actinobacteria *Rhodococcus pyridinivorans* NT2 to examine the cytotoxicity of anthraquinone-loaded ZnO NPs against HT-29 cancer cells with detectable ability to kill the cancer cell in comparison to normal peripheral blood mononuclear cells. The authors suggested that these particles have the possible application as an effective drug delivery vehicle. Nanomaterial has been employed for cancer treatment, gene delivery and gene therapy. Three-dimensional tetrapod like ZnO nanostructures has been investigated as gene vectors to deliver pEGFP-N1 DNA (which contains the gene for green fluorescent protein) to A375 human melanoma cells (Nie et al. 2006, 2007). It does not show significant cytotoxicity possibly due to three-dimensional geometry of Zn. Since a tetrahedral structure has the ability to attach another tetrapod in all directions, it is easy for ZnO to interact with any structural motif. The bond angle in such structure is 109°; therefore all distances are equal and can easily be fitted with another ZnO tetrapod. ZnO quantum dots coated with polymers are also used to condense pDNA for efficient gene delivery (Zhang and Liu 2010).

7.3.6 Anticancer Activity

Anticancer activity of ZnO NPs is associated with their ability to generate ROS and induce apoptosis. ZnO NPs have been exploited as drug carriers for loading and transporting drugs to target cells. This prevents non-targeted toxicity and also reduces quantity of the drug (Ryter et al. 2007). Interaction of cancer cells with ZNO NPs is driven by electrostatic interaction between them (Rasmussen et al. 2010). ZnO NPs have been found to be most toxic against T98G cancer cells (Wahab et al. 2013). Chemical compounds showing positive response in the Ames test (Ames et al. 1975) are most likely to be carcinogenic in mammals, and, by analogy, they may affect human somatic germ cells leading to many ailments including infertility and cancer. If the quantity and exposure time of the ZnO is in excess to cause toxicity in microorganism, it may not be toxic to human beings because the quantity (10 μg mL^{-1}) is too small (Sawai et al. 1998; Yoshida et al. 2009). Adverse symptoms appear only if an excess of a toxic substance has been administered. However, it requires comprehensive clinical studies for confirmation (Sharma et al. 2009; Kumar et al. 2011). ZnO NPs are a mixture of small and large NPs; the smaller ones may penetrate into the bacterial cell and interact with biomolecules in the cytoplasm, while the larger ones may adhere to the cell wall and inhibit their growth. Although ZnO NPs are reported to induce selective damage of cancer cells, their exact mode of transport to the target cell is still not well understood. Nonetheless, they have been shown to destroy certain types of cancer in human beings (Akhtar et al. 2012). It has been shown that the primary normal rat cells (astrocytes and hepatocytes) are unaffected by ZnO NPs (Akhtar et al. 2012). ZnO NPs selectively induce apoptosis in almost all types of cancer cells via ROS. At 5 μg ml^{-1}, the NPs were ineffective, but a significant reduction in cell viability was noted for all cancer cells at a concentration of 10 and 15 μg ml^{-1}(Akhtar et al. 2012). Ostrovsky et al. (2009) have reported that ZnO NPs are cytotoxic to human glioma cell lines but they do not affect the normal human astrocytes. This is also true that the normal living cells have the inherent tendency to undergo repair when exposed to ZnO NPs. Vimala et al. (2014) used fruit extract of *Borassus flabellifer* for fabrication of ZnO NPs and studied their effect, along with DOX, on breast cancer cell line (MCF-7) and colon cancer cell line (HT-29). They noted the biocompatibility of DOX-ZnO NPs with no significant toxicity on blood stream (in vivo) and reduction of Bcl-2 expression on MCF-7 cells (in vitro). Further, they examined the proliferation rates of cancerous cells and the cytotoxic effect of ZnO NPs with varying doses of ZnO, DOX and DOX-ZnO NPs; MTT assay was done with IC50 values of 0.5 μg mL^{-1}, 0.25 μg mL^{-1} and 0.125 μg mL^{-1} for MCF-7 cancer cells and of 0.5, 0.25 and 0.125 μg mL^{-1} for HT-29 cancer cells and demonstrated that the cytotoxicity of DOX-ZnO NPs is more prominent than that of DOX and ZnO NPs alone. Moreover, they observed a higher in vitro reduction of cancerous cell proliferation at higher doses and a minor effect on target cancer cells at lower doses in a concentration-dependent manner. Vimala et al. (2014) have suggested that NPs of varied size facilitate drug targeting and hence turn out a suitable agent to improve drug delivery for the leukaemia cancerous cells. In another study, *Aloe barbadensis* was used to produce ZnO NPs, which exhibited antibacterial and antibiofilm features and thus found suitable as nano-antibiotic or a drug carrier for delivery to target cancerous cell (Ali et al. 2016).

7.3.7 Biosensor

Biosensors are used in healthcare, biological analysis, environmental monitoring and food industry (Zhao et al. 2010). Nanomaterials generally immobilize biomolecules such as enzymes, antibodies and several other proteins. They can also allow for direct electron transfer between the active site of biomolecules and the electrode. ZnO nanomaterial-based biosensors are mainly used for the detection of small molecules such as glucose, H_2O_2, phenols, cholesterols, urea, etc. (Wang et al. 2002; Liu et al. 2005; Singh et al. 2007; Solanki et al. 2008; Ali et al. 2009; Hu et al. 2011).

7.3.8 Bioimaging

Semiconductor nanocrystals possess better photoluminescent features in comparison to the traditional organic fluorescent dyes. Moreover, because of a large variety of nanostructure, low toxicity and biodegradability of ZnO have become an indispensable substance for all mammals. Besides its application as nano optical device, energy storage, nano sensors and semi-conductor (Huang et al. 2001; Fan and Lu 2005; Wang et al. 2007; Lao et al. 2007; Yang et al. 2011; Yakimova et al. 2012), it is frequently used as an essential supplement for human system and in biomedical application such as drug delivery and gene delivery, including cellular imaging (Zvyagin et al. 2008). Transferrin-conjugated green fluorescent ZnO nanocrystals have been reported for cancer cell imaging with minimum cytotoxicity (Sudhagar et al. 2011). ZnO nanocrystals doped with cobalt, copper and nickel cations have been employed for cellular imaging in various cells (Wu et al. 2008). It has been demonstrated that Zn nanowires can be employed to targeted imaging of cancer cells (Hong et al. 2011; Shi et al. 2011).

7.3.9 Cosmetics

The inflammatory effect of ZnO NPs on eosinophils has been thoroughly studied by Donaldson and Seaton (2012) and Donaldson and Poland (2012). It has been observed that the ZnO NPs in sunscreen do not penetrate deep into the skin. Monteiro-Riviere et al. (2011) analysed the effect of ZnO and TiO_2 NP into porcine UVB (ultraviolet-B radiation)-damaged skin in vivo and in vitro. Generally, ZnO

NP penetrated into the superficial layers of epidermis, although penetration was deep into the UVB-damaged skin. ZnO NPs are frequently used in personal care products along with food additives and preservatives (Brayner et al. 2010). Application of NP in cosmetics has increased tremendously in the recent past. Gulson et al. (2010) demonstrated that sunscreen cream containing isotopic Zn^{68}, generally applied on the skin to protect against UV radiation, is negligibly absorbed. Only 0.1% of all the Zn in the blood contained isotopically labelled Zn^{68}. This isotope in the blood/urine was almost negligible as its concentration was nearly 1/1000th of the total Zn absorbed. The metal oxide NPs commonly used in sunscreen lotions are activated by the absorption of light with subsequent generation of free radicals and ROS (Nel et al. 2006). Many workers have reported the production of free radicals from TiO_2/ZnO, which damage DNA in human skin cells when exposed to UV light (Dunford et al. 1997; Nakagawa et al. 1997). According to the WHO (2006) definition, dermal absorption describes the transport of chemicals from the outer surface of the skin to the systemic circulation, and dermal penetration describes the entry of a substance into a particular layer or structure. The commercially available sunscreens contain isopropyl myristate, a chemical enhancer of skin penetration (Chan 2005), and a chelating agent, EDTA. The amount of Zn absorbed from the sunscreen is less than 0.001% of the applied dose containing ZnO NP. It is, therefore, safe to use ZnO NP in sunscreen lotion to protect the skin from the damage by UV radiation. Since EDTA is a chelating agent, it binds ZnO strongly but, being very small in quantity, remains in liquid form due to moisture or sweat. EDTA works as a carrier to transport Zn^{2+} ions into the blood stream.

7.3.10 Agriculture

ZnO NPs influence (enhanced or reduced) the growth and development of plants, when absorbed and incorporated into the plant transport system. They diffuse in the plant cells if smaller in size than the pores in the plant cell walls. Studies have shown that the absorption and translocation of various NPs in different parts of the plant depend on their bioavailability, concentration, solubility, exposure time as well as the anatomy of the plant species concerned (Rico et al. 2011; Husen and Siddiqi 2014; Zhao et al. 2014). In addition, replacing biofertilizers with bionanomaterials may sometimes be advantageous if they increase the fruit count, seed production and biomass accumulation without producing toxic effects. In an experiment, Prasad et al. (2012) used seeds of *Arachis hypogea* to examine the influence of ZnO NPs on their growth and yield parameters. Various doses of ZnO NPs (25 nm) influenced the overall plant-growth response in terms of seed germination, seedling vigour index, root growth, flowering, chlorophyll content and pod yield. The authors noted that at 1000 ppm of dose, ZnO NPs increased seed germination and seedling vigour and also increased the stem and root growth. Early germination and establishment of seeds in the soil caused early flowering and promoted leaf chlorophyll content. Foliar treatment of ZnO NPs to *Cyamopsis tetragonoloba* and

Solanum lycopersicum has shown a positive response in terms of biomass production and the chlorophyll and total soluble leaf protein contents (Raliya and Tarafdar 2013; Raliya et al. 2015). However, ZnO NPs' treatment to *Arabidopsis* seedlings at 100 mg L^{-1} concentration decreased the biomass production (Landa et al. 2012). Zafar et al. (2016) have demonstrated that ZnO NPs at 500–1500 mg L^{-1} negatively influenced the *Brassica nigra* seed germination and seedling growth. ZnO NPs also enhanced antioxidative activities and non-enzymatic antioxidant contents in *Brassica nigra* plants. In another study, ZnO NPs at 2000 mg kg^{-1} could not affect the root length and biomass production in *Cucumis sativus* grown on a loamy sand soil at pH 5.5 (Kim et al. 2012b).

Manzo et al. (2011) have demonstrated that ZnO NPs at 286 mg kg^{-1} influenced root elongation in *Lepidium sativum* sown in an artificial standard soil, whereas Du et al. (2011) noted that these NPs at 45.45 mg kg^{-1} (5 g per 110 kg soil) decreased the biomass production of *Triticum aestivum* cultivated in loamy clay soil at pH 7.36. In another study of wheat (*T. aestivum*), inhibition of root elongation was noticed in acid soil, whereas phytotoxicity was alleviated in the alkaline soil, and absorption of ZnO NP was doubled when Zn concentration in the soil was low. Soluble zinc in acidic soil was 200 times more with 10 times more Zn levels in shoots than those in the alkaline soil. Phytotoxicity was noticed in the soil spiked with humic acid, but it did not affect plant responses. The ZnO NPs' aggregation with humic acid provides bioavailable zinc. But these NPs may be distributed within the plant body only if taken up through diffusion. The plant roots are stunted in the acidic soil as the amount of soluble zinc was 100 times more in acidic soil than in alkaline soil. Nevertheless, higher concentration of Zn at 500 mg L^{-1} produced phytotoxicity to *T. aestivum* (Watson et al. 2015). In *Allium cepa* grown in hydroponic solution, Kumari et al. (2011) observed high toxicity of ZnO NPs in terms of Zn^{2+} ion, possibly due to high release of ROS. In *Vigna unguiculata* grown in amended soil with ZnO NPs and Zn^{2+}, Wang et al. (2013) found no difference in plant growth, accumulation or speciation between the treatments with zinc ions and ZnO NPs. In a recent study, ZnO NPs fabricated from *Calotropis gigantea* leaf extract (Chaudhuri and Malodia 2017) were used at the concentration of 30 mg L^{-1} to examine their effect on growth of *Azadirachta indica*, *Alstonia scholaris* and *Pongamia pinnata* at nursery stage. In comparison to the control, all treated plant populations exhibited enhanced growth in height. Among the three tested plants, *A. scholaris* exhibited maximum height. Milani et al. (2012) have reported the use of ZnO NPs as a fertilizer in soil as well as a foliar fertilizer. Studies have also revealed that ZnO NPs are useful in plant protection and in the suppression of fungal diseases. For instance, He et al. (2011) recorded the biocidal activity of ZnO NPs against post-harvest pathogenic fungi, namely, *Botrytis cinerea* and *Penicillium expansum*, which were exposed to ZnO NP of 70 ± 15 nm size, in a concentration range of 0–12 mM. Examination of plants and pathogens before and after their incubation with ZnO NPs revealed that the concentration above 3 mM significantly inhibits the growth of the above-mentioned fungi by deforming the structure of fungal hyphae. The results were supported by Raman spectral data which showed that the peaks for nucleic acids, carbohydrates, proteins and lipids undergo tremendous alterations. It was also observed that the efficacy of ZnO NPs on the inhibition of fungal growth was not uniform, possibly due to change in species and their morphology (He et al. 2011). It has been reported that integration of ZnO NPs into bacterial cell may induce continuous release of membrane lipids and proteins, which change the membrane permeability of bacterial cells (Amro et al. 2000). It is, however, certain that penetration of ZnO NPs into the microbial cell is the key to growth inhibition of microorganisms.

7.4 Conclusion

Green synthesis of ZnO NPs with the help of various plants and/or their parts and products has gained the ground because of their multifarious applications. The active compounds present in plant tissues have been shown to act as a reducing agent as well as the capping and stabilizing agents for the fabrication of ZnO NPs. The size and morphology of ZnO NPs are dependent on several factors such as incubation time, temperature, concentration and pH of the solution, among others. ZnO NPs have shown their inhibitory effect on the bacterial and fungal growth and hence are useful in food packaging and agriculture. They are also useful in drug delivery, gene delivery, bioimaging, treatment of different skin conditions, cosmetics and agricultural system. However, their toxicity towards biological systems is a matter of conjecture even though they seem to have a bright future in biomedical applications.

References

Abdul H, Sivaraj R, Venckatesh R (2014) Green synthesis and characterization of zinc oxide nanoparticles from *Ocimum basilicum* L. var. *purpurascens* Benth. – Lamiaceae leaf extract. Mater Lett 131:16–18

Akhtar MJ, Ahamed M, Kumar S, Khan MM, Ahmad J, Alrokayan SA (2012) Zinc oxide nanoparticles selectively induce apoptosis in human cancer cells through reactive oxygen species. Int J Nanomedicine 7:845–857

Aladpoosh R, Montazer M (2015) The role of cellulosic chains of cotton in biosynthesis of ZnO nanorods producing multifunctional properties: mechanism, characterizations and features. Carbohydr Polym 126:122–129

Ali A, Ansari AA, Kaushik A, Solanki PR, Barik A, Pandey MK, Malhotra BD (2009) Nanostructured zinc oxide film for urea sensor. Mater Lett 63:2473–2475

Ali K, Dwivedi S, Azam A, Saquib Q, Al-said MS, Alkhedhairy AA, Musarrat J (2016) Aloe vera extract functionalized zinc oxide nanoparticles as nanoantibiotics against multi-drug resistant clinical bacterial isolates. J Colloid Interface Sci 472:145–156

Ambika S, Sundrarajan M (2015) Green biosynthesis of ZnO nanoparticles using *Vitex negundo* L. extract: spectroscopic investigation of interaction between ZnO nanoparticles and human serum albumin. J Photochem Photobio. B Biol 149:143–148

Ames BN, McCann J, Yamasaki E (1975) Methods for detecting carcinogens and mutagens with the Salmonella/mammalian microsome mutagenicity test. Mutation Res 31:347–364

Amro NA, Kotra LP, Wadu-Mesthrige K, Bulychev A, Mobashery S, Liu GY (2000) High-resolution atomic force microscopy studies of the *Escherichia coli* outer membrane: structural basis for permeability. Langmuir 16:2789–2796

Anbuvannan M, Ramesh M, Viruthagiri G, Shanmugam N, Kannadasan N (2015a) Synthesis, characterization and photocatalytic activity of ZnO nanoparticles prepared by biological method. Spectrochim Acta A Mol Biomol Spectrosc 143:304–308

Anbuvannan M, Ramesh M, Viruthagiri G, Shanmugam N, Kannadasan N (2015b) *Anisochilus carnosus* leaf extract mediated synthesis of zinc oxide nanoparticles for antibacterial and photocatalytic activities. Mater Sci Semicond Process 39:621–628

Araujo-Lima CF, Nunes RJM, Carpes RM, Aiub FAF, Felzenszwalb I (2017) Pharmacokinetic and toxicological evaluation of a zinc gluconate-based chemical sterilant using in vitro and in silico approaches. BioMed Res Inter 2017:5746768

Auld DS (2001) Zinc coordination sphere in biochemical zinc sites. Biometals 14:271–313

Awwad AM, Albiss B, Ahmad AL (2014) Green synthesis, characterization and optical properties of zinc oxide nanosheets using *Olea europea* leaf extract. Adv Mater Lett 5:520–524

Azizi S, Mohamad R, Bahadoran A, Bayat S, Rahim RA, Ariff A, Saad WZ (2016) Effect of annealing temperature on antimicrobial and structural properties of bio-synthesized zinc oxide nanoparticles using flower extract of *Anchusa italic*. J Photochem Photobiol B Biol 161:441–449

Bala N, Saha S, Chakraborty M, Maiti M, Das S, Basu R, Nandy P (2015) Green synthesis of zinc oxide nanoparticles using *Hibiscus subdariffa* leaf extract: effect of temperature on synthesis, antibacterial activity and anti-diabetic activity. RSC Adv 5:4993–5003

Bhuyan T, Mishra K, Khanuja M, Prasad R, Varma A (2015) Biosynthesis of zinc oxide nanoparticles from *Azadirachta indica* for antibacterial and photocatalytic applications. Mater Sci Semicond Process 32:55–61

Brahms J, Mattai J, Jacoby R, Chopra S, Guenin E (2005) Dry deodorant containing a sesquiterpene alcohol and zinc oxide. U.S. Patent 20050191257 A1

Brayner R, Dahoumane SA, Yéprémian C, Djediat C, Meyer M, Couté A, Fiévet F (2010) ZnO nanoparticles: synthesis, characterization, and ecotoxicological studies. Langmuir 2010:6522–6528

Brown HE (1976) Zinc oxide: properties and applications. International Lead Zinc Research Organization, New York

Chan TCK (2005) Percutaneous penetration enhancers: an update. In: Proceedings 9th Biennial Conference of Perspectives in Percutaneous Penetration. 13 April, vol 2004. La Grande-Motte, France, pp 18–23

Chaudhuri SK, Malodia L (2017) Biosynthesis of zinc oxide nanoparticles using leaf extract of *Calotropis gigantea*: characterization and its evaluation on tree seedling growth in nursery stage. Appl Nanosci 7:501–512

Darrudi M, Oskuee RK, Kargar H (2013) Sol-gel synthesis, characterization and neurotoxicity effect of zinc oxide nanoparticles using gum tragacanth. Ceram Int 40:4827–4831

De Graaf TP, Galley E, Butcher KE (1999) Use of an antimicrobial agent. European Patent EP1079799

Deng Z, Chen M, Gu G, Wu L (2008) A facile method to fabricate ZnO hollow spheres and their photocatalytic property. J Phys Chem B 112:16–22

Ding Y, Wang ZL (2009) Structures of planar defects in ZnO nanobelts and nanowires. Micron 40:335–342

Divyapriya S, Sowmia C, Sasikala S (2014) Synthesis of zinc oxide nanoparticles and microbial activity of *Murraya koenigii*. World J Pharm Pharm Sci 12:1635–1645

Dobrucka R, Długaszewska J (2016) Biosynthesis and antibacterial activity of ZnO nanoparticles using *Trifolium pratense* flower extract. Saudi J Bio Sci 23:517–523

Donaldson K, Poland CA (2012) Inhaled nanoparticles and lung cancer – what we can learn from conventional particle toxicology. Swiss Med Wkly. https://doi.org/10.4414/smw.2012.13547

Donaldson K, Seaton A (2012) A short history of the toxicology of inhaled particles. Part Fibre Toxico, vol 9, p 13

Du WC, Sun YY, Ji R, Zhu JG, Wu JC, Guo HY (2011) TiO_2 and ZnO nanoparticles negatively affect wheat growth and soil enzyme activities in agricultural soil. J Environ Monit 13:822–828

Dunford R, Salinaro A, Cai L, Serpone N, Horikoshi S, Hidaka H, Knowland J (1997) Chemical oxidation and DNA damage catalysed by inorganic sunscreen ingredients. FEBS Lett 418:87–90

Elumalai K, Velmurugan S (2015) Green synthesis, characterization and antimicrobial activities of zinc oxide nanoparticles from the leaf extract of *Azadirachta indica*. Appl Surf Sci 345:329–336

Elumalai K, Velmurugan S, Ravi S, Kathiravan V, Adaikala Raj G (2015a) Bio-approach: plant mediated synthesis of ZnO nanoparticles and their catalytic reduction of methylene blue and antimicrobial activity. Adv Powder Technol 26:1639–1651

Elumalai K, Velmurugan S, Ravi S, Kathiravan V, Ashokkumar S (2015b) Green synthesis of zinc oxide nanoparticles using *Moringa oleifera* leaf extract and evaluation of its antimicrobial activity. Spectrochim Acta A Mol Biomol Spectrosc 143:158–164

Fan Z, Lu JG (2005) Zinc oxide nanostructures: synthesis and properties. J Nanosci Nanotechnol 5:1561–1573

Frederickson CJ, Koh JY, Bush AI (2005) The neurobiology of zinc in health and disease. Na Rev Neurosci 6:449–462

Fu L, Fu Z (2015) *Plectranthus amboinicus* leaf extract-assisted biosynthesis of ZnO nanoparticles and their photocatalytic activity. Ceram Int 41:2492–2496

Gnanasangeetha D, Thambwani DS (2013) Biogenic production of zinc oxide nanoparticles using Acalypha indica. J Chem Biol Phys Sci 1:238–246

Gulson B, McCall M, Korsch M, Gomez L, Casey P, Oytam Y, Taylor A, McCulloch M, Trotter J, Kinsley L, Greenoak G (2010) Small amounts of zinc from zinc oxide particles in sunscreens applied outdoors are absorbed through human skin. Toxicol Sci 118:140–149

Guo J, Peng C (2015) Synthesis of ZnO nanoparticles with a novel combustion method and their C_2H_5OH gas sensing properties. Ceram Int 41:2180–2186

Guo D, Wu C, Jiang H, Li Q, Wang X, Chen B (2008) Synergistic cytotoxic effect of different sized ZnO nanoparticles and daunorubicin against leukemia cancer cells under UV irradiation. J Photochem Photobiol B 93:119–126

Halioua B, Ziskind B (2005) Medicine in the days of the pharaohs. Belknap Press of Harvard University Press, Cambridge

Hanley C, Layne J, Punnoose A, Reddy KM, Coombs I, Coombs A, Feris K, Wingett D (2008) Preferential killing of cancer cells and activated human T cells using ZnO nanoparticles. Nanotechnology 19:295103

He L, Liu Y, Mustapha A, Lin M (2011) Antifungal activity of zinc oxide nanoparticles against *Botrytis cinerea* and *Penicillium expansum*. Microbiol Res 166:207–215

Hong H, Shi J, Yang Y, Zhang Y, Engle JW, Nickles RJ, Wang X, Ca W (2011) Cancer-targeted optical imaging with fluorescent zinc oxide nanowires. Nano Lett 11:3744–3750

Hu FX, Chen SH, Wang CY, Yuan R, Chai Y, Xiang Y, Wang C (2011) ZnO nanoparticle and multiwalled carbon nanotubes for glucose oxidase direct electron transfer and electrocatalytic activity investigation. J Mol Catal B Enzym 72:298–304

Huang MH, Wu Y, Feick H, Tran N, Weber E (2001) Catalytic growth of zinc oxide nanowires by vapor transport. Adv Mater 13:113–116

Hubbard NB, Culpepper ML, Howell LL (2006) Actuators for micro positioners and nano positioners. App Mech Rev 59:324–334

Husen A, Siddiqi KS (2014) Phytosynthesis of nanoparticles: concept, controversy and application. Nano Res Lett 9:229

Izu N, Shimada K, Akamatsu T, Itoh T, Shin W, Shiraishi K, Usui T (2014) Polyol synthesis of Al-doped ZnO spherical nanoparticles and their UV–vis–NIR absorption properties. Ceram Int 40:8775–8781

Jamdagni P, Khatri P, Rana JS (2018) Green synthesis of zinc oxide nanoparticles using flower extract of *Nyctanthes arbor-tristis* and their antifungal activity. J King Saud Univ – Sci 30:168–175

Jang JS, Yu CJ, Choi SH, Ji SM, Kim ES, Lee JS (2008) Topotactic synthesis of mesoporous ZnS and ZnO nanoplates and their photocatalytic activity. J Catal 254:144–155

Jansen J, Karges W, Rink L (2009) Zinc and diabetes–clinical links and molecular mechanisms. J Nutr Biochem 20:399–417

Kakiuchi K, Hosono E, Kimura T, Imai H, Fujihara S (2006) Fabrication of mesoporous ZnO nanosheets from precursor templates grown in aqueous solutions. J Sol-Gel Sci Technol 39:63–72

Karnan T, Selvakumar SAS (2016) Biosynthesis of ZnO nanoparticles using rambutan (*Nephelium lappaceum* L.) peel extract and their photocatalytic activity on methyl orange dye. J Mol Struct 1125:358–365

Kathiravan V, Ravi S, Ashokkumar S, Velmurugan S, Elumalai K, Khatiwada CP (2015) Green synthesis of silver nanoparticles using *Croton sparsiflorus* morong leaf extract and their antibacterial and antifungal activities. Spectrochim Acta A Mol Biomol Spectrosc 139:200–205

Kavitha S, Dhamodaran M, Prasad R, Ganesan M (2017) Synthesis and characterisation of zinc oxide nanoparticles using terpenoid fractions of *Andrographis paniculata* leaves. Int Nano Lett 7:141–147

Kavithaa K, Paulpandi M, Ponraj T, Murugan K, Sumathi S (2016) Induction of intrinsic apoptotic pathway in human breast cancer (MCF-7) cells through facile biosynthesized zinc oxide nanorods. Karbala Int J Mod Sci 2:46–55

Kim S, Lee S, Lee I (2012b) Alteration of phytotoxicity and oxidant stress potential by metal oxide nanoparticles in *Cucumis sativus*. Water Air Soil Pollut 223:2799–2806

Kim SW, Jung JH, Lamsal K, Kim YS, Min JS, Lee YS (2012a) Antifungal effects of silver nanoparticles (AgNPs) against various plant pathogenic fungi. Mycobiology 40:53–58

Kim MH, Seo JH, Kim HM, Jeong HJ (2014) Zinc oxide nanoparticles, a novel candidate for the treatment of allergic inflammatory diseases. Eur J Pharmacol 738:31–39

Klingshirn C (2007) ZnO: from basics towards applications. Phys Status Solidi 244:3027–3073

Krupa AND, Vimala R (2016) Evaluation of tetraethoxysilane (TEOS) sol-gel coatings, modified with green synthesized zinc oxide nanoparticles for combating microfouling. Mater Sci Eng C 61:728–735

Kumar A, Pandey AK, Sing SS, Shanker R, Dhawan A (2011) Cellular uptake and mutagenic potential of metal oxide nanoparticles in bacterial cells. Chemosphere 83:1124–1132

Kumari M, Khan SS, Pakrashi S, Mukherjee A, Chandrasekaran N (2011) Cytogenetic and genotoxic effects of zinc oxide nanoparticles on root cells of *Allium cepa*. J Hazard Mater 190:613–621

Kundu D, Hazra C, Chatterjee A, Chaudhari A, Mishra S (2014) Extracellular biosynthesis of zinc oxide nanoparticles using Rhodococcus pyridinivorans NT2: multifunctional textile finishing, biosafety evaluation and in vitro drug delivery in colon carcinoma. J Photochem Photobiol B Biol 140:194–204

Landa P, Vankova R, Andrlova J, Hodek J, Marsik P, Storchova H, White JC, Vanek T (2012) Nanoparticle-specific changes in Arabidopsis thaliana gene expression after exposure to ZnO, TiO_2, and fullerene soot. J Hazard Mater 241:55–62

Lao CS, Park MC, Kuang Q, Deng Y, Sood AK, Polla DL, Wang ZL (2007) Giant enhancement in UV response of ZnO nanobelts by polymer surface-functionalization. J Am Chem Soc 129:12096–12097

Lee HJ, Yeo SY, Jeong SH (2003) Antibacterial effect of nanosized silver colloidal solution on textile fabrics. J Mater Sci 38:2199–2204

Lingaraju K, Raja Naika H, Manjunath K, Basavaraj RB, Nagabhushana H, Nagaraju G, Suresh D (2016) Biogenic synthesis of zinc oxide nanoparticles using *Ruta graveolens* (L.) and their antibacterial and antioxidant activities. App Nanosci 6:703–710

Liu YL, Yang YH, Yang HF, Liu ZM, Shen GL, Yu RQ (2005) Nanosized flower-like ZnO synthesized by a simple hydrothermal method and applied as matrix for horseradish peroxidase immobilization for electro-biosensing. J Inorg Biochem 99:2046–2053

Long TC, Saleh N, Tilton RD, Lowry GV, Veronesi B (2006) Titanium dioxide (P25) produces reactive oxygen species in immortalized brain microglia (BV2): implications for nanoparticle neurotoxicity. Environ Sci Technol 40:4346–4352

Lovric J, Cho SJ, Winnik FM, Maysinger D (2005) Unmodified cadmium telluride quantum dots induce reactive oxygen species formation leading to multiple organelle damage and cell death. Chem Biol 12:1227–1234
Lu PJ, Huang SC, Chen YP, Chiueh LC, Shih DYC (2015) Analysis of titanium dioxide and zinc oxide nanoparticles in cosmetics. J Food Drug Anal 23:587–594
Madan HR, Sharma SC, Udayabhanu, Suresh D, Vidya YS, Nagabhushana H, Rajanaik H, Anantharaju KS, Prashantha SC, Sadananda Maiya P (2016) Facile green fabrication of nanostructure ZnO plates, bullets, flower, prismatic tip, closed pine cone: their antibacterial, antioxidant, photoluminescent and photocatalytic properties. Spectrochim Acta A Mol Biomol Spectrosc 152:404–416
Mahmud S, Abdullah MJ (2006) Nanotripods of zinc oxide, IEEE Conf. emerging technol—nanoelectron. pp 442–446
Mahmud S, Johar M, Abdullah PGA, Chong J, Mohamad AK (2006) Nanostructure of ZnO fabricated via french process and its correlation to electrical properties of semiconducting varistors. Synth React Inorg Met Org Chem Nano-Met Chem 36:155–159
Manokari M, Shekhawat MS (2016) Biogenesis of zinc oxide nanoparticles using *Couroupita guianensis* Aubl. extracts—a green approach. World Sci News 29:135–145
Manzo S, Rocco A, Carotenuto R, Picione Fde L, Miglietta ML, Rametta G, Di Francia G (2011) Investigation of ZnO nanoparticles ecotoxicological effects towards different soil organisms. Environ Sci Pollut Res Int 18:756–763
Maremanda KP, Khan S, Jena G (2014) Zinc protects cyclophosphamide-induced testicular damage in rat: Involvement of metallothionein, tesmin and Nrf2. Biochem Biophys Res Commun 445:591–596
Mehr ES, Sorbiun M, Ramazani A, Fardood ST (2018) Plant-mediated synthesis of zinc oxide and copper oxide nanoparticles by using ferulago angulata (schlecht) boiss extract and comparison of their photocatalytic degradation of Rhodamine B (RhB) under visible light irradiation. J Mater Sci: Mater Electron 29:1333–1340
Milani N, McLaughlin MJ, Stacey SP, Kirby JK, Hettiarachchi GM, Beak DG, Cornelis G (2012) Dissolution kinetics of macronutrient fertilizers coated with manufactured zinc oxide nanoparticles. J Agric Food Chem 60:3991–3998
Mirzaeia H, Darroudi M (2017) Zinc oxide nanoparticles: Biological synthesis and biomedical applications. Ceram Int 43:907–914
Mishra V, Sharma R (2015) Green synthesis of zinc oxide nanoparticles using fresh peels extract of *Punica granatum* and its antimicrobial activities. Int J Pharma Res Health Sci 3:694–699 14
Moezzi A, Cortie M, McDonagh A (2011) Aqueous pathways for the formation of zinc oxide nanoparticles. Dalton Trans 40:4871–4878
Momeni SS, Nasrollahzadeh M, Rustaiyan A (2016) Green synthesis of the Cu/ZnO nanoparticles mediated by *Euphorbia prolifera* leaf extract and investigation of their catalytic activity. J Colloid Interface Sci 472:173–179
Monteiro-Riviere NA, Wiench K, Landsiedel R, Schulte S, Inman AO, Riviere JE (2011) Safety evaluation of sunscreen formulations containing titanium dioxide and zinc oxide nanoparticles in UVB sun burned skin: an in vitro and in vivo study. Toxicol Sci 123:264–280
Nagajyothi PC, Sreekanth TVM, Tettey CO, Jun YI, Mook SH (2014) Characterization, antibacterial, antioxidant, and cytotoxic activities of ZnO nanoparticles using *Coptidis rhizome*. Bioorg Med Chem Lett 24:4298–4303
Nagajyothi PC, Cha SJ, Yang IJ, Sreekanth TVM, Kim KJ, Shin HM (2015) Antioxidant and anti-inflammatory activities of zinc oxide nanoparticles synthesized using *Polygala tenuifolia* root extract. J Photochem Photobiol B Biol 146:10–17
Nakagawa Y, Wakuri S, Sakamoto K, Tanaka N (1997) The photogenotoxicity of titanium dioxide particles. Mutat Res 394:125–132
Narendhran S, Sivaraj R (2016) Biogenic ZnO nanoparticles synthesized using *L. aculeate* leaf extract and their antifungal activity against plant fungal pathogens. Bull Mater Sci 39:1–5
Nel A, Xia T, Mädler L, Li N (2006) Toxic potential of materials at the nanolevel. Science 311:622–627

Nie L, Gao L, Feng P, Zhang J, Fu X, Liu Y, Yan X, Wang T (2006) Three-dimensional functionalized tetrapod-like ZnO nanostructures for plasmid DNA delivery. Small 2:621–625

Nie L, Gao L, Yan X, Wang T (2007) Functionalized tetrapod-like ZnO nanostructures for plasmid DNA purification, polymerase chain reaction and delivery. Nanotechnology 18:015101

Ostrovsky S, Kazimirsky G, Gedanken A, Brodie C (2009) Selective cytotoxic effect of ZnO nanoparticles on glioma cells. Nano Res 2:882–890

Oudhia A, Kulkarni P, Sharma S (2015) Green synthesis of ZnO nanotubes for bioapplications. Int J Curr Eng Technol 1:280–281

Ozgur U, Ya IA, Liu C, Teke A, Reshchikov MA, Doğan S, Avrutin V, Cho SJ, Morkoç H (2005) A comprehensive review of ZnO materials and devices. J Appl Phys 98:041301

Pandimurugan R, Thambidurai S (2016) Novel seaweed capped ZnO nanoparticles for effective dye photodegradation and antibacterial activity. Adv Powder Technol 27:1062–1072

Patnaik P (2003) Handbook of inorganic chemicals. McGraw Hill, New York

Prasad TNVKV, Sudhakar P, Sreenivasulu Y, Latha P, Munaswamy V, Raja Reddy K, Sreeprasad TS, Sajanlal PR, Pradeep T (2012) Effect of nanoscale zinc oxide particles on the germination, growth and yield of peanut. J Plant Nutr 35:905–927

Qian Y, Yao J, Russel M, Chen K, Wang X (2015) Characterization of green synthesized nano-formulation (ZnO – *A. vera*) and their antibacterial activity against pathogens. Environ Toxicol Pharmacol 39:736–746

Raja A, Ashokkumar S, Pavithra Marthandam R, Jayachandiran J, Kathiwada CP, Kaviyarasu K, Ganapathi Raman R, Swaminathan M (2018) Eco-friendly preparation of zinc oxide nanoparticles using *Tabernaemontana divaricata* and its photocatalytic and antimicrobial activity. J Photochem Photobiol B: Biol 181:53–58

Rajakumar G, Thiruvengadam M, Mydhili G, Gomathi T, IllM C (2018) Green approach for synthesis of zinc oxide nanoparticles from *Andrographis paniculata* leaf extract and evaluation of their antioxidant, anti-diabetic, and anti-inflammatory activities. Bioprocess Biosyst Eng 41:21–30

Rajalakshmi M, Sohila S, Ramya S, Divakar R, Ghosh C, Kalavathi S (2012) Blue green and UV emitting ZnO nanoparticles synthesized through a non-aqueous route. Opt Mater 34:1241–1245

Rajendran SP, Sengodan K (2017) Synthesis and characterization of zinc oxide and iron oxide nanoparticles using *Sesbania grandiflora* leaf extract as reducing agent. J Nanosci 2017:8348507

Rajiv P, Rajeshwari S, Venckatesh R (2013) Bio-Fabrication of zinc oxide nanoparticles using leaf extract of *Parthenium hysterophorus* L. and its size-dependent antifungal activity against plant fungal pathogens. Spectrochim Acta A Mol Biomol Spectrosc 112:384–387

Raliya R, Tarafdar JC (2013) ZnO nanoparticle biosynthesis and its effect on phosphorous-mobilizing enzyme secretion and gum contents in Cluster bean (*Cyamopsis tetragonoloba* L.). Agric Res 2:48–57

Raliya R, Nair R, Chavalmane S, Wang WN, Biswas P (2015) Mechanistic evaluation of translocation and physiological impact of titanium dioxide and zinc oxide nanoparticles on the tomato (*Solanum lycopersicum* L.) plant. Metallomics 7:1584–1594

Ramesh M, Anbuvannan M, Viruthagiri G (2015) Green synthesis of ZnO nanoparticles using *Solanum nigrum* leaf extract and their antibacterial activity. Spectrochim Acta A Mol Biomol Spectrosc 136:864–870

Rasmussen JW, Martinez E, Louka P, Wingett DG (2010) Zinc oxide nanoparticles for selective destruction of tumor cells and potential for drug delivery applications. Expert Opin Drug Deliv 7:1063–1077

Research and Markets (2015) Global nanotechnology market outlook 2015–2020. Available at: http://www.prnewswire.com/news-releases/globalnanotechnology-market-outlook-2015-2020%2D%2D-industry-will-grow-to-reachus-758-billion-507155671.html

Rico CM, Majumdar S, Duarte-Gardea M, Peralta-Videa JR, Gardea-Torresdey JL (2011) Interaction of nanoparticles with edible plants and their possible implications in the food chain. J Agri Food Chem 59:3485–3498

Ryter SW, Kim HP, Hoetzel A, Park JW, Nakahira K, Wang X, Choi AM (2007) Mechanisms of cell death in oxidative stress. Antioxid Redox Signal 9:49–89

Samat NA, Nor RM (2013) Sol-gel synthesis of zinc oxide nanoparticles using *Citrus aurantifolia* extracts. Ceram Int 39:S545–S548

Sawai J, Kojima H, Kano F, Igarashi H, Hashimoto A, Kawada E, Kokugan T, Shimizu M (1998) Ames assay with *Salmonella typhimurium* TA102 for mutagenicity and antimutagenicity of metallic oxide powders having antibacterial activities. World J Microbiol Biotechnol 14:773–775

Senthilkumar SR, Sivakumar T (2014) Green tea *Camellia sinensis* mediated synthesis of zinc oxide nanoparticles and studies on their antimicrobial activities. Int J Pharm Pharm Sci 6:461–465

Sharma SC (2016) ZnO nano-flowers from *Carica papaya* milk: degradation of Alizarin Red-S dye and antibacterial activity against *Pseudomonas aeruginosa* and *Staphylococcus aureus*. Opt Int J Light Electron Opt 127:6498–6512

Sharma V, Shukla RK, Saxena N, Parmar D, Das M, Dhawan A (2009) DNA damaging potential of zinc oxide nanoparticles in human epidermal cells. Toxicol Lett 185:211–218

Sharma D, Sabela MI, Kanchi S, Mdluli PS, Singh G, Stenström TA, Bisetty K (2016) Biosynthesis of ZnO nanoparticles using *Jacaranda mimosifolia* flowers extract: synergistic antibacterial activity and molecular simulated facet specific adsorption studies. J Photochem Photobiol B Biol 162:199–207

Shekhawat MS, Ravindran CP, Manokari M (2015) A green approach to synthesize the zinc oxide nanoparticles using aqueous extracts of *Ficus benghalensis* L. Int J Biosci Agric Technol 6:1–5

Shen L, Zhang H, Guo S (2009) Control on the morphologies of tetrapod ZnO nanocrystals. Mater Chem Phys 114:580–583

Shi J, Hong H, Ding Y, Yang Y, Wang F, Cai W, Wang X (2011) Evolution of zinc oxide nanostructures through kinetics control. J Mater Chem 21:9000–9008

Siddiqi KS, Husen A (2016) Fabrication of metal nanoparticles from fungi and metal salts: scope and application. Nano Res Lett 11:98

Siddiqi KS, Husen A (2017) Plant response to engineered metal oxide nanoparticles. Nano Res Lett 12:92

Siddiqi KS, Rahman A, Tajuddin, Husen A (2018) Properties of zinc oxide nanoparticles and their activity against microbes. Nano Res Lett 13:141

Sindhura KS, Prasad TN, Selvam P, Hussain OM (2014) Synthesis, characterization and evaluation of effect of phytogenic zinc nanoparticles on soil exoenzymes. Appl Nanosci 4:819–827

Singh SP, Arya SK, Pandey P, Malhotra BD (2007) Cholesterol biosensor based on rf sputtered zinc oxide nanoporous thin film. Appl Phys Lett 91:063901

Solanki PR, Kaushik A, Ansari AA, Sumana G, Malhotra BD (2008) Zinc oxide-chitosan nanobiocomposite for urea sensor. Appl Phys Lett 93:163903

Sorbiun M, Mehr ES, Ramazani A, Fardood ST (2018) Green synthesis of zinc oxide and copper oxide nanoparticles using aqueous extract of oak fruit hull (Jaft) and comparing their photocatalytic degradation of basic violet 3. Int J Environ Res 12:29–37

Speight JG (2002) Chemical and process design handbook. McGraw Hill, Inc, New York

Sudhagar S, Sathya S, Pandian K, Lakshmi BS (2011) Targeting and sensing cancer cells with ZnO nanoprobes in vitro. Biotechnol Lett 33:1891–1896

Sundrarajan M, Ambika S, Bharathi K (2015) Plant-extract mediated synthesis of ZnO nanoparticles using *Pongamia pinnata* and their activity against pathogenic bacteria. Adv Powder Technol 26:1294–1299

Suresh J, Pradheesh G, Alexramani V, Sundrarajan M, Hong SI (2018) Green synthesis and characterization of zinc oxide nanoparticle using insulin plant (*Costus pictus* D. Don) and investigation of its antimicrobial as well as anticancer activities. Adv Nat Sci: Nanosci Nanotechnol 9:015008

Suresh D, Nethravathi PC, Udayabhanu, Rajanaika H, Nagabhushana H, Sharma SC (2015) Green synthesis of multifunctional zinc oxide (ZnO) nanoparticles using Cassia fistula plant extract and their photodegradative, antioxidant and antibacterial activities. Mater Sci Semicond Process 31:446–454

Tani T, Mdler L, Pratsinis SE (2002) Homogeneous ZnO nanoparticles by flame spray pyrolysis. J Nanopart Res 4:337–343

Thema FT, Manikandan E, Dhlamini MS, Maaza M (2015) Green synthesis of ZnO nanoparticles via *Agathosma betulina* natural extract. Mater Lett 161:124–127

Vanathi P, Rajiv P, Narendhran S, Rajeshwari S, Rahman PKSM (2016) Biosynthesis and characterization of phytomediated zinc oxide nanoparticles: a green chemistry approach. Mater Lett 134:13–15

Vayssieres L, Keis K, Hagfeldt A, Lindquist SE (2001) Three-dimensional array of highly oriented crystalline ZnO microtubes. Chem Mater 13:4395–4398

Vidya C, Hiremath S, Chandraprabha MN, Venugopal I, Jain A, Bansal K (2013) Green synthesis of ZnO nanoparticle by *Calotropis gigantea*. Int J Curr Eng Technol 4:118–120

Vimala K, Sundarraj S, Paulpandi M, Vengatesan S, Kannan S (2014) Green synthesized doxorubicin loaded zinc oxide nanoparticles regulates the Bax and Bcl-2 expression in breast and colon carcinoma. Process Biochem 49:160–172

Wahab R, Kaushik NK, Kaushik N, Choi EH, Umar A, Dwivedi S, Musarrat J, Al-Khedhairy AA (2013) ZnO nanoparticles induces cell death in malignant human T98G gliomas, KB and non-malignant HEK cells. J Biomed Nanotechnol 9:1181–1189

Wang ZL (2004a) Nanostructures of zinc oxide. Mater Tod 7:26–33

Wang ZL (2004b) Zinc oxide nanostructures: growth, properties and applications. J Phys Condens Mat16:R829–R858

Wang TX, Lou TJ (2008) Solvothermal synthesis and photoluminescence properties of ZnO nanorods and nanorod assemblies from ZnO2 nanoparticles. Mater Lett 62:2329–2331

Wang L, Muhammed M (1999) Synthesis of zinc oxide nanoparticles with controlled morphology. J Mater Chem 9:2871–2878

Wang G, Xu JJ, Ye LH, Zhu JJ, Chen HY (2002) Highly sensitive sensors based on the immobilization of tyrosinase in chitosan. Bioelectrochemistry 57:33–38

Wang X, Liu J, Song J, Wang ZL (2007) Integrated nanogenerators in biofluid. Nano Lett 7:2475–2479

Wang H, Wingett D, Engelhard MH, Feris K, Reddy KM, Turner P, Layne J, Hanley C, Bell J, Tenne D, Wang C, Punnoose A (2009) Fluorescent dye encapsulated ZnO particles with cell-specific toxicity for potential use in biomedical applications. J Mater Sci Mater Med 20:11–22

Wang P, Menzies NW, Lombi E, McKenna BA, Johannessen B, Glover CJ, Kappen P, Kopittke PM (2013) Fate of ZnO nanoparticles in soils and cowpea (*Vigna unguiculata*). Environ Sci Technol 47:13822–13830

Watson JL, Fang T, Dimkpa CO, Britt DW, McLean JE, Jacobson A, Anderson AJ (2015) The phytotoxicity of ZnO nanoparticles on wheat varies with soil properties. Biometals 28:101–112

Wilson BC, Patterson MS (2008) The physics, biophysics and technology of photodynamic therapy. Phys Med Biol 53:R61–R109

World Health Organization (2006) Dermal absorption EHC 235. WHO Press, World Health Organization, Geneva

Wu YL, Fu S, Tok AI, Zeng XT, Lim CS, Kwek LC, Boey FC (2008) A dual-colored bio-marker made of doped ZnO nanocrystals. Nanotechnology 19:345605

Xia T, Kovochich M, Brant J, Hotze M, Sempf J, Oberley T, Sioutas C, Yeh JI, Wiesner MR, Nel AE (2006) Comparison of the abilities of ambient and manufactured nanoparticles to induce cellular toxicity according to an oxidative stress paradigm. Nano Lett 6:1794–1807

Xie J, Li P, Li Y, Wang Y, Wei Y (2009) Morphology control of ZnO particles via aqueous solution route at low temperature. Mater Chem Phys 114:943–947

Xie J, Cao Y, Jia D, Li Y, Wang Y (2016) Solid-state synthesis of Y-doped ZnO nanoparticles with selective-detection gas-sensing performance. Ceram Int 42:90–96

Xiong HM (2013) ZnO nanoparticles applied to bioimaging and drug delivery. Adv Mater 25:5329–3535

Xu HY, Wang H, Zhang YC, He WL, Zhu MK, Wang B, Yan H (2004) Hydrothermal synthesis of zinc oxide powders with controllable morphology. Ceram Int 30:93–97

Yakimova R, Selegard L, Khranovskyy V, Pearce R, Spetz AL, Uvdal K (2012) ZnO materials and surface tailoring for biosensing. Front Biosci (Elite Ed) 4:254–278

Yang SJ, Park CR (2008) Facile preparation of monodisperse ZnO quantum dots with high quality photoluminescence characteristics. Nanotechnology 19:035609

Yang Y, Guo W, Zhang Y, Ding Y, Wang X, Wang ZL (2011) Piezotronic effect on the output voltage of P3HT/ZnO micro/ nanowire heterojunction solar cells. Nano Lett 11:4812–4817

Yoshida R, Kitamura D, Maenosono S (2009) Mutagenicity of water-soluble ZnO nanoparticles in Ames test. J Toxicol Sci 34:119–122

Yuan Q, Hein S, Misra RD (2010) New generation of chitosan-encapsulated ZnO quantum dots loaded with drug: synthesis, characterization and in vitro drug delivery response. Acta Biomater 6:2732–2739

Yuvakkumar R, Suresh J, Nathanael AJ, Sundrarajan M, Hong SI (2014) Novel green synthetic strategy to prepare ZnO nanocrystals using rambutan (*Nephelium lappaceum* L.) peel extract and its antibacterial applications. Mater Sci Eng C 41:17–27

Zafar H, Ali A, Ali JS, Haq IU, Zia M (2016) Effect of ZnO nanoparticles on *Brassica nigra* seedlings and stem explants: growth dynamics and antioxidative response. Front Plant Sci 7:535

Zhang P, Liu W (2010) ZnO QD@PMAA-co-PDMAEMA nonviral vector for plasmid DNA delivery and bioimaging. Biomaterials 31:3087–3094

Zhang H, Chen B, Jiang H, Wang C, Wang H, Wang X (2011) A strategy for ZnO nanorod mediated multi-mode cancer treatment. Biomaterials 32:1906–1914

Zhao Z, Lei W, Zhang X, Wang B, Jiang H (2010) ZnO-based amperometric enzyme biosensors. Sensors (Basel) 10:1216–1231

Zhao L, Peralta-Videa JR, Rico CM, Hernandez-Viezcas JA, Sun Y, Niu G, Servin A, Nunez JE, Duarte-Gardea M, Gardea-Torresdey JL (2014) CeO_2 and ZnO nanoparticles change the nutritional qualities of cucumber (*Cucumis sativus*). J Agric Food Chem 62:2752–2759

Zvyagin AV, Zhao X, Gierden A, Sanchez W, Ross JA, Roberts MS (2008) Imaging of zinc oxide nanoparticle penetration in human skin in vitro and in vivo. J Biomed Opt 3:064031

Chapter 8
Plant-Mediated Synthesis of Copper Oxide Nanoparticles and Their Biological Applications

Archana Joshi, Ashutosh Sharma, Rakesh Kumar Bachheti, Azamal Husen, and Vinod Kumar Mishra

8.1 Introduction

As per the estimations of National Science Foundation (NSF), Alexandria, USA, the global market for the nanotechnology-based products would be reaching three trillion USD by the year 2020 (Roco 2011). More than a thousand commercial products containing nanoparticles (NPs), which range from 1 to 100 nm, are currently available in the market (Vance et al. 2015) with wide-ranging applications in the fields of biomedicine, wastewater treatment, environmental remediation, food processing and packaging, agriculture, horticulture, and crop protection (Husen 2017; Siddiqi and Husen 2016, 2017a, b; Siddiqi et al. 2018a, b, c).

In comparison to their bulk counterparts, NPs have a greater chemical reactivity, strength, and some novel properties due to their increased surface-to-volume ratio and quantum size effect. The surface plasmon resonance (SPR) exhibited by metal NPs is one of their most important characteristics. NPs can be produced by the breakdown (top-down) or the buildup (bottom-up) methods (Husen and Siddiqi 2014), involving various physicochemical techniques. However, these production

A. Joshi
Department of Environmental Science, Graphic Era University, Dehradun, India

A. Sharma
Department of Chemistry, Graphic Era University, Dehradun, India

R. K. Bachheti
Department of Industrial Chemistry, College of Applied Science, Addis Ababa Science and Technology University, Addis Ababa, Ethiopia

A. Husen
Department of Biology, College of Natural and Computational Sciences, University of Gondar, Gondar, Ethiopia

V. K. Mishra (✉)
Department of Biotechnology, Baba Farid Institute of Technology, Dehra Dun, Uttarakhand, India

A. Husen, M. Iqbal (eds.), *Nanomaterials and Plant Potential*,
https://doi.org/10.1007/978-3-030-05569-1_8

methods are usually expensive, labor-intensive, and potentially hazardous to the environment and living organisms. The most acceptable and effective approach for NP preparation is the bottom-up approach. Biological methods for NP synthesis utilize a bottom-up approach with the help of reducing and stabilizing agents.

Copper oxide nanoparticles (CuO NPs) are especially known for their catalytic, electric, optical, photonic, superconducting, and biological properties (Padil and Cernik 2013). However, their large-scale production has introduced risk to the environment and human health. Considering the wide-ranging applications and increasing demand of metal NPs, an alternative, cost-effective, safe, and green technology for large-scale production of CuO NPs is required. This chapter discusses the pros and cons of plant-mediated synthesis of CuO NPs, with the main focus on reduction, capping, and stabilization of NPs, and casts a cursory glance on their prospective applications.

8.2 Plant-Mediated Synthesis of Copper Oxide NPs

Extracts of various parts of different plant species have been used for synthesis of CuO NPs, as shown in Table 8.1, proving the ability of plant extracts to reduce copper ions to copper metals, yielding Cu NPs. Plant-mediated synthesis of MNPs is energetically efficient and can even be carried out at room temperature and is generally completed within few minutes.

Plant parts (leaf, root, bark, etc.) are thoroughly washed with tap water in order to remove dust particles, sun-dried for 1–2 h to remove the residual moisture, cut into small pieces, and extracted. The extract is purified by filtration and centrifugation. Different concentrations of plant extract and metal salts (e.g., cupric sulfate, cupric chloride, copper nitrate, cupric acetate) are incubated in a shaker for different time intervals, at different pH and temperature, for NP synthesis. Formation of NPs is monitored by change in color of the reaction mixture. At the end, the reaction mixture is centrifuged at low speed to remove any medium components or large particles. Finally, the NPs can be centrifuged at a high speed or with a density gradient, washed thoroughly in water/solvent (ethanol/methanol) and collected in the form of a bottom pellet.

In an experiment, CuO NPs were synthesized from black bean extract and characterized by XRD, FTIR, XPS, Raman spectroscopy, DLS, TEM, SAED, SEM, and EDX (Nagajyothi et al. 2017). XRD studies have shown that the particles were ~26.6 nm in size and spherical in shape. Rehana et al. (2017) used leaf extract of various plants (*Azadirachta indica*, *Hibiscus rosa-sinensis*, *Murraya koenigii*, *Moringa oleifera*, and *Tamarindus indica*) for synthesizing CuO NPs. UV-vis spectroscopy revealed the band centered between 220 and 235 nm, typical for CuO NPs. The SEM and TEM studies confirmed the spherical shape, with an average size of 12 nm, while SAED revealed the crystalline nature of the NPs. FTIR spectroscopy displayed bands at ~490 and ~530 cm^{-1} corresponding to metal–oxygen (Cu–O) vibration that supports the availability of monoclinic phase of CuO NPs. These NPs

Table 8.1 Plants and their parts used for biosynthesis of CuO NPs, together with their size, shape, and significance

Plant (part used)	Size (nm)	Shape	Significance	References
Acalypha indica (leaf)	29	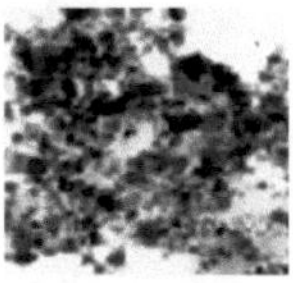 Spherical	Antibacterial and antifungal effect against *Escherichia coli*, *Pseudomonas fluorescens*, and *Candida albicans*. Anticancer activity against MCF-7 human breast cancer cell line	Sivaraj et al. (2014b)
Albizia lebbeck (leaf)	<100	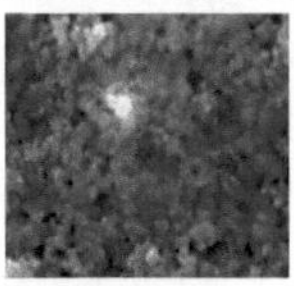 Spherical	Cheap and effective reducing agent for CuO NP production in large scale	Jayakumarai et al. (2015)
Aloe vera (leaf)	24–61	 Octahedral	Photocatalytic activities	Kerour et al. (2018)
Aloe barbadensis (leaf)	15–30	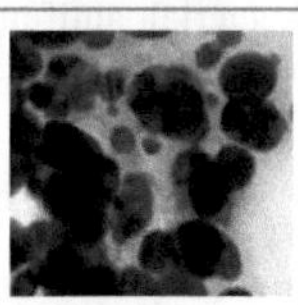 Versatile and spherical	Large-scale commercial production and health-related applications of CuO NPs	Gunalan et al. (2012)
Aloe vera (leaf)	20	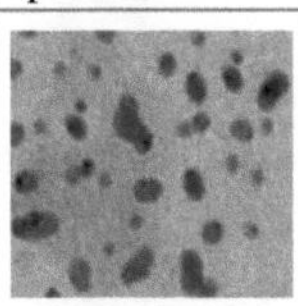 Spherical	Antibacterial activity against fish bacterial pathogens	Kumar et al. (2015)
Alternanthera sessilis (leaf)	22.6–25.2	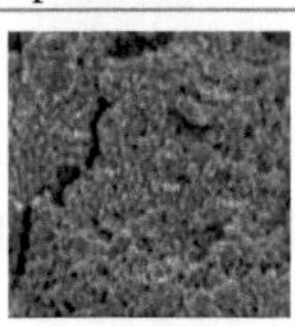 Spherical	Antimicrobial activity	Niraimathi et al. (2016)
Calotropis gigantea (leaf)	20–30	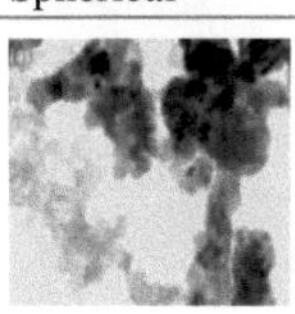 Spherical	Solar cell applications	Sharma et al. (2015)

(continued)

Table 8.1 (continued)

Plant (part used)	Size (nm)	Shape	Significance	References
Calotropis procera (leaf)	40	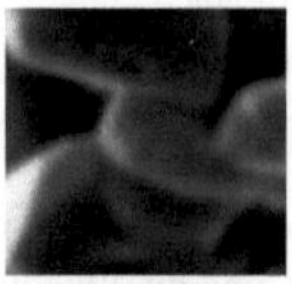 Cylindrical	Optical studies	Reddy (2017)
Camellia sinensis (leaf)	34.36	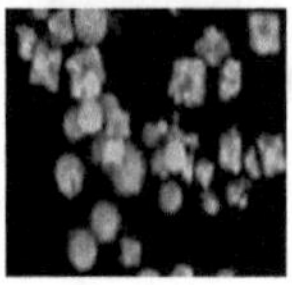 Cubes and spheres	Antibacterial activity	Riya and George (2015)
Cassia alata (flower)	110–280	 Spherical	Wide application in medicine	Jayalakshmi and Yogamoorthi (2014)
Curcuma aeruginosa (rhizome)	40	–	Antibacterial, antifungal, and antioxidant activity	George and Britto (2014)
Desmodium gangeticum (root)	28	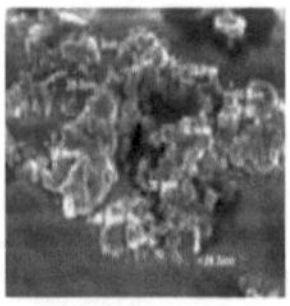 Spherical	Antioxidant activity against DPPH	Guin et al. (2015)
Ferulago angulate (aerial part)	44	 Shell-like sheet structure	Photocatalytic activity	Mehr et al. (2018)
Galeopsidis herba (Plant extract)	10	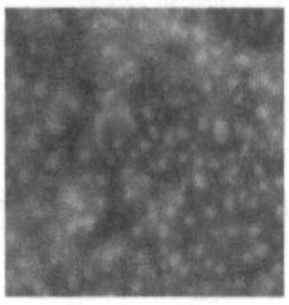 Spherical	Antioxidant and catalytic activity	Dobrucka (2018)

(continued)

Table 8.1 (continued)

Plant (part used)	Size (nm)	Shape	Significance	References
Gloriosa superba (leaves)	5–10	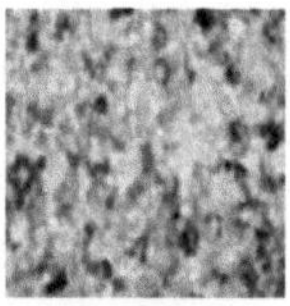 Spherical	Effective against *S. aureus* and *Klebsiella aerogenes*	Naikaa et al. (2015)
Malva sylvestris (leaf)	5–30	 Spherical	Antibacterial activity against *Shigella* and *Listeria* strains	Awwad et al. (2015)
Matricaria chamomilla (flower)	140	 Spherical	Antioxidant activity and interaction with plasmid DNA (pBR322)	Duman et al. (2016)
Pterocarpus marsupium (wood)	20–25	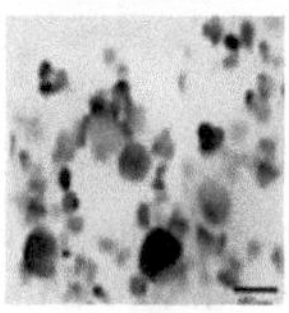 Spherical	Antibacterial activity against *E. coli*, *P. vulgaris*, *K. pneumonia*, *S. aureus*, *S. epidermidis*, and *B. cereus*	Rajgovind et al. (2015)
Pyrus pyrifolia (leaf)	24	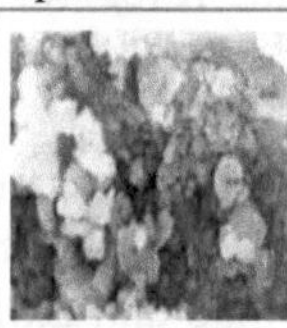 Spherical	Photocatalytic study	Sundaramurthy and Parthiban (2015)
Rubus glaucus (leaf and fruit)	43.3	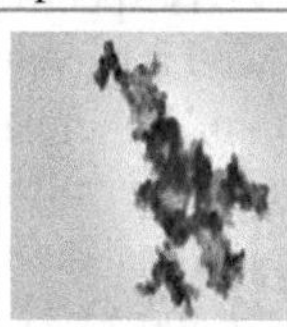 Spherical	Antioxidant activity against DPPH	Kumar et al. (2017)
Stachys lavandulifolia (flowers)	80	 Near-spherical	Antibacterial activity	Khatami et al. (2017)

(continued)

Table 8.1 (continued)

Plant (part used)	Size (nm)	Shape	Significance	References
Tabernaemontana divaricata (leaf)	48	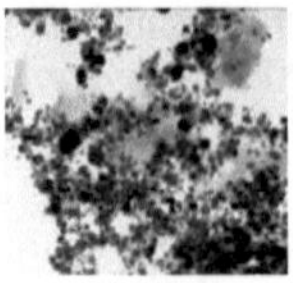 Spherical	Antibacterial activity against urinary tract pathogen	Sivaraj et al. (2014a)
Tinospora cordifolia (leaf)	6–8	 Spherical	Antioxidant, antimicrobial activity	Udayabhanu et al. (2015)
Thymus vulgaris (leaf)	30	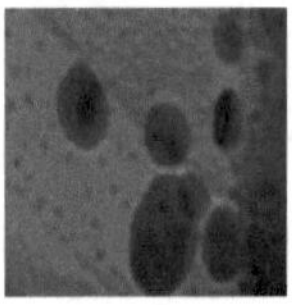 Face-centered cubic (fcc)	Catalytic activity	Nasrollahzadeh et al. (2016)

also displayed band in the region of 3442–3474 cm^{-1} due to O–H stretching vibration, whereas the band at 2370 and 2385 cm^{-1} was ascribed to the primary amines. The band at 1663 and 1674 cm^{-1} was reported due to amide I; however the band at 1530 and 1535 cm^{-1} was ascribed to the amide II region, which was characteristic of proteins and/or enzymes. Further, bands at 1030–1110 cm^{-1} corresponded to C–O stretching vibrations. Rehana et al. (2017) have suggested that the phenolic compounds and flavonoids acted as capping agents, thus preventing agglomeration, and stabilized the NP formation. Vishveshvar et al. (2018) synthesized CuO NPs using an aqueous solution of copper (II) sulfate and *Ixora coccinea* leaf extract. These particles were characterized using UV-vis spectroscopy, SEM, TEM, and FTIR spectroscopy. UV-vis spectroscopy showed a wavelength region from 200 to 300 nm. SEM images exhibited the formation of NP clusters of an average size of 300 nm. Further, TEM images of NPs, separated after ultrasonication of the dispersion, showed an average particle size of 80–110 nm. The FTIR spectroscopy peaks revealed the bonding vibrations such as Cu–O and O–H. In another recent study, Mehr et al. (2018) fabricated CuO and ZnO NPs by using the *Ferulago angulata* extract as a mild and nontoxic reducing agent and an efficient stabilizer without adding any surfactants and characterized them with the help of XRD, FTIR, and FESEM. The NPs produced were crystalline in nature, with high purity and an average size of ~ 44 nm. The FTIR spectrum of NPs showed two peaks at 912 and 620 cm^{-1}. Dobrucka (2018) synthesized CuO NPs (10 ± 5 nm in size) using the extract of *Galeopsidis herba* and characterized them by UV-vis, SEM, TEM, and

FTIR spectroscopy and EDS profile. SEM images confirmed the spherical shape of the NPs. FTIR spectrum showed bands at 417, 408, and 398 cm^{-1} indicating the formation of metal–oxygen stretching of CuO nanostructure. These NPs showed antioxidant as well as the catalytic activity in the reduction of malachite green. Octahedral and spherical CuO NPs were prepared using copper sulfate and *Aloe vera* aqueous extract (Kerour et al. 2018). The SEM images revealed octahedral and spherical agglomeration of NPs. XRD confirmed the cubic structure of NPs, which depends upon the crystallite size concentration of *Aloe vera* aqueous extract. The FTIR vibration measurements validated the presence of pure Cu_2O in the samples. The UV-vis spectra indicated that the prepared Cu_2O had a gap energy estimated from 2.5 to 2.62 eV. The photocatalytic activities enabled an improved and fast degradation of methylene blue in aqueous solution at room temperature under solar simulator irradiation (Kerour et al. 2018). Further details are given in Table 8.1.

8.2.1 Mechanism of CuO NP Synthesis

Plant extracts contain a wide range of metabolites that can act both as reducing and stabilizing agents in the metal NP synthesis. Bioreduction is a relatively complex process. As a reducing agent, biomolecules in the extract provide electrons to metal ions, thus reducing it into the elemental metal. After reduction, the atom so formed acts as the nucleation center, which is immediately followed by a period of growth when the smaller neighboring particles amalgamate to form a larger NP. In the final stage of synthesis, the plant extracts' ability to stabilize the NP ultimately determines its energetically favorable and stable morphology. In order to avoid further growth and maintain the particle in the nano-range, a substance called capping agent is added. Biomolecules present in the plant extract may act as the reducing agent, or the same molecules may function as both the reducing agent and the capping agent. Since different plant extracts contain different phytoconstituents, the NP formation mechanisms vary among different plant species, and their details are yet to be fully elucidated.

Secondary metabolites, such as terpenoids, polyphenols, flavonoids, alkaloids, phenolic acids, etc., are responsible for the reduction of metal ions to zerovalent metals or the stabilization of MNPs. Flavonoids constitute a large group of polyphenolic compounds that comprise of several classes, viz., anthocyanins, isoflavonoids, flavonols, chalcones, flavones, and flavanones. The transition of flavonoids from the enol to the keto may lead to reduction of the metal. The ability of flavonoids to chelate metal ions is well documented. Some flavonoids, such as quercetin and santin, are known to possess strong chelating activity due to the presence of hydroxyls and carbonyl functional groups (Anjum et al. 2015). Apart from the crude extract, individual pure secondary metabolites have the ability to synthesize metal NPs (Sahu et al. 2016; Kasthuri et al. 2009). It has been reported that amino acids, sugars, and fatty acids available in gum karaya could act as a reducing and capping agent for the formation of metal oxide NPs (Silva et al. 2003).

8.3 Controlling the Shape and Size of CuO NPs

Despite a significant progress in the biosynthesis of NPs, little could be achieved in controlling the shape of the metal NPs by biological routes. Polydispersity of nanoparticles still remains a challenge. During the course of biological synthesis of metal NPs, a number of physical and biological parameters, including pH, reactants' concentration, reaction time, and temperature, govern nucleation and the subsequent formation of the stabilized NP. The oxidation/reduction state of proteins and enzymes present in the cell-free extract is highly dependent upon the pH, making it a substantial factor to determine the shape and size of NPs. Further, metal ion concentration and pH, reaction time, and temperature can also affect the rate of nucleation and growth of NPs. As the reaction temperature increases, the reaction rate increases consuming metal ions to form the nuclei, thereby enhancing the biosynthesis process. Alteration of the reaction time can lead to variation in growth rate of seed particle, generating multi-shaped NPs. It is very clear that a single parameter cannot decide the fate of nuclei; rather it is the balance between all the parameters that can generate different shapes and sizes of NPs produced.

8.4 Characterization of CuO NPs

For characterization of synthesized NPs, several techniques, including ultraviolet-visible (UV-vis) spectroscopy, transmission electron microscopy (TEM), small-angle X-ray scattering (SAXS), Fourier transform infrared (FTIR) spectroscopy, X-ray fluorescence (XRF) spectroscopy, X-ray diffraction (XRD), X-ray photoelectron spectroscopy (XPS), scanning electron microscopy (SEM), field emission scanning electron microscopy (FESEM), particle size analysis (PSA), Malvern Zetasizer (MZS), energy-dispersive X-ray spectroscopy (EDX/EDS), nanoparticle tracking analysis (NTA), X-ray reflectometry (XRR), Brunauer–Emmett–Teller (BET) analysis, selected area electron diffraction (SAED), and atomic force microscopy (AFM), are used (Table 8.2).

8.5 Applications of Copper Oxide NPs

8.5.1 Biological Application

Various studies have established that CuO NPs possess potent antimicrobial, antioxidant, and anticancer activities (Table 8.3).

Table 8.2 Characterization techniques, their use, and limitations

Technique	Main role	Limitation	References
High-resonance transmission electron microscopy (HRTEM)	Used for the determination of shape and size of nanostructures	Particle size larger than 1.5 nm cannot be determined	Thakur et al. (2014)
X-ray diffraction (XRD)	Size and crystallinity of NPs can be determined by this technique	Composition of plasmon and NPs cannot be analyzed	Din et al. (2017)
Scanning electron microscopy (SEM)	Used for the determination of shape and size of nanostructures	Samples must be solid and elements with atomic number <11 cannot be determined	Kalpana et al. (2016)
Atomic force microscopy (AFM)	Particle size and characterization	Can be used for gas and liquid samples only	Daniel et al. (2013)
Brunauer–Emmett–Teller analysis (BET)	Specific surface area is measured	–	Karimi and Mohsenzadeh (2015)
X-ray fluorescence spectroscopy (XRF)	Used for the measurement of chemical composition and concentration	Limited ability to measure precisely and accurately	Karimi and Mohsenzadeh (2015)
Nanoparticle tracking analysis (NTA)	To visualize and measure particle size, concentration, and fluorescent properties of NPs	–	Cheirmadurai et al. (2014)
Particle size analysis (PSA)	To measure the distribution of size in the sample of solid or liquid particulate materials	–	Parihar and Balekar (2016); Shende et al. (2016)
Selected area electron diffraction (SAED)	Technique that can be performed inside a TEM	Cannot be recommended for quantitative identification	Karimi and Mohsenzadeh (2015)
Energy-dispersive X-ray spectroscopy (EDX/EDS)	Composition of NPs can be analyzed	Particles with size <2 nm cannot be analyzed	Nasrollahzadeh et al. (2015a); Harne et al. (2012)
X-ray photoelectron spectroscopy (XPS)	Elemental composition of NPs can be analyzed	Samples are decomposed	Lee et al. (2011)
Zeta size analyzer	Measures the size of NPs, zeta, potential, and protein mobility	In nano-range	Shende et al. (2016)

8.5.1.1 Antimicrobial Activity

The last decade introduced opportunities to investigate the bactericidal effect of metal NPs. Their small size and high surface-to-volume ratio allow them to interrelate strongly with microbial membranes which are not merely due to the release of metal ions in solution (Subhankari and Nayak 2013). Niraimathi et al. (2016) and

Table 8.3 Antimicrobial, antioxidant, anticancer, and catalytic activities of CuO NPs

Biological entity	Activity	Response	References
Aloe vera	Photocatalytic activities	Degradation of methylene blue	Kerour et al. (2018)
Alternanthera sessilis	Antimicrobial activity	Water treatment, synthetic textiles, biomedical and surgical devices, food processing and packaging	Niraimathi et al. (2016)
Andrographis paniculata	Antimicrobial activity	*Escherichia coli*, *Staphylococcus aureus*, and *Pseudomonas aeruginosa*	Devasenan et al. (2016)
Anthemis nobilis	Catalytic activity	A^3 coupling reaction	Nasrollahzadeh et al. (2015a)
Aerva lanata	Antibacterial activity	*E. coli*, *S. aureus*, *Bacillus cereus*, and *P. aeruginosa*	Hariprasad et al. (2016)
Artemisia haussknechtii	Antibacterial, antioxidant, and ion-chelating activities	*E. coil*, *S. aureus*, *S. epidermidis*, *S. marcescens*	Alavi and Karimi (2017)
Azadirachta indica	Anticancer activity	Cytotoxic against four cancer cell lines, human breast (MCF-7), cervical (HeLa), epithelioma (Hep-2), lung (A549) and one normal human dermal fibroblast (NHDF) cell line	Rehana et al. (2017)
Bifurcaria bifurcata	Antibacterial activity	*Enterobacter aerogenes* (Gram-negative) and *S. aureus* (Gram-positive)	Abboud et al. (2013)
Phaseolus vulgaris	Anticancer activity	Suppress the proliferation of HeLa cells	Nagajyothi et al. (2017)
Calotropis procera	Optical studies	Indirect optical transitions	Reddy (2017)
Centella asiatica	Catalytic activity	Photo catalytic degradation of methyl orange	Devi and Singh (2014)
Cassia auriculata	Antirheumatic activity	Arthritis	Shi et al. (2017)
Matricaria chamomilla	Antioxidant activity	Photooxidation of polymer	Duman et al. (2016)
Datura innoxia	Antimicrobial activity	Leaf blight disease of rice	Kala et al. (2016)
Desmodium gangeticum	Antioxidant activity	Inhibiting the free radical production	Guin et al. (2015)
Euphorbia nivulia	Anticancer activity	Human adenocarcinomic alveolar basal epithelial cells (A549 cells)	Valodkar et al. (2011a)
Ferulago angulata	Photocatalytic activity	Rhodamine B as organic contaminant	Mehr et al. (2018)
Galeopsidis herba	Antioxidant and catalytic activity	By DPHH method and degradation of malachite green	Dobrucka (2018)
Gloriosa superba	Antibacterial activity	*Klebsiella aerogenes*, *P. desmolyticum*, *E. coli*, and *S. aureus*	Naikaa et al. (2015)

(continued)

Table 8.3 (continued)

Biological entity	Activity	Response	References
Gundelia tournefortii	Catalytic activity	Excellent catalytic activity for reduction of 4-nitrophenol and synthesis of N-monosubstituted ureas	Nasrollahzadeh et al. (2015b)
Malus domestica	DNA-cleavage activity	pBR322 plasmid of *E. coli*	Jadhav et al. (2018)
Malus domestica	Antibacterial activity	Gram-positive and Gram-negative bacteria	Jadhav et al. (2018)
Malus domestica	Antioxidant activity	Free radical scavenging activity by the DPPH (1,1-diphenyl 2-picrylhydrazyl) method	Jadhav et al. (2018)
Magnolia	Antibacterial activity	*E. coli*	Lee et al. (2011)
Stachys lavandulifolia	Antibacterial activity	*P. aeruginosa*	Khatami et al. (2017)
Sterculia urens	Antibacterial activity	*E. coli* and *S. aureus*	Padil and Cernik (2013)
Tamarix gallica	Catalytic activity	*N*-arylation of nitrogen-containing heterocycles with aryl halides under ligand-free conditions	Nasrollahzadeh et al. (2015c)
Thymus vulgaris	Catalytic activity	*N*-arylation of indoles and amines	Nasrollahzadeh et al. (2016)

Kala et al. (2016) have reported the antimicrobial activity of copper bionanoparticles in the recent past. Acharyulu et al. (2014) studied the antimicrobial activity of biosynthesized CuO NPs from *Phyllanthus amarus* leaf extract against multidrug-resistant Gram-positive (*Bacillus subtilis* and *Staphylococcus aureus*) and Gram-negative (*Escherichia coli* and *Pseudomonas aeruginosa*) bacteria. Abboud et al. (2013) investigated that CuO NPs produced by using brown algae extract (*Bifurcaria bifurcata*) show high antibacterial activity against two different strains of *Enterobacter aerogenes* (Gram-negative) and *Staphylococcus aureus* (Gram-positive). Suleiman et al. (2013) claimed that biologically synthesized copper NPs can be used for treating several diseases; however, it requires clinical studies to ascertain their antimicrobial potential and efficacy. Padil and Cernik (2013) suggested the antimicrobial activity of CuO NPs against common pathogens *E. coli* and *S. aureus*. Das et al. (2013) and Heinlaan et al. (2008) demonstrated the antioxidant and antibacterial behavior of these NPs, whereas Hariprasad et al. (2016) observed their good antibacterial activity against *E. coli*, *S. aureus*, *Bacillus cereus*, and *Pseudomonas aeruginosa*. According to Naikaa et al. (2015), the synthesized CuO NPs were effective against the pathogenic bacteria *S. aureus* and *Klebsiella aerogenes*. Devasenan et al. (2016) also confirmed the ability of CuO NPs to inhibit the growth of various pathogens.

8.5.1.2 Antioxidant Activity

Antioxidant activity of nanomaterials is well established. Photooxidation of polymer creates aldehydes, ketones, and carboxylic acids at the end of the polymer chain. Antioxidants terminate these chain reactions by removing free radical intermediates and inhibit other oxidation reactions. The antioxidant and DNA-cleavage properties of CuO NPs biosynthesized with the help of *Chammomile* flower extract were reported by Duman et al. (2016). They suggested that CuO NPs can act as a chemical nuclease, can generate DNA cleavage, and may be useful for preventing cell proliferation. The improved antioxidant efficacy in Andean blackberry fruit than Andean blackberry leaf extract may be due to the presence of more bioactive molecules in the former than in the latter, which have a role as an encapsulating agent in CuO NPs. The highest antioxidant efficacy of CuO NPs against DPPH is probably derived through the electrostatic attraction between negatively charged bioactive compounds (COO^-, O^-) and neutral or positively charged NPs. CuO NPs bind to phytochemicals, and their bioactivity increases synergistically. Antioxidant activity of CuO NPs was measured by Das et al. (2013) using (2,2 diphenyl-1-picrylhydrazyl) DPPH method where DPPH was used as a radial source. Guin et al. (2015) found the biologically synthesized CuO NPs to be significantly effective against oxidative stress and less toxic than the precursor material.

8.5.1.3 Anticancer Activity

Copper oxide NPs also exhibit anticancer activity, as reported recently by Nagajyothi et al. (2017). Clonogenic assays have confirmed that the NPs-incubated cancer cells are not able to proliferate well. The CuO NPs can induce apoptosis (cell death) and suppress the proliferation of HeLa cells. These NPs have a high anticancer cytotoxicity on human colon cancer lines (HCT-116) with IC_{50} value of 40 μg mL^{-1}. According to Rehana et al. (2017), CuO NPs are cytotoxic against four cancer cell lines, viz., human breast (MCF-7), cervical (HeLa), epithelioma (Hep-2), and lung (A549), and one normal human dermal fibroblast (NHDF) cell line. The anticancer activity of brown algae-mediated CuO NPs was determined by MTT assay against the cell line (MCF-7) (Suleiman et al. 2013). CuO NPs synthesized by using stem latex of *Euphorbia nivulia* (common milk hedge) could be encrusted and stabilized by peptides and terpenoids present in the latex. These NPs have shown toxic effect against human adenocarcinomic alveolar basal epithelial cells (A549 cells) (Valodkar et al. 2011a, b).

8.5.1.4 Antirheumatic Activity

In a recent investigation, it was found that CuO NPs prepared using *Cassia auriculata* extract can be used as a vehicle in drug delivery as antirheumatic agent for rheumatoid arthritis treatment (Shi et al. 2017).

8.5.1.5 Catalytic Effect

According to Devi and Singh (2014), CuO NPs prepared from the leaf extract of *Centella asiatica* at room temperature show a catalytic effect. NPs have many active sites as compared to the bulk material because of their small size and large surface-to-volume ratio. These NPs could be used for the photocatalytic degradation of methyl orange. In the absence of reducing agents in aqueous medium, these NPs reduce methyl orange to its leuco form.

Nasrollahzadeh et al. (2015a) investigated the effectiveness of CuO NPs for the synthesis of propargylamines by elaborating the reaction conditions in A^3 coupling reaction between piperidine (1.2 mmol) and phenylacetylene (1.5 mmol) with benzaldehyde (1.0 mmol) as a model reaction. An assorted range of propargylamines was obtained in a superior yield. In addition, the reclaim and separation of CuO NPs were very easy, effectual, and economically viable. In another study, CuO NPs were found to be exceptionally heterogeneous catalyst for ligand-free N-arylation of indoles and amines. An excellent yield of N-arylated products was obtained, and the catalyst could be recovered and recycled for auxiliary catalytic reactions with approximately no loss in activity (Nasrollahzadeh et al. 2016).

8.6 Conclusion

Due to rich plant diversity, phytosynthesis of CuO NPs is proficient in producing superficial, ecologically safe, and economically viable NPs, in comparison to physical and chemical methods. During the bioreduction process, the biomolecules present in plant systems play a significant role. Different techniques used for the characterization of biosynthesized NPs include UV-vis spectroscopy, FTIR spectroscopy, XRD, SEM, EDX, Raman spectroscopy, DLS, TEM, SAED, etc. Applications of CuO NPs are significant especially in biomedicine and catalysis. The recent use of engineered CuO NPs in drug and gene delivery, in addition to their well-known catalytic effect and the antimicrobial, antioxidant, and anticancer activities, has attached a special prestige to them.

References

Abboud Y, Saffaj T, Chagraoui A, Bouari AE, Brouzi K, Tanane O, Ihssane B (2013) Biosynthesis, characterization and antimicrobial activity of CuO nanoparticles (CONPs) produced using brown algae extract (*Bifurcaria bifurcata*). Appl Nanosci 4:571–576

Acharyulu NPS, Dubey RS, Swaminadham V, Kalyani RL, Kollu P, SVN P (2014) Green synthesis of CuO nanoparticles using *Phyllanthus amarus* leaf extract and their antibacterial activity against multidrug resistance bacteria. Int J Eng Res Tech 3:638–641

Alavi M, Karimi N (2017) Characterization, antibacterial, total antioxidant, scavenging, reducing power and ion chelating activities of green synthesized silver, copper and titanium dioxide nanoparticles using *Artemisia haussknechtii* leaf extract. Artif Cells Nanomed Biotechnol 12:1–16

Anjum NA, Adam V, Kizek R, Duarte AC, Pereira E, Iqbal M, Lukatkin AS, Ahmad I (2015) Nanoscale copper in the soil-plant system: toxicity and underlying potential mechanisms. Environ Res 138:306–325

Awwad AM, Albiss BA, Salem NM (2015) Antibacterial activity of synthesized copper oxide nanoparticles using *Malva sylvestris* leaf extract. SMU Med J 2:91–101

Cheirmadurai K, Biswas S, Murali R, Thanikaivelan P (2014) Green synthesis of copper nanoparticles and conducting nanobiocomposites using plant and animal sources. RSC Adv 4:19507–19511

Daniel SK, Vinothini G, Subramanian N, Nehru K, Sivakumar M (2013) Biosynthesis of Cu, ZVI and Ag nanoparticles using *Dodonaea viscosa* extract for antibacterial activity against human pathogens. J Nano Res 15:1319

Das D, Nath BC, Phukon P, Dolui SK (2013) Synthesis and evaluation of antioxidant and antibacterial behavior of CuO nanoparticles. Colloids Surf B: Biointerfaces 101:430–433

Devasenan S, Beevi NH, Jayanthi SS (2016) Synthesis and characterization of copper nanoparticles using leaf extract of *Andrographis paniculata* and their antimicrobial activities. Int J ChemTech Res 9:725–730

Devi HS, Singh TD (2014) Synthesis of copper oxide nanoparticles by a novel method and its application in the degradation of methyl orange. Adv Electr Electron Eng 4:83–88

Din MI, Arshad F, Rani A, Aihetasham A, Mukhtar M, Mehmood H (2017) Single step green synthesis of stable copper oxide nanoparticles as efficient photo catalyst material. Biomed Mater 9:41–48

Dobrucka R (2018) Antioxidant and catalytic activity of biosynthesized cuo nanoparticles using extract of *Galeopsidis herba*. J Inorg Organomet Polym 28:812–819

Duman F, Ismail Ocsoy I, Kup FO (2016) Chamomile flower extract-directed CuO nanoparticle formation for its antioxidant and DNA cleavage properties. Mater Sci Eng C Mater Biol Appl 60:33–338

George M, Britto SJ (2014) Biosynthesis characterization, antimicrobial, antifungal and antioxidant activity of copper oxide nanoparticles (CONPS). Ejbps 1:199–210

Guin R, Banu S, Kurian GA (2015) Synthesis of copper oxide nanoparticles using *Desmodium gangeticum* aqueous root extract. Int J Pharmacol Pharm Sci 7:60–65

Gunalan S, Sivaraj R, Venckatesh R (2012) Aloe barbadensis miller mediated green synthesis of mono-disperse copper oxide nanoparticles: optical properties. Spectrochim Acta A Mol Biomol Spectrosc 97:1140–1144

Hariprasad S, Bai GS, Santhoshkumar J, Madhu CH, Sravani D (2016) Green synthesis of copper nanoparticles by *Areva lanata* leaves extract and their antimicrobial activates. Int J Chem TechRes 9:98–105

Harne S, Sharma A, Dhaygude M, Joglekar S, Kodam K, Hudlikar M (2012) Novel route for rapid biosynthesis of copper nanoparticles using aqueous extract of *Calotropis procera* L latex and their cytotoxicity on tumor cells. Colloids Surf B Biointerfaces 95:284–288

Heinlaan M, Ivask A, Blinova I, Dubourguier HC, Kakru A (2008) Toxicity of nanosized and bulk ZnO, CuO and TiO_2 to bacteria Vibrio fischeri and crustaceans Daphnia magna and Thamnocephalus platyurus. Chemosphere 71:1308–1316 (If journal have single word name, it should not be abbreviated, we have to write complete name)

Husen A (2017) Gold nanoparticles from plant system: synthesis, characterization and application. In: Ghorbanpourn M, Manika K, Varma A (eds) Nanoscience and plant–soil systems, vol 48. Springer International Publication, pp 455–479

Husen A, Siddiqi KS (2014) Phytosynthesis of nanoparticles: concept, controversy and application. Nanoscale Res Lett 9:229

Jadhav MS, Kulkarni S, Raikar P, Barretto DA, Vootla SK, Raikar US (2018) Green biosynthesis of CuO & Ag–CuO nanoparticles from *Malus domestica* leaf extract and evaluation of antibacterial, antioxidant and DNA cleavage activities. New J Chem 42:204–213

Jayakumarai G, Gokulpriya C, Sudhapriya R, Sharmila G, Muthukumaran C (2015) Phytofabrication and characterization of monodisperse copper oxide nanoparticles using *Albizia lebbeck* leaf extract. Appl Nanosci 5:1017–1021

Jayalakshmi, Yogamoorthi A (2014) Green synthesis of copper oxide nanoparticles using aqueous extract of flowers of *Cassia alata* and particles characterization. Int J Nanomat Biostruct 4:66–71

Kala A, Soosairaj S, Mathiyazhagan S, Raja P (2016) Green synthesis of copper bionanoparticles to control the bacterial leaf blight disease of rice. Curr Sci 110:201–2014

Kalpana VN, Chakraborthy P, Palanichamy V, Rajeswari VD (2016) Synthesis and characterization of copper nanoparticles using Tridax procumbens and its application in degradation of bismarck brown. Int J ChemTech Res 9:498–507

Karimi J, Mohsenzadeh S (2015) Rapid, Green, and eco-friendly biosynthesis of copper nanoparticles using flower extract of *Aloe Vera*. Synth React Inorg Met Org Nano-Met Chem 45:895–898

Kasthuri J, Veerapandian S, Rajendiran N (2009) Biological synthesis of silver and gold nanoparticles using apiin as reducing agent. Colloids Surf B Biointerfaces 68:55–60

Kerour A, Boudjadar S, Bourzami R, Allouche B (2018) Eco-friendly synthesis of cuprous oxide (Cu_2O) nanoparticles and improvement of their solar photocatalytic activities. J Solid State Chem 263:79–83

Khatami M, Heli H, Jahani PM, Azizi H, Nobre MAL (2017) Copper/copper oxide nanoparticles synthesis using *Stachys lavandulifolia* and its antibacterial activity. IET Nanobiotechnol 11:709–713

Kumar PPNV, Shameem U, Kollu P, Kalyani RL, Pammi SVN (2015) Green synthesis of copper oxide nanoparticles using *Aloe vera* leaf extract and its antibacterial activity against fish bacterial pathogens. BioNanoSci 5:135–139

Kumar B, Smita K, Cumbal L, Debut A, Angulo Y (2017) Biofabrication of copper oxide nanoparticles using Andean blackberry (*Rubus glaucus* Benth.) fruit and leaf. J Saudi Chem Soc 21:S475–S480

Lee HJ, Lee G, Jang NR, Yun JH, Song JY, Kim BS (2011) Biological synthesis of copper nanoparticles using plant extract. Nanotechnol 1:371–374

Mehr ES, Sorbiun M, Ramazani A, Fardood ST (2018) Plant-mediated synthesis of zinc oxide and copper oxide nanoparticles by using *Ferulago angulata* (Schlecht) Boiss extract and comparison of their photocatalytic degradation of Rhodamine B (RhB) under visible light irradiation. J Mater Sci Mater Electron 29:1333–1340

Nagajyothi PC, Muthuraman P, Sreeknath TVM, Kim DH, Shim J (2017) Anticancer activity of copper oxide nanoparticles against human cervical carcinoma cells. Arab J Chem 10:215–225

Naikaa HR, Lingarajua K, Manjunath K, Kumar D, Nagarajuc G, Suresh D, Nagabhushanae H (2015) Green synthesis of CuO nanoparticles using *Gloriosa superba* L extract and their antibacterial activity. J Taibah Univ Sci 9:7–12

Nasrollahzadeh M, Sajadi SM, Vartooni AR (2015a) Green synthesis of CuO nanoparticles by aqueous extract of *Anthemis nobilis* flowers and their catalytic activity for the A^3 coupling reaction. J Colloid Interface Sci 459:183–188

Nasrollahzadeh M, Maham M, Sajadi SM (2015b) Green synthesis of CuO nanoparticles by aqueous extract of *Gundelia tournefortii* and evaluation of their catalytic activity for the synthesis of N-monosubstituted ureas and reduction of 4-nitrophenol. J Colloid Interface Sci 455:245–253

Nasrollahzadeh M, Sajadi SM, Maham M (2015c) *Tamarix gallica* leaf extract mediated novel route for green synthesis of CuO nanoparticles and their application for *N*-arylation of nitrogen-containing heterocycles under ligand-free conditions. RSC Adv 5:40628–40635

Nasrollahzadeh M, Sajadi SM, Vartooni AR, Hussin SM (2016) Green synthesis of CuO nanoparticles using aqueous extract of *Thymus vulgaris* L. leaves and their catalytic performance for *N*-arylation of indoles and amines. J Colloid Interface Sci 466:113–119

Niraimathi KL, Lavanya R, Sudha V, Narendran R, Brindha P (2016) Bio-reductive synthesis and characterization of copper oxide nanoparticles (CuONPs) using *Alternanthera sessilis* Linn. Leaf extract. J Pharm Res 10:29–32

Padil VVT, Cernik M (2013) Green synthesis of copper oxide nanoparticles using gum karaya as a biotemplate and their antibacterial application. Int J Nanomedicine 8:889–898

Parihar G, Balekar N (2016) *Calotropis procera*: a phytochemical and pharmacological review. TJPS 40:115–131

Rajgovind SG, Kumar DG, Jasuja ND, Joshi S (2015) *Pterocarpus marsupium* derived phyto-synthesis of copper oxide nanoparticles and their antimicrobial activities. J Micro Biochem Technol 7:140–144

Reddy KR (2017) Green synthesis, morphological and optical studies of CuO nanoparticles. J Mol Struct 1150:553–557 (Check, Is it single author or two authors It is single author)

Rehana D, Mahendiran D, Kumar RS, Rahiman AK (2017) Evaluation of antioxidant and anti-cancer activity of copper oxide nanoparticles synthesized using medicinally important plant extracts. Biomed Pharmacother 89:1067–1077

Riya L, George M (2015) Green synthesis of cuprous oxide nanoparticles. Int J Adv Res Sci Eng 4:315–322

Roco MC (2011) The long view of nanotechnology development: the national nanotechnology initiative at 10 years. J Nanopart Res 13:427–445

Sahu N, Soni D, Chandrashekhar B, Satpute DB, Saravanadevi S, Sarangi BK, Pandey RA (2016) Synthesis of silver nanoparticles using flavonoids: hesperidin, naringin and diosmin, and their antibacterial effects and cytotoxicity. Int Nano Lett 6:173–181

Sharma JK, Akhtar MS, Ameen S, Srivastava P, Singh G (2015) Green synthesis of CuO nanoparticles with leaf extract of *Calotropis gigantea* and its dye-sensitized solar cells applications. J Alloys Compd 632:321–325

Shende S, Gaikwad N, Bansod S (2016) Synthesis and evaluation of antimicrobial potential of copper nanoparticle against agriculturally important Phytopathogens. Int J Bio Res 1:41–47

Shi LB, Tang PF, Zhang W, Zhao YP, Zhang LC, Zhang H (2017) Green synthesis of CuO nanoparticles using *Cassia auriculata* leaf extract and in vitro evaluation of their biocompatibility with rheumatoid arthritis macrophages (RAW 246.7). Trop J Pharm Res 16:185–192

Siddiqi KS, Husen A (2016) Engineered gold nanoparticles and plant adaptation potential. Nano Res Lett 11:400

Siddiqi KS, Husen A (2017a) Recent advances in plant-mediated engineered gold nanoparticles and their application in biological system. J Trace Elem Med Biol 40:10–23

Siddiqi KS, Husen A (2017b) Plant response to engineered metal oxide nanoparticles. Nanoscale Res Lett 12:92

Siddiqi KS, Husen A, Rao RAK (2018a) A review on biosynthesis of silver nanoparticles and their biocidal properties. J Nanobiotechnology 16:14

Siddiqi KS, Rahman A, Tajuddin HA (2018b) Properties of zinc oxide nanoparticles and their activity against microbes. Nano Res Lett 13:141

Siddiqi KS, Husen A, Sohrab SS, Osman M (2018c) Recent status of nanomaterials fabrication and their potential applications in neurological disease management. Nano Res Lett 13:231

Silva DA, Brito ACF, de Paula RCM, Feitosa JPA, Paula HCB (2003) Effect of mono and divalent salts on gelation of native, Na and deacetylated *Sterculia striata* and *Sterculia urens* polysaccharide gels. Carbohydr Polym 54:229–236

Sivaraj R, Pattanathu KSM, Rahman P, Rajiv S, Salam HA, Venckatesh R (2014a) Biogenic copper oxide nanoparticles synthesis using *Tabernaemontana divaricate* leaf extract and its antibacterial activity against urinary tract pathogen. Spectrochim Acta A Mol Biomol Spectrosc. 133:178–118

Sivaraj R, Pattanathu KSM, Rahman P, Rajiv S, Narendhran R, Venckatesh R (2014b) Biosynthesis and characterization of *Acalypha indica* mediated copper oxide nanoparticles and evaluation of its antimicrobial and anticancer activity. Spectrochim Acta A Mol Biomol Spectrosc 129:255–258

Subhankari I, Nayak PL (2013) Antimicrobial activity of copper nanoparticles synthesised by ginger (*Zingiber officinale*) extract. World J Nano Sci Technol 2:10–13

Suleiman M, Mousa M, Hussein A, Hammouti B, Hadda TB, Warad I (2013) Copper (II)- oxide nanostructures: synthesis, characterizations and their applications–review. J Mater Environ Sci 4:792–797

Sundaramurthy N, Parthiban C (2015) Biosynthesis of copper oxide nanoparticles using *Pyrus pyrifolia* leaf extract and evolve the catalytic activity. Int Res J Eng Tech 2

Thakur S, Rai R, Sharma S (2014) Study the antibacterial activity of copper nanoparticles synthesized using herbal plants leaf extracts. Int J Bio-Technol Res 4:21–34

Udayabhanu PC, Nethravathi MA, Kumar P, Suresh D, Lingaraju K, Rajanaika H, Nagabhushana H, Sharma SC (2015) *Tinospora cordifolia*, mediated facile green synthesis of cupric oxide nanoparticles and their photo catalytic, antioxidant and antibacterial properties. Mater Sci Semicond Process 33:81–88

Valodkar M, Jadeja RN, Thounaojam MC, Devkar RV, Thakore S (2011a) Biocompatible synthesis of peptide-capped copper nanoparticles and their biological effect on tumor cells. Mater Chem Phys 128:83–89

Valodkar M, Nagar PS, Jadeja RN, Thounaojam MC, Devkar RV, Thakore S (2011b) Euphorbiaceae latex induced green synthesis of non-cytotoxic metallic nanoparticle solutions: a rational approach to antimicrobial applications. Colloids Surf A Physicochem Eng Asp 384:337–344

Vance ME, Kuiken T, Vejerano EP, McGinnis SP, Hochella MF Jr, Rejeski D, Hull MS (2015) Nanotechnology in the real world: redeveloping the nanomaterial consumer products inventory. Beilstein J Nanotechol 6:1769–1780

Vishveshvar K, Aravind Krishnan MV, Haribabu K, Vishnuprasad S (2018) Green synthesis of copper oxide nanoparticles using *Ixora coccinea* plant leaves and its characterization. BioNanoScience 8:554–558

Chapter 9
Green Synthesis of Iron Oxide Nanoparticles: Cutting Edge Technology and Multifaceted Applications

Rakesh K. Bachheti, Rocktotpal Konwarh, Vartika Gupta, Azamal Husen, and Archana Joshi

9.1 Introduction

Nanotechnology represents a realm of cross-fertilized ideas and concepts. It encompasses a myriad of innovative tools and approaches that have seeded a plethora of products for a wide gamut of applications (Konwarh et al. 2009, 2012; Husen and Siddiqi 2014; Kumar et al. 2017; Siddiqi et al. 2018a, b). Be it waste-water treatment, efficient ferrying and targeted delivery of bio-cargoes including the therapeutic/bioactive molecules, tissue engineering or optoelectronics, nanotechnology has etched an indelible mark on the scientific canvas (Barua et al. 2013; Pramanik et al. 2013; Mobasser and Firoozi 2016; Reardon et al. 2017). The number of research papers and patents on applications of nanomaterials has grown by leaps and bounds over the last few years. It may be recalled that nanoscience and nanotechnology deal with the efficient coalescing of biological, chemical and engineering approaches to control/manipulate materials within the nanoscale range (1–100 nm). The unique shape-size-surface chemistry accord of various nanomaterials has endowed them

R. K. Bachheti (✉)
Department of Industrial Chemistry, Addis Ababa Science and Technology University, Addis Ababa, Ethiopia

R. Konwarh
Department of Biotechnology, Addis Ababa Science and Technology University, Addis Ababa, Ethiopia

V. Gupta
Department of Chemistry, Graphic Era University, Dehradun, Uttarakhand, India

A. Husen
Department of Biology, College of Natural and Computational Sciences, University of Gondar, Gondar, Ethiopia

A. Joshi
Department of Environment Science, Graphic Era University, Dehradun, Uttarakhand, India

A. Husen, M. Iqbal (eds.), *Nanomaterials and Plant Potential*,
https://doi.org/10.1007/978-3-030-05569-1_9

with distinct positions in diverse domains such as biotechnology, mechanics, engineering, energy and the environment, to name a few.

Iron oxide nanoparticles (NPs) have found applications in drug delivery, immunopurification, catalysis, MRI, hyperthermia, toxic-component mitigation and so on (Siddiqi et al. 2016; Dinali et al. 2017). Biocompatibility and immune acceptance of iron oxide NPs have been discussed time and again. Use of polymeric encapsulation or coating of iron oxide NPs improves their colloidal stability in water and facilitates design of 'smart' or hybrid particles (that may be used for focused delivery of therapeutic agents) as well as greater shape-size tuning that dictates the pharmacokinetics, bio-distribution and immune clearance (Ali et al. 2016). Jun et al. (2008) discussed the various nanomagnetism scaling laws of engineered nanoparticles. The size and shape concurrences along with composition, crystallographic structure, magnetic anisotropic energy, vacancies and defects are the critical parameters for determining the magnetic properties (such as coercivity, Hc and susceptibility, χ) of I iron oxide NPs.

The pros and cons of the preparation protocols of iron oxide NPs like co-precipitation of iron salts, micro-emulsion approach, thermal decomposition of iron precursors, hydrothermal methods, sonochemical method, sol-gel transition, chemical vapour deposition (CVD), flow injection, electrochemical techniques under oxidizing conditions and laser-induced pyrolysis of pentacarbonyl iron vapours have been extensively reviewed (Jun et al. 2008). Some of the bottlenecks associated with these techniques are high cost, high energy consumption, applications of toxic reagents (high ecological footprint), etc. A surmounting pressure of 'going green' amongst the scientific fraternity has dictated the researchers to explore and exploit the nature's way of preparing the nanomaterials (Thakur and Karak 2015). In this context, bio-resources like fungi and bacteria have been assessed. A wealth of informative literature on phytosynthesis of different nanomaterials, including iron oxide NPs (Hassan et al. 2018; Devi et al. 2018), can be retrieved from literature databases of SCOPUS and Web of Science, etc. (Fig. 9.1).

Abundance of various reducing and capping agents in plant extracts has been crucial in the preparation of various well-stabilized nanomaterials (Siddiqi et al. 2016; Husen 2017; Siddiqi and Husen 2017). The plant-mediated preparation of iron oxide NPs is also gaining pace (Table 9.1).

Use of phytochemicals as capping agents/stabilizing agents could be a promising stratagem to lessen the side effects and increase the biocompatibility of inorganic nanoparticles for biomedical applications. However, this may not be the complete story. Even such green materials need to undergo in vitro, in vivo and eco-toxicity assessments before being commercialized for practical applications. Nevertheless, easy and abundant availability, efficiency in terms of time utility and circumvention of the use of toxic reagents are few of the desired facets of this approach (Husen 2017; Siddiqi and Husen 2017).

In this backdrop, we compile in this chapter the recent research on plant-mediated preparation of iron oxide NPs, with special focus on their prospective applications.

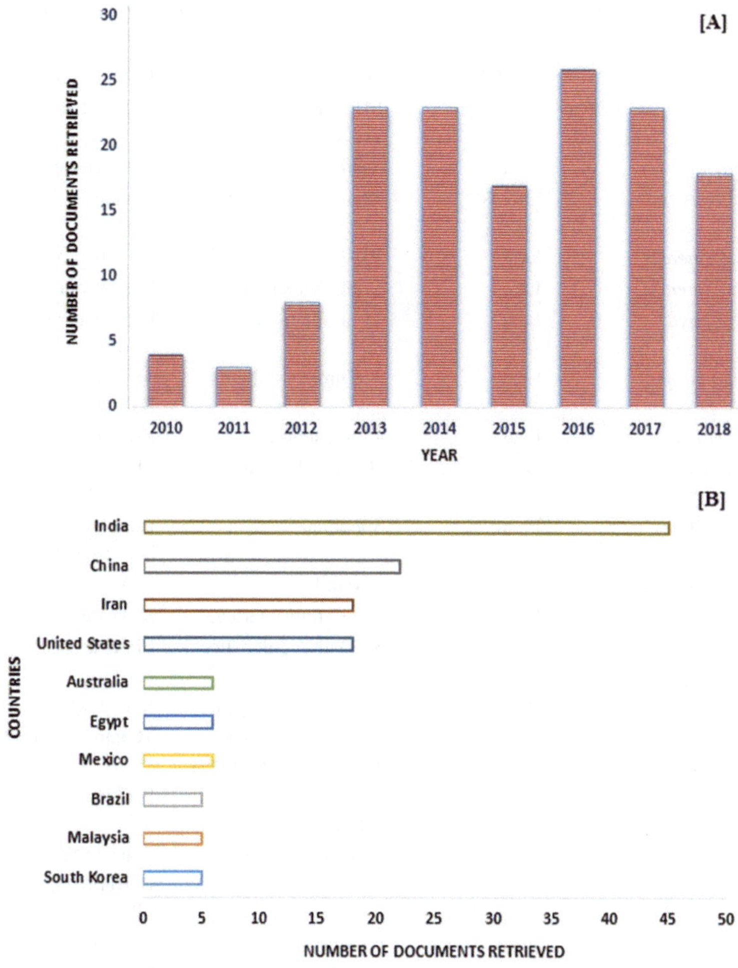

Fig. 9.1 Plots depicting the (**a**) year-wise distribution and (**b**) countries with the maximal number of scientific documents (period: 2010 to 2018) retrievable from Scopus for the search-query, 'Synthesis of iron oxide nanoparticles by plants' as on 18 June 2018

9.2 The Basics of IO-NP Phytosynthesis

9.2.1 Principle and Exemplary Illustrations

Phytosynthesis of IO-NPs represents a bottom-up strategy with oxidation/reduction being the prime reaction. The phytochemicals with antioxidant/reducing attributes mediate the reduction of various salts to their corresponding nanoparticles. A wide spectrum of plants and their active components has been explored for the synthesis of iron oxide NPs (as discussed in the subsequent sections of the chapter).

Table 9.1 Plants and their parts used for synthesis of iron oxide NPs, together with their shape and size

Plant	Plant part	Size/shape	References
Agrewia optiva and *Prunus persica*	Leaf	14–17 nm/cubic spinel structure	Mirza et al. (2018)
Aloe vera	Leaf	6–30 nm/crystalline	Phumying et al. (2013)
Cyanococcus spp.	Shoots	52.4 nm/different shape	Manquián-Cerda et al. (2017)
Camellia sinensis	Leaf	40–80 nm/spherical	Ahmmad et al. (2013)
Camellia sinensis	Leaf	5 nm ≤ crystalline	Plachtová et al. (2018)
Carica papaya	Leaf	33 nm/plate like	Latha and Gowri (2014)
Ceratonia siliqua	Leaf	4–8 nm/crystalline	Awwad and Salem (2012)
Citrus medica	Leaf	45 nm/spherical	Esam (2015)
Citrus reticulata	Peel	50 nm/spherical	Ehrampoush et al. (2015)
Curcuma longa	Plant	Spherical	Herlekar and Siddhivinayak (2015)
Cynometra ramiflora	Leaf	Crystalline	Groiss et al. (2016)
Desmodium gangeticum	Root	25–35 nm/spherical	Santoshi et al. (2015)
Excoecaria cochinchinensis	Leaves	5–20 nm/composites	Lin et al. (2018)
Hibiscus rosa-sinensis	Plant	10–15 nm/spherical	Rath et al. (2015)
Hordeum vulgare and *Rumex acetosa*	Plant	30 nm/amorphous 10–40 nm/amorphous	Makarov et al. (2014)
Juglans regia	Husk	9.7–15.4 nm/cubic shape	Izadiyan et al. (2018)
Mansoa alliacea	Leaf	30 nm/crystalline	Prasad (2016)
Medicago sativa		2–10 nm/crystalline	Herrera-Becerra et al. (2008)
Mimosa pudica	Root	67 nm/spherical and crystalline	Niraimathee et al. (2016)
Musa paradisiaca	Peel	<50 nm/polydispersed	Venkateswarlu et al. (2013)
Musa	Peel	100–200 nm/spherical	Sudha et al. (2015)
Musa acuminata and *Colocasia esculenta*	Peel/leaves	10–25 nm/cubic spinel crystal	Thakur and Karak (2014)
Ocimum sanctum	Leaf	20 nm/irregular sphere	Balamurughan et al. (2014)
Punica granatum	Peel	Amorphous	Irshad et al. (2017)
Punica granatum	Rind	Polycrystalline	Venkateswarlu et al. (2014a)
Zea mays and *Brassica rapa*	Hair of corn and cabbage	84.81 nm/crystallite sizes 48.91 nm/crystallite sizes	Patra and Baek (2017)

(continued)

Table 9.1 (continued)

Plant	Plant part	Size/shape	References
Syzygium cumini	Seed	20 nm/spherical fcc	Venkateswarlu et al. (2014b)
Terminalia chebula	Fruit pericarp	<80 nm/chain-like	Kumar et al. (2014)
Tridax procumbens	Leaf	80–100 nm/irregular spherical	Senthil and Ramesh (2012)
Vitis vinifera	Seed	30 nm/crystalline and spherical	Narayanan et al. (2012)
Vitis vinifera	Stem	<50 nm/core shell structure	Venkateswarlu et al. (2015)
Wrightia tinctoria	Plant	105–145 nm/crystalline	Sravanthi et al. (2016)

To cite for evidence, waste leaves of Oman's mango plant were used to prepare iron oxide nanorods (NRs) (Al-Ruqeishi et al. 2016). Interestingly, polyphenols like mangiferin, penta-O-galloyl-glucoside gallic acid and methyl gallate are abundant in mango (Barreto et al. 2008). These polyphenolics were envisaged to participate actively in the generation of iron oxide NRs (15 ± 2 nm in average length and 3.0 ± 0.2 nm in average diameter). As sodium hydroxide and iron sulphate interact in the aqueous medium, interaction of biomolecules of the extract of the green leaves hastens the generation of FeOOH, which dissociates yielding Fe_2O_3 nuclei. One D iron oxide NRs (the shape and size being influenced by the reaction time and temperature) are gradually built up at various places within the solution. With passage of time, Fe_2O_3 dissociates into its γ and α types. Under appropriate parameters of solution concentration, temperature and sonication, the incessant enlargement of the Fe_2O_3 nuclei leads to the attainment of the final shape of nanorod. A pivotal influence of the biomolecules of the mango leaves extract has been proposed in dictating the shape of the nanorods. Similar observations have been reported for the influence of biomolecules of plant extracts as well as bioactive molecules like enzymes on the eventual shape-size accord of various other nanomaterials as well (Konwarh et al. 2010, 2012, 2014). On the other hand, extract of *Juglans regia*'s green husk (a waste material in the walnut-production industry) is rich in phenolic compounds like vanillic acid, myricetin, coumaric acid, syringic acid, juglone and ferulic acid. The extract was efficiently exploited for the preparation of IO-NPs of high purity, with high saturation magnetization and low coercivity (Izadiyan et al. 2018). The synthesis may be summed up as underneath:

$$J.\ regia + H_2O_{(L)} + 2Fe^{+3} + Fe^{+2} \xrightarrow{\text{stirring}} \left[J.regia / 2Fe^{+3} : 1Fe^{+2}\right] \quad (9.1)$$

$$\left[J.regia / 2Fe^{+3} : 1Fe^{+2}\right] + 8OH^{-}_{(aq)} \rightarrow \left[J.regia / \text{iron oxide nanoparticles}\right]_{(s)} + 4H_2O_{(L)} \quad (9.2)$$

Similarly, polyphenols (in the form of tannins) present in the extract of dry fruit pericarp of *Terminalia chebula* were converted under mild acidic conditions to yield glucose, ellagic acid, gallic acid, etc. The conversion of phenolics to quinone was exploited for the reduction of iron salt to yield stabilized iron oxide nanoparticles (<80 nm) (Kumar et al. 2013).

9.2.2 Crucial Dictates of the Synthesis

A number of factors influence the phytosynthesis of nanoparticles. The various parameters, such as concentration of the salt, mixing ratio of the plant extract and metal salt, pH value, temperature, incubation time, etc., demand optimization to generate homogenous nanoparticles of a similar shape and size (Shah et al. 2015).

One of the prime determinants is the pH of the medium. Herrera-Becerra et al. (2008) reported the influence of pH on size distribution and nature of the iron oxides obtained by using alfalfa extract. Greater proportion of Fe_2O_3 particles (and a few Fe_3O_4) of smaller size was obtained at pH = 10, while pH = 3 and pH = 5 favoured the formation of FeO rhombohedral structure of larger dimensions. $Fe_{0.902}O$ cubic structures were obtained more or less in similar proportion for various pH values. The incubation time and storage conditions are another prime dictators of the quality of the phytosynthesized NPs. The effect of concentration of the reactant cannot be ruled out. Ehrampoush et al. (2015) reported that as the concentration of tangerine peel extract increased from 2% to 6%, the average size of the iron oxide NPs decreased from 200 nm to 50 nm. A point of pertinence is that the extraction protocol of the reducing/stabilizing agents, the age of the plants and plant parts (leaves, stems, flowers, fruits, etc. with varying compositional abundance of the active agents) used to prepare the extract might have a collective influence on the size, shape and distribution of iron oxide NPs. The presence of additional components in the reaction medium might exert additional influence. For instance, Gan et al. (2018) explored the use of the cationic surfactant CTAB as a stabilizing and capping agent for iron oxide NPs, prepared using eucalyptus leaf extract. The assembled surface monolayer of CTAB augmented the dispersion of the nanoparticles as well as dictated the phosphate removal efficiency via exerting influence over the micellar organization and accessibility of the lipophilic biomolecules of the leaf extract. Furthermore, the quality and quantity of synthesized NPs are affected by the procedure used for their purification.

9.2.3 Approaches for Physicochemical Characterization

A range of tools and techniques are routinely used for characterization of iron oxide NPs (Neamtu et al. 2018). Various spectroscopic and microscopic tools aid in comprehending the physicochemical properties like surface texture, size and shape, surface functionality, aggregation state, magnetic property, thermal and dynamic properties, surface area, etc. (Can et al. 2018). Ultraviolet-(UV)-visible spectroscopy, scanning electron microscopy (SEM), transmission electron microscopy (TEM), Fourier-transform infrared spectroscopy (FTIR), X-ray diffraction (XRD), atomic force microscopy (AFM), energy-dispersive X-ray spectroscopy (EDX), dynamic light scattering (DLS), dynamic scanning calorimetry (DSC), thermal gravimetric analysis (TGA) and differential thermal analysis (DTA), dynamic mechanical analysis (DMA), Brunauer-Emmett-Teller (BET) method and vibrating sample magnetometry (VSM) are some of the tools and techniques that are used widely. A detailed description of each characterization tool is beyond the scope of this chapter. Nevertheless, as representative physicochemical approaches to understand the properties of the iron oxide NPs, results of some of the major characterization techniques are mentioned throughout the text. Khalil et al. (2017) reported the XRD spectra of *Sageretia thea* (Osbeck.) extract-based IO-NPs (annealed at 500 °C). Single pure-phase (Fe_2O_3) tetragonal maghemite was indicated by the observed Bragg peaks. Lattice constants were calculated as $<a_{exp.}> = 8.34000$ A° and $<c_{exp.}> = 25.02000$ A°. The average size of the highly crystalline IO-NPs was found to be ~29 nm as calculated using Debye-Scherrer equation. On the other hand, FTIR spectra were illustrative of absorption peak at ~500 cm^{-1} (characteristic Fe-O vibration), broad-OH stretching at ~3400 cm^{-1} (attributed to phytophenolics) and other peaks at ~1100–1200, 1600 and 2200 cm^{-1}, ascribable to the –C–O, –C=O and –CN functional groups, respectively. Raman spectra (Argon laser, $\lambda_{exc} = 514$ nm) and EDS analysis were also used. Raman shift at ~388, ~500 and ~772 cm^{-1} could be attributed, respectively, to T_{2g}, E_g and A_{1g} Raman active phonon modes specific to maghemite. EDS analysis depicted the elemental composition of the phytogenerated iron oxide NPs (peak of carbon could be ascribed to the grid support). Furthermore, various other parameters of iron oxide NPs like cytotoxicity, antioxidant capacity, anticancer activity, biodegradability, pharmacokinetics and pharmacodynamics are assessed from the biological perspective (Feng et al. 2018).

9.2.4 Pros and Cons

Wu et al. (2015) have reviewed the pros and cons of biosynthesis of iron oxide NPs (using various plants and microbes) in comparison to other synthetic methods (e.g. co-precipitation, microemulsion, electrochemical approach, thermal decomposition, sonolysis, aerosol/vapour methods, etc.). Phytosynthesis of iron oxide NPs is considered to be a greener (involving a lesser use of hazardous chemicals) and eco-friendly approach. Furthermore, the obtained products exhibit better biocompatibility than the counterparts produced by other methods. Phytosynthesis is executed under ambient atmospheric conditions and room temperature; however, the yield is low. The reaction period may extend from few minutes or hours to days. Furthermore, the precise tuning of size and shape of iron oxide NPs prepared via phytosynthesis is a challenge and needs greater delving.

9.3 Representative Endeavours

A number of recent reports testify the application of various plant parts like stem, leaves, fruits, flowers, roots and seeds as well as various phytocompounds for the preparation and stabilization of iron oxide NPs. Tracing the yesteryears, Shahwan et al. (2011) reported the fabrication of green tea extract-based iron NPs rich in iron oxide and oxyhydroxide, behaving as Fenton-like catalyst for the removal of cationic (methylene blue) and anionic (methyl orange) dyes. On parallel lines, Ahmmad et al. (2013) synthesized photocatalytic hematite nanoparticles (average size, 60 nm, and surface area 22.5m^2 g^{-1}) using *Camellia sinensis* leaf extract. On the other hand, superparamagnetic IO/Au NPs (~30 nm) were prepared using grape seed proanthocyanidin (GSP) extract at room temperature by Narayanan et al. (2012). Awwad and Salem (2012) reported single-step *carob*-leaf-based synthesis of monodispersed iron oxide NPs (4–8 nm) with crystalline structure.

Phumying et al. (2013) synthesized *aloe vera* leaf extract-based superparamagnetic Fe_3O_4 (6–30 nm), the saturation magnetization of which rose with increase in temperature and time. Huang et al. (2014) used polyphenols and caffeine-rich oolong tea extract for the synthesis of Fe-based (40–50 nm) NPs (zero-valent iron (α-Fe), maghemite (γ-Fe_2O_3), magnetite (Fe_3O_4) and iron hydroxides), which were tested for their capacity to degrade malachite green dye. *Dodonaea viscosa* leaf extract was used by Daniel et al. (2013) to prepare iron oxide NPs (~27 nm). The extract contained flavonoids, tannin and saponins that acted as the reducing and capping agent. The NPs were reported to exhibit antibacterial activity against a number of human pathogens viz. *Bacillus subtilis*, *Klebsiella pneumonia*, *Pseudomonas fluorescens* and *Staphylococcus aureus*. In another study, Venkateswarlu et al. (2013) reported the synthesis of iron oxide NPs using the extract of plantain peel. The nearly polydispersed NPs exhibited high saturation magnetization (15.8 emu g^{-1}), particle size less than 50 nm and surface area in the range of 11.31m^2 g^{-1}. Venkateswarlu et al. (2014a) also reported the synthesis of polycrystalline, ferromagnetic iron oxide NPs, using *Punica granatum* rind extract. The nanoparticles with surface-adhered plant-derived polyphenolics, high saturation magnetization (22.7 emu g^{-1}) and surface area of 10.88m^2 g^{-1} were used for the removal of Pb (II) ions from wastewater. Venkateswarlu et al. (2014b), in another study, documented the synthesis of ferromagnetic spherical iron oxide NPs (~20 nm) with inverse spinal face-centred cubic structure, using *Syzygium cumini* seed extract that acted as the reducing and capping agent. Sodium acetate present in the seed extract served as the electrostatic stabilizing agent. On the other hand, the stem extract of *Vitis vinifera* was used for the synthesis of polycrystalline iron oxide NPs (av. particle size <50 nm) with core-shell structure and high saturation magnetization (15.74 emu g^{-1}) (Venkateswarlu et al. 2015). Similarly, López-Téllez et al. (2013) synthesized iron oxide NPs (using an ethanol-treated orange peel) with rodlike shape for the remediation of chromium ion from wastewater. Cellulosic content present in orange peel served as the reducing and capping agent.

Balamurughan et al. (2014) used an extract of *Ocimum sanctum* for the synthesis of iron oxide NPs; SEM analysis was suggestive of aggregated appearance of the NPs with irregular spherical shape and rough surface. Makarov et al. (2014) reported the application of the extracts of *Hordeum vulgare* (pH 5.8) and *Rumex acetosa* (pH 3.7) for the fabrication of 30 nm- and 10–40 nm-sized iron oxide NPs, respectively. Organic acid such as oxalic and citric acid served as the reducing and capping agent. Kumar et al. (2014) used an extract of *Passiflora tripartita* var. *mollissima* fruit for the synthesis of magnetically recyclable, spherical iron oxide NPs (22.3 ± 3 nm), exhibiting high catalytic activity for the synthesis of 2-arylbenzimidazoles with excellent yields.

Herlekar and Siddhivinayak (2015) reported the microwave-assisted synthesis of evenly dispersed spherical iron oxide NPs using the extract of *Curcuma longa* (turmeric) leaves. These NPs were used to address orthophosphate contamination and chemical oxygen demand in domestic sewage. Esam (2015) synthesized iron oxide NPs (~45 nm) using Al-Abbas's (A.S.) Hund Fruit (*Citrus medica* var. *sarcodactylis*) extract. The antimicrobial application of the NPs was assessed against the bacteria present in Al-'alqami River. On the other hand, extract of *Hibiscus* leaf was used by Rath et al. (2015) to prepare superparamagnetic, monodispersed spherical iron oxide NPs (10–15 nm). Prasad (2016) reported the synthesis of crystalline iron oxide NPs (~30 nm) using *Mansoa alliacea* (garlic vine) leaf extract abundant with polyphenol, flavonoids and other antioxidants. Niraimathee et al. (2016) resorted to *Mimosa pudica* extract for the synthesis of spherical iron oxide NPs (av. size = 67 nm). The non-protein alkaloid, mimosine (β-3-hydroxy-4 pyridone amino acid, present in the root extract), was propounded as the major active agent, responsible for reducing ferrous sulphate into iron oxide NPs. Groiss et al. (2016) used an extract of *Cynometra ramiflora* leaves for the synthesis of crystalline iron oxide NPs with surface functional groups associated with the phytochemicals. They showed antibacterial activity against *E. coli* and *S. epidermidis*, while the attribute of Fenton-like catalyst was established using Rhodamine B dye.

Thakur and Karak (2014) resorted to banana-peel-ash aqueous extract as the base source and *Colocasia esculenta* leaves' aqueous extract as the reducing agent for fabrication of superparamagnetic iron oxide NPs decorated with reduced graphene oxide (RGO). Iron oxide/RGO nanocomposite could effectively remove Pb^{2+} and Cd^{2+} within 10 min, whereas tetrabromobisphenol A within 30 min, following a pseudo-second-order adsorption kinetics.

Thus, one can easily comprehend that iron oxide NPs have been prepared using a wide varieties of plant extracts. Let us now delve into some of the documented applications of the phytogenerated iron oxide NPs.

9.4 Spectrum of Applications

A wide array of applications of iron oxide NPs has been documented in diverse realms over the years. The following section mirrors some of the novel applications of iron oxide NPs, prepared via phytosynthetic approach.

9.4.1 Biomedical Applications

The surmounting issue of antibiotic resistance has dictated researchers to explore alternative antimicrobial agents. Over the years, various nanomaterials including iron oxide NPs have been investigated for their antimicrobial potency. A number of studies vouch for the successful biomedical applications of iron oxide NPs (as testified by recent reports cited in subsequent sections). However, establishing the biocompatibility of iron oxide NPs is a prerequisite for their application in anticancer therapy or as antimicrobial agents. Investigation of interactions of NPs with the possible 'interactomes' of the prokaryotic and eukaryotic cells has been a major focus of recent research. To illustrate these aspects, Khalil et al. (2017) executed an elaborate study on iron oxide NPs prepared by using the aqueous extract of *Sageretia thea*. Five human pathogenic bacterial strains exhibited differential susceptibility towards the iron oxide NPs, *Pseudomonas aeruginosa* (MIC, 7.4 μg mL^{-1}) being the most susceptible. The compatibility at the bio-interface for prospective biomedical applications was attested by high IC_{50} value (>200 μg mL^{-1}) for human RBCs and macrophages. It is pertinent to note that 17.2 and 16.75 μg mL^{-1} were the IC_{50} values for the promastigote and amastigote cultures of *Leishmania tropica* (suggestive of plausible use of the iron oxide NPs in anti-leishmanial therapy), while the assessment of cytotoxic potential using brine shrimps (*Artemia salina*, used as standard to screen cytotoxic compounds) revealed an IC_{50} value of 16.46 μg mL^{-1}. Internalization of iron oxide NPs via receptor-mediated endocytosis and subsequent Fe^{2+} dissolution leading to interference with the stability of the various proteins, entry into mitochondria as well as genotoxicity due to ROS generation are some of the proposed routes of toxicity of iron oxide NPs. The authors further established the protein kinase (PK) (a key enzyme, targeted in cancer therapy) inhibitory potential of iron oxide NPs using *Streptomyces* 85E strain as the test organism where the enzyme plays a prime role in hyphae formation. In the absence of iron oxide NPs, the growth factors are phosphorylated (leading to their activation) by PK in response to growth stimuli. On the other hand, interference of iron oxide NPs limits the potential of PK to initiate the domino effect of the signalling cascade, thereby resulting in the inhibition of division. Thus, these preliminary investigations are suggestive of the indispensable requisite for detailed- and long-term in vitro and in vivo investigations prior to commercialization of phytogenerated iron oxide NPs. Having said that, let us peruse some of the recent reports on the biomedical applications of iron oxide NPs.

9.4.1.1 Antagonist to Microbes and Disease Vectors

In an approach to investigate the prospects of food-processing waste for nanobiotechnological application, Patra and Baek (2017) used silky hairs of corn (*Zea mays*) and outer leaves of Chinese cabbage (*Brassica rapa* L. subsp. *pekinensis*) under photo-catalyzed condition for fabricating superparamagnetic iron oxide NPs, with crystallite size of 84.81 and 48.91 nm, respectively. Apart from exhibiting anti-oxidant property, iron oxide NPs showed synergistic antibacterial action (on mixing with kanamycin and rifampicin) against pathogenic foodborne bacteria as well as anti-candidal activity (on mixing with amphotericin b) against five different *Candida* species. In a similar study, antibacterial activity of iron oxide NPs, fabricated using *Punica granatum* peel extract, had been reported against *Pseudomonas aeruginosa* (Irshad et al. 2017). The nonhaemolytic activity of the iron oxide NPs was projected as an index for establishing the biocompatibility of the iron oxide NPs. On the other hand, Fe_3O_4-Ag core-shell NPs, synthesized using *Vitis vinifera* stem extract as the green solvent, reducing and capping agent, were established as a potent antibacterial agent against gram-positive and gram-negative pathogens (Venkateswarlu et al. 2015). Antimicrobial activity of spherical iron oxide NPs (20–90 nm) prepared by ultrasound-assisted *Coriandrum sativum* leaf extract has been evaluated against *Micrococcus luteus*, *Staphylococcus aureus* and *Aspergillus niger* (Sathya et al. 2017). Researchers have also resorted to various marine plants for preparation of iron oxide NPs. El-Kassas and Ghobrial (2017) used the aqueous extract of *Halophila stipulacea*, rich in protein and polyphenols, to prepare iron oxide NPs exhibiting antagonistic action towards cyanobacteria, *Oscillatoria simplicissima* (known for the production of neurotoxins and its detrimental effects on various aquatic species). Mixed phase (α-Fe_2O_3 and γ-Fe_2O_3) spherical IO-NPs, prepared using the leaf extract of *Platanus orientalis*, were found to exhibit antifungal activity against *Aspergillus niger* and *Mucor piriformis*, the effect being more pronounced towards the latter (Devi et al. 2018).

On the other hand, in an effort to develop effective strategy to control filariasis vectors, Murugan et al. (2018) used *Ficus natalensis* aqueous extract and chemical nanosynthesis to prepare Fe_2O_3 NPs that demonstrated an LC_{50} ranging between 4.5 and 22.1 ppm against I instar larval and pupal stage of *Culex quinquefasciatus*.

9.4.1.2 Diagnostic and Curative (Anticancer) Properties

In vitro cytotoxicity of NPs prepared through green strategy (plant-mediated approach) depends on the concentration, exposure time and type of cells used for assessment (Saranya et al. 2017). Application of iron oxide NPs has been proposed for cancer therapy (Mohanasundaram et al. 2017). Iron oxide NPs, prepared using hydroethanolic leaf extract of *Annona squamosa*, were found to exhibit significant cytotoxicity on HepG2 and melanoma A375 cell lines in contrast to complete compatibility with normal liver cell line. Cubic-shaped iron oxide NPs (Fe_3O_4, 18 ± 4 nm) based on the seaweed, *Sargassum muticum*, exhibited anticancer activity (test duration, 72 h) against HepG2, MCF-7, HeLa and Jurkat cells, the corresponding IC_{50} values being 23.83 ± 1.1 μg mL^{-1}, 18.75 ± 2.1 μg mL^{-1}, 12.5 ± 1.7 μg mL^{-1} and 6.4 ± 2.3 μg mL^{-1} (Namvar et al. 2014). The reducing and capping potencies of aqueous extracts of *Vanilla planifolia* and *Cinnamomum verum* were exploited to prepare polyphenol-coated magnetic NPs with core-shell Fe_3O_4-γFe_2O_3 structure, av. size 10–14 nm, and considerably high saturation magnetization values (70.84 emu g^{-1} and 59.45 emu g^{-1} for *C. verum* and *V. planifolia* mediated samples, respectively) (Ramirez-Nuñez et al. 2018). The plausibility of NPs application in magnetic hyperthermia was vouched by their efficacy to induce in vitro death of BV2 microglial cells post 30 min at a target temperature of 46 °C. However, it is appropriate to mention here that a number of factors have posed major hurdles in clinical trials of the biogenic nanomaterials against cancer. The issues include the selection of dosage and route of administration, biodegradability and biocompatibility, uptake, retention and clearance as well as the combinatorial approach with FDA-approved anticancer drugs (Barabadi et al. 2017).

On the other hand, in an interesting study to harness the biomedical utility of 'green' iron oxide NPs, ginger (*Zingiber officinale*) rhizome extract was effectively exploited as a capping agent for self-organizing Ag and iron oxide NPs in a single-step synthesis protocol, yielding highly ordered hierarchical microstructure, dictated by the ginger polysaccharides and water molecules (Ivashchenko et al. 2017). Bactericidal and fungicidal activity as well as magnetic resonance imaging (MRI) T2 contrasting and fluorescence attributes established the multimodality of the MAg NPs. It is relevant to mention that relaxivity values (R1 and R2) are used to assess a potential MRI contrast agent. The effect of MAg was evident only on the proton T2 relaxation, the results being dictated by the Fe concentration and the kind of dispersive medium. A decrease in the R2 relaxivity was documented from the sample MAg (Ag 10%) to MAg (Ag 33%) (Table 9.2).

On the other hand, as shown in Fig. 9.2a, the fluorescence 3D image revealed the accumulation of NPs in the surrounding areas of yeast cells. The experiment was conducted without the fluorescence dye (concanavalin A). Even without the membrane contour, the cells could be visualized. The observations suggested that the MAg could be employed in the studies of cells/NP interactions without the use of extra fluorescent agent adhered to the surface of the NPs.

Table 9.2 The relaxivity data of MAg as a T_2 contrast agent

		Relaxivity data		
Ag content in the sample, Mass %	Dispersive medium	R_1 mM^{-1} S^{-1}	R_1 mM^{-1} S^{-1}	R_2/R_1
10	Water	0.33 ± 0.02	25.6 ± 0.1.3	77.52
10	BSA (1%)	0.17 ±. 0.05	37.3 ± 0.2.4	219.59
22	Water	0.36 ± 0.0.02	20 ± 2	56.08
22	BSA (1%)	0.11 ± 0.0.03	12.2 ± 0.0.7	110.55
33	Water	0.29 ± 0.0.02	6.7 ± 0.0.4	23.14
33	BSA (1%)	0.16 ± 0.0.02	7.6 ±. 0.6	47.25

Reproduced from Ivashchenko et al. (2017)

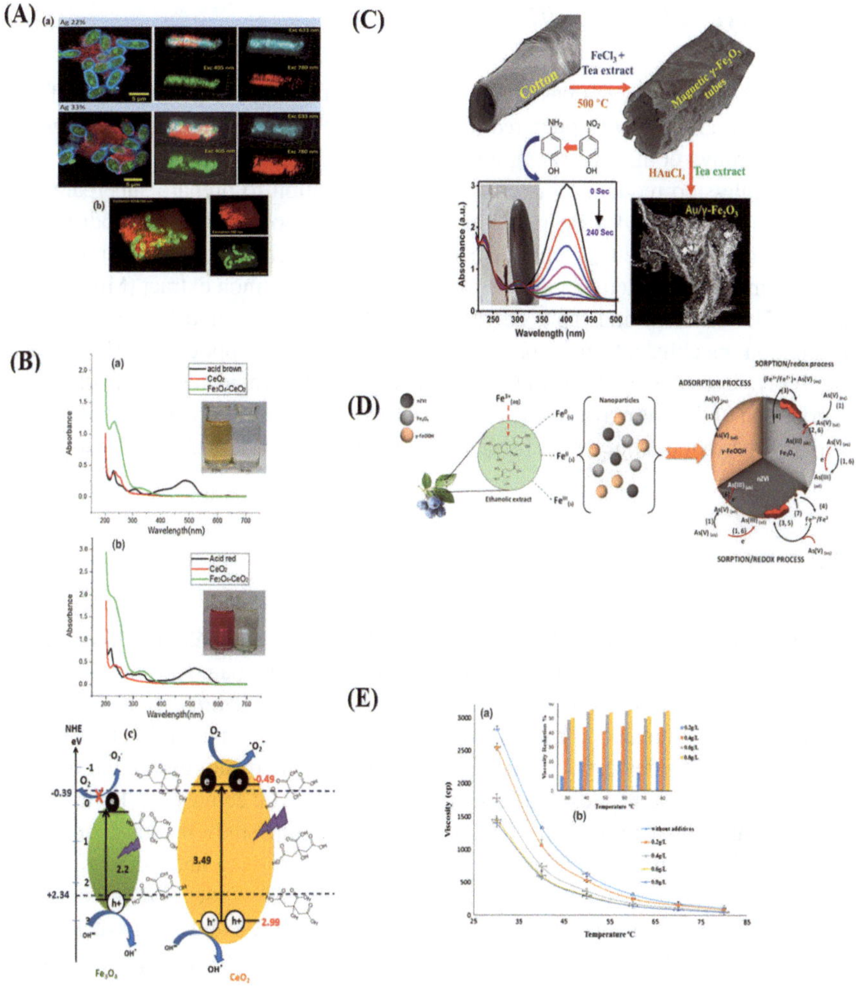

Fig. 9.2 (**a**) Fluorescence 3D images of yeast cells incubated with MAg (100 μg/mL) for 1 h (*a*) with and (*b*) without fluorescence dye. (Reproduced from Ivashchenko et al. 2017.) (**b**) Plot for change in the absorption spectrum of (*a*) acid brown and (*b*) acid red by Fe_3O_4–CeO_2. A model for the photocatalytic mechanism of the Fe_3O_4.CeO_2 nanocomposite. (Reproduced from Moradi et al. 2018.) (**c**) Metal (Au, Pd, Ag)-deposited magnetic γ-Fe_2O_3 tubular morphology was synthesized using cotton fibre as a sacrificial template. The developed material shows promising catalytic activity for the reduction of 4-nitrophenol to 4-aminophenol with a very good recyclability for the reuse of the catalyst. (Reproduced from Purbia and Paria 2018.) (**d**) Schematic representation of the possible mechanisms of As (V) removal and structural changes in lepidocrocite (γ-FeOOH), magnetite (FeO) and nZVI. Depending on the iron nanoparticles, As (V) can be removed by adsorption (*1*), absorption (*2*), precipitation (*3*), dissolution (*4*), co-precipitation (*5*) and reduction (*6*). The structure of nZVI is determined by dissolution or passivation (*4* and *7*). (Reproduced from Manquián-Cerda et al. 2017.) (**e**) (*a*) Oil viscosity vs. temperature for heavy oil before and after adding iron oxide NRs with different wt%. (*b*) An inset column chart shows viscosity reduction % at each temperature and additives wt%. (Reproduced from Al-Ruqeishi et al. 2016)

9.4.2 Dye-Adsorption/Degradation

Beheshtkhoo et al. (2018) have reported 81% dye-decolorization efficiency of iron oxide NPs (6.5–14.9 nm), prepared using the aqueous leaf extract of *Daphne mezereum*. Tharunya et al. (2017) employed ethanolic extract of *Ficus carica* fruit for a rapid, single-step reduction of ferrous sulphate solution to prepare superparamagnetic iron oxide NPs for degradation of dyes. On the other hand, monodispersed, superparamagnetic Fe_3O_4-CeO_2 nanocomposites were prepared basing on a hydrothermal method coupled to a greener protocol involving lemon extract (Moradi et al. 2018). The efficient and rapid degradation of two azo dyes under ultraviolet light irradiation established their prospective photocatalytic application (Fig. 9.2b). Similarly, Carvalho and Carvalho (2017) reported the rapid and complete decolourization of different dyes using carbon-rich catalysts based on iron (II)/(III) oxides and oxyhydroxides composites (over silica), prepared with *Camellia sinensis* extract.

9.4.3 Catalysis

Recently, Shahriary et al. (2018) have exploited the extract of *Stachys lavandulifolia* ("Chaye-e-Kohi" a native of Iran) for fabrication of a hybrid magnetic nanocomposite, with antibacterial efficacy, documented against *E. coli* and *S. aureus*. The Fe_3O_4 NPs with carbonyl and phenolic hydroxyl surface functionality, conferred by the biomolecules of the herbal tea extract, acted as the edifice for the in situ generation of silver NPs. The recyclable nanocomposite exhibited high catalytic activity for 4-nitrophenol reduction at ambient temperature. In yet another interesting study, natural cotton fibres were used as sacrificial template for the fabrication of Au, Ag and Pd nanoparticle-decorated hierarchical maghemite tubes, using green tea extract (Purbia and Paria 2018). Effective catalytic potency of the magnetically recyclable nanocatalyst was demonstrated for the hydrogenation reaction of 4-nitrophenol to aminophenol conversion (Fig. 9.2c). Fe_3O_4 nanocatalysts were prepared by Shojaee and Mahdavi Shahri (2018) using an aqueous extract of *Camellia sinensis*. The magnetically recyclable catalysts were efficiently used for the one-pot synthesis of 2-(4-chlorophenyl)-1H-benzo[*d*]imidazole that exhibited dose-dependent in vitro toxicity on MOLT-4 cells. The study vouched for the potential of iron oxide NPs, prepared by greener route for preparation of biomedically relevant therapeutic materials.

9.4.4 Eco-Hazard Mitigation/Metal-Ion Adsorption

Recalcitrant chlorobenzenes are listed as priority pollutants by the US Environmental Protection Agency (USEPA). In this context, Kuang et al. (2013) reported the synthesis of zero-valent iron (α-Fe), maghemite (γ-Fe_2O_3), magnetite (Fe_3O_4) and iron hydroxides using extracts of green tea, oolong tea and black tea. These irregular spherical NPs (with basic size: 20–40 nm and surface area of 5.82m^2 g^{-1}) could mediate Fenton-like oxidation of mono-chlorobenzene (MCB) owing to their high surface and reactivity, complemented by low leaching of Fe^{2+}. The correspondence between oxidative degradation efficacy and the associated COD removal was suggestive of the partial mineralization of MCB to carbon dioxide and water. On the other hand, Hassan et al. (2018) have demonstrated the application of iron oxide NPs, prepared using pomegranate peel extract, for adsorption of pyrene and benzo (α) pyrene. Following a pseudo-second-order mechanism of adsorption, the NPs exhibited more than 98% removal of the polycyclic aromatic hydrocarbons (PAH) from artificially contaminated water in a semi-pilot plant.

Various reports evince the use of iron oxide NPs for adsorption of metal ions. In an endeavour to develop an efficient adsorbent to remove Ni (II) ions, superparamagnetic iron oxide NPs were prepared using *Lantana camara* extract as the stabilizer and ammonia solution as the precipitating agent for Fe^{2+} and Fe^{3+} salts in water (Nithya et al. 2018). These NPs exhibited remarkably high adsorption capacity of 227.2 mg g^{-1} at a pH of 6.0 and adsorbent dose of 0.05 g. In another study, Lunge et al. (2014) reported the preparation of iron oxide NPs using an extract of tea waste at room temperature for removal of As (III) and As (V). It was observed that the maximum adsorption of As (III) and As (V) was 188.7 mg g^{-1} and 153.8 mg g^{-1} at neutral pH, respectively. Extract of leaves and shoots of blueberry (*Vaccinium corymbosum*) was used for the preparation of diverse NPs, varying in forms and associated structures, including lepidocrocite and magnetite (Manquián-Cerda et al. 2017). These were projected as an effective material for arsenic (As) (V) remediation (Fig. 9.2d). Iron oxide NPs, prepared using the extract of tangerine peels, were employed for the remediation of cadmium ions from waste water (Ehrampoush et al. 2015). Similarly, IO-NPs, synthesized using the extract of *Excoecaria cochinchinensis* leaves and subsequently modified by low-temperature calcination, were effectively used to adsorb ~98.50% of Cd (II) (10 mg L^{-1}) under optimized parameters (based on Box Behnken Design) of pH, absorbent dosage, temperature and ionic strength conditions as 8.07, 2.5 g L^{-1}, 45 °C and 0.07 mol L^{-1}, respectively (Lin et al. 2018). In an interesting study, Rao et al. (2013) obtained surface modification of *Yarrowia lipolytica* (NCIM 3589 and NCIM 3590) cells with phytoinspired Fe^0/Fe_3O_4 NPs in an effort to remove hexavalent chromium ions. The reported nanocomposite displayed higher values of Langmuir and Scatchard coefficients than untreated cells, thus suggesting that the former was a better agent for mitigating Cr (VI) toxicity.

9.4.5 Heavy-Oil Viscosity Treatment

A reduction in dynamic viscosity of crude oil results in direct microwave irradiation. On impregnation of heavy oil with phytosynthesized iron oxide NRs (Al-Ruqeishi et al. 2016), the viscosity reduction rate was enhanced. Addition of 0.2 g of iron oxide NRs to 1 L of heavy oil led to a reduction of viscosity by 10% at

30 °C. On the other hand, a further augmentation (up to 38% and 49%) in the viscosity reduction was noted for 0.4 and 0.6 g L^{-1} additives (0.8 g L^{-1} being the oil's additive saturation point) (Fig. 9.2e). Experts have forwarded the 'green' iron oxide NRs as befitting candidate for heavy crude oil cracking in the perspective of small size, distribution and thermal conversion.

9.4.6 Augmenting the Agricultural Output

Investigating the effect of nanomaterials on agricultural domain has been an interesting proposition. In this context, Sebastian et al. (2017) have recommended application of iron oxide NPs, synthesized using caffeic acid as a potential ameliorant, as a solution to Ca-induced Fe deficiency in *Oryza sativa* (rice). Bioproductivity, photosynthetic electron transport, antioxidant enzyme activity and iron accumulation under calcium stress were augmented. In an endeavour to develop green materials for environmental remediation, Sebastian et al. (2018) reported that iron oxide NPs with surface-fabricated phenolics from coconut husk could efficiently adsorb Ca and Cd. However, the interesting attribute of the study was the augmented iron accumulation in rice plants as well as tolerance towards calcium and cadmium stress. Increase of biomass and chlorophyll content attested the plant-growth-accelerating action of the iron oxide NPs.

9.5 Conclusion

Although a plethora of plants have been explored for fabrication of iron oxide NPs, it is fascinating to observe the real-time genesis of the exotic iron oxide nanostructures reported in a number of studies. Optimization of the process parameters to arrest the growth of iron oxide NRs at a particular stage could be another focus area. Use of advanced analytical and computational tools would complement our understanding on the possible evolution and structural modulations of the iron oxide NPs, mediated by the phytocompounds. Despite the fact that phytosynthesis is considered as a benign approach for preparing iron oxide NPs, it is crucial to note that nanomaterials prepared through green strategy may not be 'green' per se with respect to their action at the biological interface. The shape-size-surface chemistry-concentration accord dictates the bio-nano-interfacial outcome. Compilation of toxicity profile of the iron oxide NPs is a must prior to their commercial applications. Probing the actions at the in vitro and in vivo levels is indispensable. The experimental biocompatibility assessments may be complemented (not substituted) by computational techniques like quantitative structure-activity-relationship (QSAR) method. Execution of a 'cradle-to-grave' analysis would assist in realizing the actual application potential of iron oxide NPs. Nevertheless, the portal to several prospective applications of phytogenerated iron oxide NPs has just been opened.

References

Ahmmad B, Leonard K, Islam MS, Kurawaki J, Muruganandham M, Ohkubo T, Kuroda Y (2013) Green synthesis of mesoporous hematite (a-Fe2O3) nanoparticles and their photocatalytic activity. Adv Powder Technol 24:160–167

Ali A, Zafar H, Zia M, UL Haq I, Phull AR, Ali JS, Hussain A (2016) Synthesis, characterization, applications, and challenges of iron oxide nanoparticles. Nanotechnol Sci Appl 9:49–67

Al-Ruqeishi MS, Mohiuddin T, Al-Saadi LK (2016) Green synthesis of iron oxide nanorods from deciduous Omani mango tree leaves for heavy oil viscosity treatment. Arab J Chem. https://doi.org/10.1016/j.arabjc.2016.04.003

Awwad AM, Salem NM (2012) A green and facile approach for synthesis of magnetite nanoparticles. Nanosci Nanotechnol 2:208–213

Balamurughan MG, Mohanraj S, Kodhaiyolii S, Pugalenthi V (2014) *Ocimum sanctum* leaf extract mediated green synthesis of iron oxide nanoparticles: spectroscopic and microscopic studies. J Chem Pharma Sci Special Issue 4:201–204

Barabadi H, Ovais M, Shinwari ZK, Saravanan M (2017) Anti-cancer green bionanomaterials: present status and future prospects. Green Chem Lett Rev 10:285–314

Barreto JC, Trevisan MT, Hull WE, Erben G, Brito ESD, Pfundstein B, Würtele G, Spiegelhalder B, Owen RW (2008) Characterization and quantitation of polyphenolic compounds in bark, kernel, leaves, and peel of mango (*Mangifera indica* L). J Agric Food Chem 56:5599–5610

Barua S, Konwarh R, Mandal M, Gopalakrishnan R, Kumar D, Karak N (2013) Biomimetically prepared antibacterial, free radical scavenging poly(ethylene glycol) supported silver nanoparticles as *Aedes albopictus* larvicide. Adv Sci Eng Med 5:291–298

Beheshtkhoo N, Kouhbanani MAJ, Savardashtaki A, Amani AM, Taghizadeh S (2018) Green synthesis of iron oxide nanoparticles by aqueous leaf extract of *Daphne mezereum* as a novel dye removing material. Appl Phys A Mater Sci Process 124:363

Can HK, Kavlak S, ParviziKhosroshahi S, Güner A (2018) Preparation, characterization and dynamical mechanical properties of dextran-coated iron oxide nanoparticles (DIO-NPs). Artif Cells Nanomed Biotechnol 46:421–431

Carvalho SS, Carvalho NM (2017) Dye degradation by green heterogeneous Fenton catalysts prepared in presence of *Camellia sinensis*. J Environ Manag 187:82–88

Daniel SK, Vinothini G, Subramanian N, Nehru K, Sivakumar M (2013) Biosynthesis of Cu, ZVI, and Ag nanoparticles using *Dodonaea viscosa* extract for antibacterial activity against human pathogens. J Nanopart Res 15:1319

Devi HS, Boda MA, Shah MA, Parveen S, Wani AH (2018) Green synthesis of iron oxide nanoparticles using *Platanus orientalis* leaf extract for antifungal activity. Green Process Synth 1–8. https://doi.org/10.1515/gps-2017–0145

Dinali R, Ebrahiminezhad A, Manley-Harris M, Ghasemi Y, Berenjian A (2017) Iron oxide nanoparticles in modern microbiology and biotechnology. Crit Rev Microbiol 43:493–507

Ehrampoush MH, Miria M, Salmani MH, Mahvi AH (2015) Cadmium removal from aqueous solution by green synthesis iron oxide nanoparticles with tangerine peels extract. J Environ Health Sci Eng 13:84

El-Kassas HY, Ghobrial MG (2017) Biosynthesis of metal nanoparticles using three marine plant species: anti-algal efficiencies against *Oscillatoria simplicissima*. Environ Sci Pollut Res 24(8):7837–7849

Esam JAK (2015) Green synthesis of magnetite iron oxide nanoparticles by using Al-Abbas's (AS) hund fruit (*Citrus medica*) var. Sarcodactylis Swingle extract and used in Al-'alqami river water treatment. J Nat Sci Res 5:125–135

Feng Q, Liu Y, Huang J, Chen K, Huang J, Xiao K (2018) Uptake, distribution, clearance, and toxicity of iron oxide nanoparticles with different sizes and coatings. Sci Rep 8:2082

Gan L, Lu Z, Cao D, Chen Z (2018) Effects of cetyltrimethylammonium bromide on the morphology of green synthesized Fe_3O_4 nanoparticles used to remove phosphate. Mater Sci Eng 82:41–45

Groiss S, Selvaraj R, Thivaharan V, Vinayagam R (2016) Structural characterization, antibacterial and catalytic effect of iron oxide nanoparticles synthesised using the leaf extract of *Cynometra ramiflora*. J Mol Struct 1128:572–578

Hassan SS, Abdel-Shafy HI, Mansour MS (2018) Removal of pyrene and benzo (a) pyrene micro-pollutant from water via adsorption by green synthesized iron oxide nanoparticles. Adv Nat Sci Nanosci Nanotech 9:015006

Herlekar M, Siddhivinayak B (2015) Optimization of microwave assisted green synthesis protocol for iron oxide nanoparticles and its application for simultaneous removal of multiple pollutants from domestic sewage. Int J Adv Res 3:331–345

Herrera-Becerra R, Zorrilla C, Rius JL, Ascencio JA (2008) Electron microscopy characterization of biosynthesize iron oxide nanoparticles. Appl Phy A 91:241–246

Huang L, Weng X, Chen Z, Megharaj M, Naidu R (2014) Synthesis of iron-based nanoparticles using oolong tea extract for the degradation of malachite green. Spectrochim Acta A Mol Biomol Spectrosc 117:801–804

Husen A (2017) Gold Nanoparticles from plant system: synthesis, characterization and their application. In: Ghorbanpour M, Manika K, Varma A (eds) Nanoscience and plant–soil systems, vol 48. Springer, Cham, pp 455–479

Husen A, Siddiqi KS (2014) Phytosynthesis of nanoparticles: concept, controversy and application. Nanoscale Res Lett 9:229

Irshad R, Tahira K, Li B, Ahmada A, Siddiquia AR, Nazir S (2017) Antibacterial activity of biochemically capped iron oxide nanoparticles: a view towards green chemistry. J Photochem Photobio B 170:241–246

Ivashchenko O, Gapiński J, Peplińska B, Przysiecka Ł, Zalewski T, Nowaczyk G, Jarek M, Marcinkowska-Gapińska A, Jurga S (2017) Self-organizing silver and ultrasmall iron oxide nanoparticles prepared with ginger rhizome extract: characterization, biomedical potential and microstructure analysis of hydrocolloids. Mater Des 133:307–324

Izadiyan Z, Shameli K, Miyake M, Hara H, Mohamad SEB, Kalantari K, Taib SHM, Rasouli E (2018) Cytotoxicity assay of plant-mediated synthesized iron oxide nanoparticles using *Juglans regia* green husk extract. Arabian J Chem. https://doi.org/10.1016/j.arabjc.2018.02.019

Jun YW, Seo JW, Cheon J (2008) Nanoscaling laws of magnetic nanoparticles and their applicabilities in biomedical sciences. Acc Chem Res 41:179–189

Khalil AT, Ovais M, Ullah I, Ali M, Shinwari ZK, Maaza M (2017) Biosynthesis of iron oxide (Fe_2O_3) nanoparticles via aqueous extracts of *Sageretia thea* (Osbeck) and their pharmacognostic properties. Green Chem Lett Rev 10:186–201

Konwarh R, Karak N, Rai SK, Mukherjee AK (2009) Polymer-assisted iron oxide magnetic nanoparticle immobilized keratinase. Nanotechnology 20:225107

Konwarh R, Kalita D, Mahanta C, Mandal M, Karak N (2010) Magnetically recyclable, antimicrobial, and catalytically enhanced polymer-assisted green nanosystem-immobilized *Aspergillus niger* amyloglucosidase. Appl Microbiol Biotechnol 87:1983–1992

Konwarh R, Pramanik S, Devi KSP, Saikia N, Boruah R, Maiti TK, Deka RC, Karak N (2012) Lycopene coupled 'trifoliate' polyaniline nanofibers as multi-functional biomaterial. J Mater Chem 22:15062–15070

Konwarh R, Shail M, Medhi T, Mandal M, Karak N (2014) Sonication assisted assemblage of exotic polymer supported nanostructured bio-hybrid system and prospective application. Ultrason Sonochem 21:634–642

Kuang Y, Wang Q, Chen Z, Megharaj M, Naidu R (2013) Heterogeneous Fenton-like oxidation of monochlorobenzene using green synthesis of iron nanoparticles. J Colloid Interface Sci 410:67–73

Kumar KM, Mandal BK, Kumar KS, Reddy PS, Sreedhar B (2013) Biobased green method to synthesise palladium and iron nanoparticles using *Terminalia chebula* aqueous extract. Spectrochim Acta A Mol Biomol Spectrosc 102:128–133

Kumar B, Smita K, Cumbal L, Debut A (2014) Biogenic synthesis of iron oxide nanoparticles for 2-arylbenzimidazole fabrication. J Saudi Chem Soc 18:364–369

Kumar JP, Konwarh R, Kumar M, Gangrade A, Mandal BB (2017) Potential nanomedicine applications of multifunctional carbon nanoparticles developed using green technology. ACS Sustain Chem Eng 6:1235–1245

Latha N, Gowri M (2014) Bio synthesis and characterisation of Fe3O4 nanoparticles using *caricaya papaya* leaves extract. Syn Int J Sci Res 3:1551–1556

Lin J, Su B, Sun M, Bo C, Zuliang C (2018) Biosynthesized iron oxide nanoparticles used for optimized removal of cadmium with response surface methodology. Sci Total Environ 627:314–321

López-Téllez G, Balderas-Hernández P, Barrera-Díaz CE, Vilchis-Nestor AR, Roa-Morales G, Bilyeu B (2013) Green method to form iron oxide nanorods in orange peels for chromium (VI) reduction. J Nanosci Nanotechnol 13:2354–2361

Lunge S, Singh S, Sinha A (2014) Magnetic iron oxide (Fe_3O_4) nanoparticles from tea waste for arsenic removal. J Magn Magn Mater 356:21–31

Makarov VV, Makarova SS, Love AJ, Sinitsyna OV, Dudnik AO, Yaminsky IV, Taliansky ME, Kalinina NO (2014) Biosynthesis of stable iron oxide nanoparticles in aqueous extracts of *Hordeum vulgare* and *Rumex acetosa* plants. Langmuir 30:5982–5988

Manquián-Cerda K, Cruces E, Angélica RM, Reyes C, Arancibia-Miranda N (2017) Preparation of nanoscale iron (oxide, oxyhydroxides and zero-valent) particles derived from blueberries: reactivity, characterization and removal mechanism of arsenate. Ecotoxicol Environ Saf 145:69–77

Mirza AU, Kareem A, Nami SAA, Khan MS, Rehman S, Bhat SA, Mohammad A, Nishat N (2018) Biogenic synthesis of iron oxide nanoparticles using Agrewia optiva and Prunus persica phyto species: characterization, antibacterial and antioxidant activity. J Photochem Photobiol B 185:262–274

Mobasser S, Firoozi AA (2016) Review of nanotechnology applications in science and engineering. J Civil Eng Urban 6:84–93

Mohanasundaram S, Hemalatha S, Vanitha V, Bharathi NP, Jayalakshmi M, Pushpabharathi N (2017) Deciphering the cytotoxic activity of *Annona squamosa* iron oxide nanoparticles against selective cancer cell line. Int J Res Pharm Sci 8(2):259–263

Moradi B, Nabiyouni G, Ghanbari D (2018) Rapid photo-degradation of toxic dye pollutants: green synthesis of mono-disperse Fe_3O_4–CeO_2 nanocomposites in the presence of lemon extract. J Mater Sci Mater Electron 29(13):11065–11080

Murugan K, Dinesh D, Nataraj D, Subramaniam J, Amuthavalli P, Madhavan J, Rajasekar A, Rajan M, Thiruppathi KP, Kumar S, Higuchi A (2018) Iron and iron oxide nanoparticles are highly toxic to *Culex quinquefasciatus* with little non-target effects on larvivorous fishes. Environ Sci Pollut Res 25:10504–10514

Namvar F, Rahman HS, Mohamad R, Baharara J, Mahdavi M, Amini E, Chartrand MS, Yeap SK (2014) Cytotoxic effect of magnetic iron oxide nanoparticles synthesized via seaweed aqueous extract. Int J Nanomedicine 9:2479

Narayanan SBN, Sathy U, Mony M, Koyakutty S, Nair V, Menon D (2012) Biocompatible magnetite/gold nano hybrid contrast agents via green chemistry for MRI and CT bioimaging. ACS Appl Mater Interfaces 4:251–260

Neamtu M, Nadejde C, Hodoroaba VD, Schneider RJ, Verestiuc L, Panne U (2018) Functionalized magnetic nanoparticles: synthesis, characterization, catalytic application and assessment of toxicity. Sci Rep 8:6278

Niraimathee VA, Subha V, Ravindran RE, Renganathan S (2016) Green synthesis of iron oxide nanoparticles from Mimosa pudica root extract. Int J Environ Sustain Dev 15:227–240

Nithya K, Sathish A, Kumar PS, Ramachandran T (2018) Fast kinetics and high adsorption capacity of green extract capped superparamagnetic iron oxide nanoparticles for the adsorption of Ni (II) ions. J Ind Eng Chem 59:230–241

Patra JK, Baek KH (2017) Green biosynthesis of magnetic iron oxide (Fe3O4) nanoparticles using the aqueous extracts of food processing wastes under photo-catalyzed condition and investigation of their antimicrobial and antioxidant activity. J Photochem Photobiol B 173:291–300

Phumying S, Labuayai S, Thomas C, Amornkitbamrung V, Swatsitang E, Maensiri S (2013) *Aloe vera* plant-extracted solution hydrothermal synthesis and magnetic properties of magnetite (Fe_3O_4) nanoparticles. Appl Phys A Mater Sci Process 111(4):1187–1193

Plachtová P, Medříková Z, Zbořil R, Tuček J, Varma RS, Maršálek B (2018) Iron and Iron oxide nanoparticles synthesized with green tea extract: differences in ecotoxicological profile and ability to degrade malachite green. ACS Sustain Chem Eng 6:8679–8687

Pramanik S, Konwarh R, Deka RC, Aidew L, Barua N, Buragohain AK, Mohanta D, Karak N (2013) Microwave-assisted poly (glycidyl methacrylate)-functionalized multiwall carbon nanotubes with a 'tendrillar' nanofibrous polyaniline wrapping and their interaction at bio-interface. Carbon 55:34–43

Prasad AS (2016) Iron oxide nanoparticles synthesized by controlled bio-precipitation using leaf extract of garlic vine (*Mansoa alliacea*). Mater Sci Semicond Process 53:79–83

Purbia R, Paria S (2018) Noble metals decorated hierarchical maghemite magnetic tubes as an efficient recyclable catalyst. J Colloid Interface Sci 511:463–473

Ramirez-Nuñez AL, Jimenez-Garcia LF, Goya GF, Sanz B, Santoyo-Salazar J (2018) In vitro magnetic hyperthermia using polyphenol-coated $Fe_3O_4@\gamma Fe_2O_3$ nanoparticles from *Cinnamomun verum* and *Vanilla planifolia*: the concert of green synthesis and therapeutic possibilities. Nanotechnology 29:074001

Rao A, Bankar A, Kumar AR, Gosavi S, Zinjarde S (2013) Removal of hexavalent chromium ions by Yarrowia lipolytica cells modified with phyto-inspired Fe $0/Fe_3O_4$ nanoparticles. J Contam Hydrol 146:63–73

Rath KB, Singh A, Chandan S (2015) Biosynthesis of magnetic iron oxide nanoparticles using Hibiscus as a plant source. NanoTrends 17:1–9

Reardon PJ, Konwarh R, Knowles JC, Mandal BB (2017) Mimicking hierarchical complexity of the osteochondral interface using electrospun silk-bioactive glass composites. ACS Appl Mater Interfaces 9:8000–8013

Santoshi V, Shakila Banu A, Kurian GA (2015) Synthesis, characterization and biological evaluation of iron oxide nanoparticles prepared by *Desmodium gangeticum* root aqueous extract. Int J Pharm Pharm Sci 7(13):75–80

Saranya S, Vijayaranai K, Pavithra S, Raihana N, Kumanan K (2017) In vitro cytotoxicity of zinc oxide, iron oxide and copper nanopowders prepared by green synthesis. Toxicol Rep 4:427–430

Sathya K, Saravanathamizhan R, Baskar G (2017) Ultrasound assisted phytosynthesis of iron oxide nanoparticle. Ultrason Sonochem 39:446–451

Sebastian A, Nangia A, Prasad MNV (2017) Carbon-bound iron oxide nanoparticles prevent calcium-induced iron deficiency in *Oryza sativa* L. J AgricFood Chem 65:557–564

Sebastian A, Nangia A, Prasad MNV (2018) A green synthetic route to phenolics fabricated magnetite nanoparticles from coconut husk extract: implications to treat metal contaminated water and heavy metal stress in *Oryza sativa* L. J Clean Prod 174:355–366

Senthil M, Ramesh C (2012) Biogenic synthesis of Fe3O4 nanoparticle using Tridax procumbens leaf extract and its antibacterial activity on *Pseudomonas aeruginosa*. Dig J Nanomater Biostruct 7:1655–1661

Shah M, Fawcett D, Sharma S, Tripathy SK, Poinern GEJ (2015) Green synthesis of metallic nanoparticles via biological entities. Materials 8:7278–7308

Shahriary M, Veisi H, Hekmati M, Hemmati S (2018) In situ green synthesis of Ag nanoparticles on herbal tea extract (*Stachys lavandulifolia*)-modified magnetic iron oxide nanoparticles as antibacterial agent and their 4-nitrophenol catalytic reduction activity. Mater Sci Eng C 90:57–66

Shahwan T, Sirriah SA, Nairat M, Boyacı E, Eroğlu AE, Scott TB, Hallam KR (2011) Green synthesis of iron nanoparticles and their application as a Fenton-like catalyst for the degradation of aqueous cationic and anionic dyes. Chem Eng J 172:258–266

Shojaee S, Mahdavi Shahri M (2018) An efficient synthesis and cytotoxic activity of 2-(4-chlorophenyl)-1H–benzo [d] imidazole obtained using a magnetically recyclable Fe_3O_4 nanocatalyst-mediated white tea extract. Appl Organomet Chem 32:e3934

Siddiqi KS, Husen A (2017) Recent advances in plant-mediated engineered gold nanoparticles and their application in biological system. J Trace Elem Med Biol 40:10–23

Siddiqi KS, Rahman A, Tajuddin HA (2016) Biogenic fabrication of iron/iron oxide nanoparticles and their application. Nano Res Lett 11:498

Siddiqi KS, Husen A, Rao RAK (2018a) A review on biosynthesis of silver nanoparticles and their biocidal properties. J Nanobiotechnol 16:14

Siddiqi KS, Husen A, Sohrab SS, Osman M (2018b) Recent status of nanomaterials fabrication and their potential applications in neurological disease management. Nano Res Lett 13:231

Sravanthi M, Kumar DM, Ravichandra M, Vasu G, Hemalatha KPJ (2016) Green synthesis and characterization of iron oxide nanoparticles using *Wrightiatinctoria* leaf extract and their antibacterial studies. Int J Curr Res Aca Rev 4:30–44

Sudha K, Anitta S, Devi PM, Thejomayah G (2015) Biosynthesis of iron nanoparticle from green banana peel extract. IJSSIR 4(6):165–176

Thakur S, Karak N (2014) One-step approach to prepare magnetic iron oxide/reduced graphene oxide nanohybrid for efficient organic and inorganic pollutants removal. Mater Chem Phys 144:425–432

Thakur S, Karak N (2015) Alternative methods and nature-based reagents for the reduction of graphene oxide: a review. Carbon 94:224–242

Tharunya P, Subha V, Kirubanandan S, Sandhaya S, Renganathan S (2017) Green synthesis of superparamagnetic iron oxide nanoparticle from *Ficus carica* fruit extract, characterization studies and its application on dye degradation studies. Asian J Pharm Clin Res 10:125–128

Venkateswarlu S, Rao YS, Balaji T, Prathima B, Jyothi NVV (2013) Biogenic synthesis of Fe3O4 magnetic nanoparticles using plantain peel extract. Mater Lett 100:241–244

Venkateswarlu S, Kumar BN, Prathima B, SubbaRao Y, Jyothi NVV (2014a) A novel green synthesis of Fe3O4 magnetic nanorods using Punica granatum rind extract and its application for removal of Pb (II) from aqueous environment. Arab J Chem. https://doi.org/10.1016/j.arabjc.2014.09.006

Venkateswarlu S, Kumar BN, Prasad CH, Venkateswarlu P, Jyothi NVV (2014b) Bio-inspired green synthesis of Fe_3O_4 spherical magnetic nanoparticles using *Syzygium cumini* seed extract. Physica B Condens Matter 449:67–71

Venkateswarlu S, Kumar BN, Prathima B, Anitha K, Jyothi NVV (2015) A novel green synthesis of Fe_3O_4-Ag core shell recyclable nanoparticles using Vitis vinifera stem extract and its enhanced antibacterial performance. Physica B Condens Matter 457:30–35

Wu W, Wu Z, Yu T, Jiang C, Kim WS (2015) Recent progress on magnetic iron oxide nanoparticles: synthesis, surface functional strategies and biomedical applications. Sci Technol Adv Mater 16:023501

Chapter 10
Phytomediated Synthesis of Cerium Oxide Nanoparticles and Their Applications

Annu, Akbar Ali, Rahul Gadkari, Javed N. Sheikh, and Shakeel Ahmed

Abbreviations

AZO	Azodicarbonamide
CeO NPs	Cerium oxide nanoparticles
CeO_2	Cerium dioxide
DSC	Differential scanning calorimetry
FESEM	Field emission scanning electron microscopy
ROS	Reactive oxygen species
SAED	Selected area electron diffraction
SEM	Scanning electron microscopy
STM	Scanning tunnelling microscopy
TEM/HRTEM	Transmission electron microscopy/high-resolution transmission electron microscopy
TGA	Thermogravimetric analysis
WAXD	Wide-angle X-ray diffraction
XPS	X-ray photoelectron spectroscopy
XRD	X-ray diffraction

10.1 Introduction

Among the rare earth group elements, cerium is most abundant. Out of the 83 naturally occurring elements in the earth lithosphere, cerium ranks 28th position in its abundance. Jons Jakob Berzelius and Wilhelm Hisinger discovered cerium in

Annu · A. Ali
Department of Chemistry, Jamia Millia Islamia, New Delhi, India

R. Gadkari · J. N. Sheikh
Department of Textile Technology, Indian Institute of Technology, New Delhi, India

S. Ahmed (✉)
Department of Chemistry, Government Degree College Mendhar, Mendhar, Jammu, India

A. Husen, M. Iqbal (eds.), *Nanomaterials and Plant Potential*,
https://doi.org/10.1007/978-3-030-05569-1_10

Bastnas (Sweden) in 1803 and by Martin Heinrich Klaproth in Germany. Later, Berzelius and Hisinger proposed the name ceria to cerium in honour of the newly sighted asteroid Ceres (Weeks 1932; Dahle and Arai 2015). The main sources of cerium are allanite, bastnasite, monazite and cerite. Allanite, a silicate comprising of rare earth elements, aluminium, calcium and iron, is found extensively in Germany, Greenland, Madagascar, Scandinavia and the United States. Bastnasite, the most important commercial source of cerium and other light rare earth elements existing essentially as a rare earth fluorocarbonate, is found in Southern California. Cerite contains calcium and iron rare earth silicate and belongs mainly to Sweden. It has a high ratio of rare earth contents, but not enough to be the primary source of cerium. Monazite is a phosphate and another major source of cerium; it contains thorium and other light rare earth elements. Its main deposits are located in Australia, Brazil, the United States (Idaho and Florida), South Africa and India (Enghag 2004).

Physically, cerium is soft and ductile. In pure form, cerium has silver colour, but the commercial grade cerium is iron-grey. The pure cerium ignites, if scratched with a sharp object. Moreover, it tarnishes readily in the air, oxidizes rapidly in hot water, dissolves in acids and can burn when heated (Kilbourn 2000). Cerium dioxide (CeO_2) is a crystalline material and a major compound in the rare earth family. It has gained much attention owing to its numerous unique characteristics, such as high temperature stability, high hardness index, ultraviolet radiation-absorbing ability and its reactivity (Trovarelli et al. 1999). With the rapid progress in nanotechnology these days, nanoparticles (NPs) are now being used in various manufacturing, remedial and therapeutic processes. Cerium oxide nanoparticles (CeO NPs), also known as nanoceria, are very popular for their multifarious applications in several industries and are being synthesized in a variety of ways, which will be examined in this chapter.

10.2 Methods of Synthesis of Cerium Oxide Nanoparticle

The methods used in the synthesis of NPs play significant role in determining the morphological and structural properties of these particles. Some researchers preferred the solvothermal technique to synthesize the CeO_2 NPs so as to control conveniently their structural and morphological features. Cerium (IV) ammonium nitrate and sodium hydroxide in 1:4 molar ratio were dissolved in deionized water to form a solution of pH 12. The whole mixture was stirred for 1 h, using a magnetic stirrer with a stirring rate of 1000 rpm, and then kept in microwave oven for 30 min at 50 °C. The treatment of ammonium ceric nitrate with sodium hydroxide initiates hydrolysis, leading to the formation of products like ammonium hydroxide, cerium hydroxide and sodium nitrate. Due to the polar nature of water, a proton (H^+) gets removed from cerium hydroxide during the course of reaction, leading to the formation of hydrated CeO_2. Annealing of the hydrated CeO_2 at 800 °C for 6 h resulted in the formation of CeO_2 NPs. The particle size varies between 10 and 75 nm with an average size of 20 nm. The NPs showed good thermal stability and insulating property with a slight increase in conductivity (AC) on increasing the temperature (Kumar et al. 2013).

In another study, CeO_2 NPs were prepared from cerium nitrate as the starting material and sodium hydroxide (NaOH) as the precipitating agent, without using any stabilizer in the conventional or the sonochemical precipitation method. In the conventional method, NaOH solution (0.3 gmol) was added dropwise to 30 mL cerium nitrate hexahydrate solution (0.1 gmol) under constant stirring at 35 ± 2 °C, giving a yellowish white solution. The reaction converted the yellowish white solution to a light yellow colloidal suspension. After the completion of reaction (4 h), the solution was centrifuged to separate the product and heated at 100 °C for 3 h to get CeO_2 (light yellowish powder) from $Ce(OH)_3$ (Pinjari and Pandit 2011). The same concentrations of cerium nitrate hexahydrate and NaOH were used for the synthesis of CeO_2 by the sonochemical method, where the solutions were mixed under sonication by an ultrasonic horn (CE 22 kHz, 40% amplitude) for 2 min with a 5 s pulse and 5 s relaxation cycle at 35 ± 2 °C temperature. The addition was done in concurrence with the sonic pulse generated by the transducer, followed by the exposition of the solution to acoustic cavitation again for 18 min at constant sonication parameters. The remaining procedure was the same as in the conventional method.

The sonochemical method provides a simple, convenient, fast, economical and environmentally benign technique with more than 92% (15.15×10^{-2} kJ/g for sonochemical and 200.43×10^{-2} kJ/g for conventional method) energy saving for the synthesis of CeO_2. The crystallite size obtained in both cases was below 30 nm.

The ultrafine ceria particles of less than 5 nm size were synthesized using the conventional hydrothermal methods, employing the tetravalent cerium salts [$Ce(SO_4)_2$), $(NH_4)_2Ce(NO_3)_6$ and $Ce(NH_4)_4(SO_4)_2$] as starting materials. However, high surface energy of the nanoparticle caused agglomeration followed by precipitation, which makes it difficult to synthesize well-dispersed particles. The presence of SO_4^{2-} ions in the solution further accelerates the agglomeration processes. Protection of the NP surface will make it possible to control the size and agglomeration. The addition of citric acid molecules controls the size of cerium oxide NPs and the agglomeration process via adsorption on the surface of the particle (Hirano et al. 2000; Masui et al. 2002).

Another study suggested a new method to obtain ultrafine as well as well-dispersed CeO_2 NPs through the application of citric acid as a shielding agent against particle growth using a 1:1 molar ratio of cerium chloride (1 M aqueous solution) and citric acid adding to an excess amount of 3 M ammonium water. A dark brown transparent liquid was attained after stirring for 24 h at 323 K. The product particles were crystallized via transferring the transparent liquid and heated at 353 K for 24 h in a Teflon bottle kept in a vessel. Although the apparent status of the liquid did not change by heat treatment, the crystallinity after the treatment became higher than before. The brown liquid started showing Tyndall effect, indicating the presence of well-dispersed colloidal particles. The resulting NPs were separated by centrifugation (centrifugal force 4.1×10^4 g and 2.1×10^4 pm) for 24 h, washed with water and dried by a freeze-drying method (Masui et al. 2002).

A relatively fast "microwave hydrothermal" method has also been devised for the fast synthesis of nanoceria and nanorods. This reportedly energy-saving, high-yielding, rapid and environment-friendly method had the collective advantages of hydrothermal as well as microwave-heating techniques and produced the CeO_2 NPs of the average size of 1.6–20 nm. Besides, CeO_2 nanorods could be synthesized by altering the cerium source and regulating the quantity of the added ammonia water (Gao et al. 2006). Some researchers prepared CeO NPs by mixing the $Ce(NO_3)_3$ (0.0375 M) and hexamethylenetetramine (0.5 M) solutions in equal volumes at room temperature. Bunches of particles ranging in size between 3 and 12 nm were obtained by controlling the reaction time (Zhang et al. 2002). Nanoceria of an average particle size of 3.3 nm were also obtained sonochemically, using tetramethyl ammonium hydroxide as additive, having a molar ratio of starting materials and additives in 1:1:1. Cerium nitrate and azodicarbonamide (AZO) were used as the starting materials along with ethylenediamine or tetraalkyl ammonium hydroxide as the additive. Typically, 0.1 g AZO was added to a solution of $Ce(NO_3)_3$ (0.434 g in 50 mL distilled water) followed by the addition of varied extents of additives that showed strong influence on the size and distribution of particles. Addition of additives resulted in the production of small-sized particles of CeO_2 with a narrow range of distribution, whereas agglomerated nanoparticles were obtained without using additives (Yin et al. 2002).

10.3 Techniques Used for Characterization of Nanoparticles

10.3.1 NP Characterization by Spectroscopy

10.3.1.1 UV-Visible Spectroscopy

Ultraviolet-visible (UV-Vis) spectrophotometer consists of a source of light (xenon lamp), sample beam, reference beam, a monochromator and a detector. The UV-Vis spectra for a compound are achieved by exposing a dilute solution of the sample to UV light. Metallic nanoparticles, normally of 40–100 nm, scatter optical light elastically with significant efficiency due to a phenomenon known as surface plasmon resonance, which is a collective resonance of the conduction electrons in the metal. The intensity, peak position and spectral bandwidth of the plasmon resonance related with a nanoparticle depend on the NP composition and dimensions like shape and size. For example, Fig. 10.1a shows a typical UV-vis spectrum of CeO NPs displaying their characteristic absorption peak at a wavelength (λ max) of 345 nm. The NPs synthesized via solvothermal route were found to have the maximum absorbance at wavelength 317 nm. Phytomediated synthesized nanoceria display λmax at 345 nm generally due to the intrinsic band gap absorption, which owes to the electronic transitions from valence band [O(2p)] to the conduction band [Ce(4f)] in the nanoceria. Band gap of CeO_2 was reported to be 3.19 eV (Fig. 10.1b), whereas band gap of nanoceria was estimated around 3.59 eV. The increase (0.40 eV) in band gap might be owing to the reduction in particle size or change in oxidation state (Ce^{3+} to Ce^{4+}) (Patil and Paradeshi 2016).

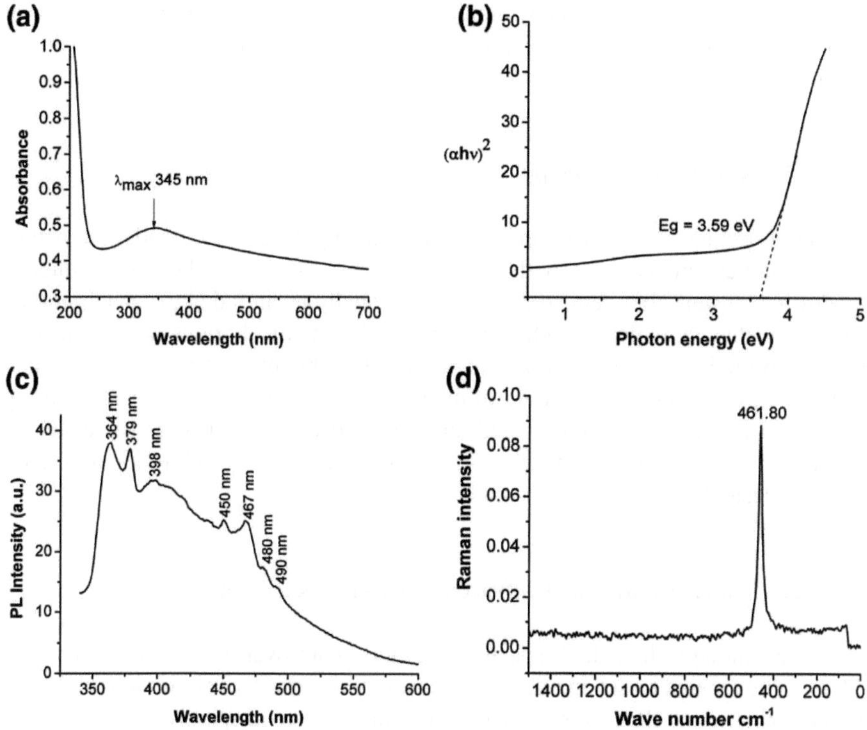

Fig. 10.1 (**a**) UV-visible spectra of CeO NPs, (**b**) photon energy level, (**c**) photoluminescence spectra recorded under light excitation wavelength 320 nm, (**d**) Raman spectra of CeO NPs synthesized using pectin (Patil and Paradeshi 2016)

10.3.1.2 Photoluminescence

Photoluminescence is the light emission from any form of matter after absorption of photons. The time periods between absorption and emission may vary ranging from femtoseconds to milliseconds. This process can be categorized on the basis of energy of exciting photon w.r.t. the emission. Green synthesized CeO NPs exhibited strong emission peaks starting from a shorter wavelength (360 nm) to a longer one (500 nm). The typical photoluminescence spectrum of CeO NPs recorded at an excitation wavelength of 320 nm is shown in Fig. 10.1c. The emission spectra displayed four bands in blue-green region (450, 467, 480 and 490 nm) and three bands in the near band edge of wavelength (364, 379 and 398 nm), respectively. The near band edge was attributed to the localized or free excitons, and blue region peaks were attributed to the bouncing of electrons from localized [Ce(4f)] state to the [O(2p)] valence band. Besides, the green peaks were ascribed to the surface defects in CeO NPs, whereas low density of oxygen vacancies caused weak intensities (Patil and Paradeshi 2016). The same was exhibited by the *Gloriosa superba* leaf extract-mediated nanoceria (Arumugam et al. 2015a). Another study reported the excitation spectra of nanoceria by monitoring the emission at 425 nm. It comprised of a sharp peak at 388 nm, attributed to Ce^{4+} intra-configuration (f–f) transitions (Malleshappa et al. 2015).

10.3.1.3 Raman Spectroscopy

It is a low-frequency spectroscopy technique, mainly used in the condensed matter physics and chemistry to study the rotational, vibrational and other low-frequency modes in a system. In chemistry, it is used normally as an indicator of vibrational frequency change, which is highly specific for the chemical bonds in molecules. It gives information in the fingerprint region (500–2000 cm^{-1}) of a molecule, which can be helpful in the identification of unknown molecules. For CeO NPs, a strong and broad Raman active mode band appears at 461.1 cm^{-1} in Fig. 10.1d, which might be ascribed to the symmetrical stretching mode of Ce-O8 vibration unit (Patil and Paradeshi 2016).

10.3.1.4 Fourier Transform Infrared (FT-IR) Spectroscopy

FT-IR technique utilizes high spectral resolution data over a wide spectral range. This can be applied to get an infrared spectrum of absorption or emission of a solid, liquid or gas. FT-IR has remarkable advantage over dispersive spectrometer. Besides, it is an analytical technique used to detect the base polymer composition, functional group and organic contaminants. It uses infrared light to probe the test samples and observe their chemical properties. The FT-IR spectra of Indian red pomelo extracted pectin-mediated CeO NPs showed the characteristic peaks as shown in Fig. 10.2 and Table 10.1. An additional peak has also been observed at 545 cm^{-1} for Ce-O stretching vibration for cubic cerium oxide (Kumar et al. 2013). On the other hand, annealed nanoceria depicted a sharp peak at 558 cm^{-1} for Ce-O stretching vibrations along with 3397, 1624 and 1471 cm^{-1}, which were due to water and CO_2 taken up from the atmosphere (Patil and Paradeshi 2016).

10.3.2 NP Characterization by Microscopy

Optical microscopy techniques are used commonly for observing particles at micron level with reasonable magnification. However, higher magnification processing is not possible through ordinary optical microscopes due to aberration and limitations in wavelength of light. Therefore, advanced imaging techniques such as scanning electron microscopy (SEM), transmission electron microscopy/high-resolution transmission electron microscopy (TEM/HRTEM) and scanning tunnelling microscopy (STM) are used to analyse the surface structure and internal morphology (size, shape and dimensions) of nanoparticles with due clarity.

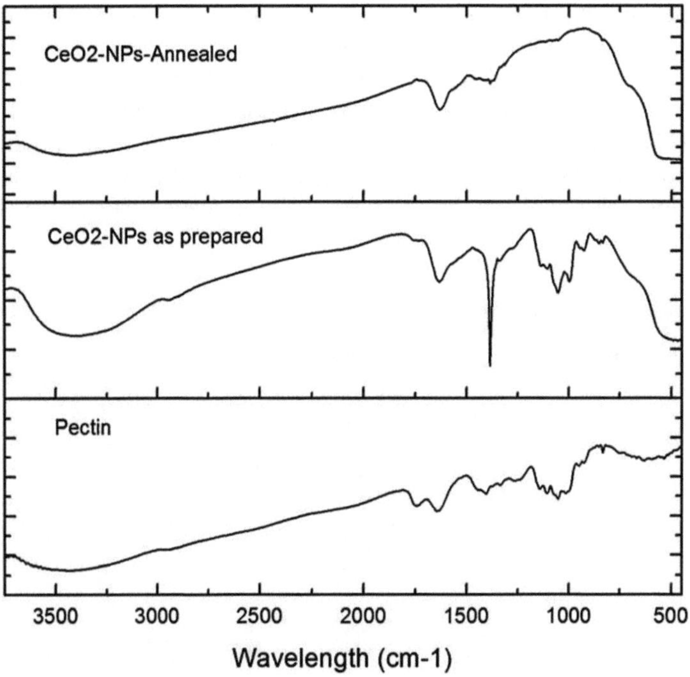

Fig. 10.2 FT-IR spectra of pectin, CeO NPs prepared using pectin and CeO NPs annealed (Patil and Paradeshi 2016)

Table 10.1 Characteristic FT-IR peak of CeO NPs (Patil and Paradeshi 2016)

S. No	IR frequency (cm^{-1})	Functional group present
1	3390	-OH, stretching
2	2939	C-H, stretching of methyl ester of galacturonic acid
3	1738	C=O, stretching vibration of methyl-esterified carboxyl group
4	1634	C=O, stretching vibration of carboxylate group
5	1402	C-O-H, in plane bending vibration
6	1295	C-O-C, asymmetric stretching vibration of $-O-CH_3$ group
7	1105	C-O-C, bending mode of acetal group
8	1051	C-O-C, bending mode of ethereal group
9	955	Glycosidic group

10.3.2.1 Scanning Electron Microscopy (SEM)

Scanning electron microscopy involves scanning of the surface of a sample with a focused beam of electrons which interacts with the atoms of the sample, producing various signals. These signals possess the sample's information about surface topography and composition. In the case of CeO NPs obtained from Indian red pomelo pectin, field emission scanning electron microscopy (FESEM) micrographs illustrated the shape of these NPs as spherical and agglomerated with their size ranging from 5 to 40 nm (Fig. 10.3), whereas the solvothermal gave the result as shown in Fig. 10.4. Additionally, the EDX pattern of CeO NPs revealed the presence of Ce, Au, O and C as the main components of the CeO NPs produced. The presence of Au and C in the EDX spectra appears due to the conductive carbon tape used to hold the sample and to the gold used for coating the sample (Patil and Paradeshi 2016).

Fig. 10.3 FESEM micrograph of CeO NPs synthesized using pectin (Patil and Paradeshi 2016)

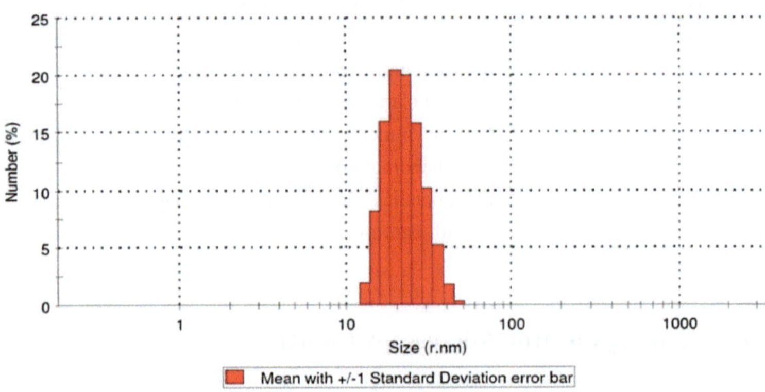

Fig. 10.4 Particle size distribution histogram of biopolymer-mediated CeO NPs (Patil and Paradeshi 2016)

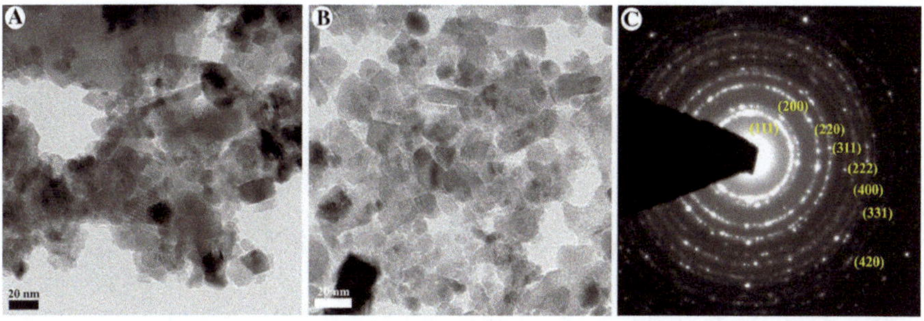

Fig. 10.5 (**a**) TEM images of mycosynthesized CeO NPs (**b**) calcined at 40 °C and (**c**) SAED of CeO NPs (Gopinath et al. 2015a, b)

10.3.2.2 Transmission Electron Microscopy (TEM)

TEM is a technique in which a beam of electrons is transmitted through a sample to form an image and the image is further magnified on a fluorescent screen. Briefly, TEM is generally used for imaging the structural details at a significantly higher resolution. The structure of CeO NPs can be observed with TEM (K. Gopinath, V. Karthika, C. Sundaravadivelan, S. Gowri 2015a). The HRTEM images showed that CeO NPs possessed spherical and cubic morphology and their size ranged between 5 and 20 nm with an average particle size of nearly 10 nm. The nanoparticles also showed fluorite cubic structure having a characteristic ring pattern in the selected area electron diffraction (SAED) and possessed a high degree of crystallinity (Fig. 10.5).

10.3.3 TGA/DTA Analysis

Thermal stability of the crystalline samples can be examined using the differential thermal analysis techniques (thermogravimetric and differential scanning calorimetry) under high temperature. The thermal stability of green synthesized nanoparticles of CeO_2 has been analysed in the temperature range of 35–1000 °C at a heating rate of 20 °C per min. A TGA/DTA curve of CeO NPs is shown in Fig. 10.6. The NPs exhibit a three-stage decomposition pattern, with the first stage accompanied by a 9.84% weight loss at 35–100 °C due to dehydration of water, moisture and extracellular fungal components, the second stage contributes a weight loss of 13.49% from 100 to 150 °C, and the third stage shows the maximum weight loss of 15.25% at the temperature range of 540–1000 °C. On DTA curve, the exothermic peaks observed at 115 °C were attributed to the combustion of organic residues, and those at 615 °C were ascribed to oxygen loss at higher temperatures and decomposition of some residual, absorbed species. Moreover, the phase change or variation in the oxidation state of cerium was shown by a slight increase in the DTA curve above 740 °C (Gopinath et al. 2015b).

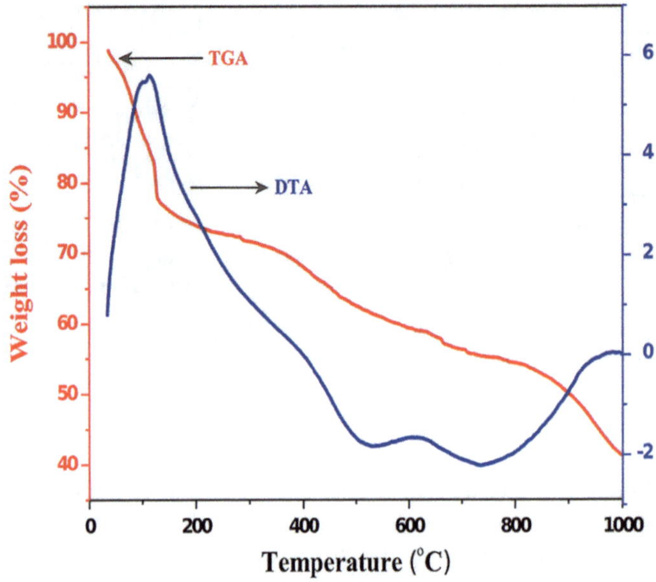

Fig. 10.6 TGA/DTA curves of green synthesized CeO_2 nanopowders (Gopinath et al. 2015a, b)

10.3.4 NP Characterization by X-Ray Technique

X-ray diffraction (XRD) technique gives information about the crystallinity, crystal size, phase composition, and crystal alignment. Two types of X-ray technique are generally used in the characterization of nanoparticles:

1. Wide-angle X-ray diffraction (WAXD): It provides information about the crystallinity, alignment of the crystal, crystallite size and the phase composition in semi-crystalline polymer.
2. X-ray photoelectron spectroscopy (XPS): This quantitative surface chemical analysis spectroscopic technique is used to evaluate the elemental composition or empirical formula, electronic state and chemical state of the elements on the surface up to 10 nm of a material. It also detects the presence of contaminants on the surface of the sample and oxidation state of each element present in the sample with the composition of the surface functionalization of the CeO NPs.

Hewer et al. (2013) prepared CeO NPs surrounded by carbon microspheres by utilizing D(+)-glucose in the one-pot green hydrothermal process. In XRD analysis, five diffraction peaks at $2\theta = 28.5, 33.0, 47.4, 56.3$ and $76.6°$ were observed and attributed to the (111), (200), (220), (311) and (331) cubic plane of CeO_2 (JCPDS 34-0394), respectively. The diffraction peak showed the face-centred cubic phase of nanoceria (Hewer et al. 2013). Another study revealed some other peaks at $2\theta = 59.09, 69.0$ and $78.99°$, attributed to (222), (400) and (420) planes of face-centred cubic nanoceria, respectively. Debye-Scherrer's formula was applied to calculate the average crystallite size of CeO NPs which came out to be 24 nm (Arumugam et al. 2015b).

10.3.5 Particle Size Analyser

Dynamic light-scattering techniques are generally coupled with various other techniques such as sieve analysis, electro-resistance counting method, laser diffraction method and optical counting method for the determination of particle size and distribution. A general particle size distribution histogram showing the CeO NPs having a hydrodynamic size (Z-average) distribution (around 29.47 nm) is presented in Fig. 10.4 (Patil and Paradeshi 2016).

10.4 Green Synthesis of Cerium Oxide Nanoparticles or Nanoceria

The term "green" is used to indicate an eco-friendly and environmentally benign use of the less energy-consuming, non-toxic chemicals and the bio-derived products for the synthesis of NPs (Husen 2017; Siddiqi and Husen 2017; Siddiqi et al. 2018). The nanoceria have been prepared using several ways and means such as solution precipitation; hydrothermal, solvothermal or thermal decomposition; thermal hydrolysis; and spray pyrolysis methods. But these methods have several disadvantages, like the use of toxic solvents and reagents, high pressure and temperature and the external additives (stabilizing or capping agents) during the reaction. Figure 10.7 displays various physical, chemical and physicochemical methods of the synthesis of nanoparticles, highlighting the green or biological approaches. Numerous bio-directed approaches involving application of natural products and other organic matrices as a stabilizer are now being used to obtain the biocompatible nanoceria. Phytosynthesis of metals and metal oxide nanoparticles is a novel subject in nanoscience engineering. Of late, phytosynthesis of nanoceria, using the extracts of plant species (such as *Aloe vera*, *Acalypha indica* and *Gloriosa superba*) as stabilizing and capping agents, has been reported (Kannan and Sundrarajan 2014; Priya et al. 2014; Arumugam et al. 2015a). Tables 10.2 and 10.3 indicate the source materials, methods for green synthesized nanoceria and also the reported advantage/disadvantage associated with these methods, respectively.

Leucas aspera leaf extract-mediated nanoceria have been characterized under diffraction technique, and the patterns (111), (200), (220) and (222) were well indexed to a pure cubic fluorite structure of CeO NPs (JCPDS Card No. 81-0792) without any impurity peak, thus indicating the single phase of CeO NPs. The study further concluded that the concentration of leaf extract strongly influences the structural parameters such as crystallite size and lattice constant (Malleshappa et al. 2015).

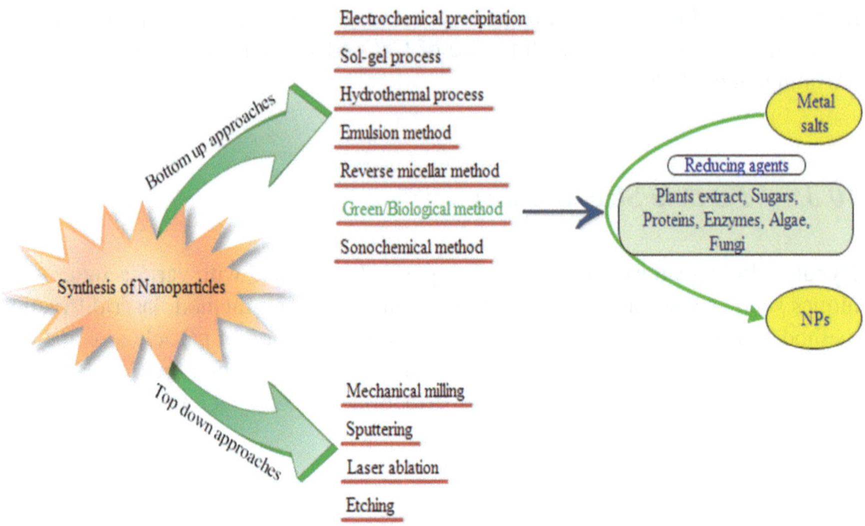

Fig. 10.7 Generalized representations of various physicochemical methods of nanoparticle synthesis, emphasizing the biological synthesis

Besides, Patil and co-workers have synthesized CeO NPs using a non-toxic biopoly-

Table 10.2 Green synthesis methods for CeO NPs

Green synthesis methods used	Materials used	Particle size (nm)	Morphology	References
Plant-mediated	*Gloriosa superba*	5	Spherical	Arumugam et al. (2015a)
Plant-mediated	*Centella asiatica*	8	Spherical and monodisperse	Sankar et al. (2015)
Plant-mediated	*Acalypha indica*	36	Spherical	Kannan and Sundrarajan (2014)
Plant-mediated	*Piper longum*	46	Spherical	Reddy Yadav et al. (2016)
Plant-mediated	China rose petal	7	Nanosheet	Qian et al. (2011)
Plant-mediated	*Aloe vera*	18.23	Spherical	Tamizhdurai et al. (2017)
Fungus-mediated	*Curvularia lunata*	5–20	Spherical	Munusamy et al. (2014)
Nutrient-mediated	Egg white	6–30	Plate like	Maensiri et al. (2007)
Nutrient-mediated	Honey	23	Spherical	Darroudi et al. (2014a)
Biopolymer-mediated	Agarose	10.5	Spherical	Mittal and Pandey (2014)
Biopolymer-mediated	Starch	6	Spherical	Darroudi et al. (2014b)
Biopolymer-mediated	Gum	10	Spherical	Charbgoo et al. (2017a)
Biopolymer-mediated	Dextran	10	Spherical	Perez et al. (2008)
Biopolymer-mediated	Pectin	≤40	Spherical	Patil and Paradeshi (2016)
Biopolymer-mediated	Chitosan	~4	Spherical	Sifontes et al. (2011)

Table 10.3 Advantages and challenges encountered by the green methods for the synthesis of CeO NPs

Green methods	Advantage	Disadvantage/challenge	Reference
Plant-mediated synthesis of CeO NPs	Spherical-shaped NPs were obtained with reduced cytotoxicity Easy process, cost-effective, energy- and time-efficient technique Capable of producing stable, water-dispersible, and highly fluorescent NPs	Non-uniformity in morphology in some cases due to which agglomeration of the individual NPs occurs Size of NPs exhibited wide distribution range from 5 to 63.6 nm, using different bio-organisms for synthesis	Gopinath et al. (2015a, b), Charbgoo et al. (2017b)
Nutrient-mediated synthesis of CeO NPs	Controllable growth and subsequent isotropic formation of small and stable CeO NPs	Lack of information about the constituents accountable for metal ions reduction	Reddy Yadav et al. (2016), Charbgoo et al. (2017b)
	Capable of providing spherical CeO NPs with narrow distribution range of particle size		
	Synthesized CeO NPs non-toxic toward human cell lines at physiological concentrations of NPs		
Biopolymer-mediated synthesis of CeO NPs	Generating NP with spherical morphology	Reproducibility of the methods need to be enhanced Abundance of structural components might affect the synthesis procedure Complete elucidation of mechanism is still a major challenge	Primo et al. (2011), Sun et al. (2012), Darroudi et al. (2014c), Kargar et al. (2015), Sathiyanarayanan et al. (2017)
	Providing NPs with no significant cytotoxic effect in human cell line at physiological concentrations		
	Capable of controlling diameter of CeO NPs		
	Providing NPs with high final purity		

mer pectin. The Braggs peaks are found at an angle (2θ) of 28.51, 33.06, 47.42, 56.30, 59.09, 69.00, 76.57 and 78.99° with Miller indices (111), (200), (220), (311), (222), (400), (331) and (420), respectively (Fig. 10.8). The synthesized CeO NPs exhibited a pure cubic fluorite structure (space group, Fm-3 m, 225), which is in agreement with the JCPDS PDF 00-033-0334. The lattice value and the unit cell volume were 5.089 Å and 131.795 Å, respectively. Furthermore, the average crystallite size of the phyto-mediated CeONPs was estimated to be 23.71 nm using the Debye-Scherrer equation (Patil and Paradeshi 2016).

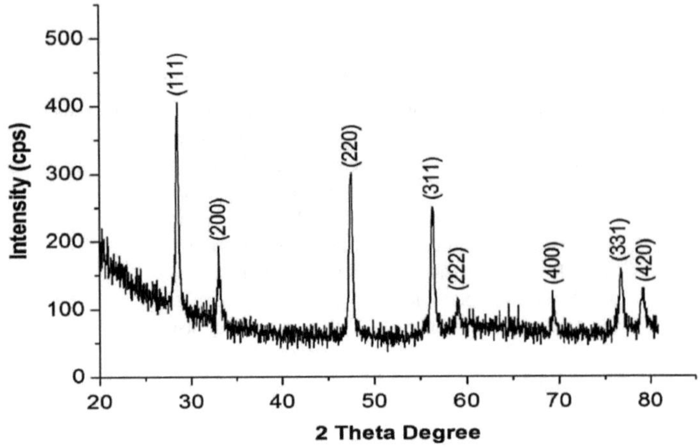

Fig. 10.8 X-ray powdered diffraction pattern of pectin-mediated synthesized CeO NPs (Patil and Paradeshi 2016)

10.5 Applications of CeO NPs

Nanoceria have the potential to be utilized in various areas of science and technology and are being used in paint industry, glass polishing, fuel additives, supercapacitors, emission control or biodiesel, biogas production, biosensing, microcosm system for soil, cosmetics, coating and most commonly biomedical sciences. However, the biomedical applications of nanoceria require a lot of care and precaution (Castano et al. 2015; Nguyen et al. 2015; Walkey et al. 2015; Oró et al. 2016; Charbgoo et al. 2017b; Li et al. 2017; Vinothkumar et al. 2017).

10.5.1 Biomedical Applications

10.5.1.1 Anti-inflammatory

Chronic inflammation, a complex immunological ailment caused by the increased levels of nitric oxide, results in irreversible organ dysfunctioning or damage. It often leads to major diseases such as cardiac disorder, rheumatoid arthritis, atherosclerosis and multiple sclerosis. Cerium oxide NPs have been used for the treatment of such chronic inflammation (Hirst et al. 2009). Khan and Ahmad have successfully synthesized CeO NPs by using the thermophilic fungus *Humicola sp.* for utilization in biomedical sciences. The fungus-secreted capping protein, involved in the capping of nanoparticles, made the particles highly stable, non-agglomerate, highly fluorescent and, most importantly, water dispersible (Khan and Ahmad 2013).

10.5.1.2 Antibacterial Activity

The nanoceria fabricated via green route exhibit good antibacterial and antifungal activity against pathogens in the oxidation states +III and +IV (Annu and Ahmed 2018). Some researchers synthesized them using the leaf extract of *Gloriosa superba* and studied the formed nanoparticles against Gram-positive and Gram-negative bacteria at varying concentrations. At high concentration of 100 mg, the zone of inhibition effect was the largest (5.33 mm) against *Staphylococcus aureus*. The *Escherichia coli* and *Shigella dysenteriae* exhibited a modulated effect on the inhibition zone of 4.00 mm and 4.33 mm, respectively. Then, *Pseudomonas aeruginosa*, *Proteus vulgaris*, *Klebsiella pneumonia* and *Streptococcus pneumoniae* showed similar inhibition zone of 4.67 mm for enhanced activity (Arumugam et al. 2015a). Similarly, the use of the leaf extract of *Acalypha indica* resulted in increased rate of antibacterial activity against Gram-positive and Gram-negative bacteria (Kannan and Sundrarajan 2014). Likewise, the nanoceria synthesized from the fungus *Aspergillus niger* showed antibacterial, larvicidal and pupicidal activities against human pathogens, viz. Gram-positive *Streptococcus pneumoniae* and *Bacillus subtilis* and Gram-negative *Proteus vulgaris* and *Escherichia coli* (Gopinath et al. 2015a, b). The authors obtained different results from different concentrations of nanoceria. For instance, 1 mg nanoceria per mL did not show any inhibition zone with any of the stains, while moderate zones of inhibition (4.67 ± 0.33 mm, 3.33 ± 0.33 mm, 3.67 ± 0.33 mm and 3.33 ± 0.33 mm) were observed in *B. subtilis*, *E. coli*, *P. vulgaris* and *S. pneumoniae* at 5 mg/mL concentration, respectively. However, at 10 mg mL^{-1} concentration of nanoceria, remarkable inhibition zones (10.33 ± 0.33 mm, 6.33 ± 0.33 mm, 8.33 ± 0.33 mm and 10.67 ± 0.33 mm) with *B. subtilis*, *E. coli*, *P. vulgaris* and *S. pneumoniae*, respectively (Figs. 10.9 and 10.10) (Gopinath et al. 2015a). Gusseme et al. (2010) reduced biogenic cerium by using the freshwater manganese-oxidizing bacteria (MOB) *Leptothrix discophora* or *Pseudomonas putida* MnB29 and demonstrated the antiviral activity of nanoceria by removing virus bacteriophage UZ1 from water due to the bacterial carrier matrix (Gusseme et al. 2010).

Patil and Paradeshi (2016) utilized pectin, extracted from Indian red pomelo fruit peels, as a renewable, non-toxic biopolymer in order to fabricate CeO NPs. They found that the biogenically synthesized nanoceria exhibited significant antibacterial property against *E. coli* and *Bacillus subtilis* at both 1 and 2 mM concentrations used (Fig. 10.11a) which was far better than one shown by the bulk and CeO_2 powder. In case of *B. subtilis*, cerium nitrate was not effective at both (1 and 2 mM) concentrations, whereas the bulk CeO_2 caused a slight reduction in survival at 2 mM concentration only (Fig. 10.11b) (Patil and Paradeshi 2016).

Causes and Mechanism

The nanoceria exhibited a higher antibacterial activity against the Gram-positive than against the Gram-negative bacteria. This is because the cell wall of Gram-positive bacteria is composed of peptidoglycan which is attached to teichoic acids.

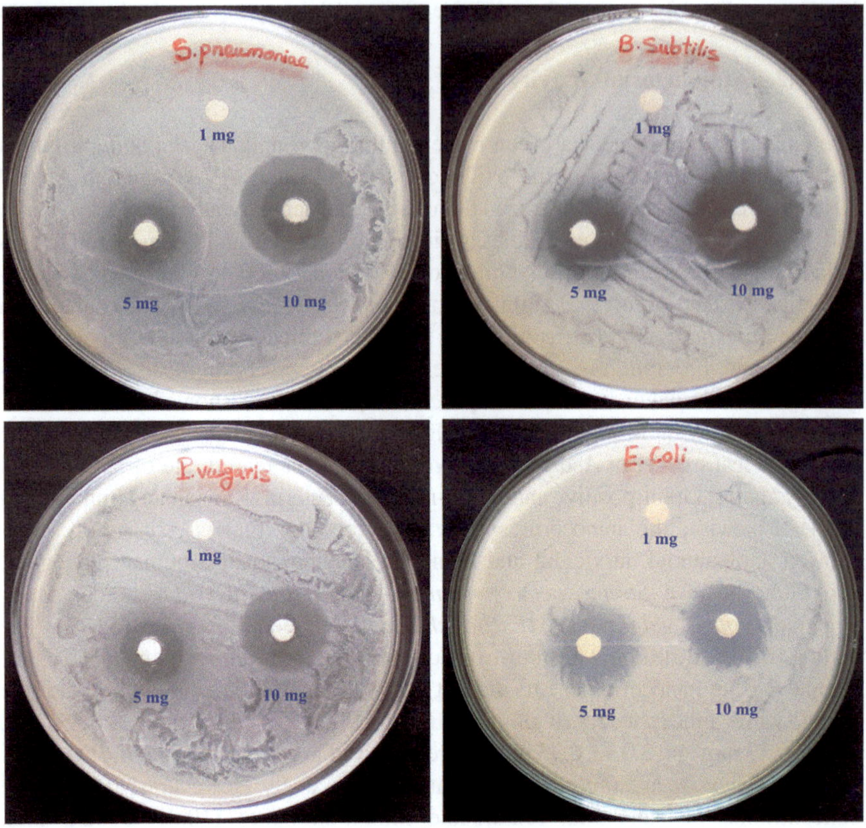

Fig. 10.9 Zones of inhibition by CeO_2 NPs at different concentrations against Gram-positive and Gram-negative bacteria (Gopinath et al. 2015a, b)

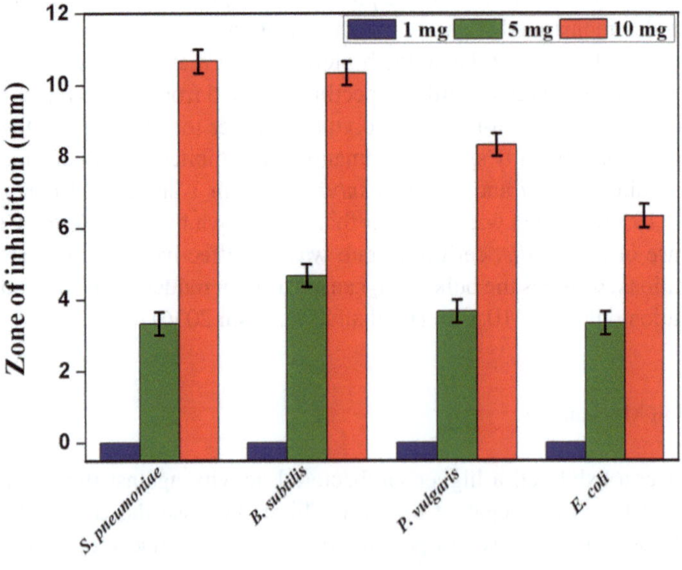

Fig. 10.10 Antibacterial activities of CeO_2 NPs against Gram-positive and Gram-negative bacteria (Gopinath et al. 2015a, b)

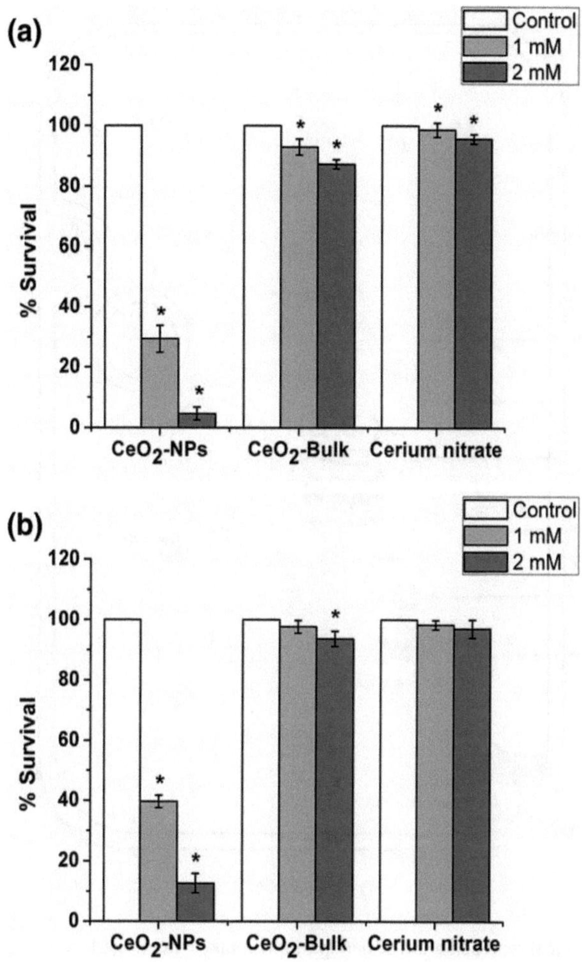

Fig. 10.11 Antibacterial activity of CeO NPs, bulk CeO_2 and cerium nitrate against (**a**) *E.coli* and (**b**) *B. subtilis*, expressed in terms of %age of the survival relative to the control group. In a group, the bars not labelled with asterisk demonstrate values (means) that are significantly different from the control ($p \leq 0.05$) by Dunnett's comparison test (Patil and Paradeshi 2016)

The electrostatic interaction between the positively charged metal or metal oxide nanoparticles and the negatively charged bacterial cell wall causes disruption of the cell wall, simultaneously generates reactive oxygen species (ROS) and ultimately leads to the death of the bacterial cell (Gopinath et al. 2015a, b). Figure 10.12 shows the schematic view of the cellular uptake of nanoparticles in general and the mechanism of inducing toxicity against bacteria (Hussain et al. 2016).

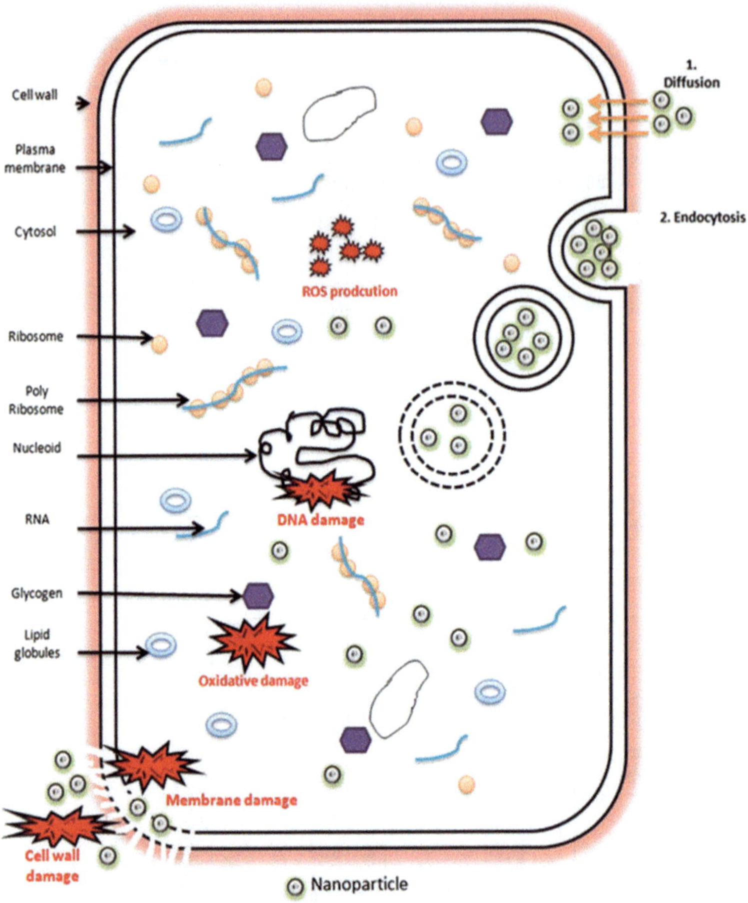

Fig. 10.12 Graphical representation of nanoparticle uptake by cells and the mechanism of toxicity induced by nanoparticles against bacteria (Hussain et al. 2016)

10.5.1.3 Cell Viability and Neurotoxicity

Agarose is a naturally occurring oxygen-rich straight-chain polysaccharide mined from red purple seaweeds. Some researchers have fabricated the CeO NPs by improved solgel route using the bioorganic agarose polymeric matrix, which acted as a stabilizing or capping agent. These biologically synthesized nanoparticles were then studied for cell viability on L929 cells at different concentrations varying from 0 to 800 μg mL^{-1}, and no significant cytotoxic effect could be observed at any concentration, suggesting that the biogenically synthesized nanoceria have the potential to be used in biomedical applications (Kargar et al. 2014). In another study, honey was used as a source to prepare CeO NPs with the help of solgel process at different

Table 10.4 Cytotoxicity of CeO NPs measured by the erythrocyte haemolysis assay (Patil and Paradeshi 2016)

S. No.	Sample	Optical density at 540 nm	% haemolysis
1	Distilled water	1.658 ± 0.028	100 ("+ve" control)
2	Physiological saline	0.034 ± 0.003	0 ("−ve" control)
3	CeO_2 NPs (0.50 mg/mL)	0.043 ± 0.003	0.55
4	CeO_2 NPs (1.00 mg/mL)	0.051 ± 0.006	1.05
5	CeO_2 NPs (2.00 mg/mL)	0.069 ± 0.005	2.16
6	CeO_2 NPs (4.00 mg/mL)	0.108 ± 0.008	4.55
7	CeO_2 NPs (4.00 mg/mL)	0.169 ± 0.008	8.31

calcinated temperatures, and their neurotoxic effect on the neuro2A cells was examined at concentrations ranging from 0 to 100 μg mL^{-1} (Darroudi et al. 2014a). It was found that the metabolic activity decreased in a concentration-dependent manner for more than 25 μg mL^{-1} after 24 h of incubation, thus proving to be more efficient in comparison to the conventionally produced nanoparticles (Darroudi et al. 2014a). In a study, based on the nanoceria synthesized with the help of pectin extracted from Indian red pomelo fruit peels, it was concluded after demonstrating the cell viability by erythrocyte haemolysis assay that the haemolysis (i.e. release of haemoglobin to plasma because of the rupture of erythrocyte membrane) increased when the concentration of nanoceria increased, as depicted in Table 10.4. This occurred at ≤4 mg mL^{-1}, which was within the permissible limit and hence biocompatible, but not much safe for human beings (Patil and Paradeshi 2016).

10.5.1.4 Antiobesity

Obesity leads to several pathologies such as type 2 diabetes, insulin resistance, cancer, hypertension and atherosclerosis. Commercially available drugs cause many side effects such as headache and insomnia. Nanoceria can strongly scavenge the ROS production for long duration and hence reduce the side effects. Rocca et al. (2015) studied the in vitro and in vivo cytotoxic effect of nanoceria on 3 T3-L1 pre-adipocytes and Wistar rats, respectively, in order to observe their interference in the lipid accumulation process and adipogenesis inhibition by reducing the mRNA transcription of genes involved in adipogenesis and by obstructing the triglyceride accumulation in 3 T3-L1 pre-adipocytes. Additionally, Wistar rats showed reduction in weight gain when injected with nanoceria; plasma level of insulin, glucose, leptin and triglycerides also declined, and no significant toxic effects appeared (Rocca et al. 2015). Another study revealed that the large surface area-to-volume ratio creates the oxygen defects as reactive sites on the nanoceria surface that can scavenge the free radicals generated in the biological system by inhibiting the ROS production and hence plays a vital role as antiobesity agent (Hirst et al. 2009).

10.5.2 Photocatalytic and Antioxidant Activity

Gogoi and Sarma (2017) biologically synthesized CeO NPs by solgel process using β-cyclodextrin in order to degrade efficiently the organic methylene blue dye from aqueous solutions at room temperature in the presence of H_2O_2 and absence of light irradiation. They found a complex pathway having different intermediates like leucomethylene blue, which is the reduced form of methylene blue and has $m/z = 286$; it is reduced further to leucomethylene blue sulfone having $m/z = 317$ amu. Thus, the colour of dye changes from blue to colourless, confirming the excellent photocatalytic activity of these nanoceria (Gogoi and Sarma 2017).

Leaf extract of *Aloe vera* was also used as stabilizing or capping agent to synthesize CeO NPs at ambient temperature (Dutta et al. 2016). The antioxidant property of nanoceria was found to be quite good which remained unaffected on being treated with H_2O_2 as the cerium was mainly present in (+III) oxidation state rather than (+IV) and the cyclic conversion took place from Ce(III)O → Ce(IV)O → Ce(III)O when reacted with H_2O_2. A gradual decrease in cell viability was noticed on increasing the H_2O_2 concentration (10–60 μM) for 24 h, and application of 4.34–78.16 μg mL^{-1} of CeO NPs remarkably enhanced cell viability, maximal at 120 μM, beyond which there was a decrease in the cell survival. Furthermore, a higher antioxidant activity and decrease of the lethal effects of H_2O_2 were observed with the higher oxygen defect in the crystal lattice. Thus, the biogenically synthesized nanoceria have the potential of being used as an antioxidant drug or a therapeutic agent for neural diseases triggered by oxidative stress and in other fields of biomedicine (Dutta et al. 2016). The catalytic properties of nanoceria, scavenging activity of nitric oxide radical, decay of peroxynitrite and their superoxide dismutase and catalase mimetic activity have been discussed recently (Walkey et al. 2015). The pectin-mediated nanoceria have been produced and evaluated for their antioxidant effect by using DPPH assay. For this assay, DPPH solution kept in the dark showed

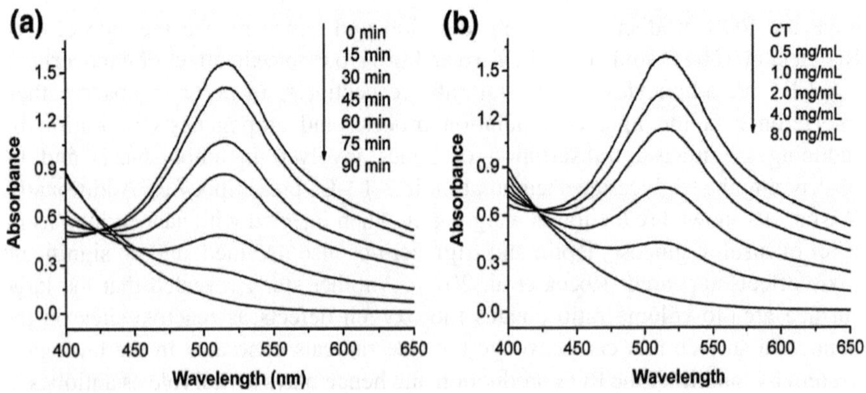

Fig. 10.13 Antioxidant activity of CeO NPs evaluated by the DPPH radical scavenging assay at (**a**) different intervals of time and (**b**) different concentrations (Patil and Paradeshi 2016)

unchanged intensity (517 nm) without any colour change, whereas in the presence of nanoceria, the absorption intensity was decreased with a colour change from deep violet to pale yellow (Fig. 10.13a). Figure 10.13b clearly indicates the decreased intensity with increase in the nanoceria concentration, and hence this decreased intensity is indicative of the free radical scavenging potential of nanoceria (Patil and Paradeshi 2016).

Nanoceria are also applicable for thermal decomposition and dye degradation. The nanoceria produced by using the *Azadirachta indica* leaf extract thermally decomposed ammonium perchlorate by dropping down the decomposition temperature of ammonium perchlorate by 130 °C and reducing the activation energy as well. This clearly indicated the excellent thermal catalytic behaviour of these NPs. On the other hand, their photocatalytic activity, having the degradation rate of 96% within 120 min, was revealed by photodegradation of Rhodamine B dye, indicating that they can be applied for dye degradation in wastewater treatment (Sharma et al. 2017).

10.6 Conclusions

Phytosynthesis of nanoparticles is an important and emerging area in nanoscience and technology. The several drawbacks of the conventional methods for NP synthesis have been overcome by the use of eco-friendly and novel green synthesis methods, which involve the use of plant extract, biopolymer and its derivatives, microbial derivatives and other natural products. Application of nanoceria is gaining ground in various fields despite many challenges. The different areas where nanoceria can be used include biosensing, plant science, coating, electronic applications, catalysis, biomedicine and nanomedicine, among others.

References

Annu AA, Ahmed S (2018) Green synthesis of metal, metal oxide nanoparticles, and their various applications. In: Martínez LMT, Kharissova OV, Kharisov BI (eds) Handbook of ecomaterials. Springer International Publishing, Cham, pp 1–45

Arumugam A, Karthikeyan C, Hameed ASH, Gopinath K, Gowri S (2015a) Synthesis of cerium oxide nanoparticles using *Gloriosa superba* L. leaf extract and their structural, optical and antibacterial properties. Mater Sci Eng C 49:408–415

Arumugam A, Karthikeyan C, Syedahamed A, Hameed ASH, Gopinath K, Gowri S, Karthika V (2015b) NU synthesis of cerium oxide nanoparticles using *Gloriosa superba*. Mater Sci Eng C Mater Biol Appl 49:408–415

Castano CE, O'Keefe MJ, Fahrenholtz WG (2015) Cerium-based oxide coatings. Curr Opin Solid State Mater Sci 19:69–76

Charbgoo F, Ahmad MB, Darroudi M (2017a) Cerium oxide nanoparticles: green synthesis and biological applications. Int J Nanomedicine 12:1401–1413

Charbgoo F, Ramezani M, Darroudi M (2017b) Bio-sensing applications of cerium oxide nanoparticles: advantages and disadvantages. Biosens Bioelectron 96:33–43

Dahle JT, Arai Y (2015) Environmental geochemistry of cerium: applications and toxicology of cerium oxide nanoparticles. Int J Environ Res Public Health 12:1253–1278

Darroudi M, Javad S, Kazemi R, Ali H (2014a) Food-directed synthesis of cerium oxide nanoparticles and their neurotoxicity effects. Ceram Int 40:7425–7430

Darroudi M, Sarani M, Oskuee RK, Zak AK, Hosseini HA, Gholami L (2014b) Green synthesis and evaluation of metabolic activity of starch mediated nanoceria. Ceram Int 40:2041–2045

Darroudi M, Sarani M, Oskuee RK, Zak AK, Amiri MS (2014c) Nanoceria: gum mediated synthesis and in vitro viability assay. Ceram Int 40:2863–2868

Dutta D, Mukherjee R, Patra M, Banik M, Dasgupta R, Mukherjee M, Basu T (2016) Green synthesized cerium oxide nanoparticle: a prospective drug against oxidative harm. Colloids Surf. B Biointerfaces 147:45–53

Enghag P (2004) The elements – origin, occurrence, discovery and names. In: Encyclopedia of the elements. Wiley-VCH Verlag GmbH & Co. KGaA, Weinheim, pp 55–78

Gao F, Lu Q, Komarneni S (2006) Fast synthesis of cerium oxide nanoparticles and nanorods. J Nanosci Nanotechnol 6:3812–3819

Gogoi A, Sarma KC (2017) Synthesis of the novel b -cyclodextrin supported CeO_2 nanoparticles for the catalytic degradation of methylene blue in aqueous suspension. Mater Chem Phys 194:327–336

Gopinath K, Karthika V, Sundaravadivelan C, Gowri S, Arumugam A (2015b) Mycogenesis of cerium oxide nanoparticles using *Aspergillus niger* culture filtrate and their applications for antibacterial and larvicidal activities. J Nanostruct Chem 5:295–303

Gusseme BDE, Laing GDU, Hennebel T, Renard P, Chidambaram D, Fitts JP, Bruneel E, Driessche IV, Verbeken K, Boon N, Verstraete W (2010) Virus removal by biogenic cerium. Environ Sci Technol 44:6350–6356

Hewer TLR, Soeira LS, Brito GES, Freire RS (2013) One-pot green synthesis of cerium oxide-carbon microspheres and their catalytic ozonation activity. J Mater Chem A 1:6169–6174

Hirano M, Fukuda Y, Iwata H, Hotta Y, Inagaki M (2000) Preparation and spherical agglomeration of crystalline cerium(IV) oxide nanoparticles by thermal hydrolysis. J Am Ceram Soc 83:1287–1289

Hirst SM, Karakoti AS, Tyler RD, Sriranganathan N, Seal S, Reilly CM (2009) Anti-inflammatory properties of cerium oxide nanoparticles. Small 5:2848–2856

Husen A (2017) Gold Nanoparticles from plant system: synthesis, characterization and their application. In: Ghorbanpour M, Manika K, Varma A (eds) Nanoscience and plant–soil systems, Springer, vol 48. Cham, Cham Switzerland, pp 455–479

Hussain I, Singh NB, Singh A, Singh H, Singh SC (2016) Green synthesis of nanoparticles and its potential application. Biotechnol Lett 38:545–560

Gopinath K, Karthika V, Sundaravadivelan C, Gowri S, Arumugam A (2015a) Mycogenesis of cerium oxide nanoparticles using *Aspergillus niger* culture filtrate and their applications for antibacterial and larvicidal activities. J Nanostruct Chem 5:295–303

Kannan SK, Sundrarajan M (2014) A green approach for the synthesis of a cerium oxide nanoparticle: characterization and antibacterial activity. Int J Nanosci 13:1450018

Kargar H, Ghasemi F, Darroudi M (2014) Bioorganic polymer-based synthesis of cerium oxide nanoparticles and their cell viability assays. Ceram Int 41:1589–1594

Kargar H, Ghasemi F, Darroudi M (2015) Bioorganic polymer-based synthesis of cerium oxide nanoparticles and their cell viability assays. Ceram Int 41:1589–1594

Khan SA, Ahmad A (2013) Fungus mediated synthesis of biomedically important cerium oxide nanoparticles. Mater Res Bull 48:4134–4138

Kilbourn BT (2000) Cerium and cerium compounds. Kirk-Othmer encyclopedia of chemical technology. Wiley, New York, In

Kumar E, Selvarajan P, Muthuraj D (2013) Synthesis and characterization of CeO_2 nanocrystals by solvothermal route. Mater Res 16:269–276

Li B, Chen Y, Liang W. Mu L, Bridges WC, Jacobson AR, Darnault CJG (2017) Influence of cerium oxide nanoparticles on the soil enzyme activities in a soil-grass microcosm system. Geoderma 299:54–62

Maensiri S, Masingboon C, Laokul P, Jareonboon W, Promarak V, Anderson PL, Seraphin S (2007) Egg white synthesis and photoluminescence of platelike clusters of CeO_2 nanoparticles. Cryst Growth Des 7:950–955

Malleshappa J, Nagabhushana H, Sharma SC, Vidya YS, Anantharaju KS, Prashantha SC, Daruka Prasad B, Raja Naika H, Lingaraju K, Surendra BS (2015) *Leucas aspera* mediated multifunctional CeO_2 nanoparticles: structural, photoluminescent, photocatalytic and antibacterial properties. Spectrochim Acta A Mol Biomol Spectrosc 149:452–462

Masui T, Hirai H, Imanaka N, Adachi G, Sakata T, Mori H (2002) Synthesis of cerium oxide nanoparticles by hydrothermal crystallization with citric acid. J Mater Sci Lett 21:489–491

Mittal S, Pandey AK (2014) Cerium oxide nanoparticles induced toxicity in human lung cells: role of ROS mediated DNA damage and apoptosis. Biomed Res Int 2014:1–14

Munusamy S, Bhakyaraj K, Vijayalakshmi L, Stephen A, Narayanan V (2014) Synthesis and characterization of cerium oxide nanoparticles using *Curvularia lunata* and their antibacterial properties. Int J Innov Res Sci Eng 2:318–323

Nguyen D, Visvanathan C, Jacob P, Jegatheesan V (2015) Effects of nano cerium (IV) oxide and zinc oxide particles on biogas production. Int Biodeter Biodegr 102:165–171

Oró D, Yudina T, Fernández-Varo G, Casals E, Reichenbach V, Casals G, González de la Presa B, Sandalinas S, Carvajal S, Puntes V, Jiménez W (2016) Cerium oxide nanoparticles reduce steatosis, portal hypertension and display anti-inflammatory properties in rats with liver fibrosis. J Hepatol 64:691–698

Patil SN, Paradeshi JS (2016) Bio-therapeutic potential and cytotoxicity assessment of pectin-mediated synthesized nanostructured cerium oxide. Appl Biochem Biotechnol 638–654

Perez JM, Asati A, Nath S, Kaittanis C (2008) Synthesis of biocompatible dextran-coated Nanoceria with pH-dependent antioxidant properties. Small 4:552–556

Pinjari DV, Pandit AB (2011) Room temperature synthesis of crystalline CeO_2 nanopowder: advantage of sonochemical method over conventional method. Ultrason Sonochem 18:1118–1123

Primo A, Marino T, Corma A, Molinari R, García H (2011) Efficient visible-light photocatalytic water splitting by minute amounts of gold supported on nanoparticulate CeO_2 obtained by a biopolymer templating method. J Am Chem Soc 133:6930–6933

Priya GS, Kanneganti A, Kumar KA, Venkateswara Rao K, Bykkam S (2014) Biosynthesis of cerium oxide nanoparticles using Aloe barbadensis miller gel. Int J Sci Res Publ 4:199–224

Qian J, Chen F, Zhao X, Chen Z (2011) China rose petal as biotemplate to produce two-dimensional ceria nanosheets. J Nanopart Res 13:7149–7158

Reddy Yadav LS, Manjunath K, Archana B, Madhu C, Raja Naika H, Nagabhushana H, Kavitha C, Nagaraju G (2016) Fruit juice extract mediated synthesis of CeO2 nanoparticles for antibacterial and photocatalytic activities. Eur Phys J Plus 131:154

Rocca A, Moscato S, Ronca F, Nitti S, Mattoli V, Giorgi M, Ciofani G (2015) Pilot in vivo investigation of cerium oxide nanoparticles as a novel anti-obesity pharmaceutical formulation. Nanomedicine: Nanotechnol Biol Med 11:1725–1734

Sankar V, SalinRaj P, Athira R, Soumya RS, Raghu KG (2015) Cerium nanoparticles synthesized using aqueous extract of *Centella asiatica*: characterization, determination of free radical scavenging activity and evaluation of efficacy against cardiomyoblast hypertrophy. RSC Adv 5:21074–21083

Sathiyanarayanan G, Dineshkumar K, Yang YH (2017) Microbial exopolysaccharide-mediated synthesis and stabilization of metal nanoparticles. Crit Rev Microbiol 43:731–752

Sharma JK, Srivastava P, Ameen S, Akhtar MS, Sengupta SK, Singh G (2017) Phytoconstituents assisted green synthesis of cerium oxide nanoparticles for thermal decomposition and dye remediation. Mater Res Bull 91:98–107

Siddiqi KS, Husen A (2017) Recent advances in plant-mediated engineered gold nanoparticles and their application in biological system. J Trace Elem Med Biol 40:10–23

Siddiqi KS, Husen A, Rao RAK (2018) A review on biosynthesis of silver nanoparticles and their biocidal properties. J Nanobiotechnology 16:14

Sifontes AB, Gonzalez G, Ochoa JL et al (2011) Chitosan as template for the synthesis of ceria nanoparticles. Mater Res Bull 46:1794–1799

Sun C, Li H, Chen L (2012) Nanostructured ceria-based materials: synthesis, properties and applications. Energy Environ Sci 5:8475–8505

Tamizhdurai P, Sakthinathan S, Chen SM, Shanthi K, Sivasanker S, Sangeeth P (2017) Environmentally friendly synthesis of CeO_2 nanoparticles for the catalytic oxidation of benzyl alcohol to benzaldehyde and selective detection of nitrite. Sci Rep 7:46372

Trovarelli A, de Leitenburg C, Boaro M, Dolcetti G (1999) The utilization of ceria in industrial catalysis. Catal Today 50:353–367

Vinothkumar G, Amalraj R, Babu KS (2017) Fuel-oxidizer ratio tuned luminescence properties of combustion synthesized Europium doped cerium oxide nanoparticles and its effect on antioxidant properties. Ceram Int 43:5457–5466

Walkey C, Das S, Seal S, Erlichman J, Heckman K, Ghibelli L, Traversa E, McGinnis JF, Self WT (2015) Catalytic properties and biomedical applications of cerium oxide nanoparticles. Environ Sci Nano 2:33–53

Weeks ME (1932) The discovery of the elements. XVI. The rare earth elements. J Chem Educ 9:1751

Yin L, Wang Y, Pang G et al (2002) Sonochemical synthesis of cerium oxide nanoparticles—effect of additives and quantum size effect. J Colloid Interface Sci 246:78–84

Zhang F, Chan S-W, Spanier JE, Apak E, Jin Q, Robinson RD, Herman IP (2002) Cerium oxide nanoparticles: size-selective formation and structure analysis. Appl Phys Lett 80:127–129

Chapter 11
Plant-Assisted Fabrication of SnO_2 and SnO_2-Based Nanostructures for Various Applications

Mohammad Mansoob Khan, Mohammad Hilni Harunsani, and Adedayo Rasak Adedeji

11.1 Introduction

Nanotechnology encompasses the understanding, manipulation, and control of matter at the nanoscale level obtained by a combination of engineering, physical, chemical, and biological approaches. It has enabled the development of advanced materials such as nanoparticles (NPs) and nanostructures with unique properties, including the novel optoelectronic, catalytic, and biological properties (Hong and Jiang 2017). The extensive practical application of NPs could be attributed to their unique characteristics that establish their superiority over their bulk counterparts (Shamaila et al. 2016).

Different metal oxides such as TiO_2, ZnO, SnO_2, and CeO_2, among others, are commonly synthesized and used widely as photocatalysts, especially in the heterogeneous photocatalysis (Kalathil et al. 2013; Khan et al. 2013, 2014, 2015a, b, Ansari et al. 2014a, b, c, d, 2016; Saravanan et al. 2015). This is due to their biocompatibility as well as exceptional stability in a variety of conditions and their capability to generate charge carriers when excited by the required amount of light energy. The favorable combination of the electronic structure, light absorption properties, charge-transport characteristics, and excited lifetimes of charge carriers in different metal oxides has made them a fine photocatalyst (Khan et al. 2015a). Among all these nanomaterials (NMs), SnO_2 NPs in particular can be utilized as gas sensors (Sun et al. 2005; Song et al. 2012; Ahamed Fazil et al. 2015; Manjula et al. 2012), photocatalysts (Haritha et al. 2016), semiconductors (Cheng et al. 2016), and antibacterial agents (Vidhu and Philip 2015a).

SnO_2 semiconductor is an important n-type oxide semiconductor with wide bandgap (3.6 eV). It has good electrical, optical, and electrochemical properties and

M. M. Khan (✉) · M. H. Harunsani · A. R. Adedeji
Chemical Sciences, Faculty of Science, Universiti Brunei Darussalam, Gadong, Brunei Darussalam

A. Husen, M. Iqbal (eds.), *Nanomaterials and Plant Potential*,
https://doi.org/10.1007/978-3-030-05569-1_11

is known as the catalytic support material for solar cells, as the solid-state chemical sensors and for its high lithium-storage capacity. The active crystal surfaces of SnO_2 have an important role in determining its interesting properties including the sensing and catalytic ones (Das and Jayaraman 2014).

The large bandgap (3.6 eV, which can only respond to UV illumination) and the high electron hole recombination rate are the main drawbacks of SnO_2 semiconductors. Doping the SnO_2 semiconductors with metal ions is one of the best methods to improve the visible light photocatalytic activity for the oxidative degradation of organic compounds. Many transition metals have been used to modify the electronic properties of SnO_2. Due to the relative ionic radius and interstitial spacing in the SnO_2 structure, the most frequently tested transition metals for this modification include inter alia gold, manganese, silver, cobalt, and iron. These transition metals have similar or lesser ionic radius, so the atoms can easily replace or fit into the interstitial sites in the SnO_2 structure (Sabergharesou et al. 2013; Cheng et al. 2016; Tomer and Duhan 2016; Sinha et al. 2017).

Gas sensors of individual SnO_2 materials are attractive due to their high sensitivity, quick response, and good stability. The hetero-structured SnO_2 with other semiconducting metal oxides will provide another promising strategy to develop novel high-performance gas sensors due to the formation of different charge carriers. This would assist to overcome effectively the challenge of high recombination rate in SnO_2 by scavenging the charge carriers (electrons and holes from the conduction and valence band, respectively) through compositing SnO_2 with other metal oxides such as n-type TiO_2, ZnO, WO_3, In_2O_3, CaO, MgO, V_2O_5, and Nb_2O_5 and p-type NiO, Co_3O_4, Sb_2O_3, La_2O_3, Cu_2O, Ag_2O, CeO_2, etc. (Cheng et al. 2016; Sudhaparimala and Vaishnavi 2016).

SnO_2 NPs have also been used widely for effective photocatalytic degradation of dye effluents and water treatment applications, using the UV or solar irradiation. This photodegradation process converts dye molecules into nonhazardous compounds and by-products. The nanostructured SnO_2 exhibits enhanced photocatalytic activity for the degradation of dyes and proves to be an effective means for the elimination of various water pollutants (Bhattacharjee et al. 2015). In addition, utility of SnO_2 NPs has been reported for antimicrobial applications (Meena Kumari and Philip 2015; Vidhu and Philip 2015a, b; Roopan et al. 2015). This report intends to elucidate the morphology and characteristics of phytosynthesized SnO_2 NPs and their diverse applications.

11.2 Synthesis of Nanoparticles

In general, there are chemical, physical, and biological methodologies (as shown in Fig. 11.1) that can be used to synthesize NPs. Chemical methods of producing the desired NPs usually involve the use of metal precursors and other chemicals to initiate the particular reaction and stabilize pH for favorable reaction condition.

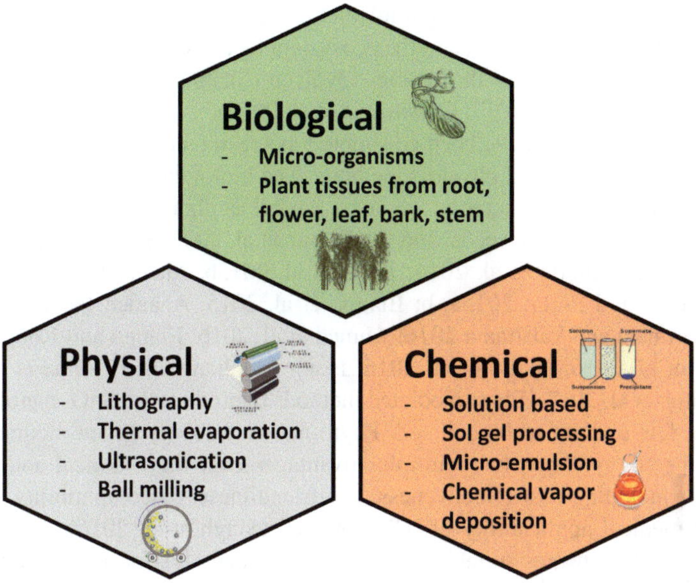

Fig. 11.1 Different methods of nanoparticle synthesis

These methods include sol-gel technique, microemulsion technique, solvo-thermal method, hydrothermal method, and chemical vapor deposition (CVD) technique, among others. In the case of physical synthesis, dry and wet mechanical grindings are used for inexpensive NP preparation. For technological applications, wet grinding is preferable because it allows more options to control the NP size (Saleh 2016). Other physical methods include ultrasonication, thermal evaporation, and lithography. Several NP preparation methods employ a combination of both physical and chemical synthesis procedures, such as the sonochemical technique based on chemical method and ultrasound technique as well as the microwave-assisted chemical syntheses. A proper combination of chemical and physical methods allows for the production of smaller NPs (Xia et al. 2008; Saleh 2016).

The expansion of these synthesis procedures to a large-scale production has several limitations such as expensive production costs resulting from high energy consumption, use of toxic organic solvents, production of hazardous intermediates, and formation of harmful waste products, leading to environmental pollution and several biological risks. Generally, wet chemical synthesis of NPs dominates the production methods. Agglomeration or aggregation of NPs usually occurs during synthesis due to the presence of attractive forces among NPs. Therefore, some capping agent is used to prevent aggregation and attain the desired morphology of the product. Besides their toxic nature, these methods also have such drawbacks as slow production rate, limited growth, and distorted structure of synthesized NPs. Size reduction may also lead to increased reactivity and toxicity of the material synthesized.

Therefore, prior to the large-scale implementation of these reactions, it is necessary to analyze the potential hazards to the ecosystem by considering the entire chemical procedure and tracking all the species involved (Shamaila et al. 2016; Hong and Jiang 2017; Sinha et al. 2017; Osuntokun et al. 2017).

Biological methods have been considered as green (eco-friendly) alternatives to the existing chemical and physical methods. The biological tools include microorganisms such as fungi, bacteria, and yeast as well as plant tissues obtained usually from roots, barks, leaves, seeds, and fruits (Sun et al. 2005; Song et al. 2012; Ansari et al. 2014d; Kamaraj et al. 2014; Khan et al. 2015b; Meena Kumari and Philip 2015; Vidhu and Philip 2015a, b; Elango et al. 2015; Ahamed Fazil et al. 2015; Sudhaparimala and Vaishnavi 2016; Ahmed et al. 2016; Elango and Roopan 2016; Haritha et al. 2016; Diallo et al. 2016; Hong and Jiang 2017; Sinha et al. 2017; Osuntokun et al. 2017). Plant-mediated methods to produce the SnO_2 nanomaterials are very few and yet to be collated. Phyto-nanotechnology (plant-mediated technique) for NP synthesis has multiple advantages over the chemical and physical methods, including cost-effectiveness, eco-friendliness, biocompatibility, scalability, and medical applicability of NPs obtained (Singh et al. 2016). The nontoxic nature of plants ensures suitability of phyto-mediated NPs for applications in biomedical and environmental areas. Extracts of various plant parts contain compounds required for the synthesis of NPs. Active compounds such as polyphenols, flavonoids, and other secondary metabolites are believed to be primarily responsible for the biological synthesis. These compounds are capable of reducing or oxidizing the ions present in the precursors by forming intermediates. The resulting NPs, such as SnO_2, are usually obtained after calcination at specified temperatures (Diallo et al. 2016; Hong and Jiang 2017; Osuntokun et al. 2017).

11.3 Plant-Assisted Fabrication of SnO_2 Nanostructures and Their Applications

Given the current chemophobia and the awareness about the hazardous effects of many chemical synthetic routes, green chemistry tends to offer an appropriate alternative by ensuring that the fabrication of materials is eco-friendly. A befitting and cost-effective means of achieving this goal in NM synthesis is the adoption of plant extracts to replace the toxic chemicals. This area of green chemistry is very promising to achieve an eco-friendly synthesis route and has gained ground with some recent reports on phytosynthesis of SnO_2 NPs (Table 11.1) and their diverse applications (Fig. 11.2) (Roopan et al. 2015; Vidhu and Philip 2015b; Elango and Roopan 2016; Diallo et al. 2016; Hong and Jiang 2017; Sinha et al. 2017; Osuntokun et al. 2017).

Table 11.1 Different SnO_2 nanostructures synthesized with plant-assisted methods and their applications

No.	Plant	Metal oxide nanoparticle	Particle size (nm)	Applications	Year	Reference
1	Cotton fiber	SnO_2 microtubes	5–15 × 10^3 (microtubes)	Gas sensing	2005	Sun et al. (2005)
2	Cotton fiber	SnO_2 nanotubules	14.2	Gas sensing	2010	Zhu et al. (2010)
3	Ramie fiber	SnO_2/C; original NaOH-treated degummed	2.0–5.0 2.5–6.3 2.6–4.7	Gas sensing	2011	He et al. (2011)
4	*Brassica campestris*	SnO_2 microreactor	40–55	Gas sensing	2012	Song et al. (2012)
5	Cotton	SnO_2/C	20	Lithium-ion batteries	2014	Li et al. (2014)
6	*Cleistanthus collinus*	SnO_2	~49.26	Antimicrobial and antioxidant agents	2014	Kamaraj et al. (2014)
7	*Saraca indica*	SnO_2	2.1–4.1	Antibacterial and antioxidant agents	2014	Vidhu and Philip (2015a)
8	*Ficus carica*	SnO_2	~132	Hg^{2+} sensor	2015	Hu J (2015)
9	*Peltophorum pterocarpum*	SnO_2 motifs	16–25	Gas sensing	2015	Ahamed Fazil et al. (2015)
10	*Trigonella foenum-graecum*	SnO_2	2.2–3.2	Nanofluids, antibacterial, and antioxidant agents	2015	Vidhu and Philip (2015b)
11	Pomegranate	SnO_2	2.5 - 2.7	Nanofluids, antibacterial, and antioxidant agents	2015	Meena Kumari and Philip (2015)
12	*Annona squamosa*	SnO_2	25 ± 5	Cytotoxicity toward HepG2	2015	Roopan et al. (2015)
13	*Persea americana*	SnO_2	~4	Photocatalytic degradation (phenol red)	2015	Elango et al. (2015)
14	*Cyphomandra betacea*	SnO_2	~21	Photocatalytic degradation (methylene blue)	2016	Elango and Roopan (2016)
15	*Catunaregam spinosa*	SnO_2	47 ± 2	Photocatalytic degradation (Congo red)	2016	Haritha et al. (2016)

(continued)

Table 11.1 (continued)

No.	Plant	Metal oxide nanoparticle	Particle size (nm)	Applications	Year	Reference
16	Aloe vera	Coupled SnO_2-ZnO	~22.27	Photocatalytic degradation (methylene orange), antimicrobial agent	2016	Sudhaparimala and Vaishnavi (2016)
17	*Aspalathus linearis*	SnO_2	2.1–19.3	Photocatalytic degradation (methylene blue, Congo red, and eosin Y.)	2016	Diallo et al. (2016)
18	*Brassica oleracea* L. var. *botrytis*	SnO_2	3.62–6.34	Photocatalytic degradation (methylene blue)	2017	Osuntokun et al. (2017)
19	*Litsea cubeba*	SnO_2	~30	Photocatalytic degradation (Congo red), antioxidant agent	2017	Hong and Jiang (2017)
20	*Saccharum officinarum*	Ag doped SnO_2	~9	Photocatalytic degradation (methylene blue, rose Bengal, methyl violet 6B, and 4-nitro phenol), antimicrobial and antioxidant agents	2017	Sinha et al. (2017)
21	*Piper nigrum*	SnO_2	8.85 ± 3.5 at 300 °C; 12.76 ± 3.9 at 500 °C; 29.29 ± 10.9 nm at 900 °C	Cytotoxicity toward cancer cells	2017	Tammina et al. (2017)

11.3.1 Synthesis of Undoped SnO_2

Vidhu and Philip (2015a) reported a cost-effective and environmentally benign method of synthesizing bioactive SnO_2 NPs of 2.1–4.1 nm, using *Saraca indica* flower. The synthesized NPs exhibited antibacterial activity against the gram-negative bacteria *Escherichia coli* and proved to be a good antioxidant and antibacterial agent. In another work, they synthesized SnO_2 NPs in the size range of 2.2–3.2 nm, using the fenugreek seeds. These particles found applications in nanofluids and biomedical field due to their unique properties such as viscosity and thermal conductivity and their antibacterial and antioxidant potential (Vidhu and Philip 2015b).

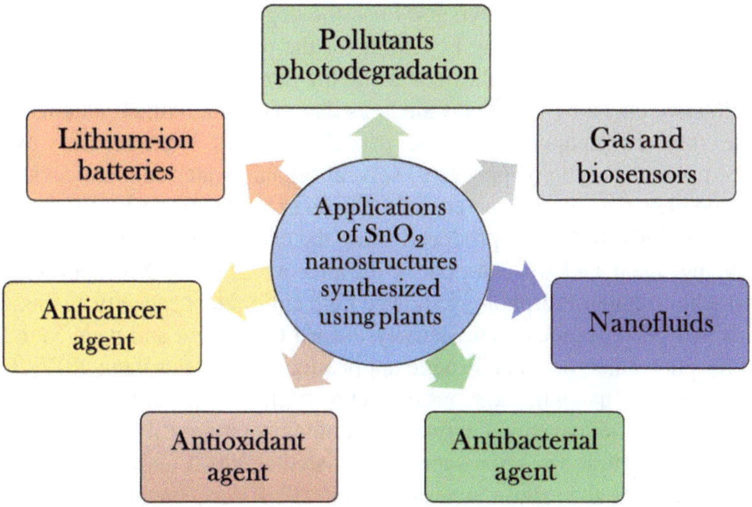

Fig. 11.2 Reported applications of SnO_2 nanostructure synthesized by using plants

Similarly, Kamaraj and coworkers prepared SnO_2 NPs using the methanol extract of *Cleistanthus collinus* plant as the reducing and capping agent. The average crystallite size of the NPs formed was 49.26 nm, and they were identified to possess antioxidant property due to their activity against DPPH radicals (Kamaraj et al. 2014). Quantum-confined spherical SnO_2 NPs have also been fabricated using different quantities of the pomegranate fruit extract as the reducing/oxidizing agent and capping agent, while tin (IV) chloride ($SnCl_4 \times H_2O$) as the precursor. The prepared NPs showed excellent antibacterial and antioxidant activity, thus emerging as a potential bioactive material (Kumari and Philip 2015).

Aqueous extract of dried peel of sugar apple (*Annona squamosa*), an agricultural waste, was used in the rapid synthesis of stable SnO_2 NPs. At specific concentrations they exhibited moderate cytotoxicity toward the hepatocellular carcinoma (HepG2) (Roopan et al. 2015). Elango et al. (2015) synthesized SnO_2 nanoparticles of 4 nm size, using the methanol extract of *Persea americana* seeds as reducing/oxidizing agents as well as capping agents and tin chloride as the precursor. These particles could cause photocatalytic degradation of phenols under ultraviolet irradiation, depicting their applicability in such processes as dye effluents pollution abatement (Elango et al. 2015). In another study, these authors used methanol extract of *Cyphomandra betacea*. The NPs produced were rod-shaped with an average size of 21 nm and showed photocatalytic ability when used for degrading methylene blue (Elango and Roopan 2016).

Likewise, Hu (2015) synthesized SnO_2 NPs from tin chloride in aqueous mixture of fig (*Ficus carica*) leaf extract. The NPs produced were about 132 nm in size and could be used as an electrode modifier for electrochemical detection of mercury in water. Its application as a Hg^{2+} sensor proved useful for water purification and pollutant detection. Haritha et al. (2016) synthesized SnO_2 NPs (47 ± 2 nm) using an aqueous extract of *Catunaregam spinosa* root bark. These particles were able to degrade the toxic Congo red dye and raise the degradation percentage up to 92% after 45 min and therefore may be useful for the purpose of pollutant abatement. SnO_2 NPs have also been synthesized using the *Aspalathus linearis* natural extract as an effective chelating agent and $SnCl_4$ as the precursor, without adding any acid or base standard compounds. These particles (2.1–19.3 nm in size) exhibited enhanced photocatalytic responses to several organic water contaminants such as methylene blue, Congo red, and eosin Y (Diallo et al. 2016).

Recently, SnO_2 NPs were produced using the extracts from *Litsea cubeba* fruits as reducing agent and tin chloride as the precursor. These NPs (approximately 30 nm) with irregular morphology showed high antioxidant capacity against DPPH radicals and good photocatalytic activity toward Congo red and were found to be useful for photocatalytic and cosmetic applications (Hong and Jiang 2017). Similar efforts were made using the aqueous extract of fresh cauliflower (*Brassica oleracea* L. var. *botrytis*) after annealing at two different temperatures (300 and 450 °C). Due to the smaller particle sizes, the samples prepared at 300 °C exhibited greater degradation efficiency than those annealed at 450 °C. It was concluded that the SnO_2 NPs so synthesized can be utilized in applications for degradation of toxic organic dyes and purification of effluent water (Osuntokun et al. 2017).

Tammina et al. (2017) synthesized tetragonal SnO_2 NPs of different sizes using *Piper nigrum* seed extract at three different calcination temperatures (300, 500, and 900 °C). Their cytotoxicity test proved them to be toxic against the colorectal (HCT116) and lung (A549) cancer cell lines, depending on their size and dose. The cytotoxicity was attributed to the formation of reactive oxygen species (ROS), which was more abundant with the smaller NPs than with the larger ones. The reported IC50 values of SnO_2 NPs with average particle sizes of 8.85 ± 3.5, 12.76 ± 3.9, and 29.29 ± 10.9 nm were 165, 174, and 208 μg L^{-1}, respectively, against HCT116, whereas 135, 157, and 187 μg L^{-1}, respectively, against the A549 carcinoma cell lines. These stable SnO_2 NPs were recommended as a potent therapeutic agent against cancerous cell lines (Tammina et al. 2017).

11.3.2 Plant-Assisted Template Synthesis of SnO_2

Different forms of NPs such as nanotubes, nanospheres, and motifs have been fabricated to enhance their convenient applications in areas of energy storage and gas sensing (Song et al. 2012; Ahamed Fazil et al. 2015; Sinha et al. 2017). Sun et al. (2005) developed a method to synthesize the biomorphic SnO_2 microtubules, using cotton fibers as templates. The fibers were infiltrated with tin alkoxide solution and subsequently sintered at high temperatures to produce the final SnO_2 microtubules of 5–15 μm diameter. The BET (Brunauer–Emmett–Teller) surface area and the pore volume of the biomorphic SnO_2 microtubules were reported to be highest ($S_{BET} = 24.2$ m^2 g^{-1}, $V_{mic} = 0.075$ mL g^{-1}) at 700 °C based on the N_2-adsorption isotherms. The fabricated porous SnO_2 NPs were recommended for gas-sensing applications (Sun et al. 2005).

SnO_2 nanotubular materials were also fabricated using a sonochemical route with the assistance of cotton fibers to obtain a desirable and porous morphology for enhanced sensing response. The nanotubules so produced consisted of nanocrystals of 14.2 nm size when calcined at 700 °C. The SnO_2 nanotubules exhibited a good selectivity for acetone at a working temperature of 350 °C, with the sensitivity to 20 ppm acetone recorded as 6.4, and a rapid response and recovery (around 9–10 s) (Zhu et al. 2010).

He et al. (2011) prepared a biotemplate, using ramie fibers for the fabrication of SnO_2/C biomorphic materials. In order to study the effect of the finishing processes on the material fabricated, the ramie fibers were given different treatments such as water washing, NaOH soaking, and oxidative bleaching. Ramie fibers and $Sn(OH)_4$ were used as the carbon and the SnO_2 precursors, respectively, for each treatment process. The crystallite size of SnO_2 in the SnO_2/C biomorphic materials was recorded to be 2.0–5.0 nm, 2.5–6.3 nm, and 2.6–4.7 nm for each template, viz., original ramie fiber, NaOH-treated ramie fiber, and degummed ramie fiber, respectively. The morphology and properties of the materials produced were reported to vary and could be effectively controlled with different finishing processes (He et al. 2011). The biomorphic SnO_2/C composites were also fabricated using the natural cotton as the structure template and the bio-carbon as the precursor (Li et al. 2014). These composites were made of the nano-sized small particles of about 20 nm. The carbon content in the composites obtained had a great impact on their electrochemical performances and could be adjusted by altering the sintering temperature. The composites prepared at 300 °C were reported to exhibit a reversible capacity of 530 mAh g^{-1} after 100 cycles at a current density of 100 mA g^{-1}. These were recommended for energy storage application such as anode materials in the lithium-ion batteries (Li et al. 2014).

Song et al. (2012) described a method of SnO_2 synthesis with morphology that mimics the bioreactors' scaffolds of pollen grains. The technique consists of a facile two-step soakage process (the NP was mobilized on the pollen grain) followed by calcinations to reconstruct the microreactors of pollen grains (*Brassica campestris*) with the SnO_2 nanoparticle as the gas-sensing materials. The resulting microreactors of SnO_2 NPs exhibited superior gas-sensing capabilities to NO_2 in terms of response value and recovery rate, compared with other SnO_2-based gas sensors (Song et al. 2012). The reported performance of other SnO_2 sensors includes the sprayed SnO_2 thin films (Leo et al. 1999), SnO_2 and WO_x–SnO_2 thick films (Chiorino et al. 2001), SnO_2 nanowires (Kim et al. 2011), nanowire-structured SnO_x–SWNT composites (Hoa et al. 2009), and ordered mesoporous SnO_2 (Hyodo et al. 2003), as shown in Table 11.2.

Ahamed Fazil and associates adopted a plant-assisted template during the synthesis of crystalline SnO_2 motifs with porous structure using the *Peltophorum pterocarpum* pollen grains. The motifs were reported to be made up of SnO_2 NPs of the size of 16–25 nm with a high BET surface area of 82.72 m^2 g^{-1} recorded when annealed at 600 °C for 2 h. Since a larger surface area of the SnO_2 motifs would aid the surface reaction with gases better, the resulting porous morphology proved to be useful for gas-sensing applications (Ahamed Fazil et al. 2015).

Table 11.2 Comparison of some SnO_2-based gas sensors

No.	SnO_2-based gas sensor	Concentration (ppm)	Response time t_1 (s)	Recovery time t_2 (s)	Temp. (°C)	Reference
1	Sprayed SnO_2 thin films	~15–50	360	300	350	Leo et al. (1999)
2	SnO_2 and WO_x–SnO_2 thick films	~2 and ~5–10	>300	>300	250	Chiorino et al. (2001)
3	SnO_2 nanowires	<30–10	20–60	>300	300	Kim et al. (2011)
4	Nanowire-structured SnO_x–SWNT composites	~25–60	~50	211	200	Hoa et al. (2009)
5	Ordered mesoporous SnO_2	<150–100	–	–	300	Hyodo et al. (2003)
6	SnO_2 microreactors	219.5–50	5	111	330	Song et al. (2012)

11.3.3 *Doped, Coupled, and Decorated SnO_2 Nanostructure Synthesis*

In order to overcome the challenge of large bandgap and high recombination rate in SnO_2, such nanostructures as decorated, doped, and coupled SnO_2 hetero-structures have been fabricated with enhanced characteristics. Only few reports have appeared so far on the synthesis of these SnO_2 nanostructures using plant tissues and extracts (Sudhaparimala and Vaishnavi 2016; Sinha et al. 2017). Sinha and coworkers have reported a phytosynthetic technique for the fabrication of sphere-shaped Ag-SnO_2 nanocomposite of average particle size of 9 nm, employing the stem extracts of *Saccharum officinarum*. The study reveals the nanocomposite efficacy as being anti-oxidant and antibacterial due to its action against *Pseudomonas aeruginosa*, *Escherichia coli*, and *Bacillus subtilis*. Also, the Ag-SnO_2 nanocomposite was found to be a good photocatalyst, and its removal efficiency for the abatement of different industrially emerging pollutants such as methylene blue, rose Bengal, methyl violet 6B, and 4-nitrophenol was recorded (Sinha et al. 2017). Further, tin (II) chloride and zinc acetate were used as the precursors and the gel of *Aloe vera* plant as the medium to fabricate the coupled nanocomposite of tin (IV) oxide and zinc oxide (SnO_2/ZnO). The average crystallite size was 22.27 nm, while application studies recorded the degradation of organic dyes, such as methyl orange, under the visible light irradiation. Moreover, the nanocomposite was effective against the growth of *Staphylococcus aureus* and *E. coli* at the microgram level (Sudhaparimala and Vaishnavi 2016).

11.4 Future Perspectives

SnO_2 NPs have been widely researched for many important applications such as energy storage, gas sensing, and photocatalytic degradation. The synthesis route determines how safely they could be adopted in biological and biomedical exercises (as nanofluids, anticancer, antioxidants, and antibacterial materials). The plant-assisted (green) synthesis method is a promising and viable technique to fabricate SnO_2 nanostructures without using harmful chemicals. With the growing acceptability of NMs in biology and medical science, new windows have opened for potential research on green synthesis routes of SnO_2 NPs, in order to render the NPs nonhazardous. Nonetheless, the complex mechanism of phytosynthesis of nanomaterials, such as SnO_2 nanostructures, is yet to be completely understood. The step reactions involved and the identification of important isolates among the components of plant extracts require in-depth studies to be taken up to adequately understand, improve, and control the reactions for the desired NM synthesis. The efficacy of different SnO_2 nanostructures as a photocatalyst in pollution control would be better appreciated if future researches on photodegradation use samples of mixed pollutants from different locations for critical analysis. These studies can facilitate an efficient and feasible field application of these photocatalysts to remove the various environmental pollutants.

In the near future, many forms of SnO_2 nanostructures with enhanced capabilities and efficiencies are likely to be phytosynthesized through doping (metal and nonmetal), co-doping, decorating, and coupling with other metals, metal oxides, chalcogenides, etc. without using the harmful chemicals. Many potential researches would also adopt plant tissues as template for synthesizing the SnO_2 nanostructures to improve their morphology and porosity and achieve increased surface area for an enhanced surface reaction, which would be suitable for such applications as gas sensing, self-cleaning, catalysis, and photocatalysis, among others.

11.5 Conclusion

By adopting the green (plant-assisted) synthesis methods, the commonly used hazardous chemicals could be avoided, and the NPs synthesized would be safe for different applications. Since there are only few reports on progress with the plant-assisted synthesis of SnO_2 nanostructures, this review has summarized the efforts hitherto made with reference to the phytosynthesis of undoped, doped, and template-assisted SnO_2 nanostructures as well as their performance and applications.

References

Ahamed Fazil A, Udaya Bhanu J, Amutha A, Joicy S, Ponpandian N, Amirthapandian S, Panigrahi BK, Thangadurai P (2015) A facile bio-replicated synthesis of SnO_2 motifs with porous surface by using pollen grains of *Peltophorum pterocarpum* as a template. Microporous Mesoporous Mater 212:91–99

Ahmed S, Ahmad M, Swami BL, Ikram S (2016) A review on plants extract mediated synthesis of silver nanoparticles for antimicrobial applications: a green expertise. J Adv Res 7:17–28

Ansari SA, Khan MM, Ansari MO, Kalathil S, Lee J, Cho MH (2014a) Band gap engineering of CeO_2 nanostructure using an electrochemically active biofilm for visible light applications. RSC Adv 4:16782–16791

Ansari SA, Khan MM, Ansari MO, Lee J, Cho MH (2014b) Visible light-driven photocatalytic and photoelectrochemical studies of Ag–SnO_2 nanocomposites synthesized using an electrochemically active biofilm. RSC Adv 4:26013

Ansari SA, Khan MM, Ansari MO, Cho MH (2016) Nitrogen-doped titanium dioxide (N-doped TiO_2) for visible light photocatalysis. New J Chem 40:3000–3009

Ansari SA, Khan MM, Lee J, Cho MH (2014c) Highly visible light active Ag@ZnO nanocomposites synthesized by gel-combustion route. J Ind Eng Chem 20:1602–1607

Ansari SA, Khan MM, Omaish Ansari M, Lee J, Cho MH (2014d) Highly photoactive SnO_2 nanostructures engineered by electrochemically active biofilm. New J Chem 38:2462–2469

Bhattacharjee A, Ahmaruzzaman M, Sinha T (2015) A novel approach for the synthesis of SnO_2 nanoparticles and its application as a catalyst in the reduction and photodegradation of organic compounds. Spectrochim Acta A Mol Biomol Spectrosc 136(Pt B):751–760

Cheng JP, Wang J, Li QQ, Liu HG, Li Y (2016) A review of recent developments in tin dioxide composites for gas sensing application. J Ind Eng Chem 44:1–22

Chiorino A, Ghiotti G, Prinetto F, Carotta MC, Malagù C, Martinelli G (2001) Preparation and characterization of SnO_2 and WO_x-SnO_2 nanosized powders and thick films for gas sensing. Sensors Actuators B Chem 78:89–97

Das S, Jayaraman V (2014) SnO_2: a comprehensive review on structures and gas sensors. Prog Mater Sci 66:112–255

Diallo A, Manikandan E, Rajendran V, Maaza M (2016) Physical & enhanced photocatalytic properties of green synthesized SnO_2 nanoparticles via *Aspalathus linearis*. J Alloys Compd 681:561–570

Elango G, Kumaran SM, Kumar SS, Muthuraja S, Roopan SM (2015) Green synthesis of SnO_2 nanoparticles and its photocatalytic activity of phenolsulfonphthalein dye. Spectrochim Acta Part A Mol Biomol Spectrosc 145:176–180

Elango G, Roopan SM (2016) Efficacy of SnO_2 nanoparticles toward photocatalytic degradation of methylene blue dye. J Photochem Photobiol B Biol 155:34–38

Haritha E, Roopan SM, Madhavi G, Elango G, Al-Dhabi NA, Arasu MV (2016) Green chemical approach towards the synthesis of SnO_2 NPs in argument with photocatalytic degradation of diazo dye and its kinetic studies. J Photochem Photobiol B Biol 162:441–447

He XH, Qi LH, Wang JB, Min-ge Y, Ming-qian S, Wei C (2011) Effect of biological template consolidation on the microstructure and properties of SnO_2/C bio-morphic materials. Xinxing Tan Cailiao/New Carbon Mater 26:375–380

Hoa ND, Van Quy N, Kim D (2009) Nanowire structured SnO_x-SWNT composites: high performance sensor for NO_x detection. Sensors Actuators B Chem 142:253–259

Hong G, Jiang C (2017) Synthesis of SnO_2 nanoparticles using extracts from *Litsea cubeba* fruits. Mater Lett 194:164–167

Hu J (2015) Biosynthesis of SnO_2 nanoparticles by Fig (*Ficus carica*) leaf extract for electrochemically determining Hg(II) in water samples. Int J Electrochem Sci 10:10668–10676

Hyodo T, Abe S, Shimizu Y, Egashira M (2003) Gas-sensing properties of ordered mesoporous SnO_2 and effects of coatings thereof. Sensors Actuators B Chem 93:590–600

Kalathil S, Khan MM, Ansari SA, Lee J, Cho MH (2013) Band gap narrowing of titanium dioxide (TiO_2) nanocrystals by electrochemically active biofilms and their visible light activity. Nanoscale 5:6323–6326

Kamaraj P, Vennila R, Arthanareeswari M, Devikala S (2014) Biological activities of tin oxide nanoparticles synthesized using plant extract. World J Pharm Pharm Sci 3:382–388

Khan MM, Adil SF, Al-Mayouf A (2015a) Metal oxides as photocatalysts. J Saudi Chem Soc 19:462–464

Khan MM, Ansari SA, Pradhan D, Ansari MO, Han DH, Lee J, Cho MH (2014) Band gap engineered TiO_2 nanoparticles for visible light induced photoelectrochemical and photocatalytic studies. J Mater Chem A 2:637–644

Khan MM, Ansari SA, Khan ME, Ansari MO, Min BK, Cho MH (2015b) Visible light-induced enhanced photoelectrochemical and photocatalytic studies of gold decorated SnO_2 nanostructures. New J Chem 39:2758–2766

Khan MM, Ansari SA, Lee J, Cho MH (2013) Enhanced optical, visible light catalytic and electrochemical properties of Au@TiO_2 nanocomposites. J Ind Eng Chem 19:1845–1850
Kim BG, Lim DG, Park JH, Choi YJ, Park JG (2011) In-situ bridging of SnO_2 nanowires between the electrodes and their NO_2 gas sensing characteristics. Appl Surf Sci 257:4715–4718
Kumari M, Philip D (2015) Synthesis of biogenic SnO_2 nanoparticles and evaluation of thermal, rheological, antibacterial and antioxidant activities. Powder Technol 270:312–319
Leo G, Rella R, Siciliano P, Capone S, Alonso JC, Pankov V, Ortiz A (1999) Sprayed SnO_2 thin films for NO_2 sensors. Sensors Actuators B Chem 58:370–374
Li B, Zai J, Xiao Y, Han Q, Qian X (2014) SnO_2/C composites fabricated by a biotemplating method from cotton and their electrochemical performances. CrystEngComm 16:3318–3322
Manjula P, Boppella R, Manorama SV (2012) A facile and green approach for the controlled synthesis of porous SnO_2 nanospheres: application as an efficient photocatalyst and an excellent gas sensing material. ACS Appl Mater Interfaces 4:6252–6260
Osuntokun J, Onwudiwe DC, Ebenso EE (2017) Biosynthesis and photocatalytic properties of SnO_2 nanoparticles prepared using aqueous extract of cauliflower. J Clust Sci 28:1883–1896
Roopan SM, Kumar SHS, Madhumitha G, Suthindhiran K (2015) Biogenic-production of SnO_2 nanoparticles and its cytotoxic effect against hepatocellular carcinoma cell line (HepG2). Appl Biochem Biotechnol 175:1567–1575
Sabergharesou T, Wang T, Ju L, Radovanovic PV (2013) Electronic structure and magnetic properties of sub-3 nm diameter Mn-doped SnO_2 nanocrystals and nanowires. Appl Phys Lett. https://doi.org/10.1063/1.4813011
Saleh TA (2016) Nanomaterials for pharmaceuticals determination. Bioenergetics 5:226
Saravanan R, Khan MM, Gupta VK, Mosquera E, Gracia F, Narayanan V, Stephen A (2015) ZnO/Ag/CdO nanocomposite for visible light-induced photocatalytic degradation of industrial textile effluents. J Colloid Interface Sci 452:126–133
Shamaila S, Sajjad AKL, Ryma N, Farooqi SA, Jabeen N, Majeed S, Farooq I (2016) Advancements in nanoparticle fabrication by hazard free eco-friendly green routes. Appl Mater Today 5:150–199
Singh P, Kim Y-J, Zhang D, Yang D-C (2016) Biological synthesis of nanoparticles from plants and microorganisms. Trends Biotechnol 34:588–599
Sinha T, Ahmaruzzaman M, Adhikari PP, Bora R (2017) Green and environmentally sustainable fabrication of Ag-SnO_2 nanocomposite and its multifunctional efficacy as photocatalyst and antibacterial and antioxidant agent. ACS Sustain Chem Eng 5:4645–4655
Song F, Su H, Han J, Lau WM, Moon WJ, Zhang D (2012) Bioinspired hierarchical tin oxide scaffolds for enhanced gas sensing properties. J Phys Chem C 116:10274–10281
Sudhaparimala S, Vaishnavi M (2016) Biological synthesis of nano composite SnO_2- ZnO – screening for efficient photocatalytic degradation and antimicrobial activity. Mater Today Proc 3:2373–2380
Sun B, Fan T, Xu J, Zhang D (2005) Biomorphic synthesis of SnO_2 microtubules on cotton fibers. Mater Lett 59:2325–2328
Tammina SK, Mandal BK, Ranjan S, Dasgupta N (2017) Cytotoxicity study of Piper nigrum seed mediated synthesized SnO_2 nanoparticles towards colorectal (HCT116) and lung cancer (A549) cell lines. J Photochem Photobiol B Biol 166:158–168
Tomer VK, Duhan S (2016) A facile nanocasting synthesis of mesoporous Ag-doped SnO_2 nanostructures with enhanced humidity sensing performance. Sensors Actuators B Chem 223:750–760
Vidhu VK, Philip D (2015a) Biogenic synthesis of SnO_2 nanoparticles: evaluation of antibacterial and antioxidant activities. Spectrochim Acta Part A Mol Biomol Spectrosc 134:372–379
Vidhu VK, Philip D (2015b) Phytosynthesis and applications of bioactive SnO_2 nanoparticles. Mater Charact 101:97–105
Xia H, Zhuang H, Zhang T, Xiao D (2008) Visible-light-activated nanocomposite photocatalyst of Fe_2O_3/SnO_2. Mater Lett 62:1126–1128
Zhu S, Zhang D, Gu J, Jiaqiang X, Junping D, Jinlong L (2010) Biotemplate fabrication of SnO_2 nanotubular materials by a sonochemical method for gas sensors. J Nanopart Res 12:1389–1400

Chapter 12
Bionanoparticles in the Treatment of Glycation-Induced Secondary Complications of Diabetes

Pamela Jha and Ahmad Ali

Abbreviations

AGE	Advanced glycation end products
CML	Carboxymethyl-lysine
GA	Glycated albumin
HbA1c	Glycated hemoglobin
MG	Methylglyoxal
NP	Nanoparticle
RAGE	Receptors for advanced glycation end products

12.1 Introduction

The European Commission has defined nanomaterial (NM) as a natural, incidental, or manufactured material containing particles in an unbound state or in an aggregate or agglomerate in which ≥50% of the particles in the number size distribution have one or more external dimensions in the size range 1–100 nm (Mu et al. 2014).

The rationale of nanoparticles (NPs) being an attractive alternative of bulk material is based on their unique features, such as their surface to mass ratio, which is much larger than that of any other particles or materials. This allows catalytic promotion of reactions as well as their ability to adsorb and carry other compounds. The reactivity of the surface originates from quantum phenomena and can make them unpredictable, immediately after their generation. NPs may have their surface modified, depending upon the presence of reactants and adsorbing compounds, which may instantaneously change with the changing compounds and thermodynamic conditions. Therefore, on one hand, NP has a large (functional) surface which

P. Jha
Amity Institute of Biotechnology, Amity University, Navi Mumbai, Maharashtra, India

A. Ali (✉)
Department of Life Sciences, University of Mumbai, Mumbai, Maharashtra, India

A. Husen, M. Iqbal (eds.), *Nanomaterials and Plant Potential*,
https://doi.org/10.1007/978-3-030-05569-1_12

is able to bind, adsorb, and carry other compounds like drugs, probes, and proteins; on the other hand, this surface might be chemically more reactive, compared to their fine analogues (Paul et al. 2004). There are different types of NPs like liposomes, nanocrystals, solid lipid NPs, polymeric NPs, dendrimers, silicon-based structures, carbon structures, and metal structures. Each of these types has its specific advantages and dedicated applications (Husen and Siddiqi 2014a, b, c; Siddiqi and Husen 2016; Husen 2017; Siddiqi et al. 2018a, b).

Diabetes has affected millions of the people all over the world (Shaw et al. 2010). Due to prolonged accumulation of glucose in the body, there is an overproduction of a group of harmful compounds commonly known as advanced glycation end products (AGEs). These products are generated as a result of nonenzymatic and covalent interaction between the carbonyl group of sugars and amino groups of proteins, nucleic acids, and lipids (Suji and Sivakami 2004). AGEs have been implicated in various secondary complications of diabetes and neurodegenerative disorders (Singh et al. 2014). A range of artificial and natural antiglycating agents have been designed to prevent the accumulation of AGEs and adverse effects of these compounds (Abbas et al. 2016; Ali et al. 2014). Recent upsurge in application of NPs in the field of medicine has led to their utilization in the management of glycation as sensors and antiglycating agents. This chapter deals with the process of glycation and application of plant-based NPs in the detection and prevention of glycated products.

12.2 Applications of Plant-Mediated NPs in Medicine and Healthcare

Nanoparticles have found many applications in the field of Science and Technology. Generally three approaches are used to synthesize NPs: chemical, physical, and biological. In the last few years, there are several reports in the literature regarding the toxicity of non-bioNPs. Accordingly focus has shifted toward the synthesis of bionanoparticles using microbes, algae, and plants. Plant-mediated NPs have found various applications in areas of medicine and healthcare and are used for diagnostics, biological imaging, biosensors, and drug development (Husen and Siddiqi 2014b; Husen 2017; Siddiqi et al. 2018a, b). The antioxidant, antimicrobial, antimalarial, and antidiabetic properties of plant-mediated NPs have made them suitable for use in medicine. The antidiabetic activity of drugs and NPs is mostly concerned with decreasing the release of glucose, increasing the utilization of glucose and insulin release. However, with increasing evidence of glucotoxicity playing major roles in secondary complications of diabetes, focus has been shifted toward the use of these NPs in diagnosis and treatment of glycation. The main theme of this chapter is to highlight the mechanism of glycation, its prevention by natural agents and plant-mediated NPs.

12.3 Glycation

Diabetes mellitus, which affects around 1–2% of the world population, has come up as a significant medical problem. Diabetic patients are prone to long-term micro- and macrovascular complications such as cardiomyopathy, atherosclerosis, retinopathy, cataract, neuropathy, and nephropathy (Suji and Sivakami 2004). Hyperglycemia has an important role in the pathogenesis of long-term complications, and the diabetic patients with poor blood glucose control are particularly at risk. Many mechanisms have been worked out to show the correlation. Production of AGEs through protein glycation reaction is one such mechanism, which depicts the role of hyperglycemia in the pathogenesis of diabetic complications. The high blood glucose nonenzymatically interacts with intracellular proteins, leading to the generation of different heterogeneous AGEs. The AGEs formed in the plasma interact with receptors for AGEs (RAGE) and activate proinflammatory response. Several lines of evidence suggest that AGE/RAGE axis could profoundly be involved in diabetic complications, cardiovascular diseases, neurodegenerative diseases, cancer, and aging (Daroux et al. 2010). Additionally, diffusion of AGEs out of the cell gives them an opportunity to react and modify the extracellular matrix molecules present in the vicinity, thereby causing cellular dysfunction since this affects the signalling between the matrix and the cell (Smit and Lutgers 2004). Elevated production of the glycation precursors, namely, the dicarbonyls methylglyoxal (MG) and glyoxal, is witnessed in hyperglycemia, aging, cancer, and neurodegeneration. This leads to increase in AGEs and the subsequent pronounced molecular glycation damage (Pun and Murphy 2012).

12.3.1 Biochemistry of Glycation

Glycation is a nonenzymatic reaction in which carbonyl groups of sugar react with amino group of proteins and nucleic acids. Research on glycation began with the discovery by Louis Camille Maillard that heating amino acids and reducing sugars together result in a color change to yellowish brown (Thorpe and Baynes 1996). For this reason, the process of glycation is also known as "Maillard reaction" (Ashraf et al. 2016). Glycation involves posttranslational modification of proteins which is responsible for various diseases such as diabetes, cataract, Alzheimer's, Parkinson's, dialysis-related amyloidosis, atherosclerosis, physiological aging, etc. (Suji and Sivakami 2004).

Glycation is initiated by the reversible formation of a Schiff base between a reducing sugar and the amino group of a protein, DNA and lipoproteins. The Schiff base, which is relatively unstable, undergoes rearrangement to form a more stable Amadori product, which in turn undergoes a series of reactions like oxidation, reduction, hydration, etc. to form AGEs. The accumulation of these AGEs in the tissues is thought to be involved in diabetic complications and aging (Ashraf et al.

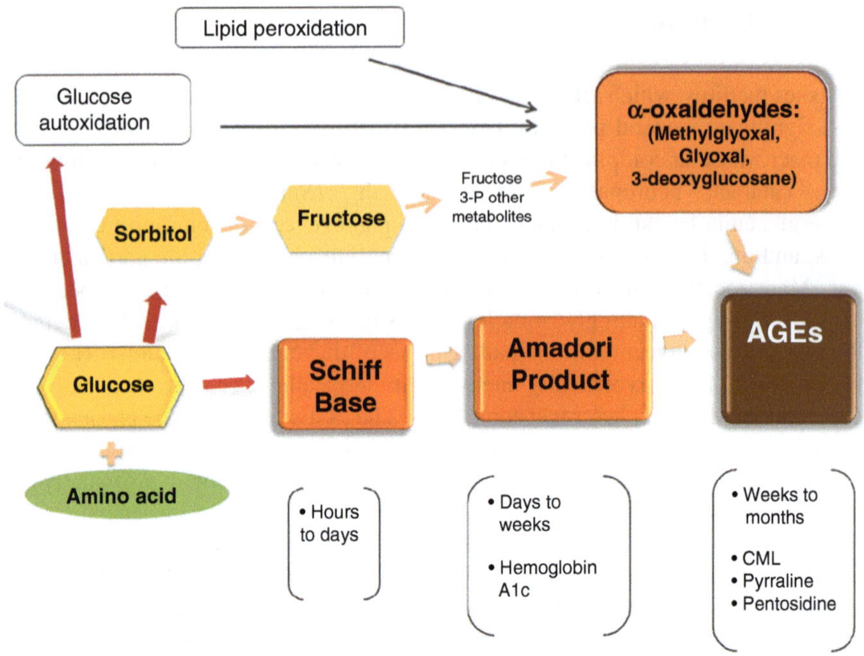

Fig. 12.1 Mechanism of the formation of advanced glycation end products. (Adapted from Luevano-Contreras and Chapman-Novakofski 2010)

2016). This process completes within few days, weeks, or months, and it is an irreversible process (Fig. 12.1). AGEs are very stable and therefore accumulate inside and outside the cells and interfere with the function of macromolecules.

The early and advanced glycation end products are continuously synthesized in the body, even at normal glucose levels. However, the deleterious effect of these products is observed due to their consequential accumulation after sometime and particularly when the blood glucose level increases above the normal range. Glycated products interfere with the homeostasis mechanism of the body, which results in various diseases (Ali and Sharma 2015). AGE formation progressively increases along with the normal process of aging even in the absence of any disease. However, they are formed at an extremely accelerated rate in diabetic condition (Ashraf et al. 2014). There are three stages of glycation: early, intermediate, and late (Fig. 12.2).

Early Stage The carbonyl group of a reducing sugar interacts in a nonenzymatic way with an amino acid to form an unstable compound known as Schiff base (Nass et al. 2007). Sugars are reactive toward lysine residues, while dicarbonyls are mainly reactive toward arginine residues.

Intermediate Stage During this phase, the Schiff base may undergo hydrolysis and produce the original sugar and amino acid, or it may undergo cyclization, and then

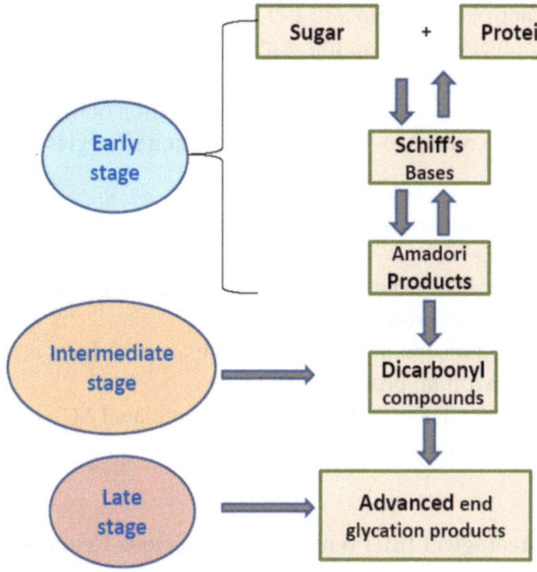

Fig. 12.2 The process of glycation comprising of three successive stages

Amadori rearranges to form Amadori products, which are relatively more stable compounds (Singh et al. 2014). However, under physiological and non-oxidative conditions, 90 percent of Amadori products may sustain a reversible reaction to the initial sugar and amino acid.

Late Stage In this phase, Amadori products can generate AGEs by oxidative or non-oxidative cleavage (Thorpe and Baynes 2003). The principal AGE produced in oxidative cleavage is carboxymethyl-lysine (CML), whereas dicarbonyl derivative 3-deoxyglucosone is produced in non-oxidative cleavage. This derivative can react with an amino acid and form CML or other AGE cross-links like pyrraline, pentosidine, imidazolone, etc.

12.3.2 Prevention of Glycation

The Amadori products and AGEs accumulate in the body even during the normoglycemic level as the process of glycation takes place continuously between the sugars and proteins. The amount of AGEs increases with an increase in the level of glucose in the blood, and, accordingly, the body is more severely affected by the deleterious effects of these products. AGEs bring about structural alteration of proteins and other biomolecules and in turn cause the functional loss. The other mechanism by which AGEs interfere with the normal functioning of cells is through the generation of reactive oxygen species. In the last few decades, efforts have been made to develop anti-AGE therapeutics. There are several stages at which AGE formation can be prevented. The simplest mechanism is to inhibit the formation of Schiff base by blocking one of the two reacting groups, carbonyl or amino. The other stages which can be interfered with are the formation of Amadori products and modification of these products to AGEs. Some common strategies for the prevention of glycation are:

(i) Use of inhibitors such as ascorbic acid, aspirin, metformin, etc., to prevent the formation of AGEs
(ii) Use of drugs, e.g., hydrazine, for deglycation and transglycation approach for Schiff bases/Amadori products
(iii) Reversal of AGE-induced modifications such as cross-links and aggregations, e.g., phenacylthiazolium bromide (PTB)
(iv) Prevention from the deleterious effects of accumulated AGEs in the body, e.g., resveratrol and curcumin

12.3.2.1 Pharmacological Intervention of Deleterious Reactions

The deleterious effects of glycation products on the human health have been discussed above. These can be prevented by inhibiting their accumulation and associated damage to the biomolecules. The AGE inhibitors share a common feature in possessing a nucleophilic group such as amine or hydrazine that can react with intermediate carbonyl compounds formed during the process of glycation and AGE formation (Suji and Sivakami 2004).

Some of the approaches commonly accepted for prevention of the damaging effects of AGEs include the use of synthetic and natural compounds to inhibit the progression of glycation and repair the damage induced by AGEs (Ali et al. 2017). There are many classes of drugs which can help in the prevention of AGE formation; some of these act by preventing the glycation-induced oxidative damage of proteins and DNA (Ali and Sharma 2015). The radical-trapping antioxidants and metal ion chelators belong to these classes of inhibitors. However, it is very difficult to reduce significantly the accumulation of AGEs by one class of inhibitors. At the same time, it can be assumed that a single inhibitor may exert its effect by more than one mechanism. The reason behind all these ambiguities is that the exact mechanism by AGEs-caused damage to the biomolecules has not been elucidated and the mechanism by which the inhibitors prevent the formation of AGEs is also not very clear. Several attempts have been made in the recent past to develop drugs, which can be used as multifunctional AGE inhibitor (Suji and Sivakami 2003).

12.3.2.2 Classes of AGE Inhibitors

Several classes of inhibitors have been identified on the basis of their mechanism of action (Abbas et al. 2016). Some of these are summarized in Table 12.1.

Table 12.1 Major class of AGE inhibitors, their mode of action, and examples

S. No.	Type of inhibitor	Mechanism of inhibition	Examples
1.	Inhibition of sugar attachment with proteins	Modification of sugar or protein molecules or ability to compete for the amino groups on the protein	Aspirin, diclofenac, pyridoxal-5-phosphate, metformin, pioglitazone, pentoxifylline, etc.
2.	Inhibition by using antioxidants (radical scavengers)	Suppression of AGE formation by attenuating glycol oxidation and preventing oxidative stress	Calcium antagonists, amlodipine, quinine, acetylsalicylic acid, ibuprofen, etc.
3.	Inhibition by metal chelators	Reduction of the metal ion catalyzed free-radical generation	Deferoxamine, DETAPAC, phytate, etc.
4.	Inhibition of dicarbonyl intermediates	Ability to scavenge both reactive carbonyls and reactive free radicals formed during glycation	Aminoguanidine, pyridoxamine, thiamine pyrophosphate, etc.
5.	Inhibition of Amadori product formation	Reaction with the sugar-derived moieties of glycated proteins and Amadori products, blocking of AGE receptors, i.e., RAGE	Tenilsetam and ethanol, antibodies against Amadori products, etc.
6.	Inhibition of AGE and protein cross-links	Breaking the cross-linking in the formed AGEs	N-Phenacylthiazolium bromide, alagebrium (ALT-711), TRC4186, etc.

12.4 Natural Inhibitors

The naturally occurring phytochemicals/products have been found to be relatively safe for human consumption, as compared to synthetic compounds. Natural compounds are relatively nontoxic and inexpensive and can be made available in an ingestible form. A large number of plants and natural biomolecules have shown antidiabetic effects. Plant extracts have been tested for antiglycating activities, but the mechanism is yet to be fully understood. It is well established that glycation and AGEs formation are accelerated and followed by oxidative stress. The antioxidant compounds may likely be promising agents for the prevention of glycation and AGE formation. The polyphenolic compounds, especially the flavonoids, have received the maximum attention with special focus on antidiabetic properties (Soumyanath 2006).

12.4.1 Natural Antiglycating Agents

There are numerous medicinal herbs and dietary plants that have been reported to possess antiglycating potential of similar or even higher order than that of aminoguanidine, an artificial AGE inhibitor (Kang et al. 2008). Plant-derived natural products possess significant antiglycating potentials (Ali et al. 2014). Studies depict that antiglycating potential is correlated with total phenolics present in plant extracts (Hsieh et al. 2007). The methanolic extracts of whole plants of *Calendula officinalis* and fruits of *Juglans regia* have shown antiglycating activity with respect to bovine serum albumin. Similarly, ethyl acetate extracts of *Erigeron annuus* inhibited glycation of BSA, prevented opacification of lenses and inhibited aldose reductase in the in vitro experiments (Jang et al. 2010). The extract of *Empetrum nigrum* L. inhibits glycation in vitro, and its antiglycating activity can be correlated with radical scavenging activity (Harris et al. 2014). In the in vitro conditions, maltol showed a significant inhibiting activity, as compared with aminoguanidine (Kang et al. 2008). Thus, antiglycating activity is exhibited by several plant species.

The polyphenolic compounds are the natural phytochemicals and are common constituents of plant-based foods which include fruits, vegetables, cereals, nuts, seeds, and chocolate and beverages like tea, coffee, and wine. The consumption of polyphenolic compounds is associated with several health benefits such as the prevention of cancer (Landis-Piwowar et al. 2007), neurodegenerative diseases (Mandel and Youdim 2004), cardiovascular diseases (Vinson et al. 2006), and diabetes (Kowluru and Kanwar 2007). Polyphenols are classified on the basis of source of origin, biological functions, and chemical structures, whereas chlorogenic acids present in *Chrysanthemum* species act as free-radical and metal scavengers and can interfere with the absorption of glucose and alter gene expression of antioxidant enzymes (Fiuza et al. 2004). The derivatives of cinnamic acid such as ferulic acid (3-methoxy-4-hydroxycinnamic acid) also display AGEs' inhibiting activity (Banan and Ali 2016; Meeprom et al. 2013). Ellagic acid also prevents glycation-mediated beta sheet formation in hemoglobin and the lysozyme that shows its ability of antiglycation (Torres-Piedra et al. 2010). It has the ability to encage MG and glyoxal and can thus inhibit AGEs formation.

12.4.2 Mechanisms for Inhibition of Glycation

Complexity of Maillard reaction is the major hurdle in identifying the mechanism behind inhibition of glycation by molecules and products of natural origin. It can be stated that AGEs are the major pathogenic culprits for diabetes and its complication. Several mechanisms have been proposed for inhibition of glycation by means of plant products and natural compounds that target essential stages of glycation.

These are certain mechanisms which can correlate antiglycating activity with antidiabetic potential of plants and their compounds (Tupe et al. 2016). The mechanisms include antiglycemic or hypoglycemic actions of plant products and their compounds (e.g., *Albizia odoratissima*, *Allium cepa*); inhibition of Amadori product formation (e.g., *Salacia chinensis*, etc.); inhibition of the formation of AGEs and its precursors (*Ilex paraguariensis*); reduction of cross-linking (green tea extract), radical scavenging and antioxidant activity (extracts of wild berries), and scavenging of dicarbonyl compounds (catechin and epicatechin, procyanidin, B2 isolated from cinnamon bark extract); etc.

12.5 Plant-Mediated NPs in Detection of Diabetes

We have discussed the significance of bioNPs in medicine and the strategies for inhibition of glycation-mediated secondary complications of diabetes. However, the natural as well as synthetic inhibitors have their own limitations. Synthetic drugs are associated with certain limitations like high cost, gastrointestinal disturbances, liver toxicity, development of hypoglycemia, fatigue, weakness, shortness of breath, nausea, dizziness, kidney toxicity, lactic acidosis, etc. (Modak et al. 2007). Renal patients are not allowed to take certain specific type of synthetic drugs. Limitations of natural inhibitors, on the other hand, include lack of dose-dependent standardization of inhibitors for their efficacy and safety (Ernst 2005). Different types of NPs, such as glucose sensors, glycated protein (Hb and albumin) sensors, protein oxidation sensors, and AGE's sensors, are used for assessment of glycation-induced complications of diabetes.

In the field of diabetes management, nanotechnology is applicable to glucose monitoring, insulin delivery, drug delivery, and wound healing. Fluorescent glucose nanosensors provide continuous glucose monitoring in contrast to the conventional finger prick tests. Whereas the conventional sensors are based on enzymes like glucose oxidase, the nanosensors incorporate the same enzyme, an oxygen-sensitive fluorescent indicator and a fluorescent dye insensitive to oxygen as a reference (Xu et al. 2002). The other enzyme, hexokinase, known to bind glucose and induce a conformational change in the protein, was used in nanosensors which caused a 25% reduction in its intrinsic fluorescence (Hussain et al. 2005). Another major issue with diabetes control is insulin delivery. The low oral bioavailability and short half-life of insulin can be overcome by encapsulating it in NPs. Insulin-loaded NPs delivered orally demonstrate a sustained effect of decreasing the blood glucose level over a longer period of time, as compared to subcutaneous injections (Lin et al. 2007). Many drugs of diabetes management have a short half-life and poor absorption characteristics. Nanodrug delivery systems, such as gliclazide-loaded Eudragit (L100 and RS), can release the drug in a controlled manner for extended periods of time (Devarajan and Sonavane 2007). One major impact of diabetes is the slow wound healing. Nanofibers have exhibited higher wound-healing rates in comparison to controls; the poly-n-acetyl glucosamine (sNAG) biodegradable nanofibers were found highly effective. Epidermal growth factor (rhEGF)-conjugated nanofibers have also been used for in vivo wound healing of diabetic ulcers in mice (Choi et al. 2008).

12.5.1 Glucose Sensors

Monitoring of glucose in an individual is the key factor in the management of diabetes. Therefore, the last four decades have observed evolution of glucose meters, noninvasive glucose monitoring (NGM) devices and continuous glucose monitoring systems (CGMS) (Rahisuddin 2018). With reference to the use of nanotechnology, the following two primary approaches have been incorporated into glucose sensors (Cash and Clark 2010).

In the first approach, the sensors can be designed inclusive of macro- or microscale components (such as electrodes and supporting hardware) but include either a nanostructured surface or a nanomaterial. The advantages of these designs are high surface area (leading to increase in current and prompt responses) and enhanced catalytic activities. These approaches with modified designs would be implemented for continuous monitoring of glucose. However, these sensors need to be studied thoroughly for their fouling and shelf life before implementation.

In the second approach, nanofabrication techniques can be used to make glucose sensors that are nanoscale in all dimensions. The advantages of these designs are that they can be used as injectables with an ease in implantation and administration. In this approach, the shelf life is potentially longer than the earlier one. However, these sensors have a limited clinical data and hence need to be researched more before implementation on a commercial scale.

12.5.2 Glycated Protein Sensors

In a recent study, Ghosh et al. (2017) have investigated an optical sensor comprising of DNA aptamer, semiconductor quantum dot, and AuNPs for the detection of glycated albumin (GA). The system "turn on," due to increase in photoluminescence intensity, was caused due to addition of GA to the sensor. This might be possible due the structure of DNA aptamer, which undergoes folding to form a hairpin loop, before the addition of analyte. In order to bind to GA, this loop is supposed to open up after the addition of the target to the sensor.

This pushes the quantum dot and the AuNPs away leading to increase in photoluminescence. A linear increase in photoluminescence intensity and quenching efficiency of the sensor is observed as the GA concentration is changed. The present work demands further studies with higher number of clinical samples to be effectively and largely employed in efficient diagnosis and monitoring of diabetes mellitus.

12.5.3 Protein Oxidation Sensors

For the electrochemical analysis of proteins, a number of sensors have been developed based on techniques including the direct as well as indirect electrochemistry following a selective reaction. Carbon nanotubes have been introduced to utilize their faster electron transfer kinetics and to provide a wire to the redox site of a protein. Applications cover a very broad range of proteins (Jacobs et al. 2010).

12.5.4 AGE's Sensors

The vanadium oxide nanoplates synthesized through microwave assistance were used as an interface material in the fabrication of modified Au working electrode for electrochemical MG (predominant precursor of AGEs) biosensor. These nanosensors showed a very high sensitivity with a linear range of 3–30 μM and a response time less than 8 s toward MG. The lifetime and percentage recovery of the sensor were found to be 25 days and 102.5–108.7%, respectively (Bhat et al. 2008). Previously, Ghosh et al. (2007) demonstrated the application of gold NPs synthesized on a protein template in the sensing of AGEs. This sensing property of gold NPs of glycated protein was confirmed using the techniques like transmission electron microscopy, surface plasmon resonance, CD, and FTIR.

12.6 Plant-Mediated NPs in Prevention of Glycation and Treatment of Diabetes

The most commonly used NPs for the treatment of glycation-induced diabetic complications are Ag, Au, and Se. The general strategy for checking the effect of NPs on glycation is to incubate NPs with the glycation system (protein + sugar) and then compare the amount of glycation products in the presence and absence of NPs. The glycation products are measured by several established methods including the measurement of browning, fructosamines, carbonyl content, total AGEs by spectrofluorimetry, HPLC, protein structural characterization by CD, and gel electrophoresis (Ali et al. 2017). The NP is classified as an antiglycating agent if there is a significant decrease in the AGEs in the glycation system.

In one report, Pickup et al. (2008) emphasized that NPs are potent therapeutic agent to control diabetes with very few side effects. They asserted that the AgNPs were efficient in control of the sugar level of 140 mg dl^{-1} in mice successfully. Manikanth et al. (2010) found that α-amylase inhibitory components are abundantly present in the ethanolic extract of *Sphaeranthus amaranthoides*. In a similar study, AgNPs synthesized by using the same plant, Swarnalatha et al. (2012) reported that these NPs inhibited a-amylase and acarbose sugar in diabetes-induced animal model. In the same year, Daisy and Saipriya (2012) found AuNPs to have high therapeutic effects against diabetic models. The AuNPs used were significantly able to reduce the level of liver enzymes such as alanine transaminase, alkaline phosphatase, serum creatinine, and uric acid in treated diabetes mice. Also, these diabetic models treated with AuNPs showed a decrease in HbA1c (glycated hemoglobin).

Collagen has been a protein of interest to study glycation because of its stability, great abundance in the body, and its application in cosmetic surgical treatments. Kim et al. (2012) reported the antiglycating effect of gold NPs on collagen. Gold NPs (nearly 20 nm) significantly decreased the level of glycation products in the glycated collagen sample. The results presented by Kim and colleagues (2012) also

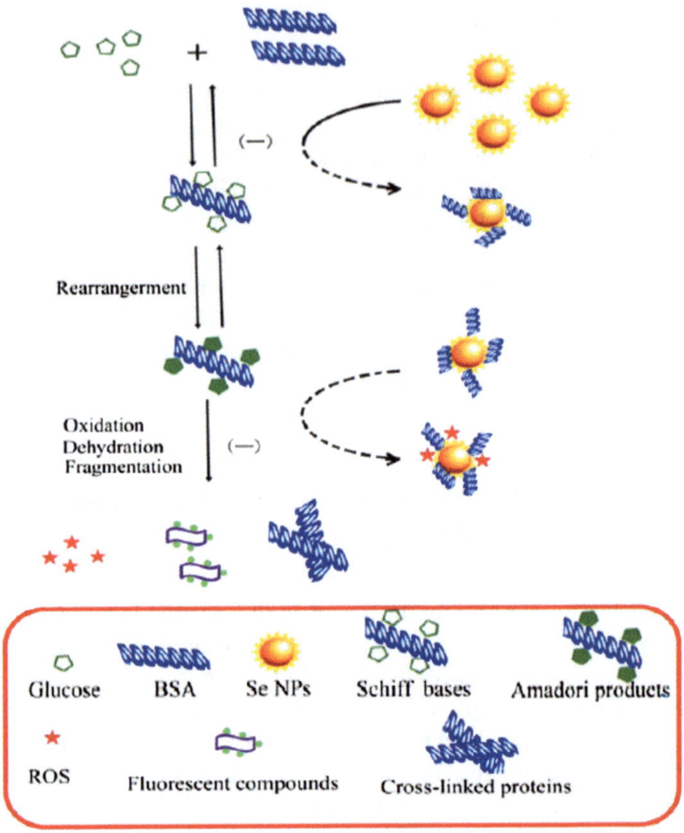

Fig. 12.3 Inhibitory effect of selenium NPs on glycated BSA. (Adapted from Yu et al. (2015)

suggest that gold NPs can be used for the prevention of glycation-induced skin aging.

In an initial effort on the use of selenium NPs in the prevention of glycation, Yu et al. (2015) studied the effect of SeNPs on BSA-glucose glycation system (Fig. 12.3) and found that SeNPs can prevent the progress of protein glycation in a concentration-dependent but time-independent manner under the specified reaction conditions (55 °C, 40 h). The mechanism inferred for the inhibitory efficacy of SeNPs might be related to NPs' (i) strong competitive activity against the available amino groups in proteins, (ii) very high scavenging activity on ROS, and (iii) inhibitory effect on the formation of α-dicarbonyl compounds. It was also proven that SeNPs protect proteins from structural modifications in the system and do not show any significant cytotoxicity toward BV-2 and BRL-3A cells up to 50 μg mL^{-1}. Thus, SeNPs may be extended to in vivo studies as the potent antiglycation agents.

Ashraf et al. (2014) observed inhibitory effect of gum arabic capped-AgNPs on AGEs formation, which proves the potential of these bioNPs to be an effective antiglycating agent. The mixtures of BSA and MG, incubated with different concentrations of NPs, caused significant reduction in AGEs, as confirmed by UV-Vis, fluorescence spectrometry, and HPLC techniques.

Leaf extract of *Solanum nigrum* was used to synthesize AgNPs, which were evaluated for antidiabetic activity in alloxan-induced diabetic rats. The AgNPs-treated diabetic rats could significantly improve the dyslipidemic condition, similar to diabetic control. Reduction in the blood glucose level was observed over the period of treatment. The body weight was also improved, showing the *S. nigrum* extract-mediated AgNPs as a potential antidiabetic agent against the alloxan-induced diabetes in rats (Sengottaiyan et al. 2016).

In another report (Ashraf et al. 2016), inhibitory strength of AgNPs, synthesized with *Aloe vera* leaf, in HSA (human serum albumin) glycation was shown. These NPs were characterized using UV-Vis spectroscopy, energy-dispersive X-ray spectroscopy, high-resolution transmission electron microscopy, X-ray diffraction, and dynamic light-scattering techniques. The inhibitory effects of AgNPs on AGEs formation were assessed by checking the degree of reactivity of free amino groups (lysine and arginine residues), protein-bound carbonyl and CML content, and the impact on protein structure. It was found that AgNPs significantly inhibited AGEs formation in a concentration-dependent manner and also had a positive effect on protein structure. It was recommended that AgNPs may play a therapeutic role in diabetes-related complications (Ashraf et al. 2016).

In a recent report, ZnONPs synthesized from aqueous extract of *Aloe vera* leaf were found to be a potent antiglycating agent, as they could inhibit the formation of AGEs and protect the protein structure from modification. This proves the therapeutic efficacy of ZnONPs in controlling the AGE-related complications (Ashraf et al. 2018).

Curcumin, an active ingredient of turmeric, has been shown to assist in wound healing in diabetic mice (Merrell et al. 2009). Curcumin treatment enhanced the biosynthesis of extracellular matrix proteins and also increased the formation of granulation tissue. However, the low in vivo stability and low bioavailability of curcumin make it difficult for oral administration. Nanofiber matrices are able to mimic the diameter of collagen fibrils in the extracellular matrix. Curcumin-loaded poly (ε-caprolactone) nanofibers were shown to reduce inflammation and enhance wound closure in vivo in a diabetic mouse model (Merrell et al. 2009). Poly-N-acetyl glucosamine (sNAG) nanofibers were shown to be effective in wound healing and are biodegradable (Scherer et al. 2009). Plant-mediated metal NPs that are used for management of secondary complications of diabetes are mentioned in Table 12.2.

In a recent report, leaf of *Stevia rebaudiana* along with chitosan was used to establish their antidiabetic potential in experimental rat model of streptozotocin (STZ)-induced diabetes mellitus. These NPs showed a significant reduction in rat's mean fasting blood glucose level, compared with the diabetic control group. Also, the serum levels of various enzymes, viz., serum glutamic oxaloacetic transaminase (SGOT), serum glutamic pyruvic transaminase (SGPT), alkaline phosphatases (ALP), lipid peroxidation, and antioxidant such as catalase (CAT), reduced glutathione (GSH), and superoxide dismutase(SOD), in these NPs-treated group were closer to normal levels than those in the diabetic control group (Perumal et al. 2016).

Table 12.2 Role of the plant-mediated NPs in glycation-induced diabetic complications

Name of the plant	Plant part	Metal used	Significance	Reference
Tephrosia tinctoria	Stem	Ag	NPs scavenged free radicals, decreased levels of enzymes that catalyze hydrolysis of complex carbohydrates (α-glucosidase and α-amylase), and increased the consumption rate of glucose	Rajaram et al. (2015)
Cassia fistula	Stem	Au	NPs-treated diabetic model showed a decrease of HbA1c level which is maintaining the normal range	Daisy and Saipriya (2012)
Sphaeranthus amaranthoides	Whole plant	Ag	Inhibitory activity on α-amylase and an IC50 was significantly lower than the standard drug, acarbose	Swarnalatha et al. (2012)
Sphaeranthus amaranthoides	Whole plant	Au	α-amylase inhibitory components are present in ethanolic extract of *S. amaranthoides*	Manikanth et al. (2010)
Acacia Senegal	Leaf	Ag	The mixtures of BSA and MG incubated with increasing concentrations of AgNPs showed significant reduction in AGEs	Ashraf et al. (2014)
Solanum nigrum	Leaf	Ag	AgNPs-treated diabetic rats showed significantly improved dyslipidemic condition as seen in the diabetic control. Blood glucose level was also reduced	Sengottaiyan et al. (2016)
Aloe vera	Leaf	Zn	Displayed strong capability of antioxidant and antiglycating agent as well as protected protein from damage by MG	Ashraf et al. (2018)

12.7 Conclusion and Future Prospective

Advanced glycation end products have been implicated in many pathophysiological conditions including diabetes, cataract, Alzheimer's, and Parkinson's diseases. Some artificial compounds have been tested during the last decade for their antiglycating potential, but a single drug, which can prevent the accumulation of AGEs, could not be developed so far. This has led to search for natural compounds and their derivatives with antidiabetic and antiglycating potential. Some compounds like phenolics, curcumin, and eugenol have shown promising results. Of late, NPs (both bio and non-bio) have been synthesized and characterized for their antidiabetic and antiglycating properties. Although the plant-mediated NPs with antiglycating potential are still very few, their advantage over the synthetic drugs makes them a suitable tool for diagnosis and prevention of glycation and glycation-mediated secondary complications of diabetes.

On the basis of in vitro and animal model studies, several plant products have been found to be effective supplements against glycation. It is therefore desirable to undertake their clinical trials on human beings in order to set their appropriate physiological concentration and understand their mode of action. Bioavailability of polyphenols is influenced by several factors such as bioaccessibility, transporters, molecular structures, metabolizing enzymes, etc. It is necessary to generate new techniques such as nanotechnology and homogenization, which can enhance bioavailability of natural inhibitors such as polyphenols. The NPs and nanoencapsulation-based albumins and polyphenols have been generated. In future more intense investigations are likely to be undertaken with regard to the long-term and acute hypoglycemic, antiglycating, and hypolipidemic effects of this homogenization, NPs, and NP encapsulation on type 2 diabetes.

References

Abbas G, Al-Harrasi AS, Hussain H, Hussain J, Rashid R, Choudhary MI (2016) Antiglycation therapy: discovery of promising antiglycation agents for the management of diabetic complications. Pharm Biol 54:198–206

Ali A, More TA, Hoonjan AK, Sivakami S (2017) Antiglycating potential of acesulfame potassium: an artificial sweetener. Appl Physiol Nutr Metab 42:1054–1063

Ali A, Sharma R (2015) A comparative study on the role of lysine and BSA in glycation-induced damage to DNA. Biosci Bioeng Commun 1:38–43

Ali A, Sharma R, Sivakami S (2014) Role of natural compounds in the prevention of DNA and proteins damage by glycation. Bionano Front 7:25–30

Ashraf JM, Ansari MA, Choi I, Khan HM, Alzohairy MA (2014) Antiglycating potential of gum arabic capped-silver nanoparticles. Appl Biochem Biotechnol 174:398–410

Ashraf JM, Ansari MA, Fatma S, Saleh MSA, Iqbal J, Madkhali A, Hamali AH, Ahmad S, Jerah A, Echeverria V, Barreto GE, Ashraf GM (2018) Inhibiting effect of zinc oxide nanoparticles on advanced glycation products and oxidative modifications: a potential tool to counteract oxidative stress in neurodegenerative diseases. Mol Neurobiol 55:7438. https://doi.org/10.1007/s12035-018-0935-x

Ashraf JM, Ansari MA, Khan HM, Alzohairy MA, Choi I (2016) Green synthesis of silver nanoparticles and characterization of their inhibitory effects on advanced glycation end products formation using biophysical techniques. Sci Rep 6:20414

Banan P, Ali A (2016) Preventive effect of phenolic acids on *in vitro* glycation. Ann Phytomed 5:97–102

Bhat BG, Younis H, Herrera J, Palacio K, Pascual B, Hur G, Jessen B, Ogilvie KM, Rejto PA (2008) Antisense inhibition of 11 betahydroxy steroid dehydrogenase type 1 improves diabetes in a novel cortisone-induced diabetic KK mouse model. Biochem Biophys Res Commun 365:740–745

Cash KJ, Clark HA (2010) Nanosensors and nanomaterials for monitoring glucose in diabetes. Trends Mol Med 16:584–593

Choi JS, Leong KW, Yoo HS (2008) *In vivo* wound healing of diabetic ulcers using electrospun nanofibers immobilized with human epidermal growth factor (EGF). Biomaterials 29:587–596

Daisy P, Saipriya K (2012) Biochemical analysis of *Cassia fistula* aqueous extract and phytochemically synthesized gold nanoparticles as hypoglycemic treatment for diabetes mellitus. Int J Nanomedicine 7:1189–1202

Daroux M, Prevost G, Maillard-Lefebvre H, Gaxatte C (2010) Advanced glycation end-products: implications for diabetic and non-diabetic nephropathies. Diabetes Metab 36:1–10

Devarajan PV, Sonavane GS (2007) Preparation and i*n vitro*/*in vivo* evaluation of gliclazide loaded eudragit nanoparticles as sustained release carriers. Drug Dev Ind Pharm 33:101–111

Ernst E (2005) The efficacy of herbal medicine - an overview. Fundam Clin Pharmacol 19:405–409

Fiuza SM, Gomes C, Teixeira LJ, Girão da Cruz MT, Cordeiro MN, Milhazes N, Borges F, Marques MP (2004) Phenolic acid derivatives with potential anticancer properties-a structure-

activity relationship study. Part 1: methyl, propyl and octyl esters of caffeic and gallic acids. Bioorg Med Chem 12:3581–3589
Ghosh S, Datta D, Cheema M, Dutta M, Stroscio MA (2017) Aptasensor based optical detection of glycated albumin for diabetes mellitus diagnosis. Nanotechnology 28:43
Ghosh MR, Bhattacharya J, Mitra CK, Basak S, Dasgupta AK (2007) Protein seeding of gold nanoparticles and mechanism of glycation sensing. Nanomedicine 3:208–214
Harris CS, Cuerrier A, Lamont E, Haddad PS, Arnason JT, Bennett SA, Johns T (2014). Investigating wild berries as a dietary approach to reducing the formation of advanced glycation endproducts: chemical correlates of in vitro antiglycation activity. Plant Foods Human Nutrition (Dordrecht, Netherlands) 69(1):71–77
Hsieh CL, Peng CH, Chyau CC, Lin YC, Wang HE, Peng RY (2007) Low-density lipoprotein, collagen, and thrombin models reveal that *Rosemarinus officinalis* L. exhibits potent antiglycative effects. J Agric Food Chem 55:2884–2891
Husen A (2017) Gold Nanoparticles from plant system: synthesis, characterization and their application. In: Ghorbanpour M, Manika K, Varma A (eds) Nanoscience and plant–soil systems, vol 48. Springer, Cham, pp 455–479
Husen A, Siddiqi KS (2014a) Carbon and fullerene nanomaterials in plant system. J Nanobiotechnol 12:16
Husen A, Siddiqi KS (2014b) Phytosynthesis of nanoparticles: concept, controversy and application. Nano Res Lett 9:229
Husen A, Siddiqi KS (2014c) Plants and microbes assisted selenium nanoparticles: characterization and application. J Nanobiotechnol 12:28
Hussain F, Birch DJS, Pickup JC (2005) Glucose sensing based on the intrinsic fluorescence of sol-gel immobilized yeast hexokinase. Anal Biochem 339:137–143
Jacobs CB, Peairs MJ, Venton BJ (2010) Carbon nanotube based electrochemical sensors for biomolecules. Anal Chim Acta 662:105–127
Jang DS, Lee YM, Jeong IH, Kim JS (2010) Constituents of the flowers of Platycodon grandiflorum with inhibitory activity on advanced glycation end products and rat lens aldose reductase in vitro. Arch Pharmacal Res 33(6):875–880
Kang KS, Yamabe N, Kim HY, Yokozawa T (2008) Role of maltol in advanced glycation end products and free radicals, *in vitro* and *in vivo* studies. J Pharm Pharmacol 60:445–452
Kim J, Hong C, Koo Y, Choi H, Lee K (2012) Anti-glycation effect of gold nanoparticles on collagen. Biol Pharm Bull 35:260–264
Kowluru RA, Kanwar M (2007) Effects of curcumin on retinal oxidative stress and inflammation in diabetes. Nutr Metab 4:8
Landis-Piwowar KR, Huo C, Chen D, Milacic V, Shi G, Chan TH, Dou QP (2007) A novel prodrug of the green tea polyphenol -epigallocatechin-3- gallate as a potential anticancer agent. Cancer Res 67:4303–4310
Lin Y-H, Chen CT, Liang HF, Kulkarni AR, Lee PW, Chen CH, Sung HW (2007) Novel nanoparticles for oral insulin delivery via the paracellular pathway. Nanotechnology 18:105102
Luevano-Contreras C, Chapman-Novakofski K (2010) Dietary advanced glycation end products and aging. Nutrients 2:1247–1265
Mandel S, Youdim MB (2004) Catechin polyphenols: neurodegeneration and neuroprotection in neurodegenerative diseases. Free Radic Biol Med 37:304–317
Manikanth SB, Kalishwaralal K, Sriram M, Pandian SRK, Hyung-seop Y, Eom SH, Gurunathan S (2010) Anti-oxidant effect of gold nanoparticles restrains hyperglycemic conditions in diabetic mice. J Nanobiotechnol 8:77–81
Meeprom A, Sompong W, Chan CB, Adisakwattana S (2013) Isoferulic acid, a new anti-glycation agent, inhibits fructose- and glucose-mediated protein glycation *in vitro*. Molecules 18:6439–6454
Merrell JG, McLaughlin SW, Tie L, Laurencin CT, Chen AF, Nair LS (2009) Curcumin loaded poly (epsilon-Caprolactone) nanofibers: diabetic wound dressing with antioxidant and anti-inflammatory properties. Clin Exp Pharmacol Physiol 36:1149–1156

Modak M, Dixit P, Londhe J, Ghaskadbi S, Paul T (2007) Indian herbs and herbal drugs used for the treatment of diabetes. J Clin Biochem Nutr 40:163–173

Mu Q, Jiang G, Chen L, Zhou H, Fourches D, Tropsha A, Yan B (2014) Chemical basis of interactions between engineered nanoparticles and biological systems. Chem Rev 114:7740–7781

Nass N, Bartling B, Navarrete Santos A, Scheubel RJ, Borgermann J, Silber RE, Simm A (2007) Advanced glycation end products, diabetes and ageing. Z Gerontol Geriatr 40:349–356

Paul J, Borm A, Kreyling W (2004) Toxicological hazards of inhaled nanoparticles -potential implications for drug delivery. J Nanosci Nanotechnol 4:1–11

Perumal V, Manickama T, Bang KS, Velmurugane P, Oh BT (2016) Antidiabetic potential of bioactive molecules coated chitosan nanoparticles in experimental rats. Int J Biol Macromol 92:63–69

Pickup JC, Zhi ZL, Khan F, Saxl T, Birch DJ (2008) Nanomedicine and its potential in diabetes research and practice. Diabetes Metab Res Rev 24:604–610

Pun PBL, Murphy MP (2012) Pathological significance of mitochondrial glycation. Inter J Cell Biol 843505:13

Rahisuddin A (2018) Metal nanoparticles as glucose sensor. In: Khan Z (ed) Nanomaterials and their applications. Advanced structured materials, vol 84. Springer, Singapore, pp 143–168

Rajaram K, Aiswarya DC, Sureshkumar P (2015) Green synthesis of silver nanoparticle using *Tephrosia tinctoria* and its antidiabetic activity. Mater Lett 138:251–254

Scherer SS, Pietramaggiori G, Matthews J, Perry S, Assmann A, Carothers A, Demcheva M, Muise-Helmericks RC, Seth A, Vournakis JN, Valeri RC, Fischer TH, Hechtman HB, Orgill DP (2009) Poly-N-acetyl glucosamine nanofibers: a new bioactive material to enhance diabetic wound healing by cell migration and angiogenesis. Ann Surg 250:322–330

Sengottaiyan A, Aravinthan A, Sudhakar C, Selvam K, Srinivasan P, Govarthanan M, Manoharan K, Selvankumar T (2016) Synthesis and characterization of *Solanum nigrum*-mediated silver nanoparticles and its protective effect on alloxan-induced diabetic rats. J Nanostruct Chem 6:41–48

Shaw JE, Sicree RA, Zimmet PZ (2010) Diabetes atlas global estimates of the prevalence of diabetes for 2010 and 2030. Diabetes Res Clin Pract 87:4–14

Siddiqi KS, Husen A (2016) Engineered gold nanoparticles and plant adaptation potential. Nano Res Lett 11:400

Siddiqi KS, Husen A (2017) Plant response to engineered metal oxide nanoparticles. Nanoscale Res Lett 12:92

Siddiqi KS, Husen A, Rao RAK (2018a) A review on biosynthesis of silver nanoparticles and their biocidal properties. J Nanobiotechnol 16:14

Siddiqi KS, Rahman A, Tajuddin, Husen A (2018b) Properties of zinc oxide nanoparticles and their activity against microbes. Nano Res Lett 13:141

Singh R, Kishore L, Kaur N (2014) Diabetic peripheral neuropathy: current perspective and future directions. Pharmacol Res 80:21–35

Smit AJ, Lutgers HL (2004) The clinical relevance of advanced glycation end products (AGE) and recent developments in pharmaceutics to reduce advanced glycation end products accumulation. Curr Med Chem 11:2767–2784

Soumyanath A (2006) Traditional medicines in modern times-anti-diabetic plants. CRC Press, Boca Raton/ London/ New York

Suji G, Sivakami S (2003) Approaches to the treatment of diabetes mellitus: an overview. Cell Mol Biol 49:635–639

Suji G, Sivakami S (2004) Glucose, glycation and aging. Biogerontology 5:365–373

Swarnalatha L, Rachel C, Ranjan S, Baradwaj P (2012) Evaluation of *in-vitro* antidiabetic activity of *Sphaeranthus amaranthoides* mediated silver nanoparticles. Int J Nanomater Biostruct 2:25–29

Thorpe SR, Baynes JW (1996) Role of the Maillard reaction in diabetes mellitus and diseases of aging. Drugs Aging 9:69–77

Thorpe SR, Baynes JW (2003) Maillard reaction products in tissue proteins: new products and new perspectives. Amino Acids 25:275–281

Torres-Piedra M, Ortiz-Andrade R, Villalobos-Molina R, Singh N, Medina-Franco JL, Webster SP, Binnie M, Navarrete-VÃ¡zquez G, Estrada-Soto S (2010) A comparative study of flavonoid analogues on streptozotocin-nicotinamide induced diabetic rats: quercetin as a potential antidiabetic agent acting via 11beta-hydroxysteroid dehydrogenase type 1 inhibition. Eur J Med Chem 45:2606–2612

Tupe RS, Kemse NG, Khaire AA, Shaikh SA (2016) Attenuation of glycation-induced multiple protein modifications by Indian antidiabetic plant extract. Pharm Biol 55:68–75

Vinson JA, Proch J, Bose P, Muchler S, Taffera P, Shuta D, Samman N, Agbor GA (2006) Chocolate is a powerful ex vivo and in vivo antioxidant, an antiatherosclerotic agent in an animal model, and a significant contributor to antioxidants in the European and American Diets. J Agric Food Chem 54:8071–8076

Xu H, Aylott JW, Kopelman R (2002) Fluorescent nano-PEBBLE sensors designed for intracellular glucose imaging. Analyst 127:1471–1477

Yu S, Zhang W, Liu W, Zhu W, Guo R, Wang Y, Zhang D, Wang J (2015) The inhibitory effect of selenium nanoparticles on protein glycation *in vitro*. Nanotechnology 10(26):145703

Chapter 13
Andrographis paniculata: From Traditional to Nano Drug for Cancer Therapy

Rabea Parveen, Bushra Parveen, Abida Parveen, and Sayeed Ahmad

13.1 Introduction

Dried stems and leaves of *Andrographis paniculata* (Burm. 128a.) Wall. ex Nees of the family Acanthaceae have long been used as a traditional herbal drug for a variety of human ailments. Its therapeutic effects are due mainly to the andrographolide, a labdane diterpenoidal compound (Chadha 1985; Chopra et al. 1956; Niranjan et al. 2010; Anonymous 2010). It is intensely bitter in taste and therefore known as "King of bitters." The plant is known as *Kalmegha* in Ayurveda, *Kiryat* in Unani, *Kalomeg* in Homeopathy systems of medicine, *Kalpnath* in Nepali medicine, *Chuan Xin Lian* in Chinese medicine, *Fah Tha Lai* in traditional Thai medicine, *Hempedu bumi* in Malaysian medicine, *Senshinren* in Japanese traditional medicine, *Green chiretta* in Scandinavian countries, and *Nain-e havandi* in Iranian folk medicine (Kumar et al. 2004; Barilla 1999; Handa and Sharma 1990).

Native to China, India, and Taiwan, *A. paniculata* is found throughout the tropical and subtropical Asia and Southeast Asia, including Cambodia, Indonesia, Laos, Malaysia, Myanmar, Sri Lanka, Thailand, Vietnam, and also the Caribbean islands (Niranjan et al. 2010; Wu et al. 1996; Benoy et al. 2012; Hossain et al. 2014). In India, it occurs largely in areas from Himachal Pradesh to West Bengal, Assam and Mizoram, and all over the South India.

The major phytoconstituents present in the herb are andrographolide, bisandrographolide A, isoandrographolide, neoandrographolide, homoandrographolide, andrographiside, andrographin, andrographan, andropanoside, andrographosterin, 14-deoxy-11,12-didehydroandrographolide, 14-deoxyandrographolide, panicolin, chlorogenic and myristic acids, and various methoxyflavones (Pholphana et al.

R. Parveen (✉) · B. Parveen · A. Parveen · S. Ahmad
School of Pharmaceutical Education and Research, Jamia Hamdard (Deemed University), New Delhi, India

A. Husen, M. Iqbal (eds.), *Nanomaterials and Plant Potential*,
https://doi.org/10.1007/978-3-030-05569-1_13

2013; Saxena et al. 2010; Chao and Lin 2010; Smith et al. 2006; Bhaskar et al. 2003; Radhika et al. 2010; Chao et al. 2010; Pholphana et al. 2004; Cheung et al. 2001; Pramanick et al. 2007). Andrographolide is the main ingredient responsible for its therapeutic effects. However, its effectiveness is hampered by its low aqueous solubility (3.29 ± 0.73 μg/mL), high lipophilicity (with log P value = 2.632 ± 0.135), and low bioavailability (Parveen et al. 2014).

13.2 Traditional Uses of *Andrographis paniculata*

Traditionally, the drug has been used as aperients, bitter tonic, and stimulant in Indian and some other traditional healthcare systems such as Chinese, Iranian, Japanese, and Thai systems of medicine (Kumar et al. 2014). *A. paniculata* has been reported to have antibacterial (Singha et al. 2003), antiviral (Calabrese et al. 2000), antiparasitic (Umar et al. 2012), and antifungal (Sule et al. 2012; Bhatnagar et al. 1961) activities and choleretic, hypoglycemic, hypocholesterolemic, adaptogenic, anti-inflammatory, emollient, astringent, diuretic, carminative, anthelmintic, antipyretic, and hepatoprotective properties (Jarukamjorn and Nemoto 2008; Parixit et al. 2012).

In China, the drug is used as a blood purifier and immunity enhancer. Because of its blood-purifying activity, it is used to treat gonorrhea, leprosy, scabies, and other dermal disorders like boils, eruptions, and patches (Deng 1978). Being bitter and cold, it is used as anti-inflammatory, antipyretic, detumescent, and detoxicant (Saxena et al. 2010) and provides a popular therapy for fevers, common cold, laryngitis, pharyngitis, tonsillitis, pneumonia, hepatitis, and respiratory infection (Chang and But 1987; Zhang 2000; Muluye et al. 2014; Chen et al. 2012).

In India, it has a traditional use in managing the liver disorders in Ayurveda (Chua 2014; Anonymous 2004; Khare 2007; Balachandran and Govindarajan 2005), against hepatic problems, leprosy, gonorrhea, scabies, boils, skin eruptions, and chronic and seasonal fevers in Unani (Kabeeruddin 1937; Dymock 1972; Kabir et al. 2014) and against general debility, dysentery dyspepsia, upper respiratory tract infection, and syphilis in Homeopathy (Rajpal et al. 2011; Carr and Nahata 2006). The leaf juice is used to relieve griping, loss of appetite, and irregular bowel habits both in infants and adults (Farooqi and Sreeramu 2001; Hossain et al. 2014).

It is used for the treatment of cough with thick sputum, carbuncle and sores, diarrhea, dysentery, pharyngolaryngitis, diabetes, hypertension, and snakebite in Malaysia (Burkill et al. 1966); for fever in Nepal (Kunwar et al. 2010); for common cold in Japan (Kumar et al. 2004); for fever, common cold, and noninfectious diarrhea in Thailand (Pholphana et al. 2004); for fever, common cold, upper lung infection, and flu in Scandinavian countries (Cáceres et al. 1997); and for skin problems, flu, respiratory disease, and snakebite in Iran (Ebrahimi and Ebrahimi 2012; Valdiani et al. 2012).

13.3 Pharmacological Potential

Apart from the above traditional uses of *A. paniculata*, recent studies on immunomodulatory and antitumor potential of andrographolide have identified it as a phytomolecule of tremendous therapeutic importance for managing a variety of disorders such as asthma, allergic reactions, atherosclerosis, cancer, common cold, diarrhea, fever, inflammation, hepatotoxicity, hyperglycemia, hypertension, intestinal disorders, leishmaniasis, microbial diseases, nervous disorders, respiratory system impairment, and urolithiasis, as illustrated in Table 13.1. Toxicological studies on animal model have proved it to be a safe phytoconstituent (Guo et al. 1988).

13.3.1 Effects on Digestive System

There is no effective therapeutic treatment for liver disorders in the modern system of medicine, only symptomatic treatment is available, and therefore some alternative or complimentary system is highly desirable. The therapeutic effect of *A. paniculata* is due to the presence of a diterpene lactone called andrographolide and some related terpenoidal analogues (Kishore et al. 2017). The hepatoprotective or hepatostimulant role of andrographolide has been discussed in the recent past (Shi et al. 2012; Singh et al. 2009; Kunwar et al. 2010). Khamphaya et al. (2016) studied its protective effect on the alpha-naphthylisothiocyanate (ANIT)-induced liver injury in a rat model and noted a decline in serum alanine aminotransferase (SGPT), aspartate aminotransferase (SGOT), alkaline phosphatase, and γ-glutamyltransferase. It also lowered the expression of α-smooth muscle actin and NF-κB, indicating the protective effect against the ANIT-induced cholestatic liver injury (Khamphaya et al. 2016).

The mechanism underlying the protective effect of andrographolide was demonstrated by Mittal et al. (2016) against the H_2O_2-induced cell death, lipid peroxidation, and reactive oxygen species in HepG2 cells. They considered the antioxidant activity to be due to activation of heme oxygenase-1, regulated transcriptionally through Nrf2, due to inhibition of GSK-3β by phosphorylation. Andrographolide also activated the p38 MAP kinase, which caused upregulation of Nrf2. Binding to adenosine A2A receptor resulted in activation of Nrf2 expression (Mittal et al. 2016).

13.3.2 Effects on Cardiac System

Andrographolide was reported to have an antiatherogenic property as well as the capability to decrease the serum biomarkers in atherogenic serum that prevents the complications of atherosclerosis caused by *Porphyromonas gingivalis* (Al Batran

Table 13.1 Therapeutic indications of *A. paniculata* on different systems of human body and against microbial infection

S. no.	Therapeutic effect	Key references
Digestive system		
1	Antidiarrheal	Deng (1978); Gupta et al. (1990); Yin and Guo (1993)
2	Hepatoprotective	Joselin and Jeeva (2014); Jayakumar et al. (2013); Niranjan et al. (2010); Maiti et al. (2006)
3	Antioxidant	Verma and Vinayak (2008); Zhang and Tan (2000); Shen et al. (2000); Lin et al. (2009)
Cardiac system		
4	Coagulant	Wang et al. (1997)
5	Antihypertensive	Huang (1987)
6	Cardioprotective	Tan & Zhang (2004); Woo et al. (2008)
7	Antiatherogenic	Al Batran et al. (2014); Lin et al. (2018)
Urinary system		
8	Antiurolithiatic	Coresh et al. (2007)
9	Renoprotective	Singh et al. (2009)
Immune system		
10	Immunomodulatory	Puri et al. (1993); Panossian et al. (1999); Iruretagoyena et al. (2005); Burgos et al. (1997); Qin et al. (2006)
Reproductive system		
11	Contraceptive	Akbarsha and Manivannan (1993); Kamal et al. (2003)
12	Antifertility	Zoha et al. (1989); Akbarsha et al. (1990); Janarthanan (1990)
Nervous system		
13	Neuroprotective	Yen et al. (2013); Varela-Nallar et al. (2015); Tapia-Rojas et al. (2015); Serrano et al. (2014)
14	Alzheimer's disease	Geng et al. (2018)
Respiratory system		
15	Bronchodilator	Li et al. (2009); Ozolua et al. (2011); Peng et al. (2016a)
16	Anti-inflammatory	Guan et al. (2013); Liao et al. (2016)
Hyperglycemia		
17	Antidiabetic	Zhang et al. (2009); Li et al. (2015)
Microbial infections		
18	Antibacterial	Rahman et al. (1999); Zaidan et al. (2005); Xu et al. (2006); Voravuthikunchai and Limsuwan (2006); Mishra et al. (2009)
19	Antifungal	Bhatnagar et al. (1961); Sule et al. (2012)
20	Antiviral	Chang et al. (1991); Wiart et al. (2005); King Spalding (2006); Lin et al. (2008); Seubsasana et al. (2011); Tang et al. (2012)

et al. 2014). Recent studies have shown that diterpene lactones from *A. paniculata* (andrographolide, 14-deoxy-11, 12-didehydroandrographolide, 14-deoxyandrographolide, and sodium 14-deoxyandrographolide-12-sulfonate) possess cardioprotective activities. Andrographolide protected the cardiomyocytes against the reoxygenation injury or hypoxia and upregulated the glutathione level and the activity of antioxidant enzymes. Andrographolide was also found to upregulate the mRNA and protein levels of GCLC (catalytic subunit of GCL) and GCLM (modifier subunit GCL) (Woo et al. 2008).

One andrographolide analogue (14-deoxy-11, 12-didehydroandrographolide) showed a lesser vasorelaxation and hypotensive activity than verapamil (a Ca^{2+} channel blocker). Yoopan et al. (2007) suggested that the smooth vascular muscle is the main site of hypotensive effect of 14-deoxy-11, 12-didehydroandrographolide. The ethyl acetate fraction of *A. paniculata* and its major phytoconstituent andrographolide have shown hypotensive effect in the anaesthetized Sprague-Dawley (SD) rats, which involved modulation of α-adrenoceptors, histaminergic receptors, and autonomic ganglion (Zhang and Tan 1997).

The hypolipidemic effect of andrographolide was demonstrated in yolk emulsion-induced hyperlipidemic mice and high-fat emulsion-induced hyperlipidemic rats. There was a significant decrease in the plasma levels of triglycerides, total cholesterol, and LDL by andrographolide at dose of 100 mg/kg body weight. A significant reduction was also found in the plasma levels of SGOT and SGPT by andrographolide when compared to that of the positive group (simvastatin). The cardioprotective effect was found to be due to upregulation of eNOS expression and downregulation of iNOS expression (Yang et al. 2013).

Due to anti-inflammatory activity of andrographolide, it inhibits the oxLDL (oxidized low-density lipoproteins)-induced macrophage foam cell formation and lowers the oxLDL-induced lipid accumulation in these cells. It also decreases protein expression of CD36 and mRNA by inducing degradation of CD36 mRNA but enhances the expression of ABCA1 and ABCG1 proteins and mRNA. It also improves the DNA-binding activity and LXR nuclear translocation. The suggested mechanism of andrographolide action in atherosclerosis is the induction of ABCG1- and ABCA1-dependent cholesterol efflux and inhibition of CD36-mediated oxLDL uptake (Lin et al. 2018).

13.3.3 Effects on Respiratory System

Andrographolide has a use in the treatment of respiratory disorders such as chronic obstructive pulmonary disease (COPD) and asthma. Andrographolide activates the redox-sensitive antioxidant transcription factor Nrf2 (nuclear factor erythroid-2-related factor 2) in COPD, which is characterized by a reduced Nrf2 activity (Mittal et al. 2016). The antioxidant effect of the drug was observed in the cigarette smoke-exposed BEAS-2B bronchial epithelial cells wherein andrographolide decreased the lavage fluid cell counts and the levels of interleukin-1β, IP-10, MCP-1, and KC. It also decreased the levels of oxidative biomarkers like 8-isoprostane, 3-nitrotyrosine, and 8-OHdG and increased the activity of glutathione peroxidase and glutathione reductase in the lungs of cigarette smoke-exposed mice (Guan et al. 2013).

The LPS and IFN-γ in combination strongly upregulate the interleukin (IL-27) production, which is related to the steroid-resistant airway hyperresponsiveness (SR-AHR). Andrographolide restored the steroid sensitivity to block IL-27 production and AHR induced by LPS and IFN-γ, due to antioxidative activity of the drug. The LPS and IFN-γ markedly decreased the level of histone deacetylase (HDAC-2), an epigenetic enzyme mediating the steroid anti-inflammatory action. The drug significantly restored the nuclear level of HDAC2 and total HDAC activity and lowered the ratio of total histone acetyltransferase/HDAC activity in the LPS and IFN-γ-exposed mouse lungs. The results indicated that andrographolide could be a potential agent in re-sensitizing the steroid action in asthma (Liao et al. 2016).

Andrographolide and its analogs have also shown anti-inflammatory effects in a variety of inflammatory disease models. Among the different signaling pathways explored, the underlying mechanism is the inhibition of NF-κB activity and improvement in the antioxidant defense caused by andrographolide via Nrf2 activation (Tan et al. 2017).

In a study by Li et al. (2012), andrographolide significantly decreased the eosinophil granulocyte count in bronchoalveolar lavage fluid and significantly downregulated the expressions of eotaxin and interleukin (IL-5) in the bronchoalveolar lavage fluid, blood, and lung tissues, thus suppressing the infiltration of eosinophil granulocytes in the ovalbumin-induced asthma in a mouse model. In another study (Guan et al. 2011), it reversed the ovalbumin-induced increases in eosinophil counts, interleukins (IL-4, IL-5, and IL-13) in lavage fluid, and the serum IgE level in an ovalbumin-induced asthma model. It decreased the ovalbumin-induced airway eosinophilia, mast cell degranulation, mucus production, pro-inflammatory biomarker expression in lung tissues, and airway hyperresponsiveness.

Andrographolide showed relaxant effects on the tracheal smooth muscles of guinea pig. *A. paniculata* was effective against histamine-induced contraction, whereas andrographolide was more effective on carbachol-induced contraction. The mechanism of action involved the inhibition of Ca^{2+} influx into the smooth muscle cells of the trachea (Ozolua et al. 2011). The anti-inflammatory activity of andrographolide studied by the allergen-induced airway inflammation and airways hyperresponsiveness indicated that it decreases the airway hyperreactivity and inhibits the extensive infiltration of inflammatory cells in the lungs (Li et al. 2009).

The effect of andrographolide sulfonate, a water-soluble form of andrographolide available under the trade name of Xiyanping injection, was studied on lipopolysaccharide-induced acute lung injury/acute respiratory distress syndrome characterized by a widespread lung inflammation (Peng et al. 2016a). The airway inflammatory cell recruitment and histological alterations of the lungs were significantly reversed, and the levels of pro-inflammatory cytokines in the serum, bronchoalveolar lavage fluid, and lung tissues declined. The drug also inhibited the MAPK-mediated inflammatory responses (Peng et al. 2016b).

In a study of cigarette smoke-induced chronic obstructive pulmonary disease and nontypeable *Haemophilus influenzae*-induced inflammation, andrographolide decreased the lung's cellular infiltrates and the levels of CXCL1/KC, IL-1β, 8-OHdG, TNF-α, and matrix metalloproteinases (MMP-8 and MMP-9). This is due to an increased Nrf2 activation and a decreased repressor function of the Kelch-like ECH-associated protein 1 (Keap1), thus indicating the potential use of andrographolide in lung disorders (Tan et al. 2016).

13.3.4 Effects on Nervous System

Andrographolide depicted neuroprotective effects against the lipopolysaccharide-induced chemokine upregulation in mouse cortex and in the cultured primary astrocytes (Wong et al. 2016). Its efficacy was assessed for psychiatric and cognitive disorders on the streptozotocin-induced diabetic Charles Foster albino rats, showing that it is effective against the diabetes-related cognitive deficits and the increased activity of acetylcholinesterase (Thakur et al. 2016).

Andrographolide is a competitive inhibitor of GSK-3β (glycogen synthase kinase-3β), a key enzyme of Wnt/β-catenin signaling cascade, suggesting the use of this drug in Alzheimer's disease. It stimulated the neurogenesis process in the adult hippocampus of the APPswe/PS1ΔE9 transgenic mouse model (Varela-Nallar et al. 2015; Tapia-Rojas et al. 2015; Serrano et al. 2014). At a dose of 0.1 mg kg^{-1} body weight, it significantly decreased the activation of microglia and reduced the production of cytokines TNF-α, IL-1β, and pro-inflammatory factor, i.e., prostaglandin (PG E2) in the ischemic brain areas in rats with permanent middle cerebral artery occlusion (pMCAO) (Chan et al. 2010). To study the effect of andrographolide sulfonate in Alzheimer's disease, APPswe/PSENΔ9 double transgenic mice were used to imitate the Alzheimer's condition. The drug was administered to mice for 7 months before the onset of Aβ plaque. Spatial memory test showed that the treatment prevented the cognitive decline. Aβ deposits were unaffected, while synapse and hippocampus damage were significantly alleviated. The mitochondrial swelling and oxidative stress were decreased, indicating that this could be a promising drug for the treatment of Alzheimer's disease via mitochondria protection (Geng et al. 2018).

13.3.5 Effects on Urinary System

Studies on the protective effect of *A. paniculata* and andrographolide against the cyclophosphamide-induced urothelial toxicity have shown that the drug could reverse the inflammation caused by cyclophosphamide in the urinary bladder, lower the urinary protein and urea levels, and increase the glutathione content in the liver and bladder of mice with urothelial toxicity induced by cyclophosphamide. It also reduced the pro-inflammatory cytokine production and increased the level of interferon-gamma (Sheeja and Kuttan 2006).

The oral administration of 200 mg/kg and 400 mg/kg of *A. paniculata* leaves extract in ethanol showed a protective effect in cisplatin-induced renal toxicity. Serum chemistry was used to determine the markers of oxidative stress, antioxidant enzymes, and serum biomarkers. All the oxidative stress markers were significantly increased in the toxic group; the condition was reversed in the treatment groups. Histopathological study supported these observations. The results of immunohistochemistry showed a higher expression of KIM-1 and lower expression of Nrf2 in the toxic group. The KIM-1 expression was significantly decreased, whereas the Nrf2 expression was increased in the treatment groups (Adeoye et al. 2018).

13.3.6 Effects on Immune System

Several plant-derived compounds, such as andrographolide, capsaicin, colchicine, curcumin, epigallocatechol-3-gallate, quercetin, resveratrol, and genistein, have shown modulatory effects on the cellular and humoral immune functions in preclinical investigations, thus proving their clinical potential (Jantan et al. 2015).

Andrographolide modulates both the innate and adaptive immune responses by producing Ag-specific antibody and regulating the macrophage activation of the phenotypic polarization (Wang et al. 2010). The immunomodulatory activity of a mixture of andrographolide, 14-deoxyandrographolide, and 14-deoxy-11, 12-didehydroandrographolide was determined in a delayed-type hypersensitivity (DTH) induced by cyclophosphamide in mouse. The mixture enhanced the immune function by increasing the total WBC count and the number of plaque-forming cells in the spleen and depressed the hemagglutination antibody titer (Naik and Hule 2009). Andrographolide could also reduce the symptoms of autoimmune encephalomyelitis by inhibition of CD4+ T-cell activation by dendritic cells and the antibody responses directed to myelin antigens (Iruretagoyena et al. 2005). Moreover, it reduced the luciferase activity of the nuclear factor of activated T cells and interfered with the nuclear distribution of the same, which leads to activation of c-jun N-terminal kinase phosphorylation. Also, it decreased the anti-CD3-induced ERK5 and ERK1 phosphorylation in T cells (Carretta et al. 2009).

Andrographolide sodium bisulfate is a water-soluble analogue of andrographolide, possessing antioxidant, anti-inflammatory, and immunomodulatory activities. Its antiphotoaging effect was studied on the ultraviolet irradiation-induced premature aging of mice. The topical application of the drug showed protective effects against the UV-induced disruption of elastic fibers and collagen fibers (Zhan et al. 2016).

13.3.7 Effects on Reproductive System

A. paniculata has shown the contraceptive (Akbarsha et al. 1990; Janarthanan 1990; Akbarsha and Manivannan 1993; Kamal et al. 2003) and antifertility (Burgos et al. 1997; Mkrtchyan et al. 2005; Sattayasai et al. 2010) effects. The contraceptive effect was found to be due to termination of spermatogenesis in male albino rats (Akbarsha et al. 1990; Kamal et al. 2003). None of the female mice that consumed the food mixed with *A. paniculata* daily could conceive after mating with the untreated male mice of potential fertility, indicating that the drug has contraceptive effect (Zoha et al. 1989).

Andrographolide inhibited the cytokinesis of the dividing spermatogenic cell lines, which consequently stopped spermatogenesis (Akbarsha and Manivannan 1993; Kamal et al. 2003). Andrographolide showed marked decreases in protein content and significant increases in the acid phosphatase, alkaline phosphatase and cholesterol levels with fructose present in the reproductive systems of rats, indicating antifertility effect of andrographolide (Janarthanan 1990). Administration of *A. paniculata* showed a negative effect on the blood progesterone content in rats, although the testosterone level was significantly increased in mice after 4 weeks of treatment (Panossian et al. 1999). Allan et al. (2009) observed no significant change in the reproductive or fertility potential of male Wistar rats on administration of an extract of *A. paniculata*. Further research is required to confirm the double stand activities of *A. paniculata* and andrographolide on the male and female sexual behaviors.

13.3.8 Hypoglycemic Effects

Some antidiabetic drugs decrease the level of blood glucose directly or indirectly, in addition to protecting both mass and insulin-secreting functions of pancreatic cells. The same dual function was shown by the andrographolide-lipoic acid conjugate in the alloxan-treated mice by decreasing the blood glucose level and increasing the insulin level. Simultaneously, it protected the pancreatic β cells by decreasing the loss of β cells and their dysfunctioning and by stimulating the glucose transport protein subtype 4 membrane translocation (Zhang et al. 2009). Andrographolide-lipoic acid conjugate also increased the insulin sensitivity and decreased the homeostasis model assessment of insulin resistance in type 2 diabetic rats (Li et al. 2015).

The ethanolic extract of *A. paniculata* and andrographolide caused delay in glucose absorption by inhibiting the α-glucosidase and α-amylase enzymes in vitro, which was confirmed by in vivo studies on diabetic rats (Subramanian et al. 2008). The purified extract of *A. paniculata* and andrographolide showed antidiabetic effect on type 1 and type 2 diabetes mellitus models of rats, resembling the effects of metformin (Nugroho et al. 2012). Andrographolide also showed antidiabetic effect in the autoimmune nonobese diabetic mice by decreasing the production of interferon-γ, IL-2, and IL-17, increasing the production of IL-10 and transforming the growth factor-β. It also showed increased expression of GATA3 mRNA and decreased expression of T-bet and RORγt mRNA (Zhang et al. 2013). The andrographolide-enriched fraction of *A. paniculata* showed hypoglycemic effect in the neonatal streptozotocin-induced diabetic rats – a model of type 2 diabetic rats (Nugroho et al. 2014). Andrographolide attenuated the diabetes-associated cognitive deficits in the streptozotocin-induced diabetic Charles Foster albino rats. It lowered the increased acetylcholinesterase activity and oxidative stress in brain tissues, the symptoms related to diabetes (Thakur et al. 2016).

13.3.9 Some Other Effects

Antibacterial Effects *A. paniculata* has shown in vitro antibacterial activity against *E. coli*, *Bordetella pertussis*, *Legionella pneumophila*, *Pseudomonas aeruginosa*, *Salmonella typhimurium*, *Shigella sonnei*, *Staphylococcus aureus*, *Streptococcus pneumonia*, and *S. pyogenes* at very low concentrations (Xu et al. 2006). The aqueous extract of leaves showed activity against gram-positive (*S. aureus* and methicillin-resistant *S. aureus*) and gram-negative (*Pseudomonas aeruginosa*) bacteria (Zaidan et al. 2005). The ethanol extract of *A. paniculata* exhibited significant activity against *E. coli* (Voravuthikunchai and Limsuwan 2006). The IC_{50} value of methanol extract was 7.2μg mL^{-1} against *Plasmodium falciparum* (Mishra et al. 2009). The antibacterial action of andrographolide was observed by measuring the minimal inhibitory concentration against gram-positive bacteria. Radiolabeled uridine, thymidine, leucine, and N-acetyl glucosamine were used to check the effect on the biosynthesis of RNA, DNA, protein, and cell wall, respectively (Banerjee et al. 2017).

Antiviral Effects The antiviral activity of andrographolide has been reported against several viruses like herpes simplex virus 1 (HSV-1) (Wiart et al. 2005; Seubsasana et al. 2011), human immunodeficiency virus (HIV) (Chang et al. 1991), pestiviruses and flaviviruses (King Spalding 2006), and Epstein-Barr virus (EBV) (Lin et al. 2008). Extract of *A. paniculata* showed cytotoxic effects against the DENV1-infected Vero E6 cells (Tang et al. 2012) and P3HR1 cells (Lin et al. 2008).

There is no approved antiviral drug against dengue virus. However, andrographolide has shown antiviral effects against a variety of viruses, giving a hope of developing new antiviral drug with diverse effects. The activity of andrographolide against the dengue virus (DENV) serotype 2 and 4 was carried out in HepG2 and serotype 2 in HeLa cell lines only. The results showed significant anti-DENV activity in HepG2 as well as HeLa cell line (Panraksa et al. 2017). In a study of dengue vector *A. aegypti*, the percent mortality of larvae increased, whereas the number of eggs decreased with increasing concentration of andrographolide (Edwin et al. 2016).

The ethanolic extract of *A. paniculata* and 5.0 μg ml^{-1} of andrographolide inhibited the expression of Epstein-Barr virus lytic proteins, Zta, Rta, and EA-D, during the lytic cycle of virus in P3HR1 cells due to inhibition of transcription of two Epstein-Barr virus immediate-early genes (BZLF1 and BRLF1) that encode Zta and Rta, respectively. This indicates that the compound is potentially useful as an anti-Epstein-Barr viral drug (Lin et al. 2018).

Antimalarial Effects The chloroform extract of the drug at 0.05 mg mL^{-1} concentration completely inhibited the growth of malarial parasitic in 24 h of incubation; similar results were obtained with methanol extract at 2.5 mg mL^{-1} concentration in 48 h. In vivo studies also showed a high antimalarial activity of *A. paniculata* (Rahman et al. 1999; Dua et al. 1999). Andrographide obtained from *A. paniculata* has shown antimalarial activity against *Plasmodium berghei* NK65 in *Mastomys natalensis* (Misra et al. 1992).

Larvicidal Effects *A. paniculata* inhibited the growth of *Anopheles stephensi* larva and caused mortality in a concentration-dependent manner (Kuppusamy and Murugan 2009). Its ethanolic extract exhibited the ovicidal activity against *Aedes stephensi* (Chenniappan and Kadarkarai 2008). The methanol, chloroform, hexane, and ethyl acetate extracts of *A. paniculata* leaves showed inhibitory action against *Culex quinquefasciatus* and *A. aegypti* (Govindarajan 2011). Another study by Sheeja et al. (2012) suggested that these extracts have provided an eco-friendly method for the control of *Culex quinquefasciatus*.

13.4 Pharmacological Validation Through Clinical Studies

Several clinical trials have been conducted on andrographolide to prove the reported therapeutic activities. A double-blind, placebo-controlled, parallel-group clinical trial was conducted on 185 individuals to study the effect of *A. paniculata* extract SHA-10 fixed combination (Kan Jang®) for the treatment of acute upper respiratory tract infections. The total score analysis revealed a significant improvement in the test group versus the placebo. The Kan Jang® had a progressive effect against these infections and also showed improvement in the inflammatory symptoms of sinusitis (Gabrielian et al. 2002). Kan Jang® tablets (*A. paniculata* extract), developed by the Swedish Herbal Institute in Gothenburg, mitigate the symptoms of common cold as well as uncomplicated sinusitis. A controlled, double-blind clinical trial conducted on 50 patients at the Health Center, Hallehälsan, in 1992, showed a significant reduction in the subjective symptoms of common cold and the duration of symptoms (Melchior et al. 1997).

Ethanol extract of *A. paniculata* cured 88.3% cases of acute bacillary dysentery and 91.3% cases of acute gastroenteritis. Administration of a tablet, containing neoandrographolide and andrographolide in the ratio of 3:7, to the patients of bacillary dysentery also gave a similar result (i.e., 91.1% cure rate), which was higher than one obtained with furazolidone or chloramphenicol (Chang and But 1987).

Another clinical trial was conducted on 13 HIV-positive patients and 5 healthy volunteers. Andrographolide was given orally at a dose of 5.0 mg kg^{-1} body weight for the first 3 weeks and then 10 mg kg^{-1} body weight for another 3 weeks. The dose was increased to 20 mg kg^{-1} body weight for the last 3 weeks. The results showed significant improvement in the CD4+ lymphocyte count (from 405 to 501 cells per mm^3) in HIV-positive patients. No significant change was observed in the mean plasma HIV-1 RNA levels (Calabrese, et al. 2000).

In a randomized, double-blind, and controlled clinical trial, *A. paniculata* was administered to 152 Thai patients of pharyngotonsillitis at a dosage of 6 g per day for 7 days, and the result was similar to one obtained by using acetaminophen (Thamlikitkul et al. 1991).

In another trial, the treatment with *A. paniculata* extract SHA-10 reduced the intensity of the symptoms of tiredness (OR = 1.28; 95% CI 1.07–1.53), sleeplessness (OR = 1.71; 95% CI 1.38–2.11), sore throat (OR = 2.3; 95% CI 1.69–3.14), and HSP (OR = 2.51; 95% CI 1.82–3.46), as compared with the placebo group in a duration-dependent manner. Intensity of all the symptoms was significantly reduced in the 4-day-treatment group (Cáceres et al. 1999).

In a double-blind, placebo-controlled, and randomized study by Saxena et al. (2010), the extract of *A. paniculata* leaves (200 mg per day) or placebo was given to the 223 patients of both sexes having upper respiratory tract infection (URTI), and 9 self-evaluated symptoms (cough, earache, expectoration, fever, headache, malaise/fatigue, nasal discharge, sleep disturbance, and sore throat) were observed. In both the groups of test and placebo, the mean scores of all symptoms exhibited a decreasing trend from day 1 to day 3. However, from the third to the fifth day, some symptoms in the placebo group remained unchanged (cough, earache, and headache) or got aggravated (sleep disturbance and sore throat), whereas all symptoms were significantly reduced in the test group (p $\leq$0.05). The overall efficacy of the test over the placebo was significant ($p \leq 0.05$) with a 2.1 times (52.7%) improvement (Saxena et al. 2010).

In a prospective, double-blind, randomized, and placebo-controlled study based on 60 patients of rheumatoid arthritis (RA), tablets of an extract of *A. paniculata* were administered three times per day for 14 weeks, followed by a 2-week washout period. At the end of the treatment, intensity of joint pain decreased in the test group, as compared to the placebo group, although the variance was not statistically significant. A reduction was also observed in rheumatoid factor, C4, and IgA (Burgos et al. 2009).

A multicenter, randomized, and double-blind, 8-week parallel-group trial was conducted on 120 patients of ulcerative colitis at five centers in China. The extract of *A. paniculata* leaves was given to patients at a dose of 1200 mg per day and compared with that of 4500 mg per day of slow-release mesalazine granules. In the test group, clinical remission was seen in 21% of patients and response in 76%, whereas 16% of remission and 82% of response were recorded in the mesalazine-treated patients (Tang et al. 2011). By colonoscopy, however, the clinical remission and response were observed in 28% and 74% of the extract-treated patients, and in 24% and 71% of the mesalazine-treated patients, respectively (Tang et al. 2011).

Two double-blind, randomized, placebo-controlled parallel, group clinical trials were conducted on 46 patients of upper respiratory tract infections in Sweden. The patients were treated with a commercial formulation consisting of the standardized extracts of *A. paniculata* and *Eleutherococcus senticosus* in a fixed combination (Kan Jang), three times a day for a minimum of 3 days and a maximum up to 8 days. Another phase III study was conducted on 179 patients for 3 days. The results of both the studies showed a highly significant relief of throat symptoms in the drug-treated groups, when compared to the placebo-treated groups (Melchior et al. 2000). In a three-arm study, the effect of Kan Jang (as an adjunct to standard treatment) was also studied in uncomplicated respiratory diseased children (4–11 years) for 10 days. Patients receiving Kan Jang showed less severe symptoms and faster recovery and a significantly lesser requirement for the standard medication (Spasov et al. 2004).

The antidiarrheal effect of andrographolide from *A. paniculata* was studied on 80 patients of acute bacterial diarrhea in China. The drug was given at a total dose of 500 mg in divided doses (three times a day) for 6 days. The results showed a cure rate of 82.5%, and the effectiveness of the therapy was confirmed by laboratory tests of stool (Yin and Guo 1993). In another study, *A. paniculata* was used to treat 1611 patients of bacterial dysentery and 955 patients of diarrhea and showed 91.3% of effectiveness (Deng 1978).

In a randomized controlled trial conducted on 60 patients of hypertriglyceridemia, the patients were treated with high and low doses of andrographolide and compared with gemfibrozil. The significant reduction caused by andrographolide in triglycerides level was comparable to one caused by gemfibrozil (Phunikhom et al. 2015).

In a trial conducted on 46 patients of hand, foot, and mouth diseases, 28 received the conventional therapy, whereas 18 received a combination of andrographolide sulfonate plus the conventional therapy. Combination therapy reduced the elevated plasma levels of myeloperoxidase, histone, S100A8/A9, and interleukin-6 in comparison to the conventional therapy. The lipopolysaccharide-stimulated neutrophil activation was inhibited in vitro by the use of andrographolide sulfonate (Wen et al. 2015).

13.5 Andrographolide in Cancer Therapy

Most of the herbal drugs used for cancer therapy in the modern medicine act either directly by inducing apoptosis or indirectly as an immunomodulatory agent by triggering the body's immune system. Preclinical and clinical studies suggest a potential use of andrographolides in cancers. The compounds that have a multifaceted mode of action and inhibit the multiple pro-cancer events are of greater interest as they are more likely to target a wider range of cancers. Andrographolide has emerged as a strong therapeutic anticancer pharmacophore that acts both directly and indirectly on the cancer cells (Kumar et al. 2004; Rajagopal et al. 2003).

A. paniculata is cited in Ayurveda as a plant with anticancer properties, and it is easy to presume the action of andrographolide on similar principles. The compound reportedly induces a G0/G1 cell-cycle arrest in various kinds of cancer cells, activates the death-receptor pathways, induces TRAIL-mediated apoptosis, activates p53 via enhanced phosphorylation, and causes inhibition of NF-κB transcriptional factors and various angiogenic factors (Fig. 13.1). It also exerts strong immunomodulatory effects against cancer cells in addition to its cytotoxic effects, a property which is similar to other anticancer agents including doxorubicin, mitomycin, cisplatin, and others. Apart from acting on various pathways to obliterate cancer cells, it also exerts a protective effect on normal cells saving them from induced toxicity, in comparison to the contrary effect against cancer cells. These characteristics make it an interesting molecule for further research (Varma et al. 2011).

The role of andrographolide in gastric cancers has been elucidated in a recent in vitro study (Dai et al. 2017), employing the gastric cell line SGC7901 and investigating the mechanisms by which andrographolide inhibits the gastric cancer cell proliferation and metastasis. It showed the survival ratio of cancer cells decreasing with increase in concentration of andrographolide in a dose-dependent manner. Andrographolide treatment blocked the G2/M2 phase of cell cycle, enhanced the proportion of cells arrested at G1/M with increase in dose, and reduced the migration and invasion of the gastric cancer cells at various concentrations of andrographolide. Therefore, it can be surmised that andrographolide inhibits cell proliferation, invasion, and migration; blocks the cell cycle; promotes apoptosis in SGC7901 cells possibly via upregulated expression of TIMP-1/TIMP-2, cyclin B1, p-Cdc2, Bax, and Bik; and downregulates the expression of MMP-2/MMP-9 and anti-apoptosis protein Bcl-2 (Dai et al. 2017).

Studies have shown that andrographolide is also effective in combination therapy with flavonoid taxifolin that has antioxidant and antiproliferative effects against different cancer cells. Combination of andrographolide and taxifolin exerted synergistic effects on HeLa cells at various concentrations. Caspase-dependent apoptosis was enhanced by the combination, while andrographolide-induced reactive oxygen species (ROS)-dependent protective autophagy was attenuated by the addition of

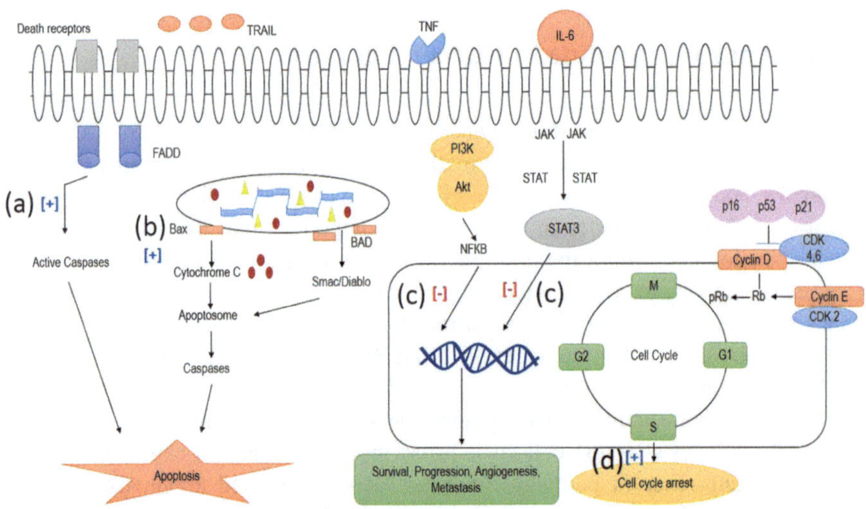

Fig. 13.1 Mechanism of the multi-target anticancer therapeutic approach of andrographolide: (**a**) activates the extrinsic death-receptor pathway and induces apoptotic cell death; (**b**) mitochondria-mediated apoptosis leading to cytochrome c release and apoptotic cell death; (**c**) inhibition of two critical transcription factors, NF-κB and STAT3, resulting in the inhibition of cellular proliferation, survival, and angiogenesis; (**d**) induces cell-cycle arrest in cancer cells at G0/G1 stage due to the induction of cell-cycle inhibitory proteins p16, p21, and p53 associated with a decreased expression of cyclin E, cyclin D, CDK4, CDK6, and CDK2, required for G1 to S transition

taxifolin. Activation of JNK was involved in the cell death of HeLa cells, while activation of p53 was involved in andrographolide-induced autophagy, which was decreased by the use of taxifolin. Mitochondrial outer membrane permeabilization (MOMP) involved in the andrographolide-induced cell death was further enhanced by the combination. This led to an increase in AIF and a cytochrome C release from mitochondria, which consequently increased the caspase-dependent and independent cell death (Alzaharna et al. 2017).

Andrographolide was recently found to reduce oncogenicity and restore radiosensitivity of ALDH1 + CD44+ oral cancer stem cells, which possess high tumorigenic and metastatic properties as well as the chemo- and radioresistance (Yang et al. 2017). Experiments with mice bearing xenograft tumors showed a reduced tumor growth after andrographolide treatment, which was attributed to the activation of miR-218/Bmil axis. This is evident in mechanistic studies showing that andrographolide treatment increased the expression of microRNA-218 (miR-218), leading to the downregulation of Bmi1, and in the knockout studies wherein deletion of miR-218 in the ALDH1-CD44-non-oral cancer stem cells enhanced the cancer stemness, while silencing of Bmi1 significantly counteracted it. This suggests that andrographolide is a valuable natural compound for anti-CSCs (cancer stem cells) treatment of oral squamous cell carcinoma (Yang et al. 2017).

Another example of combination treatment is that of andrographolide and vincristine with reference to nephroblastoma (Wilms' tumor), which is the most common malignant renal cancer in children. Wilms' tumor cells were employed to investigate the synergistic antitumor effect of andrographolide in combination with vincristine. It suppressed the SK-NEP-1 cell proliferation in vitro and inhibited the xenograft tumor growth in vivo. This was due to increased apoptosis, not autophagy. Moreover, the PI3K-AKT-p53 signaling pathway was involved in the process of combination treatment, which was confirmed when a selective AKT activator was applied (Zhang et al. 2016).

Various mechanisms are reported to be responsible for mediating the anticancer activity of andrographolides, including the inhibition of the Janus tyrosine kinases-signal transducers and activators of transcription, phosphatidylinositol 3-kinase, and NF-κB signaling pathways; the suppression of heat shock protein, cyclins, and cyclin-dependent kinases, metalloproteinases, and growth factors; and the induction of tumor-suppressor proteins p53 and p21, leading to the inhibition of cancer cell proliferation, survival, metastasis, and angiogenesis, activation of miR-218/Bmi1 axis, caspase-dependent, and independent apoptosis, upregulation of expression of TIMP-1/TIMP-2, cyclin B1, p-Cdc2, Bax, and Bik, and downregulation of expression of MMP-2/MMP-9 and anti-apoptosis protein Bcl-2. (Lim et al. 2012).

13.6 Nanoformulations of Andrographolide

Nanoformulations are the most promising drug delivery systems used to deliver herbal drugs for improvement of biopharmaceutic and pharmacokinetic properties of herbal drugs (Parveen et al. 2008). Therapeutic index of traditional drugs is enhanced via improved specificity due to targeting of drugs to a particular tissue, cell, or intracellular compartment, control over release kinetics, protection of active agent, or a combination of the above. Different nanocarrier systems are used in the novel drug delivery system (NDDS) like polymeric nanoparticles, solid lipid nanoparticles, nanocapsules, liposomes, niosomes, micelles, and dendrimers (Parveen et al. 2013). The aim of developing these formulations is to obtain systems with optimized drug loading and release properties, long shelf-life, and low toxicity.

If the herbs themselves or their purified phytopharmaceuticals are incorporated in the nano-drug delivery systems, we can get the benefits of both. Thus, it is important to incorporate the nano-drug delivery system in the Indian Ayurvedic and Unani medicines to combat serious diseases like cancer, diabetes, and hepatic damage (Parveen et al. 2010; Kamal et al. 2012). Great advances have been made in the development of NDDS for active plant metabolites and extracts in the recent past. A variety of novel herbal formulations like polymeric nanoparticles, nanocapsules, liposomes, phytosomes, nanoemulsions, microsphere, transfersomes, and ethosomes have been developed using bioactive compounds and plant extracts. These formulations have remarkable advantages over the conventional ones, including the improved solubility and bioavailability, protection from toxicity, enhanced pharmacological activity, increased stability, improved tissue macrophages distribution, sustained delivery, and protection from physical and chemical degradation (Parveen et al. 2011).

Out of several compounds isolated from *A. paniculata*, andrographolide is effective against numerous pharmacological disorders, but its effectiveness is hampered due to its low aqueous solubility ($3.29 \pm 0.73\ \mu g\ ml^{-1}$), poor cellular permeability, short biological half-life, and high lipophilicity having log p value = 2.632 ± 0.135, efflux by P-glycoprotein, and a very low oral bioavailability. Different formulation approaches have been reported to improve its biopharmaceutic properties and thereby increase its oral bioavailability (Parveen et al. 2014). Nanoformulations of andrographolide have been prepared for the treatment of cancer (Parveen et al. 2014), hepatotoxicity (Das et al. 2015), and leishmaniasis (Roy et al. 2010).

Niosomes of andrographolide were prepared by the film hydration method, and their tissue distribution was determined by liquid chromatography mass spectrometry (LCMS) in mice. It revealed the maximum absorption in the liver. The cytotoxic potential of andrographolide niosomes was analyzed by MTT assay on HepG2 cell line which showed no significant difference between the pure drug and its niosomal form (Tu et al. 2014).

Solid lipid nanoparticles of andrographolide were prepared by the solvent injection method to increase its oral bioavailability by the UPLC/Q-TOF-MS method. An improvement was observed in the pharmacokinetic parameters of the andrographolide nanoparticles in comparison to andrographolide. The antitumor activity was also studied on the Balb/c mice, showing better results with andrographolide nanoparticles than with andrographolide (Parveen et al. 2014).

Another study reports preparation of solid lipid nanoparticles to deliver andrographolide into the brain. The particles were formulated by emulsion/evaporation/solidifying method, using Compritol 888 ATO as a lipid and Brij 78 as surfactant.

Nanoparticles were spherical in shape, with narrow size distribution, 92% encapsulation efficiency, and physically and chemically stable. In vitro dissolution study at pH 7.4 showed a prolonged and sustained release of the drug from solid lipid nanoparticles. The results of in vitro permeation to blood-brain barrier using parallel artificial membrane and hCMEC/D3 cells proved that nanoparticles improve drug permeability in comparison to the free drug. After i.v. administration to healthy rats, fluorescent particles were detected in the brain parenchyma outside the vascular bed, indicating their ability to overcome the blood-brain barrier (Graverini et al. 2018).

Andrographolide was also incorporated in a poly D,L-lactide-*co*-glycolide-*b*-poly ethylene glycol-*b*-poly D,L-lactide-*co*-glycolide (PLGA-PEG-PLGA) amphiphilic triblock copolymer-based micelle formulation for enhancement of bioavailability and anticancer efficacy in vivo. The andrographolide-loaded micelles showed a higher percent proliferation inhibition, pro-apoptosis effects, and cell-cycle arrest at G2/M phase in MAD-MB-231 cell line, indicating better efficacy for intracellular transport and cellular uptake. In the in vivo pharmacokinetic studies, 2.7-fold and 2.5-fold increase was observed in plasma AUC and mean resident time, respectively, by andrographolide micelles, when compared to drug suspension (Zhang et al. 2014).

A semisynthetic analogue (19-triphenylmethyl ether andrographolide) has been reported to have the cytotoxic effect against many cancer cell lines. It was encapsulated in poly (ethylene glycol)-*b*-poly (D,L-lactide) (PEG-*b*-PLA) micelles due to its poor aqueous solubility. It increased the aqueous solubility of the drug by 280-fold. The polymeric micelles exhibited sustained drug-release pattern and excellent cytotoxicity against cholangiocarcinoma KKU-M213 cell line, with IC_{50} value of 3.33 μM (Puntawee et al. 2016).

Kim and co-workers have prepared an aqueous soluble nano-complex for evaluating the anticancer potential in vitro and in vivo. A boronic ester was formed between the cis-1, 3-diol of andrographolide with polymer. The drug release from loaded nano-complex was controlled by high ATP concentrations and low pH. The phenylboronic acid-andrographolide nano-construct showed excellent tumor targeting in vitro as well as in vivo, and a significant tumor-growth inhibition was observed in the in vivo studies (Kim et al. 2016).

Andrographolide encapsulated in mannosylated and fucosylated liposomes was used to target the macrophage-related diseases like leishmaniasis in the hamster model. Mannosylated liposomes were more potent in delivering andrographolide to phagocytic cells by reducing the parasitic burden in the spleen, and also decreased the renal and hepatic toxicity, as compared to the pure drug (Sinha et al. 2000). Likewise, Roy et al. (2010) prepared the poly D,L-lactide-co-glycolic acid-loaded andrographolide nanoparticles, using polyvinyl alcohol as stabilizer, for drug delivery into monocyte-macrophage cells infected with leishmanial amastigote. Significant antileishmanial activity was observed with one-fourth of the dose of andrographolide in nanoparticles. The IC 50 value of andrographolide in nanoparticles was found to be 160 μM with good tissue localization. Andrographolide nanoparticles may thus bring an effective low-cost alternative therapy for leishmaniasis (Roy et al. 2010).

Another analogue of andrographolide, the 14-deoxy-11-oxo-andrographolide, which shows antileishmanial activity, has been evaluated for its effect in the form of subcutaneous injection on hamster model. Different formulation approaches were used and compared with the conventional form. The conventional dosage form decreased the parasitic load of the spleen by 39%, whereas nanoformulations could cause 78% decrease by liposomes, 91% decrease by niosomes, and 59% decrease by microspheres. The reduction in spleen parasite load was inversely proportional to the size of the carrier. Because of this improved efficacy and lower toxicity, these nanocarriers offer a better approach for treating the visceral leishmaniasis (Lala et al. 2003). The available drugs for leishmaniasis are toxic in nature and have low bioavailability; their incorporation into nanocarriers will be a better option for alternative therapy of leishmaniasis. Particle size has proved to be a significant factor for efficacy of the drug delivery system (Vermelho et al. 2014).

An analogue (14-deoxy 11,12-didehydro andrographolide) of andrographolide with poor water solubility was incorporated into a biodegradable polymer polycaprolactone by Kamaraj et al. (2017). The antidiabetic activity was worked out by 2-deoxy-D-[1-3H] glucose uptake assay, which produced dose-dependent results with a maximum uptake of 108.54 ± 1.42% in L6 myoblasts. A biphasic pattern was observed in the in vitro drug release with an initial burst release followed by the sustained release pattern (Kamaraj et al., 2017).

Antileishmanial andrographolide-loaded PLGA nanoparticles were formulated using vitamin E TPGS as stabilizer for targeting the macrophage cells infected with the drug-resistant amastigotes. The nanoparticles showed the sustained drug-release pattern up to 288 h. Significant antileishmanial activity was found with a one-tenth dose of andrographolide in nanoparticles, using TPGS and one-third dose without TPGS in comparison with the pure drug. Cytotoxic potential of the nanoparticulate formulation of the drug with and without using TPGS was significantly less than that of the standard drugs (such as amphotericin B, sodium stibogluconate or paromomycin) used for leishmaniasis (Mondal et al. 2013).

Cationic modified biodegradable polylactic-co-glycolic acid (PLGA) nanoparticles of andrographolide were prepared and evaluated against hepatotoxic conditions in the experimental mouse model. The polymeric nanoparticles exhibited significant protective effect in the liver and restoration of oxidative stress markers. The better residence time in the liver and the favorable cytokine regulation of liver tissues were responsible for the rapid recovery (Roy et al. 2014).

Andrographolide-loaded PLGA nanoparticles were prepared using heparin of different densities by the emulsion solvent evaporation method to determine their hepatotoprotective effect against the paracetamol-induced hepatotoxicity in mouse. The heparin-functionalized andrographolide-loaded PLGA nanoparticles showed a hepatic localization of nanoparticles and a fast protective effect due to regeneration of antioxidant capacity of andrographolide (Roy et al. 2013).

The PLGA-loaded andrographolide nanoparticulate formulation was evaluated against the arsenic-induced liver toxicity in mice model. The nanoformulation reduced the arsenic-induced elevated serum levels of liver function markers like SGPT, SGOT, and alkaline phosphatase as well as arsenic deposition in the liver. It also increased the concentration of antioxidant enzymes such as superoxide dismutase and catalase in liver tissues. Its potency was five times greater than that of the free drug (Das et al. 2015).

The optimized and stabilized oil-in-water nanoemulsion of andrographolide prepared with the layer-by-layer technique through electrostatic deposition of chitosan over alginate by the ultra-sonication method has shown strategic drug release in vitro in the simulated biological fluids. The use of this formulation restored the markers of liver function and antioxidant capacity in the galactosamine-lipopolysaccharide-induced liver toxicity in mice (Mishra et al. 2017).

Qiao et al. (2017) prepared a nanosuspension of andrographolide using D-α-tocopheryl polyethylene glycol 1000 succinate (TPGS) as the surfactant and sodium lauryl sulfate as stabilizer by the wet media milling method followed by freeze-drying. The optimized product by Box-Behnken design was amorphous and showed a faster dissolution rate in vitro than the pure drug suspension. The membrane-permeability studies on the Caco-2 cell monolayer showed a significantly higher permeability rate of andrographolide nanosuspension with TPGS than the nanosuspension without TPGS and the pure suspension. The pharmacokinetic study revealed a significant improvement in the area under plasma-response curve and C_{max} (maximum plasma concentration). The anti-inflammatory activity was evaluated on the carr-induced paw edema in rat model. Serum levels of anti-inflammatory markers were significantly restored after the treatment with andrographolide nanosuspension, and the formulation was more efficacious than the pure drug (Qiao et al. 2017).

A solid self-nanodispersion of andrographolide was formulated by homogenization technique to increase its solubility and oral bioavailability. It was then converted to dried nanocrystals using the spray-drying technology. Dissolution rate of nanocrystals was faster (85.87%) than that of the pure drug or a physical mixture of drug with stabilizer. The maximum plasma concentration and the area under the plasma-response curve were significantly higher, compared to the pure drug (Xu et al.). The nanocrystals of andrographolide were also prepared from nanosuspension using different cryoprotectants (Xie et al. 2016).

An injectable form of andrographolide nanosuspension was also prepared using poloxamer 188 as the nonionic stabilizer and sodium deoxycholate/sodium taurursodeoxycholate as the ionic stabilizers by wet milling method, followed by freeze-drying using mannitol as a cryoprotectant. The nanosuspension showed a significantly faster release rate, as compared to the bulk drug. It also brought improvement in the pharmacokinetic parameters and tissue distribution (Guo et al. 2017). Similarly, the andrographolide nanosuspension was developed by Chen et al. (2017), using the natural triterpenoidal saponin (i.e., glycyrrhizin) as stabilizer, by the homogenization technique followed by freeze-drying to convert the drug into a solid powdered form. Glycyrrhizin prevents aggregation between the nanoparticles, an easy re-dispersibility in the vehicle. This natural saponin possessed the interfacial property and electrostatic effect, which help in entrapment of nanoparticles into a network-like structure. The dissolution rate was significantly increased up to 99.87% in comparison to the free drug (42.35%) (Chen et al. 2017).

Zinc oxide nanoparticles of *A. paniculata* leaf extract were synthesized using its reducing and capping potential. The nanoparticles were characterized using different analytical techniques such as UV-visible spectrometry, X-ray diffraction, FT-infrared spectrometry, and the scanning and transmission electron microscopy. The results showed that phenolics, terpenoids, and proteins of *A. paniculata* were involved in stability and nucleation of nanoparticles. The particles were spherical in shape and in the size range of 96–115 nm (Rajakumar et al. 2018).

13.7 Conclusion

Herbal drugs are believed to boost the body's immune system without producing harmful side effects and therefore provide an effective means of treating the chronic diseases. To overcome problems relating to the delivery of herbal drugs, novel pharmaceutical formulations such as nanoemulsion, nanoparticles, nanocrystals, nanodispersion liposomes and niosomes, etc. are being introduced and have dramatically increased the oral bioavailability as well as the therapeutic effectiveness of the drug. The synergy of potential herbal drugs and contemporary nanotechniques must lead to increased therapeutic indices, reduced toxicities/side effects, improved bioavailability, and localized drug targeting. Andrographolide, obtained from the traditional hepatoprotective herb *A. paniculata,* is a diterpenoidal labdane responsible for the therapeutic properties of the herb, including the anticancer activity. Its effectiveness was hindered due to its poor water solubility, short biological half-life, and low oral bioavailability. Nanoformulations have the potential to improve the biopharmaceutic and pharmacokinetic properties of andrographolide and thus open up new perspectives for the poorly water-soluble herbal drugs. Nanotechniques used for the incorporation of andrographolide have increased its oral bioavailability, dissolution rate, and degree of efficacy, compared to the pure andrographolide suspension.

References

Adeoye BO, Asenuga ER, Oyagbemi AA, Omobowale TO, Adedapo AA (2018) The protective effect of the ethanol leaf extract of *Andrographis paniculata* on cisplatin-induced acute kidney injury in rats through nrf2/KIM-1 signaling pathway. Drug Res (Stuttg) 68:23–32

Akbarsha MA, Manivannan B (1993) Biochemical changes in the testis and male accessory organs of albino rats on treatment with *Andrographis paniculata* Nees. Indian J Comp Anim Physiol 11:103–108

Akbarsha MA, Manivannan B, Hamid KS, Vijayan B (1990) Antifertility effect of *Andrographis paniculata* (Nees) in male albino rat. Indian J Exp Biol 28:421–426

Al Batran R, Al-Bayaty F, Al-Obaidi MM, Ashrafi A (2014) Insights into the antiatherogenic molecular mechanisms of andrographolide against *Porphyromonas gingivalis*-induced atherosclerosis in rabbits. Naunyn Schmiedeberg's Arch Pharmacol 387:1141–1152

Allan JJ, Pore MP, Deepak M, Murali B, Mayachari AS, Agarwal A (2009) Reproductive and fertility effects of an extract of *Andrographis paniculata* in male wistar rats. Int J Toxicol 28:308–317

Alzaharna M, Alqouqa I, Cheung HY (2017) Taxifolin synergizes andrographolide-induced cell death by attenuation of autophagy and augmentation of caspase dependent and independent cell death in HeLa cells. PLoS One 12:e0171325

Anonymous IP (2004) WHO monographs on selected medicinal plants, vol 2. Ministry of Health & Family Welfare, Government of India

Anonymous IP (2010) Herbs and herbal products. Ministry of Health & Family Welfare, Government of India

Balachandran P, Govindarajan R (2005) Cancer: an ayurvedic perspective. Pharmacol Res 51:19–30

Banerjee M, Parai D, Chattopadhyay S (2017) Andrographolide: antibacterial activity against common bacteria of human health concern and possible mechanism of action. Folia Microbiol (Praha) 62:237–244

Barilla J (1999) *Andrographis paniculata*. Keats Publishing, Los Angeles, CA, USA

Benoy GK, Animesh DK, Aninda M, Priyanka DK, Sandip H (2012) An overview on *Andrographis paniculata* (burm. F.) Nees. Int J Res Ayurved Pharm 3:752–760

Bhaskar Reddy MV, Kishore PH, Rao CV, Gunasekar D, Caux C, Bodo B (2003) New 2′-oxygenated flavonoids from *Andrographis affinis*. J Nat Prod 66:295–297

Bhatnagar SS, Santapau H, Desa JD, Maniar AC, Ghadially NC, Solomon MJ, Yellore S, Rao TN (1961) Biological activity of Indian medicinal plants. I. Antibacterial, antitubercular and antifungal action. Indian J Med Res 49:799–813

Burgos RA, Caballero EE, S'anchez NS, Schroeder RA, Wikman GK, Hancke JL (1997) Testicular toxicity assessment of *Andrographis paniculata* dried extract in rats. J Ethnopharmacol 58:219–224

Burgos RA, Hancke JL, Bertoglio JC, Aguirre V, Arriagada S, Calvo M, Caceres DD (2009) Efficacy of an *Andrographis paniculata* composition for the relief of rheumatoid arthritis symptoms: a prospective randomized placebo-controlled trial. Clin Rheumatol 28:931–946

Burkill IH, Birtwistle W, Foxworthy F, Scrivenor J, Watson J (1966) A dictionary of the economic plants of the Malay Peninsula, vol I. Ministry of Agriculture and Cooperative, Kuala Lumpur, Malaysia

C'aceres DD, Hancke JL, Burgos RA, Sandberg F, Wikman GK (1999) Use of visual analogue scale measurements (VAS) to assess the effectiveness of standardized *Andrographis paniculata* extract SHA-10 in reducing the symptoms of common cold. A randomized double blind-placebo study. Phytomedicine 6:217–223

Cáceres DD, Hancke JL, Burgos RZ, Wikman GK (1997) Prevention of common colds with *Andrographis paniculata* dried extract. A pilot double-blind study. Phytomedicine 4:101–104

Calabrese C, Berman SH, Babish, Ma X, Shinto L, Dorr M, Wells K, Wenner CA, Standish LJ (2000) A phase I trial of andrographolide in HIV positive patients and normal volunteers. Phytother Res 14:333–338

Carr RR, Nahata MC (2006) Complementary and alternative medicine for upper-respiratory-tract infection in children. Am J Health Syst Pharm 63:33–39

Chadha YR (1985) The wealth of India: raw materials, vol 1A. CSIR, New Delhi

Carretta MD, Alarcón P, Jara E, Solis L, Hancke JL, Concha II, Hidalgo MA, Burgos RA (2009) Andrographolide reduces IL-2 production in T-cells by interfering with NFAT and MAPK activation. Eur J Pharmacol 602:413–421

Chan SJ, Wong WS, Wong PT, Bian JS (2010) Neuroprotective effects of andrographolide in a rat model of permanent cerebral ischaemia. Br J Pharmacol 161:668–679

Chang HM, But PPH (1987) Pharmacology and applications of Chinese Materia Medica, Vol 2, vol 2. World Scientific Publishing Co. Pvt. Ltd, Singapore, pp 918–928

Chang RS, Ding L, Chen GQ, Pan QC, Zhao ZL, Smith KM (1991) Dehydroandrographolide succinic acid monoester as an inhibitor against the human immunodeficiency virus (43225). Proc Soc Exp Biol Med 197:59–66

Chao WW, Kuo YH, Lin BF (2010) Anti-inflammatory activity of new compounds from *Andrographis paniculata* by NF-κB transactivation inhibition. J Agric Food Chem 58:2505–2512

Chao WW, Lin BF (2010) Isolation and identification of bioactive compounds in *Andrographis paniculata* (Chuanxinlian). Chin Med 5:17

Chen Q, Liu Y, Liu YM, Liu GY, Zhang MQ, Jia JY, Lu C, Yu C (2012) Pharmacokinetics and tolerance of dehydroandrographolide succinate injection after intravenous administration in healthy Chinese volunteers. Acta Pharmacol Sin 33:1332–1336

Chen Y, Liu Y, Xu J, Xie Y, Zheng Q, Yue P, Yang M (2017) A natural triterpenoid saponin as multifunctional stabilizer for drug nanosuspension powder. AAPS PharmSciTech 18:2744–2753

Chenniappan K, Kadarkarai M (2008) Oviposition deterrent, ovicidal and gravid mortality effects of ethanolic extract of *Andrographis paniculata* Nees against the malarial vector *Anopheles stephensi* Liston (Diptera: Culicidae). Entomol Res 38:119–125

Cheung HY, Cheung CS, Kong CK (2001) Determination of bioactive diterpenoids from *Andrographis paniculata* by micellar electrokinetic chromatography. J Chromatogr A 930:171–176

Chopra RN, Nayar SL, Chopra IC, Asolkar LV, Kakkar KK (1956) Glossary of Indian medicinal plants. Council of Scientific & Industrial Research, New Delhi

Chua LS (2014) Review on liver inflammation and antiinflammatory activity of *Andrographis paniculata* for hepatoprotection. Phytother Res 28:1589–1598

Coresh J, Selvin E, Stevens LA, Manzi J, Kusek JW, Eggers P, Van Lente F, Levey AS (2007) Prevalence of chronic kidney disease in the United States. JAMA 298(17):2038–2047

Dai L, Wang G, Pan W (2017) Andrographolide inhibits proliferation and metastasis of SGC7901 gastric cancer cells. Biomed Res Int 2017:1

Das S, Pradhan GK, Das S, Nath D, Das Saha K (2015) Enhanced protective activity of nano formulated andrographolide against arsenic induced liver damage. Chem Biol Interact 242:281–289

Deng WL (1978) Outline of current clinical and pharmacological research on *Andrographis paniculata* in China. Newsl Chin Herb Med 10:27–31

Dua VK, Ojha VP, Biswas S, Valecha N, Singh N, Sharma VP (1999) Antimalarial activity of different fractions isolated from the leaves of *Andrographis paniculata*. J Med Aromat Plant Sci 21:1069–1073

Dymock W (1972) Pharmacographia Indica. e Institute of Health and Tibbi Research, Hamdard National Foundation, Karachi

Ebrahimi E, Ebrahimi S (2012) Application of mathematical modeling in production of solvent extraction of a medicinal plant. World Appl Sci J 17:296–300

Edwin ES, Vasantha-Srinivasan P, Senthil-Nathan S, Thanigaivel A, Ponsankar A, Pradeepa V, Selin-Rani S, Kalaivani K, Hunter WB, Abdel-Megeed A, Duraipandiyan V, Al-Dhabi NA (2016) Anti-dengue efficacy of bioactive andrographolide from *Andrographis paniculata* (Lamiales: Acanthaceae) against the primary dengue vector *Aedes aegypti* (Diptera: Culicidae). Acta Trop 163:167–178

Farooqi AA, Sreeramu BS (2001) Cultivation of aromatic and medicinal crop. Universities Press Ltd, Hyderabad

Gabrielian ES, Shukarian AK, Goukasova GI, Chandanian GL, Panossian AG, Wikman G, Wagner H (2002) A double blind, placebo-controlled study of *Andrographis paniculata* fixed combination Kan Jang in the treatment of acute upper respiratory tract infections including sinusitis. Phytomedicine 9:589–597

Geng J, Liu W, Xiong Y, Ding H, Jiang C, Yang X, Li X, Elgehama A, Sun Y, Xu Q, Guo W, Gao J (2018) Andrographolide sulfonate improves Alzheimer-associated phenotypes and mitochondrial dysfunction in APP/PS1 transgenic mice. Biomed Pharmaco 97:1032–1039

Govindarajan M (2011) Evaluation of *Andrographis paniculata* Burm. f. (Family: Acanthaceae) extracts against *Culex quinquefasciatus* (Say.) and *Aedes aegypti* (Linn.) (Diptera: Culicidae). Asian Pac J Trop Med 4:176–181

Graverini G, Piazzini V, Landucci E, Pantano D, Nardiello P, Casamenti F, Pellegrini-Giampietro DE, Bilia AR, Bergonzi MC (2018) Solid lipid nanoparticles for delivery of andrographolide across the blood-brain barrier: *in vitro* and *in vivo* evaluation. Colloids Surf B Biointerfaces 161:302–313

Guan SP, Kong LR, Cheng C, Lim JC, Wong WS (2011) Protective role of 14-deoxy-11,12-didehydroandrographolide, a noncytotoxic analogue of andrographolide, in allergic airway inflammation. J Nat Prod 74:1484–1490

Guan SP, Tee W, Ng DS, Chan TK, Peh HY, Ho WE, Cheng C, Mak JC, Wong WS (2013) Andrographolide protects against cigarette smoke-induced oxidative lung injury via augmentation of Nrf2 activity. Br J Pharmacol 168:1707–1718

Guo L, Kang L, Liu X, Lin X, Di D, Wu Y, Kong D, Deng Y, Song Y (2017) A novel nanosuspension of andrographolide: preparation, characterization and passive liver target evaluation in rats. Eur J Pharm Sci 104:13–22

Guo SY, Li DZ, Li WS, Fu AH, Zhang LF (1988) Study of the toxicity of andrographolide in rabbits. J Beijing Med Univ 5:422–428

Gupta S, Choudhary MA, Yadava JNS, Srivastava V, Tandon JS (1990) Antidiarrhoeal activity of diterpenes of *Andrographis paniculata* (kalmegh) against *Escherichia coli* enterotoxin in in *vivo* models. Int J Crude Drug Res 28:273–283

Handa SS, Sharma A (1990) Hepatoprotective activity of andrographolide from *Andrographis paniculata* against carbon tetrachloride. Indian J Med Res 92:284–292

Hossain MS, Urbi Z, Sule A, Rahman KMH (2014) *Andrographis paniculata* (Burm. f.) Wall. ex Nees: a review of ethnobotany, phytochemistry, and pharmacology. Sci World J 2014:274905

Huang LY (1987) The effects of andrographolide on experimental blood deficiency of cardiac muscle. Chin Herb Med 18:26–28

Iruretagoyena MI, Tobar JA, Gonz'alez PA, Sepúlveda SE, Figueroa CA, Burgos RA, Hancke JL, Kalergis AM (2005) Andrographolide interferes with T cell activation and reduces experimental autoimmune encephalomyelitis in the mouse. J Pharmacol Exp Ther 312:366–372

Janarthanan S (1990) Antifertility effects of andrographolide in rats. J Ecobiol 2:325–329

Jantan I, Ahmad W, Bukhari SN (2015) Plant-derived immunomodulators: an insight on their preclinical evaluation and clinical trials. Front Plant Sci 6:655

Jarukamjorn K, Nemoto N (2008) Pharmacological aspects of *Andrographis paniculata* on health and its major diterpenoid constituent andrographolide. J Health Sci 54:370–381

Jayakumar T, Hsieh CY, Lee J, Sheu JR (2013) Experimental and clinical pharmacology of *Andrographis paniculata* and its major bioactive phytoconstituent andrographolide. Evid Based Complement Altern Med 2013:846740

Joselin J, Jeeva S (2014) *Andrographis paniculata*: a review of its traditional uses, phytochemistry and pharmacology. Med Aromat Plants 3:169

Kabeeruddin M (1937) Kitabul Advia, vol 2. Aligarh Barqi Press, Delhi

Kabir MH, Hasan N, Rahman MM, Rahman MA, Khan JA, Hoque NT, Bhuiyan MR, Mou SM, Jahan R, Rahmatullah M (2014) A survey of medicinal plants used by the Deb barma clan of the Tripura tribe of Moulvibazar district. Bangladesh J Ethnobiol Ethnomed 10:19

Kamal R, Gupta RS, Lohiya NK (2003) Plants for male fertility regulation. Phytother Res 17:579–590

Kamal YT, Musthaba SM, Singh M, Parveen R, Ahmad S, Baboota S, Ali I, Siddiqui KM, Zaidi SMA (2012) Development and validation of HPLC method for simultaneous estimation of piperine and guggulsterones in compound Unani formulation (tablets) and a nanoreservoir system. Biomed Chromatogr 26:1183–1190

Kamaraj N, Rajaguru PY, Issac PK, Sundaresan S (2017) Fabrication, characterization, *in vitro* drug release and glucose uptake activity of 14-deoxy, 11, 12-didehydroandrographolide loaded polycaprolactone nanoparticles. Asian J Pharm Sci 12:353–362

Khamphaya T, Chansela P, Piyachaturawat P, Suksamrarn A, Nathanson MH, Weerachayaphorn J (2016) Effects of andrographolide on intrahepatic cholestasis induced by alpha-naphthylisothiocyanate in rats. Eur J Pharmacol 789:254–264

Khare CP (2007) Indian medicinal plants: an illustrated dictionary. Springer, Berlin

Kim J, Lee J, Lee YM, Pramanick S, Im S, Kim WJ (2016) Andrographolide-loaded polymerized phenylboronic acid nanoconstruct for stimuli-responsive chemotherapy. J Control Release S0168-3659:31082–31083

King Spalding LLP (2006) Andrographolide derivatives to treat viral infections. US patent 20060333785

Kishore V, Nagendra Y, Anupam B, Gunasekar D, Caux C, Bodo B (2017) Multi-targeting Andrographolide and its Natural Analogs as Potential Therapeutic Agents. Current Topics in Medicinal Chemistry 17:845–857

Kumar RA, Sridevi K, Kumar NV, Nanduri S, Rajagopal S (2004) Anticancer and immunostimulatory compounds from *Aandrographis paniculata*. J Ethnopharmacol 92(2–3):291–295

Kumar V, Thakur AK, Chatterjee SS (2014) Perspective of *Andrographis paniculata* in neurological disorders. Clin Pharmacol Biopharm S2:005

Kunwar RM, Shrestha KP, Bussmann RW (2010) Traditional herbal medicine in Far-west Nepal: a pharmacological appraisal. Journal of Ethnobiology and Ethnomedicine 6:35

Kuppusamy C, Murugan K (2009) Mosquitocidal effect of *Andographis paniculata* Nees against the malaria vector, *Anopheles stephensi* Liston (Diptera: culicidae). Int J Integr Biol 5:75–81

Lala S, Nandy AK, Mahato SB, Basu MK (2003) Delivery *in vivo* of 14-deoxy-11-oxoan- drographolide, an antileishmanial agent, by different drug carriers. Indian J Biochem Biophys 40:169–174

Li J, Luo L, Wang X, Liao B, Li G (2009) Inhibition of NF-kappaB expression and allergen-induced airway inflammation in a mouse allergic asthma model by andrographolide. Cell Mol Immunol 6:381–385

Li Y, Yan H, Zhang Z, Zhang G, Sun Y, Yu P, Wang Y, Xu L (2015) Andrographolide derivative AL-1 improves insulin resistance through down-regulation of NF-κB signalling pathway. Br J Pharmacol 172:3151–3158

Liao W, Tan WS, Wong WS (2016) Andrographolide restores steroid sensitivity to block lipopolysaccharide/ifn-γ-induced il-27 and airway hyperresponsiveness in mice. J Immunol 196:4706–4712

Lim JC, Chan TK, Ng DS, Sagineedu SR, Stanslas J, Wong WS (2012) Andrographolide and its analogues: versatile bioactive molecules for combating inflammation and cancer. Clin Exp Pharmacol Physiol 39:300–310

Lin FL, Wu SJ, Lee SC, Ng LT (2009) Antioxidant, antioedema and analgesic activities of *Andrographis paniculata* extracts and their active constituent andrographolide. Phytother Res 23:958–964

Lin HC, Lii CK, Chen HC, Lin AH, Yang YC, Chen HW (2018) Andrographolide inhibits oxidized LDL-induced cholesterol accumulation and foam cell formation in macrophages. Am J Chin Med 46:87–106

Lin TP, Chen SY, Duh PD, Chang LK, Liu YN (2008) Inhibition of the Epstein-Barr virus lytic cycle by andrographolide. Biol Pharm Bull 31:2018–2023

Li YH, Wang MY, Jin R, Guo S, Fan XY, Ma H, Wu LX, Zhang JH (2012) Effects of andrographolide on the expression of eosinophil granulocytes and possible mechanisms. Zhongguo Dang Dai Er Ke Za Zhi. 14:371–374

Maiti K, Gantait A, Mukherjee K, Saha BP, Mukherjee PK (2006) Therapeutic potentials of andrographolide from *Andrographis paniculata*: a review. J Nat Remed 6:1–13

Melchior J, Palm S, Wikman G (1997) Controlled clinical study of standardized *Andrographis paniculata* extract in common cold- a pilot trial. Phytomedicine 3:315–318

Melchior J, Spasov AA, Ostrovskij OV, Bulanov AE, Wikman G (2000) Double-blind, placebo-controlled pilot and phase III study of activity of standardized *Andrographis paniculata* Herba Nees extract fixed combination (Kan Jang) in the treatment of uncomplicated upper-respiratory tract infection. Phytomedicine 7:341–350

Mishra K, Dash AP, Swain BK et al (2009) Antimalarial activities of *Andrographis paniculata* and *Hedyotis corymbosa* extracts and their combination with curcumin. Malar J 8:26

Mishra N, Yadav KS, Rai VK, Yadav NP (2017) Polysaccharide encrusted multilayered nano-colloidal system of andrographolide for improved hepatoprotection. AAPS PharmSciTech 18:381–392

Misra P, Pal NL, Guru PY, Katiyar JC, Srivastava V, Tandon JS (1992) Antimalarial activity of *Andrographis paniculata* (Kalmegh) against *Plasmodium berghei* NK 65 in *Mastomys natalensis*. Int J Pharmacogn 30:263–274

Mittal SP, Khole S, Jagadish N, Ghosh D, Gadgil V, Sinkar V, Ghaskadbi SS (2016) Andrographolide protects liver cells from H_2O_2 induced cell death by upregulation of Nrf-2/HO-1 mediated via adenosine A2a receptor signalling. Biochim Biophys Acta 1860:2377–2390

Mkrtchyan A, Panosyan V, Panossian A, Wikman G, Wagner H (2005) A phase I clinical study of *Andrographis paniculata* fixed combination Kan Jang versus ginseng and valerian on the semen quality of healthy male subjects. Phytomedicine 12:403–409

Mondal S, Roy P, Das S, Halder A, Mukherjee A, Bera T (2013) *In vitro* susceptibilities of wild and drug resistant *Leishmania donovani* amastigote stages to andrographolide nanoparticle: role of vitamin E derivative TPGS for nanoparticle efficacy. PLoS One 8:e81492

Muluye RA, Bian Y, Alemu PN (2014) Anti-inflammatory and antimicrobial effects of heat-clearing chinese herbs: a current review. J Tradit Complement Med 4:93–98

Naik SR, Hule A (2009) Evaluation of immunomodulatory activity of an extract of andrographolides from *Andographis paniculata*. Planta Med 75:785–791

Niranjan A, Tewari SK, Lehri A (2010) Biological activities of kalmegh (*Andrographis paniculata* Nees) and its active principles- a review. Indian J Nat Prod Res 1:125–135

Nugroho AE, Andrie M, Warditiani NK, Siswanto E, Pramono S, Lukitaningsih E (2012) Antidiabetic and antihiperlipidemic effect of *Andrographis paniculata* (Burm. f.) Nees and andrographolide in high-fructose-fat-fed rats. Indian J Pharmacol 44:377–381

Nugroho AE, Rais IR, Setiawan I, Pratiwi PY, Hadibarata T, Tegar M, Pramono S (2014) Pancreatic effect of andrographolide isolated from *Andrographis paniculata* (Burm. f.) Nees. Pak J Biol Sci 17:22–31

Ozolua RI, Adejayan A, Aigbe OP, Uwaya DO, Argawal A (2011) Some characteristic relaxant effects of aqueous leaf extract of *Andrographis paniculata* and andrographolide on Guinea pig tracheal rings. Niger J Physiol Sci 26:119–124

Panossian A, Kochikian A, Gabrielian E, Muradian R, Stepanian H, Arsenian F, Wagner H (1999) Effect of *Andrographis paniculata* extract on progesterone in blood plasma of pregnant rats. Phytomedicine 6:157–162

Panraksa P, Ramphan S, Khongwichit S, Smith DR (2017) Activity of andrographolide against dengue virus. Antivir Res 139:69–78

Parixit B, Bharath C, Rajarajeshwari N, Ganapaty S (2012) The genus Andrographis- a review. Int J Pharm Sci 4:1835–1856

Parveen R, Ahmad FJ, Iqbal Z (2013) Potential botanicals for the treatment of breast cancer- pharmaceutical approaches used to increase the absorption of herbal drugs. In: Iqbal M, Ahmad A (eds) Recent trends in medicinal botany. IK International Publishing Company, New Delhi, pp 110–130

Parveen R, Ahmad FJ, Iqbal Z, Samim M, Ahmad S (2014) Solid lipid nanoparticles of anticancer drug andrographolide: formulation, *in vitro* and *in vivo* studies. Drug Dev Ind Pharm 40:1206–1212

Parveen R, Baboota S, Ahmad S, Ali J, Ahuja A (2008) Approaches used for oral bioavailability enhancement of herbal drugs. Online J Pharmacol Pharmacokinet 4:18–31

Parveen R, Baboota S, Ali J, Ahuja A, Vasudev SS, Ahmad S (2011) Oil based nanocarrier for improved oral delivery of silymarin: *In vitro* and *in vivo* studies. Int J Pharm 413:245–253

Parveen R, Baboota S, Ali J, Ahuja A, Vasudev SS, Ahmad S (2010) Nanoemulsion as a novel carrier for delivery of silymarin against carbon tetrarchloride-induced hepatic damage. Arch Pharmcal Res 34:767–774

Peng S, Gao J, Liu W, Guo W, Jiang C, Yang X, Xu Q, Sun Y (2016a) Andrographolide ameliorates OVA-induced lung injury in mice by suppressing ROS-mediated NF-κB signaling and NLRP3 inflammasome activation. Oncotarget 7:80262–80274

Peng S, Hang N, Liu W, Guo W, Jiang C, Yang X, Xu Q, Sun Y (2016b) Andrographolide sulfonate ameliorates lipopolysaccharide-induced acute lung injury in mice by down-regulating MAPK and NF-κB pathways. Acta Pharm Sin B 6:205–211

Pholphana N, Rangkadilok N, Saehun J, Ritruechai S, Satayavivad J (2013) Changes in the contents of four active diterpenoids at different growth stages in *Andrographis paniculata* (Burm.f.) Nees (Chuanxinlian). Chin Med 8:2

Pholphana N, Rangkadilok N, Thongnest S, Ruchirawat S, Ruchirawat M, Satayavivad J (2004) Determination and variation of three active diterpenoids in *Andrographis paniculata* (Burm.f.) Nees. Phytochem Anal 15:365–371

Phunikhom K, Khampitak K, Aromdee C, Arkaravichien T, Sattayasai J (2015) Effect of *Andrographis paniculata* extract on triglyceride levels of the patients with hypertriglyceridemia: a randomized controlled trial. J Med Assoc Thail 98:S41–S47

Pramanick S, Banerjee S, Achari B, Mukhopadhyay S (2007) Phytochemicals from the genus Andrographis. In: Govil JN, Singh VK, Bhardwaj R (eds) Recent progress in medicinal plants: phytomedicines. Studium Press LLC, Houston, USA, pp 339–387

Puntawee S, Theerasilp M, Reabroi S, Saeeng R, Piyachaturawat P, Chairoungdua A (2016) Solubility enhancement and *in vitro* evaluation of PEG-b-PLA micelles as nanocarrier of semi-synthetic andrographolide analogue for cholangiocarcinoma chemotherapy. Pharm Dev Technol 21:437–444

Puri A, Saxena R, Saxena RP, Saxena KC, Srivastava V, Tandon JS (1993) Immunostimulant agents from *Andrographis paniculata*. J Nat Prod 56:995–999

Qiao H, Chen L, Rui T, Wang J, Chen T, Fu T, Li J, Di L (2017) Fabrication and *in vitro/in vivo* evaluation of amorphous andrographolide nanosuspensions stabilized by d-α-tocopheryl polyethylene glycol 1000 succinate/sodium lauryl sulfate. Int J Nanomed 12:1033–1046

Qin LH, Kong L, Shi GJ, Wang ZT, Ge BX (2006) Andrographolide inhibits the production of TNF-ũFC; and interleukin-12 in lipopolysaccharide-stimulated macrophages: role of mitogen-activated protein kinases. Biol Pharm Bull 29:220–224

Radhika P, Prasad YR, Lakshmi KR (2010) Flavones from the stem of *Andrographis paniculata* Nees. Nat Prod Commun 5:59–60

Rahman NNNA, Furuta T, Kojima S, Takane K, Ali M (1999) Antimalarial activity of extracts of Malaysian medicinal plants. J Ethnopharmacol 64:249–254

Rajagopal S, Kumar RA, Deevi DS, Satyanarayana C, Rajagopalan R (2003) Andrographolide, a potential cancer therapeutic agent isolated from *Andrographis paniculata*. J Exp Ther Oncol 3:147–158

Rajakumar G, Thiruvengadam M, Mydhili G, Gomathi T, Chung IM (2018) Green approach for synthesis of zinc oxide nanoparticles from *Andrographis paniculata* leaf extract and evaluation of their antioxidant, anti-diabetic, and anti-inflammatory activities. Bioprocess Biosyst Eng 41:21–30

Rajpal, Singh VK, Dutta BN (2011) *Andrographis paniculata*: a multicentric, randomized, double-blind homoeopathic pathogenetic trial. Indian J Res Homoeopath 5:8–14

Roy P, Das S, Auddy RG, Saha A, Mukherjee A (2013) Engineered andrographolide nanoparticles mitigate paracetamol hepatotoxicity in mice. Pharm Res 30:1252–1262

Roy P, Das S, Auddy RG, Mukherjee A (2014) Engineered andrographolides systems for smart recovery in hepatotoxic conditions. Int J Nanomed 9:4723–4735

Roy P, Das S, Bera T, Mondol S, Mukherjee A (2010) Andrographolide nanoparticles in leishmaniasis: characterization and *in vitro* evaluations. Int J Nanomedicine 5:1113–1121

Sattayasai J, Srisuwan S, Arkaravichien T, Aromdee C (2010) Effects of andrographolide on sexual functions, vascular reactivity and serum testosterone level in rodents. Food Chem Toxicol 48:1934–1938

Saxena RC, Singh R, Kumar P, Yadav SC, Negi MP, Saxena VS, Joshua AJ, Vijayabalaji V, Goudar KS, Venkateshwarlu K, Amit A (2010) A randomized double blind placebo controlled clinical evaluation of extract of *Andrographis paniculata* (KalmCold) in patients with uncomplicated upper respiratory tract infection. Phytomedicine 17:178–185

Serrano FG, Tapia-Rojas C, Carvajal FJ, Hancke J, Cerpa W, Inestrosa NC (2014) Andrographolide reduces cognitive impairment in young and mature AβPPswe/PS-1 mice. Mol Neurodegener 9:61

Seubsasana S, Pientong C, Ekalaksananan T, Thongchai S, Aromdee C (2011) A potential andrographolide analogue against the replication of herpes simplex virus type 1 in vero cells. Med Chem 7:237–244

Sheeja BD, Sindhu D, Ebanasar J, Jeeva S (2012) The larvicidal activity of *Andrographis paniculata* (Burm. f)Nees against *Culex quinquefasciatus* Say (Insecta: Diptera-Culicidae), a filarial vector. Asian Pac J Trop Dis 2:S574–S578

Sheeja K, Kuttan G (2006) Protective effect of *Andrographis paniculata* and andrographolide on cyclophosphamide-induced urothelial toxicity. Integr Cancer Ther 5:244–251

Shen YC, Chen CF, Chiou WF (2000) Suppression of rat neutrophil reactive oxygen species production and adhesion by the diterpenoid lactone andrographolide. Planta Med 66:314–317

Shi G, Zhang Z, Zhang R, Zhang X, Lu Y, Yang J, Zhang D, Zhang Z, Li X, Ning G (2012) Protective effect of andrographolide against concanavalin A-induced liver injury. Naunyn-Schmiedeberg's Arch Pharmacol 385:69–79

Singh P, Srivastava MM, Khemani LD (2009) Renoprotective effects of *Andrographis paniculata* (Burm. f.) Nees in rats. Ups J Med Sci 114:136–139

Singha PK, Roy S, Dey S (2003) Antimicrobial activity of *Andrographis paniculata*. Fitoterapia 74:692–694

Sinha J, Mukhopadhyay S, Das N, Basu MK (2000) Targeting of liposomal andrographolide to l. donovani infected macrophages *in vivo*. Drug Deliv 7:209–213

Smith PL, Maloney KN, Pothen RG, Clardy J, Clapham DE (2006) Bisandrographolide from *Andrographis paniculata* activates TRPV4 channels. J Biol Chem 281:29897–29904

Spasov AA, Ostrovskij OV, Chernikov MV, Wikman G (2004) Comparative controlled study of *Andrographis paniculata* fixed combination, Kan Jang and an Echinacea preparation as adjuvant, in the treatment of uncomplicated respiratory disease in children. Phytother Res 18:47–53

Subramanian R, Asmawi MZ, Sadikun A (2008) *In vitro* alpha-glucosidase and alpha-amylase enzyme inhibitory effects of *Andrographis paniculata* extract and andrographolide. Acta Biochim Pol 55:391–398

Sule A, Ahmed QU, Latip J, Samah OA, Omar MN, Umar A, Dogarai BB (2012) Antifungal activity of *Andrographis paniculata* extracts and active principles against skin pathogenic fungal strains *in vitro*. Pharm Biol 50:850–856

Tan BK-H, Zhang ACY (2004) *Andrographis paniculata* and the cardiovascular system. Oxid Stress Dis 14:441–455

Tan WS, Liao W, Zhou S, Wong WS (2017) Is there a future for andrographolide to be an anti-inflammatory drug? Deciphering its major mechanisms of action. Biochem Pharmacol 139:71–81

Tan WS, Peh HY, Liao W, Pang CH, Chan TK, Lau SH, Chow VT, Wong WS (2016) Cigarette smoke-induced lung disease predisposes to more severe infection with nontypeable *Haemophilus influenzae*: protective effects of andrographolide. J Nat Prod 79:1308–1315

Tang LIC, Ling APK, Koh RY, Chye SM, Voon KGL (2012) Screening of anti-dengue activity in methanolic extracts of medicinal plants. BMC Complement Altern Med 12:1–10

Tang T, Targan SR, Li Z.-S, Xu C, Byers VS, Sandborn WJ (2011) Randomised clinical trial: herbal extract HMPL-004 in active ulcerative colitis - a double-blind comparison with sustained release mesalazine. Alimen Pharmacol Therap 33:194–202

Tapia-Rojas C, Schüller A, Lindsay CB, Ureta RC, Mejías-Reyes C, Hancke J, Melo F, Inestrosa NC (2015) Andrographolide activates the canonical Wnt signalling pathway by a mechanism that implicates the non-ATP competitive inhibition of GSK-3β: autoregulation of GSK-3β *in vivo*. Biochem J 466:415–430

Thakur AK, Rai G, Chatterjee SS, Kumar V (2016) Beneficial effects of an *Andrographis paniculata* extract and andrographolide on cognitive functions in streptozotocin-induced diabetic rats. Pharm Biol 54:1528–1538

Thamlikitkul V, Theerapong S, Boonroj P, Chantrakul C, Boonroj P, Punkrut W, Ekpalakorn W, Boontaeng N, Taechaiya S, Petcharoen S (1991) Efficacy of *Andrographis paniculata* nees for pharyngotonsillitis in adults. J Med Assoc Thail 74:437–442

Tu YS, Sun DM, Zhang JJ, Jiang ZQ, Chen YX, Zeng XH, Huang DE, Yao N (2014) Preparation and characterisation of andrographolide niosomes and its anti-hepatocellular carcinoma activity. J Microencapsul 31:307–316

Umar S, Sareer O, Ahad A (2012) Prophylactic and lenitive effects of *Andrographis paniculata* against common human ailments: an exhaustive and comprehensive reappraisal. J Pharm Res Opin 2:138–162

Valdiani A, Kadir MA, Tan SG, Talei D, Abdullah MP, Nikzad S (2012) Nain-e Havandi *Andrographis paniculata* present yesterday, absent today: a plenary review on underutilized herb of Iran's pharmaceutical plants. Mol Biol Rep 39:5409–5424

Varela-Nallar L, Arredondo SB, Tapia-Rojas C, Hancke J, Inestrosa NC (2015) Andrographolide stimulates neurogenesis in the adult hippocampus. Neural Plast 2015:1

Varma A, Padh H, Shrivastava N (2011) Andrographolide: a new plant-derived antineoplastic entity on horizon. Evid Based Complement Alternat Med 2011:1

Verma N, Vinayak M (2008) Antioxidant action of *Andrographis paniculata* on lymphoma. Mol Biol Rep 35:535–540

Vermelho AB, Supuran CT, Cardoso V, Menezes D, Silva JRA, Ferreira JLP, Amaral ACF, Rodrigues IA (2014) Leishmaniasis: possible new strategies for treatment. In: Claborn DM (ed) Leishmaniasis- trends in epidemiology, diagnosis and treatment. InTech Publisher. https://doi.org/10.5772/57067

Voravuthikunchai SP, Limsuwan S (2006) Medicinal plant extracts as anti-*Escherichia coli* O157:H7 agents and their effects on bacterial cell aggregation. J Food Prot 69:2336–2341

Wang HW, Zhao HY, Xiang SQ (1997) Effects of *Andrographis paniculata* component on nitric oxide, endothelin and lipid peroxidation in experimental atherosclerotic rabbits. Zhongguo Zhong Xi Yi Jie He Za Zhi 17:547–549

Wang W, Wang J, Dong SF, Liu CH, Italiani P, Sun SH, Xu J, Boraschi D, Ma SP, Qu D (2010) Immunomodulatory activity of andrographolide on macrophage activation and specific antibody response. Acta Pharmacol Sin 31:191–201

Wen T, Xu W, Liang L, Li J, Ding X, Chen X, Hu J, Lv A, Li X (2015) Clinical efficacy of andrographolide sulfonate in the treatment of severe hand, foot, and mouth disease (HFMD) is dependent upon inhibition of neutrophil activation. Phytother Res 29:1161–1167

Wiart C, Kumar K, Yusof MY, Hamimah H, Fauzi ZM, Sulaiman M (2005) Antiviral properties of ent-labdene diterpenes of *Andrographis paniculata* Nees, inhibitors of herpes simplex virus type 1. Phytother Res 19:1069–1070

Wong SY, Tan MG, Banks WA, Wong WS, Wong PT, Lai MK (2016) Andrographolide attenuates LPS-stimulated up-regulation of C-C and C-X-C motif chemokines in rodent cortex and primary astrocytes. J Neuroinflamm 13:34

Woo AY, Waye MM, Tsui SK, Yeung ST, Cheng CH (2008) Andrographolide up-regulates cellular-reduced glutathione level and protects cardiomyocytes against hypoxia/reoxygenation injury. J Pharmacol Exp Ther 325:226–235

Wu Z, Raven PH, Hong DY, Garden MB (1996) Flora of China: cucurbitaceae through valerianaceae with annonaceae and berberidaceae. Science Press, Beijing

Xie Y, Ma Y, Xu J, Dan J, Yue P, Wu Z, Yang M, Zheng Q (2016) Roles of cryo/thermal strength for redispersibility of drug nanocrystals: a representative study with andrographolide. Arch Pharm Res 39:1404–1417

Xu J, Ma Y, Xie Y, Chen Y, Liu Y, Yue P, Yang M (2016) Design and evaluation of novel solid self-nanodispersion delivery system for andrographolide. AAPS PharmSciTech 18:1572–1584

Xu Y, Marshall RL, Mukkur TKS (2006) An investigation on the antimicrobial activity of *Andrographis paniculata* extracts and andrographolide *in vitro*. Asian J Plant Sci 5:527–530

Yang PY, Hsieh PL, Wang TH, Yu CC, Lu MY, Liao YW, Lee TH, Peng CY (2017) Andrographolide impedes cancer stemness and enhances radio-sensitivity in oral carcinomas via miR-218 activation. Oncotarget 8:4196–4207

Yang T, Shi HX, Wang ZT, Wang CH (2013) Hypolipidemic effects of andrographolide and neoandrographolide in mice and rats. Phytother Res 27:618–623

Yen TL, Hsu WH, Huang SK, Lu WJ, Chang CC, Lien LM, Hsiao G, Sheu JR, Lin KH (2013) A novel bioactivity of andrographolide from *Andrographis paniculata* on cerebral ischemia/reperfusion-induced brain injury through induction of cerebral endothelial cell apoptosis. Pharm Biol 51:1150–1157

Yin J, Guo L (1993) Contemporary traditional Chinese medicine. Xie Yuan, Beijing

Yoopan N, Thisoda P, Rangkadilok N, Sahasitiwat S, Pholphana N, Ruchirawat S, Satayavivad J (2007) Cardiovascular effects of 14-deoxy-11, 12-didehydroandrographolide and *Andrographis paniculata* extracts. Planta Med 73:503–511

Zaidan MR, Rain AN, Badrul AR, Adlin A, Norazah A, Zakiah I (2005) *In vitro* screening of five local medicinal plants for antibacterial activity using disc diffusion method. Trop Biomed 22:165–170

Zhan JY, Wang XF, Liu YH, Zhang ZB, Wang L, Chen JN, Huang S, Zeng HF, Lai XP (2016) Andrographolide sodium bisulfate prevents uv-induced skin photoaging through inhibiting oxidative stress and inflammation. Mediat Inflamm 2016:1

Zhang C, Gui L, Xu Y, Wu T, Liu D (2013) Preventive effects of andrographolide on the development of diabetes in autoimmune diabetic NOD mice by inducing immune tolerance. Int Immunopharmacol 16:451–456

Zhang CY, Tan BK (1997) Mechanisms of cardiovascular activity of *Andrographis paniculata* in the anaesthetized rat. J Ethnopharmacol 56:97–101

Zhang J, Li Y, Gao W, Repka MA, Wang Y, Chen M (2014) Andrographolide-loaded PLGA-PEG-PLGA micelles to improve its bioavailability and anticancer efficacy. Expert Opin Drug Deliv 11:1367–1380

Zhang T (2000) Advances in the study of *Andrographis paniculata* (Burm.f.) Nees. Zhong Yao Cai 23:366–368

Zhang XF, Tan BKH (2000) Anti-diabetic property of ethanolic extract of *Andrographis paniculata* in streptozotocin diabetic rats. Acta Pharmacol Sin 21:1157–1164

Zhang Z, Jiang J, Yu P et al (2009) Hypoglycemic and beta cell protective effects of andrographolide analogue for diabetes treatment. J Transl Med 7:62

Zhang M, Xue E, Shao W (2016) Andrographolide promotes vincristine-induced SK-NEP-1 tumor cell death via PI3K-AKT-p53 signaling pathway. Drug Des Devel Ther 10:3143–3152

Zoha MS, Hussain AH, Choudhury SA (1989) Antifertility effect of *Andrographis paniculata* in mice. Bangladesh Med Res Counc Bull 15:34–37

Part II
Interaction of Plants and Nanomaterials

Chapter 14
Impact of Nanomaterials on Plant Physiology and Functions

Rubbel Singla, Avnesh Kumari, and Sudesh Kumar Yadav

14.1 Introduction

Nanotechnology has generated a variety of nanoparticles (NPs) with differences in size, shape, surface charge, and surface chemistry (Albanese et al. 2012). Among the various types of nanomaterials synthesized and released in the environment, metallic NPs are found in greater amount. The metal-based NPs have been extensively used in different applications (Husen and Siddiqi 2014a; Siddiqi et al. 2016, 2018; Husen 2017) which are directly or indirectly related to humans (Zhu and Njuguna 2014) and the environment (Remédios et al. 2012). The systematic design and engineering of NPs pose risks through inhalation, dermal penetration, and environmental persistence to humans as well as environment (Maurer-Jones et al. 2013). Small size and large surface area of NPs make them more prone for causing toxicity in the ecosystem (Ostiguy and IRSST (Québec) 2006). Plants, the most abundant species in the ecosystem, possess the highest likelihood to experience adverse or positive effects of NPs (Maiti et al. 2015). The presence of NPs in the air, water, and soil shows inevitable effects on plants. NPs are absorbed directly or indirectly by rootless or rooted surface by any of the physical or chemical processes. NPs are transported to leaves and other parts of plants through stem and may get accumulated in seeds (Maiti et al. 2015). Plants originating from seeds having NP concentration above bioconcentration may develop toxicity. NPs also play vital role in the protection of plants against different abiotic stresses as these can mimic the role of antioxidant enzymes (Rico et al. 2013a, b; Wei and Wang 2013; Siddiqi and Husen 2016, 2017). They may also enhance the photosynthesis rate by suppressing the

R. Singla · A. Kumari
Biotechnology Division, CSIR-Institute of Himalayan Bioresource Technology, Palampur, HP, India

S. K. Yadav (✉)
Center of Innovative and Applied Bioprocessing (CIAB), Mohali, PB, India

A. Husen, M. Iqbal (eds.), *Nanomaterials and Plant Potential*,
https://doi.org/10.1007/978-3-030-05569-1_14

osmotic and oxidative stresses (Haghighi and Pessarakli 2013; Qi et al. 2013). On the other hand, plants also show toxic effects depending upon the plant species and the type and concentration of NPs used (Begum et al. 2011; Slomberg and Schoenfisch 2012). For instance, the presence of NPs in growth media inhibits the seed germination rate, root and shoot length, plant biomass, and even the level of nutrients (Barhoumi et al. 2015; Da Costa and Sharma 2016; Wang et al. 2015). Exposure of plants to NPs has detrimental effects on cell biosynthesis, cell organization, electron transport, and energy pathways of the plant system (Landa et al. 2012; Van Aken 2015). Extensive studies have been carried out to investigate the interaction of NPs with plants, but the role of NPs in plant physiology and functions still remains ambiguous and unclear.

This chapter addresses the role of metallic NPs in plant physiology affecting the various functions of plant systems. The mechanism of NPs uptake, transport, and accumulation in the plant and plant parts and the role of NPs associated factors in relation to major physiological processes such as photosynthesis, water and nutrient uptake, respiration, transpiration, and seed germination have been highlighted. Both the negative and positive effects of metallic NPs on the activation of plant growth and development have been discussed with special focus on the NP-mediated gene delivery to plants as well as the phytotoxicity of NPs.

14.2 NPs Interactions with Plants

As already mentioned, the wide applicability and use of metal NPs in our daily life are directly proportional to the release of NPs in the environment (Maurer-Jones et al. 2013). NPs discharged into the environment possess a great tendency to interact with the terrestrial or aquatic plants through the atmosphere, soil, or water (Rico et al. 2011). The mechanism of NPs interactions with plants covers mainly three phases: uptake, translocation, and accumulation. The interactions of NPs with plants depends upon the type of plant species, type of NPs, the chemical composition, surface functionality, properties, shape and size of NPs, and many more factors (Rico et al. 2011). The uptake of nanomaterials by plants is a current topic of research in the scientific world, as most of the data available till date provides a view merely in the formative stage. This is because the protocols followed to measure the quantity of NPs entering in the plant tissues are not yet well-defined (Remédios et al. 2012). The lack of proper information regarding how the NPs enter the plant tissues, their transport, and then accumulation in various food chains leads to a defective understanding of the phenomena. The metal-based nanomaterials include Ag, Au, Al_2O_3, Cu, Zn, ZnO, TiO_2, CeO_2, SiO_2, Fe, Fe_2O_3, CdS, ZnSe, etc. Several reports are available on the uptake of these different kinds of NP by the plant systems and their effect on the physiology and growth of plants.

14.3 Uptake, Translocation, and Bio-distribution of Metallic NPs in Plants

In general, uptake of NPs occurs when a particle penetrates the cell walls of plants. The plant cell wall acts as a semipermeable barrier that regulates the trafficking of materials across the membrane through various pores to the plant cells. Several routes have been proposed in the literature describing the uptake of NPs by the plants. Pore sizes of the plant cell wall vary depending upon the plant species but generally range from 5 to 20 nm (Miralles et al. 2012; Fleischer et al. 1999). To pass through these pores, the size, orientation, and properties of NPs play a critical role. NPs or their aggregates having a size smaller than the diameter of the cell wall pores could easily penetrate the cell wall and reach the plasma membrane (Navarro et al. 2008). NPs may enter into plant cells by binding to carrier proteins and ion channels, through endocytosis, by binding to organic acids present in the culture media, or by forming complexes with the transporters of the plasma membrane (Kurepa et al. 2010; Rico et al. 2013a, b). Once the NPs reach the cell cytoplasm, these may combine with various cellular organelles and obstruct the metabolic reactions at that particular site (Jia et al. 2005). Although in vitro studies have been performed using isolated plant cells to determine the NPs uptake and interactions, these studies did not prove enough to explain the interactions of NPs with the whole plant. Most of the studies have suggested that NPs uptake by plants is associated with the absorption of nutrients from the culture media in which plants are grown (Lin et al. 2009). The uptake of NPs by the plant mainly occurs through seed coat, roots, and leaves. It takes place through the root tips in many species, though NPs can also enter the plant tissues through the stomatal openings present on the leaf surface. The mucilage and exudates excreted by root caps and root hairs also assist in the NP uptake in various plant species (Chichiriccò and Poma 2015; Schwab et al. 2016). Details of the uptake through different routes are described below.

Metallic NPs are being rapidly discharged into the environment and used extensively in agriculture system over the past few years. Studies on the uptake, translocation, and accumulation of iron oxide (Fe_2O_3) NPs in pumpkin plant grown in the aqueous media showed that NPs were taken up by the roots and then transported to leaves through the stem of the plant (Zhu et al. 2008). In wheat plant (*Triticum aestivum*) grown in the hydroponic culture, uptake of the citric acid-coated Fe_2O_3 NPs (20 mg L^{-1}) took place by roots through the apoplastic pathway (Iannone et al. 2016). Leaves and roots of arugula (*Eruca sativa*) and escarole (*Cichorium endivia*) were exposed to Pt NPs to study root-to-leaf and leaf-to-root translocation. Inductively coupled plasma mass spectrometry results showed that for both plants, the proportion of Pt translocated from roots to leaves (99% for arugula and 28% for escarole) was higher than that from leaves to roots (<1% for both plants) (Kranjc et al. 2018). It was depicted that foliar surface free energy affects Pt NPs adhesion, uptake, and translocation from leaves to roots in arugula and escarole. Similarly, the fluorescence and transmission electron microscopy images revealed that the magnetic Fe_2O_3 NPs penetrated the corn (*Zea mays*) root epidermis and migrated from the epidermal layer to endodermal cells through the apoplastic route (Li et al. 2016a, b). It was hypothesized that the macromolecular exudates excreted by root cells might be responsible for the NPs accumulation in the root epidermis (Li et al. 2016a, b). A different study regarding the uptake and accumulation of the anatase titanium dioxide TiO_2 NPs (12 nm) has shown the uptake of these NPs by wheat plant roots from the hydroponics media. After 7 days of exposure, TiO_2 NPs were observed in the parenchyma cells of the root and also in the vascular cylinder, indicating the transfer of NPs from media to vegetal tissues of the wheat (Larue et al. 2011). In another study, the uptake and transport of TiO_2 NPs were studied in two crops (*Phaseolus vulgaris* and *T. aestivum*), a wetland species (*Rumex crispus*), and a floating aquatic plant (*Elodea canadensis*). All the rooted plants showed higher concentrations of Ti in roots. Ti from these NPs was translocated from roots to shoots in *R. crispus*. It was proposed that this species might increase the availability of metal NPs by producing siderophores (which have the tendency to bind metals) or by altering the rhizosphere pH (Jacob et al. 2013).

The uptake and transport of silver nanoparticles (AgNPs) of 20 nm size in *Arabidopsis thaliana* seedlings grown in gel media have been studied by Geisler-Lee et al. (2014), who noticed the presence of AgNPs on root surface and root tips and also in the root hair cells 14 days after planting. By the 17th day, AgNPs reached the vascular tissues, viz., xylem and phloem, and got transported through them to the whole plant. The AgNPs accumulated in cotyledons quite early, entered the stomatal pores at day 14, and then heavily accumulated in stomata and pavement cells by the 17th day of planting. The study thus showed the accumulation and a long-distance transport of AgNPs in *A. thaliana* (Geisler-Lee et al. 2014). Another study revealed the uptake of tannate and citrate-coated gold (Au) NPs of 10, 30, and 50 nm size in tobacco *Nicotiana* and *Triticum aestivum* plants. The uptake behavior of Au NPs was different in the two species. Au NPs were found in the leaf midrib of tobacco plant in detectable concentrations, whereas the wheat leaf showed no evidence of Au NPs accumulation. The authors suggested that there is no requirement of passive movement through the pores of the cell wall for NPs uptake inside the plants (Judy et al. 2012). Similarly, polyvinylpyrrollidone (PVP)-coated Au NPs of diameter 40 nm were applied to tomato seeds at concentrations of 0.2 mg L^{-1} and 5 mg L^{-1}, and the tomato plants obtained were capable of absorbing the 40 nm-sized Au NPs and transporting them as intact NPs to the plant shoots (Dan et al. 2015).

A recent study (Antisari et al. 2015) described the uptake and translocation of metal oxides CeO_2, Fe_3O_4, SnO_2, and TiO_2 and metallic Ag, Co, and Ni NPs in tomato plants grown in a soil and peat mixture. Fe_3O_4 and TiO_2 did not show any uptake by plant organs. Ag, Co, and Ni accumulated in leaves, and Fe, Ce, and Sn were accumulated in roots, while Ni was found in stems. Ag and Co were translocated from roots to the aboveground organs, whereas no translocation was observed in the case of Ce, Sn, and Ti. The translocation of metal NPs could be associated with the absorption behavior of water and nutrients in plant roots (Antisari et al. 2015). Similarly, CuO NPs exposed to *Z. mays* were studied for their transport and bio-distribution in the plant. The results demonstrated that CuO NPs were taken up by the plant roots. The evidence of translocation of NPs was provided by the presence of NPs in the xylem saps which might be responsible for the transport of particles from roots to shoots (Wang et al. 2012). NPs could also be translocated back to roots from shoots by the phloem. In general, CuO NPs existed in the cell wall of epidermal cells, intracellular spaces, and the cytoplasm of cortical cells. Their presence was also observed in the nucleus at 100 mg L^{-1} concentration. The study suggested that NPs may pass through epidermis and cortex to enter the intracellular space via the apoplastic pathway; further, the presence of endosome confirmed that CuO NPs entered the plant cells by endocytosis (Wang et al. 2012). Another study of zinc oxide (ZnO) NPs (30 nm) in relation to maize plants has shown that ZnO NPs were present in the epidermal cells, cortex, and root tips of the plant. Few NPs also entered the vascular cylinder of tap roots through the primary root-lateral root junction. No NPs were observed in shoots, thus indicating a lack of NPs translocation to the aerial plant parts (Lv et al. 2015) (Table 14.1).

Table 14.1 Role of metal and metal oxide nanoparticles in plant growth and development

Type of nanoparticles (NPs)	Plant	Concentration of NPs used	Effect on plants	Reference
AgNPs	*A. thaliana*	300 mg L^{-1}	Inhibition of root elongation and leaf expansion, decreased photosynthetic efficiency	Sosan et al. (2016)
AgNPs	Wheat, cowpea, and *Brassica*	50 and 75 ppm	Improvement in growth, root nodulation and shoot parameters	Mehta et al. (2016)
AgNPs	*Lycopersicon esculentum*		Drastic decrease in germination index, root and shoot length	Mehrian et al. (2016)
AgNPs	Maize	100 mg kg^{-1}	Increase in biomass	Sillen et al. (2015)
Cu NPs and AgNPs	*Cucurbita pepo*	–	Decrease in biomass and transpiration	Musante and White (2012)
AgNPs	*Raphanus*	500 mg L^{-1}	Decrease in root and shoot length, decrease in water content	Zuverza-Mena et al. (2016)
AgNPs	*Brassica*	50 ppm	Positive effect on root and shoot length and seedling index	Sharma et al. (2012)
AgNPs	Rice	10 or 100 mg L^{-1}	Decrease in growth and germination of seedlings	Thuesombat et al. (2014)
CeO_2	*Arabidopsis thaliana*	250 ppm	Increase in biomass	Ma et al. (2013)
Fe NPs	Barley	300 mg L^{-1}	Reduction in germination rate	El-Temsah and Joner (2012)
Alumina NPs	*Lemna minor*	0.3 g L^{-1}	Increase in root length and accumulation of biomass	Juhel et al. (2011)

(continued)

Table 14.1 (continued)

Type of nanoparticles (NPs)	Plant	Concentration of NPs used	Effect on plants	Reference
ZnO	*Arachis hypogea*	1000 ppm	Enhancement in stem and root growth, high yield	Prasad et al. (2012)
ZnO	Soybean	500 mg L^{-1}	Increase in root growth of the plant	Lopez-Moreno et al. (2010)
Fe, Co, and, Cu NPs	Soybean	<300 mg $hectare^{-1}$	Increase in yield of the crop plant	Ngo et al. (2014)
Zerovalent Fe	Peanut	40 and 80 μmol L^{-1}	Stimulation of seedling development and growth	Li et al. (2015a, b)
ZnO, FeO, and ZnCuFe-oxide NPs	*Vigna radiata*	–	Improvement in shoot growth of seedlings	Dhoke et al. (2013)
Cu NPs	Lettuce	0.013% (w/w)	Improvement in shoot/root ratio	Shah and Belozerova (2009)
CuO	*Triticum aestivum*	500 mg kg^{-1}	Increase in plant biomass	Dimkpa et al. (2012)
TiO_2	Wheat	2 and 10 ppm	Promotion of seed germination	Feizi et al. (2012)
Au NPs	*Gloriosa superba*	1000 μM	Enhancement in seed yield	Gopinath et al. (2014)
TiO_2	Fennel	60 ppm	Positive effect on seed germination and shoot dry weight	Feizi et al. (2013)
ZnO	*Vigna radiata*	20, 40, 60 and 100 mg	Increase in germination rate, fresh and dry weights of roots and shoots	Jayarambabu et al. (2014)
Au NPs	*Brassica*	10–25 ppm	Increase in seed yield as well as increase in total sugar content	Arora et al. (2012)
Al NPs	Raddish, rape	2000 mg L^{-1}	Improvement in growth of plant roots	Lin and Xing (2007)

Quantum dots (QDs) are another important kind of metal-based NPs used in various applications. The impact of these NPs after their release in the environment is necessary to understand. Thereby, few recent studies have explored the uptake, translocation, and bioaccumulation of CdSe/CdZnS QDs in plants. In a study, *A. thaliana* leaf petiole was exposed to three types of CdSe/CdZnS coated with anionic, cationic, and neutral coatings. It was found that the anionic NPs were taken up by the leaf petiole and roots and distributed uniformly in the plant leaves. The cationic and neutral QDs showed destabilization in the plant growth media. The cationic QDs failed to translocate readily to the distal parts of plant leaves possibly due to their aggregation. Confocal microscopic studies demonstrated that the anionic and neutral charged QDs get accumulated within cells of the main root, root hairs, and leaf veins (Koo et al. 2014). In another study of *A. thaliana* exposed to CdSe/CdZnS QDs, intact QDs were not internalized by the plants. Analysis of Cd and Se in the roots and leaves showed that QDs were mainly present on the external surfaces of roots and were not transported to the plant leaves (Navarro et al. 2012). The strong adsorption of QDs on the root surface was supposed to be due possibly to the mechanism of van der Waals interactions as well as the cross-linking of CO_2 groups on the plant roots and QDs. The uptake of QDs occurs when QDs are conjugated to nitrogenous organic compounds like arginine, glycine, and chitosan (Silver and Ou 2005).

14.4 Effect of NPs Uptake on Plant Physiology and Functions

The main processes that constitute plant physiology are water uptake, transpiration, photosynthesis, plant nutrition, respiration, seed dormancy and germination, and plant growth and movements. Since, the NPs are widely used in various products that are released into the environment, it is necessary to study the impact of these NPs on plant physiology. Studies have shown both the positive and negative effects of NPs on the physiological parameters of plants.

14.4.1 Seed Germination

Water is important for seed germination, a process by which a plant grows from the seed through the emergence of radical and plumule. NPs present in the plant growth media highly regulate the water imbibitions by the seed coat and thus affect seed germination. In a recent study (Andersen et al. 2016), TiO_2 NPs and CeO_2 NPs were found to affect seed germination as well as the timing of cotyledon development in ten different plant species. Exposure of TiO_2 NPs significantly altered the germination rate in five species, although the germination was enhanced in some species and suppressed in some others. CeO_2 NPs did not show alteration of germination in any of the ten species (Andersen et al. 2016). The mechanism behind alteration in germination rate may be related to the photocatalytic activity of these NPs (Ma et al. 2012). Another study investigated the influence of ZrO_2, SiO_2, Al_2O_3, and TiO_2 NPs on maize seed germination. Al_2O_3 and TiO_2 reduced the germination percentage, while SiO_2 has enhanced it under all the growth conditions. The uptake of metal NPs by seeds was the maximum for SiO_2, followed by TiO_2, Al_2O_3, and then ZrO_2 (Karunakaran et al. 2016). Corn seedlings exposed to 400 mg L^{-1} concentration of ZnO NPs (24 ± 3 nm) for 15 days showed a significant reduction of 40 and 53% in seed germination at exposure temperature of 20 and 25 °C, respectively (López-Moreno et al. 2017). The effect of AgNPs on the germination (%) of pea, rice, and maize plants varied with the size of the NPs. AgNPs of all sizes (77.5, 111.7, 68.6, and 98.9 nm) showed a 100% higher germination in peanut, whereas AgNPs of 77.5 and 111.7 nm showed lesser germination in maize and rice, as compared to controls (Prasad et al. 2016). Seed germination was greatly influenced upon the application of metallic NPs like Ag, Au, and Cu. The seed germination was 98.6% for Ag, 69.6% for Cu, and 56.5% for Au NPs suspensions, compared to the control, after 35 days of inoculation (Hussain et al. 2017). Similarly, effect of AgNPs at concentration (0, 10, 20, 30, and 40 $\mu g\ mL^{-1}$) was investigated on the rate of seed germination in lentils. At low concentrations (10 $\mu g\ mL^{-1}$), AgNPs promoted seed germination, while adverse effects were seen at higher concentrations (Hojjat and Hojjat 2016). Influence of metal NPs (Si, Pd, Cu, and Au) has also been studied on seed germination in lettuce plants; the Pd and Au NPs at low concentrations whereas Cu and Si NPs at high concentrations could enhance seed germination after a 15-day exposure and ultimately affected the root and shoot growth of the plants (Shah and Belozerova 2009). Another study focused on the germination of lettuce seeds exposed to low concentrations of four metal NPs (CuO, Fe_2O_3, ZnO, and MnO_2) showed that CuO NPs inhibited seed germination and the MnO_2 and Fe_2O_3 NPs enhanced the growth of seedlings, whereas no alteration was caused by ZnO NPs, as compared to their respective controls (Liu et al. 2016). Spinach seeds treated with 50 $\mu g\ mL^{-1}$ solutions of graphene oxide NPs prompted early and extensive seed germination and increased the mass of spinach, compared to seeds treated with 0 as well as 200 $\mu g\ mL^{-1}$ NP concentration (He et al. 2018).

14.4.2 Water Absorption

Water uptake by plant roots is an important process to carry out the various metabolic functions in the plant. Plants absorb water from the media (e.g., the soil) in which they are grown. The nanomaterials present in the plant growth media or soil greatly alter the water absorption capacity of plant roots by regulating the gene expression of water channel proteins involved in uptake process. In a recent study, effect of a short-term exposure of the nano zerovalent iron and Fe_2O_3 NPs on the water uptake capacity and root water content of tomato plants was investigated (Martinez-Fernandez and Komárek 2016). Fe_2O_3 NPs at a concentration of 100 $mg\ L^{-1}$ have significantly reduced water uptake by plant roots (40% less water content than in the control), whereas at the same concentration, the nanoscale zerovalent iron did not show any effect on the water uptake. Thus, the potential adherence of Fe_2O_3 NPs to the root surface could be responsible for the observed effects. Hence the difference in relative effect of two types of NPs could be due to their degree of aggregation on the surface of roots, as well as their oxidation products. It was presumed that at this concentration, Fe_2O_3 NPs got accumulated in the epithelial root cell wall causing blockage of water uptake (Martinez-Fernandez and Komárek 2016). Similarly, TiO_2 NPs also decreased the water transport of maize roots because the aggregates of NPs formed in the cell wall of roots blocked the water uptake (Asli and Neumann 2009). Also, TiO_2 NPs could interfere with the pores of the root cell wall leading to a reduction in the water uptake and ultimately inhibited the growth of tobacco roots. Similarly, cobalt oxide NPs affected the water uptake by blocking the water channels through adsorption, whereas ZnO NPs permeated the onion root cells and damaged their morphology (Ghodake et al. 2011).

14.4.3 Mineral Uptake

Since the NPs are released frequently in the soil, it is a prerequisite to study the impact of NPs on the nutrient uptake by the plants. Micro- and macronutrients absorbed by the plant roots from the soil have a great role in various physiological processes such as photosynthesis, nitrogen fixation, and nitrogen metabolism. The *Raphanus sativus* sprouts were exposed to AgNPs at a concentration of 500 mg L^{-1} to study the impact of NPs on the nutrient uptake. The Ca and Mg uptake was found to decline by 33% and 19%, respectively. Even the uptake of micronutrients (B, Zn, Mn, and Cu) was found impaired at this concentration of AgNPs. It was believed that the accumulation of AgNPs at high concentration might block the diffusion pathways or Ca and Mg channels, affecting their active absorption by the plants (Zuverza-Mena et al. 2016). Similarly, the pumpkin plants exposed to the neodymium oxide (Nd_2O_3) NPs at 100 mg L^{-1} showed significantly reduced levels of S, Ca, K, and Mg. On the other hand, when tannic acid was used in addition to Nd_2O_3 NPs, plants restored the levels of these nutrients, because the tannic acid aided in altering the surface charge of Nd_2O_3 NPs, which reduced the likelihood of NPs binding onto the roots of the plants (Chen et al. 2016).

Another study (Peralta-Videa et al. 2014) conducted on the soil-cultivated soybean has shown that CeO_2 (1000 mg kg^{-1}) exposure increased the P and Cu contents but decreased the Ca content of pods, as compared to controls. However, plant exposure to ZnO NPs resulted in a greater Zn, Mn, and Cu accumulation. Both these NPs altered the nutritional uptake in soybean to a different extent. It was suggested that CeO_2 interferes with the uptake of those nutrients which play a role in nitrogen metabolism and photosynthesis of plants (Peralta-Videa et al. 2014). In contrast, application of TiO_2 NPs (500 mg kg^{-1}) to cucumber plants grown in sandy loam soil caused accumulation of about 35% more K and 34% more P than in the control plants (Servin et al. 2013). The authors reported that TiO_2 NPs have a similar positive effect as do the plant hormones cytokinins and gibberellins that favor K and P uptake by plants. Likewise, exposure to three types (unmodified, hydrophilic, and hydrophobic) of TiO_2 NPs has been found to affect the homeostasis of essential elements in *Ocimum basilicum* plant. At 500 mg kg^{-1} concentration, the unmodified NPs increased the Cu (104%) and Fe (90%) contents, and hydrophilic NPs increased Fe content (90%), while hydrophobic ones increased Mn (339%) but reduced Ca (71%), Cu (58%), and P (40%) contents, with reference to the control (Tan et al. 2017). Another study described the effect of Cu/CuO NPs (10 and 20 mg L^{-1}) on nutrient uptake in lettuce plants showing that these NPs significantly increased the uptake of Al, Zn, S, and Fe whereas decreased that of Mg, Mn, and P (Trujillo-Reyes et al. 2014). Lettuce plants exposed to the other kind of core/shell Fe/Fe_2O_3 NPs (10 mg L^{-1}) were found to increase the uptake of Al. The uptake was less affected by the Fe/Fe_2O_3 NPs than by the Cu/CuO NPs, which might be due to less damage of the cellular membrane by the former NPs and consequently a lesser reduction in selectivity for the nutrient uptake (Trujillo-Reyes et al. 2014). Leaf of *Capsicum annum* plants exposed to 125 mg kg^{-1} of CuO NPs was found to show lower concentration of P by 42% than in plants exposed to the bulk CuO. In fruit samples upon 500 mg kg^{-1} CuO NPs exposure, the P concentration was significantly lower than the bulk CuO at 62.5 and 125 mg kg^{-1} exposure. CuO NPs at 500 mg kg^{-1} also reduced the amount of Zn in leaves compared to bulk CuO treatment (Rawat et al. 2018). The effect of citrate-coated Au NPs has also been studied on the uptake of micro- and macronutrients by the barley plants. Au NPs slightly increased the uptake of Ca when at 10 mg L^{-1} concentration and significantly increased the uptake of K when at 5 mgL^{-1} concentration. The uptake of micronutrients Zn and Fe increased, whereas that of Mn decreased after the plant's exposure to Au NPs (Feichtmeier et al. 2015).

14.4.4 Root Water Transport and Transpiration

Maintenance of the water transport capacity of roots is essential to meet the water utilization for evaporation and leaf growth in plants. Reduction in water supply by the roots may cause xylem tension, leaf growth inhibition, stomatal closure, reduced transpiration, and, ultimately, plant wilting (Neumann 2008). Imbalance in the root water transport may occur as a result of environmental stress. Any material present in the external water sources has to pass through the cell wall pores of the epidermal layer of roots to reach the root xylem through the parallel transport of molecules via both the symplastic and apoplastic pathways. Plant cell walls act as semipermeable membranes to allow the passage of water or only the small molecules through them and block the movement of the larger ones (Steudle and Peterson 1988). Therefore, the presence of NPs in the soil water sources may affect the root hydraulic conductivity and ultimately the transpiration. In a study, TiO_2 NPs applied to *Z. mays* caused a concentration-dependent inhibitory effect on the root hydraulics in a progressive manner and ultimately decreased the transpiration rate. Physical interaction between the NPs and the root cell wall, buildup of the interfacial viscosity, and further decrease in the pore diameter were considered to be reasons behind the inhibition of the root hydraulic conductivity (Asli and Neumann 2009). AgNPs and Cu NPs have reduced the transpiration rate of *Cucurbita pepo* by 66–84% and 60–70%, respectively, compared to untreated controls (Musante and White 2012). In other study, treatments of 500 and 100 mg L^{-1} of AgNPs decreased the transpiration rate of *Cucurbita pepo* by 41% (Stampoulis et al. 2009).

14.4.5 Stomatal Conductance and Gas Exchange

In a study, exposure of *A. thaliana* plants to ZnO NPs (300 mg L^{-1}) reduced the stomatal conductance and intracellular CO_2 as compared to controls (Wang et al. 2015). CeO_2 and TiO_2 NPs applied to soil-grown *Clarkia unguiculata* plants inhibited CO_2 assimilation in plants (Conway et al. 2015). Similarly, application of CeO_2 and ZnO NPs (at 0, 400 and 800 mg kg^{-1} dosage) on maize (*Zea mays*) plants revealed that CeO_2 NPs did not affect the stomatal conductance and gaseous exchange parameters, while ZnO NPs (at 800 mg kg^{-1}) reduced the stomatal conductance by 15%, compared to controls, 20 days after exposure (Zhang et al. 2015). In contrast, *Cucurbita pepo* plants grown under saline (NaCl) stress showed a decreased gas exchange and stomatal conductance due to excessive accumulation of Na^+ and Cl^- ions. However, addition of SiO_2 NPs alleviated the adverse effects of NaCl treatment by enhancing the gas exchange parameters in plants (Siddiqui et al. 2014). A recent study has described the effect of positive-charged and negative-charged CeO_2 NPs on the physiology of soybean. At 100% field capacity, both types of NPs significantly increased the stomatal conductance to a similar extent (Cao et al. 2018).

14.4.6 Photosynthetic Parameters

The effect of metal and metal oxide NPs on the chlorophyll content and photosynthesis rate of plants has been widely studied. Of late, the effect of mesoporous SiO_2 NPs on the chlorophyll and photosynthetic activity of treated wheat and lupin plants was investigated, showing an increase in chlorophyll (both *a* and *b* pigments) and the maximum photosynthetic activity when plants were exposed to 500 mg L^{-1} of NPs (Sun et al. 2016). It was believed that the application of Si increased the expression of genes (*HemD* and *PsbY*) that might be related to chlorophyll biosynthesis. Enhanced expression of these genes might result in greater activity of photosystem II and the electron transfer rate, which led to enhanced chlorophyll content (Song et al. 2014; Li et al. 2015a, b). Application of 0.1, 0.2 and 0.4% anatase TiO_2 NPs on *Ulmus elongata* seedlings lowered the net photosynthetic activity, as compared to control plants. It also led to reduction and blockage of electron flow from Q_a to Q_b (Gao et al. 2013). At 60 mg kg^{-1} concentration, TiO_2 NPs of less than 20 nm increased the chlorophyll content in wheat (*Triticum aestivum*) plants by 32.3% in comparison to the control. However, there was a decrease of 11.1% at 100 mg kg^{-1} concentration, as the plants were not able to tolerate NPs at concentrations beyond 60 mg kg^{-1} (Rafique et al. 2018). In contrast, TiO_2 NPs increased the regulated photosystem II (PS II) energy dissipation and decreased the nonregulated PS II energy dissipation during heat stress in tomato plants and then promoted photosynthesis (Qi et al. 2013).

In a similar way, application of ZnO NPs (300 and 400 mg L^{-1}) on *A. thaliana* showed that ZnO NPs significantly lowered the chlorophyll *a* and *b* content by 50%, as compared to controls. The net photosynthetic activity was also reduced by more than 50% at 300 mg L^{-1} concentration of these NPs. Even the expression of chlorophyll synthesis genes and photosystem structure genes was reduced fivefold (Wang et al. 2015). CeO_2 NPs at a concentration of 400 or 800 mg kg^{-1} showed no effect on the net photosynthetic activity after 10, 15, and 20 days of germination in maize (Zhao et al. 2012a, b). In a similar manner, CeO_2 and ZnO NPs did not affect the chlorophyll content and net photosynthetic rate of cucumber plants (Zhao et al. 2013). On the other hand, CuO NPs of <50 nm size when applied to barley leaves at concentrations of 0.5, 1.0, and 1.5 mM showed no noticeable change in chlorophyll contents on the 10th day of growth. A sudden loss of chlorophyll contents was, however, observed on the 20th day of growth (Shaw et al. 2014). In another study, exposure of the aquatic plant *Lemna gibba* L. to CuO NPs for 48 h inhibited the photosynthetic rate due to inactivation of reaction centers of PS II and a decline in electron transport (Perreault et al. 2014). AgNPs showed a concentration-dependent fluorescence quenching of chlorophyll in soybean plants and also reduced the photosynthetic pigments. The fluorescence quenching of chlorophyll may be attributed to the process of electron transport from excited chlorophyll molecules to AgNPs (Falco et al. 2015). A recent study reveals that at 100% field capacity, the content of chlorophyll *a* was increased by 18 and 20% in soybean exposed to uncoated CeO_2 NPs and PVP-coated CeO_2 NPs, respectively. However, the uncoated CeO_2 NPs reduced chlorophyll *b* content by 12% and 21% at 55% and 100% field capacity, respectively, whereas the reduction caused by PVP-coated CeO_2 NPs at 55 and 100% field capacity was up to 15 and 12%, respectively. The net photosynthetic rate was significantly higher in the CeO_2 NP-treated plants than in their respective controls (Cao et al. 2018).

14.5 Factors Affecting Behavior of NPs

It is clear from the above discussion that the use of metal-based NPs has a direct impact on physiology and growth of plants. Let us now examine the important factors that decide the behavior of NPs toward plant physiology.

14.5.1 *Size, Shape, and Type of NPs*

Size and shape of NPs are the most important factors which have noticeable impact on plant performance (Siddiqi and Husen 2016, 2017). NPs of the same metal having different sizes and shapes may have different behaviors toward the plant physiological processes. In a study of rice plants exposed to AgNPs of different sizes (20, 30–60, 70–120, and 150 nm), the uptake was increased when the seeds were treated with small (20 nm)-sized AgNPs. Treatment with AgNPs of 150 nm diameter resulted in leaf cell deformation (Thuesombat et al. 2014). *Nicotiana tabacum* L. cv Xanthi (tobacco) and *Triticum aestivum* (wheat) have been exposed to tannate (T-MNPs)- or citrate (C-MNPs)-coated Au NPs of 10, 30, and 50 nm sizes. All these MNPs bioaccumulated in tobacco, but no bioaccumulation of MNPs was observed in wheat (Judy et al. 2012). In another study, ZnO and CeO_2 NPs were exposed to soybean plants. Presence of CeO_2 NPs was observed in plant roots, but no ZnO NPs were present in roots, demonstrating the differential effect of different types of metal NPs exposed to plants (Lopez-Moreno et al. 2010). Similarly, the effect of four different metallic NPs (Al_2O_3, SiO_2, Fe_3O_4, and ZnO) was different on the development of *Arobidopsis thaliana*; *ZnO NPs were most phytotoxic, followed by* Fe_3O_4, SiO_2, and Al_2O_3. ZnO NPs inhibited the germination of seeds, and the effect was size dependent (Lee et al. 2010). A similar study evaluated the effect of anatase and rutile TiO_2 on flax seeds. The anatase TiO_2 were more toxic than the rutile TiO_2 due to differences in their crystalline behavior (Clement et al. 2013). Different types of carbon nanomaterials, namely, fullerenes, single-walled carbon nanotube (CNT), multi-walled CNT, and graphene oxide NPs, showed different effects on the physiological parameters, viz., concentration of phytohormones, activity of antioxidant enzymes, and extension growth of plant axis in rice (*Oryza sativa*) grown in loamy soil (Hao et al. 2018).

14.5.2 Surface Coating or Surface Functionality of NPs

Surface characteristics of NPs are important with reference to phytotoxicity of NPs. Five plant species, *Zea mays* (corn), *Cucumis sativus* (cucumber), *Glycine max* (soybean), *Brassica oleracea* (cabbage), and *Daucus carota* (carrot), treated with monomolecular layers of alumina NPs, showed that phytotoxicity of alumina NPs was reduced appreciably by the loading of 10% monomolecular layers of phenanthrene. Alumina NPs loaded with phenanthrene were less toxic than the non-loaded NPs (Yang and Watts 2005). Likewise, corn plants treated with bare and alginate-coated cerium oxide (CeO_2) NPs in soil exhibited that alginate coating on NPs surface increased the uptake of Ce in plants (Zhao et al. 2012a, b). In another study, the surface of ultra-small TiO_2 NPs was modified with alizarin red S and sucrose and the effect was seen in the intact *Arabidopsis thaliana*. The nano-conjugates traversed the cell walls to enter plant cells and accumulated in specific subcellular locations (Kurepa et al. 2010). One more study demonstrated the effect of surface ligands, i.e., cysteamine, cysteine, and thioglycolic acid, on the uptake of Au NPs in tomato and rice. The negatively charged cysteine-coated Au NPs were more easily internalized by roots and then transferred to plant shoots as compared to the thioglycolic acid-coated Au NPs of similar size and having similar charge. It was noted that the uptake and translocation of NPs depend more upon the type of surface ligand rather than the surface charge (Li et al. 2016a, b). Similarly, citric acid coated and uncoated CeO_2 NPs could increase the shoot length and total chlorophyll content in tomato plants, whereas the same were reduced after incubation of plants with the bulk cerium oxide and cerium acetate. The ligand coating on the surface of NPs reduced the Ce uptake by plants and did not have any effect on the translocation of NPs (Barrios et al. 2016).

14.5.3 Surface Charge of NPs

Surface charge (cationic, anionic, or neutral) of NPs is a major deciding factor for their uptake by the plant parts. The positively charged Au NPs were most readily taken up by plant roots, while the negatively charged NPs were most efficiently translocated from roots to shoots (Zhu et al. 2012). CeO_2 NPs functionalized with positive-charged, negative-charged, and neutral dextran coatings were assessed for their uptake by wheat seedlings. A 15–20% reduction from Ce(IV) to Ce(III) was found in both roots and leaves. Due to their high affinity with the negative-charged cell walls, $CeO_2(+)$ NPs adhered in a strongest way to the plant roots. After 34 h, plants exposed to $CeO_2(-)$ and $CeO_2(0)$ NPs showed higher leaf Ce concentrations than the plants exposed to $CeO_2(+)$ NPs (Spielman-Sun et al. 2017).

14.5.4 Concentration/Dose of NPs

Concentration of NPs has a significant effect on the growth and development of plants (Siddiqi and Husen 2016, 2017). Exposure to 24 nm-sized green synthesized Au NPs at concentrations of 10 and 80 $\mu g\ mL^{-1}$ enhanced the shoot length of *A. thaliana* seedlings by 1.42- and 1.64-folds in comparison to the control (Kumar et al. 2013). Cucumber plants exposed to CeO_2 or ZnO NPs at concentrations of 0, 400, and 800 $mg\ kg^{-1}$ have indicated that remained unaffected with reference to plant growth, gas exchange, and chlorophyll content. However, at 800 $mg\ kg^{-1}$ concentration, CeO_2 NPs reduced the yield, compared to the control (Zhao et al. 2013). The hydroponically grown lettuce treated with core–shell (Fe/Fe_3O_4, Cu/CuO NPs) at 10 and 20 $mg\ L^{-1}$ and $FeSO_4{\cdot}7H_2O$ and $CuSO_4{\cdot}5H_2O$ at 10 $mg\ L^{-1}$ indicated no effect of iron ions/NPs on the physiological parameters. Conversely, Cu ions/NPs decreased the water content, root length, and dry biomass of the lettuce plants (Trujillo-Reyes et al. 2014). Treatment with 2000 $mg\ L^{-1}$ and 4000 $mg\ L^{-1}$ CeO_2 NPs has induced genotoxic effects in soybean plants (Lopez-Moreno et al. 2010). *Lemna minor* exposed to media with various concentrations of TiO_2 NPs showed that superoxide dismutase activity was decreased when TiO_2 NPs concentration was higher than 200 $mg\ L^{-1}$ and the plant cell membrane experienced serious damage at 500 $mg\ L^{-1}$ TiO_2 NP concentration (Song et al. 2012). Similarly, TiO_2 NPs have a significant effect on seed germination, plant growth, and development of switchgrass (*Panicum virgatum*) in a dose-dependent manner (Boykov et al. 2018).

The well-known model plant, *A. thaliana*, when exposed to different concentrations of CeO_2 and indium oxide (In_2O_3) NPs, indicated that CeO_2 NPs at 250 ppm can significantly increase the plant biomass, but at 500–2000 ppm the plant growth was reduced by 85%. CeO_2 NPs at 1000 and 2000 ppm reduced the chlorophyll production by nearly 60% and 85%, respectively, whereas anthocyanin production was enhanced three- to fivefold. Malondialdehyde (MDA) production was unaffected by the exposure to 250–500 ppm CeO_2 NPs, but it increased by 2.5-fold at 1000 ppm (Ma et al. 2013). *Application of* CeO_2 NPs at 0–1000 mg kg^{-1} and ZnO NPs at 0–500 mg kg^{-1} reduced the nutritional value of soybean plants. Compared to the control, CeO_2 NPs at 1000 mg kg^{-1} could cause significantly less accumulation of Ca and more of P and Cu in pods, while 100 mg kg^{-1} ZnO NPs led to a higher accumulation of Zn, Mn, and Cu (Peralta-Videa et al. 2014). The higher concentration of NPs had a more negative effect on seedling growth (Thuesombat et al. 2014). Growth of roots of buckwheat treated with ZnO NPs and CuO NPs at 2000 and 4000 mg L^{-1}, respectively, was reduced considerably in comparison to the control (Lee et al. 2013).

14.5.5 Type of Plant Species Exposed

NPs of the same type often have different impact on different plant species (Husen and Siddiqi 2014b, c; Siddiqi and Husen 2016, 2017). A study revealed that radish and ryegrass roots generally accumulated more amounts of Au NPs (14–900 ng mg^{-1}) than rice and pumpkin roots (7–59 ng mg^{-1}). Accumulation was statistically significant in rice shoots (1.1–2.9 ng mg^{-1}), while none of the Au NPs accumulated in the shoots of radish and pumpkins (Zhu et al. 2012). NPs bioaccumulation has also been reported to be plant species-dependent possibly due to interspecific differences in the nature of chemical exudations from roots (Judy et al. 2012). In another study, effect of Cu NPs on the growth of *Polyboroides radiatus* and *Triticum aestivum* was studied and found that *P. radiatus* was more sensitive to Cu NPs than *T. aestivum* (Lee et al. 2008).

14.5.6 Nature of Growth Media

The medium used for plant cultivation also has an effect on the phytotoxicity of NPs. The toxicity and bioavailability of AgNPs to *Polyboroides radiatus* and *Sorghum bicolor* were measured in the agar and soil media. Growth of *S. bicolor* was reduced in a concentration-dependent manner in the agar medium. In the case of soil, growth of *P. radiatus* was not significantly affected, but *S. bicolor* showed a slightly reduced growth rate (Lee et al. 2012). In another study, it was shown that the soil organic matter plays a vital role in the mobility and bioavailability of CeO_2 NPs in the soil solution (Zhao et al. 2012a, b). The pumpkin and wheat plants were exposed to CeO_2 NPs dissolved in media containing gum arabic or fulvic acid. None of the plants showed a reduced growth or any toxic response. CeO_2 NPs were found to be translocated shoots in pumpkin but not in wheat plants. The presence of fulvic acid and gum arabic acid also affected the amount of NPs associated with roots (Schwabe et al. 2013).

14.6 Effect of NPs-Mediated Gene Delivery on Plant Growth

Metallic NPs, nanofibers, and nanocapsules offer a new set of tools to manipulate the genes. Appropriately functionalized NPs serve as a platform to transport a large number of genes as well as chemicals that trigger gene expression in plants (McKnight et al. 2003; Radu et al. 2004; Roy et al. 2005; Torney et al. 2007; Xia et al. 2009). Fluorescent-labeled starch NPs were used for transport of genes across the plant cell wall. Integration of different genes on NPs and, at the same time, imaging of the fluorescent NPs were possible with fluorescence microscope so as to monitor the movement of exterior genes along with the expression of genes (Liu et al. 2008). SiO_2 NPs capped with Au NPs have been used for particle bombardment in plant cells. The results have demonstrated that the plasmid DNA transferred by the gene gun method, using the gold-capped SiO_2 NPs, was successfully expressed in the intact tobacco and maize tissues. The main benefit was the simultaneous delivery of both DNA and effector molecules to the specific sites that resulted in the site-targeted delivery and expression of chemicals and genes, respectively (Torney et al. 2007). In another study, plasmids were covalently bound to magnetic NPs and successfully delivered to canola cells. The expression of plasmid delivery after 48 h in suspension culture was confirmed by the appearance of blue-color expression of GUS gene due to staining with 5-bromo-4-chloro-3-indolyl-β-d-glucuronic acid (X-Gluc) in the plant protoplasts (Hao et al. 2013). Likewise, SiO_2 NPs served as a vehicle for the delivery of foreign DNA into the roots of *Arabidopsis thaliana*. The expression of DNA was noticed in the epidermal layer and also in the inner most cortical and endodermal regions of plant roots (Chang et al. 2013). Another study has described the role of gold-coated SiO_2 NPs for co-delivery of protein and DNA into the onion plant cells by the particle bombardment method. The expression of marker gene (enhanced green fluorescent protein) and fluorescently labeled BSA protein was detected in the onion epidermal cells (Martin-Ortigosa et al. 2012). Thus, the metallic NPs serve as candidates suitable for the plasmid/DNA delivery to plant cells.

14.7 Phytotoxicity of NPs

NPs are used in a wide variety of consumer products such as cosmetics, wound dressings, textiles, and so on. After the end usage of NPs, these are discharged into the environment inevitably, leading ultimately to the pollution of water bodies and soil. Plants also get exposed to NPs in the aquatic and terrestrial environments (Choudhury et al. 2016). Airborne NPs possess the tendency to attach to leaves and other aerial parts of plants, whereas roots interact with NPs through the waterborne or soil materials. Once present on the leaf surface, NPs penetrate the plants via the bases of trichomes or through stomata and then get translocated to tissues of different plant systems. The main damage to the ecosystem due to NP deposition is related to the alteration of competition pattern among the species, which results in a drastic effect on plant biodiversity (Choudhury et al. 2016). More sensitive species may be eliminated, while the growth, flowering, and fructification of other species may be favored. This situation has brought the phytotoxicity issues of NPs to the fore.

Higher plants strongly interact with their atmospheric and terrestrial environments. These interactions are expected to be affected by their exposure to NPs (Navarro et al. 2008). Most of the studies undertaken on the effects of NPs on higher plants have been confined to seed germination and root elongation test (Lin and Xing 2007; Yang et al. 2006; Zheng et al. 2005). TiO_2 NPs have shown positive effects on germination of aged spinach seeds as well as on the growth of seedlings (Zheng et al. 2005). Similarly, TiO_2 NPs significantly promoted the growth of spinach and accelerated the nitrogen assimilation (Yang et al. 2006). TiO_2 NPs were not much toxic to willow trees, and the toxic effects did not follow a clear dose-dependent relationship, probably due to the formation of aggregates and their subsequent sedimentation (Seeger et al. 2009). Similarly, effects of TiO_2 NPs on the photochemical reaction of chloroplasts of *Spinacia oleracea* were analyzed (Hong et al. 2005), which showed that the nano-TiO_2 treatment could enhance the Hill reaction and the activity of chloroplasts, which further accelerated the FeCy reduction and oxygen evolution. Moreover, noncyclic photophosphorylation activity was higher than the cyclic photophosphorylation activity during the TiO_2 treatment, which also increased the germination rate and vigor indexes (Zheng et al. 2005). The plant dry weight also increased due to increase in the chlorophyll formation and the ribulose bisphosphate carboxylase/oxygenase activity and, consequently, in the photosynthetic rate. Gao et al. (2006) have shown that Rubisco carboxylase activity was 2.67 times higher in the nano-anatase TiO_2-treated *Spinacia oleracea* than that in the control. The molecular mechanism of the carbon reaction promoted by nano-anatase TiO_2 during photosynthesis is still not clearly understood. Reverse transcription PCR and northern blotting experiments have evidenced that mRNAs encoding the small and large subunit of Rubisco were promoted in the NPs-treated plants (Xuming et al. 2008). Accordingly, the protein expression of Rubisco from the nano-anatase-treated spinach was increased by 40%, as compared to the control.

The phytotoxicity of AgNPs was investigated in crop plants *Phaseolus radiatus* and *Sorghum bicolor* grown in the agar and soil media. AgNPs showed a concentration-dependent growth inhibition of *P. radiatus* and *S. bicolor* seeds in agar media and also caused browning and necrosis of root tips. The phytotoxicity was relatively less in soil media due to the reduced bioavailability of AgNPs in soil than in agar (Lee et al. 2012). In a similar study, exposure of glucoxylan-mediated green synthesized AgNPs (40 nm) and Au NPs (6 nm) did not show any significant effect on germination of radish seeds. These two types of NPs behaved differently toward the root growth of radish seedlings, where AgNPs induced a stimulatory effect on root length in a concentration-dependent manner, while Au NP exposure caused no significant change in the root length (Iram et al. 2014). Zn and ZnO NPs have shown effect on the root and shoot growth in radish, rape, and ryegrass. The phytotoxicity of ZnO NPs was not directly related with their limited dissolution in the bulk nutrient solution or rhizosphere (Lin and Xing 2007). In the aquatic plant *Hydrilla verticillata*, growth in early stages was inhibited after exposure to ZnO NPs at a high concentration of 1000 mg mL^{-1}, whereas *Phragmites australis* indicated a decline in growth rate after few weeks of exposure. It was concluded that high dosage of ZnO NPs caused significant phytotoxicity in these aquatic plants (Song and Lee 2016). In another study to determine the toxicity of ZnO NPs in the corn and cucumber plants, root length was found to reduce by 17% and 21%, respectively, while seed germination remained unaffected by the ZnO NPs at a concentration of 1000 mg mL^{-1} (Zhang et al. 2015).

In a study of the phytotoxicity and bioavailability of Cu NPs to *Phaseolus radiatus* and *Triticum aestivum*, growth rate of both the crop plants was found to drop as a result of exposure to NPs. Bioaccumulation was concentration-dependent, and the contents of NPs in plant tissues increased with increasing NPs concentration in the growth media. *T. aestivum* showed a relatively greater accumulation of Cu NPs in roots due to its peculiar root morphology (Lee et al. 2008). Of late, toxicity of Cu NPs was evaluated in 10–15 days old alfalfa and lettuce seedlings grown in hydroponics. Plant size and the nutrient content of tissues got reduced, whereas enzyme activity was altered in both the plants (Hong et al. 2015). In *A. thaliana* grown in agar media, CeO_2 NPs caused stimulatory effect at a low dose, while high NP concentrations led to adverse effects on antioxidant system, photosystem, and the overall plant growth (Yang et al. 2016). The toxicity effects of Al_2O_3 NPs (13 nm) were investigated in a time- and dose-dependent manner on the root growth and development of *T. aestivum*. There was a decrease in root elongation with increase in the concentration of NPs applied. Moreover, a decrease in the total protein content, an increase in peroxidase activity and accumulation of lignin, and a damage to root cortex were also recorded (Yanik and Vardar 2015).

The magnetic NPs coated with tetramethylammonium hydroxide had a magnetic influence on the enzyme structures implied at different stages of photosynthesis during the early growth of *Zea mays* plants. Small concentrations of aqueous ferrofluid solution added to the culture medium showed a stimulating effect on growth, while the larger amounts induced an inhibitory effect (Racuciu and Creanga 2007). Toxic effects of semiconductor L-cysteine capped CdS NPs (15–20 nm) were examined on the aquatic plant *Spirodela polyrrhiza*, which experienced significant reduction in plant growth as well as changes in the activity of antioxidant enzymes (peroxidase and superoxide dismutase), demonstrating the phytotoxicity of the synthesized NPs (Khataee et al. 2014). Earlier, phytotoxicity and biotransformation of La_2O_3 NPs were assessed in cucumber roots. The NPs over a concentration of 200 mg L^{-1} affected the root elongation and growth rate. The organic acids secreted from the root cells could possibly cause dissolution of the NPs applied. Moreover, the biotransformation of La_2O_3 NPs from spherical shape to needlelike form was observed in the intercellular spaces and middle lamella of the root tissues (Ma et al. 2011).

14.8 NPs Exposure Causes Oxidative Stress in Plants

There are limited reports in the literature regarding the oxidative stress caused or altered by NPs in the plant tissues. Reactive oxygen species (ROS) are free radicals which contain atoms of oxygen and are generally formed inside several plant structures due to various metabolic pathways (Anjum et al. 2015a, b; Aref et al. 2016). Various environmental factors are responsible for increasing the ROS in the cellular milieu of plants, leading to stressful conditions. Increase in ROS inside plant cell membranes may cause several damages and disturb normal activity of plants (Gill and Tuteja 2010; Anjum et al. 2012). Heavy metal ions present in soil or water may affect the plant metabolism and cause oxidative stress (Lopez-Moreno et al. 2010). Excessive increase in ROS levels may elicit detoxification mechanisms involving both enzymatic and nonenzymatic systems that tend to prevent oxidation of biological molecules like proteins or lipids (Gill and Tuteja 2010). A difference in balance of both the ROS production and detoxification leads to oxidative stress. Oxidative stress caused in plants by metal NPs has been mentioned in several studies (Anjum et al. 2015a, b).

In a study, wheat (*T. aestivum*) plants exposed to Fe_2O_3 NPs were evaluated for any oxidative damage by monitoring the activity of antioxidant enzymes. It came out that the activity of enzymes such as ascorbate peroxidase, superoxide dismutase (SOD), guaiacol peroxidase, and catalase (CAT) was increased significantly in the roots and aerial parts of NPs-treated plants (Iannone et al. 2016). Similarly, *R. sativus* exposed to Fe_2O_3 NPs (1 mg mL^{-1} concentration) showed a 219.5% increase in the ROS inside the cells. Consequently, activities of CAT, SOD, and the glutathione content also increased, showing that Fe_2O_3 NPs generated heavy oxidative stress which was countered by the antioxidant enzymes in order to develop the defense system (Saquib et al. 2016). AgNPs of size 20 nm (20 and 50 mg mL^{-1}) applied to *Vigna radiata* L. (mung bean) for 21 days produced ROS and increased the formation of hydrogen peroxide and lipid peroxidation, which caused cellular damage (Nair and Chung 2015). In contrast, AgNPs synthesized by synthetic route caused a decrease in SOD activity in *E. crassipes* at all the doses used, whereas activity of CAT was increased. In water hyacinth plant, the activity of catalase (CAT) and peroxidase (POD) was decreased with increase in the concentration of biosynthesized AgNPs applied. The study suggested that synthetic AgNPs could produce more oxidative stress in comparison to the biosynthesized AgNPs, which was combated by modulation of antioxidant enzymes accordingly (Rani et al. 2016).

CuO and ZnO NPs in *Cucumis sativus* significantly increased the activity of SOD, CAT, and POD enzymes. The activity of SOD and POD in root cells was increased by more than 50% than in the control (Kim et al. 2012). Similarly, oxidative stress in the sand-grown wheat plants treated with CuO and ZnO NPs was evidenced by increased lipid peroxidation and oxidized glutathione in roots, leading to increased ROS production (Dimkpa et al. 2012). A significant increase in the levels of SOD and POD was found in response to ROS generation in cotton plants caused by the low dose of ZnO NPs carrying phycomolecule ligands present in the extract of algae *Halimeda tuna* (Venkatachalam et al. 2017). On the other hand, ZnO NPs at higher concentration induced the oxidative stress in *S. lycopersicum* through the generation of ROS. The activity of SOD was increased at lower concentrations of ZnO NPs, whereas lipid peroxidation measured in terms of malondialdehyde increased with increase in the ZnO NP concentration (Singh et al. 2016). In rice plants, H_2O_2 content increased by 162% over the control on exposure to 500 mg L^{-1} CeO_2. At 125 mg L^{-1}, the activity of SOD, glutathione reductase, and dehydroascorbate reductase was significantly reduced, causing damage to cellular membranes (Rico et al. 2013a, b). Similarly, radish plants grown with 125 mg of CeO_2 NPs showed enhanced CAT and ascorbate peroxidase activity in root tubers but a reduced one in leaves (Corral-Diaz et al. 2014). *Medicago sativa* cells exposed to MPA-CdSe/ZnS QDs at 10, 50, and 100 nM showed increased activity of SOD, CAT, and glutathione reductase enzymes, possibly due to the release of Cd from the degradation of CdSe/ZnS QDs. The ROS generated in the plants after the treatment with QDs was thus overcome by mobilizing the ROS-scavenging mechanisms through enzymatic actions (Santos et al. 2013).

14.9 Conclusions

A variety of metallic NPs released in our environment have a great impact on the living beings such as humans, plants, and animals. Uptake, translocation, and accumulation of metallic NPs occur in plant roots, stem, leaves, and other plant parts. Uptake of NPs affects plant physiological processes such as photosynthesis, respiration, mineral and water uptake, and transpiration. This leads to variations in seed germination rate, plant growth, and biomass. Plants' exposure to NPs has both positive and negative effects on plant growth and development, depending on several factors such as size and shape of NPs, plant species exposed, concentration of NPs, and duration of exposure. Delivery of plasmids/DNA to plant cells mediated by NPs carriers and also the phytotoxicity and oxidative stress caused in plants by application of NPs are of utmost importance. It is a prerequisite to decide the various parameters related to NPs (size, shape, concentration of use, exposure duration, etc.) before their applicability in the plant systems so as to avoid toxicity effects and get the beneficial response.

Acknowledgments RS is thankful to UGC for awarding a Senior Research Fellowship. Financial support from the Department of Biotechnology (DBT), Government of India, and the Council of Scientific and Industrial Research (CSIR), New Delhi, is also acknowledged.

References

Albanese A, Tang PS, Chan WC (2012) The effect of nanoparticle size, shape, and surface chemistry on biological systems. Annu Rev Biomed Eng 14:1–16

Andersen CP, King G, Plocher M, Storm M, Pokhrel LR, Johnson MG, Rygiewicz PT (2016) Germination and early plant development of ten plant species exposed to TiO_2 and CeO_2 nanoparticles. Environ Toxicol Chem 35:2223–2229

Anjum NA, Ahmad I, Mohmood I, Pacheco M, Duarte AC, Pereira E, Umar S, Ahmad A, Khan NA, Iqbal M, Prasad MNV (2012) Modulation of glutathione and its related enzymes in plants' responses to toxic metals and metalloids: a review. Environ Exp Bot 75:307–324

Anjum NA, Adam V, Kizek R, Duarte AC, Pereira E, Iqbal M, Lukatkin AS, Ahmad I (2015a) Nanoscale copper in the soil-plant system: toxicity and underlying potential mechanisms. Environ Res 138:306–325

Anjum NA, Sofo A, Scopam A, Roychoudhury A, Gill SS, Iqbal M, Lukatkin AS, Pereira E, Duarte AC, Ahmad I (2015b) Lipids and proteins – major targets of oxidative modifications in abiotic stressed plants. Environ Sci Pollut Res 22:4099–4121

Antisari LV, Carbone S, Gatti A, Vianello G, Nannipieri P (2015) Uptake and translocation of metals and nutrients in tomato grown in soil polluted with metal oxide (CeO_2, Fe_3O_4, SnO_2, TiO_2) or metallic (Ag, Co, Ni) engineered nanoparticles. Environ Sci Pollut Res 22:1841–1853

Aref IM, Khan PR, Khan S, El-Atta H, Ahmed AI, Iqbal M (2016) Modulation of antioxidant enzymes in *Juniperus procera* needles in relation to habitat environment and dieback incidence. Trees Struct Funct 30:1669–1681

Arora S, Sharma P, Kumar S (2012) Gold-nanoparticle induced enhancement in growth and seed yield of *Brassica juncea*. Plant Growth Regul 66:303–310

Asli S, Neumann PM (2009) Colloidal suspensions of clay or titanium dioxide nanoparticles can inhibit leaf growth and transpiration via physical effects on root water transport. Plant Cell Environ 32:577–584

Barhoumi L, Oukarroum A, Taher LB, Smiri LS, Abdelmelek H, Dewez D (2015) Effects of superparamagnetic iron oxide nanoparticles on photosynthesis and growth of the aquatic plant *Lemna gibba*. Arch Environ Contam Toxicol 68(3):510–520

Barrios AC, Rico CM, Trujillo-Reyes J, Medina-Velo IA, Peralta-Videa JR, Gardea-Torresdey JL (2016) Effects of uncoated and citric acid coated cerium oxide nanoparticles, bulk cerium oxide, cerium acetate, and citric acid on tomato plants. Sci Total Environ 563:956–964

Begum P, Ikhtiari R, Fugetsu B (2011) Graphene phytotoxicity in the seedling stage of cabbage, tomato, red spinach, and lettuce. Carbon 49:3907–3919

Boykov IN, Shuford E, Zhang B (2018) Nanoparticle titanium dioxide affects the growth and microRNA expression of switchgrass (*Panicum virgatum*). Genomics. https://doi.org/10.1016/j.ygeno.2018.03.002

Cao Z, Rossi L, Stowers C, Zhang W, Lombardini L, Ma X (2018) The impact of cerium oxide nanoparticles on the physiology of soybean [*Glycine max* (L.) Merr.] under different soil moisture conditions. Environ Sci Pollut Res 25:930–939

Chang FP, Kuang LY, Huang CA, Jane WN, Hung Y, Yue-ie CH, Mou CY (2013) A simple plant gene delivery system using mesoporous silica nanoparticles as carriers. J Mater Chem B 1:5279–5287

Chen G, Ma C, Mukherjee A, Musante C, Zhang J, White JC, Dhankher OP, Xing B (2016) Tannic acid alleviates bulk and nanoparticle Nd_2O_3 toxicity in pumpkin: a physiological and molecular response. Nanotoxicology 10:1–11

Chichiriccò G, Poma A (2015) Penetration and toxicity of nanomaterials in higher plants. Nanomater 5(2):851–873

Choudhury R, Majumder M, Roy DN, Basumallick S, Misra TK (2016) Phytotoxicity of Ag nanoparticles prepared by biogenic and chemical methods. Int Nano Lett 6:153–159

Clement L, Hurel C, Marmier N (2013) Toxicity of TiO_2 nanoparticles to cladocerans, algae, rotifers and plants–effects of size and crystalline structure. Chemosphere 90:1083–1090

Conway JR, Beaulieu AL, Beaulieu NL, Mazer SJ, Keller AA (2015) Environmental stresses increase photosynthetic disruption by metal oxide nanomaterials in a soil-grown plant. ACS Nano 9(12):11737–11749

Corral-Diaz B, Peralta-Videa JR, Alvarez-Parrilla E, Rodrigo-García J, Morales MI, Osuna-Avila P, Niu G, Hernandez-Viezcas JA, Gardea-Torresdey JL (2014) Cerium oxide nanoparticles alter the antioxidant capacity but do not impact tuber ionome in *Raphanus sativus* (L). Plant Physiol Biochem 84:277–285

Da Costa MVJ, Sharma PK (2016) Effect of copper oxide nanoparticles on growth, morphology, photosynthesis, and antioxidant response in *Oryza sativa*. Photosynthetica 54(1):110–119

Dan Y, Zhang W, Xue R, Ma X, Stephan C, Shi H (2015) Characterization of gold nanoparticle uptake by tomato plants using enzymatic extraction followed by single-particle inductively coupled plasma–mass spectrometry analysis. Environ Sci Technol 49:3007–3014

Dhoke SK, Mahajan P, Kamble R, Khanna A (2013) Effect of nanoparticles suspension on the growth of mung (*Vigna radiata*) seedlings by foliar spray method. Nanotechnol Dev 3:1

Dimkpa CO, McLean JE, Latta DE, Manangon E, Britt DW, Johnson WP, Boyanov MI, Anderson AJ (2012) CuO and ZnO nanoparticles: phytotoxicity, metal speciation, and induction of oxidative stress in sand-grown wheat. J Nanopart Res 14:1–15

El-Temsah YS, Joner EJ (2012) Impact of Fe and Ag nanoparticles on seed germination and differences in bioavailability during exposure in aqueous suspension and soil. Environ Toxicol 27:42–49

Falco WF, Queiroz AM, Fernandes J, Botero ER, Falcão EA, Guimarães FE, M'Peko JC, Oliveira SL, Colbeck I, Caires AR (2015) Interaction between chlorophyll and silver nanoparticles: a close analysis of chlorophyll fluorescence quenching. J Photochem Photobiol A Chem 299:203–209

Feichtmeier NS, Walther P, Leopold K (2015) Uptake, effects, and regeneration of barley plants exposed to gold nanoparticles. Environ Sci Pollut Res 22:8549–8558

Feizi H, Moghaddam PR, Shahtahmassebi N, Fotovat A (2012) Impact of bulk and nanosized titanium dioxide (TiO_2) on wheat seed germination and seedling growth. Biol Trace Elem Res 146:101–106

Feizi H, Kamali M, Jafari L, Moghaddam PR (2013) Phytotoxicity and stimulatory impacts of nanosized and bulk titanium dioxide on fennel (*Foeniculum vulgare* Mill). Chemosphere 91:506–511

Fleischer A, O'Neill MA, Ehwald R (1999) The pore size of non-graminaceous plant cell walls is rapidly decreased by borate ester cross-linking of the pectic polysaccharide rhamnogalacturonan II. Plant Physiol 121:829–838

Gao F, Hong F, Liu C, Zheng L, Su M, Wu X, Yang F, Wu C, Yang P (2006) Mechanism of nano-anatase TiO_2 on promoting photosynthetic carbon reaction of spinach. Biol Trace Elem Res 111:239–253

Gao J, Xu G, Qian H, Liu P, Zhao P, Hu Y (2013) Effects of nano-TiO_2 on photosynthetic characteristics of *Ulmus elongata* seedlings. Environ Pollut 176:63–70

Geisler-Lee J, Brooks M, Gerfen JR, Wang Q, Fotis C, Sparer A, Ma X, Berg RH, Geisler M (2014) Reproductive toxicity and life history study of silver nanoparticle effect, uptake and transport in *Arabidopsis thaliana*. Nanomaterials 4:301–318

Ghodake G, Seo YD, Lee DS (2011) Hazardous phytotoxic nature of cobalt and zinc oxide nanoparticles assessed using *Allium cepa*. J Hazard Mater 186:952–955

Gill SS, Tuteja N (2010) Reactive oxygen species and antioxidant machinery in abiotic stress tolerance in crop plants. Plant Physiol Biochem 48:909–930

Gopinath K, Gowri S, Karthika V, Arumugam A (2014) Green synthesis of gold nanoparticles from fruit extract of *Terminalia arjuna* for the enhanced seed germination activity of *Gloriosa superba*. J Nanostruct Chem 4:1–11

Haghighi M, Pessarakli M (2013) Influence of silicon and nano-silicon on salinity tolerance of cherry tomatoes (*Solanum lycopersicum* L.) at early growth stage. Sci Hortic 161:111–117

Hao Y, Yang X, Shi Y, Song S, Xing J, Marowitch J, Chen J, Chen J (2013) Magnetic gold nanoparticles as a vehicle for fluorescein isothiocyanate and DNA delivery into plant cells. Botany 91:457–466

Hao Y, Ma C, Zhang Z, Song Y, Cao W, Guo J, Zhou G, Rui Y, Liu L, Xing B (2018) Carbon nanomaterials alter plant physiology and soil bacterial community composition in a rice-soil-bacterial ecosystem. Environ Pollut 232:123–136

He Y, Hu R, Zhong Y, Zhao X, Chen Q, Zhu H (2018) Graphene oxide as a water transporter promoting germination of plants in soil. Nano Res 11:1928–1937

Hojjat SS, Hojjat H (2016) Effects of silver nanoparticle exposure on germination of Lentil (*Lens culinaris* Medik.). Int J Farm Allied Sci 5:248–252

Hong F, Yang F, Liu C, Gao Q, Wan Z, Gu F, Wu C, Ma Z, Zhou J, Yang P (2005) Influences of nano-TiO_2 on the chloroplast aging of spinach under light. Biol Trace Elem Res 104:249–260

Hong J, Rico CM, Zhao L (2015) Toxic effects of copper-based nanoparticles or compounds to lettuce (*Lactuca sativa*) and alfalfa (*Medicago sativa*). Environ Sci Processes Impacts 17:177–185

Husen A (2017) Gold nanoparticles from plant system: synthesis, characterization and application. In: Ghorbanpourn M, Manika K, Varma A (eds) Nanoscience and plant–soil systems, vol 48. Springer International Publication, Cham, pp 455–479

Husen A, Siddiqi KS (2014a) Plants and microbes assisted selenium nanoparticles: characterization and application. J Nanobiotechnol 12:28

Husen A, Siddiqi KS (2014b) Carbon and fullerene nanomaterials in plant system. J Nanobiotechnol 12:16

Husen A, Siddiqi KS (2014c) Phytosynthesis of nanoparticles: concept, controversy and application. Nano Res Lett 9:229

Hussain M, Raja NI, Iqbal M, Sabir S, Yasmeen F (2017) *In vitro* seed germination and biochemical profiling of *Artemisia absinthium* exposed to various metallic nanoparticles. 3 Biotech 7:101

Iannone MF, Groppa MD, de Sousa ME, van Raap MB, Benavides MP (2016) Impact of magnetite iron oxide nanoparticles on wheat (*Triticum aestivum* L.) development: evaluation of oxidative damage. Environ Exp Bot 131:77–88

Iram F, Iqbal MS, Athar MM (2014) Glucoxylan-mediated green synthesis of gold and silver nanoparticles and their phyto-toxicity study. Carbohydr Polym 104:29–33

Jacob DL, Borchardt JD, Navaratnam L, Otte ML, Bezbaruah AN (2013) Uptake and translocation of Ti from nanoparticles in crops and wetland plants. Int J Phytoremediation 15:142–153

Jayarambabu N, Kumari BS, Rao KV, Prabhu YT (2014) Germination and growth characteristics of mungbean seeds (*Vigna radiata* L.) affected by synthesized zinc oxide nanoparticles. Int J Curr Eng Technol 4:3411–3416

Jia G, Wang H, Yan L, Wang X, Pei R, Yan T, Zhao Y, Guo X (2005) Cytotoxicity of carbon nanomaterials: single-wall nanotube, multi-wall nanotube, and fullerene. Environ Sci Technol 39:1378–1383

Judy JD, Unrine JM, Rao W, Wirick S, Bertsch PM (2012) Bioavailability of gold nanomaterials to plants: importance of particle size and surface coating. Environ Sci Technol 46:8467–8474

Juhel G, Batisse E, Hugues Q, Daly D, van Pelt FN, O'Halloran J, Jansen MA (2011) Alumina nanoparticles enhance growth of *Lemna minor*. Aquat Toxicol 105:328–336

Karunakaran G, Suriyaprabha R, Rajendran V, Kannan N (2016) Influence of ZrO_2, SiO_2, Al_2O_3 and TiO_2 nanoparticles on maize seed germination under different growth conditions. IET Nanobiotechnol 10:171–177

Khataee A, Movafeghi A, Nazari F, Vafaei F, Dadpour MR, Hanifehpour Y, Joo SW (2014) The toxic effects of L-cysteine-capped cadmium sulfide nanoparticles on the aquatic plant *Spirodela polyrrhiza*. J Nanopart Res 16:1–10

Kim S, Lee S, Lee I (2012) Alteration of phytotoxicity and oxidant stress potential by metal oxide nanoparticles in *Cucumis sativus*. Water Air Soil Pollut 223:2799–2806

Koo Y, Wang J, Zhang Q, Zhu H, Chehab EW, Colvin VL, Alvarez PJ, Braam J (2014) Fluorescence reports intact quantum dot uptake into roots and translocation to leaves of *Arabidopsis thaliana* and subsequent ingestion by insect herbivores. Environ Sci Technol 49:626–632

Kranjc E, Mazej D, Regvar M, Drobne D, Remskar M (2018) Foliar surface free energy affects platinum nanoparticle adhesion, uptake, and translocation from leaves to roots in arugula and escarole. Environ Sci Nano 5:520–532

Kumar V, Guleria P, Kumar V, Yadav SK (2013) Gold nanoparticle exposure induces growth and yield enhancement in *Arabidopsis thaliana*. Sci Total Environ 461:462–468

Kurepa J, Paunesku T, Vogt S, Arora H, Rabatic BM, Lu J, Wanzer MB, Woloschak GE, Smalle JA (2010) Uptake and distribution of ultrasmall anatase TiO_2 alizarin red S nanoconjugates in *Arabidopsis thaliana*. Nano Lett 10:2296–2302

Landa P, Vankova R, Andrlova J, Hodek J, Marsik P, Storchova H, White JC, Vanek T (2012) Nanoparticle-specific changes in *Arabidopsis thaliana* gene expression after exposure to ZnO, TiO_2, and fullerene soot. J Hazard Mater 241:55–62

Larue C, Khodja H, Herlin-Boime N, Brisset F, Flank AM, Fayard B, Chaillou S, Carrière M (2011) Investigation of titanium dioxide nanoparticles toxicity and uptake by plants. Journal of physics: conference series, IOP Publishing, 304(1), 012057

Lee WM, An YJ, Yoon H, Kweon HS (2008) Toxicity and bioavailability of copper nanoparticles to the terrestrial plants mung bean (*Phaseolus radiatus*) and wheat (*Triticum aestivum*): plant agar test for water-insoluble nanoparticles. Environ Toxicol Chem 27:1915–1921

Lee CW, Mahendra S, Zodrow K, Li D, Tsai YC, Braam J, Alvarez PJ (2010) Developmental phytotoxicity of metal oxide nanoparticles to *Arabidopsis thaliana*. Environ Toxicol Chem 29:669–675

Lee WM, Kwak JI, An YJ (2012) Effect of silver nanoparticles in crop plants *Phaseolus radiatus* and *Sorghum bicolor*: media effect on phytotoxicity. Chemosphere 86:491–499

Lee S, Chung H, Kim S, Lee I (2013) The genotoxic effect of ZnO and CuO nanoparticles on early growth of buckwheat, *Fagopyrum esculentum*. Water Air Soil Pollut 224:1–11

Li P, Song A, Li ZJ, Fan F, Liang Y (2015a) Silicon ameliorates manganese toxicity by regulating both physiological processes and expression of genes associated with photosynthesis in rice (*Oryza sativa* L.). Plant Soil 397:289e301

Li X, Yang Y, Gao B, Zhang M (2015b) Stimulation of peanut seedling development and growth by zero-valent iron nanoparticles at low concentrations. PLoS One 10:e0122884

Li H, Ye X, Guo X, Geng Z, Wang G (2016a) Effects of surface ligands on the uptake and transport of gold nanoparticles in rice and tomato. J Hazard Mater 314:188–196

Li J, Hu J, Ma C, Wang Y, Wu C, Huang J, Xing B (2016b) Uptake, translocation and physiological effects of magnetic iron oxide (γ-Fe 2 O 3) nanoparticles in corn (*Zea mays* L.). Chemosphere 159:326–334

Lin D, Xing B (2007) Phytotoxicity of nanoparticles: inhibition of seed germination and root growth. Environ Pollut 150:243–250

Lin S, Reppert J, Hu Q, Hudson JS, Reid ML, Ratnikova TA, Rao AM, Luo H, Ke PC (2009) Uptake, translocation, and transmission of carbon nanomaterials in rice plants. Small 5:1128–1132

Liu J, Wang FH, Wang LL, Xiao SY, Tong CY, Tang DY, Liu XM (2008) Preparation of fluorescence starch-nanoparticle and its application as plant transgenic vehicle. J Cent S Univ Technol 15:768–773

Liu R, Zhang H, Lal R (2016) Effects of stabilized nanoparticles of copper, zinc, manganese, and iron oxides in low concentrations on lettuce (*Lactuca sativa*) seed germination: nanotoxicants or nanonutrients? Water Air Soil Pollut 227:1–14

Lopez-Moreno ML, de la Rosa G, Hernández-Viezcas JÁ, Castillo-Michel H, Botez CE, Peralta-Videa JR, Gardea-Torresdey JL (2010) Evidence of the differential biotransformation and genotoxicity of ZnO and CeO_2 nanoparticles on soybean (*Glycine max*) plants. Environ Sci Technol 44:7315–7320

López-Moreno ML, de la Rosa G, Cruz-Jiménez G, Castellano L, Peralta-Videa JR, Gardea-Torresdey JL (2017) Effect of ZnO nanoparticles on corn seedlings at different temperatures; X-ray absorption spectroscopy and ICP/OES studies. Microchem J 134:54–61

Lv J, Zhang S, Luo L (2015) Accumulation, speciation and uptake pathway of ZnO nanoparticles in maize. Environ Sci Nano 2:68–77

Ma Y, He X, Zhang P, Zhang Z, Guo Z, Tai R, Xu Z, Zhang L, Ding Y, Zhao Y, Chai Z (2011) Phytotoxicity and biotransformation of La_2O_3 nanoparticles in a terrestrial plant cucumber (*Cucumis sativus*). Nanotoxicology 5:743–753

Ma H, Brennan A, Diamond SA (2012) Photocatalytic reactive oxygen species production and phototoxicity of titanium dioxide nanoparticles are dependent on the solar ultraviolet radiation spectrum. Environ Toxicol Chem 31(9):2099–2107

Ma C, Chhikara S, Xing B, Musante C, White JC, Dhankher OP (2013) Physiological and molecular response of *Arabidopsis thaliana* (L.) to nanoparticle cerium and indium oxide exposure. ACS Sustain Chem Eng 1:768–778

Maiti S, El Fahime E, Benaissa M, Brar SK (2015) Nano-ecotoxicology of natural and engineered nanoparticles for plants. In: Brar SK, Zhang TC, Verma M, Surampalli RY, Tyagi RD (eds) Nanomaterials in the environment. American Society of Civil Engineers, USA, pp 469–485

Martinez-Fernandez D, Komárek M (2016) Comparative effects of nanoscale zero-valent iron (nZVI) and Fe_2O_3 nanoparticles on root hydraulic conductivity of *Solanum lycopersicum* L. Environ Exp Bot 131:128–136

Martin-Ortigosa S, Valenstein JS, Lin VSY, Trewyn BG, Wang K (2012) Gold functionalized mesoporous silica nanoparticle mediated protein and DNA codelivery to plant cells via the biolistic method. Adv Funct Mater 22:3576–3582

Maurer-Jones MA, Gunsolus IL, Murphy CJ, Haynes CL (2013) Toxicity of engineered nanoparticles in the environment. Anal Chem 85:3036–3049

McKnight TE, Melechko AV, Griffin GD, Guillorn MA, Merkulov VI, Serna F, Hensley DK, Doktycz MJ, Lowndes DH, Simpson ML (2003) Intracellular integration of synthetic nanostructures with viable cells for controlled biochemical manipulation. Nanotechnology 14:551

Mehrian SK, Heidari R, Rahmani F, Najafi S (2016) Effect of chemical synthesis silver nanoparticles on germination indices and seedlings growth in seven varieties of *Lycopersicon esculentum* Mill (tomato) plants. J Clust Sci 27:327–340

Mehta CM, Srivastava R, Arora S, Sharma AK (2016) Impact assessment of silver nanoparticles on plant growth and soil bacterial diversity. 3 Biotech 6:254

Miralles P, Church TL, Harris AT (2012) Toxicity, uptake, and translocation of engineered nanomaterials in vascular plants. Environ Sci Technol 46:9224–9239

Musante C, White JC (2012) Toxicity of silver and copper to *Cucurbita pepo*: differential effects of nano and bulk-size particles. Environ Toxicol 27(9):510–517

Nair PMG, Chung M (2015) Study on the correlation between copper oxide nanoparticles induced growth suppression and enhanced lignification in Indian mustard (*Brassica juncea* L.). Ecotoxicol Environ Saf 113:302–313

Navarro E, Baun A, Behra R, Hartmann NB, Filser J, Miao AJ, Quigg A, Santschi PH, Sigg L (2008) Environmental behavior and ecotoxicity of engineered nanoparticles to algae, plants, and fungi. Ecotoxicology 17:372–386

Navarro DA, Bisson MA, Aga DS (2012) Investigating uptake of water-dispersible CdSe/ZnS quantum dot nanoparticles by *Arabidopsis thaliana* plants. J Hazard Mater 211:427–435

Neumann PM (2008) Coping mechanisms for crop plants in drought-prone environments. Ann Bot 101:901–907

Ngo QB, Dao TH, Nguyen HC, Tran XT, Van Nguyen T, Khuu TD, Huynh TH (2014) Effects of nanocrystalline powders (Fe, Co and Cu) on the germination, growth, crop yield and product quality of soybean (Vietnamese species DT-51). Adv Nat Sci Nanosci Nanotechnol 5:015016

Ostiguy C, IRSST (Québec) (2006) Les effets à la santé reliés aux nanoparticules. Rapp Tech R451. IRSST

Peralta-Videa JR, Hernandez-Viezcas JA, Zhao L, Diaz BC, Ge Y, Priester JH, Holden PA, Gardea-Torresdey JL (2014) Cerium dioxide and zinc oxide nanoparticles alter the nutritional value of soil cultivated soybean plants. Plant Physiol Biochem 80:128–135

Perreault F, Samadani M, Dewez D (2014) Effect of soluble copper released from copper oxide nanoparticles solubilisation on growth and photosynthetic processes of *Lemna gibba* L. Nanotoxicology 8:374–382

Prasad TNVKV, Sudhakar P, Sreenivasulu Y, Latha P, Munaswamy V, Reddy KR, Sreeprasad TS, Sajanlal PR, Pradeep T (2012) Effect of nanoscale zinc oxide particles on the germination, growth and yield of peanut. J Plant Nutr 35:905–927

Prasad TN, Adam S, Rao PV, Reddy BR, Krishna TG (2016) Size dependent effects of antifungal phytogenic silver nanoparticles on germination, growth and biochemical parameters of rice (*Oryza sativa* L), maize (*Zea mays* L) and peanut (*Arachis hypogaea* L). IET Nanobiotechnol 11:277–285

Qi M, Liu Y, Li T (2013) Nano-TiO_2 improve the photosynthesis of tomato leaves under mild heat stress. Biol Trace Elem Res 156:323–328

Racuciu M, Creanga DE (2007) TMA-OH coated magnetic nanoparticles internalized in vegetal tissues. Rom J Phys 52:395–402

Radu DR, Lai CY, Jeftinija K, Rowe EW, Jeftinija S, Lin VS (2004) A polyamidoamine dendrimer-capped mesoporous silica nanosphere-based gene transfection reagent. J Am Chem Soc 126:13216–13217

Rafique R, Zahra Z, Virk N, Shahid M, Pinelli E, Park TJ, Kallerhoff J, Arshad M (2018) Dose-dependent physiological responses of *Triticum aestivum* L. to soil applied TiO_2 nanoparticles: alterations in chlorophyll content, H_2O_2 production, and genotoxicity. Agric Ecosyst Environ 255:95–101

Rani PU, Yasur J, Loke KS, Dutta D (2016) Effect of synthetic and biosynthesized silver nanoparticles on growth, physiology and oxidative stress of water hyacinth: *Eichhornia crassipes* (Mart) Solms. Acta Physiol Plant 38:1–9

Rawat S, Pullagurala VL, Hernandez-Molina M, Sun Y, Niu G, Hernandez-Viezcas JA, Peralta-Videa JR, Gardea-Torresdey JL (2018) Impacts of copper oxide nanoparticles on bell pepper (*Capsicum annum* L.) plants: a full life cycle study. Environ Sci Nano 5:83–95

Remédios C, Rosário F, Bastos V (2012) Environmental nanoparticles interactions with plants: morphological, physiological, and genotoxic aspects. J Bot 2012.: Article ID 751686:8. https://doi.org/10.1155/2012/751686

Rico CM, Majumdar S, Duarte-Gardea M, Peralta-Videa JR, Gardea-Torresdey JL (2011) Interaction of nanoparticles with edible plants and their possible implications in the food chain. J Agric Food Chem 59:3485–3498

Rico CM, Hong J, Morales MI, Zhao L, Barrios AC, Zhang JY, Peralta-Videa JR, Gardea-Torresdey JL (2013a) Effect of cerium oxide nanoparticles on rice: a study involving the antioxidant defense system and *in vivo* fluorescence imaging. Environ Sci Technol 47:5635–5642

Rico CM, Morales MI, McCreary R, Castillo-Michel H, Barrios AC, Hong J, Tafoya A, Lee WY, Varela-Ramirez A, Peralta-Videa JR, Gardea-Torresdey JL (2013b) Cerium oxide nanoparticles modify the antioxidative stress enzyme activities and macromolecule composition in rice seedlings. Environ Sci Technol 47:14110–14118

Roy I, Ohulchanskyy TY, Bharali DJ (2005) Optical tracking of organically modified silica nanoparticles as DNA carriers: a nonviral, nanomedicine approach for gene delivery. Proc Natl Acad Sci U S A 102:279–284

Santos AR, Miguel AS, Macovei A, Maycock C, Balestrazzi A, Oliva A, Fevereiro P (2013) CdSe/ZnS quantum dots trigger DNA repair and antioxidant enzyme systems in *Medicago sativa* cells in suspension culture. BMC Biotechnol 13:111

Saquib Q, Faisal M, Alatar AA, Al-Khedhairy AA, Ahmed M, Ansari SM, Alwathnani HA, Okla MK, Dwivedi S, Musarrat J, Praveen S (2016) Genotoxicity of ferric oxide nanoparticles in *Raphanus sativus*: deciphering the role of signaling factors, oxidative stress and cell death. J Environ Sci 47:49–62

Schwab F, Zhai G, Kern M, Turner A, Schnoor JL, Wiesner MR (2016) Barriers, pathways and processes for uptake, translocation and accumulation of nanomaterials in plants–critical review. Nanotoxicology 10:257–278

Schwabe F, Schulin R, Limbach LK, Stark W, Bürge D, Nowack B (2013) Influence of two types of organic matter on interaction of CeO_2 nanoparticles with plants in hydroponic culture. Chemosphere 91:512–520

Seeger EM, Baun A, Kästner M, Trapp S (2009) Insignificant acute toxicity of TiO_2 nanoparticles to willow trees. J Soils Sediments 9:46–53

Servin AD, Morales MI, Castillo-Michel H, Hernandez-Viezcas JA, Munoz B, Zhao L, Nunez JE, Peralta-Videa JR, Gardea-Torresdey JL (2013) Synchrotron verification of TiO_2 accumulation

in cucumber fruit: a possible pathway of TiO_2 nanoparticle transfer from soil into the food chain. Environ Sci Technol 47:11592–11598
Shah V, Belozerova I (2009) Influence of metal nanoparticles on the soil microbial community and germination of lettuce seeds. Water Air Soil Pollut 197:143–148
Sharma P, Bhatt D, Zaidi MGH, Saradhi PP, Khanna PK, Arora S (2012) Silver nanoparticle-mediated enhancement in growth and antioxidant status of *Brassica juncea*. Appl Biochem Biotechnol 167:2225–2233
Shaw AK, Ghosh S, Kalaji HM, Bosa K, Brestic M, Zivcak M, Hossain Z (2014) Nano-CuO stress induced modulation of antioxidative defense and photosynthetic performance of Syrian barley (*Hordeum vulgare* L.). Environ Exp Bot 102:37–47
Siddiqi KS, Husen A (2016) Engineered gold nanoparticles and plant adaptation potential. Nano Res Lett 11:400
Siddiqi KS, Husen A (2017) Plant response to engineered metal oxide nanoparticles. Nano Res Lett 12:92
Siddiqi KS, Rahman A, Tajuddin, Husen A (2016) Biogenic fabrication of iron/iron oxide nanoparticles and their application. Nano Res Lett 11:498
Siddiqi KS, Husen A, Rao RAK (2018) A review on biosynthesis of silver nanoparticles and their biocidal properties. J Nanobiotechnol 16:14
Siddiqui MH, Al-Whaibi MH, Faisal M, Al Sahli AA (2014) Nano-silicon dioxide mitigates the adverse effects of salt stress on *Cucurbita pepo* L. Environ Toxicol Chem 33:2429–2437
Sillen WM, Thijs S, Abbamondi GR, Janssen J, Weyens N, White JC, Vangronsveld J (2015) Effects of silver nanoparticles on soil microorganisms and maize biomass are linked in the rhizosphere. Soil Biol Biochem 91:14–22
Silver J, Ou W (2005) Photoactivation of quantum dot fluorescence following endocytosis. Nano Lett 5:1445–1449
Singh A, Singh NB, Hussain I, Singh H, Yadav V, Singh SC (2016) Green synthesis of nano zinc oxide and evaluation of its impact on germination and metabolic activity of *Solanum lycopersicum*. J Biotechnol 233:84–94
Slomberg DL, Schoenfisch MH (2012) Silica nanoparticle phytotoxicity to *Arabidopsis thaliana*. Environ Sci Technol 46(18):10247–10254
Song U, Lee S (2016) Phytotoxicity and accumulation of zinc oxide nanoparticles on the aquatic plants *Hydrilla verticillata* and *Phragmites Australis*: leaf-type-dependent responses. Environ Sci Pollut Res 23:8539–8545
Song G, Gao Y, Wu H, Hou W, Zhang C, Ma H (2012) Physiological effect of anatase TiO_2 nanoparticles on *Lemna minor*. Environ Toxicol Chem 31:2147–2152
Song A, Li P, Fan FL, Li Z, Liang Y (2014) The effect of silicon on photosynthesis and expression of its relevant genes in rice (*Oryza sativa* L.) under high zinc stress. PLoS One 9:e113782
Sosan A, Svistunenko D, Straltsova D, Tsiurkina K, Smolich I, Lawson T, Subramaniam S, Golovko V, Anderson D, Sokolik A, Colbeck I (2016) Engineered silver nanoparticles are sensed at the plasma membrane and dramatically modify the physiology of *Arabidopsis thaliana* plants. Plant J 85:245–257
Spielman-Sun E, Lombi E, Donner E, Howard D, Unrine JM, Lowry GV (2017) Impact of surface charge on cerium oxide nanoparticle uptake and translocation by wheat (*Triticum aestivum*). Environ Sci Technol 51:7361–7368
Stampoulis D, Sinha SK, White JC (2009) Assay-dependent phytotoxicity of nanoparticles to plants. Environ Sci Technol 43:9473–9479
Steudle E, Peterson CA (1988) How does water get through roots? J Exp Bot 49:775–788
Sun D, Hussain HI, Yi Z, Rookes JE, Kong L, Cahill DM (2016) Mesoporous silica nanoparticles enhance seedling growth and photosynthesis in wheat and lupin. Chemosphere 152:81–91
Tan W, Du W, Barrios AC, Armendariz R, Zuverza-Mena N, Ji Z, Chang CH, Zink JI, Hernandez-Viezcas JA, Peralta-Videa JR, Gardea-Torresdey JL (2017) Surface coating changes the physiological and biochemical impacts of nano-TiO_2 in basil (*Ocimum basilicum*) plants. Environ Pollut 222:64–72

Thuesombat P, Hannongbua S, Akasit S, Chadchawan S (2014) Effect of silver nanoparticles on rice (*Oryza sativa* L. cv. KDML 105) seed germination and seedling growth. Ecotoxicol Environ Saf 104:302–309

Torney F, Trewyn BG, Lin VSY, Wang K (2007) Mesoporous silica nanoparticles deliver DNA and chemicals into plants. Nat Nanotechnol 2:295–300

Trujillo-Reyes J, Majumdar S, Botez CE, Peralta-Videa JR, Gardea-Torresdey JL (2014) Exposure studies of core–shell Fe/Fe_3O_4 and Cu/CuO NPs to lettuce (*Lactuca sativa*) plants: are they a potential physiological and nutritional hazard? J Hazard Mater 267:255–263

Van Aken B (2015) Gene expression changes in plants and microorganisms exposed to nanomaterials. Curr Opin Biotechnol 33:206–219

Venkatachalam P, Priyanka N, Manikandan K, Ganeshbabu I, Indiraarulselvi P, Geetha N, Muralikrishna K, Bhattacharya RC, Tiwari M, Sharma N, Sahi SV (2017) Enhanced plant growth promoting role of phycomolecules coated zinc oxide nanoparticles with P supplementation in cotton (*Gossypium hirsutum* L.). Plant Physiol Biochem 110:118–127

Wang Z, Xie X, Zhao J, Liu X, Feng W, White JC, Xing B (2012) Xylem-and phloem-based transport of CuO nanoparticles in maize (*Zea mays* L.). Environ Sci Technol 46:4434–4441

Wang X, Yang X, Chen S, Li Q, Wang W, Hou C, Gao X, Wang L, Wang S (2015) Zinc oxide nanoparticles affect biomass accumulation and photosynthesis in arabidopsis. Front Plant Sci 6:1243

Wei H, Wang E (2013) Nanomaterials with enzyme-like characteristics (nanozymes): next-generation artificial enzymes. Chem Soc Rev 42:6060–6093

Xia T, Kovochich M, Liong M (2009) Polyethyleneimine coating enhances the cellular uptake of mesoporous silica nanoparticles and allows safe delivery of siRNA and DNA constructs. ACS Nano 3:3273–3286

Xuming W, Fengqing G, Linglan M, Jie L, Sitao Y, Ping Y, Fashui H (2008) Effects of nano-anatase on ribulose-1, 5-bisphosphate carboxylase/oxygenase mRNA expression in spinach. Biol Trace Elem Res 126:280–289

Yang L, Watts DJ (2005) Particle surface characteristics may play an important role in phytotoxicity of alumina nanoparticles. Toxicol Lett 158:122–132

Yang F, Hong F, You W, Liu C, Gao F, Wu C, Yang P (2006) Influence of nano-anatase TiO_2 on the nitrogen metabolism of growing spinach. Biol Trace Elem Res 110:179–190

Yang X, Pan H, Wang P, Zhao FJ (2016) Particle-specific toxicity and bioavailability of cerium oxide (CeO_2) nanoparticles to *Arabidopsis thaliana*. J Hazard Mater 322:292–300

Yanik F, Vardar F (2015) Toxic effects of aluminum oxide (Al_2O_3) nanoparticles on root growth and development in *Triticum aestivum*. Water Air Soil Pollut 226:1–13

Zhang R, Zhang H, Tu C, Hu X, Li L, Luo Y, Christie P (2015) Phytotoxicity of ZnO nanoparticles and the released Zn (II) ion to corn (*Zea mays* L.) and cucumber (*Cucumis sativus* L.) during germination. Environ Sci Pollut Res 22:11109–11117

Zhao L, Peng B, Hernandez-Viezcas JA, Rico C, Sun Y, Peralta-Videa JR, Tang X, Niu G, Jin L, Varela-Ramirez A, Zhang JY (2012a) Stress response and tolerance of *Zea mays* to CeO_2 nanoparticles: cross talk among H_2O_2, heat shock protein, and lipid peroxidation. ACS Nano 6:9615–9622

Zhao L, Peralta-Videa JR, Varela-Ramirez A, Castillo-Michel H, Li C, Zhang J, Aguilera RJ, Keller AA, Gardea-Torresdey JL (2012b) Effect of surface coating and organic matter on the uptake of CeO_2 NPs by corn plants grown in soil: insight into the uptake mechanism. J Hazard Mater 225:131–138

Zhao L, Sun Y, Hernandez-Viezcas JA, Servin AD, Hong J, Niu G, Peralta-Videa JR, Duarte-Gardea M, Gardea-Torresdey JL (2013) Influence of CeO_2 and ZnO nanoparticles on cucumber physiological markers and bioaccumulation of Ce and Zn: a life cycle study. J Agric Food Chem 61:11945–11951

Zheng L, Hong F, Lu S, Liu C (2005) Effect of nano-TiO_2 on strength of naturally aged seeds and growth of spinach. Biol Trace Elem Res 104:83–91

Zhu H, Njuguna J (2014) Nanolayered silicates/clay minerals: uses and effects on health. Health Environ Saf Nanomater, Woodhead Publishing Limited:133–146

Zhu H, Han J, Xiao JQ, Jin Y (2008) Uptake, translocation, and accumulation of manufactured iron oxide nanoparticles by pumpkin plants. J Environ Monit 10:713–717

Zhu ZJ, Wang H, Yan B, Zheng H, Jiang Y, Miranda OR, Rotello VM, Xing B, Vachet RW (2012) Effect of surface charge on the uptake and distribution of gold nanoparticles in four plant species. Environ Sci Technol 46:12391–12398

Zuverza-Mena N, Armendariz R, Peralta-Videa JR, Gardea-Torresdey JL (2016) Effects of silver nanoparticles on radish sprouts: root growth reduction and modifications in the nutritional value. Front Plant Sci 7:90

Chapter 15
Impacts of Metal and Metal Oxide Nanoparticles on Plant Growth and Productivity

Mukesh Kumar Kanwar, Shuchang Sun, Xianyao Chu, and Jie Zhou

15.1 Introduction

The art of manipulating matter on an atomic, molecular, or supramolecular level is known as nanotechnology (Banerjee and Kole 2016). Usually, materials lesser than 100 nm in one dimension, are treated as nanomaterial. Therefore, they can be one dimensional (rod-shaped), two dimensional (films), three dimensional (any shape), or zero dimensional (all dimensions are at nanoscales), based on the modification of matter (Bernhardt et al. 2010; Tiwari et al. 2014; Banerjee and Kole 2016). Because of this, they possess some characteristic features, including the physical and chemical properties that have drawn a general attention to their pivotal application in plant sciences with reference to plant growth and development.

In general, nanoparticles (NPs) can be distinguished into three main groups, viz., natural, incidental, and engineered or manufactured NPs (Nowack and Bucheli 2007; Monica and Cremonini 2009). NPs of the first type are present since the beginning of the earth and are released through natural processes like volcanic eruptions, forest fires, dust storms, and photochemical reactions. The second form of NPs is anthropogenic in nature, which emanates usually from petrol/diesel exhaust, burning of coal,

M. K. Kanwar
Department of Horticulture, Zijingang Campus, Zhejiang University,
Hangzhou, People's Republic of China

Department of Botany and Environmental Sciences, Sri Guru Granth Sahib World University,
Fatehgarh Sahib, Punjab, India

S. Sun · J. Zhou (✉)
Department of Horticulture, Zijingang Campus, Zhejiang University,
Hangzhou, People's Republic of China

X. Chu
Zhejiang Institute of Geological Survey/Geological Research Center for Agricultural
Applications, China Geological Survey, Hangzhou, People's Republic of China

A. Husen, M. Iqbal (eds.), *Nanomaterials and Plant Potential*,
https://doi.org/10.1007/978-3-030-05569-1_15

and industrial exhausts (Buzea et al. 2007). The engineered NPs (ENPs) can be categorized as carbon-based NPs (CB-NPs), metal-based NPs (MB-NPs), magnetic NPs, dendrimers, and composite NPs. The metal and metal oxide-based NPs are purposely produced by humans from different metals like gold (Au), silver (Ag), zinc (Zn), nickel (Ni), ferrum (Fe), and copper (Cu) and from metal oxides like titanium dioxide (TiO_2), ferroferric oxide (Fe_3O_4), silicon dioxide (SiO_2), cerium oxide (CeO_2), aluminum oxide (Al_2O_3), etc. (Fedlheim and Foss 2001).

During the last two decades, a significant amount of research has been conducted on metal NPs (MNPs) and metal oxide NPS (MONPs) particularly from agricultural perspective, because these NPs can easily slip into the plant system (Tripathi et al. 2011; Husen and Siddiqi 2014a, b; Raliya et al. 2015). Because of their unique properties, NPs are reported to boost plant metabolism (Nair et al. 2011). Excessive use of NPs in the industrial sector, in food and agricultural products, and in remediation technologies has conjured the issue of contamination of ecosystems (Gardea-Torresdey et al. 2014; Nair 2016). This has dragged the attention of many research groups to explore the potential effects of NPs on plants (Monica and Cremonini 2009; Nair et al. 2011; Li et al. 2014; Rico et al. 2015). Being sessile in nature, plants are frequently exposed to NPs. Plants exposed to MNPs and MONPs subsequently accumulate them in their underground and aerial parts. When present into the rhizosphere, NPs can easily enter the epidermis and cortex via the apoplastic route (Rizwan et al. 2016). Translocation to aerial parts is much dependent on exposure time, plant species, and the shape and size of NPs (Li et al. 2014; Rico et al. 2015). On finding their entry into crop plants, NPs also generate a threat to human population through contamination of the food chain. Potential toxicity of NPs toward living organisms is well established. Given this, it becomes imperative to study the interactions among plants and NPs, which determine the mode of the NP uptake and accumulation, and subsequently their fate in the environment (Nair 2016). This chapter provides the latest information related to the interaction of plants with NPs and elucidates the consequent effects on plant growth and development.

15.2 Metal and Metal Oxide Nanoparticles

Being a new field, nanotechnology has a potential to provide a platform for researchers to design and incorporate new tools for studying the key functioning of NPs into the plant system (Cossins 2014). Metallic NPs are simple to synthesize because of their tunable features like size, shape, composition, structure, and encapsulation, out of different reported NPs (Subbenaik 2016). Of the synthesized NPs, Au and Ag NPs are most frequently used because of their simplicity in preparation, bio-conjugation, and appealing results under various tested systems. Limited size and high density of corner gave exclusive properties to metal oxide NPs also (Picó and Blasco 2012; Raliya and Tarafdar 2013; Subbenaik 2016). Different metals (Au, Ag, Zn, Ni, Fe, and Cu) and metal oxides (TiO_2, Fe_2O_4, SiO_2, CeO_2, Al_2O_3, etc.) have been used to design NPs (Fedlheim and Foss 2001) that suit different plant-related processes including protection and fertilization (Gogos et al. 2012). For example, the use of SiO_2 and Al_2O_3 nanoparticles reportedly increases the germination percentage and growth of roots in plants (Lin and Xing 2007; Siddiqui and Al-Whaibi 2014).

However, the rapidly increasing use of MNPs and MONPs in various operations and their presence in the environment has raised the issues of environmental health. Regulation of their optimum levels within the soil as nutrients/facilitators/pollutants for sustainable agriculture and crop production is a tedious task. As plant development is regulated by diverse environmental factors like nutrient availability, temperature, soil morphology, and light intensity, it is important to explore whether MNPs/MONPs also have a potential to influence the plant growth and development and/or create toxicity in the plant system.

15.3 Effects of NPs on Plant Growth and Development

Plant growth and development is a holistic term that starts from the initial stages of seed germination and extends up to the senescence. The effects of MNPs/MONPs on the overall growth process are found to be both positive and negative, possibly depending upon the size, composition, concentration, physical and chemical properties of NPs, and also the nature of plant species (Khodakovskaya et al. 2012; Husen and Siddiqi 2014b; Nair 2016; Siddiqi and Husen 2016, 2017a, b; Husen 2017; Fig. 15.1). This chapter is planned to discuss the plausible role of NPs on overall plant growth and productivity.

15.3.1 Seed Germination

Effects of NPs on seed germination are both positive and negative (Hong et al. 2015). Nanoparticles synthesized from lead, palladium, gold, and copper have considerably swayed the growth of lettuce (*Lactuca sativa*) seeds (Shah and Belozerova 2009). The activity of nitrate reductase was increased by the exogenous treatment of nano-SiO_2 and nano-TiO_2 which results in the better germination of soybean seeds (Lu et al. 2002). Improved seed germination was also noticed in lettuce and cucumber (*Cucumis sativus*) (Barrena et al. 2009), Indian mustard (*Brassica juncea*) (Arora et al. 2012), *Boswellia ovalifoliolata* (Savithramma et al. 2012), and *Gloriosa superba* (Gopinath et al. 2014) plants when given Au NP treatment. Similarly, application of nano-SiO_2 improves seed germination and nutrient's availability to maize plants (Suriyaprabha et al. 2012). Better germination of seeds was noticed in peanut (*Arachis hypogaea*) (Prasad et al. 2012), soybean (*Glycine max*) (Sedghi et al. 2013), wheat (*Triticum aestivum*) (Ramesh et al. 2014), and onion (*Allium cepa*) (Raskar and Laware 2014) by the application of Zn nanoparticle.

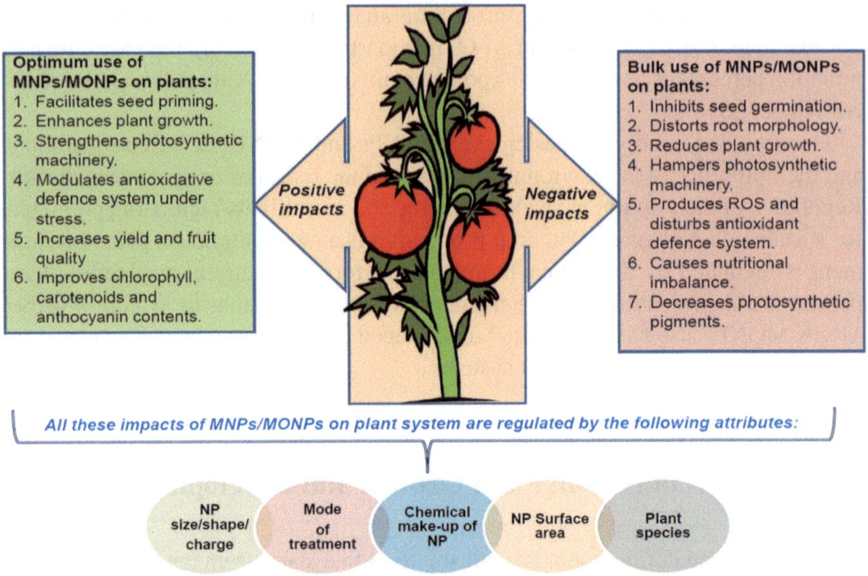

Fig. 15.1 Schematic summary depicting the potential impacts of metal and metal oxide nanoparticles on plant growth and productivity. These impacts are affected by size, composition, concentration, physical and chemical properties of NPs, mode of their application, and also on the nature of plant species

Negative impacts of NPs on seed germination have also been recorded. Rice, barley, faba bean, and turnip have shown a dose-dependent decrease when treated with Ag NPs (El-Temsah and Joner 2012; Thuesombat et al. 2014; Thiruvengadam et al. 2014). Similarly, NPs of copper oxide (CuO), nickel oxide (NiO), TiO_2, iron oxide (Fe_2O_3), and cobaltosic oxide (Co_3O_4) also reduced the seed germination of lettuce, radish, and cucumber plants (Wu et al. 2014). Plants growing in soils under natural environment somehow behave differently from those in the lab or greenhouse. Seed germination varied with different soils and was less affected in rye grass, barley, and flax plants when supplied with differential doses of Fe or Ag NPs in soil as compared to water. For instance, it was less obvious in clay soil than in sandy ones (El-Temsah and Joner 2012). In another supporting study, germination of lettuce and radish was less pretentious in soil than in water when treated with Ag NPs (Gruyer et al. 2013).

It is also known that a lower dose of NPs may serve as a seed-priming agent, whereas a higher dose causes phytotoxicity in crops. Moreover, plant response varies significantly with the NPs tested and can be correlated to their dose and size (De Rosa et al. 2013). Although the positive or negative interaction between NPs and seed germination is voraciously studied in plants, the mechanism operative behind the scene is still obscure and needs a comprehensive evaluation.

15.3.2 Mineral Uptake

Nutrients play a significant role in the process of plant growth, and their deficiency or nonavailability may lead to devastating effects such as stunting, deformity, discoloration, distress, and even death of the plant. The toxic metal ions considerably hamper the uptake of minerals in plants (Zaheer et al. 2015; Rizwan et al. 2016). Foliar spray of Ag NPs at a differential dose range reduced the mineral uptake in tomato seedlings, which led to nutrient deficiency (Shams et al. 2013). Treatment of CeO_2 and SiO_2 NPs altered the nutrient supply in shoots and roots of transgenic cotton(*Gossypium*)plant (Le et al. 2014; Li et al. 2014). Similarly, lettuce plants were challenged for the nutrients on application of CuO NPs (Trujillo-Reyes et al. 2014). Zhao et al. (2014) found an increase in the uptake of micronutrients in cucumber plants when treated with ZnO NPs. However, 5, 10, and 20 mg L^{-1} doses of CuO NPs increased the content of various nutrients, like Cu, P, and S in alfalfa (*Medicago sativa*) shoots, and lowered the uptake of P and Fe in lettuce shoots (Hong et al. 2015). Such a decrease in mineral uptake by plants on the application of NPs might be because of the release of toxic metal ions from NPs (Dimkpa et al. 2012; Mahmoodzadeh et al. 2013). However, more research is required to know the details of the mechanism involved.

15.3.3 Photosynthetic Machinery

Photosynthesis is a vital process by which plants convert light energy into chemical energy that can later be used for their growth and development. Of the total solar radiation energy falling on the surface of earth, approximately 2–4% is converted by plants for their growth and development (Kirschbaum 2011). It is important to increase the efficiency of this profit-yielding process for the better growth of plants. One essential modification for increasing the efficiency of photosynthesis in plants is modifying the rubisco activity, an enzyme that catalyzes the conversion of carbon dioxide (CO_2) into the biomolecules. Recently, genes of cyanobacterium *Synechococcus elongatus* were incorporated in tobacco (*Nicotiana tabacum*) plants by replacing the Rubisco gene for carbon fixation in tobacco plant (Lin et al. 2014). These new engineered plants showed more photosynthetic efficiency than the native ones. It has been reported that the application of SiO_2 NPs to plants improves the photosynthetic rate by improving the activity of carbonic anhydrase enzyme (that supplies CO_2 to the Rubisco) and synthesis of photosynthetic pigments (Xie et al., 2012, Siddiqui and Al-Whaibi 2014; Siddiqui et al. 2014). Similarly, the use of nano-anatase TiO_2 enhances photosynthesis by stimulating the Rubisco activity that could eventually increase the growth rate of plants (Gao et al. 2006; Chand and Siddiqui, 2012). It is also noteworthy that the bulk use of MNPs/MONPs generates the oxidative burst in plants by releasing their metal ions (Rizwan et al. 2016). This oxidative stress may result in the production of ROS that might interfere with many biochemical reactions and reduce the photosynthesis and gas exchange capacity of plants (Adrees et al. 2015; Das et al. 2015). The toxic effect of NPs on photosynthesis and gas exchange in food crops has been widely studied (Mirzajani et al. 2013; Abouzeid and Moustafa 2014; Rao and Shekhawat 2014; Shaw et al. 2014).

The use of Ag NPs as seed-priming agent in wheat, soybean, and barley significantly hampered the content of photosynthetic pigments and also quenched the chlorophyll fluorescence (Zhao et al. 2013; Abouzeid and Moustafa 2014; Gorczyca et al. 2015). The excessive application of MNPs/MONPs caused a significant reduction in the total chlorophyll content of some crop plants including Indian mustard, pea, and soybean (Pradhan et al. 2013; Mukherjee et al. 2014; Rao and Shekhawat 2014). Therefore, it is quite now known that the toxic effects of MNPs/MONPs on the photosynthetic machinery depend on the duration, type, and dose of the NPs. Plants have the in-built mechanism to withstand against NP toxicity for certain period of time, and the extended exposure to NPs could hamper the photosynthetic system.

15.3.4 Plant Morphology

Plants flourishing in suited environment are characterized with better morphology based on the shoot and root lengths, the shoot and root biomass, and the leaf area. Stressful environment has adverse effects on these characters. Asli and Neumann (2009) reported that application of $nTiO_2$ NPs repressed leaf growth and leaf functions in maize seedlings via affecting the water uptake. Implications of the ZnO, Fe_2O_3, Al_2O_3, and CuO NPs in modifying the morphological parameters have been assessed in various crop systems (Dimkpa et al. 2012; Mahmoodzadeh et al. 2013; Nair and Chung 2014). It is suggested that the possible reason for the negative effect on plant growth parameters is the plausible release of toxic metal ions from these manufactured NPs. NPs synthesized from Ag significantly inhibited the root growth and biomass of different tested crops such as wheat, rice, sorghum, and tomato (Mazumdar and Ahmed 2011; Dimkpa et al. 2012; Song et al. 2013; Vannini et al. 2014). Exposure of Ag NPs, at a dose of 0.2, 0.5, and 1 mg L^{-1} for 1 week, remarkably reduced the shoot and root biomass as well as the root elongation in rice seedlings (Nair et al. 2014). The length of wheat seedlings as well as the root growth in soybean and chickpea was noticeably hampered on application of CuO NPs (Adhikari et al. 2012; Dimkpa et al. 2012). Similarly, a dose-dependent decrease in plant height and in the shoot and root biomass of cotton seedlings was caused by the CeO_2 and SiO_2 NPs added to the growth media (Le et al. 2014). Additionally, occurrence of silver nanoparticles was detected in plasmodesmata and cell wall (Geisler-Lee et al. 2013), which would certainly result in wall seepage and snarl-up of intercellular communication (Geisler-Lee et al. 2014), sequentially distorting the performance of nutrient transporter proteins and intercellular transport and thereby affecting the overall growth of the plants. The anatomical and ultrastructural responses of *Capsicum annuum* toward Fe NPs have been studied recently by Yuan et al. (2018) who found the responses to be dose-dependent. For instance, improvement in leaf growth, chloroplast number, grana stacking, and development of vascular bundles was observed by light and electron microscopes at low concentrations of Fe NPs. In contrast, at elevated doses, Fe NPs appeared to be aggregated in cell walls and entered into the roots via the apoplastic pathway, thereby blocking the movement of iron.

In addition, many studies have shown the stimulating effects of different NPs toward plant morphology (Wang et al. 2012; Wang et al. 2013a; Antisari et al. 2015). Exposure of tomato seedlings to CeO_2 NPs (0.1-10 mg L^{-1}) caused a little increase in plant height and biomass (Wang et al. 2012). Application of $nTiO_2$ to spinach (*Spinacia oleracea*) significantly improved its growth (Hong et al. 2015); treatment with ZnO at different doses (125, 250, and 500 mg^{-1} kg^{-1} soil) enhanced the root length of green peas (*Pisum sativum*), which was almost doubled in comparison to the control (Mukherjee et al. 2014). Treatment of TiO_2 NPs improves the growth and yield of wheat plants in water-deficit condition (Jaberzadeh et al. 2013). Similarly, SiO_2 NP application (at a dose of 5–20 kg ha^{-1}) in sandy loam soil appreciably increased the shoot and root length and leaf area of 20-day-old maize seedlings (Suriyaprabha et al. 2012). Tobacco plants exposed to different concentrations (0.1%, 0.5%, and 1%) of Al_2O_3 NPs exhibited increase in root length and biomass but a drastic decrease in leaf count (Burklew et al. 2012; Verma et al. 2018). Likewise, physiological changes in watermelon were evaluated in vitro after exposure to γ-Fe_2O_3 NPs (Wang et al. 2016). An optimum dose of NPs was able to recover chlorosis and iron deficiency and promote plant growth.

Interaction between NPs and rhizosphere is immensely important because of the probable impact of NPs on the root-bacteria symbiosis. ZnO NPs proved perilous to rhizobium-legume symbiosis, as they disrupted early communication between rhizobia and the plant along with the nodule development, and subsequently delayed the onset of nitrogen fixation, an important factor in relation to plant growth and productivity (Huang et al. 2014). Additionally, the presence of NPs also brought other molecular changes in terms of hormonal imbalance in plants. For instance, CuO NP application on cotton and *Bt*-cotton caused momentous alterations in the intrinsic levels of indole-3-acetic acid (IAA) and abscisic acid (ABA) (Nhan Le et al. 2016). Although the MNPs/MONPs may have synergistic and antagonistic effects on the morphology of plants, their impact depends primarily on size and dose of NPs, duration of exposure, and experimental setup (Pokhrel and Dubey 2013; Thuesombat et al. 2014). It is germane to mention that most of the studies performed on NPs with respect to plant morphology have been of short duration and were conducted in controlled conditions. Therefore, for a better understanding of NP-plant interaction, long-term studies have to be carried out under natural environmental conditions.

15.3.5 Grain Yield and Quality

Higher yield and excellent quality of seeds signify the best growth of plants. Researchers have applied diverse nanotechnology inputs to get these desired outcomes in plants. It is now well established that plant exposure to NPs alters the uptake of nutrients and ensues biological activities, causing variation in growth and yield in different plant species. For instance, plants raised in the soil with unmodified or modified nano-TiO_2 (hydrophobic or hydrophilic coating) for 65 days showed that plant growth, inorganic nutrient uptake (Ca, Mg, P, Cu, Fe, Mn, and Se), enzyme activity, chlorophyll, and carbohydrate production were augmented with coated NPs (Tan et al. 2017). Application of Ag, Fe_2O_3, and CeO_2 NPs increased fruit yield and biomass in cucumber, soybean, and tomato plants (Sheykhbaglou et al. 2010; Wang et al. 2012; Shams et al. 2013). An increased vigor index was observed in fennel seeds exposed to TiO_2 NPs, as compared to the control and bulk TiO_2-treated plants (Feizi et al. 2013; Verma et al. 2018). Likewise, increased pod biomass and kernel, together with shelling percentage, was observed in peanut, when treated with ZnO NPs (Prasad et al. 2012). Additionally, NPs also affect the nutritional components and dietary value of fruits, and the impact may be trans-generational as observed in tomato plants treated with cerium oxide NPs (Wang et al. 2013b). However, some researchers have encountered contrary effects in other crops like barley (*Hordeum vulgare*), where exposure of CeO_2 NPs prevented seed setting (Rico et al. 2015). Application of the same NP (CeO_2) noticeably decreased Fe, sulfur (S), starch, and amino acid content in rice seedlings (Rico et al. 2013). Significant decrease in potassium (K) and phosphorus (P) was observed with the application of TiO_2 in cucumber fruits (Servin et al. 2012). The ZnO and CeO_2treatments significantly reduce overall yield of maize plants, remarkably altering the quality of corn by disrupting the mineral elements in cobs and kernels (Zhao et al. 2015). In the nutshell, application of different MNPs/MONPs affects the yield and quality of fruits and seeds, which entirely depend on the type and size of NPs and on the mode of treatment. Further illustrative studies need to be conducted to find the optimum dose of NPs. In-depth analysis in terms of long-term exposure, dose-dependent experiments, and molecular studies like proteomics and metabolics can be a handy tool in deciding the exact role of NPs on grain yield and quality.

15.4 Conclusion and Future Directions

Nanotechnology has evinced immense potential for the growth of agriculture sector and hence used excessively. However, bulk production and inadvertent discharge of NPs into the environment have dragged attention toward the contamination of ecosystems and food supply (Medina-Velo et al. 2016). Being nano in dimension, NPs can easily be inserted into the plant system and translocated to different organs of the plant. Therefore, it becomes imperative to evaluate their interaction on the plant system, irrespective of benefits or hazards. Furthermore, the kinetic studies have revealed the rapid and highly reactive nature of NPs and showed them to be inherently interactive with impurities (Subbenaik 2016). Recently, phytotoxicity and beneficial aspects of these xenobiotic compounds in the plant system have been discussed by several research groups (Sheikh Mohamed and Sakthi Kumar 2016). But the exact picture of their roles and interaction with plants is still blurred. Comprehensive evidence of the toxicity/benefits of different MNPs/MONPs has been presented and discussed in this chapter (Fig. 15.1). It is clear that, from their entry to accumulation, NPs depend upon the plant species, growth stage, NP size, and the mode of treatment (Kole et al. 2013; Raliya et al. 2015).

It is germane to mention that the presence of NPs in plants also poses a threat of human exposure to NPs through the food chain. Therefore, the effects of NPs in plants need to be evaluated from a wider perspective and over several generations of plants to get the better insights on their fate in the ecosystem. A comprehensive research should also be conducted at the molecular level to assess the precise roles of NPs in plants, which could be used as a platform for designing the future research in the field of nano-agriculture.

References

Abou-Zeid HM, Moustafa Y (2014) Physiological and cytogenetic responses of wheat and barley to silver nanopriming treatment. Int J Appl Biol PharmTechnol 5:265–278

Adhikari T, Kundu S, Biswas AK, Tarafdar JC, Rao AS (2012) Effect of copper oxide nanoparticle on seed germination of selected crops. J Agric Sci Technol 2:815–823

Adrees M, Ali S, Rizwan M, Ibrahim M, Abbas F, Farid M, Rehman MZ, Irshad MK, Bharwana SA (2015) The effect of excess copper on growth and physiology of important food crops: a review. Environ Sci Pollut Res 22:8148–8162

Antisari LV, Carbone S, Gatti A, Vianello G, Nannipieri P (2015) Uptake and translocation of metals and nutrients in tomato grown in soil polluted with metal oxide (CeO_2, Fe_3O_4, SnO_2, TiO_2) or metallic (Ag, Co Ni) engineered nanoparticles. Environ Sci Pollut Res 22(3):1841–1853

Arora S, Sharma P, Kumar S, Nayan R, Khanna PK, Zaidi MGH (2012) Gold-nanoparticle induced enhancement in growth and seed yield of *Brassica juncea*. Plant Growth Regul 66:303–310

Asli S, Neumann PM (2009) Colloidal suspensions of clay or titanium dioxide nanoparticles can inhibit leaf growth and transpiration via physical effects on root water transport. Plant Cell Environ 32:577–584

Banerjee J,Kole C (2016) Plant nanotechnology: an overview on concepts strategies and tools. Chittaranjan, Kole, D. Sakthi Kumar and Mariya V. Khodakovskaya, Plant Nanotechnology-principals and practices, Springer International Publishing Switzerland, pp 1–14

Barrena R, Casals E, Colon J, Font X, Sanchez A, Puntes V (2009) Evaluation of the ecotoxicity of model nanoparticles. Chemosphere 75:850–857

Bernhardt ES, Colman BP, Hochella JMF, Cardinale BJ, Nisbet RM, Richardson CJ, Yin L (2010) An ecological perspective on nanomaterial impacts in the environment. J Environ Qual 39:1954–1965

Burklew CE, Ashlock J, Winfrey WB, Zhang B (2012) Effects of aluminum oxide nanoparticles on the growth, development, and microRNA expression of tobacco (*Nicotiana tabacum*). PLoS One 7:e34783

Buzea C, Pacheco II, Robbie K (2007) Nanomaterials and nanoparticles: sources and toxicity. Biointerphases 2:Mr17–Mr71

Chand N, Siddiqui N (2012) Improvement in thermo mechanical and optical properties of in situ synthesized PMMA/TiO_2 nanocomposite. Compos Interfaces 19:51–58

Cossins D (2014) Next generation: nanoparticles augment plant functions. The incorporation of synthetic nanoparticles into plants can enhance photosynthesis and transform leaves into biochemical sensors. The Scientist, News & Opinion. http://www.the-scientist.com/?articles.view/articleNo/39440/title/Next-Generation–Nanoparticles-Augment-Plant-Functions/

Das S, Wolfson BP, Tetard L, Tharkur J, Bazata J, Santra S (2015) Effect of N-acetyl cysteine coated CdS:Mn/ZnS quantum dots on seed germination and seedling growth of snow pea (*Pisum sativum* L.): imaging and spectroscopic studies. Environ Sci Nano 2(2):203–212

De Rosa G, Lopez-Moreno ML, De Haro D, Botez CE, Peralta-Videa JR, Gardea-Torresdey J (2013) Effects of ZnO nanoparticles in alfalfa, tomato, and cucumber at the germination stage: root development and X-ray absorption spectroscopy studies. Pure Appl Chem 85(12):2161–2174

Dimkpa CO, McLean JE, Latta DE, Manangon E, Britt DW, Johnson WP et al (2012) CuO and ZnO nanoparticles: phytotoxicity, metal speciation, and induction of oxidative stress in sand-grown wheat. J Nanopart Res 14:1–15

El-Temsah YS, Joner EJ (2012) Impact of Fe and Ag nanoparticles on seed germination and differences in bioavailability during exposure in aqueous suspension and soil. Environ Toxicol 27:42–49

Feizi H, Kamali M, Jafari L, Rezvani MP (2013) Phytotoxicity and stimulatory impacts of nanosized and bulk titanium dioxide on fennel (*Foeniculum vulgare* Mill). Chemosphere 91:506–511

Fedlheim DL, Foss CA (2001) Metal Nanoparticles: Synthesis, Characterization, and Applications. CRC Press, Boca Raton, FL, USA

Gao FQ, Hong FH, Liu C, Zheng L, Su MY, Wu X, Yang F, Wu Yang P (2006) Mechanism of nano-anatase TiO2 on promoting photosynthetic carbon reaction of spinach–including complex of Rubisco-Rubisco activase. Biol Trace Elem Res 111:239–253

Gardea-Torresdey JL, Rico CM, White JC (2014) Trophic transfer, transformation, and impact of engineered nanomaterials in terrestrial environments. Environ Sci Technol 48:2526–2540

Geisler-Lee J, Wang Q, Yao Y, Zhang W, Geisler M, Li K, Huang Y, Chen Y, Kolmakov A, Ma X (2013) Phytotoxicity, accumulation and transport of silver nanoparticles by *Arabidopsis thaliana*. Nanotoxicology 7:323–337

Geisler-Lee J, Brooks M, Gerfen J, Wang Q, Fotis C, Sparer A, Ma X, Berg R, Geisler M (2014) Reproductive toxicity and life history study of silver nanoparticle effect, uptake and transport in *Arabidopsis thaliana*. Nano 4:301–318

Gogos A, Knauer K, Bucheli TD (2012) Nanomaterials in plant protection and fertilization: current state, foreseen applications, and research priorities. J Agric Food Chem 60:9781–9792

Gopinath K, Karthika V, Gowri S, Senthil Kumar V, Kumaresan S, Arumugam A (2014) Antibacterial activity of ruthenium nanoparticles synthesized using *Gloriosa superba L.* leaf extract. J Nanostruct Chem 4:83

Gorczyca A, Pociecha E, Kasprowicz M, Niemiec M (2015) Effect of nanosilver in wheat seedlings and *Fusarium culmorum* culture systems. Eur J Plant Pathol 142(2):251–261

Gruyer N, Dorais M, Bastien C, Dassylva N, Triffault-Bouchet G (2013) Interaction between silver nanoparticles and plant growth. In: International symposium on new technologies for environment control, energy-saving and crop production in green house and Plant 1037:795–800

Hong J, Rico CM, Zhao L, Adeleye AS, Keller AA, Videa JRP, Gardea-Torresdey JL (2015) Toxic effects of copper-based nanoparticles or compounds to lettuce (*Lactuca sativa*) and alfalfa (*Medicago sativa*). Environ Sci Processes Impacts 17:177–185

Huang YC, Fan R, Grusak MA, Sherrier JD, Huang CP (2014) Effects of nano-ZnO on the agronomically relevant Rhizobium-legume symbiosis. Sci Total Environ 466-467:503–512

Husen A, Siddiqi KS (2014a) Carbon and fullerene nanomaterials in plant system. J Nanobiotechnol 12:16

Husen A, Siddiqi KS (2014b) Phytosynthesis of nanoparticles: concept, controversy and application. Nano Res Lett 9:229

Husen A (2017) Gold nanoparticles from plant system: synthesis, characterization and application. In: Ghorbanpourn M, Manika K, Varma A (eds) Nanoscience and plant–soil systems, vol 48. Springer International Publishing AG, Cham, Switzerland, pp 455–479

Jaberzadeh A, Moaveni P, Moghadam HRT, Zahedi H (2013) Influence of bulk and nanoparticles titanium foliar application on some agronomic traits, seed gluten and starch contents of wheat subjected to water deficit stress. Notulae Bot Hort Agrobo 41:201–207

Khodakovskaya M, Dervishi E, Mahmood M, Xu Y, Li ZR, Watanabe F, Biris AS (2012) Carbon nanotubes are able to penetrate plant seed coat and dramatically affect seed germination and plant growth (retraction of vol 3, pg 3221, 2009). ACS Nano 6:7541–7541

Kirschbaum MUF (2011) Does enhanced photosynthesis enhance growth? Lessons learned from CO_2 enrichment studies. Plant Physiol 155:117–124

Kole C, Kole P, Randunu KM, Choudhary P, Podila R, Ke PC, Rao AM, Marcus RK (2013) Nanobiotechnology can boost crop production and quality: first evidence from increased plant biomass, fruit yield and phytomedicine content in bitter melon (*Momordica charantia*). BMC Biotechnol 13:37

Le VN, Rui Y, Gui X, Li X, Liu S, Han Y (2014) Uptake, transport, distribution and bio-effects of SiO_2 nanoparticles in Bt-transgenic cotton. J Nanobiotechnol 12:50

Li X, Gui X, Rui Y, Ji W, Yu Z, Peng S (2014) Bt-transgenic cotton is more sensitive to CeO_2 nanoparticles than its parental non-transgenic cotton. J Hazard Mater 274:173–180

Lin D, Xing B (2007) Phytotoxicity of nanoparticles: inhibition of seed germination and root growth. Environ Pollut 150(2):243–250

Lin MT, Occhialini A, Andralojc PJ, Parry MAJ, Hanson MR (2014) A faster Rubisco with potential to increase photosynthesis in crops. Nature 513:547–550

Lu CM, Zhang CY, Wen JQ, Wu GR, Tao MX (2002) Research on the effect of nanometer materials on germination and growth enhancement of *Glycine max* and its mechanism. Soybean Sci 21:68–172

Mahmoodzadeh H, Aghili R, Nabavi M (2013) Physiological effects of TiO_2 nanoparticles on wheat (*Triticum aestivum*). Tech J Eng Appl Sci 3:1365–1370

Mazumdar H, Ahmed GU (2011) Phytotoxicity effect of silver nanoparticles on *Oryza sativa*. Int J Chem Technol Res 3:1494–1500

Medina-Velo IA, Zuverza-Mena N, Tan W, Hernandez-Viezcas JA, Peralta-Videa JR, Gardea-Torresdey JL (2016) Biophysical methods of detection and quantification of uptake, translocation, and accumulation of nanoparticles. In: Chittaranjan K, Sakthi Kumar D, Khodakovskaya MV (eds) Plant nanotechnology- principals and practices. Springer International Publishing, Switzerland, pp 29–64. https://doi.org/10.1007/978-3-319-42154-4

Mirzajani F, Askari H, Hamzelou S, Farzaneh M, Ghassempour A (2013) Effect of silver nanoparticles on *Oryza sativa* L. and its rhizosphere bacteria. Ecotoxicol Environ Saf 88:48–54

Monica RC, Cremonini R (2009) Nanoparticles and higher plants. Caryologia Int J Cytol Cytosyst Cytogenet 62(2):161–165

Mukherjee A, Pokhrel S, Bandyopadhyay S, Mädler L, Peralta-Videa JR, Gardea-Torresdey JL (2014) A soil mediated phyto-toxicological study of iron doped zinc oxide nanoparticles (Fe@ZnO) in green peas (*Pisum sativum* L.). Chem Eng J 258:394–401

Nair PMG, Chung IM (2014) Impact of copper oxide nanoparticles exposure on *Arabidopsis thaliana* growth, root system development, root lignification, and molecular level changes. Environ Sci Pollut Res 21(22):12709–12722

Nair PMG, Kim SH, Chung IM (2014) Copper oxide nanoparticle toxicity in mung bean (*Vigna radiata* L.) seedlings: physiological and molecular level responses of in vitro grown plants. Acta Physiol Plant 36:2947–2958

Nair R (2016) Effects of nanoparticles on plant growth and development. In: Chittaranjan K, Sakthi Kumar D, Khodakovskaya MV (eds) Plant Nanotechnology- principals and practices. Springer International Publishing, Switzerland, pp 95–118

Nair R, Poulose AC, Nagaoka Y, Yoshida Y, Maekawa T, Kumar DS (2011) Uptake of FITC labeled silica nanoparticles and quantum dots by rice seedlings: effects on seed germination and their potential as biolabels for plants. J Fluoresc 21:2057–2068

Nhan LV, Ma C, Shang J, Rui Y, Liu S, Xing B (2016) Effects of CuO nanoparticles on insecticidal activity and phytotoxicity in conventional and transgenic cotton. Chemosphere 144:661–670

Nowack B, Bucheli TD (2007) Occurrence, behavior and effects of nanoparticles in the environment. Environ Pollut 150:5–22

Picó Y, Blasco C (2012) Nanomaterials in food, which way forward? Analys Risk Nanomater Environ Food Samp 59:305

Pokhrel LR, Dubey B (2013) Evaluation of developmental responses of two crop plants exposed to silver and zinc oxide nanoparticles. Sci Total Environ 452:321–332

Pradhan S, Patra P, Das S, Chandra S, Mitra S, Dey KK, Goswami A (2013) Photochemical modulation of biosafe manganese nanoparticles on *Vigna radiata*: a detailed molecular biochemical, and biophysical study. Environ Sci Technol 47:13122–13131

Prasad TNVK, Sudhakar P, Sreenivasulu Y, Latha P, Munaswamy V, Reddy KR, Sreeprasad TS, Sajanlal PR, Pradeep T (2012) Effect of nanoscale zinc oxide particles on the germination, growth and yield of peanut. J Plant Nutr 35:905–927

Raliya R, Nair R, Chavalmane S, Wang W-N, Biswas P (2015) Mechanistic evaluation of translocation and physiological impact of titanium dioxide and zinc oxide nanoparticles on the tomato (*Solanum lycopersicum* L.) plant. Metallomics 7:1584–1594

Raliya R, Tarafdar J (2013) ZnO nanoparticle biosynthesis and its effect on phosphorous-mobilizing enzyme secretion and gum contents in Cluster bean (*Cyamopsis tetragonoloba* L.). Agric Res 2:48–57

Ramesh M, Palanisamy K, Babu K, Sharma NK (2014) Effects of bulk and nano-titanium dioxide and zinc oxide on physio-morphological changes in *Triticum aestivum* L. J Glob Biosci 3:415

Rao S, Shekhawat GS (2014) Toxicity of ZnO engineered nanoparticles and evaluation of their effect on growth, metabolism and tissue specific accumulation in *Brassica juncea*. J Environ Chem Eng 2:105–114

Raskar SV, Laware SL (2014) Effect of zinc oxide nanoparticles on cytology and seed germination in onion. Int J Curr Microbiol App Sci 3:467–473

Rico CM, Barrios AC, Tan W, Rubenecia R, Lee SC, Ramirez AV (2015) Physiological and biochemical response of soil-grown barley (*Hordeum vulgare* L.) to cerium oxide nanoparticles. Environ Sci Pollut Res 22(1):10551–10558

Rico CM, Morales MI, Barrios AC, McCreary R, Hong J, Lee WY, Nunez J, Peralta-Videa JR, Gardea-Torresdey JL (2013) Effect of cerium oxide nanoparticles on the quality of rice (*Oryza sativa* L.) grains. J Agric Food Chem 67:11278–11285

Rizwan M, Ali S, Adrees M, Rizvi H, Rehman MZ, Hannan F, Qayyum MF, Hafeez F, Ok YS (2016) Cadmium stress in rice: toxic effects, tolerance mechanisms and management: a critical review. Environ Sci Pollut Res 23(18):17859–17879

Savithramma N, Ankanna S, Bhumi G (2012) Effect of nanoparticles on seed germination and seedling growth of *Boswellia ovalifoliolata* – an endemic and endangered medicinal tree taxon. Nano Vis 2:61–68

Sedghi M, Hadi M, Toluie SG (2013) Effect of nano zinc oxide on the germination of soybean seeds under drought stress. Ann West Univ Timisoara 16(2):73–78

Servin AD, Michel HC, Viezcas JAH, Diaz BC, Videa JRP, Torresdey JLG (2012) Synchrotron micro-XRF and micro-XANES confirmation of the uptake and translocation of TiO(2) nanoparticles in cucumber (*Cucumis sativus*) plants. Environ Sci Technol 46:7637–7643

Shah V, Belozerova I (2009) Influence of metal nanoparticles on the soil microbial community and germination of lettuce seeds. Water Air Soil Pollut 197:143–148

Shams G, Ranjbar M, Amiri AA, Khodarahmpour Z (2013) The effect of 35 nm silver nanoparticles on antagonistic and synergistic mineral elements in leaves and fruit of tomato (*Lycopersicon esculentum* Mill.). Int J Agric Crop Sci 5:439–500

Shaw AK, Ghosh S, Kalaji HM, Bosa K, Brestic M, Zivcak M, Hossain Z (2014) Nano-CuO stress induced modulation of antioxidative defense and photosynthetic performance of Syrian barley (*Hordeum vulgare* L.). Environ Exp Bot 102:37–47

Sheikh Mohamed M, Sakthi Kumar D (2016) Methods of using nanoparticles. In: Chittaranjan K, Sakthi Kumar D, Khodakovskaya MV (eds) Plant Nanotechnology: Principles and Practices. Springer International Publishing, Cham

Sheykhbaglou R, Sedghi M, Shishevan MT, Sharifi RS (2010) Effects of nano-iron oxide particles on agronomic traits of soybean. Notulae Scientia Biologicae 2:112–113

Siddiqui MH, Al-Whaibi MH (2014) Role of nano-SiO_2 in germination of tomato (*Lycopersicum esculentum* Mill.) seeds. Saud J Biol Sci 21:13–17

Siddiqui MH, Al-Whaibi MH, Faisal M, Al Sahli AA (2014) Nano-silicon dioxide mitigates the adverse effects of salt stress on *Cucurbita pepo* L. Environ Toxicol Chem 33:2429–2437

Siddiqi KS, Husen A (2016) Engineered gold nanoparticles and plant adaptation potential. Nano Res Lett 11:400

Siddiqi KS, Husen A (2017a) Recent advances in plant-mediated engineered gold nanoparticles and their application in biological system. J Trace Elem Med Biol 40:10–23

Siddiqi KS, Husen A (2017b) Plant response to engineered metal oxide nanoparticles. Nano Res Lett 12:92

Song U, Jun H, Waldman B, Roh J, Kim Y, Yi J, Lee EJ (2013) Functional analysis of nanoparticle toxicity: a comparative study of the effects of TiO_2 and Ag on tomatoes (*Lycopersicon esculentum*). Ecotoxicol Environ Saf 93:60–67

Subbenaik SC (2016) Physical and chemical nature of nanoparticles. In: Chittaranjan K, Sakthi Kumar D, Khodakovskaya MV (eds) Plant nanotechnology- principals and practices. Springer International Publishing, Switzerland, pp 15–28

Suriyaprabha R, Karunakaran G, Yuvakkumar R, Rajendran V, Kannan N (2012) Silica nanoparticles for increased silica availability in maize (*Zea mays* L.) seeds under hydroponic conditions. Curr Nanosci 8:902–908

Tan W, Du W, Barrios AC, Jr AR, Zuverzamena N, Ji Z, Chang CH, Zink JI, Hernandezviezcas JA, Peraltavidea JR (2017) Surface coating changes the physiological and biochemical impacts of nano-TiO2 in basil (*Ocimum basilicum*) plants. Environ Pollut 222:64–72

Thiruvengadam M, Gurunathan S, Chung IM (2014) Physiological, metabolic, and transcriptional effects of biologically-synthesized silver nanoparticles in turnip (*Brassica rapa* ssp. *rapa* L.). Protoplasma 252(4):1031–1046

Thuesombat P, Hannongbua S, Akasit S, Chadchawan S (2014) Effect of silver nanoparticles on rice (*Oryza sativa* L cv. KDML 105) seed germination and seedling growth. Ecotoxicol Environ Saf 104:302–309

Tiwari DK, Dasgupta-Schubert N, Cendejas LMV, Villegas J, Montoya LC, Garcia SEB (2014) Interfacing carbon nanotubes (CNT) with plants: enhancement of growth, water and ionic nutrient uptake in maize (*Zea mays*) and implications for nanoagriculture. Appl Nanosci 4:577–591

Tripathi S, Sonkar SK, Sarkar S (2011) Growth stimulation of gram (*Cicer arietinum*) plant by water soluble carbon nanotubes. Nanoscale 3:1176–1181

Trujillo-Reyes J, Majumdar S, Botez CE, Videa JRP, Gardea-Torresdey JL (2014) Exposure studies of core–shell Fe/Fe_3O_4 and Cu/CuO NPs to lettuce (*Lactuca sativa*) plants: are they a potential physiological and nutritional hazard. J Hazard Mater 267:255–263

Vannini C, Domingo G, Onelli E, De Mattia F, Bruni I, Marsoni M, Bracale M (2014) Phytotoxic and genotoxic effects of silver nanoparticles exposure on germinating wheat seedlings. J Plant Physiol 171:1142–1148

Verma SK, Das AK, Patel MK, Shah A, Kumar V, Gantait S (2018) Engineered nanomaterials for plant growth and development: a perspective analysis. Sci Total Environ 630:1413–1435

Wang P, Menzies NW, Lombi E, McKenna BA, Johannessen B, Glover CJ et al (2013a) Fate of ZnO nanoparticles in soils and cowpea (*Vigna unguiculata*). Environ Sci Technol 47:13822–13830

Wang Q, Ebbs SD, Chen Y, Ma X (2013b) Trans-generational impact of cerium oxide nanoparticles on tomato plants. Metallomics 5:753–759

Wang Q, Ma X, Zhang W, Pei H, Chen Y (2012) The impact of cerium oxide nanoparticles on tomato (*Solanum lycopersicum* L.) and its implications for food safety. Metallomics 4:1105–1112

Wang Z, Xu L, Zhao J, Wang X, White JC, Xing B (2016) CuO nanoparticle interaction with *Arabidopsis thaliana*: toxicity, parent-progeny transfer, and gene expression. Environ Sci Technol 50:6008

Wu SG, Huang L, Head J, Ball M, Tang YJ, Chen DR (2014) Electrospray facilitates the germination of plant seeds. Aerosol Air Qual Res 14:632–641

Xie Y, Li B, Zhang Q, Zhang C (2012) Effects of nano-silicon dioxide on photosynthetic fluorescence characteristics of *Indocalamus barbatus* McClure. J Nanjing For Univ (Natl Sci Ed) 2:59

Yuan J, Chen Y, Li H, Lu J, Zhao H, Liu M, Nechitaylo GS, Glushchenko NN (2018) New insights into the cellular responses to iron nanoparticles in *Capsicum annuum*. Sci Rep 8:3228

Zaheer IE, Ali S, Rizwan M, Farid M, Shakoor MB, Gill RA, Najeeb U, Iqbal N, Ahmad R (2015) Citric acid assisted phytoremediation of copper by *Brassica napus* L. Ecotoxicol Environ Saf 120:310–317

Zhao L, Hernandez-Viezcas JA, Videa JRP, Bandyopadhyay S, Peng B, Munoz B (2013) ZnO nanoparticle fate in soil and zinc bioaccumulation in corn plants (*Zea mays*) influenced by alginate. Environ Sci Processes Impacts 15:260–266

Zhao L, Peralta-Videa JR, Rico CM, Hernandez-Viezcas JA, Sun Y, Niu G, Servin A, Nunez JE, Duarte-Gardea M, Gardea-Torresdey JL (2014) CeO2 and ZnO nanoparticles change the nutritional qualities of cucumber (*Cucumis sativus*). J Agric Food Chem 62(13):2752–2759

Zhao L, Sun Y, Viezcas JRP, Hong J, Majumdar S (2015) Monitoring the environmental effects of CeO_2 and ZnO nanoparticles through the life cycle of corn (*Zea mays*) plants and in situ -XRF mapping of nutrients in kernels. Environ Sci Technol 49(5):2921–2928

Chapter 16
Ecotoxicological Effects of Nanomaterials on Growth, Metabolism, and Toxicity of Nonvascular Plants

Sophia Mavrikou and Spyridon Kintzios

16.1 Introduction

Nanomaterials (NMs), defined as materials with one or more dimensions of the size of 1–100 nm (ASTM/E2456-06 2012; Reiners 2013), have gained increasing attention due to their unique properties, relative to their bulk counterparts, which impart them beneficial characteristics including a high specific surface area and reaction activity (Laurent et al. 2008) and the quantum confinement effects (Amelia et al. 2012). Nowadays with the increasing insertion of nanotechnology in our daily life, nanomaterials of different shapes and diameters have been developed and used in various consumer products, pharmaceuticals, cosmetics (Melo et al. 2015), and other commodities (Bradley et al. 2011; Ghasemzadeh et al. 2014; Singh 2017).

Nanoparticles (NPs), a subgroup of nanomaterials, are classified into various categories depending on their size, morphology, and chemical properties. The primary focus of this chapter is metal, metal oxide, carbon-based NPs, and quantum dots. Metal NPs, i.e., Cu, Ag, and Au, in nanometer range possess unique optical, electrical, and magnetic properties due to their localized surface plasmon resonance (LSPR) characteristics (Dreaden et al. 2012). They are considered as the potential candidates for catalysis due to their large surface area per volume or weight unit, compared to their bulk counterparts, typically functioning on metal surfaces (Roldan Cuenya 2010). Metal oxide nanoparticles represent a class of engineered nanomaterials that can be synthesized via several routes (Lang et al. 2011). The usual practices for manufacturing metal oxide nanoparticles are chemical vapor synthesis (Stankic et al. 2016) and the addition of oxidizing/precipitating agents during their synthesis (Sanchez-Dominguez et al. 2009). Metal oxide nanoparticles include both individual (e.g., TiO_2, CeO_2, CrO_2, ZnO, Bi_2O_3, and MoO_3) and binary oxides (e.g., $BaTiO_2$, InSnO, and $LiCoO_2$). They too have wide applications in industry.

S. Mavrikou (✉) · S. Kintzios
Laboratory of Cell Technology, Department of Biotechnology, School of Food, Biotechnology and Development, Agricultural University of Athens, Athens, Greece

A. Husen, M. Iqbal (eds.), *Nanomaterials and Plant Potential*,
https://doi.org/10.1007/978-3-030-05569-1_16

Carbon-based NPs comprise mainly of two major groups: fullerenes and carbon nanotubes (CNTs). Fullerenes are globular cage-like structures with various numbers of carbon atoms (e.g., C_{60}, C_{70}). They have created noteworthy commercial interest due to their remarkable electronic and structural properties, high strength, and versatility (Astefanei et al. 2015). The fullerene-C_{60} is the most commercially attractive carbon-based NP due to its ability to increase the efficiency of drugs, cosmetics, and electronics (Bianco and Da Ros 2011). Carbon nanotubes (CNTs) have been listed at the third position of the most important engineered nanoparticles (ENPs) found in the consumer product inventories (Vance et al. 2015). They are elongated, tubular structure with a large length/diameter ratio (Ibrahim 2013). They can be metallic or semiconducting reliant on a chiral vector value (the way they are rolled up) (Aqel et al. 2012). Structurally, they resemble graphite sheet rolling upon itself. Depending on the number of carbon sheets, they can either be single-walled (SWNTs), double-walled (DWNTs), or multiwalled carbon nanotubes (MWNTs), respectively (Elliott et al. 2013). CNTs have many potential applications, e.g., in plastics, batteries, paints, composites, touch screens, and drug delivery (De Volder et al. 2013).

Semiconductor materials possess properties between metals and nonmetals and therefore have found various applications (Ali et al. 2017). Semiconductor NPs or quantum dots (QDs) possess wide bandgaps and therefore show significant alteration in their properties with bandgap tuning by exhibiting particle size-dependent tunable photoluminescence (Dybiec et al. 2007). These properties render them very useful materials in photocatalysis, photo optics, electronic devices (Chow and Jahnke 2013), biology, and medicine (Zhou et al. 2015). The most common commercially available QDs are CdSe/ZnS-QDs due to their bright and unique emission with wide excitation spectra and narrow emission bandwidths (Deerinck 2008).

The increased use of nanomaterials in several industrial applications and consumer products ranging from cosmetics to medicine (e.g., odor-resistant textiles, household appliances, including wound dressings) (Rai et al. 2009; Namasivayam et al. 2010; Chaudhari et al. 2012) during the past decade has led to a rise in concerns about the potential toxic effects of accidentally or incidentally released NPs into the environment and their likely access into ambient aquatic systems (Colvin 2003; Service 2008). Scientists have expressed concerns about the potential adverse effects of NPs to beneficial bacteria in the environment, especially in soil and water. Although toxic effects of NPs on bacterial, fungal, and mammalian cells have been well investigated (Shrivastava et al. 2007, 2009; Kim et al. 2009), their impact on the growth and biology of algae and nonvascular lower plants has not been sufficiently documented (Lee et al. 2005).

"Lower plants" represent a heterogeneous group of plants and plantlike organisms, including algae, bryophytes, and lichens, which are characterized primarily by their lack of vascular tissues (which circulate water and nutrients in higher plants) (Eddy et al. 1992). However, pteridophytes, which are also included in lower plants due to absence of seeds, have the vasculature (phloem and xylem tissues), like the seed-bearing "higher plants." This chapter is focused on algae and bryophytes, the nonvascular lower plants, because there is hardly any information on responses of sister groups (like lichens, pteridophytes) to NPs.

Algae represent a large and diverse group of photosynthetic eukaryotes. They comprise of many ancient and miscellaneous lineages, including various symbiotic relationships with animals and fungi. Moreover, they display many degrees of organismal complexity: they range from microscopic unicells to macroscopic bodies and also possess multicellular thalli more than a meter in length (De Clerck et al. 2012). Further, algal cell wall surface has an additional layer of rigid, porous cell wall for modulating the entry of foreign materials, ions, and particles (Chen et al. 2012). Algae are commercially important since they can be used as biofertilizers, pollution control agents (algae bioreactors) (Pimratcha et al. 2015), biofuels (Hannon et al. 2010), stabilizers of casein, and source of nutrition (B complex vitamins and minerals) and can be incorporated to cosmetics (Spolaore et al. 2006). They form a critical component of almost all aquatic and many terrestrial ecosystems. As primary producers in the aquatic ecosystem, they are important indicators for environmental pollution monitoring and therefore constitute widely used model organisms in ecotoxicity studies of nanomaterials (Ma and Lin 2013; Quigg et al. 2013). The observed toxicities of NPs to algae have been attributed mainly to three mechanisms (Schwab et al. 2011; Long et al. 2012): (1) reduction of photosynthetic rate due to inhibition of light transmittance (shading effect) (Miazek et al. 2015); (2) NP agglomeration and physical interaction with algal cells, leading to the internalization of NPs and the disruption of the cell membrane (García-Cambero et al. 2013); and (3) induction of intracellular reactive oxygen species (ROS) formation leading to membrane lipid peroxidation and changes in the concentration of nonenzymatic antioxidants and in the activity of antioxidant enzymes (Chen et al. 2012).

Like algae, bryophytes are also nonvascular plants that can grow on the surface of tree trunks or rocks and generally absorb water and nutrients direct through leaf surfaces from their immediate environment (Dymytrova 2009; Schröder et al. 2010; Harmens et al. 2011). Some species play an important role in the colonization of bare or degraded soil and facilitate the installation and maintenance of higher plants, a significant fact for a healthy plant-soil dynamics and sustainable ecosystem development. Moreover, these lower plants are recognized as good accumulators of pollutants, especially for metal trace elements (Garrec and Van Haluwyn 2002; Faburé et al. 2010). Furthermore, these organisms lack a root system ensuring their exposure to atmospheric pollutants and are excellent sensors of air quality for different contexts and pollutants (Faburé et al. 2010; Meyer et al. 2010; Lodenius 2013). Owing to these properties, bryophytes are considered excellent models for evaluating atmospheric pollutant impact on the environment in many parts of the world (Oishi 2013; Agnan et al. 2015; Vuković et al. 2015). The observed toxicities of NPs to bryophytes have been mainly attributed to (a) NP agglomeration that leads to the NP uptake by leaves (Canivet et al. 2014) and (b) overproduction of ROS/RNS and induction of glutathione status modulation (Canivet et al. 2015).

Since the last decade, the use of NPs in our daily life has considerably increased, and the ecotoxicity studies about NP effects on plants and animals come up rapidly. However, the published data on NP toxicity on algae and plants of sister groups are insufficient, and the experimental designs and testing conditions are inconsistent across the studies. This chapter provides an outline of the NP ecotoxicological effects on these plants based on the data available in the published literature and focuses on the underlying mechanisms of NP toxicity.

16.2 Effects of Metal Nanoparticles

16.2.1 Gold Nanoparticles (Au-NPs)

A recent study explored the impacts of amine-coated 10 nm gold NP contaminations on the green alga *Scenedesmus subspicatus* (Renault et al. 2008). The growth/mortality effects were determined by the algal cell numerations. The lethal dose for 50% of the population was reached within a 24 h exposure at 1.6×10^5 Au-NPs/cell, while mortality for the lowest contamination condition was 20%. TEM examination revealed that Au-NPs were strongly adsorbed by the cell wall of algae, leading to progressive intracellular and wall disturbances. However, bare and hyaluronic acid-capped Au-NPs (particle size 12.5 nm) were found to be harmless to *S. subspicatus* in contrast to bulk soluble gold that gave an EC_{50} value 1.91 mg L^{-1} (García-Cambero et al. 2013). The carbonate- and citrate-coated Au-NPs did not cause any significant toxicity to the green algae *Chlamydomonas reinhardtii* (Behra et al. 2015).

The effects of Au-NPs with amphiphilic polymer coating (AP) or amphiphilic coating to which 10 kDa polyethylene glycol chains were attached (AP-PEG) were also assessed on the green algae *Pseudokirchneriella subcapitata* (Van Hoecke et al. 2013). The AP Au-NPs were more toxic than the AP-PEG after a 72 h exposure with EC_{50} values 7.5 and 39 mg Au-NPs L^{-1}, respectively. In another study Au-NP toxicity effects on *P. subcapitata* (formerly known as *Selenastrum capricornutum*) were evaluated by three biomass measuring techniques (coulter counting, cell counting in hemocytometer, and fluorescence of pigment extracts) (Hartmann et al. 2013). The coulter counting method gave unreliable results. Therefore, it was not considered suitable for biomass quantification. At 48 h, algal cultures showed follow the growth of the control sample – both with respect to cell number and pigment content – whereas at 72 h a leveling off was noticed in pigment content of all algal cultures exposed to Au-NP dispersions accompanied by a decrease in EC10. On the contrary, cell number growth rates followed continued exponential trends. The results also indicated that pigment (such as chlorophyll and carotenoid) synthesis was affected by the exposure to Au-NP dispersions despite the continued exponential cell growth. Although the EC_{50} values were higher than the highest tested concentration, subsequent tests of the effects of the "Starch Control" (starch/glucose/MES solution in concentrations identical to that of the Au nanoparticle dispersions) revealed that the dispersion constituents, and not Au-NPs themselves, were largely responsible for inhibitory effects as well as the characteristic leveling off in pigment content after 48 h.

A recent study suggested that the naturally existing ions like zinc ions could modulate the toxicity of Au-NPs (Iswarya et al. 2017). The effects of Au-NPs with two surface cappings (citrate and PVP) and three different sizes (16, 27, and 37 nm) were explored on a predominant freshwater alga *Scenedesmus obliquus* in the sterile freshwater matrix. Among the different-sized Au-NPs, the highest toxicity (54%) was observed at 1 mg L^{-1} of citrate-capped Au-NPs with particle size 37 nm, whereas PVP-capped Au-NPs showed 42% toxicity. A statistically significant reduction in the Au-NP (both citrate-capped/37 nm and PVP-capped/37 nm) toxicity was observed when Zn^{2+} (5 mg L^{-1}) was added to the growth medium (12% for citrate-capped −37 and 11% for of PVP-capped −37). All the above information on the Au-NP toxicity on lower plants is summarized in Table 16.1, specifying the NP size, the algal species used, the half maximal effective concentration, the exposure time, and the effects observed.

Table 16.1 Summarized results from Au-NP toxicity studies performed on the lower plants

Species	Particle size (nm)	EC_{50} (mg Au-NPs L^{-1}) – incubation time	Effects observed	Reference
Scenedesmus subspicatus (Chodat)	10	1.6 × 10^5 Au-NPs per cell at 24 h	Agglutination, intracellular and wall disturbances	Renault et al. (2008)
S. subspicatus (Chodat)	12.8	1.91 at 72 h	Agglomeration	García-Cambero et al. (2013)
Pseudokirchneriella subcapitata (Korshikov) F. Hindák	51 by dynamic light scattering (DLS) measurements 46 by NTA measurements	36 at 48 h 38 at 72 h fluorescence of pigments/ 83 at 72 h hemocytometer counts	Formation of larger agglomerates/ aggregate during time, reduction of pigment (chlorophyll and carotenoid) synthesis	Hartmann et al. (2013)
P. subcapitata (Korshikov) F. Hindák	4–5	AP:7.5 at 72 h AP-PEG:39 at 72 h	Growth inhibition	Van Hoecke et al. (2013)
Chlamydomonas reinhardtii P.A. Dangeard (strain CC 125 and wall-free strain CC-400)	5	–	No effects observed	Behra et al. (2015)
Scenedesmus obliquus (Turpin) Kützing	16	PVP-capped: >1	Aggregation, cytotoxicity	Iswarya et al. (2017)
		Citrate-capped: >1		
	27	PVP-capped: >1		
	37	Citrate-capped: >0.1 and <1		

16.2.2 Silver Nanoparticles (Ag-NPs)

Ag-NPs have shown extensive adverse effects on growth and morphology of the green algae *Pithophora oedogonia* and *Chara vulgaris* in a dose-dependent manner (Dash et al. 2012). Exposure of algal thalli to increasing concentrations of Ag-NPs resulted in progressive chromosome instability, mitotic disturbance, depletion of chlorophyll content, and the associated morphological malformations in algal filaments. SEM micrographs revealed dramatic alterations in cell wall, characterized with cell wall rupture and degradation in the NP-treated *Pithophora*. Discoloration of filaments due to chloroplast contraction followed by disintegration, regional bulging of filaments, thinning and disruption of cell wall permitting exclusion of the chlorophyll pigments, adsorption of Ag-NPs on cell surface and organellar membranes, mitostatic effect, induction of chromosomal anomalies, and irreversible genetic damage were the significant detrimental effects of nanosilver recorded in the tested algae. Ag-NPs also caused growth inhibition on the green alga*Pseudokirchneriella subcapitata* with an EC_{50} value 0.19 mg Ag-NPs L^{-1} after 96 h incubation (Griffitt et al. 2008).

The short-term toxicity of citrate-stabilized Ag-NPs and ionic silver Ag(I) to the ichthyotoxic marine raphidophyte *Chattonella marina* has also been investigated (He et al. 2012). The addition of Ag-NPs to GSe medium caused aggregation and dissolution of Ag-NPs. Cellular uptake of dissolved Ag(I) was observed, and toxicity effects were much higher for Ag(I) than for Ag-NPs. However, these inhibitory effects of Ag(I) and Ag-NPs were completely removed by the addition of cysteine, a strong Ag(I) ligand, suggesting that the toxicity of Ag-NPs was due to the release of Ag(I). The growth inhibition effects of Ag-NPs have been studied also on the bloom-forming cyanobacterial *Microcystis aeruginosa* strain after a 10-day exposure (Duong et al. 2016). A dose-dependent reduction of the cell growth was observed by increasing Ag-NP concentrations. The EC_{50} value based on the cell growth was 0.0075 mg L^{-1}, and the inhibition efficiency at the highest concentration of Ag-NPs (1 mg L^{-1}) was 98.7%. SEM and TEM images indicated shrunken and damaged cell wall attributed to toxicity of Ag-NPs.

The toxic effects of large-sized Ag-NPs (50 nm) were investigated on the freshwater microalga *Chlorella vulgaris* and the marine microalga *Dunaliella tertiolecta* after 24 h exposure (Oukarroum et al. 2012). Ag-NPs interacted directly with the *Chlorella vulgaris* cell surface forming large aggregates and caused significant decrease in chlorophyll contents and algal viability, while they induced ROS formation and lipid peroxidation in both algae, showing a variability in sensitivity (1 mg L^{-1} Ag-NPs induced a 44% decrease of viable cells for *D. tertiolecta* and 33% for *C. vulgaris*). In another study, aggregation and dissolution behavior of gum arabic (GA)- and polyvinylpyrrolidone (PVP)-coated Ag-NPs were compared in a mixture of aquatic plants *Potamogeton diversifolius* and *Egeria densa* (Unrine et al. 2012). Plants released dissolved organic matter (DOM) into the water column either through active or passive processes in response to Ag exposure that bound Ag ions.

As a result, the plant-derived DOM stabilized PVP-Ag-NPs as the primary particles but removed GA-Ag-NPs from the water column, possibly by dissolution and binding of the released Ag ions on sediment and plant surfaces.

The extent and mechanisms of toxicity of two Ag-NPs with differing size distributions (AG1 and AG2) and capping agents were investigated on two model organisms, a green alga (*Chlamydomonas reinhardtii*) and a cyanobacterium (*Synechococcus leopoliensis*) (Taylor et al. 2016a). Their effects on the production of extracellular polymeric substances (EPS) were also assessed. Both silver forms had a significant effect on viability and membrane integrity in *C. reinhardtii* but hardly affected ROS production, whereas no toxicity effects were observed in *S. leopoliensis*. The levels of EPS produced by both the species were similar for all the treatments. The EPS composition was affected from AG1 in a concentration-dependent manner and conversely from AG2. Higher levels of lower molecular weight material were produced by *C. reinhardtii* in the presence of all silver forms. Reduction in growth rate was observed for *S. leopoliensis*, but the impact on viability and ROS was lower than for *C. reinhardtii* probably due to differences in relevant biological properties (e.g., algal cell size and cell wall composition).

The abovementioned toxicity data of Ag-NPs on lower plants are presented in Table 16.2, specifying the particle size of Ag-NPs, the algal species used, the half maximal effective concentration and exposure time, and the effects observed.

16.2.3 Platinum Nanoparticles (Pt-NPs)

Pt-NP toxicity toward green microalgae *Pseudokirchneriella subcapitata* and *Chlamydomonas reinhardtii* was assessed (Sørensen et al. 2016), using the standard ISO tests for estimation of growth rate inhibition (EC_{50} values of 15–200 mg Pt-NPs L^{-1}). By using a double-vial setup, cells were separated from Pt-NPs, which indicated that shading is an important artifact for Pt-NP toxicity. Membrane damage was not severe, but substantial oxidative stress was detected at 0.1–80 mg Pt-NPs L^{-1} in both the algal species. Pt-NPs caused a growth rate inhibition and oxidative stress in *P. subcapitata*, in a low concentration of dissolved Pt, indicating the NP-specific toxicity of Pt. In addition, higher body burdens were measured in this species, possibly due to a favored binding of Pt to the polysaccharide-rich cell wall. However, in a previous study, the Pt-NP concentration causing total inhibition of algal growth was 22.2 mg L^{-1} (Książyk et al. 2015). Similar results were obtained by analyzing the levels of photosynthetic pigments in *P. subcapitata* exposed to nanoparticles. In another study, where the acute toxicity of $PtCl_4$ and Pt-NPs was investigated, EC_{50} values were 14 mg L^{-1} after 48 h exposure and 28 mg L^{-1} after 2 h, indicating that the toxicity was dependent on the exposure duration (Delgado et al. 2013).

Table 16.2 Summarized results from Ag-NP toxicity studies performed on lower plants

Species	Particle size (nm)	EC_{50} (mg Ag-NPs/L^{-1}) – incubation time	Effects observed	Reference
Pseudokirchneriella subcapitata (Korshikov) F. Hindák	20–30	0.19 at 96 h	Growth inhibition	Griffitt et al. (2008)
Pithophora oedogonia (Mont.) Wittrock *Chara vulgaris* Linn.	10–15	n.m.[a]	Depletion in algal chlorophyll content, chromosome instability, mitotic disturbance, morphological malformations in algal filaments	Dash et al. (2012)
Chattonella marina (Subrahmanyan) Hara et Chihara (strain CMPL01)	56	10–20 uM at 1 h	Growth inhibition, aggregation, cellular uptake	He et al. (2012)
Chlorella vulgaris Beyerinck *Dunaliella tertiolecta* Butcher (CPCC-420)	50	n.m.[a] >1 at 24 h	Aggregation, decrease in chlorophyll content and algal viability, induction of ROS formation and lipid peroxidation	Oukarroum et al. (2012)
Potamogeton diversifolius Raf.	49.3 (PVP)	Not assayed	Aggregation	Unrine et al. (2012)
Egeria densa Planch.	12.0 (GA)			
Microcystis aeruginosa (Kützing) Kützing	10–15	0.0075	Growth inhibition, changes in cell structure and morphology	Duong et al. (2016)
Chlamydomonas reinhardtii P.A. Dangeard (CCAP strain 11/32c) *Synechococcus leopoliensis* (Raciborski) Komárek	3–8 (AG1) 50 (AG2: PVP)	0.038 (AG1) and 0.031 (AG2) at 72 h0.028 (AG1) and 0.01 (AG2) at 72 h	Growth inhibition only, variation on the composition of the produced EPS	Taylor et al. (2016a, b)

[a]Not mentioned

16.3 Effects of Metal Oxide Nanoparticles

16.3.1 Alumina Nanoparticles

The toxicological impact of Al_2O_3-NPs to lower plants was demonstrated on algal species *Scenedesmus* sp. and *Chlorella* sp. (Aruoja et al. 2015). The observed EC_{50} value of Al_2O_3-NPs with particle size <50 nm after 72 h was 45.4 mg L^{-1} for *Chlorella* sp. and 39.35 mg L^{-1} for *Scenedesmus* sp. Bulk alumina (particle size <5 um) also showed toxicity in a lower range ($EC_{50} = 110.2$ mg L^{-1} for *Chlorella* sp.; 100.4 mg L^{-1} for *Scenedesmus* sp.). Additionally, chlorophyll content declined, and the cell surface was also affected (Sadiq et al. 2011b). Earlier, Al_2O_3-NPs with a particle size of 51 nm showed an EC_{50} value 8.30 mg L^{-1} at 96 h for the algal species *Pseudokirchneriella subcapitata* (Griffitt et al. 2008), whereas smaller Al_2O_3-NPs (particle size 8–21 nm) showed an EC_{50} value >10–100 mg Al_2O_3 L^{-1} at 72 h.

In addition, toxic effects of binary compounds of aluminum oxide alpha-forms (7 and 70 nm) and macro form (4um) on the growth of unicellular algae *Chlorella vulgaris* (Gosteva et al. 2015) were found to be concentration-dependent. A selective dependence of Al_2O_3-NPs toxicity on the size, concentration, and chemical nature of NPs was revealed. The EC_{50} values for the small-sized Al_2O_3-NPs (7 and 70 nm) were around 1 mg L^{-1}, whereas no toxicity was observed for the macro form.

16.3.2 Cerium Oxide Nanoparticles (CeO_2-NPs)

An assessment of the molecular and phenotypic effects of CeO_2-NPs was conducted with the unicellular green alga, *Chlamydomonas reinhardtii*, by using well-characterized monodispersed NPs (particle size 4– nm) based on the hypothesis that their toxicity is likely to be higher than the macro form (Taylor et al. 2016a, b). The potential toxicity of NPs was investigated by transcriptomics and metabolomics approaches in a wide range of exposure concentrations in order to provide insight into molecular toxicity pathways. Even though CeO_2-NPs inserted the intracellular vesicles within *C. reinhardtii*, they did not cause significant changes in the algal growth at any exposure concentration. At supra-environmental CeO_2-NPs concentrations, downregulation of photosynthesis and the carbon fixation perturbations were detected with further effects on energy metabolism.

In a previous study, on a short-term exposure, dissolved Ce^{3+} decreased the photosynthetic yield in a concentration-dependent manner with EC_{50} values of 7.5 uM for the wild type and 6.3 uM for a cell wall-free mutant strain of *C. reinhardtii*, whereas precipitated $CePO_4$(s) was not bioavailable and, hence, not toxic (Röhder et al. 2014). The intracellular ROS levels increased upon exposure to Ce^{3+} with the effective concentrations being similar to those inhibiting photosynthesis. Moreover, CeO_2-NPs agglomerated in exposure media and caused a slight inhibition of photosynthesis and reduction of intracellular ATP content upon a short-term exposure at the highest (100 uM) concentrations possibly due to Ce^{3+} ions co-occurring in the nanoparticle suspension, whereas no effect was observed for dispersed CeO_2-NPs as the dissolved Ce^{3+} got precipitated with phosphate and, hence, was not bioavailable. Moreover, flocculation of algal cells upon exposure to agglomerated CeO_2-NPs and Ce^{3+} was observed. The cell wall-free mutant and wild type of *C. reinhardtii* showed the same sensitivity to either Ce^{3+} or CeO_2-NPs toxicity indicating that the cell wall does not have a protective effect against CeO_2-NPs or Ce^{3+}.

16.3.3 Titanium Dioxide Nanoparticles

The studies on a potential impact of TiO_2-NPs on the environment have been conducted on green alga *Desmodesmus subspicatus* by determining its growth during a 72 h incubation period (Hund-Rinke and Simon 2006). Twenty-five nm-sized TiO_2-NPs (EC_{50} = 44 mg TiO_2-NPs L^{-1}) were more toxic to *D. subspicatus* than 100 nm TiO_2-NPs (no toxicity observed). It was demonstrated that the smaller particles caused a clear dose-dependent reduction in the algal growth, whereas the larger ones showed less toxicity ($C < 50$ mg TiO_2-NPs L^{-1}). On similar lines, TiO_2-NPs (10, 30, 300 nm) were applied to the algal species *Pseudokirchneriella subcapitata*, and their effects assessed after 72 h incubation (Hartmann et al. 2010). Smaller TiO_2-NPs (<10 nm) showed inhibition (21% reduction) in growth rate at the concentration of 2 mg L^{-1}, whereas, 30 and 300 nm TiO_2-NPs showed a slight stimulation of algal growth. In earlier studies, TiO_2-NPs with particle size of 30 nm (Griffitt et al. 2008) and ~ 100 nm (Blaise et al. 2008) were shown not to be toxic to this alga.

In another study, *P. subcapitata* was used for toxicity assessment of not readily soluble NPs with standardized algal growth inhibition tests (Hartmann et al. 2013). TiO_2-NPs formed large (micron-sized) agglomerates/aggregates in a dose-dependent manner. Three biomass surrogate measuring techniques (coulter counting, cell counting in hemocytometer, and fluorescence of pigment extracts) were evaluated. The results showed a concentration-dependent reduction in algal growth by both the biomass quantification techniques, yielding an EC_{50} value of 160 mg TiO_2-NPs L^{-1} (by hemocytometer). The EC_{50} value based on measurements of pigment fluorescence was found to be 200 mg L^{-1} (the highest tested concentration was 560 TiO_2-NPs L^{-1}). *P. subcapitata* was also used for toxicity assessment of fine (140 nm) and ultrafine (~140 nm) TiO_2 (uf-TiO_2) particles after incubation for 72 h (Warheit et al. 2007). EC_{50} values (95% fiducial limits) based on inhibition of growth and healthy average cell counts were 16 mg L^{-1} for fine TiO_2-NPs and 21 mg L^{-1} for uf-TiO_2-NPs. Aruoja et al. (2009) found that bulk TiO_2 (EC_{50} = 35.9 mg Ti L^{-1}) were less toxic to this algal species than their nano formulations (EC_{50} = 5.83 mg TiO_2-NPs L^{-1}). TiO_2-NPs formed characteristic aggregates entrapping the algal cells, thus contributing therefore to the toxic effect of TiO_2-NPs to algae. In a later study, it was indicated that the agglomerates entrapped nearly all algal cells so that the cells could mostly be seen inside the agglomerates and rarely in the surrounding medium (Aruoja et al. 2015). The high variability in the observed toxicity of TiO_2-NPs has been discussed by Menard et al. (2011), but no discernable correlation between primary particle size and toxic effect could be proved because the existing data were insufficient for confirmation (Menard et al. 2011).

Manier et al. (2016) indicated that the type of exposure system also affects the toxicity of TiO_2-NPs. Different exposure systems including the Erlenmeyer flasks and 24-well microplates (both using an orbital shake system) and an alternative system using cylindrical vials and magnetic stirring were used. After a 72 h exposure of *P. subcapitata* to two different types of TiO_2-NPs (particle size <10 and 20 nm), the authors found that the exposure systems applied to achieve the test can substantially affect the ecotoxicological results and the subsequent calculated EC_{50} values. The selected systems influenced both the interaction between algal cells and TiO_2-NPs as well as the growth inhibition level (Manier et al. 2016). The cytotoxicity potential of TiO_2-NPs was also assessed toward freshwater algal isolate *Scenedesmus obliquus* under dark and UV conditions at low exposure levels (≤ 1 μg mL^{-1}) (Dalai et al. 2013). Statistically significant reduction in cell viability and photosynthetic pigment content and increase in ROS production and membrane permeability (light vs. dark) were observed. Cell viability at 1 μg mL^{-1} concentration under UV illumination and dark conditions was 59.1% and 69.46%, respectively, for 72 h exposure period. Cellular uptake of NPs was indicated in electron micrographs, whereas fluorescence micrographs and images from confocal laser scanning microscopy (CLSM) brought out their probable genotoxic effects (Dalai et al. 2013).

In addition, a comparative study was conducted by Sadiq et al. (2011a, b) to demonstrate the toxic effects caused by TiO_2-NPs toward the freshwater algae (*Scenedesmus* sp. and *Chlorella* sp.) isolated from freshwater environment after 72 h incubation. The particles had a growth-inhibiting effect for both species (EC_{50} = 16.12 mg TiO_2-NPs L^{-1} for *Chlorella* sp.; EC_{50} = 21.2 mg TiO_2-NPs L^{-1} for *Scenedesmus* sp.). Bulk (micron-sized) TiO_2 also showed toxicity though to a lesser extent (EC_{50} = 35.50 mg TiO_2-NPs L^{-1} for *Chlorella* sp.; EC_{50} = 44.40 mg TiO_2-NPs L^{-1} for *Scenedesmus* sp.). A concentration-dependent reduction in the fluorescence of chlorophyll content was also observed (Sadiq et al. 2011a, b). These species were also used in a comparative study of the photocatalytic activity of P25 TiO_2-NPs under dark, visible light and UVA conditions (Roy et al. 2016). *Chlorella* was more sensitive toward the toxicity effects than *Scenedesmus*. Furthermore, at the highest exposure concentration, ROS generation was found correlated with inactivation of the antioxidant enzymes (SOD and GSH) for both the algae under visible light and UVA conditions. Additionally, TiO_2-NPs increased catalase activity and LPO release, indicating the membrane damage, particularly high in *Chlorella*, which is a single-cell algae and therefore is more susceptible to TiO_2-NP uptake in comparison to *Scenedesmus*, which shows a high colonization tendency.

Moreover, toxicity of NPs of binary compounds of titanium dioxides (with particle size 5, 50, 90, and 350 nm) was studied again on the unicellular alga *Chlorella vulgaris* (Gosteva et al. 2015). Substantiating the findings of Hartmann et al. (2010), this study revealed a selective dependence of TiO_2-NP toxicity on size and concentration of NPs. TiO_2-NPs with particle size 5 and 90 nm were classified to the category "acute toxicity 1," whereas no acute toxicity was registered for particle size 50 nm. Physiological, biochemical, and molecular genetic levels were assessed on the unicellular green alga*Chlamydomonas reinhardtii* after application of TiO_2-NPs (Wang et al. 2008). Growth inhibition was observed during the first 2–3 days of incubation with TiO_2-NPs, but later a dose-dependent recovery was observed. Oxidative stress occurred within the cell after exposure to TiO_2-NPs, which caused an increase in malondialdehyde levels, while four stress response genes (sod1, gpx, cat, and ptox2) were upregulated in cultures containing even 1 mg L^{-1} of TiO_2-NPs. The maximum transcripts of cat, sod1, gpx, and ptox2 occurred at 1.5, 3, 3, and 6 h, respectively, proportional to the initial concentration of the NPs.

Kulacki and Cardinale (2012) examined how TiO_2-NPs (ranging from 0 to 300 mg TiO_2-NPs L^{-1}) affect the population dynamics and production of biomass across a range of the North American freshwater algae (*Anabaena* spp., *Navicula subminuscula*, *Nitzschia pusilla*, *Oscillatoria* spp., *Planothidium lanceolatum*, *Scenedesmus quadricauda*, *Selenastrum minutum*, *Spirogyra communis*, *Stigeoclonium tenue*, *Tabularia fasciculata*). The effects of TiO_2-NPs on the population growth rate of each algal species over a period of 25 days were not significant ($p = 0.376$), even though there was a considerable species-specific differentiation in responses (strong inhibition of maximum growth rate for *Spirogyra communis*, whereas strong stimulation of maximum growth rate for *Stigeoclonium tenue*). On the contrary, exposure to TiO_2-NPs tended to increase the maximum biomass achieved by species in culture ($p = 0.06$).

Finally, the effects of TiO_2-NPs and bulk particles on the marine microalga *Nitzschia closterium* were evaluated with reference to growth, oxidative stress, and cellular uptake after 96 h incubation (Xia et al. 2015). Toxicity of TiO_2-NPs to algal cells significantly increased with decreasing nominal particle size, and the EC_{50} values were 88.78, 118.80, and 179.05 mg L^{-1} for 21, 60, and 400 nm NPs, respectively. The growth was significantly inhibited on exposure to 5 mg L^{-1} of 21 nm TiO_2 NPs. Activities of antioxidant enzymes, viz., peroxidase (POD), superoxide dismutase (SOD), and catalase (CAT), were induced at the beginning and thereupon were inhibited, whereas malondialdehyde (MDA) levels and reactive oxygen species (ROS) increased following the exposure to 5 mg L^{-1} TiO_2 NPs, indicating damages on the cell membrane. Flow cytometry and TEM studies and Ti content measurements indicated that TiO_2-NPs were internalized in *N. closterium* cells. The level of extracellular ROS was negligible, as compared to the intracellular ROS level, suggesting that the elevated TiO_2 toxicity in marine environments is related to increased ROS levels caused by internalization of TiO_2-NPs.

The abovementioned toxicity data of TiO_2-NPs are presented in Table 16.3, specifying the particle size, the algal species used, the half maximal effective concentration, and the exposure time used.

Table 16.3 Summarized results from TiO_2-NP toxicity studies performed on algae

Species	Particle size (nm)	EC_{50} (mg TiO_2-NPs L^{-1}) – incubation time	Effects observed	Reference
Desmodesmus subspicatus Chodat	25 100	44 at 72 h	Reduction of fluorescence intensity	Hund-Rinke and Simon (2006)
Pseudokirchneriella subcapitata (Korshikov) F. Hindák	140	16 (fine TiO_2-NPs) at 72 h	Growth inhibition	Warheit et al. (2007)
	Mean~ 140	21 (uf-C TiO_2-NPs) at 72 h		

(continued)

Table 16.3 (continued)

Species	Particle size (nm)	EC_{50} (mg TiO_2-NPs L^{-1}) – incubation time	Effects observed	Reference
P. subcapitata (Korshikov) F. Hindák	30	n.m.[a]	Growth inhibition	Griffitt et al. (2008)
Chlamydomonas reinhardtii P.A. Dangeard	21	n.m.[a] >10 at 72 h	Aggregate during time, growth inhibition, lipid peroxidation, increase of malondialdehyde level, upregulation of sod1, gpx, cat, and ptox2 gene transcripts	Wang et al. (2008)
Pseudokirchneriella subcapitata (Korshikov) F. Hindák	<100	n.m.[a] at 72 h	No toxicity effects mentioned	Blaise et al. (2008)
P. subcapitata (Korshikov) F. Hindák	25 70	5.83 at 72 h 35.9 at 72 h	Aggregate during time, growth inhibition	Aruoja et al. (2009)
P. subcapitata (Korshikov) F. Hindák	10 30 300	241 at 72 h, 71.1 at 72 h, 145 at 72 h	Concentration-dependent aggregation, reduction of fluorescence intensity	Hartmann et al. (2010)
Scenedesmus sp.	25	21.2 (TiO_2-NPs) at 72 h,	Reduction of fluorescence intensity	Sadiq et al. (2011a, b)
		35.5 (bulk TiO_2-NPs) at 72 h		
		16.12 (TiO_2-NPs) at 72 h,		
Chlorella sp.		44.4 (bulk TiO_2-mPs) at 72 h		
Anabaena spp., *Navicula subminuscula* Manguin, *Nitzschia pusilla* Grunow, *Oscillatoria* spp., *Planothidium lanceolatum* (Brébisson ex Kützing) Lange-Bertalot, *Scenedesmus quadricauda* (Turpin) Brébisson, *Selenastrum minutum* (Nägeli) Collins, *Spirogyra communis* (Hassall) Kützing, *Stigeoclonium tenue* (C.Agardh) Kützing, *Tabularia fasciculata* (C. Agardh) D.M.Williams and Round	27	n.m.[a]	Moderate growth inhibition across all species, increase of maximum biomass	Kulacki and Cardinale (2012)

(continued)

Table 16.3 (continued)

Species	Particle size (nm)	EC_{50} (mg TiO_2-NPs L^{-1}) – incubation time	Effects observed	Reference
Pseudokirchneriella subcapitata (Korshikov) F. Hindák	21	200 at 72 h (fluorescence of pigments), 160 at 72 h (hemocytometer counts)	Formation of larger agglomerates/aggregate during time, reduction of pigment (chlorophyll and carotenoid) synthesis	Hartmann et al. (2013)
Scenedesmus obliquus (Turpin) Kützing	25	n.m.[a] >1 at 72 h	Agglomeration, ROS generation, genotoxicity, increased membrane permeability and TiO_2-NP uptake	Dalai et al. (2013)
Pseudokirchneriella subcapitata (Korshikov) F. Hindák	8–21	>1–10 at 72 h	Growth inhibition, agglomeration	Aruoja et al. (2015)
Chlorella vulgaris Beyerinck	5 50 90 350	1.153 at 72 h, >100 at 72 h, 7.7 at 72 h, >100 at 72 h	Growth inhibition	Gosteva et al. (2015)
Nitzschia closterium (Ehrenberg) W.Smith	21 60 400	88.78 118.80 179.05	Aggregation, growth inhibition, decrease in the activities of SOD, CAT, and POD, increase in MDA and ROS levels, damages in cell membrane, NP internalization	Xia et al. (2015)
Pseudokirchneriella subcapitata (Korshikov) F. Hindák	<10	8.5 at 72 h (24-well microplates), 2.7 at 72 h (in cylindrical vial systems), >50 at 72 h (in Erlenmeyer flasks)	Growth inhibition	Manier et al. (2016)
	20	>50 at 72 h (24-well microplates), 39 at 72 h (in cylindrical vial systems), >50 at 72 h (in Erlenmeyer flasks)		
Scenedesmus sp.	21	7.6 at 72 h under dark conditions	Growth inhibition under UVA, ROS generation, decreased SOD activity and GSH levels especially under UVA, increased catalase levels, LPO release and TiO_2-NP uptake	Roy et al. (2016)
		4.1 at 72 h under visible light		
		2.7 at 72 h under UVA		
Chlorella sp.		5.9 at 72 h under dark conditions		
		2.16 at 72 h under visible light		
		1.5 at 72 h under UVA		

[a]Not mentioned

16.3.4 Zinc Oxide Nanoparticles

Toxicity of ZnO-NPs to the algae *Pseudokirchneriella subcapitata* was determined by the OECD 201 algal growth inhibition test (Aruoja et al. 2009). EC_{50} values of bulk and nano ZnO particles were both similar to that of $ZnSO_4$ (at 72 h EC_{50} ~ 0.04 mg ZnO-NPs L^{-1}), and no aggregation formation was observed. This was attributed to the dissolved Zn possibly because most of the ZnO is dissolved at these low concentrations (Franklin et al. 2007). The toxicity data were close to those obtained by Franklin et al. (2007) who showed EC_{50} values for the same algal species to be 0.063 mg Zn L^{-1} for bulk ZnO and 0.068 mg Zn L^{-1} for nano ZnO after 72 h exposure. In another study performed on *P. subcapitata*, EC_{50} value was between 0.1 and 1 mg ZnO-NPs L^{-1} with particle size 8–21 nm (Aruoja et al. 2015).

Toxicity of ZnO-NPs with particle size 20 nm was also evaluated on the unicellular alga *Chlorella vulgaris* (Morgalev et al. 2015). A concentration-dependent reduction in the fluorescence of growth rate was observed after application of ZnO-NPs. The detected value of EC_{50} was 0.17 mg ZnO-NPs L^{-1}. In assessing the maximum effect of ZnO-NPs on *Chlorella* and other organisms according to GHS and EU Directive 93/67/ EEC, they were assigned to dangerous substances with a high-degree toxicity "acute toxicity 1."

16.3.5 Iron Oxide and Zerovalent Iron Nanoparticles

Engineered zerovalent nano-iron particles (Fe-NPs) hold promise for remediation of several pollutants, but their impact on the environment is not completely clear. The effects of three types of nZVI, (a) Nanofer 25 (uncoated), (b) Nanofer 25S (surface coated with a Na-acrylic copolymer), and (c) Nanofer STAR (Surface stabilized Transportable Air-stable Reactive) powder with an inorganic coating, were assessed on the growth, cell morphology, and metabolic status of marine microalgae *Pavlova lutheri*, *Isochrysis galbana*, and *Tetraselmis suecica* after 23-day exposure (Kadar et al. 2012). The algal growth rate, size distribution, and cellular structure were not altered significantly in any of the three species. The total cellular lipid content increased in *T. suecica* grown on media enriched with uncoated Nanofer 25 and in *P. lutheri* with Nanofer STAR, when compared at equimolar exposures. Furthermore, there occurred a significant change in fatty acid composition complementing the Nanofer STAR-mediated increase in lipid content of *P. lutheri*. Likewise, Zehnder medium fortified with zerovalent Fe-NPs (Nanofer 25 and Nanofer 25S) boosted the growth of four green algae (*Desmodesmus subspicatus*, *Dunaliella salina*, *Parachlorella kessleri*, and *Raphidocelis subcapitata*), two eustigmatophycean algae (*Nannochloropsis limnetica* and *Trachydiscus minutus*), and the cyanobacterium *Arthrospira maxima* (Pádrová et al. 2015). In all the species studied, zerovalent Fe-NPs induced lipid accumulation, the saturated and monounsaturated fatty acid (except palmitoleic acid) and polyunsaturated fatty acid contents in cells. The authors suggested that these particles may provide a source of iron that increases cell growth and enhances metabolic changes leading to higher lipid production and changes in the composition of fatty acids.

In another study, toxicities of four zerovalent Fe-NPs of different sizes (20, 50, 80, and 100 nm), Fe_2O_3-NPs of two sizes (30 and 20 nm) of different crystal phases (α, γ), and Fe_3O_4-NPs of one size (20 nm) were assessed with green alga *Chlorella pyrenoidosa*, focusing on the effects of particle size, crystal phase, oxidation state, and environmental aging (Lei et al. 2016). The results indicated a significant increase in toxicity as particle size decreased. The algal growth inhibition decreased with oxidation of the NPs with an order of zerovalent Fe-NPs > Fe_3O_4 NPs > Fe_2O_3 NPs, while α-Fe_2O_3 NPs ($EC_{50} = 71$ mg L^{-1}) presented significantly higher toxicity than γ-Fe_2O_3 NPs ($EC_{50} = 132$ mg L^{-1}). The EC_{50} values after a 96 h exposure to zerovalent Fe-NPs were for 100 nm (91.3 mg L^{-1}) > 80 nm (81.2 mg L^{-1}) > 50 nm (74.1 mg L^{-1}) > 20 nm (19.8 mg L^{-1}). The NP-induced oxidative stress was the main toxic mechanism, which could give a possible explanation in the difference in algal toxicity caused by NPs with the contribution of agglomeration and physical interactions.

The effects of Fe-NPs have also been assessed on the bryophyte *Physcomitrella patens* subsp. *patens* after foliar exposure (Canivet et al. 2015). The effects (cytotoxicity, oxidative stress, lipid peroxidation of membrane) of Fe-NPs from industrial emissions of metallurgical industries were determined through the axenic culture of *P. patens* exposed at five different concentrations (5 ng, 50 ng, 500 ng, 5 mg, and 50 mg per plant). At concentrations tested over a short period (24 h, 72 h), the levels of ROS, MDA, and glutathione were not significantly disturbed, but after internalization (168 h) the Fe-NPs could interact with the intracellular medium and cause cytotoxic effects and/or oxidative stress. Additionally, confocal microscopy experiments revealed that Fe-NPs (particle size 20–80 nm) penetrated the leaves of *Aphanorrhegma patens* when applied as mineral water suspensions (Canivet et al. 2014). By the way, this was the first demonstration of NP uptake by a bryophyte, and the actual penetration mechanism remains mysterious as also in higher plants.

16.3.6 Copper Oxide Nanoparticles

In a recent study, toxicity of CuO-NPs toward the algal species *Pseudokirchneriella subcapitata* was estimated by the OECD 201 algal growth inhibition test (Aruoja et al. 2009). CuO-NPs with mean particle size 30 nm were found to cause higher toxicity (72 h $EC_{50} = 0.71$ mg CuO-NPs L^{-1}) than one caused by bulk CuO (72 h $EC_{50} = 11.55$ mg CuO-NPs L^{-1}). No aggregates were observed in the growth medium. These findings are in agreement with those of a previous study where 15–45 nm CuO-NPs gave 0.54 mg CuO-NPs L^{-1} EC_{50} value at 96 h (Griffitt et al. 2008). The toxicity of CuO-NPs to algae has also been assessed in the presence of dissolved organic matter (DOM). One of the main fractions of DOM (Suwannee river fulvic acid) (20 mg L^{-1}) was added alone or in the presence of CuO-NPs in the culture medium of the prokaryotic alga *Microcystis aeruginosa* (Wang et al. 2011). Internalization of CuO-NPs was observed in the intact algal cells at certain locations (e.g., thylakoids and granules), and the cell uptake was enhanced by Suwannee river fulvic acid (SRFA). The main form of intracellular NPs observed was Cu_2O, indicating that intracellular environment may reduce CuO into Cu_2O. The increased CuO nanotoxicity observed in the presence of SRFA was related to the decreased rate of aggregation formation, the higher Cu^{2+} release, and the induction in the internalization of CuO-NPs.

In another study, the short-term effects of core-shell copper oxide NPs (CS-CuO-NPs) in two different agglomeration states on the green alga *Chlamydomonas reinhardtii* were examined, and toxicity was investigated with regard to change in cellular population structure, primary photochemistry of photosystem II, and the ROS formation (Saison et al. 2010). CS-CuO-NPs induced cellular aggregation processes and reduced chlorophyll levels by inhibiting photosystem II. This process (inhibition of photosynthetic electron transport) induced a strong energy dissipation process via non-photochemical pathways indicating the formation of reactive oxygen ROS. However, no ROS formation was observed when *C. reinhardtii* was exposed to the core without the shell or to the shell only. The toxicity of carbon-coated copper nanoparticles (Cu-NPs) was also investigated using the alga *C. reinhardtii* and compared with effects of dissolved Cu^{2+} (provided as $CuSO_4$) (Müller et al. 2015). The Cu-NPs agglomerated in the medium from original size of 6–7 nm to average particle sizes of 140–200 nm possibly due to the hydrophobic properties of the carbon coating. Cu-NPs strongly decreased the photosynthetic yield of *C. reinhardtii* after 1–2 h exposure to dissolved Cu^{II} in a concentration range 1–100uM, whereas this decrease occurred in a concentration range of 0.1–10 uM for $CuSO_4$. Cu-NP effects on photosynthetic yield were similar for the same concentration of dissolved Cu^{2+} for 1 h exposure and slightly stronger after longer exposure times. After the addition of EDTA as a strong ligand for Cu^{II}, toxicity of both dissolved Cu^{II} and of Cu-NPs was completely suppressed.

The toxicity effects of sonicated and non-sonicated CuO suspensions (<50 nm) were elucidated on macrophytic (*Nitellopsis obtusa*) and microphytic (*Chlorella* spp.) algae cells (Manusadžianas et al. 2012). Cell lethality and resting potential depolarization were used to measure the NP effects on *N. obtusa*, whereas photosynthetic efficiency was assessed on *Chlorella* spp. There were no substantial differences between the effects of non-sonicated and sonicated CuO-NP suspensions. The particles rapidly reagglomerated within 5 min after sonication. The lethal concentrations of CuO-NPs did not evoke a rapid cell membrane depolarization in *N. obtusa* within the initial 90 min period, indicating that charophyte cell wall might have delayed the NP toxic effects. Significant cell membrane depolarization could be observed only after a 6 h exposure. In addition, rewash lethality tests revealed that 5 min exposure in 100 mg CuO-NPs L^{-1} concentration induced algal cells mortality by 70% after 8 days, whereas 6 h exposure at 0.64 mg L^{-1} of Cu^{2+} evoked less than 40% cell mortality. The observed lethal effects of algae cells as well as delayed cell membrane depolarization were evoked by nanoparticles or their agglomerates per se, but not by dissolved Cu, as neither chemical analysis nor biological testing confirmed the presence of Cu^{2+} in toxic amounts.

The macrophyte *Lemna gibba* was also used to evaluate CuO-NP toxicity. On exposure to CuO-NPs or soluble copper for 48 h, photosynthetic activity was inhibited due to inactivation of photosystem II reaction centers, causing a decrease in electron transport rate and an increase of thermal energy dissipation (Perreault et al. 2014). Toxicity of CuO-NPs was mainly driven by copper ions released from particles due to the NPs' tendency to agglomerate in the culture medium.

The data on CuO-NPs versus algae are presented in Table 16.4.

Table 16.4 Summarized results from CuO-NP toxicity studies performed on lower plants

Species	Particle size (nm)	EC_{50} (mg CuO-NPs L^{-1}) – incubation time	Effects observed	Reference
Pseudokirchneriella subcapitata (Korshikov) F. Hindák	15–45	0.54 at 96 h	Growth inhibition	Griffitt et al. (2008)
P. subcapitata (Korshikov) F. Hindák	30	0.71at 72 h	Growth inhibition	Aruoja et al. (2009)
Chlamydomonas reinhardtii P.A. Dangeard (wall-free strain CC-400)	81	n.m.[a]	Formation of aggregates, inhibition of photosynthesis, ROS formation	Saison et al. (2010)
Microcystis aeruginosa (Kützing) Kützing	<10	0.47 at 72 h	Growth inhibition, ROS formation, DNA damage, membrane integrity damage	Wang et al. (2011)
Nitellopsis obtusa (N.A. Desvaux) J. Groves	~30	4.3 for non-sonicated NPs at 96 h	Growth inhibition, agglomeration, inhibition of photosynthesis, cell membrane depolarization	Manusadžianas et al. (2012)
		2.8 for sonicated NPs at 96 h		
Chlorella sp.		57 for non-sonicated NPs at 30 min		
		47 for sonicated NPs at 30 min		
Lemna gibba L.	30–40	n.m.[a]	Growth inhibition, inactivation of photosystem II reaction centers, decrease in electron transport rate, increase of thermal energy dissipation	Perreault et al. (2014)
C. reinhardtii P.A. Dangeard	6–7	1.87 ± 3 at 2 h	Agglomeration, inhibition of photosynthesis	Müller et al. (2015)

[a]Not mentioned

16.3.7 Nickel Oxide Nanoparticles

NiO-NPs (20 nm average size) were found to provoke a severe growth inhibition on a marine microalga strain of *Chlorella vulgaris* in a sterilized enriched seawater medium (f/2 medium) when treated with 40–50 mg L^{-1} during 72–120 h of exposure, with EC_{50} being 32.28 mg NiO L^{-1} at 72 h and 44.33 mg NiO L^{-1} at 120 h (Gong et al. 2011). The observed inhibitory effect was accompanied by cellular structural alterations such as cytomembrane breakage (detached or degraded plasma membrane), plasmolysis (leak of cytosol), and disorder of thylakoids. At the same time, living algae showed a tendency to increase the agglomeration-deposition capacity of NiO-NPs as well as to reduce them for zero valence nickel.

In another study, the aquatic plant *Lemna gibba* was used to investigate and compare the toxicity induced by 30 nm NiO-NPs and nickel(II) oxide as bulk (NiO Bulk) (Oukarroum et al. 2015). Plants were exposed for 24 h to NiO-NPs, or NiO Bulk caused agglomerations of NiO-NPs in culture medium, due to ionic strength. Both NPs and bulk enhanced ROS formation, especially at 1000 mg L^{-1} (five times compared to control), indicating the cellular oxidative stress. Both types of NiO induced a strong inhibitory effect on the PSII quantum yield, indicating a reduction of the photosynthetic electron transport performance due to damage to the structural and functional properties of PSII (Oukarroum et al. 2015).

Likewise, *Chlorella vulgaris* exposed to NiO-NPs for 96 h showed cellular alterations, which were related to NiO-NP concentration (EC_{50} of 13.7 mg L^{-1}). They particularly inhibited cell division (relative cell size and granularity), deteriorated the photosynthetic apparatus (chlorophyll synthesis and photochemical reactions of photosynthesis), and induced oxidative stress (ROS formation). The TEM and X-ray analysis indicated that NiO-NPs were able to cross biological membranes and accumulate inside the algal cells (Oukarroum et al. 2017). In addition, 20 nm NiO-NPs displayed severe inhibitory effect on the growth of *C. vulgaris* after 96 h exposure with EC_{50} value of 31.4 mg L^{-1} (Li et al. 2017). The changes observed were plasmolysis with a shriveled cell shape, disruption of plasma membrane, cytosol leakage, and disorders in thylakoid grana lamella. Moreover, NP aggregation as well as partial reduction to Ni^0 could be observed, suggesting a possible remediation strategy of aquatic pollution (Li et al. 2017). These results for 20 nm NiO-NPs are comparable to the 0.35 mg NiO L^{-1} EC_{50} value obtained in a previous study for *Pseudokirchneriella subcapitata* (Griffitt et al. 2008). Data on toxicity effects of NiO-NPs are presented in Table 16.5.

16.3.8 Silica Oxide Nanoparticles

In order to test the hypothesis that the ecotoxicity of nanoparticles is related to their surface area and not to their mass, toxicity of silica (SiO_2) nanoparticles was monitored on the growth of *Pseudokirchneriella subcapitata* exposed to stable silica suspensions (Van Hoecke et al. 2008). Commercial Ludox suspensions of NPs with diameters 12.5 and 27 nm were toxic, with 20% effect concentration (EC_{20}) values on growth rate of 20.0 and 28.8 mg L^{-1}, respectively, at 72 h. Because no aggregation was

Table 16.5 Summarized results from NiO-NP toxicity studies performed on lower plants

Species	Particle size (nm)	EC_{50} (mg NiO-NPs L^{-1}) – incubation time	Effects observed	Reference
Pseudokirchneriella subcapitata (Korshikov) F.Hindák	5–20	0.35 at 96 h	Growth inhibition	Griffitt et al. (2008)
Chlorella vulgaris Beyerinck	20	32.28 at 72 h 44.33 at 120 h	Plasmolysis, cytomembrane breakage, and thylakoid disorder	Gong et al. (2011)
Lemna gibba L.	30	n.m.[a]/at 24 h	Increase in ROS formation, inhibition of the photochemical activity of the photosystem II (PSII), reduction in the quantum yield of PSII electron transport	Oukarroum et al. (2015)
Chlorella vulgaris Beyerinck	20	31.4 at 96 h	Growth inhibition, plasmolysis	Li et al. (2017)
C. vulgaris Beyerinck	30	13.7 at 96 h	Agglomeration formation, inhibition in cell division, deterioration of photosynthetic apparatus, ROS formation	Oukarroum et al. (2017)

[a]Not mentioned

observed and the dissolution of NPs was negligible, the toxicity was attributable to the solid nanospheres. There was no significant difference in toxicity on expressing the concentration as a surface area. The 72 h EC_{20} values were 4.7 and 3.9 mg L^{-1}. Silica bulk material was found to be nontoxic up to 1 g L^{-1}. TEM studies with 100 mg SiO_2-NPs L^{-1} (particle size 12.5 and 27 nm) elucidated no evidence of particle uptake even though the particles clearly adhered to the cell wall surface (Van Hoecke et al. 2008).

In a different study, 96 h exposure of *Scenedesmus obliquus* to SiO_2-NPs (10–20 nm, 25–200 mg L^{-1}) resulted in a significant concentration-dependent decrease in chlorophyll content, whereas the carotenoid content was unaffected. EC_{50} value could not be determined since SiO_2-NPs were not toxic probably due to shading on the cell surface (Wei et al. 2010a).

16.4 Carbon-Based Nanoparticles

16.4.1 Fullerene

The effects of carbon fullerene C_{60} were investigated on *Chlamydomonas reinhardtii*. The assays included also a bioaccumulation test to observe whether the algae accumulate nanomaterials and whether they have negative effects on *Daphnia magna* (water flea), a small planktonic crustacean belonging to the subclass Phyllopoda and consecutively on the whole trophic chain (Luo 2007). Population changes were measured over an initial period of 48 h, which was then extended to 480 h to estimate the long-term effects. The effects were long lasting, as the algal population treated with 10 mg L^{-1} of C_{60}-NPs was unable to recover within a 20-day period. C_{60} treatment caused color changes, cell lysis, and difficulties in reproduction. More algae died on 1 mg L^{-1} of C_{60}, compared to 10 mg L^{-1} of C_{12}. C_{60} NPs were also more toxic to *D. magna* than C_{12}. Bioaccumulation studies indicate the relocation of nanomaterials from the alga to *Daphnia*, primarily through water but also through the alga, but the trend is not conclusive. Dynamic light scattering studies indicated aggregate formation, when the particles were introduced in an aquatic environment that led to the induction of oxidative stress.

The nanocrystalline fullerene (nC_{60}) uptake amounts and trophic transfer efficiency to the predator (*Daphnia magna*) through dietary exposure to algae or algal subcellular fractions (*Scenedesmus obliquus*) have also been investigated (Chen et al. 2016). The nC_{60}-contaminated algae were separated into the cell wall (CW), cell organelle (CO), and cell membrane (CM) fractions. The highest nC_{60} distribution was in CW, followed by CO and CM subparts. Further, the sublethal concentration for *S. obliquus* has been determined as 0.09 mg L^{-1} after 72 h exposure (Tao et al. 2015). During a sublethal experiment of C_{60} that was carried out for 60 days, the photosynthesis processes, the photosynthetic polysaccharide products, soluble protein, and total lipids in *S. obliquus* were decreased. Additionally, chlorophyll *a* and chlorophyll *b* were negatively impacted possibly due to the 40% algal Mg^{2+} decline at the sublethal concentration (0.09 mg L^{-1}) of C_{60}. The decline was due to inhibition of Mg^{2+}-ATPase activity caused by nC_{60} aggregates. On the other hand, on comparing the highest and lowest sensitivity responses of a bioassay on *Pseudokirchneriella subcapitata*, Blaise et al. (2008) found fullerene C_{60} to be less toxic than other NPs and placed it in the "not toxic" category (>100 mg L^{-1}) (Blaise et al. 2008).

Pseudokirchneriella subcapitata as well as the crustacean *D. magna* were used in a series of toxicity tests for studying the influence of C_{60} aggregates on toxicity and bioaccumulation (Baun et al. 2008). C_{60} powder was stirred in water over 2 months, and the aggregates formed were mixed with four environmental contaminants (atrazine, methyl parathion, pentachlorophenol-PCP, and phenanthrene) with different physicochemical properties and toxic modes of action, 5 days prior to testing. The sorption to C_{60} aggregates was 85% for phenanthrene and 10% for the rest of the compounds. In the presence of C_{60} suspensions, the toxicity of phenanthrene increased (from 720 μg L^{-1} to 430 μg L^{-1}), and the toxicity of PCP decreased (from 36 μg L^{-1} to 70 μg L^{-1}), and a consequent increase in toxicity was found for phenanthrene after addition of C_{60} to the aqueous solution. Addition of C_{60} suspensions reduced the toxicity of PCP. Finally, no enhanced bioaccumulation of phenanthrene was observed in the presence of C_{60} (Baun et al. 2008).

16.4.2 Carbon Nanotubes (CNTs)

In a recent study, multiwalled CNT (MWNT) material was carboxylated by microwave-assisted acid oxidations (f-MWNTs) and was examined for potential toxicity effects, using the unicellular marine chlorophyte alga*Dunaliella tertiolecta* (Wei et al. 2010a, b). Concentrations 5 and 10 mg f-MWNTs L^{-1} caused substantial growth lag phase, and the EC_{50} value at 96 h was 0.82 mg L^{-1}. This impact is in line with the 72 h IC_{25} value (1.04 mg L^{-1}) of single-walled carbon nanotubes (SWCNTs) on the growth of green alga *Pseudokirchneriella subcapitata* (Blaise et al. 2008). Especially, at 10 mg L^{-1} f-MWNTs, 36% reduction in exponential growth rate was observed indicating the presence of oxidative stress and 22% reduction in photosystem II (PSII) quantum yield (Wei et al. 2010a, b). The results differed in a later study, where oxidized SWCNTs (f-SWCNTs) caused 30% growth inhibition, 18% decrease in the photosynthetic yield, and 95% reduction in the intracellular glutathione levels of *Dunaliella tertiolecta* (Thakkar et al. 2016).

In a study of growth inhibition and photosynthetic activity in *Chlorella vulgaris* and *P. subcapitata*, EC_{50} values were 1.8 mg CNTs L^{-1} and 20 mg CNTs L^{-1}, respectively, in well-dispersed suspension whereas 24 mg CNTs L^{-1} and 36 mg CNTs L^{-1}, respectively, in agglomerated suspension (Schwab et al. 2011). The photosynthetic activity was not affected, whereas growth inhibition was correlated to the shading of CNTs and the agglomeration of algal cells, suggesting that the growth might be affected by the shading caused by the CNTs and by alga-CNT agglomerates. However, in another study the toxicological effects of MWCNTs were not dose-dependent (Pereira et al. 2014). Exposure of *C. vulgaris* to MWCNTs induced SOD activity, decreased intracellular ATP levels, and further induced ultrastructural cell damage. Uptake of MWCNTs was observed when cells were cultured in BB medium, but this internalization was not repeated when cells were cultured in Seine river water. The toxicity of MWCNTs was also investigated in *Chlorella* sp. focusing on the four possible mechanisms for the algal growth inhibition (i.e., oxidative stress, agglomeration, physical interactions, and shading effects) and their correlation to the MWCNT size and concentration. At MWCNT concentrations near EC_{50} at 96 h, the oxidative stress accounted for approximately 50% of the algal growth inhibition, whereas 25% of it owes to agglomeration-physical interactions and 25% to the shading effects (Long et al. 2012). Moreover, toxicity of MWCNTs toward *Chlorella pyrenoidosa* was investigated in the presence of different dissolved organic matters, i.e., a natural originated humic acid (HA) and two synthetic surfactants [sodium dodecylbenzenesulfonate (SDBS) and octyl phenoxy polyethoxyethanol (TX100)] (Zhang et al. 2015). Cell internalization of MWCNTs and induction of oxidative stress were promoted by SDBS and TX100, while HA alleviated the MWCNT toxicity by limiting the cell internalization of MWCNTs and reducing the oxidative stress.

The acute aquatic toxicity of SWCNTs (~20 um in length and 1 ~ 1.2 nm in diameter) has been evaluated toward two freshwater microalgae (*Raphidocelis sub-*

capitata and *Chlorella vulgaris*) after a 72 h incubation period (Sohn et al. 2015). The SWCNTs inhibited the growth of *R. subcapitata* and *C. vulgaris* with EC_{50} values of 29.99 and 30.96 mg CNTs L^{-1}, respectively, and were classified as "acute category 3" in the Globally Harmonized System (GHS) of classification and labeling of chemicals. A study of the effects of SWCNTs was undertaken on a population of microalga *Chromochloris zofingiensis* with a special focus on the profile and production of pigments and fatty acids (Wang and Yang 2013). The alga after a 6-day incubation with SWCNTs showed biomass enhancement at low concentrations (40–160 mg L^{-1}) and inhibition at high concentrations (320 mg L^{-1}). By contrast, fatty acids and pigments accumulation decreased over the range of the tested concentrations indicating an increasing sensitivity of the inhibitive toxicity markers as follows: biomass and fatty acids < primary carotenoids < chlorophylls < secondary carotenoids. The data recorded for toxicity of CNTs are shown in Table 16.6.

16.5 Quantum Dots

Potential toxicity of quantum dots (QDs) was assessed by using *Chlamydomonas reinhardtii* as a model system (Wang et al. 2008). The response of the organism to QDs was initially assessed by growth kinetics that showed growth inhibition and formation of aggregates during the first 2–3 days of cultivation (EC_{50} value = 5 mg QDs L^{-1}), followed by a rapid recovery and reduction of cell aggregation as the culture proceeded. Cellular oxidative stress occurred 6 h after exposure to QDs, as confirmed by the transcriptional expression profiling of three stress response genes (*SOD1*, *GPX*, and *CAT*). The expression of these genes was temporarily enhanced in cultures containing 0.1 mg QDs L^{-1}, with the maximum transcripts of *SOD1*, *GPX*, and *CAT* occurring after 3 h of treatment, proportionally to the initial concentration of QDs. As the cultures continued, recovery in growth was observed, and the extent of recovery, as indicated by the final cell concentration, was dosage-dependent. In another study, the adsorption of carboxyl-functionalized polymer-coated QDs (CdSe/ZnS-QDs) and their effects on *C. reinhardtii* photosynthesis were examined (Lin et al. 2009). The amount of QDs adsorbed onto algae logarithmically depends upon the equilibrium concentration of the QDs with Freundlich constants determined as $k = 0.588$ ppm^{1-n} and $n = 0.629$. Furthermore, CO_2 depletion and O_2 production assays showed a significantly inhibited photosynthetic activity of the alga exposed to QDs in concentrations above 100 ppm and 5 ppm, respectively, suggesting the potential impact of NP adsorption on the obstruction of gas flow and nutrients uptake for the algae.

In addition, when *C. reinhardtii* was exposed to increasing QD concentrations, dissolution increased with decreasing pH (Domingos et al. 2011). QDs were accumulated by the algal cells (in a size-dependent manner), though the particles may have been dissolved upon entry into the cells. Whole transcriptome screening using

Table 16.6 Summarized results from CNT toxicity studies performed on algae

Species	Type	Length (μm) and outer diameter (nm)	EC_{50} (mg CNTs L^{-1}) – incubation time	Effects observed	Reference
Pseudokirchneriella subcapitata (Korshikov) F. Hindák	SWCNTs	2–5 and 1.2–1.5	n.m.[a] >1.04 at 72 h	Growth inhibition	Blaise et al. (2008)
Dunaliella tertiolecta Butcher	Oxidized MWCNTs	50 and 20–30	0.82 at 96 h	Aggregation, growth inhibition, reduction of photosynthesis, oxidative stress indication	Wei et al. (2010a, b)
Chlorella vulgaris Beyerinck *Pseudokirchneriella subcapitata* (Korshikov) F. Hindák	CNTs	2–5 and 5–15	1.8 mg CNTs L^{-1} and 20 mg CNTs L^{-1} in well-dispersed suspensions at 96 h/24 mg CNTL^{-1} and 36 mg CNT L^{-1} in agglomerated suspensions at 96 h	Aggregation, growth inhibition	Schwab et al. (2011)
Chlorella sp.	MWCNT10	0.8 and <10	38.7 (light) and 70.8 (dark) at 96 h	Agglomeration, growth inhibition, increase in MDA and ROS levels	Long et al. (2012)
	MWCNT40	1.7 and 20–40	12.6 (light) and 40.7(dark) at 96 h		
	MWCNT100	3.2 and 60–100	10.8 (light) and 45 (dark) at 96 h		
Chromochloris zofingiensis Dönz	SWCNTs	n.m.[a]	Not assayed	Biomass enhancement at concentrations 40–160 mg L^{-1} and inhibition at 320 mg L^{-1}, decrease in fatty acids and pigment accumulation	Wang and Yang (2013)
C. vulgaris Beyerinck	MWCNTs	40–60 and 20–40	Nonconclusive	Growth inhibition, increase in SOD activity, decrease of intracellular ATP levels, internalization of MWCNTs, inhibition of photosynthesis at 96 h	Pereira et al. (2014)

Raphidocelis subcapitata (Korshikov) Nygaard, Komárek, J.Kristiansen and O.M.Skulberg *Chlorella vulgaris* Beyerinck	SWCNTs	20 and 1–1.2	29.99 at 72 h 30.96 at 72 h	Growth inhibition	Sohn et al. (2015)
Chlorella pyrenoidosa H. Chick	MWCNTs	3.2 and 70	14.5 at 96 h 10.0 (with SDBS) at 96 h 12.6 (with TX100) at 96 h 23.1 (with HA) at 96 h	Growth inhibition, oxidative stress induction, agglomeration, internalization of MWCNTs	Zhang et al. (2015)
Dunaliella tertiolecta Butcher	Oxidized SWCNTs	5–30 and 1.1	n.m.[a] >20 at 120 h	Growth inhibition, decrease in photosynthesis activity, reduction of the GSH levels	Thakkar et al. (2016)

[a]Not mentioned

RNA-Seq analysis identified 174 transcripts that were specifically upregulated by QDs. Moreover, pathways linked to transmembrane activity, proteolysis involving proteasome activation, and ubiquitin-mediated processes were observed. In a different study, effects of CdSe/ZnS-QDs were studied on the availability of Cu and Pb on two strains of *C. reinhardtii* (a wall-less and a walled strain containing glycoproteins) and a microalga (*Chlorella kesslerii*) that possesses a cellulosic cell wall (Worms et al. 2012). The results indicated that QDs decreased the intracellular Cu and Pb contents (non-extractable by EDTA) to almost half in *C. kesslerii* and in the walled strain of *C. reinhardtii* but increased them about 3.5–4 times in the wall-less strain, suggesting that CdSe/ZnS-QDs could influence metal bioavailability due to the interactions of QDs with the cell wall.

As the stability of NPs in seawater is an important requisite for efficient interactions with living organisms, the effects of water-soluble CdSe-QDs were assessed on the marine microalga *Phaeodactylum tricornutum* (Morelli et al. 2012). High QD concentrations (>0.5 nM) caused a dose-dependent inhibition of growth rate and induced ROS formation as well as modulation of SOD and CAT activities. Similarly, functionalized CdSe/ZnS-QDs (amine- and carboxyl-) showed limited toxicity to the marine diatom *Thalassiosira pseudonana* after a 5-day exposure under varied nutrient conditions (enriched versus nitrogen-limited media) (Zhang et al. 2013). Production of proteins in *T. pseudonana* was induced suggesting that these extracellular proteins might be involved in the detoxification of QDs by this alga via the Cd release of QDs.

The bioaccumulation kinetics of thioglycolic acid-stabilized CdTe quantum dots (TGA-CdTe-QDs) was investigated in a freshwater alga *Ochromonas danica* (Wang et al. 2013). Flow cytometry measurements showed high photoluminescent intensity in cells during the exposure time (1, 5, 10, 15, 20, 30, 40, 50, 60 min) suggesting internalization of TGA-CdTe-QDs, while a significant NP uptake was observed through the mechanism of micropinocytosis. The intracellular TGA-CdTe-QDs had negligibly direct acute effects on the algae, and their toxicity was mainly caused by Cd ion liberation into the bulk medium. Quick elimination in the photoluminescent intensity of cellular TGA-CdTe-QDs was also observed, and it was correlated to QD dissolution, surface modification, or expulsion out of the cells. In another report, cytotoxicity of two types of QDs, i.e., carbon QDs (N,S-doped CQDs, N-doped CQDs, no-doped CQDs) and metal QDs (CdTe-QDs, CdS-QDs, $CuInS_2$/ZnS-QDs), was investigated on *Chlorella pyrenoidosa* (Xiao et al. 2016). On treating *C. Pyrenoidosa* with various concentrations of QDs, the total protein and chlorophyll *a* contents were reduced in a dose-response manner. The EC_{50} values (mg L^{-1}) of CQDs and MQDs (shown in Table 16.7) were determined by a growth inhibition biotest (algal cells counting) for 96 h, and their toxicity order was $CuInS_2$/ZnS-QDs < no-doped CQDs < N-doped CQDs < N,S-doped CQDs < CdS-QDs < CdTe-QDs. QDs enhanced the activity of antioxidant enzyme superoxide dismutase (SOD) and decreased the reduced glutathione (GSH) level in a dose-dependent manner. Additionally, QDs enhanced the accumulation of malondialdehyde (MDA). Finally, the toxicity of CQDs was smaller than MQDs, with the toxicity of $CuInS_2$/ZnS-QDs being the smallest one (Xiao et al. 2016).

Table 16.7 Summarized results from QD toxicity studies performed on algae

Species	Type of QDs	EC_{50} (mg QDs L^{-1}) – incubation time	Effects	Reference
Chlamydomonas reinhardtii P.A. Dangeard	CdTe-QDs	5 at 72 h	Cell aggregation, growth inhibition, lipid peroxidation gene expression, temporary enhancement of sod1, gpx, and cat genes	Wang et al. (2008)
C. reinhardtii P.A. Dangeard	CdSe/ZnS-QDs	n.m.[a]	Inhibition of photosynthesis, reduction of CO_2 depletions, reduction of oxygen production	Lin et al. (2009)
C. reinhardtii P.A. Dangeard	CdTe/CdS-QDs	n.m.[a]	Increase in dissolution, QD uptake, upregulation of transmembrane activity pathway proteolysis involving proteasome activation and ubiquitin-mediated processes	Domingos et al. (2011)
Phaeodactylum tricornutum Bohlin	CdSe-QDs CdSe/ZnS-QDs	n.m.[a] >0.5 nM	ROS formation, increase in SOD and CAT activities	Morelli et al. (2012)
Chlamydomonas reinhardtii P.A. Dangeard (wall-less and walled strain containing glycoproteins) *Chlorella kessleri* Fott and Nováková	CdSe/ZnS-QDs	n.m.[a]	Twofold decrease in the intracellular contents of Cu and Pb in *C. kesslerii* and in the walled strain of *C. reinhardtii*, 3.5–4 times increase in the intracellular contents of Cu and Pb in the wall-less strain of *C. reinhardtii*	Worms et al. (2012)
Ochromonas Danica E.G. Pringsheim	TGA-CdTe-QDs	n.m.[a]	QD uptake through micropinocytosis, limited exocytosis/expulsion and dissolution. Quick photoluminescence elimination	Wang et al. (2013)
Thalassiosira pseudonana Hasle and Heimdal	CdSe/ZnS-QDs (amine- and carboxyl-)	n.m.[a]	Aggregation of amine-functionalized QDs, increase in the production of extracellular proteins	Zhang et al. (2013)
Chlorella pyrenoidosa H. Chick	N, S-doped CQDs	38.56 at 96 h	Reduction of total proteins and chlorophyll *a* contents, induction of SOD activity and MDA accumulation, decrease in GSH levels	Xiao et al. (2016)
	N-doped CQDs	185.83 at 96 h		
	No-doped CQDs	232.47 at 96 h		
	CdTe-QDs	0.015 at 96 h		
	CdS-QDs	4.88 at 96 h		
	$CuInS_2$/ZnS-QDs	459.5 at 96 h		

[a]Not mentioned

16.6 Conclusions

The quick growth of nanotechnology over the years has led to rapid development of its commercial applications, which involves the use of a great variety of manufactured NPs. The use of these organic and inorganic nanomaterials may result in the surreptitious discharge of these materials into the environment through soil, sediment, and biosolids from wastewater treatment. Algae and other nonvascular plants constitute an important component of our ecosystem, and toxic effects of nanoparticles on their growth attract serious concerns. Most of the studies conducted on NP toxicity to nonvascular plants have been focused on algae; only a few could encompass bryophytes. The toxic action of NPs can involve some distinct mechanisms, but drawing a general conclusion regarding factors that determine the toxicological effects of NPs is not possible, because toxicity data generated thus far are conflicting and inconsistent. There are indications that NPs might interact directly with algae due to secondary particle size and/or specific surface area or indirectly through release of toxic substances into the exposure media. Further, the duration of exposure to NPs may be an important parameter for the assessment of their toxicity potential (even at low concentrations), which may be more representative of real environmental conditions. Additional comprehensive investigations are urgently and immensely required to examine the impact of NPs on the food chain and the environment and also to be able to reach at logical conclusions for establishing regulations over the use, confinement, and disposal of NPs for the protection of the ecosystem and humankind.

References

Agnan Y, Séjalon-Delmas N, Claustres A, Probst A (2015) Investigation of spatial and temporal metal atmospheric deposition in France through lichen and moss bioaccumulation over one century. Sci Total Environ 529:285–296

Ali S, Khan I, Khan S, Sohail M, Ahmed R, Rehman A, Ansari MS, Morsy MA (2017) Electrocatalytic performance of Ni@Pt core–shell nanoparticles supported on carbon nanotubes for methanol oxidation reaction. J Electroanal Chem 795:17–25

Amelia M, Lincheneau C, Silvi S, Credi A (2012) Electrochemical properties of CdSe and CdTe quantum dots. Chem Soc Rev 41:5728–5743

Aqel A, El-Nour K, Ammar R, Al-Warthan A (2012) Carbon nanotubes, science and technology part (I) structure, synthesis and characterisation. Arab J Geosci 5:1–23

Aruoja V, Dubourguier H, Kasemets K, Kahru A (2009) Toxicity of nanoparticles of CuO, ZnO and TiO_2 to microalgae *Pseudokirchneriella subcapitata*. Sci Total Environ 407:1461–1468

Aruoja V, Pokhrel S, Sihtmäe M, Mortimer M, Mädlerb L, Kahrua A (2015) Toxicity of 12 metal-based nanoparticles to algae, bacteria and protozoa. Environ Sci Nano 2:630–644

Astefanei A, Núñez O, Galceran M (2015) Characterisation and determination of fullerenes: a critical review. Anal Chim Acta 882:1–21

ASTM/E2456-06 (2012) Standard terminology relating to nanotechnology. American Society for Testing and Materials, West Conshohocken Retrieved from www.astm.org

Baun A, Sørensen S, Rasmussen R, Hartmann N, Koch C (2008) Toxicity and bioaccumulation of xenobiotic organic compounds in the presence of aqueous suspensions of aggregates of nano-C_{60}. Aquat Toxicol 86:379–387

Behra R, Wagner B, Sgier L, Kistler D (2015) Colloidal stability and toxicity of gold nanoparticles and gold chloride on *Chlamydomonas reinhardtii*. Aquat Geochem 21:331–342

Bianco A, Da Ros T (2011) Biological applications of fullerenes. In: Langa F, Nierengarten J (eds) Fullerenes: principles and applications, 2nd edn. RSC Publishing, Cambridge, pp 507–545

Blaise C, Gagné F, Férard J, Eullaffroy P (2008) Ecotoxicity of selected nanomaterials to aquatic organisms. Environ Toxicol 23:591–598

Bradley E, Castle L, Chaudhry Q (2011) Applications of nanomaterials in food packaging with a consideration of opportunities for developing countries. Trends Food Sci Technol 22:604–610

Canivet L, Dubot P, Denayer F (2014) Uptake of iron nanoparticles by *Aphanorrhegma patens* (Hedw.) Lindb. J Bryol 36:104–109

Canivet L, Dubot P, Garçon G, Denayer F (2015) Effects of engineered iron nanoparticles on the bryophyte, *Physcomitrella patens* (Hedw.) Bruch & Schimp, after foliar exposure. Ecotoxicol Environ Saf 113:499–505

Chaudhari P, Masurkar S, Shidore V, Kamble S (2012) Effect of biosynthesized silver nanoparticles on *Staphylococcus aureus* biofilm quenching and prevention of biofilm formation. Nano-Micro Lett 4:34–39

Chen P, Powell B, Mortimer M, Ke P (2012) Adaptive interactions between zinc oxide nanoparticles and Chlorella sp. Environ Sci Technol 46:12178–12185

Chen Q, Hu X, Yin D, Wang R (2016) Effect of subcellular distribution on nC_{60} uptake and transfer efficiency from *Scenedesmus obliquus* to *Daphnia magna*. Ecotoxicol Environ Saf 128:213–221

Chow W, Jahnke F (2013) On the physics of semiconductor quantum dots for applications in lasers and quantum optics. Prog Quantum Electron 37:109–184

Colvin V (2003) The potential environmental impact of engineered nanomaterials. Nat Biotechnol 21:1166–1170

Dalai S, Pakrashi S, Nirmala M, Chaudhri A, Chandrasekaran N, Mandal A, Mukherjee A (2013) Cytotoxicity of TiO_2 nanoparticles and their detoxification in a freshwater system. Aquat Toxicol 138:139: 1–139:11

Dash A, Singh A, Chaudhary B, Singh S, Dash D (2012) Effect of silver nanoparticles on growth of eukaryotic green algae. Nano-Micro Lett 4:158–165

De Clerck O, Bogaert K, Leliaert F (2012) Diversity and evolution of algae: primary endosymbiosis. Adv Bot Res 64:55–86

De Volder M, Tawfick S, Baughman R, Hart A (2013) Carbon nanotubes: present and future commercial applications. Science 339:535–539

Deerinck T (2008) The application of fluorescent quantum dots to confocal, multiphoton, and electron microscopic imaging. Toxicol Pathol 36:112–116

Delgado C, Sørensen S, Engelbrekt C, Baun A (2013) Toxicity of platinum nanoparticles to freshwater algae and crustaceans. In: SETAC North America 34th annual meeting, Nashville

Domingos R, Simon D, Hauser C, Wilkinson K (2011) Bioaccumulation and effects of CdTe/CdS quantum dots on Chlamydomonas reinhardtii—nanoparticles or the free ions? Environ Sci Technol 45:7664–7669

Dreaden E, Alkilany A, Huang X, Murphy C, El-Sayed M (2012) The golden age: gold nanoparticles for biomedicine. Chem Soc Rev 41:2740–2779

Duong T, Le T, Huong Tran T, Nguyen T, Ho C, Dao T, Quynh Le T, Nguyen H, Dang D, Huong Le T, Ha P (2016) Inhibition effect of engineered silver nanoparticles to bloom forming cyanobacteria. Adv Nat Sci Nanosci Nanotechnol 7:035018

Dybiec M, Chomokur G, Ostapenko S, Wolcott A, Zhang J, Zajac A, Phelan C, Sellers T, Gerion G (2007) Photoluminescence spectroscopy of bioconjugated CdSe/ZnS quantum dots. Appl Phys Lett 90:263112

Dymytrova L (2009) Epiphytic lichens and bryophytes as indicators of air pollution in Kyiv city (Ukraine). Folia Cryptogam Est 46:33–44

Eddy A, Galloway D, John D, Tittley I (1992) Lower plant diversity. In: Groombridge B (ed) Global biodiversity. Springer, Dordrecht, pp 55–57

Elliott J, Shibuta Y, Amara H, Bichara C, Neyts E (2013) Atomistic modelling of CVD synthesis of carbon nanotubes and graphene. Nanoscale 5:6662–6676

Faburé J, Meyer C, Denayer F, Gaudry A, Gilbert D, Bernard N (2010) Accumulation capacities of particulate matter in an acrocarpous and a pleurocarpous moss exposed at three differently polluted sites (industrial, urban and rural). Water Air Soil Pollut 212:205–217

Franklin N, Rogers N, Apte S, Batley G, Gadd G, Casey P (2007) Comparative toxicity of nanoparticulate ZnO, bulk ZnO, and ZnCl2 to a freshwater microalga (*Pseudokirchneriella subcapitata*): the importance of particle solubility. Environ Sci Technol 41:8484–8490

García-Cambero J, García M, Díaz López G, López Herranz A, Cuevas L, Pérez-Pastrana E, Sendra Cuadal J, Ramis Castelltort M, Castaño Calvo A (2013) Converging hazard assessment of gold nanoparticles to aquatic organisms. Chemosphere 93:1194–1200

Garrec J, Van Haluwyn C (2002) Biosurveillance végétale de la qualité de l'air. Tech & Doc, Lavoisier, Paris

Ghasemzadeh G, Momenpour M, Omidi F, Hosseini M, Ahani M, Barzegari A (2014) Applications of nanomaterials in water treatment and environmental remediation. Front Environ Sci Eng 8:471–482

Gong N, Shao K, Feng W, Lin Z, Liang C, Sun Y (2011) Biotoxicity of nickel oxide nanoparticles and bio-remediation by microalgae *Chlorella vulgaris*. Chemosphere 83:510–516

Gosteva I, Morgalev Y, Morgaleva T, Morgalev S (2015) Effect of AL_2O_3 and TiO_2 nanoparticles on aquatic organisms. In: IOP conference series: materials science and engineering, vol 98. Curran Associates, Inc., Tambov, p 012007

Griffitt R, Luo J, Gao J, Bonzongo J, Barber D (2008) Effects of particle composition and species on toxicity of metallic nanomaterials in aquatic organisms. Environ Toxicol Chem 27:1972–1978

Hannon M, Gimpel J, Tran M, Rasala B, Mayfield S (2010) Biofuels from algae: challenges and potential. Biofuels 1:763–784

Harmens H, Norris D, Cooper D, Mills G, Steinnes E, Kubin E, Thöni L, Aboal J, Alber R, Carballeira A, Coşkun M, De Temmerman L, Frolova M, González-Miqueo L, Jeran Z, Leblond S, Liiv S, Zechmeister H (2011) Nitrogen concentrations in mosses indicate the spatial distribution of atmospheric nitrogen deposition in Europe. Environ Pollut 59:2852–2860

Hartmann N, von der Kammer F, Hofmann T, Baalousha M, Ottofuelling S (2010) Algal testing of titanium dioxide nanoparticles – testing considerations, inhibitory effects and modification of cadmium bioavailability. Toxicology 269:190–197

Hartmann N, Engelbrekt C, Zhang J, Ulstrup J, Kusk K, Baun A (2013) The challenges of testing metal and metal oxide nanoparticles in algal bioassays: titanium dioxide and gold nanoparticles as case studies. Nanotoxicology 7:1082–1094

He D, Dorantes-Aranda J, Waite T (2012) Silver nanoparticle-algae interactions: oxidative dissolution, reactive oxygen species generation and synergistic toxic effects. Environ Sci Technol 46:8731–8738

Hund-Rinke K, Simon M (2006) Ecotoxic effect of photocatalytic active nanoparticles (TiO_2) on algae and daphnids. Environ Sci Pollut Res 13:225–232

Ibrahim K (2013) Carbon nanotubes-properties and applications: a review. Carbon Lett 14:131–144

Iswarya V, Johnson J, Parashar A, Pulimi M, Chandrasekaran N, Mukherjee A (2017) Modulatory effects of Zn^{2+} ions on the toxicity of citrate- and PVP-capped gold nanoparticles towards freshwater algae, *Scenedesmus obliquus*. Environ Sci Pollut Res Int 24:3790–3801

Kadar E, Rooks P, Lakey C, White D (2012) The effect of engineered iron nanoparticles on growth and metabolic status of marine microalgae cultures. Sci Total Environ 439:8–17

Kim K, Sung W, Suh B, Moon S, Choi J, Kim J, Lee D (2009) Antifungal activity and mode of action of silver nano-particles on *Candida albicans*. Biometals 22:235–242

Książyk M, Asztemborska M, Stęborowski R, Bystrzejewska-Piotrowska G (2015) Toxic effect of silver and platinum nanoparticles toward the freshwater microalga *Pseudokirchneriella subcapitata*. Bull Environ Contam Toxicol 94:554–558

Kulacki K, Cardinale B (2012) Effects of nano-titanium dioxide on freshwater algal population dynamics. PLoS One 7:e47130

Lang X, Hirata A, Fujita T, Chen M (2011) Nanoporous metal/oxide hybrid electrodes for electrochemical supercapacitors. Nat Nanotechnol 6:232–236

Laurent S, Forge D, Port M, Roch A, Robic C, van der Elst L, Muller R (2008) Magnetic iron oxide nanoparticles: synthesis, stabilization, vectorization, physicochemical characterizations, and biological applications. Chem Rev 108:2064–2110

Lee D, Fortin C, Campbell P (2005) Contrasting effects of chloride on the toxicity of silver to two green algae, Pseudokirchneriella subcapitata and *Chlamydomonas reinhardtii*. Aquat Toxicol 75:127–135

Lei C, Zhang L, Yang K, Zhu L, Lin D (2016) Toxicity of iron-based nanoparticles to green algae: effects of particle size, crystal phase, oxidation state and environmental aging. Environ Pollut 218:505–512

Li Y, Xiao R, Liu Z, Liang X, Feng W (2017) Cytotoxicity of NiO nanoparticles and its conversion inside *Chlorella vulgaris*. Chem Res Chin Univ 33:107–111

Lin S, Bhattacharya P, Rajapakse N, Brune D, Ke P (2009) Effects of quantum dots adsorption on algal photosynthesis. J Phys Chem C 113:10962–10966

Lodenius M (2013) Use of plants for biomonitoring of air borne mercury in contaminated areas. Environ Res 125:113–123

Long Z, Ji J, Yang K, Lin D, Wu F (2012) Systematic and quantitative investigation of the mechanism of carbon nanotubes' toxicity towards algae. Environ Sci Technol 46:8458–8466

Luo J (2007) Toxicity and bioaccumulation of nanomaterial in aquatic species. J US SJWP 2:1–16

Ma S, Lin D (2013) The biophysicochemical interactions at the interfaces between nanoparticles and aquatic organisms: adsorption and internalization. Environ Sci Processes Impacts 15:145–160

Manier N, Le Manach S, Bado-Nilles A, Pandard P (2016) Effect of two TiO_2 nanoparticles on the growth of unicellular green algae using the OECD 201 test guideline: influence of the exposure system. Toxicol Environ Chem 98:860–876

Manusadžianas L, Caillet C, Fachetti L, Gylyte B, Grigutyte R, Jurkoniene S, Karitonas R, Sadauskas K, Thomas F, Vitkus R, Ferard J (2012) Toxicity of copper oxide nanoparticle suspensions to aquatic biota. Environ Toxicol Chem 31:108–114

Melo A, Amadeu M, Lancellotti M, de Hollanda LM, Machado D (2015) The role of nanomaterials in cosmetics: national and international legislative aspects. Quim Nova 38:599–603

Menard A, Drobne D, Jemec A (2011) Ecotoxicity of nanosized TiO_2. Review of in vivo data. Environ Pollut 159:677–684

Meyer C, Gilbert D, Gaudry A, Franchi M, Nguyen-Viet H, Fabure J, Bernard N (2010) Relationship of atmospheric pollution characterized by gas (NO_2) and particles (PM_{10}) to microbial communities living in bryophytes at three differently polluted sites (rural, urban, and industrial). Microb Ecol 59:324–334

Miazek K, Iwanek W, Remacle C, Richel A, Goffin D (2015) Effect of metals, metalloids and metallic nanoparticles on microalgae growth and industrial product biosynthesis: a review. Int J Mol Sci 16:23929–23969

Morelli E, Cioni P, Posarelli M, Gabellieri E (2012) Chemical stability of CdSe quantum dots in seawater and their effects on a marine microalga. Aquat Toxicol 122–123:153–162

Morgalev Y, Morgaleva T, Gosteva I, Morgalev S, Kulizhskiy S, Astafurova T (2015) Effect of zinc oxide nanoparticles on the test function of water organisms of different trophic levels. In: IOP conference series: materials science and engineering, vol 98. Curran Associates, Inc., Tambov, p 012005

Müller E, Behra R, Sigg L (2015) Toxicity of engineered copper (Cu0) nanoparticles to the green alga *Chlamydomonas reinhardtii*. Environ Chem 13:457–463

Namasivayam S, Gnanendra K, Reepika R (2010) Synthesis of silver nanoparticles by *Lactobacillus acidophilus* 01 strain and evaluation of its in vitro genomic DNA toxicity. Nano-Micro Lett 2:160–163

Oishi Y (2013) Comparison of pine needles and mosses as bio-indicators for polycyclic aromatic hydrocarbons. J Environ Prot 4:106–113

Oukarroum A, Bras S, Perreault F, Popovic R (2012) Inhibitory effects of silver nanoparticles in two green algae, *Chlorella vulgaris* and *Dunaliella tertiolecta*. Ecotoxicol Environ Saf 78:80–85

Oukarroum A, Barhoumi L, Samadani M, Dewez D (2015) Toxic effects of nickel oxide bulk and nanoparticles on the aquatic plant *Lemna gibba* L. Biomed Res Int 2015:501326

Oukarroum A, Zaidi W, Samadani M, Dewez D (2017) Toxicity of nickel oxide nanoparticles on a freshwater green algal strain of *Chlorella vulgaris*. Biomed Res Int 2017:9528180

Pádrová K, Lukavský J, Nedbalová L, Čejková A, Cajthaml T, Sigler K, Vítová M, Řezanka T (2015) Trace concentrations of iron nanoparticles cause overproduction of biomass and lipids during cultivation of cyanobacteria and microalgae. J Appl Phycol 27:1443–1451

Pereira M, Mouton L, Yéprémian C, Couté A, Lo J, Marconcini J, Ladeira LO, Raposo N, Brandão H, Brayner R (2014) Ecotoxicological effects of carbon nanotubes and cellulose nanofibers in *Chlorella vulgaris*. J Nanobiotechnol 12:15

Perreault F, Samadani M, Dewez D (2014) Effect of soluble copper released from copper oxide nanoparticles solubilisation on growth and photosynthetic processes of *Lemna gibba* L. Nanotoxicology 8:374–382

Pimratcha S, Butsat S, Kesmala T (2015) Application of blue-green algae and mineral fertilizers. ScienceAsia 41:305–314

Quigg A, Chin W, Chen C, Zhang S, Jiang Y, Miao A (2013) Direct and indirect toxic effects of engineered nanoparticles on algae: role of natural organic matter. ACS Sustain Chem Eng 1:686–702

Rai M, Yadav A, Gade A (2009) Silver nanoparticles as a new generation of antimicrobials. Biotechnol Adv 27:76–83

Reiners R (2013) Definition and standardization of nanomaterials. In: Wolfgang Luther AZ (ed) Safety aspects of engineered nanomaterials. Pan Stanford Publishing Pte. Ltd, Singapore, pp 1–27

Renault S, Baudrimont M, Mesmer-Dudons N, Gonzalez P, Mornet S, Brisson A (2008) Impacts of gold nanoparticle exposure on two freshwater species: a phytoplanktonic alga (*Scenedesmus subspicatus*) and a benthic bivalve (*Corbicula fluminea*). Gold Bull 41:116–126

Röhder L, Brandt T, Sigg L, Behra R (2014) Influence of agglomeration of cerium oxide nanoparticles and speciation of cerium(III) on short term effects to the green algae *Chlamydomonas reinhardtii*. Aquat Toxicol 152:121–130

Roldan Cuenya B (2010) Synthesis and catalytic properties of metal nanoparticles: size, shape, support, composition, and oxidation state effects. Thin Solid Films 518:3127–3150

Roy R, Parashar A, Bhuvaneshwari M, Chandrasekaran N, Mukherjee A (2016) Differential effects of P_{25} TiO_2 nanoparticles on freshwater green microalgae: *Chlorella* and *Scenedesmus* species. Aquat Toxicol 176:161–171

Sadiq I, Dalai S, Chandrasekaran N, Mukherjee A (2011a) Ecotoxicity study of titania (TiO_2) NPs on two microalgae species: *Scenedesmus* sp. and *Chlorella* sp. Ecotoxicol Environ Saf 74:1180–1187

Sadiq I, Pakrashi S, Chandrasekaran N, Mukherjee A (2011b) Studies on toxicity of aluminum oxide (Al_2O_3) nanoparticles to microalgae species: *Scenedesmus* sp. and *Chlorella* sp. J Nanopart Res 13:3287–3299

Saison C, Perreault F, Daigle J, Fortin C, Claverie J, Morin M, Popovic R (2010) Effect of core-shell copper oxide nanoparticles on cell culture morphology and photosynthesis (photosystem II energy distribution) in the green alga, *Chlamydomonas reinhardtii*. Aquat Toxicol 96:109–114

Sanchez-Dominguez M, Boutonnet M, Solans C (2009) A novel approach to metal and metal oxide nanoparticle synthesis: the oil-in-water microemulsion reaction method. J Nanopart Res 11:1823

Schröder W, Holy M, Pesch R, Harmens H, Fagerli H, Alber R, Coşkun M, De Temmerman L, Frolova M, González-Miqueo L, Jeran Z, Kubin E, Leblond S, Liiv S, Maňkovská B, Piispanen J, Santamaría H, Simonèièn P, Suchara I, Yurukova L, Thöni L, Zechmeister H (2010) First europe-wide correlation analysis identifying factors best explaining the total nitrogen concentration in mosses. Atmos Environ 44:3485–3491

Schwab F, Bucheli T, Lukhele L, Magrez A, Nowack B (2011) Are carbon nanotube effects on green algae caused by shading and agglomeration? Environ Sci Technol 4:6136–6144

Service R (2008) Report faults U.S. strategy for nanotoxicology research. Science 322:1779

Shrivastava S, Bera T, Roy A, Singh G, Ramachandrarao P, Dash D (2007) Characterization of enhanced antibacterial effects of novel silver nanoparticles. Nanotechnology 18:225103

Shrivastava S, Bera T, Singh S, Singh G, Ramachandrarao P, Dash D (2009) Characterization of antiplatelet properties of silver nanoparticles. ACS Nano 3:1357–1364

Singh N (2017) Nanotechnology innovations, industrial applications and patents. Environ Chem Lett 15:185–191

Sohn E, Chung Y, Johari S, Kim T, Kim J, Lee J, Lee Y, Kang S, Yu I (2015) Acute toxicity comparison of single-walled carbon nanotubes in various freshwater organisms. Biomed Res Int 2015:323090

Sørensen S, Engelbrekt C, Lützhøft H, Jiménez-Lamana J, Noori J, Alatraktchi F, Delgado C, Slaveykova V, Baun A (2016) A multimethod approach for investigating algal toxicity of platinum nanoparticles. Environ Sci Technol Lett 50:10635–10643

Spolaore P, Joannis-Cassan C, Duran E, Isambert A (2006) Commercial applications of microalgae. J Biosci Bioeng 101:87–96

Stankic S, Suman S, Haque F, Vidic J (2016) Pure and multi metal oxide nanoparticles: synthesis, antibacterial and cytotoxic properties. J Nanobiotechnol 14:73

Tao X, Yu Y, Fortner J, He Y, Chen Y, Hughes J (2015) Effects of aqueous stable fullerene nanocrystal (nC60) on *Scenedesmus obliquus*: evaluation of the sub-lethal photosynthetic responses and inhibition mechanism. Chemosphere 122:162–167

Taylor C, Matzke M, Kroll A, Read D, Svendsen C, Crossley A (2016a) Toxic interactions of different silver forms with freshwater green algae and cyanobacteria and their effects on mechanistic endpoints and the production of extracellular polymeric substances. Environ Sci Nano 3:396–408

Taylor N, Merrifield R, Williams T, Chipman J, Lead J, Viant M (2016b) Molecular toxicity of cerium oxide nanoparticles to the freshwater alga *Chlamydomonas reinhardtii* is associated with supra-environmental exposure concentrations. Nanotoxicology 10:32–41

Thakkar M, Mitra S, Wei L (2016) Effect on growth, photosynthesis, and oxidative stress of single walled carbon nanotubes exposure to marine alga *Dunaliella tertiolecta*. J Nanomater 2016:8380491

Unrine J, Colman B, Bone A, Gondikas A, Matson C (2012) Biotic and abiotic interactions in aquatic microcosms determine fate and toxicity of ag nanoparticles. Part 1. Aggregation and dissolution. Environ Sci Technol 46:6915–6924

Van Hoecke K, De Schamphelaere K, Van Der Meeren P, Lucas S, Janssen C (2008) Ecotoxicity of silica nanoparticles to the green alga *Pseudokirchneriella subcapitata*: importance of surface area. Environ Toxicol Chem 27:1948–1957

Van Hoecke K, De Schamphelaere K, Ali Z, Zhang F, Elsaesser A, Rivera-Gil P, Parak W, Smagghe G, Howard C, Janssen C (2013) Ecotoxicity and uptake of polymer coated gold nanoparticles. Nanotoxicology 7:37–47

Vance M, Kuiken T, Vejerano E, McGinnis S, Hochella M, Rejeski D, Hull M (2015) Nanotechnology in the real world: redeveloping the nanomaterial consumer products inventory. Beilstein J Nanotechnol 6:1769–1780

Vuković G, Urošević M, Goryainova Z, Pergal M, Škrivanj S, Samson R, Popović A (2015) Active moss biomonitoring for extensive screening of urban air pollution: magnetic and chemical analyses. Sci Total Environ 521–522:200–210

Wang Y, Yang K (2013) Toxicity of single-walled carbon nanotubes on green microalga *Chromochloris zofingiensis*. Chin J Oceanol Limnol 31:306–311

Wang J, Zhang X, Chen Y, Sommerfeld M, Hu Q (2008) Toxicity assessment of manufactured nanomaterials using the unicellular green alga *Chlamydomonas reinhardtii*. Chemosphere 73:1121–1128

Wang Z, Li J, Zhao J, Xing B (2011) Toxicity and internalization of CuO nanoparticles to prokaryotic alga *Microcystis aeruginosa* as affected by dissolved organic matter. Environ Sci Technol 45:6032–6040

Wang Y, Miao A, Luo J, Wei Z, Zhu J, Yang L (2013) Bioaccumulation of CdTe quantum dots in a freshwater alga *Ochromonas danica*: a kinetics study. Environ Sci Technol 47:10601–10610

Warheit D, Hoke R, Finlay C, Donner E, Reed K, Sayes C (2007) Development of a base set of toxicity tests using ultrafine TiO_2 particles as a component of nanoparticle risk management. Toxicol Lett 171:99–110

Wei C, Zhang Y, Guo J, Han B, Yang X, Yuan J (2010a) Effects of silica nanoparticles on growth and photosynthetic pigment contents of *Scenedesmus obliquus*. J Environ Sci 22:155–160

Wei L, Thakkar M, Chen Y, Ntim S, Mitra S, Zhang X (2010b) Cytotoxicity effects of water dispersible oxidized multiwalled carbon nanotubes on marine alga, *Dunaliella tertiolecta*. Aquat Toxicol 100:194–201

Worms I, Boltzman J, Garcia M, Slaveykova V (2012) Cell-wall-dependent effect of carboxyl-CdSe/ZnS quantum dots on lead and copper availability to green microalgae. Environ Pollut 167:27–33

Xia B, Chen B, Sun X, Qu K, Ma F, Du M (2015) Interaction of TiO_2 nanoparticles with the marine microalga *Nitzschia closterium*: growth inhibition, oxidative stress and internalization. Sci Total Environ 508:525–533

Xiao A, Wang C, Chen J, Guo R, Yan Z, Chen J (2016) Carbon and metal quantum dots toxicity on the microalgae *Chlorella pyrenoidosa*. Ecotoxicol Environ Saf 133:211–217

Zhang S, Jiang Y, Chen C, Creeley D, Schwehr K, Quigg A, Chin W, Santschi P (2013) Ameliorating effects of extracellular polymeric substances excreted by *Thalassiosira pseudonana* on algal toxicity of CdSe quantum dots. Aquat Toxicol 126:214–223

Zhang L, Lei C, Chen J, Yang K, Zhu L, Lin D (2015) Effect of natural and synthetic surface coatings on the toxicity of multiwalled carbon nanotubes toward green algae. Carbon 83:198–207

Zhou J, Yang Y, Zhang C (2015) Toward biocompatible semiconductor quantum dots: from biosynthesis and bioconjugation to biomedical application. Chem Rev 115:11669–11717

Chapter 17
Oxidative Stress Biomarkers and Antioxidant Defense in Plants Exposed to Metallic Nanoparticles

Naser A. Anjum, Sarvajeet Singh Gill, Armando C. Duarte, and Eduarda Pereira

17.1 Introduction

Nanoparticles (NPs) are being engineered to at least one dimension ≤100 nm in order to achieve twofold goals: first, to enhance their technologically interesting properties and, second, to explore their use in multiple disciplines such as electronics, textiles, cosmetics, environment, and health/medicines (Royal Society 2004; Elder et al. 2009; Bhatt and Tripathi 2011; Husen and Siddiqi 2014; Husen 2017; Siddiqi et al. 2018). Engineered NPs may end up in various environmental compartments (such as air, water, and soils), after their multidisciplinary and/or indiscriminate use. However, information is meager on the fate and persistence of engineered NPs and also on the outcome of their possible interaction with living systems and the subsequent consequences therein (Handy et al. 2008; Mueller and Nowack 2008; Ma et al. 2010). Among the engineered NPs, metallic NPs (such as Ag, CuO, ZnO, TiO_2, CeO_2, CrO_2, MoO_3, Bi_2O_3) are being increasingly used in a variety of disciplines such as electronics, optics, textiles, medical applications, cosmetics, food packaging, water treatment technology, fuel cells, catalysts, biosensors, and environmental remediation. Environmental contamination due to these NPs is therefore obvious. Given this, systematic investigations on the potential impact of NPs on the health of environment and biota (plant/animal/human) have been strongly advocated (Handy et al. 2008; Thakkar et al. 2010; Gerloff et al. 2012; Anjum et al. 2016; Siddiqi and Husen 2016).

N. A. Anjum (✉)
Department of Botany, Aligarh Muslim University, Aligarh, UP, India

S. S. Gill
Stress Physiology and Molecular Biology Laboratory, Centre for Biotechnology, MD University, Rohtak, India

A. C. Duarte · E. Pereira
Department of Chemistry and CESAM-Centre for Environmental and Marine Studies, University of Aveiro, Aveiro, Portugal

A. Husen, M. Iqbal (eds.), *Nanomaterials and Plant Potential*,
https://doi.org/10.1007/978-3-030-05569-1_17

17.2 Reactive Oxygen Species, Oxidative Stress, and Its Major Biomarkers

Reactive oxygen species (ROS) are produced in living systems, including plants, as the by-products of a normal metabolism. In plants, specific organelles such as chloroplasts, mitochondria, or peroxisomes are the major source of the cellular ROS, as these organelles have a highly oxidizing metabolic activity or intense rate of electron flow (Gill and Tuteja 2010). A myriad of biotic and abiotic stress factors elevate the generation of ROS in plants, where the failure of ROS-metabolizing system gives rise to the oxidative stress, which in turn damages biomolecules and cellular organelles and disrupts the cellular redox homeostasis (Anjum et al. 2010, 2012a, b; Gill and Tuteja 2010; Gill et al. 2013). ROS-mediated mechanism has been considered as one of the major mechanisms underlying the metallic NP-caused toxicity in living organisms including plants (Hou et al. 2018). Ample literature is now available on the ROS-caused oxidative stress in plants exposed to the environmental NPs (He et al. 2011; Panda et al. 2011; Zhao et al. 2012; Rico et al. 2013; Shaw and Hossain 2013; Anjum et al. 2013a, b; Zhang et al. 2013; Yasur and Rani 2013; Marslin et al. 2017; Yanık and Vardar 2018). The organ/tissue or cellular levels of the ROS ($O_2^{\bullet -}$, OH•, and H_2O_2), and the ROS-led protein and membrane lipid products, reactive/protein carbonyls, and thiobarbituric acid reactive substances (TBARS)/malondialdehyde (MDA), are considered to be the major biomarkers of oxidative stress in plants (Anjum et al. 2010; 2012a, b; Gill and Tuteja 2010). Recently, distribution of metallic NPs in root/shoot of a number of plants and their impact on plant performance have been reviewed, together with the role of oxidative stress in suppressing the plant growth (Siddiqi and Husen 2017). However, literature is inconclusive on whether oxidative stresses caused by NPs could activate defense mechanisms in order to counteract the nanotoxicity in plants (Morales et al. 2013; Rico et al. 2013; Li et al. 2016).

17.3 Membrane Lipid Peroxidation Versus Metallic Nanoparticles

As mentioned above, owing to their unique physical (size, shape, crystallinity, surface charge) and chemical (surface coating, elemental composition and solubility) attributes, NPs can produce chemical conditions conducive for the generation of ROS (Carlson et al. 2008). Elevated or unmetabolized ROS may alter the cellular redox homeostasis and cause adverse biological consequences (Anjum et al. 2010, 2012a, b; Gill and Tuteja 2010). Moreover, an elevated production of ROS and the subsequent oxidative stress, lipid peroxidation, proteins, and DNA damage in plants are the major causes for the metallic NP's phytotoxicity (reviewed by Arruda et al. 2015; Ma et al. 2015a, b; Tripathi et al. 2017) (Fig. 17.1). Biosolids disposal from wastewater treatment plant can add to the CeO_2 metal-oxide NPs in soils. A

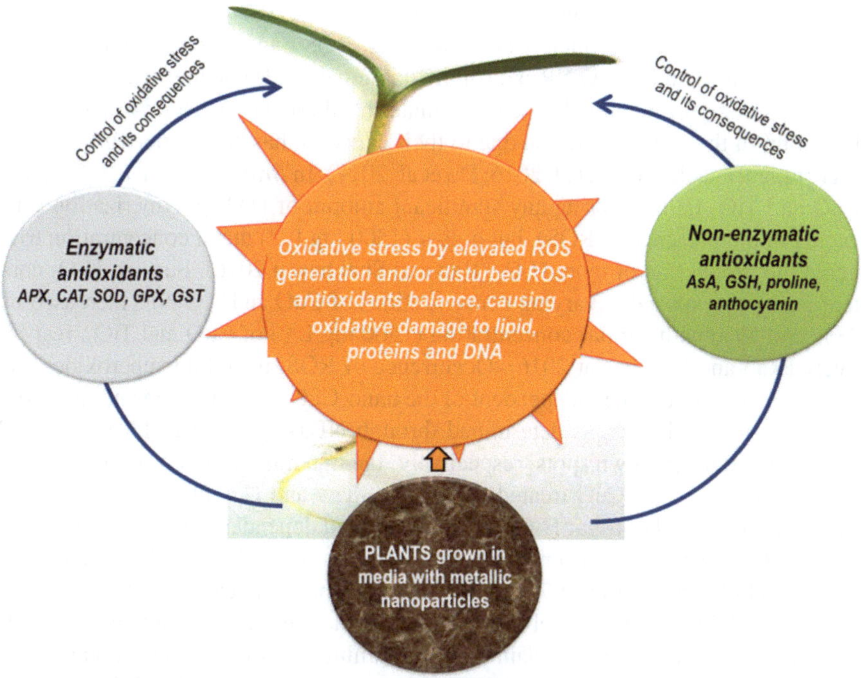

Fig. 17.1 Schematic presentation highlighting the interaction of nanoparticle exposure and oxidative stress and their impact on plant growth and metabolism via antioxidant defense system

concentration-dependent increase in oxidative stress via elevated levels of cellular H_2O_2 and lipid peroxidation has been reported in maize (*Zea mays*) exposed to CeO_2 NPs (Zhao et al. 2012). Elevated generation of ROS, including H_2O_2, was evidenced in duckweed (*Spirodela punctata*) exposed to Ag and ZnO NPs (Thwala et al. 2013). Increasing use of ZnO NPs in consumer products may enhance their release into the environment. In *Allium cepa*, ZnO NP-mediated high toxicity due to Zn^{2+} ion in hydroponic solution was attributed to a higher release of ROS (Kumari et al. 2011). The effect of ZnO NPs on lipid peroxidation, as examined by measuring TBARS concentration, was evident at all the concentrations (25, 50, 75, and 100 μg ml^{-1}), compared to bulk ZnO. ROS-sensitive organic compounds such as dye 3, 3′-diaminobenzidine (DAB) polymerize in the presence of ROS and yield deep greenish-brown polymerization product that could be visualized by naked eyes. By the in vivo detection of H_2O_2 through DAB staining, Shaw et al. (2014) found high accumulation of H_2O_2 in leaves of nano-CuO-stressed *Hordeum vulgare* seedlings after 20 days of stress treatment. Shaw and Hossain (2013) also reported a similar increasing trend of H_2O_2 (and MDA levels) in leaves of rice (*Oryza sativa*) seedlings subjected to nano-CuO stress. Priester et al. (2017) reported a differential level of leaf ROS in soybean (*Glycine max*) grown in the soil amended with either nano-CeO_2 (0.1, 0.5, or 1 g kg^{-1} soil) or nano-ZnO (0.05, 0.1, or 0.5 g kg^{-1} soil). Compared to the control, total leaf ROS concentrations (measured as fluorescence intensity units (FIU) g^{-1} dry leaf) were higher with medium (0.5 g kg^{-1} soil) and high (1.0 g kg^{-1} soil) doses of nano-CeO_2. However, ROS concentrations did not differ from the control under ZnO NP exposure. Titanium dioxide nanoparticles (TiO_2NPs) are also among the maximally used nanomaterials these days (Piccinno et al. 2012). Elevation in the ROS generation due to their exposure has been reported in several higher plant species (Cox et al. 2016; Du et al. 2017). In *Brassica juncea*, 200 mg L^{-1} CuO and TiO_2 failed to bring any significant amount of DAB polymerization (that occurs in the presence of ROS), but higher (1500 mg L^{-1}) metal concentration triggered the production of more H_2O_2 (Rao and Shekhawat 2016). Further, H_2O_2 content in shoots exposed to varying concentrations of CuO and TiO_2NPs increased by 28% and 19% with similar concentration (1500 mg L^{-1}) of CuO and TiO_2, respectively (Rao and Shekhawat 2016). Occurrence of ROS, including superoxides and H_2O_2, was reported to be independent of the nano-CuO concentration and the exposure time in rice leaves, as determined through NBT-stained dark blue spots and DAB-stained deep brown spots, respectively; the maximum spots were observed in the leaves of 1.5 mM CuO-treated seedlings (Shaw and Hossain 2013). However, the increasing trend of foliar H_2O_2 content exhibited dependence on time period and nano-CuO concentration; even a low concentration (0.5 mM) could markedly elevate the H_2O_2content after a 7-day treatment. A sharp increase (1.7-fold over the control) in the foliar H_2O_2 level was recorded with a high CuO treatment (1.5 mM) after 14 days (Shaw and Hossain 2013). Staining of stressed plant organs with 2′,7′-dichlorofluorescin diacetate (DCFH-DA) and subsequent DCF fluorescence have also been used to measure the status of oxidative stress caused by ROS generation in plant organs (Wang and Joseph 1999; Faisal et al. 2013). Compared to the untreated control, Faisal et al. (2013) reported a sharp increase of DCF fluorescence in roots of tomato exposed to higher concentrations (1.0, 1.5, and 2 mg ml^{-1}) of NiO, which was further confirmed with the help of flow cytometry (Faisal et al. 2013). Recently, Yanık and Vardar (2018) reported a dose-dependent increase in H_2O_2 content and lipid peroxidation in *Triticum aestivum* exposed to varying concentrations (5.0, 25, and 50 mg ml^{-1}; 96 h) of 13-nm-sized Al_2O_3 NPs for 96 h. Significant decreases in growth traits and damages in the photosynthetic apparatus of barley due to exposure to NiO NP (120 mg kg^{-1}; 14 d) were found to be accompanied by an overproduction of superoxide anion ($O_2^{\bullet-}$) and the enhancement of lipid peroxidation Soares et al. (2018a, b). In their earlier study also, these authors observed the occurence of significant increases in lipid peroxidation, $O_2^{\bullet-}$ generation, and subsequent cell death in barley plants exposed even to the lowest (87.8 mg kg^{-1}) of the applied dose of NiO NPs (Soares et al. 2016).

Thiobarbituric acid reactive substances (TBARS) and malondialdehyde (MDA) have been used as indicators of membrane lipid peroxidation. Cellular status of lipid peroxidation (LPO) (and also electrolyte leakage) is among the major traits to measure the NP-caused oxidative stress in plants (Rico et al. 2015). The cellular status of major oxidative stress biomarkers, including H_2O_2 generation, lipid peroxidation, and electrolyte leakage, responded differently to the CeO_2 NP concentrations in *Oryza sativa* (Rico et al. 2013). In particular, 500 mg $nCeO_2$ L^{-1} increased the H_2O_2

content by 162%, whereas 125 and 250 mg $nCeO_2$ L^{-1} did not change the H_2O_2 content in roots. However, NP concentration of 125 mg L^{-1} caused significant increases in lipid peroxidation (by 19–31%) and electrolyte leakage (by 4–18%) in comparison with the other $nCeO_2$ treatments (Rico et al. 2013). In the case of shoots, the highest lipid peroxidation was recorded at 500 mg nCeO2 L^{-1} treatment. Likewise, high levels of H_2O_2 and $O_2^{\bullet-}$ have been reported in AgNP-exposed *Allium cepa* (Panda et al. 2011). All concentrations of ZnO NPs and ZnO bulk caused increase in TBARS formation in *A. cepa*, the degree of increase being higher with ZnO NPs than their bulk counterparts (Kumari et al. 2011). Notably, a rapid generation of ROS was supposed to be the possible mechanism for higher intrinsic cyto-/genotoxicity of ZnO NPs to *A. cepa*, where ROS presumably converted fatty acids to toxic lipid peroxides (such as TBARS formation), subsequently damaging the membrane permeability. The highest concentration of ZnO NP (0.5 g kg^{-1} dry soil) caused a higher leaf lipid peroxidation in *Pisum sativum* (Mukherjee et al. 2014) and *Glycine max* (Priester et al. (2017). In contrast to no increase in ROS in the ZnO NP-exposed *G. max*, the occurrence of H_2O_2 along with the elevated lipid peroxidation in *P. sativum* indicates the differential response of plant species to Zn (Mukherjee et al. 2014; Priester et al. 2012).

The MDA level indicates changes in the cell membrane integrity due to oxidative stress. In addition, the cellular level of MDA may also represent plant senescence and injury under the pollution stress (Song et al. 2012; Nair and Chung 2014b). Li et al. (2016) determined the oxidative stress indicated by the five- to sevenfold higher (vs. control) MDA levels in corn roots treated with g-Fe_2O_3 NPs (20–100 mg L^{-1}). Application of silver nanoparticles (Ag NPs) at 200 and 250 mg L^{-1} could significantly elevate the MDA content in shoot and root of the wild-type (WT) *Crambe abyssinica* (Ma et al. 2015b). Exposure of *Arabidopsis thaliana* to 1000 ppm CeO_2 NPs caused fourfold higher MDA levels, compared to the control (Ma et al. 2013). However, exposure to In_2O_3 NPs (up to 1000 ppm) did not affect the MDA level and lipid peroxidation in *A. thaliana* (Ma et al. 2013). It can be inferred that induction in the ROS formation even at high NP exposure concentrations depends on the NP types and concentration and/or the ROS-scavenging efficiency of the plant's detoxification pathways. In *A. thaliana*, Nair and Chung (2014b) reported a differential impact of 0.2 mg L^{-1} of AgNPs and $AgNO_3$ on the modulation of lipid peroxidation, where a significantly increased MDA level was observed with the former. In the CuO NP-exposed Syrian barley (*Hordeum vulgare*), Shaw et al. (2014) reported a nano-CuO concentration-dependent gradual increase in the foliar MDA level; the maximum increase (~1.8-fold) was noticed with 1.5 mM CuO treatment, irrespective of the stress period involved. Earlier, CeO_2NPs were reported to cause the oxidative stress and modulate the antioxidant defense system in germinating rice seeds (Rico et al. 2013). In *Glycine max*, grown in soil amended with either nano-CeO_2 (0.1, 0.5, or 1.0 g kg^{-1} soil) or nano-ZnO (0.05, 0.1, or 0.5 g kg^{-1} soil), low and medium nano-CeO_2 treatments significantly elevated lipid peroxidation (as MDA concentration) by approximately 50%; but the nano-ZnO treatments could cause no significant differences in MDA concentrations relative to the controls (Priester et al. 2017). Increase in H_2O_2 (and also MDA level) may be accompanied by increased leaf damage, necrotic leaf spots, and reduced photosynthesis in *Glycine max* and *Hordeum vulgare* exposed to CeO_2 NPs (Priester et al. 2017) and CuO NPs, respectively (Shaw et al. 2014). Nano-Cu (0.5, 1.0, and 1.5 mM; <50 nm)-mediated elevated level of foliar MDA, a general marker of oxidative damage to lipid membranes, was evidenced in *Oryza sativa* seedlings, where the maximum increase (1.5-fold) was noticed with 1.5 mM CuO treatment at 7-day exposure, while at 14-day exposure, the highest MDA level appeared with 1.0 mM CuO treatment (Shaw and Hossain 2013). To confirm the occurrence of oxidative stress in the ZnO NP-treated tomato seedlings, Li et al. (2016) biochemically quantified the concentration of H_2O_2 and MDA, the important marker of lipid peroxidation, and correlated the observed growth inhibition with the excessive levels H_2O_2 and MDA. Earlier, metallic NPs were extensively shown to cause plant growth inhibition by inducing oxidative stress in a range of plant species (Dietz and Herth 2011; Chichiriccò and Poma 2015). CuO NPs (1000 and 1500 mg L^{-1}) caused significant increases over the control in MDA level in the roots of *Brassica juncea* (Rao and Shekhawat 2016), while the increase caused by 1500 mg TiO_2NPs L^{-1} was almost 50% and 9% less in the shoot and roots, respectively. In another study, Cu NPs caused about 180% increase in lipid peroxidation in *Elodea densa* when compared to Cu^{2+} (Nekrasova et al. 2011).

17.4 Antioxidant Metabolism Versus Metallic Nanoparticles

Much of the toxicity research on nanomaterials has focused mammals (including mammalian cell cultures) and aquatic species, and comparatively little has been done with the terrestrial plants (USEPA 2007; Klaine et al. 2008). Notably, unmetabolized ROS and their reaction products may bring irreparable changes in biomolecules and cellular organelles, disrupt the cellular redox homeostasis, and lead to cell death (He et al. 2011; Rico et al. 2013; Shaw and Hossain 2013; Anjum et al. 2013a, b; Yasur and Rani 2013; Zhang et al. 2013). To avoid the elevated/non-metabolized ROS-mediated consequences, plants possess efficient antioxidant defense system, where a synchronous action of enzymatic (such as superoxide dismutase, SOD; catalase, CAT; guaiacol peroxidase, GPX; glutathione sulfotransferase, GST; ascorbate peroxidase, APX; monodehydroascorbate reductase, MDHAR; dehydroascorbate reductase, DHAR; glutathione reductase, GR) and nonenzymatic (such as ascorbate, AsA; glutathione, GSH; carotenoids; tocopherols; phenolics) antioxidants metabolizes the ROS and their reaction products and nullifies their toxic consequences (Anjum et al. 2010, 2012a, b; Gill and Tuteja 2010; Gill et al. 2013).

The following text is focused on the modulation of major enzymatic and nonenzymatic antioxidants in the metallic NP-exposed plants.

17.4.1 Enzymatic Antioxidants

Reduced activity of DHAR and MDHAR in *Hordeum vulgare* plants exposed to nano-CuO (0.5, 1.0, and 1.5 mM; size, <50 nm) was considered to cause severe decreases in the recycling of AsA pool (Shaw et al. 2014). In the CeO NP-exposed *Raphanus sativus*, CAT activity was higher in tubers but lower in leaves with reference to the control, as observed at 125 mg $nCeO_2$ dose (Corral-Diaz et al. 2014). The enhanced CAT activity can reduce the excess of H_2O_2 likely to be generated by external stressors and enhanced cellular activity during the tuber formation (Marcelis et al. 1997). Increased CAT and APX activity has been reported in the $nCeO_2$ (125 mg kg^{-1})-treated shoots of cilantro plants (Morales et al. 2013) and in the Ce^{3+}-exposed rice seedlings (Liu et al. 2012). In addition, Rico et al. (2013) considered the triggering of increments in CAT and APX activity as an outcome of the $nCeO_2$ (500 mg L^{-1})-mediated enhancement in the electrolyte leakage and lipid peroxidation. Significant reduction in CAT activity was noticed in *P. sativum* leaves with all ZnO NP concentrations, whereas APOX was reduced in both roots and leaves (Mukherjee et al. 2014). However, APOX activity was downregulated in the root and leaf, and CAT remained unchanged under the bulk ZnO exposure. A high activity of H_2O_2-metabolizing enzymes (such as peroxidase and CAT) was observed in the CuO and ZnO NP-exposed wheat grown in sandculture (Dimkpa et al. 2012a). The elevated APX activity was not sufficient to control the H_2O_2 level in the nano-CuO (0.5, 1.0, and 1.5 mM; size, <50 nm)-exposed *Hordeum vulgare* (Shaw et al. 2014). In the nano-CuO-exposed *O. sativa* and *H. vulgare*, Shaw and Hossain (2013) and Shaw et al. (2014) reported the failure of APX-mediated ROS (such as H_2O_2)-scavenging system, where high levels of H_2O_2 occurred despite an elevated APX activity. Nano-CuO-caused inhibition in the cellular H_2O_2-metabolizing potential of APX and CAT was evidenced by Trujillo-Reyes et al. (2014), where they observed a decreased APX activity (at 10 mg L^{-1}) and CAT activity (at 20 mg L^{-1}) in *Lactuca sativa* roots and leaves. Plant organs (root and shoot) may respond differently to metallic NPs in terms of the stress enzyme behavior therein. To this end, Hong et al. (2015) noted a differential ROS-generating potential of nano-CuO and nano-Cu and the ROS-metabolizing capacity in root cells of the test plants. They observed a downregulated CAT activity and an upregulated APX activity in the roots of both *Lactuca sativa* and *Medicago sativa* when exposed to either nano-CuO or nano-Cu concentrations (5, 10, and 20 mg L^{-1}) (Hong et al. 2015).

The expression patterns of genes of H_2O_2-metabolizing enzymes such as APX (APX1 and APX2) and CAT (CAT2 and CAT3) can be differently modulated in plants when exposed to different (0.5, 1, 2, 5, 10, 20, 50, and 100 mg L^{-1}) nano-CuO concentrations (Nair and Chung 2014a). These nano-CuO concentrations can also differently modulate the expression patterns of genes of SODs (MnSOD gene, MSD; CuZnSOD genes, CSD1 and CSD2) and control the dismutation of $O_2^{\bullet-}$ into H_2O_2 (Nair and Chung 2014a). In *Ricinus communis*, AgNP and $AgNO_3$ (0, 500, 1000, 2000, and 4000 mg L^{-1}) enhanced the activities of SOD and peroxidase (Yasur and Rani 2013). This was considered to be due to the AgNP-mediated elevation in the ROS level and subsequent oxidative stress. In CuO and TiO_2 NP-exposed *Brassica juncea*, APX exhibited a much higher affinity to H_2O_2 than CAT. Decline in the CAT activity due to increasing CuO NP concentration increased the level of H_2O_2 content. It was inferred that APX could maintain the low levels of H_2O_2 and CAT can control the excess of H_2O_2. Metallic NPs may also play a protective role against the ROS in plants. To this end, ROS (H_2O_2)-metabolizing enzymes such as CAT and APX were reported to rise in velvet mesquite (*Prosopis juliflora-velutina*) with increase in nano-ZnO doses (Hernandez-Viezcas et al. 2011) and in corn (*Zea mays*) with increase in nano-CeO_2 doses (Zhao et al. 2012). Later, Rico et al. (2013) explained the protective response of rice seedlings to nano-CeO_2 at certain concentrations. At 125 mg L^{-1}, $nCeO_2$ caused a low H_2O_2 content with a greatly enhanced lipid peroxidation and electrolyte leakage. It was advocated that the decreased activity of SOD (and also that of DHAR and GR) resulted in an inefficient removal of H_2O_2, which led to enhancement in the membrane damage. In the NiO NP (2 mg ml^{-1})-exposed tomato plants, activity of CAT and SOD increased by 6.8- and 1.7-fold, compared to the control (Faisal et al. 2013).

17.4.2 Nonenzymatic Antioxidants

Let us now examine the modulation of nonenzymatic antioxidants such as AsA, GSH, amino acid (such as proline), phenols and phenolic acids, and anthocyanin in plants exposed to different metallic NPs. Being a major free thiol in most living cells, reduced GSH is involved in the scavenging of ROS such as H_2O_2 and in detoxification of xenobiotics. In stressed cells, the status of intracellular GSH is indicative of the overall cellular health (Akerboom and Sies 1981). ZnO NP (under 50 mg L^{-1}) brought tomato seedlings to oxidized redox state by significantly decreasing the GSH/GSSG ratio (Li et al. 2016). Increased accumulation of oxidized glutathione (GSSG, indicative of oxidative stress) was reported in AgNP-exposed *Triticum aestivum* (Van Aken 2015). Nano-CuO can bring a significant increase in GSH oxidation (Dimkpa et al. 2012a) and also a decreased GSH and the GSH/GSSG ratio (Shaw and Hossain 2013; Shaw et al. 2014). The use of GSH or GSSG quantification as an alternative method for demonstrating the role of the GSH biosynthesis pathway in NP detoxification has been focused recently (reviewed by Ma et al. 2015a). Metallic NPs such as CeO_2 and In_2O_3 have the potential to induce upregulation of genes related to GSH biosynthesis pathways (Ma et al. 2013). It has also been pointed out that the depletion of GSH and elevation of GSSG levels could not be directly considered as the indictor of ROS conversion into H_2O (Ma et al. 2013). The AgNP and Ag^+ ion treatments elevated the GSSG level in wheat (Dimkpa et al. 2012b). CuO and ZnO NP exposure could also elevate the GSSG level in wheat (Dimkpa et al. 2012a). GSH level responded differently to NiO NPs in tomato plants in a dose-dependent manner (Faisal et al. 2013). Moderate concentration of NiO NPs (1000 mg L^{-1}) caused elevation in GSH levels, whereas the higher ones (1500 and 2000 mg L^{-1}) brought a decline. Very little is known about AsA and its redox couple (AsA, AsA/DHA) in relation to metallic NPs. In Syrian barley

(*Hordeum vulgare*), Shaw et al. (2014) found a significant impact of CuO NPs on the foliar AsA pool. Marked changes in the reduced ascorbate (AsA) contents were observed 20 days after treatment. The gradual increase was in accordance with the increase in CuO concentration applied. The highest AsA accumulation (tenfold over the control) was observed in the leaves of the 1.5 mM CuO-exposed seedlings. Similar increasing trends were recorded in the total as well as oxidized (DHA) ascorbate contents. Nevertheless, the changes observed in the AsA/DHA ratio in stressed leaves, in comparison with the control, were mostly nonsignificant except at the 1.5 mM nano-copper treatment for 20 days. Rico et al. (2013) reported the highest AsA in the rice roots because of the 500 mg $nCeO_2$ L^{-1} treatment.

In addition to the major antioxidant metabolites (such as AsA and GSH), the potential of amino acids (such as proline), phenols and phenolic acids, and anthocyanin with reference to the protection of plants is affected by exposure to metallic NPs (Winkel-Shirley 2002; Krishnaraj et al. 2012; Ma et al. 2013, 2015a; Yasur and Rani 2013; Nair and Chung 2014a, b; Rao and Shekhawat 2016). *Arabidopsis thaliana* was reported to respond to the abiotic stress caused by CeO_2 and In_2O_3 NPs by producing anthocyanin at the physiological level (Ma et al. 2013). Proline, which can scavenge ROS, maintain redox homeostasis, and improve plant resistance to oxidative stress, was found to be affected by nano-CuO concentrations (10 and 20 mg L^{-1}) in *Arabidopsis thaliana*, as the proline biosynthesis genes (such as P5CS1 and P5CS2) upregulated significantly under the nano-CuO exposure (Nair and Chung 2014a, b). In the CuO- and TiO_2-exposed *Brassica juncea*, Rao and Shekhawat (2016) reported a sharp rise in proline content, with the highest content (1500 mg L^{-1}) detected in roots. CuO- and TiO_2-treated roots had 65 and 66% higher proline content (vs. control). Phenols and phenolic acids may protect plants against Ag^+- and AgNP-induced oxidative stress (Winkel-Shirley 2002; Krishnaraj et al. 2012; Yasur and Rani 2013). Anthocyanin, a type of flavonoid located in the vacuole system (Mourato et al. 2012) and acting as a superoxide radical scavenger, hydrogen donor, and metal chelator, responded to the abiotic stress caused by the metal-oxide NPs (CeO_2 and In_2O_3) in *A. thaliana* (Ma et al. 2013, 2015a). Anthocyanin production was unaffected by In_2O_3 NPs, but a treatment of 1000 ppm CeO_2 NPs clearly promoted the anthocyanin production; however, its production decreased at higher doses of CeO_2 NPs. It was argued that the higher (2000 ppm) dose of CeO_2 NPs might have inhibited the activity of phenylalanine ammonium-lyase (PAL) (a catalytic enzyme involved in the biosynthesis of anthocyanin via a phenylpropanoid pathway) that caused elevation in the ROS (H_2O_2) level and subsequently deceased the anthocyanin biosynthesis.

17.5 Conclusions and Prospects

The field of "plants-nanoparticles interaction" is emerging with diverse avenues. Particularly, exploration of physiological/biochemical and molecular insights into plant nanotoxicology has been the focus of recent researches on the subject. It is clear from the literature appraisal that metallic NPs have the potential to provoke oxidative stress via elevating the cellular ROS and thereby halting the cellular metabolism and function probably as a result of a dual toxicity of metallic NPs through metal and NPs. Further efforts need to be made to explore the impact of NPs (and also that of NP-NP interaction) on individual enzymatic and nonenzymatic components of antioxidant defense system in plants and also to understand the outcomes of the cross talk between these antioxidants in order to get rid of ROS and their consequences in plants exposed to metallic NPs.

References

Akerboom TP, Sies H (1981) Assay of glutathione, glutathione disulfide, and glutathione mixed disulfides in biological samples. Methods Enzymol 77:373–382

Anjum NA, Umar S, Chan MT (2010) Ascorbate–glutathione pathway and stress tolerance in plants. Springer, Dordrecht

Anjum NA, Ahmad I, Mohmood I, Pacheco M, Duarte AC, Pereira E, Umar S, Ahmad A, Khan NA, Iqbal M, Prasad MNV (2012a) Modulation of glutathione and its related enzymes in plants' responses to toxic metals and metalloids – a review. Environ Exp Bot 75:307–324

Anjum NA, Umar S, Ahmad A (2012b) Oxidative stress in plants: causes, consequences and tolerance. IK International Publishing House, New Delhi

Anjum NA, Gill SS, Duarte AC, Pereira E, Ahmad I (2013a) Silver nanoparticles in soil-plant systems. J Nanopart Res 15:1896. https://doi.org/10.1007/s11051-013-1896-7

Anjum NA, Singh N, Singh MK, Shah ZA, Duarte AC, Pereira E, Ahmad I (2013b) Single-bilayer graphene oxide sheet tolerance and glutathione redox system significance assessment in faba bean (*Vicia faba* L.). J Nanopart Res 15:1770. https://doi.org/10.1007/s11051-013-1770-7

Anjum NA, Rodrigo MAM, Moulick A, Heger Z, Kopel P, Zitka O, Adam V, Lukatkin AS, Duarte AC, Pereira E, Kizek R (2016) Transport phenomena of nanoparticles in plants and animals/humans. Environ Res 151:233–243

Arruda SCC, Silva ALD, Galazzi RM, Azevedo RA, Arruda MAZ (2015) Nanoparticles applied to plant science: a review. Talanta 131:693–705

Bhatt I, Tripathi BN (2011) Interaction of engineered nanoparticles with various components of the environment and possible strategies for their risk assessment. Chemosphere 82:308–317

Carlson C, Hussain SM, Schrand AM, K. Braydich-Stolle L, Hess KL, Jones RL, Schlager JJ (2008) Unique cellular interaction of silver nanoparticles: size-dependent generation of reactive oxygen species. J Phys Chem B 112(43):13608–13619

Chichiriccò G, Poma A (2015) Penetration and toxicity of nanomaterials in higher plants. Nanomaterials 5:851–873

Corral-Diaz B, Peralta-Videa JR, Alvarez-Parrilla E, Rodrigo-García J, Morales MI, Osuna-Avila P, Niu G, Hernandez-Viezcas JA, Gardea-Torresdey JL (2014) Cerium oxide nanoparticles alter the antioxidant capacity but do not impact tuber ionome in *Raphanus sativus* (L). Plant Physiol Biochem 84:277–285

Cox A, Venkatachalam P, Sahi S, Sharma N (2016) Silver and titanium dioxide nanoparticle toxicity in plants: a review of current research. Plant Physiol Biochem 107:147–163

Dietz KJ, Herth S (2011) Plant nanotoxicology. Trends Plant Sci 16:582–589

Dimkpa CO, McLean JE, Latta DE, Manangón E, Britt DW, Johnson WP, Boyanov MI, Anderson AJ (2012a) CuO and ZnO nanoparticles: phytotoxicity, metal speciation, and induction of oxidative stress in sand-grown wheat. J Nanopart Res 14:1125

Dimkpa CO, McLean JE, Martineau N, Britt DW, Haverkamp R, Anderson AJ (2012b) Silver nanoparticles disrupt wheat (*Triticum aestivum* L.) growth in a sand matrix. Environ Sci Technol 47:1082–1090

Du W, Tan W, Peralta-Videa JR, Gardea-Torresdey JL, Ji R, Yin Y, Guo H (2017) Interaction of metal oxide nanoparticles with higher terrestrial plants: physiological and biochemical aspects. Plant Physiol Biochem 110:210–225

Elder A, Lynch I, Grieger K, Chan-Remillard S, Gatti A, Gnewuch H, Kenawy E, Korenstein R, Kuhlbusch T, Linker F, Matias S (2009) Human health risks of engineered nanomaterials: critical knowledge gaps in nanomaterials risk assessment. In: Linkov I, Steevens J (eds) Nanomaterials: risks and benefits. Springer, Dordrecht, pp 3–29

Faisal M, Saquib Q, Alatar AA, Al-Khedhairy AA, Hegazy AK, Musarrat J (2013) Phytotoxic hazards of NiO-nanoparticles in tomato: A study on mechanism of cell death. J Hazard Mater 250–251:318–332

Gerloff K, Fenoglio I, Carella E, Kolling J, Albrecht C, Boots AW, Förster I, Schins RP (2012) Distinctive toxicity of TiO_2 rutile/anatase mixed phase nanoparticles on Caco-2 cells. Chem Res Toxicol 25:646–655

Gill SS, Tuteja N (2010) Reactive oxygen species and antioxidant machinery in abiotic stress tolerance in crop plants. Plant Physiol Biochem 48:909–930

Gill SS, Anjum NA, Hasanuzzaman M, Gill R, Trivedi DK, Ahmad I, Pereira E, Tuteja N (2013) Glutathione and glutathione reductase: a boon in disguise for plant abiotic stress defense operations. Plant Physiol Biochem 70:204–212

Handy RD, Owen R, Valsami-Jones E (2008) The ecotoxicology of nanoparticles and nanomaterials: current status, knowledge gaps, challenges, and future needs. Ecotoxicology 17:315–325

He D, Jones AM, Garg S, Pham AN, Waite TD (2011) Silver nanoparticle–reactive oxygen species interactions: application of a charging–discharging model. J Phys Chem C 115:5461–5468

Hernandez-Viezcas J, Castillo-Michel H, Servin A, Peralta-Videa J, Gardea-Torresdey J (2011) Spectroscopic verification of zinc absorption and distribution in the desert plant *Prosopis juliflora-velutina* (velvet mesquite) treated with ZnO nanoparticles. Chem Eng J 170:346–352

Hong J, Rico CM, Zhao L, Adeleye AS, Keller AA, Peralta-Videa JR, Gardea-Torresdey JL (2015) Toxic effects of copper-based nanoparticles or compounds to lettuce (*Lactuca sativa*) and alfalfa (*Medicago sativa*). Environ Sci Processes Impacts 17:177–185

Hou J, Wu Y, Li X, Wei B, Li S, Wang X (2018) Toxic effects of different types of zinc oxide nanoparticles on algae, plants, invertebrates, vertebrates and microorganisms. Chemosphere 193:852–860

Husen A (2017) Gold nanoparticles from plant system: synthesis, characterization and application. In: Ghorbanpourn M, Manika K, Varma A (eds) Nanoscience and plant–soil systems, vol 48. Springer International Publication, Cham, pp 455–479

Husen A, Siddiqi KS (2014) Phytosynthesis of nanoparticles: concept, controversy and application. Nanoscale Res Lett 9:229

Klaine SJ, Alvarez PJJ, Batley GE, Fernandes TF, Handy RD, Lyon DY, Mahendra S, McLaughlin MJ, Lead JR (2008) Nanomaterials in the environment: behavior, fate, bioavailability and effects. Environ Toxicol Chem 27:1825–1851

Krishnaraj C, Jagan G, Ramachandran R, Abirami SM, Mohan N, Kalaichelvan PT (2012) Effect of biologically synthesized silver nanoparticles on *Bacopa monnieri* L. Wettst. plant growth metabolism. Process Biochem 47:651–658

Kumari M, Khan SS, Pakrashi S, Mukherjee A, Chandrasekaran N (2011) Cytogenetic and genotoxic effects of zinc oxide nanoparticles on root cells of *Allium cepa*. J Hazard Mater 190:613–621

Li M, Ahammed GJ, Li C, Bao X, Yu J, Huang C, Yin H, Zhou J (2016) Brassinosteroid ameliorates zinc oxide nanoparticles-induced oxidative stress by improving antioxidant potential and redox homeostasis in tomato seedling. Front Plant Sci 7:615. https://doi.org/10.3389/fpls.2016.00615

Liu D, Wang X, Lin Y, Chen Z, Xu H, Wang L (2012) The effects of cerium on the growth and some antioxidant metabolisms in rice seedlings. Environ Sci Pollut Res 19:3282–3291

Ma X, Geiser-Lee J, Deng Y, Kolmakov A (2010) Interactions between engineered nanoparticles (ENPs) and plants: phytotoxicity, uptake and accumulation. Sci Total Environ 408:3053–3061

Ma C, Chhikara S, Xing B, Musante C, White JC, Dhankher OP (2013) Physiological and molecular response of *Arabidopsis thaliana* (L.) to nanoparticle cerium and indium oxide exposure. ACS Sustain Chem Eng 1:768–778

Ma C, White JC, Dhankher OP, Xing B (2015a) Metal-based nanotoxicity and detoxification pathways in higher plants. Environ Sci Technol 49:7109–7122

Ma C, Chhikara S, Minocha R, Long S, Musante C, White JC, Xing B, Dhankher OP (2015b) Reduced silver nanoparticle phytotoxicity in *Crambe abyssinica* with enhanced glutathione production by overexpressing bacterial γ-glutamylcysteine synthase. Environ Sci Technol 49:10117–10126

Marcelis LFM, Hauvelink E, van Dijk D (1997) Pithiness and growth of radish tubers as affected by irradiance and plant density. Ann Bot 79:397–402

Marslin G, Sheeba CJ, Franklin G (2017) Nanoparticles alter secondary metabolism in plants via ROS burst. Front Plant Sci 8:832. https://doi.org/10.3389/fpls.2017.00832

Morales MI, Rico CM, Hernandez-Viezcas JA, Nunez JE, Barrios AC, Flores Marges JP, Peralta-Videa JR, Gardea-Torresdey JL (2013) Toxicity assessment of cerium oxide nanoparticles in cilantro (*Coriandrum sativum*) plants grown in organic soil. J Agric Food Chem 61:6224–6230

Mourato M, Reis R, Martins ML (2012) Chapter 2: Characterization of plant antioxidative system in response to abiotic stresses: a focus on heavy metal toxicity. In: Montanaro G, Dichio B (eds) Advances in selected plant physiology aspects. InTech, Rijeka ISBN 978-953-51-0557.

Mueller NC, Nowack B (2008) Exposure modeling of engineered nanoparticles in the environment. Environ Sci Technol 42:4447–4453

Mukherjee A, Peralta-Videa JR, Bandyopadhyay S, Rico CM, Zhao LJ, Gardea-Torresdey JL (2014) Physiological effects of nanoparticulate ZnO in green peas (*Pisum sativum* L.) cultivated in soil. Metallomics 6:132–138

Nair PG, Chung I (2014a) Impact of copper oxide nanoparticles exposure on *Arabidopsis thaliana* growth, root system development, root lignification, and molecular level changes. Environ Sci Pollut Res 21:12709–12722

Nair PMG, Chung IM (2014b) Assessment of silver nanoparticle-induced physiological and molecular changes in *Arabidopsis thaliana*. Environ Sci Pollut Res 21:8858–8869

Nekrasova GF, Ushakova OS, Ermakov AE, Uimin MA (2011) Effects of copper (II) ions and copper oxide nanoparticles on *Elodea densa* Planch. Russ J Ecol 42:458–463

Panda KK, Achary VMM, Krishnaveni R, Padhi BK, Sarangi SN, Sahu SN, Panda BB (2011) *In vitro* biosynthesis and genotoxicity bioassay of silver nanoparticles using plants. Toxicol In Vitro 25:1097–1105

Piccinno F, Gottschalk F, Seeger S, Nowack B (2012) Industrial production quantities and uses of ten engineered nanomaterials in Europe and the world. J Nanopart Res 14(9):1109

Priester JH, Ge Y, Mielke RE, Horst AM, Moritz SC, Espinosa K, Gelb J, Walker SL, Nisbet RM, An Y-J, Schimel JP, Palmer RG, Hernandez-Viezcas JA, Zhao L, Gardea-Torresdey JL, Holden PA (2012) Soybean susceptibility to manufactured nanomaterials with evidence for food quality and soil fertility interruption. Proc Natl Acad Sci 109:E2451–E2456

Priester JH, Moritz SC, Espinosa K, Ge Y, Wang Y, Nisbet RM, Schimel JP, Goggi AS, Gardea-Torresdey JL, Holden PA (2017) Damage assessment for soybean cultivated in soil with either CeO_2 or ZnO manufactured nanomaterials. Sci Total Environ 579:1756–1768

Rao S, Shekhawat GS (2016) Phytotoxicity and oxidative stress perspective of two selected nanoparticles in *Brassica juncea*. 3 Biotech 6(2):244. https://doi.org/10.1007/s13205-016-0550-3

Rico C, Hong J, Morales MI, Zhao L, Barrios AC, Zhang J-Y, Peralta-Videa JR, Gardea-Torresdey JL (2013) Effect of cerium oxide nanoparticles on rice: a study involving the antioxidant defense system and *in vivo* fluorescence imaging. Environ Sci Technol 47:5635–5642

Rico CM, Peralta-Videa JR, Gardea-Torresdey JL (2015) Chemistry, biochemistry of nanoparticles, and their role in antioxidant defense system in plants. In: Siddiqui MH, Al-Whaibi MH, Mohammad F (eds) Nanotechnology and plant sciences. Springer International Publishing, Cham, pp 1–17

Royal Society (2004) Nanoscience and nanotechnologies: opportunities and uncertainties. http://www.nanotec.org.uk/finalReport.htm.

Shaw AK, Hossain Z (2013) Impact of nano-CuO stress on rice (*Oryza sativa* L.) seedlings. Chemosphere 93:906–915

Shaw AK, Ghosh S, Kalaji HM, Bosa K, Brestic M, Zivcak M, Hossain Z (2014) Nano-CuO stress induced modulation of antioxidative defense and photosynthetic performance of Syrian barley (*Hordeum vulgare* L.). Environ Exp Bot 102:37–47

Siddiqi KS, Husen A (2016) Engineered gold nanoparticles and plant adaptation potential. Nanoscale Res Lett 11:400

Siddiqi KS, Husen A (2017) Plant response to engineered metal oxide nanoparticles. Nanoscale Res Lett 12:92

Siddiqi KS, Husen A, Rao RAK (2018) A review on biosynthesis of silver nanoparticles and their biocidal properties. J Nanobiotechnol 16:14

Soares C, Branco-Neves S, de Sousa A, Pereira R, Fidalgo F (2016) Ecotoxicological relevance of nano-NiO and acetaminophen to Hordeum vulgare L.: combining standardized procedures and physiological endpoints. Chemosphere 165:442–452

Soares C, Branco-Neves S, de Sousa A, Pereira R, Fidalgo F (2018a) Ecotoxicological relevance of nano-NiO and acetaminophen to *Hordeum vulgare* L.: combining standardized procedures and physiological endpoints. Chemosphere 165:442–452

Soares C, Branco-Neves S, de Sousa A, Azenha M, Cunha A, Pereira R, Fidalgo F (2018b) SiO_2 nanomaterial as a tool to improve *Hordeum vulgare* L. tolerance to nano-NiO stress. Sci Total Environ 622-623:517–525

Song G, Gao Y, Wu H, Hou W, Zhang C, Ma H (2012) Physiological effect of anatase TiO_2 nanoparticles on *Lemna minor*. Environ Toxicol Chem 31:2147–2152

Thakkar KN, Mhatre SS, Parikh RY (2010) Biological synthesis of metallic nanoparticles. Nanomed Nanotechnol Biol Med 6:257–262

Thwala M, Musee N, Sikhwivhilu L, Wepener V (2013) The oxidative toxicity of Ag and ZnO nanoparticles towards the aquatic plant *Spirodela punctata* and the role of testing media parameters. Environ Sci Processes Impacts 15:1830–1843

Tripathi DK, Tripathi A, Shweta Singh S, Singh Y, Vishwakarma K, Yadav G, Sharma S, Singh VK, Mishra RK, Upadhyay RG, Dubey NK (2017) Uptake, accumulation and toxicity of silver nanoparticle in autotrophic plants, and heterotrophic microbes: a concentric review. Front Microbiol 8:07. https://doi.org/10.3389/fmicb.2017.00007

Trujillo-Reyes J, Majumdar S, Botez CE, Peralta-Videa JR, Gardea-Torresdey JL (2014) Exposure studies of core–shell Fe/Fe_3O_4 and Cu/CuO NPs to lettuce (*Lactuca sativa*) plants: Are they a potential physiological and nutritional hazard? J Hazard Mater 267:255–263

US-Environmental Protection Agency (2007) Nanotechnology white paper. US Environmental Protection Agency, Washington, DC, p 132

Van Aken B (2015) Gene expression changes in plants and microorganisms exposed to nanomaterials. Curr Opin Biotechnol 33:206–219

Wang H, Joseph JA (1999) Quantifying cellular oxidative stress by dichlorofluorescein assay using microplate reader. Free Radic Biol Med 27:612–616

Winkel-Shirley B (2002) Biosynthesis of flavonoids and effects of stress. Curr Opin Plant Biol 5:218–222

Yanık F, Vardar F (2018) Oxidative stress response to aluminum oxide (Al_2O_3) nanoparticles in *Triticum aestivum*. Biologia 73:129–135

Yasur J, Rani P (2013) Environmental effects of nanosilver: impact on castor seed germination, seedling growth, and plant physiology. Environ Sci Pollut Res 20:8636–8648

Zhang H, Peng C, Yang J-J, Shi J-Y (2013) Eco-toxicological effect of metal-based nanoparticles on plants: research progress. Chin J Appl Ecol 24:885–892

Zhao L, Peng B, Hernandez-Viezcas JA, Rico C, Sun Y, Peralta-Videa JR, Tang X, Niu G, Jin L, Varela-Ramirez A, Zhang JY (2012) Stress response and tolerance of *Zea mays* to CeO_2 nanoparticles: cross talk among H_2O_2, heat-shock protein, and lipid peroxidation. ACS Nano 6:9615–9622

Chapter 18
Role of Nanomaterials in the Mitigation of Abiotic Stress in Plants

Sanjay Singh and Azamal Husen

18.1 Introduction

Nanotechnology is a novel scientific approach that involves the use of materials and equipments capable of manipulating the physical and chemical properties of a substance at molecular levels to develop products and services used in diverse fields ranging from medicine to agriculture (Siddiqi and Husen 2016, 2017). We know that agriculture is the backbone of developing countries, with more than 60% of their population depending on it for their livelihood (Brock et al. 2011). Nanotechnology has the potential to revolutionize the agricultural and food industry by providing novel tools for the molecular management of plant diseases, and by enhancing the ability of plants to absorb nutrients, and withstand the inhospitable environmental conditions (Khan et al. 2016). Thus, it can potentially enhance crop yields and their nutritional values, develop improved systems for monitoring the environmental conditions, and enhance plant resistance/tolerance to environmental stress (Tarafdar et al. 2013).

Nanotechnology promises to provide smart sensors and delivery systems that will help the agricultural industry immensely. In the near future, nanostructured catalysts are likely to be available to increase the efficiency of input utilization, allowing lower doses of agricultural inputs and managing the crop production systems more efficiently (Liu and Lal 2015). Nanotechnology will also protect the environment indirectly through the use of alternative (renewable) energy supplies and filters or catalysts to reduce pollution and clean up the existing pollutants in soil and water (Adeleye et al. 2016; Mohamed 2017). In the agricultural sector, nanotech

S. Singh (✉)
Department of Plant Science, College of Agriculture and Natural Resources, Mizan-Tepi University, Mizan, Ethiopia

A. Husen
Department of Biology, College of Natural and Computational Sciences, University of Gondar, Gondar, Ethiopia

A. Husen, M. Iqbal (eds.), *Nanomaterials and Plant Potential*,
https://doi.org/10.1007/978-3-030-05569-1_18

research and development may frame the next level of expansion of genetically modified crops, animal production inputs, chemical pesticides, and precision farming techniques (Prasad et al. 2012). While nanochemical pesticides are already in use, other applications related mainly to the fine-tuning and more precise micromanagement of problem soils, the more efficient and targeted use of inputs in the wake of climate change and global warming, and the new toxin formulations for pest control are yet to come in full swing (Aragay et al. 2012; Kah 2015).

With the emergence of nanotechnology and its wide-spectrum applications in diverse fields, more attention is being paid on the synthesis of nanomaterials (NMs) from metals like Ag, Au, Pb, etc. or metal oxides, e.g., TiO_2, SiO_2, ZnO, etc. There are several physical, chemical, or biological methods through which nanoparticles (NPs) can be synthesized (Husen and Siddiqi 2014; Tulinski and Jurczyk 2017). Fabrication of metallic NPs using plants or their extracts is becoming more popular now. All plants contain a variety of chemical compounds such as enzymes, sugars, proteins, and other phytochemicals like flavanoids, latex, phenolics, terpenoids, alcohols, amines, cofactors, etc., which act as the reducing and stabilizing agents during the synthesis of metal NPs from metal salts with well-defined size and shape without creating any atmospheric pollution (Husen 2017). It is also observed that under certain conditions plants are capable of producing natural mineralized NMs necessary for their growth (Wang et al. 2001). However, designing a material from the atomic level to achieve a tailored response in extreme conditions is a grand challenge in material research. Controlling the radiation-induced defects via interfaces is the key factor in reducing the damage and imparting stability to certain NMs under conditions where bulk materials exhibit void swelling and/or embrittlement (Tulinski and Jurczyk 2017). The recovery of radiation-induced point defects at free surfaces and grain boundaries and stabilization of helium bubbles at interphase boundaries have been reviewed recently, and an approach for processing the bulk nanocomposites containing interfaces that are stable under irradiation has been suggested (Beyerlein et al. 2013). As the understanding of nanotechnology deepens, it will hopefully become a major economic driving force that will benefit consumers as well as farmers with no adverse effect on humans and their environment (Lutz and Steevens 2009). This chapter is focused on strategies adopted for mitigating the harmful effects of abiotic stresses by using the potential NMs to achieve the optimum plant growth and yield.

18.2 NMs to Mitigate the Harmful Effects of Abiotic Stress

NPs, also termed as nanoscale particles (NSPs), are small molecular aggregates having dimensions between 1 and 100 nm. Due to their extremely small size, such particles acquire some peculiar and diverse physicochemical properties in comparison to their standard or bulk form, as described in detail in Chaps. 1, 2, and 3 of this book. Nanostructured metals and composites provide huge interfaces that attract, absorb, and annihilate point and line defects. These interfaces recover and control defects produced in materials subjected to extremes of displacement damage, impurity implantation, stress, and temperature (Tulinski and Jurczyk 2017). Among the abiotic stresses, drought, salinity, flooding, mineral deficiencies, and various pollutants including radiations are considered as the major factors that affect crop growth and productivity (Andy 2016; Husen et al. 2014, 2016). Plants, being the sessile organism in general, have no choice but to face various environmental stresses throughout their life cycle. They, therefore, develop their defense against environmental stresses at various levels by modulating the molecular, biochemical, and physiological pathways (Husen et al. 2017, 2018). There are several studies indicating that NP-mediated effects on plant growth and development are concentration-dependent. NPs are involved in upregulating the activities of antioxidative enzymes like catalase, superoxide dismutase, and peroxidase under oxidative stress (Laware and Raskar 2014; Farhangi-Abriz and Torabian 2018). The following text summarizes the efforts being made to use various NPs for mitigating the harmful effects of abiotic stresses on plant species.

18.2.1 *Salinity Stress*

Salinity, which is caused normally by accumulation of chloride (Cl^-) and sulfate ($SO4^{2-}$) anions of predominantly sodium (Na^+) cation but sometimes of calcium (Ca^{2+}) and magnesium (Mg^{2+}) cations as well in soils of arid, semiarid, and coastal areas, is a major abiotic stress factor limiting food production. It is estimated that more than 20% of cultivated land worldwide is experiencing salinity stress and the area is increasing day by day. Since the majority of the major crop species belong to the glycophyte category, they are susceptible to salt stress (Munns and Tester 2008). Salinity stress causes negative impact on various biochemical and physiological processes associated with plant growth and yield. Lowering of soil osmotic potential, creation of nutritional imbalance, enhancing specific ionic toxicity (salt stress), and/or the combination(s) of these factors are some of the common implications of salinity stress experienced by plants. Some other vital processes of plant system, like photosynthesis, protein synthesis, and lipid metabolism, are also affected markedly (Qureshi et al. 2013; Yousuf et al. 2016, 2017). Recent studies on the application of multi-walled carbon nanotubes (MWCNTs) in broccoli (Martinez-Ballesta et al. 2016), chitosan NPs in maize (Bruna et al. 2016) and tomato (Hernandez-Hernandez et al. 2018), and silver NPs in wheat (Mohamed et al. 2017; Abou-Zeid and Ismail 2018) have clearly revealed their alleviating effect on salinity stress.

Studies have indicated that NPs of silica (SiO_2) and titanium dioxide (TiO_2) improve growth and yield in plants under stress. This tolerance may be attributed to the absorption of silicon NPs by the root where they develop a fine layer in the cell wall which helps plant to resist various stresses and maintain the yield (Derosa et al. 2010; Latef et al. 2018). Suriyaprabha et al. (2012) have reported that nanoSiO_2 particles are absorbed better and faster than the bulk or micro-SiO_2, Na_2SiO_3, and H_4SiO_4 when applied to maize roots and seeds. Because of their fast absorption,

NPs can be immediately utilized by plants to fulfill their growth needs (Suriyaprabha et al. 2012). In tomato (*Lycopersicon esculentum*) and squash (*Cucurbita pepo*) plants grown under salinity stress, seed germination and the antioxidant system have shown improvement when treated with nanoSiO_2 particles (Haghighi and Pourkhaloee 2013).

The effect of AgNPs application was concentration-dependent, with the optimum at 50 ppm, for growth amelioration in *Brassica juncea* under salinity stress (Sharma et al. 2012b). Similarly, silicon NPs and silicon fertilizer exhibited promising effects on physiological and morphological traits of basil (*Ocimum basilicum*) under salinity stress, as was evident from significant increase in growth and development indices, chlorophyll content (chl-a) and proline level (Kalteh et al. 2014). This could be due to tolerance induction in plants, thereby mitigating the effect of salinity stress (Haghighi et al. 2012). Application of SiO_2NPs has shown potential increase in chlorophyll content, fresh weight and dry weight of leaves, proline accumulation, and antioxidant enzymes' activity in some other plants also grown under salinity stress (Almutairi 2016; Mahmoud et al. 2017). Application of silicon NPs revealed significant increase in seed germination percentage and seedling growth of lentil (*Lens culinaris*) genotypes, which had suffered a significant reduction in these parameters due to salinity stress (Sabaghnia and Janmohammadi 2014). Salinity stress reduced the crop growth and yield due to Na^+ ion toxicity in tomato plants, but the use of SiO_2NPs reduced the ionic toxicity leading to enhanced crop growth and yield (Savvasd et al. 2009). Some similar results were obtained in maize also (Gao et al. 2006). These NPs apparently reduce Na^+ ion concentration, perhaps by reducing Na^+ ion absorption by plant tissues.

Treatment of squash seeds with NaCl reduced the seed germination percentage and the vigor, length, and biomass of roots and shoots (Siddiqui et al. 2014). Application of nanoSiO_2 improved these growth parameters by reducing the levels of malondialdehyde (MDA), hydrogen peroxide (H_2O_2), and electrolyte leakage. In addition, it reduced chlorophyll degradation and enhanced the net photosynthetic rate (P_n), stomatal conductance (g_s), transpiration rate, and water-use efficiency. The increase in plant germination and growth characteristics through application of nanoSiO_2 might reflect an enhanced K/Na ratio (Alsaeedi et al. 2018) and a reduction in oxidative damage due to expression of antioxidant enzymes (Torabian et al. 2016; Farhangi-Abriz and Torabian 2018). Askary et al. (2017) reported that Fe_2O_3NPs caused increases in foliar fresh and dry weights and in the phosphorus, potassium, iron, zinc, and calcium contents of peppermint (*Mentha piperita*) under salinity stress, but had no effect on sodium content. Lipid peroxidation and proline contents decreased significantly. The maximum activities of antioxidant enzymes (catalase, superoxide dismutase, and guaiacol peroxidase) were observed in plants treated with 150 mM of NaCl, but application of Fe_2O_3NPs suppressed these activities. Further, seedlings of cv. California wonder sweet pepper (*Capsicum annuum*) plants were irrigated with saline water with an EC (electrical conductivity) of 5.47 dS m^{-1} and silicon in regular (25%) as well as nanosilicon (25%) forms. All aspects of plant growth and yield were improved under all silicon treatments compared to non-treated plants. Treatment with 1 mL L^{-1} of nanosilicon recorded the maximal and significant success in mitigating the negative effects of salinity. Thus, nanosilicon was more effective and efficient in mitigating the salt-stress impact on sweet pepper plants (Tantawy et al. 2015).

Haghighi and Pessarakli (2013) studied the influence of silicon and nanosilicon on salinity tolerance of cherry tomatoes (*Lycopersicon esculentum*) at early growth stage and reported that salinity caused deleterious effects on plant-growth parameters. Conversely, electrolyte leakage increased by increasing the salinity levels. High salinity levels diminished the sub-stomatal CO_2 content, photosynthetic rate, mesophyll conductance, and water-use efficiency. Si application alleviated the effect of salinity stress on chlorophyll concentration, photosynthetic rate, leaf water content, and, ultimately, on fresh weight of the plant. In short, application of Si was beneficial in improving the salt tolerance of tomato plants; the effect of Si and nanoSi was not significantly different. Torabian et al. (2016) have reported increase in growth, net CO_2 assimilation rate, sub-stomatal CO_2 concentration, chlorophyll content, Fv/Fm and Zn contents, and a decreased Na content in the leaves of sunflower (*Helianthus annuus*) under the influence of nanoZnO. Exposure of wheat plants to CuO/ZnONPs improved the growth of both root and shoot, and this was attributed to lower solubility of CuONPs. It has been suggested that due to alkaline soil, Cu and Fe may precipitate as hydroxides and therefore may not be available for absorption by plants. However, chelated Cu and FeNPs may be an answer to this problem (Fathi et al. 2017). The above discussion shows that the NMs of differrent compounds are capable of mitigating the ill-effects of salinity stress and enhancing the growth of several plant species.

18.2.2 Drought Stress

Climate change and global warming have paved the way for several problems, the global water crisis being the most prominent among them. The ever-increasing scarcity of agricultural water is affecting the agricultural production adversely (FAO 2012) and destroying the green belts in the arid and semiarid regions worldwide (Aref et al. 2013, 2016). Therefore, identification of tolerant plant material or augmentation of drought tolerance in plants is always a prime concern related to sustainable agriculture and crop production. Efforts have been made to mitigate the harmful effects of limited water supply or drought stress on crop plants by using NMs. Use of nanosensors in global positioning systems yielding satellite images of fields might enable farmers to detect the impact of stress in plants at an early stage (Khot et al. 2012). In addition, due to their porous properties and the capillary suction they exert, nanozeolites can be used for enhanced water retention in soils (Boroghani et al. 2011). Improving water-retention capacity of soils could result in increased crop production in areas prone to drought (Gururaj and Krishna 2016).

Application of different fractions of silica NPs improved the tolerance of hawthorn (*Crataegus* sp.) toward drought stress (Ashkavand et al. 2015). Silicon NPs seem to be involved in maintaining the critical physiological and biochemical attributes in order to induce drought tolerance in hawthorn seedlings, but the exact mechanism is yet to be understood. Sedghi et al. (2013) demonstrated that nanozinc oxide can increase seed germination in soybean subjected to water stress by facilitating the rapid use of seed reservoirs for seedling growth. Absorption of microelement iron in plants under drought stress may also have a pivotal role in drought tolerance. Foliar application of iron NPs exhibited drought stress-mitigating effects on yield components and oil percentage of Goldasht spring safflower cultivars (Davar et al. 2014). *Martinez-Fernandez* et al. *(2015) reported that meghemite (Fe_2O_3)* counteracted drought stress with no effect on proline, total amino acids, and mobilization of trace elements in sunflower (*Helianthus annuus*). Further, foliar application of 0.02% TiO_2NPs improved various agronomic traits such as plant height, ear weight, ear number, seed number, 1000-seed weight, final yield, biomass, harvest index, and the gluten and starch contents of wheat under drought stress (Jaberzadeh et al. 2013).

In the case of dragonhead (*Dracocephalum moldavica*) grown under water-deficit condition, plants treated with 10 ppm TiO_2NPs had more proline and much less H_2O_2 and malondialdehyde contents, as compared to untreated plants (Mohammadi et al. 2014). It was concluded that water-deficit stress-induced damages such as oxidative stress and membrane damage can be ameliorated by foliar application of TiO_2NPs at appropriate concentrations. Similarly, AgNPs application reduced the negative effects of drought stress induced by polyethylene glycol on germination rate, germination percentage, and the length as well as fresh and dry weights of roots in lentil (*Lens culinaris*) (Hojjat 2016). The drought impact was alleviated by foliar application of fullerenol NPs in sugarbeet (*Beta vulgaris*) (Borisev et al. 2016) and TiO_2NPs in linseed (*Linum usitatissimum*) (Aghdam et al. 2016). Multi-walled carbon nanotubes (MWCNTs) and sodium nitroprusside (SN) NPs could improve tolerance of barley against drought and salinity by boosting seed water absorption and increasing seedling water content (Karami and Sepehri 2017). Recently, a group of Iranian scientists has revealed the beneficial effects of nanographine oxide at low concentrations and its phytotoxicity at higher concentrations on callus cells of *Plantago major* (a medicinal plant) grown under the polyethylene glycol-induced drought (Ghorbanpoura et al. 2018). In soybean, The use of composite micronutrient NPs (Dimkpa et al. 2017) and cerium oxide (CeO_2) NPs could enhance crop performance of soybean under drought stress (Cao et al. 2018). In vitro application of iron NPs along with salicylic acid proved to be a useful method for enhancing adaptation of strawberry plants to drought before their transplantation in the field (Mozafari et al. 2018).

A complete understanding of the mechanism of ameliorating action of NPs at cellular and metabolic levels will likely help scientists to develop resistant crop cultivars and manage them well for better yield not only in regions of limited water availability but also under other abiotic stresses (Singh et al. 2017). Some studies have confirmed the synthesis of new proteins and enzymes called dehydrins whose enhanced accumulation is related to acquisition of drought tolerance (Lopez et al. 2003). These enzymes, among others, lead to the synthesis of compatible solutes like proline, betaine, polyamines, etc., which protect cells from excessive loss of water and maintain the membranes and cell integrity (Paleg et al. 1984). It would be interesting to know at which point(s) of metabolic pathways the NPs interact with abiotic stresses for enahncing the stress tolerance of plants. It is, however, clear that NPs can be used to enhance crop yield under limited water-supply conditions.

18.2.3 Flooding Stress

Heavy rainfall can cause flooding of fields, if proper drainage is not available. Overflow of water in rivers and rise of groundwater table also cause flooding, which may destroy the standing crops. In addition, excessive irrigation also results in flooding or temporary waterlogging. All these situations may cause either partial anaerobiosis (hypoxia) or complete anaerobic condition (anoxia) in the soil, driving out air from the soil (Ricard et al. 1994; Vartapetian et al. 2014). Flooding causes not only the deficiency of O_2 whose diffusion rate is 10^4-fold slower in water than in air (Armstrong and Drew 2002) but also produces toxic compounds, which retard plant growth and often kill the plants. Hypoxia and anoxia result in energy-deficit conditions, inhibition of respiration, and upregulation of ethylene and abscisic acid synthesis-related genes as strategies for adaptation to waterlogging (Komatsu et al. 2009). Other adaptive strategies include the formation of nodal roots at the air-water interphase and the development of aerenchymatous cells in the root cortex in order to facilitate diffusion of oxygen (Yamauchi et al. 2013). It is, therefore, imperative to know whether NPs enhance the flooding tolerance or mitigate the ill-effects of flooding in plants (Bailey-Serres and Colmer 2014). A study of saffron (*Crocus sativus*), a medicinal and aromatic plant species (Rezvani et al. 2012), revealed a reduction in the number of roots, root length, root biomass, and leaf biomass caused by a 10-day flooding stress, whereas soaking the saffron corms with 40 or 80 ppm concentration of nanosilver mitigated the ill-effects of flooding stress, as indicated by increased production of roots. The effect of Al_2O_3NPs of 30–60 nm was studied on soybean plant under flooding condition with the result that the root length increased while mitochondrial proteins related to glycolysis were suppressed (Mustafa and Komatsu 2016). Also, Al_2O_3NPs of varying size and shape modulated the scavenging activity of cells by regulating the ascorbate/glutathione pathway (Mustafa et al. 2015a). In addition, some other physiological and biochemical strategies include (i) shifting of carbohydrate metabolism toward fermentation (Banti et al. 2013) and (ii) downregulation of alcohol dehydrogenase enzyme in tolerant species (Mustafa et al. 2015b). The apparent paucity of information on NPs-plants interaction warrants further well-planned investigations. However, based on the available information, it can be postulated that NMs play a vital role in alleviating the harmful effects of flooding stress through modulating the metabolism and gene expression.

18.2.4 Mineral Nutrient Stress

Application of fertilizers is a common practice in agriculture to increase the crop productivity. Efforts have been made to identify NMs that may not only increase the crop production and yield but also minimize the nutrient loss of fertilizers, causing nutrient deficiency, and augment their effective availability to plants (Liu and Lal 2015). Development of nanofertilizers (NFs) or nano-encapsulated nutrients that may ensure constant availability and smooth release of nutrients could be an effective tool toward sustainable agriculture. The NFs can be classified under four categories: macronutrient NFs, micronutrient NFs, nutrient-loaded NFs, and plant growth-enhancing NMs (Liu and Lal 2015). Application of NFs could be a potential approach to address issues of soil toxicity and associated stress problems on one hand and enhance the nutrient-use efficiency, on the other. Xiumei et al. (2005) demonstrated that application of nano$CaCO_3$ coordinated with humic acid and organic manures could improve growth and development of peanut plants significantly, and the contents of soluble sugar and protein also increased notably in the stems and leaves accompanied by enhanced absorbability of nutrient elements (calcium, nitrogen, phosphorus, and potassium). The effect of synthesized zinc NPs, characterized for size, shape, surface structure, crystalline nature, and elemental proportion, was studied in pearl millet (*Pennisetum americanum* L.) cv. HHB 67 (Tarafdar et al. 2014). Results indicated that synthesized NPs of 15–25 nm brought significant improvement in root and shoot growth, chlorophyll content, total soluble leaf protein, plant dry biomass, and enzyme activities of acid phosphatase, alkaline phosphatase, phytase, and dehydrogenase in 6-week-old plants. Moreover, grain yield was improved by 37.7% due to application of zinc NF.

Silicon NPs have distinctive physicochemical characteristics and are able to enter into plants, influence the plant metabolism, and improve plant growth and yield under unfavorable environmental conditions. The effect of nanosilicon foliar application on safflower growth under organic and inorganic fertilizer regimes was also investigated (Janmohammadi et al. 2016). Safflower plants were treated with SiO_2 suspension at leaf development, branching, and capitulum emergence stages. The findings suggested that application of organic fertilizers with foliar spray of SiO_2NPs can improve safflower production. Further, the response of two Iranian rice cultivars to nitrogen and nanopotassium fertilizer was studied by Lemraski et al. (2017) who found that NPs consumption by rice cultivars resulted in improved yields. In brief, the use of NFs with a view to reducing nutrient deficiencies in the soil may be quite effective and enhance the growth and yield of different crops.

18.2.5 Temperature Stress

Since the industrial revolution, the atmospheric CO_2 concentration is consistently increasing, giving rise to "global warming." Adverse effects of rising temperatures on crop production have been predicted (Asseng et al. 2015; Kumar 2016), and efforts are being made world over to arrest the rise of temperature to make agriculture sustainable. Both high temperature (heat stress) and low temperature (cold stress or chilling/freezing stress) become injurious to plants on crossing the threshold level.

18.2.5.1 Heat Stress

High temperature stress or heat stress is the rise in temperature beyond a critical threshold for a period of time sufficient to cause irreversible damage to plant growth, development, and yield (Wahid 2007). Heat stress accelerates overproduction of ROS and creates oxidative stress, which causes disintegration of membrane lipids, leakage of electrolytes, and denaturation of biomolecules (Savicka and Skute 2010; Karuppanapandian et al. 2011; Aref et al. 2016) and decreases the chlorophyll content and photosynthesis rate (Prasad et al. 2011). Iqbal et al. (2017) conducted experiments to investigate the effect of AgNPs on the regulation of growth of wheat under heat stress. Plant extract of *Moringa oleifera* was used for AgNPs synthesis followed by its characterization through UV-Vis spectroscopy, X-ray diffractrometry (XRD), scanning electron microscopy (SEM), and atomic force microscopy (AFM). Different concentrations of AgNPs were applied to wheat plants at three-leaf stage under heat stress of 35–40 °C for 3 h day^{-1} for 3 days. Exposure of heat stress alone reduced several plant-growth parameters including dry biomass, whereas application of AgNPs at 50 and 75 mg L^{-1} protected wheat plants against heat stress and improved their growth significantly. Application of low concentration of selenium (Se) NPs also increased chlorophyll content, hydration level, and growth of tomato and sorghum plants because of their antioxidative properties (Haghighi et al. 2014; Djanaguiraman et al. 2018). Heat shock proteins (HSPs), the molecular chaperons, are synthesized in response to heat stress (Schulze et al. 2005). HSPs and compatible solutes help other proteins in maintaining their stability under heat stress (Paleg et al. 1981; Wahid et al. 2007) and play an important role in thermotolerance. Khodakovkazya et al. (2011), using genetic, photothermal, and photoacoustic methods, demonstrated that MWCNTs upregulated the expression of various stress-related genes including *HSP90* (Table 18.1). These findings are also supported by Zhao et al. (2012) who observed that cerium oxide NPs (CeO_2) caused stress response in corn plants resulting in overproduction of H_2O_2 and upregulation of *HSP70*. Qi et al. (2013) found that TiO_2NPs alleviated heat stress through the regulation of stomatal opening. These interesting findings, inter alia, are encouraging especially for those working on aspects related to global warming versus agricultural productivity and food security.

18.2.5.2 Cold Stress

The frequent and unusual occurrence of low temperature is also a cause of concern to farmers. Cold stress (0–15 °C) is caused by the temperature cool enough to produce injury without forming ice crystals in plant tissues, while freezing stress

Table 18.1 Alleviating effects of frequently used nanomaterials on abiotic stresses in plants

Nanomaterial	Abiotic stress	Plant species	Effects	References
Ag	Post-harvest	Orchid (*Dendrobium* sp.), Peruvian lily (*Alstroemeria* sp.)	Suppressed the abscission of flowers and flower buds	Wagstaff et al. (2005) and Uthaichay et al. (2007)
		Chrysanthemums (*Chrysanthemum morifolium*)	Increased survival and succulence of cut flowers, decreased fresh weight loss, reduced stem bacterial colonies	Kazemipour et al. (2013)
	Flooding	Saffron (*Crocus sativus*)	Blocked ethylene signaling, promoted root growth	Rezvani et al. (2012)
		Soybean (*Glycine max*)	Reduced generation of cytotoxic by-products of glycolysis, increased the abundance of stress-related proteins, enhanced seedling growth	Mustafa et al. (2015b)
	Dark	Horse-shoe pelargonium (*Pelargonium zonale*)	Increased antioxidative enzyme activities, leaf chl and carotenoid content, reduced lipid peroxidation, increased petal longevity,decreased petal abscission	Hatami and Ghorbanpour (2013, 2014)
	Cold	Arabidopsis (*Arabidopsis thaliana*)	Activated and enriched antioxidant genes (*MeCu/ZnSOD* and *MeAPX2*), 35% of similar genes were regulated by both AgNPs and cold stress	Kohan-Baghkheirati and Geisler-Lee (2015)
	Drought	Lentil (*Lens culinaris*)	Reduced negative effects on germination rate and germination percentage, root length, root fresh and dry weights	Hojjat (2016)
	Heat	Wheat (*Triticum aestivum*)	Protected plants against heat stress and improved plant growth significantly	Iqbal et al. (2017)
		Wheat (*Triticum aestivum*)	Alleviated the harmful effects of salinity stress	Mohamed et al. (2017) and Abou-Zeid and Ismail (2018)
Al_2O_3	Flooding	Soybean (*Glycine max*)	Regulated energy metabolism and cell death, improved growth	Mustafa et al. (2015a)
			Increased root length, suppressed mitochondrial proteins related to glycolysis	Mustafa and Komatsu (2016)

Anatase-TiO_2	UV-B radiation	Spinach (*Spinacia oleracea*)	Decreased reactive oxygen species (ROS) and MDA content, increased activities of antioxidative enzymes, improved the rate of oxygen evolution	Lei et al. (2007, 2008)
Au^{3+}	Au^{3+}	Cowpea (*Vigna unguiculata*)	Increased generation of Au-nanoparticles and decreased phenolic-induced reduction of toxic Au^{3+} to form non-/less toxic AuNPs	Shabnam et al. (2014)
$CaCO_3$	Mineral nutrient	Peanut (*Arachis hypogaea*)	Improved the growth and development of peanut significantly, also increased the contents of soluble sugar and protein notably in the stems and leaves	Xiumei et al. (2005)
Carboxyl-CdSe/ZnS quantum dots	Pb and Cu	*Chlorella kesslerii* (cell-walled strain), *Chlamydomonas reinhardtii* (walled and wall-less strain)	Decreased intracellular Cu and Pb contents in walled strains, increased Cu and Pb in wall-less strains	Worms et al. (2012)
Cd-telluride quantum	UV-B radiation	Wheat (*Triticum aestivum*)	Programmed cell death and DNA laddering, inhibited root and shoot growth	Chen et al. (2011)
CeO_2	UV-B radiation	Green alga (*Chlorella vulgaris*)	Facilitated efficient absorption of UV radiation without scattering the useful visible light, limited oxidative stress damage	Sicard et al. (2011)
	Heat	Maize (*Zea mays*)	Enhanced production of H_2O_2 and upregulation of *HSP70*	Zhao et al. (2012)
	Drought	Soybean (*Glycine max*)	Enhanced growth, development, and yield resulting in the overall better performance of crop	Cao et al. (2018)
Chitosan	Salinity	Maize (*Zea mays*)	Alleviated the harmful effects of salinity stress	Bruna et al. (2016)
		Tomato (*Lycopersicon esculentum*)	Alleviated the harmful effects of salinity stress	Hernandez-Hernandez et al. (2018)
Cu	Post-harvest	Parsley (*Petroselinum crispum*)	Prevented weight loss, reduced lipid peroxidation, and maintained higher ascorbic acid concentration	Ouzounidou and Gaitis (2011)
CuO	UV radiation	Waterweed (*Elodea nuttallii*)	Increased Cu accumulation, reduced chlorophyll content and photosynthetic capacity, enhanced peroxidase activity	Regier et al. (2015)
	Salinity	Wheat (*Triticum aestivum*)	Improved the growth of both root and shoot	Fathi et al. (2017)

(continued)

Table 18.1 (continued)

Nanomaterial	Abiotic stress	Plant species	Effects	References
Fe	Drought	Safflower (*Carthamus tinctorius*)	Reduced impact of drought and improved yield due to foliar application	Davar et al. (2014)
		Strawberry (*Fragaria ananassa*)	Enahanced acclimation and ressitance of plants to drought before their transplantation in the field	Mozafari et al. (2018)
Fullerenol	Drought	Sugar beet (*Beta vulgaris*)	Injurious effects of drought dissapeared by foliar application of fullerenol NPs	Borisev et al. (2016)
Maghemite (y-Fe_2O_3)	Drought	Sunflower (*Helianthus annuus*)	Counteracted drought stress with no effect on proline, total amino acids, and mobilization of trace elements	Martinez-Fernandez et al. (2015)
	Salinity	Peppermint (*Mentha piperita*)	Increased leaf fresh and dry weights, phosphorus, potassium, iron, zinc, and calcium contents	Askary et al. (2017)
MWCNTs (multi-walled carbon nanotubes)	Heat	Tomato (*Lycopersicon esculentum*)	Upregulated the expression of various stress-related genes including *HSP90*	Khodakovskaya et al. (2011)
	Salinity	Cabbage (*Brassica oleracea*)	Alleviated the harmful effects of salinity stress	Martinez-Ballesta et al. (2016)
	Drought and salinity	Barley (*Hordeum vulgare*)	Improved drought and salinity tolerance by boosting seed water absorption and increasing seedling water content	Karami and Sepehri (2017)
Na_2SeO_4	Heat and cold	Tomato (*Lycopersicon esculentum*)	Improved plant growth, chlorophyll, and leaf-relative water contents	Haghighi et al. (2014)
Na_2SiO_3	Cr (VI)	Pea (*Pisum sativum*)	Protected pea seedlings against Cr(VI) phytotoxicity, reduced uptake of Cr(VI) and oxidative stress, upregulated antioxidative defense systems, and enhanced accumulation of nutrient elements leading to improved growth	Tripathi et al. (2015)
Nanopotassium	Mineral nutrient	Rice (*Oryza sativa*)	Enhanced total tillers and fertile tillers per hill	Lemraski et al. (2017)
Se	Heat	Tomato (*Lycopersicon esculentum*)	Increased chlorophyll content, hydration of plants, and growth	Haghighi et al. (2014) and Djanaguiraman et al. (2018)
SiO_2	Salinity	Tomato (*Lycopersicon esculentum*)	Enhanced seed germination potential, root length, and dry weight	Haghighi et al. (2012)

			Alleviated the effect of salinity on fresh weight, chlorophyll concentration, photosynthetic rate, and leaf water content	Haghighi and Pessarakli (2013) and Haghighi and Pourkhaloee (2013)
			Upregulated the expression profile of four salt-stress genes (*AREB*, *TAS14*, *NCED3*, and *CRK1*), and downregulated six genes (*RBOH1*, *APX2*, *MAPK2*, *ERF5*, *MAPK3*, and *DDF2*), suppressed the effect of salinity on seed germination rate, root length, and fresh weight	Almutairi (2016)
		Squash (*Cucurbita pepo*)	Improved seed germination and growth characteristics; reduced levels of MDA, H_2O_2, and electrolyte leakage; reduced chlorophyll degradation and oxidative damage; enhanced photosynthetic parameters and antioxidative enzymes	Haghighi and Pourkhaloee (2013) and Siddiqui et al. (2014)
		Basil (*Ocimum basilicum*)	Increased fresh and dry weights, chlorophyll, and proline contents	Kalteh et al. (2014) and
		Lentil (*Lens culinaris*)	Enhanced seed germination and seedling growth	Sabaghnia and Janmohammadi (2014)
		Cucumber (*Cucumis sativus*)	Increased plant germination and growth characteristics	Alsaeedi et al. (2018)
		Soybean (*Glycine max*)	Reduced oxidative damage due to expression of antioxidative enzymes	Farhangi-Abriz and Torabian (2018)
	Cold	Wheatgrass (*Agropyron elongatum*)	Overcame seed dormancy, enhanced seed germination and seedling weight	Azimi et al. (2014)
	Drought	Hawthorn (*Crataegus* sp.)	A significant positive effect on photosynthetic rate, stomatal conductance, and plant biomass, nonsignificant effect on chlorophyll and carotenoid contents	Ashkavand et al. (2015)
	Cr	Pea (*Pisum sativum*)	Alleviated metal-induced toxicity of plants	Tripathi et al. (2015)
	Mineral nutrient	Safflower (*Carthamus tinctorius*)	Improved safflower production	Janmohammadi et al. (2016)
	UV-B radiation	Wheat (*Triticum aestivum*)	Alleviated harmful effects of UV-B radiation in seedlings	Tripathi et al. (2017)

(continued)

Table 18.1 (continued)

Nanomaterial	Abiotic stress	Plant species	Effects	References
TiO_2	Excessive light	Spinach (*Spinacia oleracea*)	Increased activities of antioxidative enzymes, decreased accumulation of reactive oxygen free radicals and the level of MDA, membrane stability and structure of chloroplast remained intact	Hong et al. (2005a, b)
	Heat	Tomato (*Lycopersicon esculentum*)	Enhanced photosynthesis by regulating energy dissipation, caused cooling of leaves through inducing stomatal opening	Qi et al. (2013)
	Drought	*Wheat (Triticum aestivum)*	Increased growth, yield, gluten, and starch content	Jaberzadeh et al. (2013)
		Moldavian dragonhead *(Dracocephalum moldavica)*	More proline and much less H_2O_2 and MDA contents, less oxidative stress and membrane damage	Mohammadi et al. (2014)
		Lin seed (*Linum usitatissimum*)	Enhanced chlorophyll and carotenoid content, improved growth and yield attributes, decreased hydrogen peroxide (H_2O_2) and malondialdehyde (MDA) contents	Aghdam et al. (2015)
	Cold	Chickpea (*Cicer arietinum*)	Enhanced activities of antioxidative enzymes, decreased H_2O_2 content and electrolyte leakage, enhanced accumulation of TiO_2 in sensitive genotype than in tolerant one	Mohammadi et al. (2013a, b)
			Enhanced expression of Rubisco- and chlorophyll-binding protein genes, decreased H_2O_2 content, enhanced activity of phosphoenolpyruvate carboxylase	Hasanpour et al. (2015)
Zn	Mineral nutrient	Pearl millet (*Pennisetum americanum*)	Improved shoot length, root length, root area, chlorophyll content, total soluble leaf protein, plant dry biomass, and enzyme activities significantly	Tarafdar et al. (2014)
ZnO	Drought	*Soybean* (*Glycine max*)	Increased germination percentage and germination rate, decrease in seed residual fresh and dry weights	Sedghi et al. (2013)
	Salinity	Sunflower (*Helianthus annuus*)	Increased growth, net CO_2 assimilation rate, sub-stomatal CO_2 concentration, chlorophyll content, Fv/Fm and Zn content, and decreased Na content in leaves	Torabian et al. (2016)
		Wheat (Triticum aestivum)	Improved the growth of both root and shoot	Fathi et al. (2017)

(<0 °C) results in the formation of ice crystals in plant tissues causing frost-killing of crops (Hasanuzzaman et al. 2013). Loss of fluidity of membranes and leakage of electrolytes are the distinct effects of cold stress. Plants exposed to cold stress show poor germination, retarded growth, and reduced crop yield (Welti et al. 2002; Suzuki et al. 2008). However, susceptibility to cold stress varies among species and cultivars. Plants with higher tolerance levels show less membrane injury than the sensitive species (Amiri et al. 2010; Heidarvand et al. 2011).

With the objective of how to diminish the ill-effects of cold stress on crop plants, Mohammadi et al. (2013a) assessed the effect of TiO_2NPs on changes in membrane damage indices like electrolyte leakage index and malondialdehyde content during cold stress of 4 °C in a sensitive (ILC 533) and a tolerant (Sel 11439) chickpea (*Cicer arietinum*) genotypes. Bioaccumulation of NPs within the vacuole and chloroplast showed that, under thermal treatments, the sensitive genotype had more permeability to NPs compared to the tolerant one, and TiO_2 content was higher during cold stress than at the optimum temperature. Obviously, physiological indices were positively affected by NPs during thermal treatments. TiO_2 treatments not only prevented oxidative damage in chickpea genotypes but also alleviated membrane damage under cold stress. It was suggested for the first time that TiO_2NPs improved the redox status of the genotypes under thermal treatments. Extending their studies, Mohammadi et al. (2013b) confirmed the earlier results and suggested that TiO_2NPs confer an increased tolerance of chickpea plants to cold stress by decreasing the level of injuries and increasing the capacity of defense systems. Further research may likely endorse the usefulness of NPs in general and TiO_2NPs in particular for increasing the crop tolerance against cold stress.

Photosynthesis, a vital process, is prone to cold stress. Plants exposed to cold stress have less chlorophyll content, reduced CO_2 assimilation, and degradation of Rubisco enzyme (Yordanova and Popova 2007; Liu et al. 2012). Positive effects of NMs on photosynthesis enhance carboxylation of Rubisco (Gao et al. 2006), light absorption capacity of chloroplast (Ze et al. 2011), electron transport rate, and inhibition of ROS generation in chloroplast (Giraldo et al. 2014). Application of nano-TiO_2 enhances the expression level of Rubisco- and chlorophyll-binding protein genes (Hasanpour et al. 2015) and activities of catalase, superoxide dismutase, and ascorbate peroxidase (Mohammadi et al. 2014), maintains the stability of chlorophyll and carotenoid contents, and increases plant tolerance to cold stress (Table 18.1). Exposure of plants to chilling stress causes upregulation of *MeCu/ZnSOD* and *MeAPX2* genes and enhances the activities of monodehydroascorbate reductase, dehydroascorbate reductase, and glutathione reductase that improves ROS scavenging, leading to suppressed oxidative stress parameters such as lipid peroxidation, chlorophyll degradation, and H_2O_2 synthesis that ultimately causes stress resistance (Xu et al. 2014). On the other hand, application of NMs alone or along with a short-term chilling stress treatment has been shown to improve growth and the physiological and biochemical attributes of cold-stressed plants (Azimi et al. 2014; Haghighi et al. 2014; Kohan-Baghkheirati and Geisler-Lee 2015, Table 18.1). More work is certainly required to arrive at a concrete conclusion regarding the role of NPs in enhancing cold tolerance of different agricultural and horticultural crops.

18.2.6 Ultraviolet (UV) Radiation Stress

Exposure of plants to ultraviolet-B (UV-B, 280–315 nm), photosynthetically inactive and non-ionizing radiation, leads to accumulation of enhanced level of ROS in cells (Mackerness et al. 2001), which damages DNA, chloroplast structure, and several cellular processes including photosynthesis (Chen et al. 2011; Wang et al. 2012; Hideg et al. 2013). DNA damage results from the formation of dimer of pyrimidine bases leading to the loss of biological activity. RNA, proteins, ABA, and IAA also absorb UV radiation and, consequently, lose their biological activity. Besides activating their enzymatic and nonenzymatic antioxidant defense system, plants tend to adapt to the stressful situation by accumulating phenolic compounds such as flavonoids and flavones that absorb harmful UV radiations (Shen et al. 2010a, b). The use of NMs protects photosynthetic systems from UV-B stress (Table 18.1) by enhancing chl content and Rubisco activity and suppressing oxidative stress (Sicard et al. 2011). It is interesting to note that nanoanatase TiO_2 with a photocatalyzed characteristic under light could cause an oxidation-reduction reaction. Hong et al. (2005a, b) observed that nanoTiO_2 could promote photosynthesis and greatly improve spinach growth. However, the mechanism of the action of nanoTiO_2 in promoting conversion from light energy to electron energy and from electron energy to active chemical energy remains largely unclear. Along these findings, Lei et al. (2007, 2008) reported that during the photosynthesis, electron transfer, photoreduction activity of photosystem II, oxygen evolution, and photophosphorylation of chloroplast in the nanoanatase TiO_2-treated spinach were greatly increased under both visible and ultraviolet light illumination. Likewise, the energy-enriched electron from nanoanatase TiO_2 could enter chloroplast under ultraviolet light, got transferred to photosynthetic electron transport chain, and made NADP+ be reduced to NADPH and coupled to photophosphorylation, making electron energy be transformed to ATP. Moreover, nanoanatase h+, which photogenerated electron holes, captured an electron from water, which accelerated water photolysis and O_2 evolution. However, designing a material from the atomic level to achieve a tailored response in extreme conditions is a grand challenge in material research. NMs for the control of radiation-induced defects via interfaces are the key factors in reducing damage and providing stability in camparison to bulk materials, which display void swelling or embrittlement (Beyerlein et al. 2013; Tulinski and Jurczyk 2017).

The mechanisms by which nanoanatase promotes antioxidant stress in spinach chloroplasts under UV-B radiation are not fully understood (Lei et al. 2008). However, these authors showed that nanoanatase treatment could significantly decrease the accumulation of superoxide radicals (O· −2), hydrogen peroxide (H_2O_2), and malondialdehyde (MDA) contents; increase activities of catalase (CAT), superoxide dismutase (SOD), ascorbate peroxidase (APX), and guaiacol peroxidase (GPX); and elevate the oxygen evolution rate in spinach chloroplasts under UV-B radiation, thus mitigating the oxidative stress caused by UV-B radiation. Lately, silicon NPs have also been found to alleviate effectively the harmful effects of UV-B radiation in wheat seedlings (Tripathi et al. 2017). Such studies are opening new vistas in research that may prove revolutionary in terms of food production in the future under conditions of abiotic stress.

18.2.7 Irradiance Stress

Enhancing radiation-use efficiency is a prime consideration for realizing the maximum yield of crops. However, when the energy absorbed by the photosystems exceeds the energy that can be used in the photochemistry, it causes photodamage (photooxidation, i.e., bleaching of chl) to the plants (Havaux et al. 2000). Moreover, plants develop mechanisms to avoid photodamage by losing excess light energy as heat due to induced acidification of thylakoid lumen resulting from the formation of a proton-motive force that involves xanthophyll cycle. This leads to the enzymatic conversion of carotene violaxanthin into zeaxanthin which loses the energy as heat (Taiz and Zeiger 2010). Toward this end, the effects of nanoTiO_2 (rutile) on the aging of spinach chloroplast under light were studied (Hong et al. 2005a), and the results showed that when the chloroplasts were illuminated for relatively short periods with 500 μmol cm^{-2} min^{-1} light intensity, oxygen evolution rate rapidly increased; when the chloroplasts were treated for longer periods with similar light intensity, the rate gradually decreased. When spinach was treated with 0.25% nano-TiO_2, the rate of oxygen evolution in chloroplasts with different illumination times (1, 5, 10, 20, 30, and 40 min) was higher than that of the control. It suggested that nanoTiO_2 treatment could protect chloroplasts from aging due to longtime illumination. The results also indicated that nanoTiO_2 treatment could significantly increase the activities of SOD, CAT, and POD, decrease accumulation of reactive oxygen free radicals and the level of MDA, and maintain the stability of membrane structure of chloroplasts under light. Hong et al. (2005b) extended their studies on the effects of nanoTiO_2 (rutile) on the photochemical reaction of chloroplasts of spinach and demonstrated that when spinach was treated with 0.25% nanoTiO_2, the Hill reaction such as the reduction rate of Fe-cytochrome and the rate of oxygen evolution of chloroplasts was accelerated, and noncyclic photophosphorylation (nc-PSP) activity of chloroplasts was higher than cyclic photophosphorylation (c-PSP) activity; the chloroplast coupling was improved, and activities of Mg^{2+}-ATPase and chloroplast coupling factor I (CF_1)-ATPase on thylakoid membranes were activated. These findings suggest that the promotion of photosynthesis by nanoTiO_2 might be related to activation of photochemical reaction of the chloroplasts. On the whole, very little information is available on the effects of NPs on photosynthesis, the sole source of chemical energy for living organisms on the Earth, under long-term illumination with a range of high and low levels of irradiance for different agricultural crops. This is a productive area of basic research which might prove beneficial in enhancing crop yields, especially under low-light stress conditions often resulting from mutual shading of plants (Li et al. 2014).

18.2.8 Post-harvest Stress

The long-distance commercial shipping and handling of horticultural commodities are subjected to various post-harvest stresses. Dark stress, among them, is known to suppress chl content, photosynthesis, and antioxidant enzymes' acitivities, to increase ethylene production and ROS generation and induce membrane damage (Prochazkova and Wilhelmova 2007). These changes in the cellular system result in senescence and abscission that affect the shelf life and commercial value of plants and plant products. Post-harvest stress in horticultural products is attributed to increased water loss, respiration, oxidative stress, and lipid peroxidation, which affect the weight and nutritional value of the commodities concerned (Ouzounidou and Gaitis 2011). Although several approaches have been adopted to counteract these stresses, recent involvement of NMs in modulating the physiological and biochemical processes of plants has gained special attention of experts. Application of AgNPs suppresses abscission of flowers and flower buds, as observed in *Alstroemeria* and *Dendrobium* species (Wagstaff et al. 2005; Uthaichay et al. 2007); AgNPs are more promising in alleviating dark stress than the routine Ag salts. Treatment with CuNPs and AgNPs also improved shelf life of parsley leaves and longevity of cut flowers of chrysanthemum, respectively (Ouzounidou and Gaitis 2011; Kazemipour et al. 2013). Alleviating effect of AgNPs has also been reported with respect to alleviation of the dark stress-induced oxidative stress, and the consequent increase in the longevity of petals due to AgNP treatment has been reported in *Pelargonium zonale* (Hatami and Ghorbanpour 2013, 2014). These authors also observed decreased petal abscission in geranium cultivars treated with AgNPs and thidiazuron during storage in the dark (Table 18.1).

18.2.9 Pollutant Stress

Rapid industrialization and urbanization in the recent past have significantly contributed to the man-made pollution of air, land, and water. Air pollutants include O_3, SO_2, NO, NO_2, NH_3, CFCs, peroxyacylnitrate (PAN), and volatile organic compounds (VOC) apart from high concentration of CO_2 (Iqbal et al. 2000), whereas heavy metals such as As, Cd, Cr, Hg, Pb, etc., in addition to Na, Cl, SO_4, CO_3, HCO_3, and NO_3 ions, are the major land and water pollutants (Umar et al. 2005). Industrial solid waste products and effluents mostly consisting of the abovementioned toxic heavy metals and other chemical ions contaminate the food chain also. Detoxification or remediation of harmful pollutants using synthetic clay nanomineral is quite common (Prasad et al. 2014). The water to be filtered is percolated through a column of hydrotalcite (synthetic clay mineral), which can also be coupled with leaching through porous pots or filter candles (Gilman 2006). Zinc oxide NPs can be used to remove arsenic, using a point-of-source purification device. Nanoscale zero-valent iron is the most widely used NM that could be deployed to remediate pollutants in soil or groundwater (Chibuike and Obiora 2014). Metal toxicity often suppresses enzyme activities, disrupts uptake of essential elements causing deficiency symptoms (Capuana 2011), and generates ROS leading to oxidative stress coupled with denaturation of cell structure, biomembrane, and macromolecules (Rascio and Navari-Izzo 2011; Sharma et al. 2012a). However, plants evolve a diverse range of defense systems such as biophysical barriers serving as the first line of defense against metal stress. Further they accumulate metal chelates, organic acids, and polyphosphates, which restrict the uptake and promote endogenous sequestration of metals and the activation of antioxidant defense system to scavenge ROS. The timely and target-oriented activation of these defense systems is crucial to counter the ill-effects of metal stress.

Engineered NMs are highly effective in alleviating the metal-induced toxicity of plants (Gunjan et al. 2014; Siddiqi and Husen 2017). As NPs are smaller in size with large surface area, they easily penetrate into the contamination zone and show strong affinity to metals. Availability of Cu and Pb may be reduced due to metal binding to quantum dots (QDs) with the cell wall acting as an additional barrier (Worms et al. 2012). In case the metals pass the biophysical barriers and enter the cell, plants build up resistance to counteract their harmful effects by accumulating biomolecules and nutrients and activating their defense systems. The role of nano-TiO_2, among several NPs, has been most extensively studied. TiO_2NPs restricted Cd uptake in pea seedlings, whereas SiNPs protected them against Cr phytotoxicity (Tripathi et al. 2015). Exposure of cowpea seeds to gold (Au) ion stress provides reduction of Au^{3+} to non-/less toxic AuNPs by phenolics released by the seed coat of germinating seeds (Shabnam et al. 2014, Table 18.1). Other NMs that could be used in remediation include nanoscale zeolites, metal oxides, carbon nanotubes and fibers, and various noble metals (mainly as bimetallic NPs). Titanium dioxide nanoparticle filters can be used to remove organic particles and pesticides, for example, dichlorodiphenyltrichloroethane (DDT), endosulfan, malathion, and chlorpyrifos from water. A variety of nanoparticle filters have been used in remediation of waste sites in the developed countries (Karn et al. 2009).

Thus, NMs are useful in getting the environment and ecosystems freed from the clutches of pollutant stress and protecting plants from the ill-effects of other abiotic stresses. The mode and mechanism of NPs action within the cell to make the cellular machinery more effective and render them less toxic to plants facing abiotic stresses is depicted in Fig. 18.1.

18.2.10 Stress Caused by NMs and Their Impact on Ecosystems

With the rapid advancement in nanotechnology, release of nanoscale materials into the environment is inevitable. This may negatively influence the functioning of the ecosystems. Many manufactured NPs contain heavy metals, which can cause soil and water contamination (Sect. 18.2.9). Proteomic techniques have contributed substantially in understanding the molecular mechanisms of plant responses against

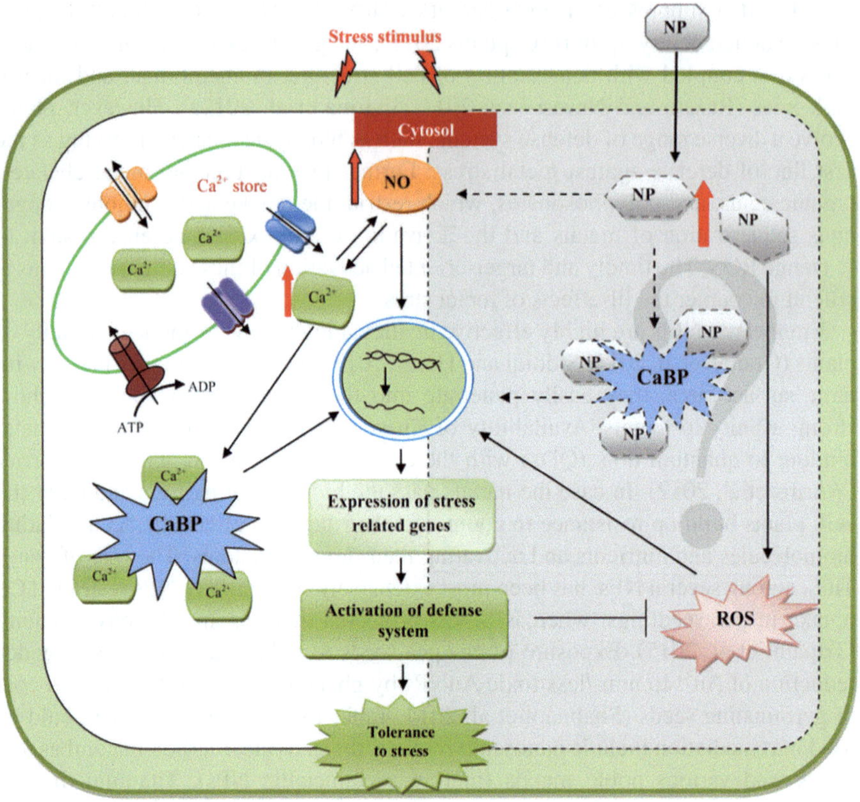

Fig. 18.1 Depiction of the role of NPs in cellular defense against abiotic stress. Gray half of the diagram at right-hand side shows the possible mechanism of NPs action. Red arrows indicate increase, whereas solid arrows and dashed arrows show established pathway and possible pathway, respectively. Perception of stress stimulus elevates the cytosolic Ca^{2+} concentration ($[Ca^{2+}]_{cyt}$), which is sensed by Ca^{2+}-binding protein (CaBP). Binding of Ca^{2+} activates CaBP that directly binds to the promoters of specific genes and induces or represses their expression, causing plant tolerance to the stress. The NPs cause overexpression of CaBPs and bind to CaBPs that trigger downstream signaling and finally the expression of stress-related genes and activation of plant's defense system. Similar to $[Ca^{2+}]_{cyt}$, NPs also induce nitric oxide (NO) synthesis that acts as stress-signaling molecule and induces expression of stress-related genes that activate the defense system and ultimately generate tolerance to stress. Optimum concentration of NPs also maintains reactive oxygen species (ROS), via regulating antioxidant defense system, at the level sufficient enough for stress signaling leading to activation of plant's defense system and development of tolerance to stress. (Source: Khan et al. 2016 with slight modification)

various stresses by providing a link between gene expression and cell metabolism (Hossain et al. 2015). Because TiO_2 generates reactive oxygen species (ROS) when exposed to ultraviolet radiation (UV-R), nanoparticulate TiO_2 has been used in antibacterial coatings and wastewater disinfection and has been investigated as an anticancer agent (Charpentier et al. 2012). In vivo tests of TiO_2 toxicity with aquatic organisms have typically shown low toxicity, though results across studies have been variable. No work has demonstrated that photoactivity causes environmental toxicity of TiO_2 under natural levels of UV-R. Relatively low levels of ultraviolet light, consistent with those found in nature, can induce toxicity of TiO_2NPs to marine phytoplankton, the most important primary producers on the Earth. No effect of TiO_2 on phytoplankton was found in treatments where UV light was blocked. Under low-intensity UV-R, ROS in seawater increased with increasing nanoTiO_2 concentration. These increases may lead to increased oxidative stress in seawater contaminated by TiO_2, and cause decreased resiliency of marine ecosystems (Miller et al. 2012). Further, the solar UV radiation impacts on toxicity of ZnONPs through photocatalytic ROS generation and photo-induced dissolution. Toxicity of ZnONPs to *Daphnia magna* was examined by Ma et al. (2014) who found enhanced ZnONPs toxicity under simulated solar UV radiation, being parallel to those of photocatalytic ROS generation and increased particle dissolution. These findings demonstrate the importance of considering the environmental solar UV radiation when assessing ZnO and other NPs' toxicity and risk in the aquatic systems.

A comprehensive study of the uptake and toxic effects of CuONPs and Cu^{2+} ions, alone and in combination with UV radiation, was undertaken using the aquatic macrophyte, *Elodea nuttallii,* as the test plant (Regier et al. 2015). Its growth was inhibited when treated with $CuSO_4$ or CuONPs with a lower amount of accumulated copper in $CuSO_4$-treated plants than in those treated with CuONPs. The difference has been attributed to the solubility of Cu^{2+} in CuO nanoparticle medium. Surprisingly, the relation between accumulated Cu and dissolved Cu^{2+} was higher in plants exposed to 256 μg L^{-1} Cu^{2+} than those exposed to 10 mg L^{-1} CuONPs containing nearly 2 mg L^{-1} dissolved Cu, as the formation of large agglomerates prevents the dissolution of Cu. Interestingly, Cu accumulation was enhanced under UV radiation in shoots of *Pisum sativum* after 4 h, but there was no direct evidence of enhanced solubility of Cu^{2+} in CuO nanoparticle suspension (Agrawal and Mishra 2009). Rai et al. (1995) observed an altered membrane permeability due to lipid peroxidation in cell membranes of UV-exposed cells of cyanobacteria. In plants, under UV radiation, photosynthetic capacity is strongly reduced. When higher quantities of Cu accumulate in plants, the response of oxidative stress-related enzymes (peroxidase and superoxide dismutase) is also high. Therefore, phototoxicity must be considered while evaluating the environmental impacts of NMs, as many of them are photoactive.

The cells in higher plants act against the toxic effects of NPs, and in doing so, certain metals are absorbed, and certain others are precipitated. Phytotoxicity of commercial CuO (<50 nm) and ZnNPs (<100 nm) against sand-grown wheat (*T. aestivum*) has been investigated (Watson et al. 2014). Oxidative stress in the nanoparticle-treated plants was reflected by an increase in lipid peroxidation and oxidized glutathione and higher peroxidase and catalase activities in roots. Also, CuONPs have been shown to induce DNA damage in plants (Atha et al. 2012). Growth inhibition in radish (*Raphanus sativus*), perennial ryegrass (*Lolium perenne*), and annual ryegrass (*Lolium rigidum*) under laboratory conditions has been reported. Germination of radish seeds in the presence of CuONPs induces substantial accumulation of mutagenic DNA lesions. Radish and similar other plants produce oxygen-derived species (O· −2, H_2O_2,·OH) during germination (Schopfer et al. 2001). H_2O_2 enhances seed germination but, in the presence of peroxidase or transition metal ions, such as iron or copper, produces an excess of ·OH via the Fenton reaction (Sharma et al. 2012a). It is therefore suggested that copper ions produced from CuONPs may catalyze the formation of ·OH which probably inhibited radish root growth to the extent of 79%, which is much more than the effect of Cu^{2+} ions alone.

Morales-Diaz et al. (2017) studied the application of nanoelements in plant nutrition and its impact on ecosystems and consumers. As discussed earlier, nanofertilizers are more effective and efficient than traditional fertilizers due to their impact on crop nutritional quality and stress tolerance in plants. However, there are virtually no studies on the potential environmental impact of NMs when used in agriculture. Such studies are highly desirable because NPs or NMs can be transferred to ecosystems by various pathways where they can cause toxicity to organisms, affecting the biodiversity in ecosystems and also causing risks to human health.

18.3 Conclusion

NPs, being small in size and having large surface area, penetrate into plant cells easily, and their absorption and targeted delivery to the sites and substances are also facilitated in relatively shorter period of time, as compared to their standard or bulk chemical forms. Several have been used for mitigating the harmful effects of various abiotic stresses on fruits, flowers, and vegetables. Some of the frequently used NMs like TiO_2, SiO_2, AgNPs, etc. have been found to have a multitude of beneficial effects, compared to their bulk chemical forms, on the expression of morphological, physiological, and biochemical attributes and enhance crop yields under a variety of abiotic stresses. Their application could have a positive impact on seed germination, plant growth, quality and quantity of yield, chlorophyll and caretonoid contents, photosynthesis rate, expression of Rubisco- and chlorophyll-binding protein genes, stomatal opening, WUE, ability to dissipate excess light energy in the form of heat, etc., thus causing less oxidative damage and enhancing tolerance to abiotic stresses in a range of crops including cereals, pulses, oil seeds, vegetables, and various ornamental, medicinal, and fruit plants.

On the contrary, NMs may generate ROS and cause several toxic effects on plants. However, the enhanced ROS level brought about by NMs could be associated with the amplification of a stress signal that perhaps activates defense systems of plants more effeciently. Information available on the mode and mechanism of action of NMs is still insufficient and inconsistent. It seems, however, that induction of an array of signaling molecules triggers gene expression leading to de novo synthesis of new proteins and enzymes, thus enhancing the tolerance of plants to abiotic stresses, resulting in optimum growth and yield of crop plants (Fig. 18.1). It may be concluded that NMs alleviate the abiotic stress-caused damage through activating the defense system in plants.

So far, there have not been enough studies at field level to allow commercialization of the large-scale use of NMs for enhancing food production. It is essential to understand plant-NP interaction and optimization of NP size and concentration before their practical applications can be taken to farmers' fields. Moreover, their possible negative impact on natural environment and crops should also be assessed and minimized. The prospective research on NM-plant interaction must address especially the effect of hormones in combination with various NPs, seed priming with penetrable (NPs of inorganic salt) and non-penetrable solutes (NPs of polyethylene glycol of higher molecular wt.) and hormones, DNA photolyase, a unique class of flavoenzymes that use blue light to catalyze repair of UV-induced damage to plant DNA, and the effect of soil acidity on the efficiency of NMs. While nanochemical pesticides are already in use, other applications that include mainly the fine-tuning and more precise micromanagement of problem soils, the more efficient and targeted use of inputs in the wake of climate change and global warming, and new toxin formulations for pest control are yet to come in full swing. Also, it would be interesting to know at which point(s) of metabolic pathways the NPs interact with abiotic stresses for enhancing the stress tolerance of plants.

References

Abou-Zeid HM, Ismail GSM (2018) The role of priming with biosynthesized silver nanoparticles in the response of *Triticum aestivum* L. to salt stress. Egypt J Bot 58:73–85

Adeleye AS, Conway JR, Garner K, Huang Y, Su Y, Kell AA (2016) Engineered nanomaterials for water treatment and remediation: costs, benefits, and applicability. Chem Eng J 286:640–662

Aghdam MTB, Mohammadi H, Ghorbanpour M (2016) Effects of nanoparticulate anatase titanium dioxide on physiological and biochemical performance of *Linum usitatissimum* (Linaceae) under well-watered and drought stress conditions. Braz J Bot 39:139–146

Agrawal SB, Mishra S (2009) Effects of supplemental ultraviolet-B and cadmium on growth, antioxidants and yield of *Pisum sativum* L. Ecotoxicol Environ Saf 72:610–618

Almutairi ZM (2016) Effect of nano-silicon application on the expression of salt tolerance genes in germinating tomato (*Solanum lycopersicum* L.) seedlings under salt stress. POJ 9:106–114

Alsaeedi A, El-Ramady A, Alshaal T, El-Garawani M, Elhawat N, Al-Otaibi A (2018) Exogenous nanosilica improves germination and growth of cucumber by maintaining K^+/Na^+ ratio under elevated Na^+ stress. Plant Physiol Biochem 125:164–171

Amiri RM, Yur'eva NO, Shimshilashvili KR, Goldenkova-Pavlova IV, Pchelkin VP, Kuznitsova EI, Tsydendambaev VD, Trunova TI, Los DA, Salehi Jouzani G, Nosov AM (2010) Expression of acyl-lipid D 12-desaturase gene in prokaryotic and eukaryotic cells and its effect on cold stress tolerance of potato. J Integr Plant Biol 52:289–297

Andy P (2016) Plant abiotic stress challenges from the changing environment. Front Plant Sci 7:1123

Aragay G, Pino F, Merkoci A (2012) Nanomaterials for sensing and destroying pesticides. Chem Rev 112:5317–5338

Aref MI, El-Atta H, El-Obeid M, Ahmed AI, Khan PR, Iqbal M (2013) Effect of water stress on relative water and chlorophyll contents of *Juniperus procera* Hochst. ex Endlicher in Saudi Arabia. Life Sci J 10(4):681–685

Aref IM, Khan PR, Khan S, El-Atta H, Ahmed AI, Iqbal M (2016) Modulation of antioxidant enzymes in *Juniperus procera* needles in relation to habitat environment and dieback incidence. Trees 30:1669–1681

Armstrong W, Drew MC (2002) Root growth and metabolism under oxygen deficiency. In: Waisel Y, Eshel A, Kafkafi U (eds) Plant roots: the hidden half. Marcel Dekker, New York, pp 729–761

Ashkavand P, Tabari M, Zarafshar M, Tomaskova I, Struve D (2015) Effect of SiO_2 nanoparticles on drought resistance in hawthorn seedlings. For Res Pap 76:350–359

Askary M, Talebi SM, Amini F, Bangan ADB (2017) Effects of iron nanoparticles on *Mentha piperita* under salinity stress. Biologija 63:65–75

Asseng S, Zhu Y, Wang E, Zhang W (2015) Crop modeling for climate change impact and application. In: Sadras VO, Calderini DF (eds) Crop physiology: applications for genetic improvement and agronomy, 2nd edn. Academic Press: San Diego, CA, USA, pp 505–546

Atha DH, Wang H, Petersen EJ, Cleveland D, Holbrook RD, Jaruga P, Dizdaroglu M, Xing B, Nelson BC (2012) Copper oxide nanoparticle mediated DNA damage in terrestrial plant models. Environ Sci Technol 46:1819–1827

Azimi R, Borzelabad MJ, Feizi H, Azimi A (2014) Interaction of SiO_2 nanoparticles with seed prechilling on germination and early seedling growth of tall wheatgrass (*Agropyron elongatum* L.). Pol J Chem Tech 16:25–29

Bailey-Serres J, Colmer TD (2014) Plant tolerance of flooding stress – recent advances. Plant Cell Environ 37:2211–2215

Banti V, Giuntoli B, Gonzali S, Loreti E, Magneschi L, Novi G, Paparelli E, Parlanti S, Pucciariello C, Santaniello A, Perata P (2013) Low oxygen response mechanisms in green organisms. Int J Mol Sci 14:4734–4761

Beyerlein I, Caro A, Demkowicz M, Mara N, Misra A, Uberuaga B (2013) Radiation damage tolerant nanomaterials. Mater Today 16:443–449

Borisev M, Borisev I, Zupunski M, Arsenov D, Pajevic S (2016) Drought impact is alleviated in sugar beets (*beta vulgaris*) by foliar application of fullerenol nanoparticles. PLoS One 11:e0166248

Boroghani M, Mirnia SK, Vahhabi J, Ahmadi SJ, Charkhi A (2011) Nanozeolite synthesis and the effect of on the runoff and erosion control under rainfall simulator. Aust J Basic Appl Sci 5:1156–1164

Brock DA, Douglas TE, Queller DC, Strassmann JE (2011) Primitive agriculture in a social amoeba. Nature 469:393–396

Bruna HCO, Gomes CR, Milena T, Pelegrino A, Seabra B (2016) Nitric oxide-releasing chitosan nanoparticles alleviate the effects of salt stress in maize plants. Nitric Oxide 61:10–19

Cao Z, Rossi L, Stowers C, Zhang W, Lombardini L, Ma X (2018) The impact of cerium oxide nanoparticles on the physiology of soybean (*Glycine max* (L.) Merr.) under different soil moisture conditions. Environ Sci Pollut Res 25:930–939

Capuana M (2011) Heavy metals and woody plants biotechnologies for phytoremediation. J Biogeo Sci For 4:7–15

Charpentier PA, Burgess K, Wang L, Chowdhury RR, Lotus AF, Moula G (2012) Nano-TiO2/polyurethane composites for antibacterial and self-cleaning coatings. Nanotechnol 23:425606

Chen H-ZZ-J, Du M-T, Han R (2011) Influence of enhanced UV-B radiation on factin in wheat division cells. Plant Diver Resour 33:306–310

Chibuike GU, Obiora SC (2014) Heavy metal polluted soils: effect on plants and bioremediation methods. Appl Environ Soil Sci 2014:1–13

Davar F, Zareii AR, Amir H (2014) Evaluation the effect of water stress and foliar application of Fe nanoparticles on yield, yield components and oil percentage of safflower (*Carthamus tinctorious* L.). Int J Adv Biol Biomed Res 2:1150–1159

Derosa MR, Monreal C, Schmitzer M, Walsh R, Sultan Y (2010) Nanotechnology in fertilizers. Nat Nanotechnol 1:193–225

Dimkpa CO, Bindraban PS, Fugice J, Agyin-Birikorang S, Singh U, Hellums D (2017) Composite micronutrient nanoparticles and salts decrease drought stress in soybean. Agron Sustain Dev 37:5

Djanaguiraman M, Belliraj N, Bossmann SH, Prasad PVV (2018) High-temperature stress alleviation by selenium nanoparticle treatment in grain sorghum. ACS Omega 3:2479–2491
FAO (2012) Coping with water scarcity- an action framework for agriculture and food security. FAO water reports. FAO Publication Division, Rome
Farhangi-Abriz S, Torabian S (2018) Nano-silicon alters antioxidant activities of soybean seedlings under salt toxicity. Protoplasma 255:953–962
Fathi A, Zahedi M, Torabian S, Khoshgoftar A (2017) Response of wheat genotypes to foliar spray of ZnO and Fe_2O_3 nanoparticles under salt stress. J Plant Nutr 40:1376–1385
Gao FQ, Hong FS, Liu C, Zheng L, Su MY, Wu X, Yang F, Wu C, Yang P (2006) Mechanism of nanoanatase TiO_2 on promoting photosynthetic carbon reaction of spinach: inducing complex of rubisco-rubisco activase. Biol Trace Elem Res 11:239–254
Gao X, Zou CH, Wang L, Zhang F (2006) Silicon decreases transpiration rate and conductance from stomata of maize plants. J Plant Nutr 29:1637–1647
Ghorbanpoura M, Farahani AHK, Hadian J (2018) Potential toxicity of nano-graphene oxide on callus cell of *Plantago major* L. under polyethylene glycol-induced dehydration. Ecotoxicol Environ Saf 148:910–922
Gilman GP (2006) A simple device for arsenic removal from drinking water using hydrotalcite. Sci Total Environ 366:926–931
Giraldo JP, Landry MP, Faltermeier SM, McNicholas TP, Iverson NM, Boghossian AA, Reuel NF, Hilmer AJ, Sen F, Brew JA, Strano MS (2014) Plant nanobionics approach to augment photosynthesis and biochemical sensing. Nat Mater 13:400–408
Gunjan B, Zaidi MGH, Sandeep A (2014) Impact of gold nanoparticles on physiological and biochemical characteristics of *Brassica juncea*. J Plant Biochem Physiol 2:133
Gururaj SB, Krishna BSVSR (2016) Water retention capacity of biochar blended soils. J Chem Pharm Sci 9:1438–1441
Haghighi M, Afifipour Z, Mozafariyan M (2012). The effect of N-Si on tomato seed germination under salinity levels. J Biol Environ Sci 6:87–90
Haghighi M, Pessarakli M (2013) Influence of silicon and nano-silicon on salinity tolerance of cherry tomatoes (*Solanum lycopersicum* L.) at early growth stage. Sci Hortic 161:111–117
Haghighi M, Pourkhaloee A (2013) Nanoparticles in agricultural soils: their risks and benefits for seed germination. Minerva Biotecnol 25:123–132
Haghighi M, Abolghasemi R, Teixeira da Silva JA (2014) Low and high temperature stress affect the growth characteristics of tomato in hydroponic culture with Se and nano-Se amendment. Sci Hort 178:231–240
Hasanpour H, Maali-Amiri R, Zeinali H (2015) Effect of TiO_2 nanoparticles on metabolic limitations to photosynthesis under cold in chickpea. Russ J Plant Physiol 62:779–787
Hasanuzzaman M, Nahar K, Fujita M (2013) Extreme temperature responses, oxidative stress and antioxidant defense in plants. In: Vahdati K, Leslie C (eds) Abiotic stress- plant responses and applications in agriculture. InTech Open Access Publisher, London, UK, pp 169–205
Hatami M, Ghorbanpour M (2013) Effect of nanosilver on physiological performance of pelargonium plants exposed to dark storage. J Hort Res 21:15–20
Hatami M, Ghorbanpour M (2014) Defense enzyme activities and biochemical variations of *Pelargonium zonale* in response to nanosilver application and dark storage. Turk J Biol 38:130–139
Havaux M, Bonfils JP, Lutz C, Niyogi KK (2000) Photodamage of the photosynthetic apparatus and its dependence on the leaf developmental stage in the *npq1* Arabidopsis mutant deficient in the xanthophyll cycle enzyme violaxanthin de-epoxidase. Plant Physiol 24:273–284
Heidarvand L, Maali-Amiri R, Naghavi MR, Farayedi Y, Sadeghzadeh B, Alizadeh KH (2011) Physiological and morphological characteristics of chickpea accessions under low temperature stress. Russ J Plant Physiol 58:157–163
Hernandez-Hernandez H, Gonzalez-Morales S, Benavides-Mendoza A, Ortega-Ortiz H, Cadenas-Pliego G, Juarez-Maldonado A (2018) Effects of chitosan–PVA and cu nanoparticles on the growth and antioxidant capacity of tomato under saline stress. Molecules 23:178

Hideg E, Jansen MA, Strid A (2013) UV-B exposure, ROS, and stress: inseparable companions or loosely linked associates? Trends Plant Sci 18:107–115

Hojjat (2016) The effect of silver nanoparticle on lentil seed germination under drought stress. Int J Farm Allied Sci 5:208–212

Hong F, Yang F, Liu C, Gao Q, Wan Z, Gu F, Wu C, Ma Z, Zhou J, Yang P (2005a) Influence of nano-TiO_2 on the chloroplast aging of spinach under light. Biol Trace Elem Res 104:249–260

Hong F, Zhou J, Liu C, Yang F, Wu C, Zheng L, Yang P (2005b) Effect of nano-TiO_2 on photo chemical reaction of chloroplasts of spinach. Biol Trace Elem Res 105:269–279

Hossain Z, Mustafa G, Komatsu S (2015) Plant responses to nanoparticle stress. Int J Mol Sci 16:26644–26653

Husen A (2017) Gold nanoparticles from plant system: synthesis, characterization and application. In: Ghorbanpourn M, Manika K, Varma A (eds) Nanoscience and plant–soil systems, vol 48. Springer International Publication, Switzerland, pp 455–479

Husen A, Siddiqi KS (2014) Phytosynthesis of nanoparticles: concept, controversy and application. Nano Res Lett 9:229

Husen A, Iqbal M, Aref MI (2014) Growth, water status and leaf characteristics of *Brassica carinata* under drought stress and rehydration conditions. Braz J Bot 37:217–227

Husen A, Iqbal M, Aref IM (2016) IAA-induced alteration in growth and photosynthesis of pea (*Pisum sativum* L.) plants grown under salt stress. J Environ Biol 37:421–429

Husen A, Iqbal M, Aref IM (2017) Plant growth and foliar characteristics of faba bean (*Vicia faba* L.) as affected by indole-acetic acid under water-sufficient and water-deficient conditions. J Environ Biol 38:179–186

Husen A, Iqbal M, Sohrab SS, Ansari MKA (2018) Salicylic acid alleviates salinity-caused damage to foliar functions, plant growth and antioxidant system in Ethiopian mustard (*Brassica carinata* A. Br.). Agri Food Security. 7:44

Iqbal M, Srivastava PS, Siddiqi TO (2000) Anthropogenic stresses in the environment and their consequences. In: Iqbal M, Srivastava PS, Siddiqi TO (eds) Environmental hazards: plants and people. CBS Publishers, New Delhi, pp 1–38

Iqbal M, Raja NI, Mashwani ZR, Hussain M, Ejaz M, Yasmeen F (2017) Effect of silver nanoparticles on growth of wheat under heat stress. Iran J Sci Technol Trans A Sci https://doi.org/10.1007/s40995-017-0417-4

Jaberzadeh A, Payam M, Hamid R, Tohidi M, Hossein Z (2013) Influence of bulk and nanoparticles titanium foliar application on some agronomic traits, seed gluten and starch contents of wheat subjected to water deficit stress. Not Bot Horti Agrobot Cluj-Na 41:201–207

Janmohammadi M, Amanzadeh T, Sabaghnia N, Ion V (2016) Effect of nano-silicon foliar application on safflower growth under organic and inorganic fertilizer regimes. Bot Lith 22:53–64

Kah M (2015) Nanopesticides and nanofertilizers: emerging contaminants or opportunities for risk mitigation? Front Chem 3:64

Kalteh M, Alipour ZT, Ashraf S, Aliabadi MM, Nosratabadi AF (2014) Effect of silica nanoparticles on basil (*ocimum basilicum*) under salinity stress. J Chem Health Risk 4:49–55

Karami A, Sepehri A (2017) Multiwalled carbon nanotubes and nitric oxide modulate the germination and early seedling growth of barley under drought and salinity. Agric Conspec Sci 82:331–339

Karn B, Kuiken T, Otto M (2009) Nanotechnology and in situ remediation: a review of benefits and potential risks. Environ Health Perspect 117:1823–1831

Karuppanapandian T, Wang HW, Prabakaran N, Jeyalakshmi K, Kwon M, Manoharan K, Kim W (2011) 2,4-dichlorophenoxyacetic acid-induced leafsenescence in mung bean (*Vigna radiata* (L.) Wilczek) and senescence inhibition by co-treatment with silver nanoparticles. Plant Physiol Biochem 49:168–217

Kazemipour S, Hashemabadi D, Kaviani B (2013) Effect of silver nanoparticles on the vase life and quality of cut chrysanthemum (*Chrysanthemum morifolium* L.) flower. Eur J Exp Biol 3:298–302

Khan MN, Mobin M, Abbas ZK, Almutairi KA, Siddiqui ZH (2016) Role of nanomaterials in plants under challenging environments. Plant Physiol Biochem 110:194–2019

Khodakovskaya MV, de Silva K, Nedosekin DA, Dervishi E, Biris AS, Shashkov EV, Ekaterina IG, Zharov VP (2011) Complex genetic, photo thermal, and photo acoustic analysis of nanoparticle-plant interactions. Proc Natl Acad Sci U S A 108:1028–1033

Khot LR, Sankaran S, Maja JM, Ehsani R, Schust EW (2012) Applications of nanomaterials in agricultural production and crop protection: a review. Crop Prot 35:64–70

Kohan-Baghkheirati E, Geisler-Lee J (2015) Gene expression, protein function and pathways of *Arabidopsis thaliana* responding to silver nanoparticles in comparison to silver ions, cold, salt, drought, and heat. Nano 5:436–467

Komatsu S, Yamamoto R, Nanjo Y, Mikami Y, Yunokawa H, Sakata K (2009) A comprehensive analysis of the soybean genes and proteins expressed under flooding stress using transcriptome and proteome techniques. J Proteome Res 8:4766–4778

Kumar M (2016) Impact of climate change on crop yield and role of model for achieving food security. Environ Monit Assess 188:465

Latef AAHA, Srivastava AK, El-Sadek MSA, Kordrostami M, Tran LP (2018) Titanium dioxide nanoparticles improve growth and enhance tolerance of broad bean plants under saline soil conditions. Land Degrad Dev 29:1065–1073

Laware SL, Raskar S (2014) Effect of titanium dioxide nanoparticles on hydrolytic and antioxidant enzymes during seed germination in onion. Int J Curr Microbiol Appl Sci 3:749–760

Lei Z, Mingyu S, Chao L, Liang C, Hao H, Xiao W, Xiaoqing L, Fan Y, Fengqing G, Fashui H (2007) Effects of nanoanatase TiO_2 on photosynthesis of spinach chloroplasts under different light illumination. Biol Trace Elem Res 119:68–76

Lei Z, Mingyu S, Xiao W, Chao L, Chunxiang Q, Liang C, Hao H, Xiaoqing L, Fashui H (2008) Antioxidant stress is promoted by nano-anatase in spinach chloroplasts under UV-B radiation. Biol Trace Elem Res 121:69–79

Lemraski MG, Normohamadi G, Madani H, Abad HHS, Mobasser HR (2017) Two Iranian rice cultivars' response to nitrogen and nano-fertilizer. Open J Ecol 7:591–603

Li T, Liu LN, Jiang CD, Liu YJ, Shi L (2014) Effects of mutual shading on the regulation of photosynthesis in field-grown sorghum. J Photochem Photobiol B Biol 137:31–38

Liu R, Lal R (2015) Potentials of engineered nanoparticles as fertilizers for increasing agronomic productions. Sci Total Environ 514:131–139

Liu YF, Qi MF, Li TL (2012) Photosynthesis, photoinhibition, and antioxidant system in tomato leaves stressed by low night temperature and their subsequent recovery. Plant Sci 196:8–17

Lopez CJ, Banowetz GM, Peterson CJ, Kronstad WE (2003) Dehydrin expression and drought tolerance in seven wheat cultivars. Crop Sci 43:577–582

Lutz C, Steevens JA (2009) Nanomaterials: risks and benefits. Springer, Dordrecht

Ma H, Wallis L, Diamond S, Li S, Canas J, Cano A (2014) Impact of solar UV radiation on toxicity of ZnO nanoparticles through photocatalytic reactive oxygen species (ROS) generation and photo-induced dissolution. Environ Pollut 193:165–172

Mackerness SAH, John CF, Jordan B, Thomas B (2001) Early signalling components in ultraviolet-B responses: distinct roles for different reactive oxygen species and nitric oxide. FEBS Lett 489:237–242

Mahmoud EF, Abdel-Haliem HS, Hegazy NS, Hassan DMN (2017) Effect of silica ions and nano silica on rice plants under salinity stress. Ecol Eng 99:282–289

Martinez-Ballesta MC, Zapata L, Chalbi N, Carvajal M (2016) Multiwalled carbon nanotubes enter broccoli cells enhancing growth and water uptake of plants exposed to salinity. J Nanobiotech 14:42

Martinez-Fernandez D, Vítkova M, Bernal MP, Komarek M (2015) Effects of nano-maghemite on trace element accumulation and drought response of *Helianthus annuus* L. in a contaminated mine soil. Water Air Soil Pollut 226(101). https://doi.org/10.1007/s11270-015-2365-y

Miller RJ, Bennett S, Keller AA, Pease S, Lenihan HS (2012) TiO_2 nanoparticles are phototoxic to marine phytoplankton. PLoS One 7:e30321

Mohamed EF (2017) Nanotechnology: future of environmental air pollution control. Environ Mgmt Sust Dev 6:429–454

Mohamed AKSH, Qayyum MF, Abdel-Hadi AM, Rehman RA, Ali S, Rizwan M (2017) Interactive effect of salinity and silver nanoparticles on photosynthetic and biochemical parameters of wheat. Arch Agron Soil Sci 63(12):1736–1747

Mohammadi R, Amiri NM, Mantri L (2013a) Effect of TiO_2 nanoparticles on oxidative damage and antioxidant defense systems in chickpea seedlings during cold stress. Russ J Plant Physiol 61:768–775

Mohammadi R, Amiri RM, Abbasi A (2013b) Effect of TiO_2 nanoparticles on chickpea response to cold stress. Biol Trace Elem Res 152:403–410

Mohammadi H, Esmailpour M, Gheranpaye A (2014) Effects of TiO_2 nanoparticles and water-deficit stress on morpho-physiological characteristics of dragonhead (*Dracocephalum moldavica* L.) plants. Environ Toxicol Chem 33:2429–2437

Morales-Diaz AB, Ortega-Ortiz H, Juarez-Maldonado A, Cadenas-Pliego G, Gonzalez-Morales S, Benavides-Mendoza A (2017) Application of nanoelements in plant nutrition and its impact in ecosystems. Adv Nat Sci Nanosci Nanotechnol 013001:1–11

Mozafari AA, Havas F, Ghaderi N (2018) Application of iron nanoparticles and salicylic acid in in vitro culture of strawberries (*Fragaria × ananassa* Duch.) to cope with drought stress. Plant Cell Tissue Organ Cult 132:511–523

Munns R, Tester M (2008) Mechanisms of salinity tolerance. Annu Rev Plant Biol 59:651–681

Mustafa G, Komatsu S (2016) Insights into the response of soybean mitochondrial proteins to various sizes of aluminum oxide nanoparticles under flooding stress. J Proteome Res 15:4464–4475

Mustafa G, Sakata K, Hossain Z, Komatsu S (2015a) Proteomic analysis of flooded soybean root exposed to aluminum oxide nanoparticles. J Proteome 128:280–297

Mustafa G, Sakata K, Hossain Z, Komatsu S (2015b) Proteomic study on the effects of silver nanoparticles on soybean under flooding stress. J Proteome 122:100–118

Ouzounidou G, Gaitis F (2011) The use of nano-technology in shelf life extension of green vegetables. J Innov Econ Manag 2:163–171

Paleg LG, Douglas TJ, van Daal A, Keech DB (1981) Proline, betaine and other organic solutes protect enzymes against heat inactivation. Aust J Plant Physiol 8:107–114

Paleg LG, Stewart GR, Bradbeer JW (1984) Proline and glycine betaine influence protein solvation. Plant Physiol 75:974–978

Prasad PVV, Pisipati SR, Mom I, Ristic Z (2011) Independent and combined effects of high temperature and drought stress during grain filling on plant yield and chloroplast EF-Tu expression in spring wheat. J Agron Crop Sci 197:430–441

Prasad R, Bagde US, Varma A (2012) Intellectual property rights and agricultural biotechnology: an overview. Afr J Biotechnol 11:13746–13752

Prasad R, Kumar V, Prasad KS (2014) Nanotechnology in sustainable agriculture: present concerns and future aspects. Afr J Biotechnol 13:705–713

Prochazkova D, Wilhelmova N (2007) Leaf senescence and activities of the antioxidant enzymes. Biol Plant 51:401–406

Qi M, Liu Y, Li T (2013) Nano-TiO_2 improve the photosynthesis of tomato leaves under mild heat stress. Biol Trace Elem Res 156:323–328

Qureshi MI, Abdin MZ, Ahmad J, Iqbal M (2013) Effect of long-term salinity on cellular antioxidants, compatible solute and fatty acid profile of sweet annie (*Artemisia annua* L.). Phytochemistry 95:215–223

Rai LC, Tyagi B, Mallick N, Rai PK (1995) Interactive effects of UV-B and copper on photosynthetic activity of the cyanobacterium *Anabaena doliolum*. Environ Exp Bot 35:177–185

Rascio N, Navari-Izzo F (2011) Heavy metal hyperaccumulating plants: how and why do they do it? And what makes them so interesting? Plant Sci 180:169–181

Regier N, Cosio C, von Moos N, Slaveykova VI (2015) Effects of copper-oxide nanoparticles, dissolved copper and ultraviolet radiation on copper bioaccumulation, photosynthesis and oxidative stress in the aquatic macrophyte *Elodea nuttallii*. Chemosphere 128:56–61

Rezvani N, Sorooshzadeh A, Farhadi N (2012) Effect of nano-silver on growth of saffron in flooding stress. Int J Biol Biomol Agri Food Biotechnol Engr 6:11–16

Ricard B, Couee I, Raymond P, Saglio H, Saint-Ges P, Veronique B, Pradet A (1994) Plant metabolism under hypoxia and anoxia. Plant Physiol Biochem 32:1–10

Sabaghnia N, Janmohammadi M (2014) Effect of nano-silicon particles application on salinity tolerance in early growth of some lentil genotypes. Ann UMCS Biol 69:39–55

Savicka M, Skute N (2010) Effects of high temperature on malondialdehyde content, superoxide production and growth changes in wheat seedlings (*Triticum aestivum* L.). Ekologija 56:26–33

Savvasd G, Giotes D, Chatzieustratiou E, Bakea M, Patakioutad G (2009) Silicon supply in soilless cultivation of Zucchini alleviates stressinduced by salinity and powdery mildew infection. Environ Exp Bot 65:11–17

Schopfer P, Plachy C, Frahry G (2001) Release of reactive oxygen intermediates (superoxide radicals, hydrogen peroxide, and hydroxyl radicals) and peroxidase in germinating radish seeds controlled by light, gibberellin, and abscisic acid. Plant Physiol 125:1591–1602

Schulze ED, Beck E, Muller-Hohenstein K (2005) Plant ecology. Springer, Berlin

Sedghi M, Hadi M, Toluie SG (2013) Effect of nano zinc oxide on the germination parameters of soybean seeds under drought stress. Ann West Uni Timisoara 16:73–78

Shabnam N, Pardha-Saradhi P, Sharmila P (2014) Phenolics impart Au^{3+}-stress tolerance to cowpea by generating nanoparticles. PLoS One 9:e85242

Sharma P, Jha AB, Dubey RS, Pessarakli M (2012a) Reactive oxygen species, oxidative damage, and antioxidative defense mechanism in plants under stressful conditions. J Bot 2012:1–26

Sharma P, Bhatt D, Zaidi MG, Saradhi PP, Khanna PK, Arora S (2012b) Silver nanoparticle-mediated enhancement in growth and antioxidant status of *Brassica juncea*. Appl Biochem Biotechnol 167:2225–2233

Shen CX, Zhang QF, Li J, Bi FC, Yao N (2010a) Induction of programmed cell death in Arabidopsis and rice by single-wall carbon nanotubes. Am J Bot 97:1–8

Shen X, Zhou Y, Duan L, Li Z, Eneji AE, Li J (2010b) Silicon effects on photosynthesis and antioxidant parameters of soybean seedlings under drought and ultraviolet-B radiation. J Plant Physiol 167:1248–1252

Sicard C, Perullini M, Spedalieri C, Coradin T, Brayner R, Livage J, Jobbagy M, Bilmes SA (2011) CeO_2 nanoparticles for the protection of photosynthetic organisms immobilized in silica gels. Chem Mater 23:1374–1378

Siddiqi KS, Husen A (2016) Engineered gold nanoparticles and plant adaptation potential. Nano Res Lett 11:400

Siddiqi KS, Husen A (2017) Plant response to engineered metal oxide nanoparticles. Nanoscale Res Lett 12:92

Siddiqui MH, Al-Whaibi MH, Faisal M, Al Sahli AA (2014) Nano-silicon dioxide mitigates the adverse effects of salt stress on *Cucurbita pepo* L. Environ Toxicol Chem 33:2429–2437

Singh S, Vishwakarma K, Singh S, Sharma S, Dubey NK, Singh VK, Liu S, Tripathi DK, Chauhan DK (2017) Understanding the plant and nanoparticle interface at transcriptomic and proteomic level: a concentric overview. Plant Gene 11:265–272

Suriyaprabha R, Karunakaran G, Yuvakkumar R, Prabu P, Rajendran V, Kannan N (2012) Growth and physiological responses of maize (*Zea mays* L.) to porous silica nanoparticles in soil. J Nanopart Res 14:1–14

Suzuki K, Nagasuga K, Okada M (2008) The chilling injury induced by high root temperature in the leaves of rice seedlings. Plant Cell Physiol 49:433–442

Taiz L, Zeiger E (2010) Plant physiology, 5th edn, Sinauer Associates Inc. Publishers, Sunderland, Massachusetts, USA

Tantawy AS, Salama YAM, El-Nemr MA, Abdel-Mawgoud AMR (2015) Nano silicon application improves salinity tolerance of sweet pepper plants. Int J ChemTech Res 8:11–17

Tarafdar JC, Sharma S, Raliya R (2013) Nanotechnology: interdisciplinary science of applications. Afr J Biotechnol 12:219–226

Tarafdar JC, Raliya R, Mahawar H, Rathore I (2014) Development of zinc nanofertilizer to enhance crop production in pearl millet (*Pennisetum americanum*). Agric Res 3:257–262

Torabian S, Zahedi M, Khoshgoftar AH (2016) Effects of foliar spray of two kinds of zinc oxide on the growth and ion concentration of sunflower cultivars under salt stress. J Plant Nutr 39:172–180

Tripathi DK, Singh VP, Prasad SM, Chauhan DK, Dubey NK (2015) Silicon nanoparticles (SiNp) alleviate chromium (VI) phytotoxicity in *Pisum sativum* (L.) seedlings. Plant Physiol Biochem 96:189–198

Tripathi DK, Singh S, Singh VP, Mohan PS, Dubey NK, Chauhan DK (2017) Silicon nanoparticles more effectively alleviated UV-B stress than silicon in wheat (*Triticum aestivum*) seedlings. Plant Physiol Biochem 110:70–81

Tulinski M, Jurczyk M (2017) Nanomaterials synthesis methods. In: Mansfield E, Kaiser DL, Fujita D, Van de Voorde M (eds) Metrology and standardization of nanotechnology: protocols and industrial innovations. Wiley-VCH Verlag GmbH & Co KGaA, Weinheim, pp 75–98

Umar S, Moinuddin, Iqbal M (2005) Heavy metal availability, accumulation and toxicity in plants. In: Dwivedi P, Dwivedi RS (eds) Physiology of abiotic stress in plants. Agrobios (India), Jodhpur, pp 325–348

Uthaichay N, Ketso S, Van Doorn WG (2007) 1-MCP pretreatment prevents bud and flower abscission in *Dendrobium* orchids. Postharvest Biol Technol 43:374–380

Vartapetian BB, Dolgikh YI, Polyakova LI, Chichkova NV, Vartapetian AB (2014) Biotechnological approaches to creation of hypoxia and anoxia tolerant plants. Acta Nat 6:19–30

Wagstaff C, Chanasut U, Harren FJM, Laarhoven LJ, Thomas B, Rogers HJ, Stead AD (2005) Ethylene and flower longevity in *Alstroemeria*: relationship between petal senescence, abscission and ethylene biosynthesis. J Exp Bot 56:1007–1016

Wahid A (2007) Physiological implications of metabolites biosynthesis in net assimilation and heat stress tolerance of sugarcane (*Saccharum officinarum*) sprouts. J Plant Res 120:219–228

Wahid A, Gelani S, Ashraf M, Foolad M (2007) Heat tolerance in plants: an overview. Environ Exp Bot 61(3):199–223

Wang LJ, Guo ZM, Li TJ, Li M (2001) The nano structure SiO_2 in the plants. Chin Sci Bull 46:625–631

Wang X, Han H, Liu X, Gu X, Chen K, Lu D (2012) Multi-walled carbon nanotubes can enhance root elongation of wheat (*Triticum aestivum*) plants. J Nanopart Res 14:1–10

Watson JL, Fang T, Dimkpa C, Britt D, Mclean J, Jacobson A, Anderson A (2014) The phytotoxicity of ZnO nanoparticles on wheat varies with soil properties. Biometals Int J Role Metal Ions Biol Biochem Med 28:101–112

Welti R, Li W, Li M, Sang Y, Biesiada H, Zhou HE, Rajashekar CB, Williams TD, Wang X (2002) Profiling membrane lipids in plant stress responses. Role of phospholipase D alpha in freezing-induced lipid changes in Arabidopsis. J Biol Chem 277:31994–32002

Worms IAM, Boltzman J, Garcia M, Slaveykova VI (2012) Cell-wall-dependent effect of carboxyl-CdSe/ZnS quantum dots on lead and copper availability to green microalgae. Environ Pollut 167:27–33

Xiumei L, Fudao Z, Shuqing Z, Xusheng H, Rufang W, Zhaobin F, Yujun W (2005) Responses of peanut to nano-calcium carbonate. Plant Nutr Fert Sci 11:385–389

Xu J, Yang J, Duan X, Jiang Y, Zhang P (2014) Increased expression of native cytosolic cu/Zn superoxide dismutase and ascorbate peroxidase improves tolerance to oxidative and chilling stresses in cassava (*Manihot esculenta* Crantz). BMC Plant Biol 14:208

Yamauchi T, Shimamura S, Nakazono M, Mochizuki T (2013) Aerenchyma formation in crop species: a review. Field Crops Res 152:8–16

Yordanova R, Popova L (2007) Effect of exogenous treatment with salicylic acid on photosynthetic activity and antioxidant capacity of chilled wheat plants. Gen Appl Plant Physiol 33:155–170

Yousuf PY, Ahmad A, Aref IM, Ozturk M, Hemant GAH, Iqbal M (2016) Salt-stress-responsive chloroplast proteins in *Brassica juncea* genotypes with contrasting salt tolerance and their quantitative PCR analysis. Protoplasma 253:1565–1575

Yousuf PY, Ahmad A, Ganie AH, Sareer O, Krishnapriya V, Aref IM, Iqbal M (2017) Antioxidant response and proteomic modulations in Indian mustard grown under salt stress. Plant Growth Regul 81:31–50

Ze Y, Liu C, Wang L, Hong M, Hong F (2011) The regulation of TiO_2 nanoparticles on the expression of light-harvesting complex II and photosynthesis of chloroplasts of *Arabidopsis thaliana*. Biol Trace Elem Res 143:1131–1141

Zhao L, Peng B, Hernandez-Viezcas JA, Rico C, Sun Y, Peralta-Videa JR, Tang X, Niu G, Jin L, Ramirez AV, Zhang JY, Gardea-Torresdey JL (2012) Stress response and tolerance of *Zea mays* to CeO_2 nanoparticles: cross talk among H_2O_2, heat shock protein and lipid peroxidation. ACS Nano 6:9615–9622

Chapter 19
Nano-fertilization to Enhance Nutrient Use Efficiency and Productivity of Crop Plants

Muhammad Iqbal, Shahid Umar, and Mahmooduzzafar

19.1 Introduction

For generations, farmers have understood the importance of nitrogen (N) for the growth of their crops and struggled consistently to boost up N levels in the soil, but the fact remains that only a small extent (30%) of the conventional fertilizer is utilized by crops, as the bulk (70%) of it is destroyed by water or wind before being used by plants (Crawshaw 2018). Fertilizers of N, P, and K are quite popular, and a balanced use of these chemical boosters helps in protecting plants from several risks and improving their growth and yield. Today, when nanotechnology research heralds the beginning of a new era of advancement in human life, it opens promising avenues concerning agriculture also. In nanotechnology, which deals with the nanoscale particles/structures, when the crystallite size of inorganic materials is reduced to nanoscale, two different phenomena occur. One is the quantum size effect, which exhibits radical changes in the physicochemical properties of the material; the other is the huge ratio of surface area to volume, which endows nanoparticles (NPs) with remarkably good transduction properties (Prasad et al. 2017; Chap. 2 of this book). Nanomaterials (NMs) have a special role in managing nutrient availability for plants and their protection from a variety of risks through health improvement, much of which could be assigned to optimized utilization of the nutrients required.

Nano-fertilizers (NFs) are nutrient carriers developed by using the substrates with nano-dimensions (1–100 nm), which have extensive surface area and can hold abundance of nutrients to be released slowly and steadily. In the case of conventional fertilizers, nutrient use efficiency hardly exceeds 30–35%, 18–20%, and 35–40% for N, P, and K, respectively, and these estimates remain constant for the past several decade (Preetha and Balakrishnan 2017), whereas the nano-clay-based

M. Iqbal (✉) · S. Umar · Mahmooduzzafar
Department of Botany, Faculty of Science, Jamia Hamdard (Deemed University), Hamdard Nagar, New Delhi, India

A. Husen, M. Iqbal (eds.), *Nanomaterials and Plant Potential*,
https://doi.org/10.1007/978-3-030-05569-1_19

fertilizer formulations (zeolite and montmorillonite with a dimension of 30–40 nm) are capable of releasing N for a much longer period of time (>1000 h) than the conventional fertilizers (<500 h) (Rahale 2011). Nano-fertilizers may contain NPs of zinc, silica, iron and titanium dioxides, ZnCdSe/ZnS core-shell QDs, InP/ZnS core-shell QDs, Mn/ZnSe QDs, gold nanorods, core-shell QDs, etc. (Sadeghzadeh 2013; Prasad et al. 2017). Since they provide a larger surface area for reaction and a prolonged availability of nutrients to the crop plant, this situation favors quality parameters, such as protein, oil, and sugar contents, by enhancing the rate of reaction or synthesis process in the plant system, as observed in various crops (Singh et al. 2017).

The use of NFs not only enhances the nutrient use efficiency but also reduces the frequency of fertilizer application and consequently the soil toxicity and other potential negative effects associated with excessive use of chemicals (Qureshi et al. 2018). A critical evaluation of nano-agrochemicals versus conventional analogues has shown that the gain in efficacy relative to the conventional products is about 20–30% (Kah et al. 2018). Nano-fertilizers motivate early seed germination and a superior plant growth and yield. The uptake of NFs relies mainly on the type of the plant and the type and size of particles of the engineered NF. A current collaborative research project undertaken by the University of Alberta, Carleton University, Agrium (a fertilizer company), and Nanogrande (a nanotechnology firm) is likely to give rise to smart products, which would be able to release nitrogen in the soil only when the plant needs it and in the amount as required by the plant. This will increase the fertilizer efficiency from 30% to more than 80% (Crawshaw 2018). The component biosensors of these smart products detect the chemical signals sent by plant roots regarding the N requirement of the plant and then allow microbes' access to the fertilizer-N inside the polymer-protected particle so that this N may be converted to forms utilizable by plants. As the different crop species send out different chemical signals, an intelligent NF product should be able to respond aptly to the varying needs of different crops (Crawshaw 2018).

Nonetheless, in some cases plants as well as soil microflora suffer from toxic effects of the nano-forms of fertilizers and other chemicals (e.g., pesticides and herbicides), which are normally concentration-dependent (Dimkpa and Bindraban 2018; Kah et al. 2018). The NMs may have adverse effects on environmental health also. The changes brought about in nano-formulations of agrochemicals may not necessarily reduce the environmental damage. Unfortunately, most of the NM studies conducted so far have lacked the nano-specific quality assurance and adequate controls, and no comprehensive study is yet available on the efficacy and environmental impact of nano-agrochemicals under field conditions (Kah et al. 2018). Several concerns related to (a) toxicity of nanoscale materials, (b) scanty nano-research with key crop nutrients, (c) inadequacy of soil- or field-based studies with NFs, (d) types of NM to produce fertilizers, (e) ways of effective NF application at the field scale, and (f) the economics of NFs are yet to be satisfied (Dimkpa and Bindraban 2018; Raliya et al. 2018).

On the whole, nanotechnology has made a positive contribution in agriculture sector by offering farmers better solutions of their problems, improving the crop production and ensuring ecological sustainability and economic stability. Here is an effort to evaluate, on the basis of information available, the role of NFs in meeting the current challenges in areas of crop production and protection.

19.2 Management of Crop's Nutritional Requirement

Soil is a natural body of finely divided rocks, minerals, and organic matter, in which sand, silt, clay, and organic matter facilitate the maintenance of necessary aeration and favorable water-intake rates. However, these components seldom have adequate nutrition to maintain soil fertility and sustain a continuous healthy plant growth. Loss of essential elements through leaching, volatilization, and/or erosion is a common phenomenon, leading to reduced soil fertility. Deficiency of macro- and micronutrients in the soil, topsoil erosion, depleted amount of water, sanitation problems, and industry effluent are some other factors that adversely affect arable lands (El-Ladan et al. 2014; IFA 2016). Sometimes, adequate nutrients are present in the soil, but these do not become bioavailable to plants due to inappropriate soil pH. The widespread nutrient deficiency in soils thus causes a decline in quantity and nutritional quality of grains for humans and livestock. The large-scale application of chemical fertilizers has increased crop production markedly but also disturbed the soil mineral balance and reduced the soil fertility, besides having adverse effects on food chains across ecosystems, often leading to heritable mutations in future generations of consumers (Solanki et al. 2015).

Eighteen elements are supposed to be essential for maintaining plant growth and development (Table 19.1). These elements are supplied in a readily available form for plant use by certain chemical materials called fertilizers. Of these required elements, three non-mineral nutrients (carbon, hydrogen, oxygen) are obtained by plants from air and water, whereas 15 mineral nutrients, including three primary nutrients or "macronutrients" (nitrogen, phosphorus, potassium), three "secondary nutrients" (calcium, magnesium, sulfur), and nine "micronutrients" (manganese, molybdenum, copper, zinc, iron, chlorine, boron, cobalt, nickel), come from the soil or added fertilizers (Savoy 1999; White and Brown 2010). Carbon dioxide (CO_2) enters the plant through small leaf openings called stomata, whereas water is taken from the soil system into the plant through absorption by roots. When chlorophyll (green pigments) of plants are exposed to light, carbon (C), hydrogen (H), and oxygen (O) obtained from air and water are combined in a process called photosynthesis, which results into the formation of carbohydrates (the food and the building material for plants) with a subsequent release of oxygen. The water and nutritional status of plants markedly affect the rate of photosynthesis (Iqbal et al. 2000; Kirschbaum 2011).

Limitation of nutrients results in poor growth of plants and often makes them susceptible to a variety of diseases and stresses. Soil productivity can be maintained by well-planned applications of multiple-element fertilizers, the primary function of which is to correct the deficiency of micro- and macronutrients of the soil (Umar

Table 19.1 Plant nutrients, their function in plant cells, and their form in which taken up by plants from the soil solution

Nutrient element	Function in plants	Form to be taken up by plants
Nitrogen	Promotes rapid growth, chlorophyll formation, and protein synthesis	Anion and cation
Phosphorus	Stimulates early root growth. Hastens maturity. Stimulates blooming and seed formation	Anion
Potassium	Increases resistance to drought and disease. Enhances stalk and straw strength. Improves quality of grain and seed	Cation
Calcium	Improves root formation, stiffness of straw, and vigor. Increases resistance to seedling diseases	Cation
Magnesium	Facilitates chlorophyll formation and phosphorus metabolism. Helps in uptake of other nutrients	Cation
Sulfur	Imparts dark green color; stimulates seed production; forms part of amino acids, vitamins	Anion
Boron	Aids carbohydrate transport and cell division	Anion
Copper	Forms part of enzymes, light reactions	Cation[a]
Iron	Facilitates chlorophyll formation	Cation[a]
Manganese	Oxidation-reduction reactions. Hastens seed germination and maturation	Cation[a]
Zinc	Stimulates auxins, enzymes	Cation[a]
Molybdenum	Aids nitrogen fixation and nitrate assimilation	Anion
Cobalt	Facilitates nitrogen fixation	Cation
Nickel	Facilitates grain filling, seed viability	Cation
Chlorine	Helps in water-use efficiency	Anion
Oxygen,	Component of most plant compounds	Obtained from air and water
Hydrogen,	-do-	
Carbon	-do-	

[a]Also available to plants in chelate form (a nutrient form having the essential nutrient linked to an organic compound so that it stays available for plant use within certain range of soil pH)

et al. 2011; IFA 2016). Spray of inorganic fertilizers on the leaves and soil is a better way of supplying N, P, and K in equal amounts. One reason of the low (about 30–40%) utilization of the applied fertilizer by plants is the chemical nature of the nutrient materials. For instance, plants can't use elemental nitrogen (N); they take up N only when it is in the NO_3 or NH_4 form. This means that each part of nitrogen has three associated parts of oxygen with (NO_3) or four parts of hydrogen with (NH_4). Thus, N forms only a small part of the bioavailable compounds. Similarly, phosphorus (P) is absorbed by plants as $H_2PO_4^-$, HPO_4^-, or PO_4^- depending upon the soil pH. Application of NFs increases the chances of bioavailability of nutrient elements as well as compact nutrients or vitamins and even hormones.

Recent reports indicate that the world human population, which is around seven billion now, will be almost ten billion in 2050 (Kah et al. 2018). In view of this tremendous increase in population and a consistent reduction of agricultural land, the pressure to extract more amount of food from less area of cultivable land is inevitable. Moreover, approximately 40% of the world's agricultural land has degraded seriously and lost its fertility due to several reasons including intensive farming practice (Kale and Gawade 2016). As a result, huge amounts of fertilizer need to be applied to improve soil fertility and crop production, because very little of the added nutrients reaches the target site due to loss caused by leaching, drift, runoff, hydrolysis, evaporation, and the photolytic or microbial degradation of the nutrients. The enormous fertilizer load, on the other hand, pollutes the air and contaminates rivers and water reservoirs (Sabir et al. 2014a; Solanki et al. 2015; Dubey and Mailapalli 2016). While only 27 kg NPK ha^{-1} was required to produce 1 ton of grain in the early 1970s, as much as 109 kg of NPK ha^{-1} was consumed to obtain the same amount of grain in 2008. As per the estimates of International Fertilizer Industry Association (IFIA), the world fertilizer consumption now hovers around 193 Mt (Manjunatha et al. 2016). Urea is a common nitrogen source in fertilizers, but it quickly breaks down into ammonia, which is rapidly flushed away by rainfall. To account for that loss, farmers need to apply extra fertilizer to crops, which increases the cost involved. Ammonia may also lead to harmful algal blooms in waterways and/or enter the atmosphere as nitrogen dioxide, a potent greenhouse gas.

Biofertilizers have also been tried in place of chemical fertilizers in order to avoid the hazards associated with the use of agrochemicals and improve the quality and quantity of crop yield. Biofertilizers are the natural carrier or liquid-based products containing living or dormant microbes (i.e., bacteria, fungi, algae, actinomycetes) alone or in combination, which (a) increase soil fertility by fixing atmospheric N or solubilizing different soil nutrients, (b) stimulate plant growth through synthesis of growth-promoting substances, and/or (c) activate the biological process to facilitate nutrients availability for plants (Simarmata et al. 2017; Dineshkumar et al. 2018). Different biofertilizers may be identified as N_2-fixing biofertilizers, P-solubilizing biofertilizers, P-mobilizing biofertilizers, plant growth-promoting rhizobacteria, and nutrient-carrying biofertilizers. They can improve soil health as well as the nutrient uptake efficiency, growth, and yield of plants (El-Ghamry et al. 2018).

Of late, smart delivery of fertilizers through nanoscale carriers has initiated the era of NFs, which turned out highly efficient and compatible to the environment. The use of NFs increases the bioavailability of poorly bioavailable nutrients (such as P and Zn) and reduces the loss of runaway nutrients (such as nitrate) to the surrounding environment, thus ensuring a full nutrient use efficiency of crops. It also excludes environmental deterioration, including the eutrophication (Mukherjee et al. 2016). Some NMs (such as TiO_2 or carbon nanotubes) also act as alternative growth promoters without any nutrients attached. Introduction of NF is likely to remove the macro- and micronutrient deficiency of the soil and help plants achieve their full nutrient use efficiency by regulating a sustained release of nutrients according to the need of the crop concerned (Prasad et al. 2014). Nano-fertilizers are produced mainly by encapsulating nutrients inside nanoporous materials or coating with thin polymer film or are delivered as NPs or emulsions as in the case of cationic (NH_4^+, K^+, Ca_2^+, Mg_2^+) nutrients or after surface modification as for anionic (NO_3^-,

PO_4^-, SO_4^-) nutrients (Subramanian et al. 2015; Panpatte et al. 2016). Based on their mode of action, NFs are identified as "controlled release fertilizers," "slow release fertilizers," "controlled-loss fertilizers," "magnetic fertilizers," or "nanocomposite fertilizers" (Lateef et al. 2016; Liu et al. 2016a; Chhowalla 2017). Besides, depending on the type of the nutrient and the role of the nanomaterial involved, three types of NF are broadly identified, i.e., (a) NFs made of macronutrients, (b) NFs made of micronutrients, and (c) NFs acting as carriers for macronutrients. In the first two cases, the NM itself is nutrient, while in the third case, it acts as an additive (Kah et al. 2018) and is often referred to as "nutrient-loaded nanofertilizer" or "NM-enhanced fertilizer" (Liu and Lal 2015). All these "plant growth-enhancing NMs" and "NMs acting as carriers for micronutrients" are either a nanoparticle or a nanostructured entity.

19.3 Nanomaterials Used for Improving Fertilizers

In general, all NFs provide a slow, steady, and time-dependent release of essential nutrients to ensure delivery of essential elements to the plant in a balanced and need-oriented form. The nanostructured nutrient carriers may be categorized as nanoclays, hydroxyapatite NPs, polymeric NPs, carbon-based NMs, mesoporous silica, and miscellaneous NMs.

19.3.1 Nanoclays

Nanoclays, which are defined as layered silicates with bidimensional platelets of nanoscale thickness (frequently ~ 1 nm) and a length of several micrometers, are the most frequently used nanocarriers of fertilizers and have the widest range of materials both in anionic and cationic forms. Anionic clays, also called layered double hydroxides (LDH), have a unique anion-exchange capacity and hence are highly suitable for transport of nitrate, phosphate, and borate (Everaert et al. 2016; Bernardo et al. 2018; Songkhum et al. 2018). Zeolites, montmorillonite, and kaolinite are among the most common cationic nutrient carriers. The quality of nanoclays as nutrient carriers is generally determined by (a) their ability to protect nutrient molecules through physical barriers provided by their structural components and (b) intercalation of nutrients into the layers of nanoclays through ion-exchange or non-electrostatic interactions such as hydrogen bonding (Guo et al. 2018).

Zeolites are natural porous crystalline aluminosilicates, composed of SiO_4 and AlO_4 tetrahedral linked by oxygen atoms. Each aluminum atom in the zeolite contributes one negative charge, which is balanced by an exchangeable cation (Ca^{2+}, Mg^{2+}, Na^+, K^+, etc.) located in the channels and cavities throughout the zeolite structure. These cations, bound to the aluminosilicate structure by weak electrostatic bonds, have a pivotal role in determining the adsorption capacity and thermal properties of the zeolite (Elizondo-Villarreal et al. 2016). This solid crystal with a relatively open, three-dimensional structure built from the combination of aluminum, oxygen, and silicon with alkali or alkaline-earth metals such as sodium, potassium, and magnesium, plus water molecules trapped in the gaps among these components, may act as a sustained release NF, ensuring that the whole amount of nutrient is taken up by plants due to continuous release in small portions. Made up of many different crystalline structures with large open pores arranged in a regular way, zeolite has a large surface area and can accommodate many organic molecules, which are delivered on receiving signal from the plant (Guo et al. 2018).

Clinoptilolite, one of the most commonly occurring natural minerals, is a weathering product of volcanic glass. It is microporous in structure and may have the surface area as large as 500 m^2 g^{-1}. It is a very good adsorbent for cations (notably ammonium) and non-charged but polar organic compounds and is known for high ion exchange and excellent molecular sieving (Elizondo-Villarreal et al. 2016). It belongs to a large zeolite family called heulandites. A single-phase clinoptilolite can be hydrothermally synthesized in an autoclave from various silica, alumina, and alkali sources with an initial Si/Al ratio from 3.0 to 5.0 at 120–195 °C. Its crystallization rate and crystallinity can be improved by seeding (Ambrozova et al. 2017). The clinoptilolite particles can be reduced to nanosize with the help of ball mill and modified chemically without disturbing its crystalline structure but incorporating adequate surface moieties for various applications (Elizondo-Villarreal et al. 2016). Bortolin et al. (2013) synthesized a novel series of hydrogels (gels in which the liquid component is water) consisting of polyacrylamide, methylcellulose, and calcic montmorillonite, which displayed a synergistic effect for controlled release of fertilizers, giving a very high fertilizer loading in their structure. The presence of montmorillonite in the hydrogel caused the system to liberate the nutrient in a more controlled manner than with the pure hydrogel in different pH ranges. The hydrolyzed hydrogels containing 50% calcic montmorillonite gave the best desorption performance, releasing larger amounts of nutrient at about 200 times slower rate than by the pure urea, i.e., without hydrogel.

19.3.2 Hydroxyapatite Nanoparticles

Hydroxyapatite [$(Ca_{10}(PO_4)_6(OH)_2$], also called "bone mineral," is a biocompatible material naturally present in human and animal hard tissues. It has a high surface area to volume ratio and holds potential to deliver both Ca and P. Kottegoda et al. (2011) used urea fertilizer modified with NPs of hydroxyapatite to achieve an extended slow release of N in three sets of soil having three different pH values. They could obtain a slow release of nitrogen from modified urea up to 60 days after an initial burst, while the commercial urea supplied nitrogen for 30 days only. The initial burst of N was inversely related to pH of the soil, i.e., the lower the pH, the greater was the initial burst of nitrogen. Also, Kottegoda et al. (2017) synthesized urea-laden hydroxyapatite nanohybrids as a fertilizer to enact slow release of nitrogen. These urea-laden HA nanohybrids displayed a much slower release profile (up to 1 week) than pure urea did (within minutes) and increased rice (*Oryza sativa*) yield by 8.2%, while having only half the N demand of pure urea. HA NPs combined with a secondary protective material to form a hybrid nutrient carrier gave even better results than the pure urea or urea-HA NPs (Giroto et al. 2017). However, no evidence was found that HA NPs were internalized by the plant (Guo et al. 2018).

19.3.3 Polymeric Nanoparticles

Polymeric materials used as fertilizer carriers are supposed to be biodegradable. Chitosan, as a natural and biodegradable biopolymer, exhibits sorbent and bactericidal properties, rendering it a potential agrochemical carrier. Suspensions of chitosan NPs containing N, P, and K fertilizers and urea-modified hydroxyapatite (HA) NPs have proved quite useful for gradual release of nutrients (Corradini et al. 2010; Kottegoda et al. 2011; Malebra and Cerana 2018). These NFs exhibit an initial burst followed by a gradual slow release of nutrient for a much longer period than commercial fertilizers. Foliar spray of NPK-loaded chitosan NPs developed through polymerization of methacrylic acid in a chitosan solution followed by further loading of NPK, accelerated growth, and yield of wheat (*Triticum aestivum*) and markedly raised the harvest index, crop index, and mobilization index of yield variables, in comparison to the conventional NPK fertilizer as well as the unfertilized control (Abdel-Aziz et al. 2016), though the release rates of NPK from the chitosan were not known. Moreover, it is not known if the enhancement was due to a controlled release of NPK from the chitosan NPs, to a direct NP internalization with subsequent nutrient release, or to both (Guo et al. 2018). Further, polymers that are not nanostructured can be used as bonding agents or as secondary protective layers for nano-enabled fertilizers (Roshanravan et al. 2014, 2015). Such polymers could enhance the mechanical strength of fertilizers and play a significant role in reducing emission of N_2O, one of the contributors to climate change (Kundu et al. 2016).

19.3.4 Carbon-Based Nanomaterials

Ashfaq et al. (2017) have shown that Cu nanoparticle (NPs)-loaded carbon nanofibers (CNFs) yielded slower release of Cu in water than Cu-loaded activated carbon microfibers (ACFs) did. These nanofibers enhanced the water-uptake capacity, seed germination rate, shoot and root lengths, and chlorophyll and protein contents of chickpea (*Cicer arietinum*). The authors suggested that a substrate or coating material such as a biodegradable polymer to encapsulate the NF would promote the nano-formulation stability. Kumar et al. (2018) investigated a combination of polymer film (PVAc starch) with carbon nanofibers as delivery vehicle for Cu-Zn nanoparticles and found that the polymeric formulation protected the Cu-Zn NPs from rapid release into the soil, showed scavenging effects on reactive oxygen species (ROS), and enhanced the growth of chickpea plants. Likewise, application of carbon NPs together with fertilizer increased grain yields of rice (10.29%), spring maize (10.93%), soybean (16.74%), winter wheat (28.81%) and vegetables (12.34–19.76%) (Liu et al. 2009). However, none of these studies could work out the mode of action (i.e., controlled nutrient release or plant uptake of nutrient-loaded nanocarriers) responsible for the enhanced crop productivity (Guo et al. 2018).

19.3.5 Mesoporous Silica

Formation of mesoporous silica (MS) involves more demanding synthesis methods. Wanyika et al. (2012) produced urea-loaded MS NPs, which had a high capacity to adsorb urea (up to 80% (w/w)) and yielded a slow release profile (fivefold increase in release period as compared with pure urea) into both water and soil. Earlier, Hossain et al. (2008) used MCM-41, another common type of mesoporous silica, as a support for urease, a nickel-based large metalloenzyme for urea hydrolysis. The pore-expanded MCM-41 silica had a higher adsorption capacity for urease (102 mg g^{-1}) than regular MCM-41 (56 mg g^{-1}) or silica gel adsorbent (SGA, 21 mg g^{-1}), causing a significantly slower rate of urea hydrolysis. The impact of these nanostructures on N-use efficiency and crop growth and productivity has not been evaluated and confirmed.

19.3.6 Miscellaneous Nanomaterials

Many other NMs have also been used as the fertilizer or fertilizer carrier. Experiment with a nanosized Mn carbonate hollow core-shell loaded with $ZnSO_4$ demonstrated that the release of Zn from this NM to a soil column (inceptisol) was slower than from conventional $ZnSO_4$ and in consonance with the plants' demand (Yuvaraj and Subramanian 2014). As plant roots absorbed the nutrients from growth media, the dissolution and ion-exchange reactions in the core-shell replenished the Zn in the soil solution to meet the nutritional requirement of the plant. This could improve the NUE, as the Zn release was extended to 29 days in comparison to 17 days with the conventional $ZnSO_4$. The core-shell-loaded Zn increased the rice grain yield by 36% and 27% under aerobic and submerged soil conditions, respectively, relative to the conventional $ZnSO_4$. It was, however, difficult to conclude whether the benefits were induced by the slow release of nutrients or by the nanocarriers themselves. Application of pine oleoresin (POR; 5%)-coated urea to soil (0.92 g N kg^{-1} soil) reduced N_2O emissions by 20.3%, whereas addition of 2% nanoscale zinc oxide (ZnO) and 35% nanoscale rock phosphate (RP) particles lowered the N_2O emission by 44.95% and 40.15%, respectively (Kundu et al. 2016). Application of nanocalcite $CaCO_3$ (40%) with nano-SiO_2 (4%), MgO (1%), and Fe_2O_3 (1%) markedly improved the uptake of not only Ca, Mg, and Fe but also of P, Zn, and Mn (Sabir et al. 2014b). Several metal- and metal-oxide NPs have the potential of acting as NF. Berahmand et al. (2012) reported enhanced yield of maize, using Ag NPs as NFs.

19.4 Intelligent Nano-fertilizers and Their Application

Nanostructured fertilizers are capable of improving the nutrient use efficiency through such mechanisms as targeted delivery and slow or controlled release of their active ingredients in response to environmental triggers and biological demands. They may increase N-use efficiency threefolds and enable plants to tolerate environmental stress. They improve crop productivity by promoting various biological phenomena through stimulation of seed germination, seedling growth, nitrogen metabolism, photosynthetic activity, protein synthesis, and antioxidant system (Siddiqi and Husen 2017; Sohair et al. 2018). Supplementations of nanoformulated or nano-entrapped micronutrients maintain the soil health and vigor and ultimately promote crop growth and productivity. Nano-fertilizers have been tried with several crops in different agroecosystems (soil and water), as enumerated by El-Ramady et al. (2018), and their toxicity analyzed in both soil microbiota and plants (Anjum et al. 2015). The toxicity level was found to be comparable with that of the conventional counterparts; e.g., CuO NPs were slightly more toxic than other Cu ions; ZnO NPs were similar to Zn ions, but manganese and iron oxide NPs caused less toxicity than their ionic counterparts and significantly enhanced the growth of lettuce seedlings (Liu et al. 2016b). NF application promoted growth, development, total phenolic content, and antioxidant activity in rice (Benzon et al. 2015). Carbon dots (CDs) of ~5 nm with different oxygen contents could penetrate into all parts of rice plants, including the cell nuclei, enhanced RuBisCO activity (by 42%), loosened the DNA structure, and increased the thionin (*Os06g32600*) gene expression, which resulted in enhanced potential of rice plant to resist diseases (Li et al. 2018). The CDs applied were degraded by the plant to form plant-hormone analogues and CO_2 and then the former promoted plant growth, while the latter got converted into carbohydrates through the Calvin cycle of photosynthesis. In consequence, the total rice yield was increased by 14.8% (Li et al. 2018).

In order to prevent their uncontrolled release in the environment, NFs are associated with such materials as hydrogels, films, or other biopolymers like chitosan, which aggregate the fertilizers in complexes with mineral NPs in the soil or other types of ceramic materials so that they respond to environmental stimuli (such as temperature or irradiance) by modifying the release of nutrients according to the plants' need (Morales-Díaz et al. 2017). Another way to control the NF supply is foliar spray, especially for elements with limited bioavailability in the soil such as Fe, Cu, and Ni (El-Kereti et al. 2013). Emulsions or encapsulated organic NPs can be useful for this purpose. However, the best way of reducing the release of NMs in the environment is perhaps to match their quantities with the stage of crop growth with maximal response. Application of small amounts of NMs to seeds at a pre-germinative stage (seed priming) proves highly beneficial, leading to a higher germination rate in senescent seeds or in seeds germinating in stressful environment (Nair et al. 2012). This is suggestive of NM application by seed priming to enhance stress tolerance and/or growth and productivity of crops (Azimi et al. 2013). An ideal NF must be responsive to any chemical or physical stimuli indicative of the plant's need of nutrition, such as rhizosphere acidification or ethylene production by roots under stress. However, these signals may sometimes be modified by the presence of NPs themselves. It is on record that Ag NPs interfered with the perception and synthesis of ethylene in *Arabidopsis*, thereby affecting other metabolic pathways also (Syu et al. 2014). Inhibition of ethylene synthesis may be accompanied by root elongation. Therefore, it is important to analyze whether such a growth enhancement is because of the improved nutritional status of the plant or due to blocking of perception and synthesis of ethylene under stress (Mirzajani et al. 2013).

While developing the nano-enabled fertilizers, evaluation system should include such criteria as release kinetics, crop productivity, nutrient use efficiency, environmental compatibility, and economic feasibility. Moreover, search for newer and safer nanocarriers, such as zein NPs made from a maize protein (Xu et al. 2011; Oliveira et al. 2018), should also continue. Multifunctional nanocarriers may also be developed; for instance, carbon nanotubes (CNTs) serving as plant growth regulators (Khodakovskaya et al. 2013) may be loaded with fertilizers to achieve plant growth regulation and nutrient delivery simultaneously. Some nanocomposites have been used as carriers for pesticides, fertilizers, and growth regulators, indicating that multiple objectives can be achieved through a single formulation (Guo et al. 2018). Efficacy of beneficial soil microbes with reference to crop plants can also be enhanced by co-application of NMs. For instance, integration of beneficial bacteria with nano-titania NPs was found to increase adherence of bacteria to plant roots, which improved crop growth and stress management (Palmqvist et al. 2015).

Various groups of NFs are identified on the basis of the nutrient they carry for utilization by plants, the major ones of which are described below.

19.4.1 Nitrogen-Based Nano-fertilizers

Enormous absorption of essential nutrients by plants, on one hand, and the natural phenomena of leaching, volatilization, and denitrification, on the other, often lead to depletion of soil nutrients. Such a situation necessitates an efficient availability of nitrogen fertilizers to the growing crops, as inorganic nitrogen is the most essential requirement of plants for their handsome growth and yield. This can be obtained from ammonia (NH_3) and its derivatives including ammonium nitrate ($N_2H_4O_3$), urea (CH_4N_2O), and diammonium phosphate [$(NH_4)_2HPO_4$] as well as the calcium products such as calcium nitrate [$Ca(NO_3)_2$] and calcium cyanamide ($CaCN_2$). N-use efficiency of plants becomes typically low when fertilizer release rate is higher than the rate of its absorption by plants and/or when fertilizers/nutrients are converted to forms that are not bioavailable to crops. A sustained N availability for a low but continuous N-feeding of crops, which is too difficult to manage manually, is now possible using a variety of N-based NFs. In order to raise the level of N-use efficiency of crops, which normally varies from 30% to 50%, modified N fertilizers, termed as "intelligent nano-fertilizers," have been put to use. This has raised the N-use efficiency of plants to 80% (Solanki et al. 2015).

One of the drawbacks of nitrogen NFs is their highly soluble nature, which may cause acute damage to crops and their surroundings. To overcome this problem, nano- and microporous zeolites are designed such that in combination with urea, their controlled release behavior ensures an improved uptake of urea N by plants. They help in retaining the maximum possible moisture to store soil nutrients and prevent their loss (Ramesh and Reddy 2011). The N-based NFs (e.g., porous nanomaterials such as zeolites, clay, or chitosan) could synchronize the release of fertilizer-N with its demand by the crop, thus preventing the undue loss of N and enhancing the plant uptake process (Abdel-Aziz et al. 2016; Panpatte et al. 2016). Similarly, ammonium-charged zeolites could improve the solubility of phosphate minerals, thus enhancing the availability of P for uptake by the crop (Dwivedi et al. 2016). Graphene oxide film (a carbon-based NP) could prolong the process of potassium nitrate release, thus extending the time of physiological action and avoiding the loss of nutrient by leaching (Zaytseva and Neumann 2016). Manikandan and Subramanian (2016) analyzed the impact of zeolite-based N fertilizers on maize plants grown in inceptisol (clay loam) and alfisol (sandy loam) soil textures. The nutrient uptake, plant growth and yield, and grain N content were consistently higher for treatments with nano-zeo-urea than with the conventional urea, the response being more pronounced in alfisol than in inceptisol soil. The mechanism of how the applied nanoclays modify soil properties and later affect plant growth is yet to be clearly understood.

19.4.2 Phosphorus-Based Nano-fertilizers

Phosphorus (P), the second most important nutrient requirement of plants after nitrogen, functions as an energy transfer molecule in plants and has a vital role in processes like respiration, photosynthesis, biosynthesis of nucleic acids, and energy generation. It is an integral part of several plant cell components such as phospholipids (Soliman et al. 2016). However, in most of the tropical agricultural soils, P is the least accessible macronutrient, meagerly available to plants for uptake mainly because of its fixation with Ca in alkaline soils and its poor recovery from fertilizers applied (Marschner 1995). Supply of P at an early stage of crop is essential for the development of reproductive structures. In general, the commercially available P fertilizers are water-soluble phosphate salts, which are easily dissolved in in the soil solution and are available to be taken up by plants. However, being highly mobile in the soil, they undergo leaching and runoff and give rise to eutrophication in water bodies. P fertilizers in solid form, on the other hand, are less effective in providing P to the plant. Allen et al. (1996) concluded from their experiments that mixtures of zeolite and phosphate rock had the potential to provide slow release fertilization to plants in synthetic soils by means of dissolution and ion-exchange reactions. Malhi et al. (2002) reported that zeolites (clinoptilolite), when saturated with monovalent nutrient cations, such as NH_4^+ and K^+, can increase the solubility of phosphate rock (PR). P-use efficiency of crops ranged from 18% to 20% during the year of application; the remaining 78–80% became part of the soil P pool to be released to the crop over the following months and years. Bansiwal et al. (2006) found that P supply from fertilizer-loaded surface-modified zeolite (SMZ) was available even after 1080 h of continuous percolation, whereas P from solid KH_2PO_4 was exhausted within 264 h. Thus, being a good sorbent for PO_4^{3-}, SMZ showed a great potential to act as a fertilizer carrier for slow release of P, which was duly confirmed by the subsequent research. Rahale (2011) studied the PO_4^- release pattern, using various nanoclays and zeolites in a percolation reactor. Nano-formulations could release phosphate for an extended period of 40–50 days, whereas the conventional fertilizer let out nutrients only up to 10–12 days. Liu and Lal (2014) noted that application of nanosized hydroxyapatite (nHA) increased the growth rate and seed yield of soybean plants by 32.6% and 20.4%, respectively, over those treated with a regular P fertilizers [$Ca(H_2PO_4)_2$]. Improvement in biomass production was 18.2% in shoots and 41.2% in roots. Preetha and Balakrishnan (2017) suggest that the use of SMZ could be a potential strategy to promote P-use efficiency, which hardly exceeds 20% in the conventional system. Benício et al. (2017) showed that layered double hydroxide (LDH)-P (15, 30, 45, and 60 mg kg^{-1}) increased soil pH in both sandy and clayey soils on harvesting maize (*Zea mays*) plants 25 days after sowing. They speculated that pH increase might facilitate the adsorption of P by plants. Hydroxyapatite particles were also used in agronomic applications for systematic release of P. Soliman et al. (2016) evaluated the effectiveness of foliar spray of different sources of P on *Adansonia digitata* and found that hydroxyapatite nanoparticles (nHA) caused a significant increase not only in plant growth parameters but also in chemical contents and the anticancer activity of leaves against Ehrlich ascites carcinoma cells (EACC), compared to controls. A group of Danish scientists led by Søren Husted is engaged in encapsulating P in biodegradable NPs that plants can absorb direct through leaves, avoiding the need to bind P to the soil before absorption. These researchers believe that plants may absorb P more effectively through their leaves than roots. Normally, farmlands are overfertilized with P, the excess of which accumulates in the soil without benefitting the crop. The objective of this research is to replace the 25 kg per hectare of P currently used on an annual basis, with merely 5–8 kg of P that will enter plants directly (Husted 2018).

Arbuscular mycorrhizal fungi have a crucial role in mobilization of soil P to roots (Das et al. 2013). *Piriformospora indica*, a newly discovered arbuscular mycorrhiza-like root-colonizing fungus, showed growth-promoting potential on being inoculated into *Bacopa monnieri* (Prasad et al. 2013). Rane et al. (2015) assessed the effect of calcium phosphate (CaP) NPs on *Zea mays* inoculated with *P. indica* and *Glomus mosseae* and found that CaP NPs in combination with both *G.*

mosseae and *P. indica* together were more potent plant growth promoter than alone or in combination with any of these fungi individually. The CaP NPs alone or in combination with *P. indica* could improve chlorophyll *a* content and performance index of the treated maize plants. The authors concluded that CaP NPs exhibited synergistic growth-promoting effect with the endosymbiotic and arbuscular mycorrhizal fungi used.

19.4.3 Potassium-Based Nano-fertilizers

While potassium (K) is not a constituent of any plant structure or compound, it has an important regulatory role in nearly all processes needed to sustain plant growth and reproduction, e.g., in photosynthesis, protein synthesis, photosynthate translocation, ionic balance, stomatal function, water-use efficiency, and activation of more than 60 enzymes. Plants with sufficient K are more resistant to drought, to flood, and to high and low temperatures. Umar et al. (2003) asserted that a proper K fertilization can even abolish the need of pesticide application on crops. Some natural zeolites contain considerable amounts of exchangeable K^+ that can enhance plant growth (Mazur et al. 1986). Natural zeolites are highly selective for K^+ than for Na^+ or divalent cations, such as calcium and magnesium, due to location and density of negative charge in the structure and dimensions of interior channels (Ming and Mumpton 1989). Potassium has a very high ion-exchange capacity of 216 cmol kg^{-1} (Dakovic et al. 2007) and hence is easily released from the crystal zeolite structure into the soil solution, eventually increasing its total content in the soil. Zeolites dominated by exchangeable K may be well-suited for plant growth-promoting applications. Zhou and Huang (2007) reported a slow and steady release of K from nano-zeolites possibly due to their ion exchangeability with selected nutrient cations. Li et al. (2010) used K^+-loaded zeolite as a slow release fertilizer and studied the growth features of hot pepper and the changes in the N and K contents of the soil. Subbarao et al. (2013) studied the dissolution rate of water-soluble polymer-coated potash prepared in the form of a cylindrical pellet and found that the lower the quantity of fertilizer and water, the slower was the release of K. The strength of the pellet had a vital role in determining the replenishment time.

Plants use the mineral nutrients of NFs either (a) as nanostructured elements incorporated in a carrier complex that may or may not be a nanomaterial, e.g., NPs of essential elements incorporated by absorption or adsorption in a matrix such as chitosan, polyacrylic acid, clay, or zeolite or (b) as the element per se in a nanostructured form (in suspension or encapsulated), such as NPs of Fe or Zn (Ghahremani et al. 2014). Treatments with 6/1000 nano-K concentration were found most effective in increasing the leaf area, harvest index, grain yield, biological yield, potassium percentage, and chlorophyll content in *Ocimum basilicum* (Ghahremani et al. 2014). In order to achieve a slow release of fertilizer, Khalifa and Hasaneen (2018) obtained chitosan (CS) NPs by polymerizing methacrylic acid (PMAA) for the entrapment of N, P, and K NPs and evaluated the impact of this CS-PMAA-NPK

NPs complex on garden pea (*Pisum sativum* var. Master B) plants by treating 5-day-old seedlings through root system. They noted a reduction in root-elongation rate and in starch accumulation at the root tip in a dose-dependent manner. Low concentrations induced mitotic cell division and upregulated some major proteins such as convicilin, vicilin, and legumin β. However, all the concentrations used exhibited genotoxic effect on DNA, based on the comet assay data 48 h after treatment. Foliar spray of different concentrations of nano-K fertilizer on *Cucurbita pepo* caused a significant increase in the number of leaves, fresh and dry weights, and product quality, by improving the nutrient absorption. Disease and pest resistance as well as drought tolerance were also enhanced (Gerdini 2016).

19.4.4 Zinc-Based Nano-fertilizers

Zinc deficiency is considered to be a major factor in limiting agricultural productivity in the alkaline soils (Sadeghzadeh 2013). Normally, Zn-use efficiency does not exceed 2–3%, and the major portion of added Zn gets fixed in the soil. However, the zinc-based NFs have shown a great promise (Wang et al. 2016b). Biogenic synthesis of zinc oxide (ZnO) NPs, by using different plant extracts (Sabir et al. 2014b; Chap. 7 of this book), and their application in agriculture are quite common these days. Of the various methods to synthesize ZnO nanocrystals, solution-evaporation method is more popular, as it may be carried out at ambient temperature, with a good control over particle morphology. ZnO NPs may be applied to crop plants by foliar spray, soil mixing method, or seed priming method, the last one being more effective, simple, and cost-effective (Narendhran et al. 2016; Sharifi et al. 2016; Khanm et al. 2018; Munir et al. 2018). Application of low concentrations ($\leq$100 mg kg^{-1}) of ZnO NPs to the soil increased Zn uptake by cucumber plant in comparison to the application of their bulk counterparts, but higher concentrations (1000 mg kg^{-1}) inhibited plant growth (Moghaddasi et al. 2017). The natural zeolite (clinoptilolite), ball milled to achieve nano-dimension (90–110 nm) and fortified with Zn by loading zinc sulfate ($ZnSO_4$), was found to prolong the duration of Zn release from the substrate into the soil solution 5.44 times, compared to the Zn release from ordinary $ZnSO_4$ (Yuvaraj and Subramanian 2017), serving therefore as a slow release Zn fertilizer to improve the nutrient use efficiency of crops.

Application of nano-zinc oxide to Zn-deficient soil improved the overall growth and the fresh and dry weights of sesame seedlings (Narendhran et al. 2016). In similar conditions, application of ZnO together with other fertilizer increased barley yield up to 91%, while the traditional bulk $ZnSO_4$ caused 31% increase over the control (Kale and Gawade 2016). Tarafdar et al. (2014) biosynthesized and characterized zinc NPs (15 and 25 nm) and used them as NF to enhance production of pearl millet (*Pennisetum americanum* L.) cv. HHB 67. A significant improvement in shoot length (15.1%), root length (4.2%), root area (24.2%), chlorophyll content (24.4%), total soluble leaf protein (38.7%), plant dry biomass (12.5%), and enzyme activities of acid phosphatase (76.9%), alkaline phosphatase (61.7%), phytase

(322.2%), and dehydrogenase (21%) was observed over the control in 6-week-old plants. The grain yield at crop maturity was found to increase by 37.7%. Davarpanah et al. (2016) tried foliar application of zinc (Zn) and boron (B) NFs on pomegranate (*Punica granatum* cv. Ardestani) before full bloom. It increased the leaf concentrations of both microelements and the pomegranate fruit yield. Fertilization with higher doses led to significant improvements in fruit quality, including 4.4–7.6% increase in total soluble solids (TSS), 9.5–29.1% decline in titratable acidity (TA), 20.6–46.1% increase in maturity index (TSS:TA ratio), and 0.28–0.62 pH unit increase in juice pH, whereas the physical fruit characteristics remained unaffected. Changes in total sugars and total phenolic compounds in the juice were minor, while the total anthocyanins and antioxidant activity stayed unaltered (Davarpanah et al. 2016).

Dapkekar et al. (2018) used zinc-complexed chitosan nanoparticles (Zn-C NPs) for biofortification of durum wheat in field-scale experiments. They analyzed the efficacy of Zn-C NPs in comparison to conventional $ZnSO_4$ (0.2%; 400 mg L^{-1} zinc) fertilizer in MACS-3125 and UC-1114 genotypes of durum wheat. Grain zinc enrichment on using Zn-CNP nanocarrier (~36%) and conventional $ZnSO_4$ (~50%) was comparable, although the zinc used in the former case was ten times less. By all accounts, NF application proved effective in enhancing grain zinc content (without affecting grain yield, protein content, spikelets per spike, kernel weight, etc.) in 4-year field trials conducted on plots differing in the soil zinc content, thus confirming the usefulness of Zn-CNP in ferti-fortification of wheat crop (Dapkekar et al. 2018). However, while showing that zinc in the form of NPs (nano ZnO) gets into roots of sunflower (*Helianthus annuus*) more effectively than the common zinc source ($ZnSO_4 \cdot 7H_2O$), Sturikova et al. (2017) have asserted that it is important to deal with the toxicity of zinc NPs, because zinc in nano-form is often more toxic to plants than its equivalent in the form of simple inorganic salts.

19.5 Mechanism of Nano-fertilizer Uptake and Translocation in Plants

The entrapped fertilizer nutrients enter the soil network via hydrogen bonds, surface tension, molecular force, or viscous force, thus enlarging their spatial scale so that they are easily prevented from soil filtration and remain fixed in the soil around the crop roots. This also reduces the migration rate of N elements in the environment (Cai et al. 2014). Liu et al. (2016a) observed that application of controlled-loss fertilizer reduced the nitrogen runoff and leaching loss by 21.6% and 24.5%, respectively, and caused a 9.8% increase in the residual mineral N of the soil, ultimately leading to 5.5% more wheat production than the production with traditional fertilizers.

Plant cell walls in the surface cell layer, which act as a barrier between inner tissues and outer environment, are porous and let the nutrients enter the plant cell. The

NF particles enter the plant tissue either through roots or the aboveground parts including root junctions and wounds. They first interact with plant cell wall, which is made up basically of a cellulosic framework that permits the entry of small particles, restricting the larger ones. The size exclusion limit for the plant cell wall is between 5 and 20 nm (Dietz and Herth 2011). Functionalized NPs promote enlargement of pore size or induction of new pores in the cell wall, thus facilitating the entry of large NPs (Kurepa et al. 2010). NPs may also enter into the cell via ion channels or endocytosis, binding to carrier proteins through aquaporin or forming complexes with membrane transporters or root exudates (Nair et al. 2010). Kurepa et al. (2010) noted that mucilage released by *Arabidopsis thaliana*roots develops pectin hydrogel complex around the roots, which paves the way for the entry of NP-dye complex. The NF particles absorbed by the root move through the apoplastic and symplastic pathways to reach the xylem and then spread to different plant parts through the vascular bundles, as has been reported for the mesoporous silica NPs and SiO_2 NPs (Sun et al. 2014). However, ZnO NPs, which follow the same transport pattern, hardly move beyond the endodermis (Ma et al. 2010; Morales-Díaz et al. 2017). Having entered into the plant tissue, NPs can move from one cell to the other through plasmodesmata (Rico et al. 2011). Figure 19.1 explains the process of uptake and transport of nutrients in plant tissues. Some authors consider the cytoplasmic symplastic pathway as a better option for their entry into plants, because NPs facilitate a rapid cellular internalization and movement through aquaporins and ion channels, compared to their bulk counterpart (Hemraj 2017). While studying the uptake and accumulation of ZnO NPs in *Glycine max* seedlings

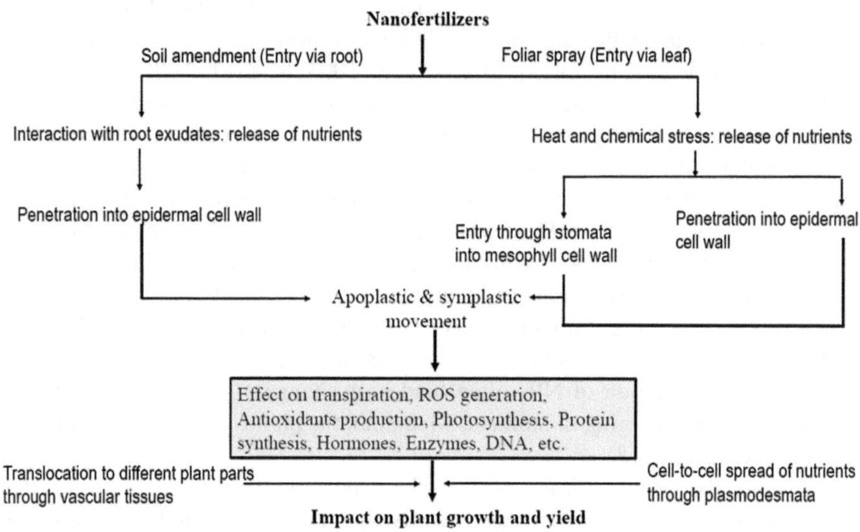

Fig. 19.1 Mechanism of action of the nutrients supplied through nano-fertilizer. Entry may occur through root and/or leaf tissues, and translocation takes place both by apoplastic and symplastic routes. The nutrients influence various cell organelles and their functions, which have a direct bearing on growth and productivity of the plant

raised from seeds treated with different concentrations (500–4000 ppm) of ZnO NPs, Lopez-Moreno et al. (2010) observed a higher Zn uptake at low concentrations. They concluded that NPs are agglomerated at high concentrations, which inhibit their entry into the seeds via cell wall pores. Further, the presence of Zn_2^+ ions, instead of Zno, suggested a role of roots in ZnO ionization on their surface. Wong et al. (2016) proposed a scheme, based on lipid exchange mechanism, for NP transport within the plant cells, showing that size, magnitude, and zeta potential of NPs primarily determine the efficiency of the process.

Nanomaterials may affect plant metabolism by providing micronutrients (Liu and Lal 2015), regulating genes (Nair and Chung 2014), or interfering with different oxidative processes in plants (Hossain et al. 2015). If accumulated in the cell in heavy concentration, NMs can interfere with electron transport chain of mitochondria and chloroplasts, affecting the formation of ROS (Rastogi et al. 2017), which may influence protein modifications, lipid peroxidation, and DNA health (Van Breusegem and Dat 2006; Chap. 17 of this book). Whether the ROS have a destructive or a signaling role depends on the equilibrium between ROS production and their scavenging. In order to combat the effect of ROS, cells have a strong antioxidant mechanism comprising of enzymatic (superoxide dismutase, catalase, guaiacol peroxidase, etc.) and nonenzymatic (ascorbate, glutathione, carotenoids, tocopherols, and phenolics) molecules (Sharma et al. 2012; Aref et al. 2016), and the NMs often promote the production of antioxidant molecules in plant cells (Faisal et al. 2013; Jiang et al. 2014; Costa and Sharma 2016; Chap. 18 of this book). Different hormonal pathways may also be upregulated or downregulated in response to different NMs (Rastogi et al. 2017). Studies have shown that NMs also influence the photosynthetic pigment and activity in plants (Hong et al. 2005; Perreault et al. 2014; Tripathi et al. 2017), and an altered photosynthesis then affects all aspects of plant metabolism and physiology (Ahmad et al. 2003; Ruhil et al. 2015).

Plant roots and leaf surfaces are highly porous on nanometer scale and constitute the main nutrient gateways for NMs entry into plants (Eichert and Goldbach 2008). The NF uptake through these pores could be facilitated by their complexation with molecular transporters or root exudates, creation of new pores, or exploitation of endocytosis or ion channels (Mastronardi et al. 2015). Size reduction of NMs especially helps increasing the surface-mass ratio of particles, which facilitates adsorption of many more nutrient ions and then their slow and steady desorption over an extended period of time (Subramanian et al. 2015; Monreal et al. 2016). However, if NPs accumulate at stomata-bearing surfaces of leaves, foliar heat is produced. Moreover, accumulation of NPs often hinders exchange of gases between leaves and the air due to clogging of stomata, which inhibits photosynthesis and transpiration (Gruyer et al. 2013).

NPs may also enter plant leaves through stomata quite rapidly. It is known that magnetic NPs penetrate more conveniently and are translocated to xylem and phloem more efficiently (Wang et al. 2012b). Wang et al. (2013) examined the leaf-to-root translocation of titanium (Ti), magnesium (Mg), and zinc (Zn) NPs and found a higher NP concentration in leaves (1.87, 8.13, and 5.74%, respectively) than in roots (5.45, 21.2, and 13.9%, respectively). They noted that aerosolized NPs of smaller diameter, such as TiO_2 (27 nm), MgO (35 nm), and ZnO (45 nm), have an easy passage through stomata, but NPs of large diameter obstruct the stomatal functioning, thus inhibiting transpiration and photosynthesis. This obviously indicates a shortcoming of the foliar application of NFs. However, it is argued that the foliar application is important for protection from infectious agents (Gogos et al. 2012; Giannousi et al. 2013). Further, the NP uptake is affected by the type of growth medium, e.g., it was higher in hydroponic medium, relatively low in sand medium, and nil in the soil, as recorded for roots and shoots of *Cucurbita maxima* (Zhu et al. 2008).

19.6 Nano-biosensors in Agriculture

Nanosensors (or nano-biosensors) have come up as promising tools in agriculture and food production sectors for determining microbes, contaminants, pollutants, toxins, moisture content, and food freshness. These are extremely small devices that can bind to entities to be detected and send back a signal. Compared with the traditional sensors, nanosensors are superior in having high sensitivity and selectivity, near real-time detection, low cost, and easy portability (Lu and Bowles 2013). These devices can be developed by using top-down lithography, molecular self-assembly, and bottom-up assembly approaches and may occur in the form of NPs, nanostructured materials (e.g., porous silicon), nanoprobes, nanowires, cantilevers, and nanoelectromechanical systems (NEMS). Nanosensors are normally categorized as NP-based nanosensors, electrochemical nanosensors, and optical nanosensors (Omanović-Mikličanin and Maksimović 2016). Details about the nano-biosensors employed for monitoring soil conditions and status of plant growth hormones and for detecting plant pathogens and pesticide residue have been summarized by Kaushal and Wani (2017).

Although nitrogen is an essential and basic requirement of plants, the unusual and excess uptake of N becomes harmful and disturbs the ideal crop production. There are sensors that may detect and accurately calculate the amounts of N present in the soil and/or taken up by the plant. Certain biosensors can also monitor the presence of biological substances in the field. They detect the excess of chemicals or the toxicity level in the soil, the variation factors such as pH level, development of herbs, the presence of moisture, and the growth condition of different plant organs (Kaushal and Wani 2017). Fertilizers can be coated by a polymer containing nano-biosensors that ensure a time-bound release of nutrients in exact quantities. In the case of N depletion, these nano-biosensors receive signals in the form of chemicals released by the plant roots and manage to release specified amount of N (Khodakovskaya et al. 2012).

Aptamers are used extensively as recognition elements in the fabrication of apta-integrated sensors, which consist of aptamers (the target-recognition element) and nanomaterial (the signal transducers and/or signal enhancers). Aptamers of a natural or synthetic origin are single-stranded nucleic acid or peptide molecules of a size less than 25 kDa. They are highly specific and selective toward their target entities (ions, proteins, toxins, microbes, viruses) due to their precise and well-defined three-dimensional structures. These are also termed as synthetic antibodies due to their selection and generation through an in vitro combinatorial molecular process called systematic evolution of ligands by exponential enrichment (SELEX). Dissociation constants of aptamers are in nanomolar or picomolar range. Based on the detection systems, aptamers are classified into optical and electrochemical systems. A variety of nanomaterials (metal NPs and nanoclusters, semiconductor NPs, carbon NPs, magnetic NPs, etc.) can be used in apta-sensors (Sharma et al. 2015). Without disturbing the cell functioning, these sensors help monitoring the signaling pathways and are used specifically for detecting plant pathology. Various transducing systems have been employed in apta-sensors for food quality assessment and safety. These devices, processed with photoluminescence, monitor the toxicity level in the foodstuff efficiently (Yasmin et al. 2016).

Carbon nanotubes (CNTs), with their typical construction, have proved useful in improving water utilization, nutrient uptake, seed germination, and the rate of plant growth (Husen and Siddiqi 2014). CNTs are hollow cylinders formed of carbon atoms. These appear like coiled tube graphite ducts, and their walls look like hexadic carbon rings. These are produced normally in huge bundles and categorized as single-walled carbon nanotube (SWCNT) and multi-walled carbon nanotube (MWCNT), the former consisting of a sole, cylindrical graphene coat, whereas the latter possessing multiple graphene coats. Having a high tensile strength, CNTs are considered to be 100 times tougher than steel and are good conductors of electricity. CNT-based nanodevices are popular because of their good electrical conductivity, elevated electrochemical catalytic efficiency, bio-suitability, and non-toxicity (Kaushal and Wani 2017). Very strong van der Waals forces bind carbon atoms in their specified position due to which these tubes are especially useful for monitoring of microenvironments, energy storage, fertilizers packaging, and also in crop improvement (Tripathi and Chauhan 2017).

Rodrigues et al. (2013) determined the impact of carboxyl-functionalized single-walled carbon nanotubes (SWCNTs) on fungal and bacterial communities in the soil through culture-dependent and -independent methods. The bacterial soil community was transiently affected by the presence of SWCNTs. The major impact figured 3 days after exposure, but the community recovered fully after 14 days. However, fungal community could not recover during the period of experiment. Physiological and DNA analyses suggested that fungi and bacteria involved in biogeochemical cycles of C and P nutrients could suffer by the presence of SWCNTs. In a similar study by Shrestha et al. (2013), application of 10, 100, and 1000 mg kg^{-1} concentrations of multi-walled carbon nanotubes (MWCNTs) of high purity caused no change in soil respiration, enzymatic activity, and microbial community composition, but the highest concentration (10,000 mg kg^{-1}) increased the fungal fatty acid methyl ester markers. Further, pyrosequencing revealed a decline in abundance of some bacterial genera like *Derxia*, *Holophaga*, *Opitutus*, and *Waddlia* at the highest treatment, while other genera like *Cellulomonas*, *Nocardioides*, *Pseudomonas*, and *Rhodococcus*, regarded as potential degraders of recalcitrant contaminants such as polycyclic aromatic hydrocarbons, increased in population. This was indicative of a shift in the soil microbial community composition to more tolerant microbial populations due to very high MWCNT concentration (Shrestha et al. 2013). Oxidized MWCNTs have been found effective in increasing root cell elongation and facilitating root growth and biomass production by augmenting dehydrogenase activity in plants such as wheat (Wang et al. 2012a; Arif et al. 2016). Tomato production was remarkably enhanced when treated with fullerene (a spherical pure carbon molecule having a hollow network of at least 60 carbon atoms) (Khodakovskaya et al. 2013).

Quantum dots (QD) are very small semiconductor particles, only several nanometers in size, having optical and electronic properties different from those of larger particles. Spherical in shape and 2–8 nm in diameter, QDs fluoresce on being stimulated by an excitation light source. These are composed of a core, which decides the color to be emitted, and a shell, both inorganic. Further, there is an aqueous organic film where biomolecules get joined to target different biomarkers having size-dependent fluorescent properties (Patolsky et al. 2006). A single wavelength can excite QDs of different sizes, and emissions governed at varying dots are utilized as high-resolution biological fluorescent probes. QDs are being used in agriculture for sensing several chemical/biochemical compounds outside and within plants (Kaushal and Wani 2017). In a study of the impact of QDs and superparamagnetic NPs on *Fusarium oxysporum*, a fungal plant pathogen, Rispail et al. (2014) observed that both these nanomaterials rapidly interacted with the fungal hypha, labeling the presence of the pathogenic fungus, and caused low toxicity to the fungal germination, growth, and viability. Whereas magnetic NPs appeared to be on the cell surface, QDs were significantly taken up by the fungal hyphae showing their potential for a possible control of the pathogen following appropriate functionalization (Rispail et al. 2014).

19.7 Toxicity Assessment of Nano-fertilizers

Based on the information available, we often conclude that the use of NFs has several positive effects on plant metabolism, growth, and yield. However, it may also have adverse effects causing a slower plant growth, increased oxidative stress (Dimkpa et al. 2013), chromosomal abnormalities (Raskar and Laware 2014), decreased photosynthetic rate (Wang et al. 2016a), disturbed water transport and water status of the plant (Martínez-Fernández et al. 2016), lower concentration of growth hormones (Le et al. 2014), metabolic disorders or necrosis (Lin and Xing 2008), changes in the transcriptional profile of many genes (Khodakovskaya et al. 2012; García-Sánchez et al. 2015; Van Aken 2015), or increased susceptibility to natural toxins such as As (Hu et al. 2014). Phytotoxicity issues of metal/metal-oxide NPs were examined by Ruttkay-Nedecky et al. (2017), who concluded that while assessing the NM phytotoxicity, we must build a connection between the characteristics of NPs applied (surface area, particle size, surface tension) and the toxicity symptoms and also keep it in view that our studies conducted in a hydroponic system and under controlled lab conditions do not reflect the interaction of NMs with soil and soil microorganisms. However, they opine that, on the whole, NPs of essential metals and metal oxides favor crop production and that the NPs of iron oxides and manganese oxides seem to be the least phytotoxic (Ruttkay-Nedecky et al. 2017).

Plant responses to nanomaterials vary depending on the chemical elements that constitute the respective NMs, their concentration and aggregation state, duration of exposure, metabolic potential of plant species, and the local environmental conditions that often extensively modify the NP properties (Pan and Xing 2012; Mrakovcic et al. 2013; Schultz et al. 2015). The surface charge of NPs, the formation of ions from NMs, and the agglomeration or aggregation of NPs are modified by the presence of dissolved organic material that forms organic coatings on the NPs (El-Badawy et al. 2011; Romih et al. 2015). The organic matter also interacts with environmental factors such as irradiance, UV radiation, temperature, the presence of dissolved elements, and biotic activity, which in turn affect their reactivity with NMs (Glenn and Klaine 2013; Rodrigues et al. 2016). The pH and presence of complexing agents such as citrate or ascorbate in water or soil solution and the bacterial or root exopolysaccharides may significantly modify NP toxicity by way of increasing the stability of NPs and limiting their aggregation, which would enhance their toxicity. Depending on the experimental conditions or the habitat environment, the same NMs may behave differently in terms of their positive or negative impact on plants (El-Temsah and Joner 2012; Parveen and Rao 2015). Their impact may also differ on different plant species that differ in their capacity of absorbing the NMs or their ability to metabolize these materials (Morales-Díaz et al. 2017). Given these complex interactions and variation in conditions, it is not easy to predict the environmental fate of NMs in the long term. Further, NMs may cause toxicity to plants either directly via the absorption or adsorption of NPs by the plant or indirectly through release of ions due to decomposition of NMs or by inducing amplified responses of other environmental toxins (Aruoja et al. 2015; Wang et al. 2016a; Massalha et al. 2017). There is, however, paucity of information on interactions between NPs and the organic/inorganic abiotic soil components.

The ethical and safety issues regarding the use of NPs in crop production are very many (Solanki et al. 2015), and exposure to nano-fertilizers and nanopesticides can contribute to health hazards (Berekaa 2015). It needs to be investigated deeply whether NFs are fully transformed into ionic forms in the plant and later incorporated into proteins and different metabolites or some residue remains intact and is transferred to consumers through food chain. The NFs in crop fields may be transferred to the soil, water, and air by contaminated leachate, runoff due to rain, transport by wind or trophic transfer through harvested organs or agricultural waste, or by some other means. NPs are also generated extrasomatically by natural processes including volcanism and meteoric dust, weathering, and nucleation and crystallization of minerals and by the action of microbes or organic matter in soil and water (Nowack and Bucheli 2007; Hochella et al. 2008). Given this, natural soil may have NPs of clay, organic matter, and Fe oxides, but their concentration in the ecosystem remains insignificant (Vittori et al. 2015). Only the occasional catastrophic events such as volcanic eruptions and perhaps some interactive happenings in the space may result in sudden bulk production of NMs (Gabrielli et al. 2004, Tepe and Bau 2014). The picture of the postharvest phase of the crop also remains blurred, and we are little informed about the fate of NFs in terms of their transfer to the environment, to different trophic levels, and to the end consumers of the crop, although some studies have suggested their transfer to different trophic levels (Zhu et al. 2010; Werlin et al. 2011; Bielmyer-Fraser et al. 2014; De la Torre et al. 2015; Kim et al. 2016).

Another apprehension regarding our understanding of the NM-plant interaction is that our conclusions, often drawn from experiments conducted under controlled laboratory conditions normally with low irradiance, temperature control, and absence of stresses, may not fit accurately to the natural conditions of open agricultural fields. Plant response to nano-enabled fertilizers under the harsh or less favorable field conditions may be markedly different depending on the habitat environment, which is modulated often by the plant's own concentration of NPs, dynamic equilibrium based on release and aggregation of ions, degree of irradiance, pH level, temperature range, the presence of other metal ions, oxidation-reduction potential (ORP), and the amount of organic material present in the soil (Cheloni et al. 2016; Mashock et al. 2016; Read et al. 2016). Extreme care must, therefore, be taken before concluding about the adverse, neutral, or beneficial effects of NMs at the ecosystem scale. Although several NFs are now freely available in the market (Table 19.2), major chemical companies still seem to be cautious in launching the NF production (Prasad et al. 2017).

Table 19.2 Some common commercial products of nano-fertilizers

Commercial product	Composition	Manufacturing company
Nano-Gro™	Plant growth regulator and immunity enhancer	Agro Nanotechnology Corp., FL, United States
Nano Green	Extracts of corn, grain, soybeans, potatoes, coconut, and palm	Nano Green Sciences, Inc., India
Nano-Ag Answer®	Microorganisms, sea kelp, and mineral electrolyte	Urth Agriculture, CA, United States
Biozar Nano-Fertilizer	Combination of organic materials, micronutrients, and macromolecules	Fanavar Nano-Pazhoohesh Markazi Company, Iran
Nano Max NPK Fertilizer	Multiple organic acids chelated with major nutrients, amino acids, organic carbon, organic micronutrients/trace elements, vitamins, and probiotic	JU Agri Sciences Pvt. Ltd, New Delhi, India
Master Nano Chitosan Organic Fertilizer	Water-soluble liquid chitosan, organic acid and salicylic acids, phenolic compounds	Pannaraj Intertrade, Thailand
TAG NANO (NPK, phos, zinc, cal, etc.) fertilizers	Proteino-lacto-gluconate chelated with micronutrients, vitamins, probiotics, seaweed extracts, humic acid	Tropical Agrosystem India (P) Ltd, India

After Prasad et al. (2017)

19.8 Conclusions

Having gone through the relevant literature, as cited in this chapter too, we have reached the conclusions that:

(a) Most of the studies hitherto published have compared the nutrient-release rate of the nano-enabled fertilizers with that of conventional fertilizers but ignored the release mechanisms and kinetic models, which are important for determining the dose and frequency of nutrient supply to meet the crop demand.
(b) The major objective of developing nano-enabled fertilizers is to improve NUE and crop productivity, but most of the published reports have analyzed merely the properties of the nano-fertilizers, whereas investigations on NUE and crop productivity have been relative few and that too considering diverse criteria for assessing these ultimate parameters, making the comparison of the efficacy of different formulations too difficult.
(c) Environmental compatibility and economic feasibility of the nanostructured material should be given due attention. These aspects need to be studied even for naturally occurring materials such as chitosan and nanoclays.
(d) Assessment of the potential negative effects of nanocarrier design and nutrient encapsulation on crops, consumers, and the environment, though highly essential, has been unduly ignored in the case of NFs unlike the case of nanopesticides.
(e) Phosphorus concentration can be increased by using the phosphate-solubilizing bacteria. On receiving stimuli from the unique signal molecules released by plants under nutrient deficiency or stress, stimuli-responsive nanocarriers can possibly provide more active and intelligent delivery systems than the normally employed diffusion and/or ion-exchange pathways. Release of P can perhaps be harmonized with the need of crops by designing nanocarriers that may be degraded or dissolved by P deficiency-related enzymes or organic acids.
(f) NFs are prone to be leached away by irrigation water and can enter direct into the food chain when taken up by plants. Therefore, limits of concentration of NPs need to be defined. Moreover, stability of NPs in food, their physicochemical properties, and toxicokinetics in the consumer's body also merit special attention.

References

Abdel-Aziz HM, Hasaneen MN, Omer AM (2016) Nano chitosan-NPK fertilizer enhances the growth and productivity of wheat plants grown in sandy soil. Spanish J Agri Res 14:0902

Ahmad A, Abrol YP, Iqbal M (2003) Photosynthetic nitrogen use efficiency under variable environments. In: Pant RC, Ghildiyal MC (eds) Sustainable plant productivity under changing environment, Souvenir, 2nd international congress of plant physiology. IARI, New Delhi, pp 59–66

Allen ER, Hossner LR, Ming DW, Henninger DL (1996) Release rates of phosphorus, ammonium and potassium in clinoptilolite-phosphate rock systems. Soil Sci Soc Amer J 60:1467–1472

Ambrozova P, Kynicky J, Urubek T, Nguyen VD (2017) Synthesis and modification of clinoptilolite. Molecules 22:1107

Anjum NA, Adam V, Kizek R, Duarte AC, Pereira E, Iqbal M, Lukatkin AS, Ahmad I (2015) Nanoscale copper in the soil-plant system – toxicity and underlying potential mechanisms. Environ Res 138:306–325

Aref IM, Khan PR, Khan S, El-Atta H, Ahmed AI, Iqbal M (2016) Modulation of antioxidant enzymes in *Juniperus procera* needles in relation to habitat environment and dieback incidence. Trees Struc Func 30:1669–1681

Arif N, Yadav V, Singh S, Singh S, Mishra RK, Sharma S, Dubey NK, Tripathi DK, Chauhan DK (2016) Current trends of engineered nanoparticles (ENPs) in sustainable agriculture: an overview. J Environ Anal Toxicol 6:5

Aruoja V, Pokhrel S, Sihtmäe M, Mortimer M, Mädler L, Kahru A (2015) Toxicity of 12 metal-based nanoparticles to algae, bacteria and protozoa. Environ Sci Nano 6:13

Ashfaq M, Verma N, Khan S (2017) Carbon nanofibers as a micronutrient carrier in plants: efficient translocation and controlled release of Cu nanoparticles. Environ Sci Nano 4:138–148

Azimi R, Feizi H, Hosseini MK (2013) Nanosized titanium dioxide particles improve seed germination features of wheatgrass (*Agropyron desertorum*). Not Sci Biol 5:325–331

Bansiwal AK, Rayalu SS, Labhasetwar NK, Juwarkar AA, Devotta S (2006) Surfactant modified zeolite as a slow release fertilizer for phosphorus. J Agric Food Chem 54:4773–4779

Benício LPF, Constantino VRL, Pinto FG, Vergütz L, Tronto J, da Costa LM (2017) Layered double hydroxides: new technology in phosphate fertilizers based on nanostructured materials. ACS Sustain Chem Eng 5:399–409

Benzon HRL, Rubenecia MRU, Ultra VU Jr, Lee SC (2015) Nano-fertilizer affects the growth, development, and chemical properties of rice. Int J Agron Agri Res 7:105–117

Berekaa M (2015) Nanotechnology in food industry: advances in food processing, packaging and food safety. Int J Curr Microbiol App Sci 4:345–357

Bernardo MP, Guimarães GGF, Majaron VF, Ribeiro C (2018) Controlled release of phosphate from layered double hydroxide structures: dynamics in soil and application as smart fertilizer. ACS Sustain Chem Eng 6:5152–5161

Bielmyer-Fraser GK, Jarvis TA, Lenihan HS, Miller RJ (2014) Cellular partitioning of nanoparticulate versus dissolved metals in marine phytoplankton. Environ Sci Technol 48:13443–13450

Bortolin A, Aouada F, Mattoso LHC, Ribeiro C (2013) Nanocomposite PAAm/methyl cellulose/montmorillonite hydrogel: evidence of synergistic effects for the slow release of fertilizers. J Agri Food Chem 61:7431–7439

Cai D, Wu Z, Jiang J, Wu Y, Feng H, Brown IG, Chu PK, Yu Z (2014) Controlling nitrogen migration through micro-nano networks. Sci Rep 4:3665

Cheloni G, Marti E, Slaveykova VI (2016) Interactive effects of copper oxide nanoparticles and light to green alga *Chlamydomonas reinhardtii*. Aquat Toxicol 170:120–128

Chhowalla M (2017) Slow-release nano-fertilizers for bumper crops. ACS Cent Sci 3:156–157

Corradini E, De Moura MR, Mattoso LHC (2010) A preliminary study of the incorporation of NPK fertilizer into chitosan nanoparticles. Express Polym Lett 4:509–515

Costa MVJD, Sharma PK (2016) Effect of copper oxide nanoparticles on growth, morphology, photosynthesis, and antioxidant response in *Oryza sativa*. Photosynthetica 54:110–119

Crawshaw C (2018) Intelligent nano-fertilizers herald the future. Alberta Barley Blog. http://www.albertabarley.com

Dakovic A, Tomasevic M, Rottinghaus EG, Matijasevic S, Sekulic Z (2007) Fumonisin B1 adsorption to octadecyldimetylbenzyl ammonium-modified clinoptilolite-rich zeolitic tuff. Micropor Mesopor Mat 105:285–290

Dapkekar A, Deshpande P, Oak MD, Paknikar KM, Jyutika M, Rajwade JM (2018) Zinc use efficiency is enhanced in wheat through nanofertilization. Sci Rep 8:6832

Das A, Prasad R, Srivastava RB, Deshmukh S, Rai MK, Varma A (2013) Co-cultivation of *Piriformospora indica* with medicinal plants: case studies. In: Varma A, Kost G, Oelmüller R (eds) *Piriformospora indica*: Sebacinales and their biotechnological applications. Springer, Berlin, Heidelberg, pp 149–171

Davarpanah S, Tehranifar A, Davarynejad G, Abadía J, Khorasani R (2016) Effects of foliar applications of zinc and boron nano-fertilizers on pomegranate (*Punica granatum* cv. Ardestani) fruit yield and quality. Sci Hort 210:57–64

De la Torre RR, Servin A, Hawthorne J, Xing B, Newman LA, Ma X, Chen G, White JG (2015) Terrestrial trophic transfer of bulk and nanoparticle La_2O_3 does not depend on particle size. Environ Sci Technol 49:11866–11874

Dietz K-J, Herth S (2011) Plant nanotoxicology. Trends Plant Sci 16:582–589

Dimkpa CO, Bindraban PS (2018) Nanofertilizers: new products for the industry? J Agric Food Chem 66:6462–6473

Dimkpa CO, McLean JE, Martineau N, Britt DW, Haverkamp R, Anderson AJ (2013) Silver nanoparticles disrupt wheat (*Triticum aestivum* L.) growth in a sand matrix. Environ Sci Technol 47:1082–1090

Dineshkumar R, Kumaravel R, Gopalsamy J, Sikder MNA, Sampathkumar P (2018) Microalgae as bio-fertilizers for rice growth and seed yield productivity. Waste Biomass Valor 9:793–800

Dubey A, Mailapalli DR (2016) Nanofertilisers, nanopesticides, nanosensors of pest and nanotoxicity in agriculture. In: Lichtfouse E (ed) Sustainable agriculture reviews, vol 22. Springer International Publishing, Dordrecht, pp 307–330

Dwivedi S, Saquib Q, Al-Khedhairy AA, Musarrat J (2016) Understanding the role of nanomaterials in agriculture. In: Singh DP, Singh HB, Prabha R (eds) Microbial inoculants in sustainable agricultural productivity. Springer, India, pp 271–288

Eichert T, Goldbach HE (2008) Equivalent pore radii of hydrophilic foliar uptake routes in stomatous and astomatous leaf surfaces: further evidence for a stomatal pathway. Physiol Plant 132:491–502

El-Badawy AM, Silva RG, Morris B, Scheckel KG, Suidan MT, Tolaymat TM (2011) Surface charge-dependent toxicity of silver nanoparticles. Environ Sci Technol 45:283–287

El-Ghamry AM, Mosa AA, Alshall TA, El-Ramady HR (2018) Nanofertilizers vs. biofertilizers: new insights. Environ Biodiv Soil Secur 2:22

Elizondo-Villarreal N, Obregón-Guerra R, García-Méndez M, Sánchez-Espinoza A-P, Alcorta-García M-A, Torres-Barrera RO, Coello V, Castaño VM (2016) Nanomodification of a natural clinoptilolite zeolite. Rev Adv Mater Sci 47:74–78

El-Kereti MA, El-Feky SA, Khater MS, Osman YA, El-Sherbini EA (2013) ZnO nanofertilizer and He Ne laser irradiation for promoting growth and yield of sweet basil plant. Recent Pat Food Nutr Agric 5:169–181

El-Ladan IY, Maiwada NA, Rumah AA (2014) Factors affecting soil quality maintenance in northern Katsina State, Nigeria. Sci World J 9:39–45

El-Ramady H, Abdalla N, Alshaal T, El-Henawy A, Elmahrouk M, Bayoumi Y, Shalaby T, Amer M, Shehata S, Fari M, Domokos-Szabolcsy E, Sztrik A, Prokisch J, Pilon-Smits EAH, Pilon M, Selmar D, Haneklaus S, Schnug E (2018) Plant nano-nutrition: perspectives and challenges. In: Gothandam K, Ranjan S, Dasgupta N, Ramalingam C, Lichtfouse E (eds) Nanotechnology, food security and water treatment, Environmental chemistry for a sustainable world series. Springer International Publishing, Cham, pp 129–161

El-Temsah YS, Joner EJ (2012) Impact of Fe and Ag nanoparticles on seed germination and differences in bioavailability during exposure in aqueous suspension and soil. Environ Toxicol 27:42–49

Everaert M, Warrinnier R, Baken S, Gustafsson JP, De Vos D, Smolders E (2016) Phosphate-exchanged Mg-Al layered double hydroxides: a new slow release phosphate fertilizer. ACS Sustain Chem Eng 4:4280–4287

Faisal M, Saquib Q, Alatar AA, Al-Khedhairy AA, Hegazy AK, Musarrat J (2013) Phytotoxic hazards of NiO-nanoparticles in tomato: a study on mechanism of cell death. J Hazard Mater 250–251:318–332

Gabrielli P, Barbante C, Plane JM, Varga A, Hong S, Cozzi G, Gaspari V, Planchon FA, Cairns W, Ferrari C, Crutzen P, Cescon P, Boutron CF (2004) Meteoric smoke fallout over the Holocene epoch revealed by iridium and platinum in Greenland ice. Nature 432:1011–1014

García-Sánchez S, Bernales I, Cristobal S (2015) Early response to nanoparticles in the *Arabidopsis* transcriptome compromises plant defence and root-hair development through salicylic acid signalling. BMC Genomics 16:341

Gerdini FS (2016) Effect of nano potassium fertilizer on some parchment pumpkin (*Cucurbita pepo*) morphological and physiological characteristics under drought conditions. Int J Farm Allied Sci 5:367–371

Ghahremani A, Akbari K, Yousefpour M, Ardalani H (2014) Effects of nano-potassium and nano-calcium chelated fertilizers on qualitative and quantitative characteristics of *Ocimum basilicum*. Int J Pharm Res Schol 3:00167

Giannousi K, Avramidis I, Dendrinou-Samara C (2013) Synthesis, characterization and evaluation of copper based nanoparticles as agrochemicals against *Phytophthora infestans*. RSC Adv 3:21743–21752

Giroto AS, Guimarães GGF, Foschini M, Ribeiro C (2017) Role of slow-release nanocomposite fertilizers on nitrogen and phosphate availability in soil. Sci Rep 7:46032

Glenn JB, Klaine SJ (2013) Abiotic and biotic factors that influence the bioavailability of gold nanoparticles to aquatic macrophytes. Environ Sci Technol 47:10223–10230

Gogos A, Knauer K, Bucheli TD (2012) Nanomaterials in plant protection and fertilization: current state, foreseen applications, and research priorities. J Agric Food Chem 60:9781–9792

Gruyer N, Dorais M, Bastien C, Dassylva N, Triffault-Bouchet G (2013) Interaction between sliver nanoparticles and plant growth. In: Proceedings of the international symposium on new technologies for environment control, energy-saving and crop production in greenhouse and plant factory. Greensys-2013, vol 6, pp 225–227

Guo H, White JC, Wang Z, Xing B (2018) Nano-enabled fertilizers to control the release and use efficiency of nutrients. Curr Opin Environ Sci Health. 6:77–83

Hemraj C (2017) Nanofertilizers and nanopesticides for agriculture. Environ Chem Lett 15:15–22

Hochella MF Jr, Lower SK, Maurice PA, Penn RL, Sahai N, Sparks DL, Twining BS (2008) Nanominerals, mineral nanoparticles, and earth systems. Science 319:1631–1635

Hong F, Zhou J, Liu C, Yang F, Wu C, Zheng L, Yang P (2005) Effect of nano-TiO_2 on photochemical reaction of chloroplasts of spinach. Biol Trace Elem Res 105:269–279

Hossain KZ, Monreal CM, Sayari A (2008) Adsorption of urease on PE-MCM-41 and its catalytic effect on hydrolysis of urea. Coll Surf Biointerf 62:42–50

Hossain Z, Mustafa G, Komatsu S (2015) Plant responses to nanoparticle stress. Int J Mol Sci 16:26644–26653

Hu X, Kang J, Lu K, Ruiren Zhou R, Mu L, Zhou Q (2014) Graphene oxide amplifies the phytotoxicity of arsenic in wheat. Sci Rep 4:6122

Husen A, Siddiqi KS (2014) Carbon and fullerene nanomaterials in plant system. J Nanobiotechnol 12:16

Husted S (2018) Innovative approach taken to phosphorus nanofertilizer research. AG Chemi Group. https://www.agchemigroup.eu/

IFA (2016) Nutrient management handbook, International Fertilizer Association (IFA). https://www.fertilizer.org/images/Library_Downloads/2016_Nutrient_Management_Handbook.pdf

Iqbal M, Ali ST, Mahmooduzzafar (2000) Photosynthetic performance of certain dicotyledonous tropical plants under degraded environment. In: Khan MA, Farooq S (eds) Environment, biodiversity and conservation. APH Publishing Corporation, New Delhi, pp 408–427

Jiang HS, Qiu XN, Li GB, Li W, Yin LY (2014) Silver nanoparticles induced accumulation of reactive oxygen species and alteration of antioxidant systems in the aquatic plant *Spirodela polyrhiza*. Environ Toxicol Chem 33:1398–1405

Kah M, Kookana RS, Gogos A, Bucheli TD (2018) A critical evaluation of nanopesticides and nanofertilizers against their conventional analogues. Nature Nanotechnol 13:677–684

Kale AP, Gawade SN (2016) Studies on nanoparticle induced nutrient use efficiency of fertilizer and crop productivity. Green Chem Tech Letter 2:88–92

Kaushal M, Wani SP (2017) Nanosensors: frontiers in precision agriculture. In: Prasad R, Kumar M, Kumar V (eds) Nanotechnology: an agricultural paradigm. Springer Nature, Singapore, pp 279–291

Khalifa NS, Hasaneen MN (2018) The effect of chitosan-PMAA-NPK nanofertilizer on *Pisum sativum* plants. 3 Biotech 8:193

Khanm H, Vaishnavi BA, Shankar AG (2018) Rise of nano-fertilizer era: effect of nano scale zinc oxide particles on the germination, growth and yield of tomato (*Solanum lycopersicum*). Int J Curr Microbiol App Sci 7:1861–1871

Khodakovskaya MV, de Silva K, Biris AS, Dervishi E, Villagarcia H (2012) Carbon nanotubes induce growth enhancement of tobacco cells. ACS Nano 6:2128–2135

Khodakovskaya MV, Kim BS, Kim JN, Alimohammadi M, Dervishi E, Mustafa T, Cernigla CE (2013) Carbon nanotubes as plant growth regulators. Effects on tomato growth, reproductive system, and soil microbial community. Small 9:115–123

Kim JI, Park HG, Chang KH, Nam DH, Yeo MK (2016) Trophic transfer of nano-TiO_2 in a paddy microcosm: a comparison of single-dose versus sequential multi-dose exposures. Environ Pollut 212:316–324

Kirschbaum MUF (2011) Does enhanced photosynthesis enhance growth? Lessons learned from CO_2 enrichment studies. Plant Physiol 155:117–124

Kottegoda N, Munaweera I, Madusanka N, Karunaratne V (2011) A green slow-release fertilizer composition based on urea-modified hydroxyapatite nanoparticles encapsulated wood. Curr Sci 101:73–78

Kottegoda N, Sandaruwan C, Priyadarshana G, Siriwardhana A, Rathnayake UA, Arachchige BDM, Kumarasinghe AR, Dahanayake D, Karunaratne V, Amaratunga GAJ (2017) Urea-hydroxyapatite nanohybrids for slow release of nitrogen. ACS Nano 11:1214–1221

Kumar R, Ashfaq M, Verma N (2018) Synthesis of novel PVA–starch formulation-supported Cu–Zn nanoparticle carrying carbon nanofibers as a nanofertilizer: controlled release of micronutrients. J Mater Sci 53:7150–7164

Kundu S, Adhikari T, Mohanty SR, Rajendiran S, Vassanda Coumar M, Saha JK, Patra AK (2016) Reduction in nitrous oxide emission from nano zinc oxide and nano rock phosphate coated urea. Agrochimica 60:59–70

Kurepa J, Paunesku T, Vogt S, Arora H, Rabatic BM, Lu J, Wanzer MB, Woloschak GE, Smalle JA (2010) Uptake and distribution of ultrasmall anatase TiO_2 Alizarin red S nano-conjugates in *Arabidopsis thaliana*. Nano Lett 10:2296–2302

Lateef A, Nazir R, Jamil N, Alam S, Shah R, Khan MN, Saleem M (2016) Synthesis and characterization of zeolite based nano–composite: an environment friendly slow release fertilizer. Micropor Mesopor Mat 232:174–183

Le VN, Rui Y, Gui X, Li X, Liu S, Han Y (2014) Uptake, transport, distribution and bio-effects of SiO_2 nanoparticles in Bt-transgenic cotton. J Nanobiotechnol 12:50

Li JX, Wee CD, Sohn BK (2010) Growth response of hot pepper applied with ammonium ($NH4^+$) and potassium (K^+)-loaded zeolite. Korean J Soil Sci Fert 43:619–625

Li H, Huang J, Lu F, Liu Y, Song Y, Sun Y, Zhong J, Huang H, Wang Y, Li S, Lifshitz Y, Lee S-T, Kang Z (2018) Impacts of carbon dots on rice plants: boosting the growth and improving the disease resistance. ACS Appl Bio Mat 1:663–672

Lin D, Xing B (2008) Root uptake and phytotoxicity of ZnO nanoparticles. Environ Sci Technol 42:5580–5585

Liu R, Lal R (2014) Synthetic apatite nanoparticles as a phosphorus fertilizer for soybean (*Glycine max*). Sci Rep 4:5686

Liu R, Lal R (2015) Potentials of engineered nanoparticles as fertilizers for increasing agronomic productions. Sci Total Environ 514:131–139

Liu J, Zhang YD, Zhang ZM (2009) The application research of nano-biotechnology to promote increasing of vegetable production. Hubei Agril Sci 1:041

Liu R, Kang Y, Pie L, Wan S, Liu S, Liu S (2016a) Use of a new controlled-loss fertilizer to reduce nitrogen losses during winter wheat cultivation in the Danjiangkou Reservoir area of China. Commun Soil Sci Plant Anal 47:1137–1147

Liu R, Zhang H, Lal R (2016b) Effects of stabilized nanoparticles of copper, zinc, manganese, and iron oxides in low concentrations on Lettuce (*Lactuca sativa*) seed germination: Nanotoxicants or nanonutrients? Water Air Soil Pollut 227:42

Lopez-Moreno ML, de la Rosa G, Hernández-Viezcas JA, Castillo-Michel H, Botez CE, Peralta-Videa JR, Gardea-Torresdey JL (2010) Evidence of the differential biotransformation and genotoxicity of ZnO and CeO_2 nanoparticles on soybean (*Glycine max*) plants. Environ Sci Technol 44:7315–7320

Lu J, Bowles M (2013) How will nanotechnology affect agricultural supply chains? Int Food Agribus Manag Rev 16:21–42

Ma X, Geisler-Lee J, Deng Y, Kolmakov A (2010) Interactions between engineered nanoparticles (ENPs) and plants: phytotoxicity, uptake and accumulation. Sci Total Environ 408:3053–3061

Malebra M, Cerana R (2018) Recent advances of chitosan applications in plants. Polymers 10:118

Malhi SS, Haderlin LK, Pauly DG, AM J (2002) Improving fertiliser use efficiency. Better Crops 86:22–25

Manikandan A, Subramanian KS (2016) Evaluation of zeolite based nitrogen nano-fertilizers on maize growth, yield and quality on inceptisols and alfisols. Int J Plant Soil Sci 9:1–9

Manjunatha SB, Biradar DP, Aladakatti YR (2016) Nanotechnology and its applications in agriculture: a review. J Farm Sci 29:1–13

Marschner H (1995) Mineral nutrition of higher plants, 2nd edn. Academic Press, London 889 pp

Martínez-Fernández D, Barroso D, Komárek M (2016) Root water transport of *Helianthus annuus* L. under iron oxide nanoparticle exposure. Environ Sci Pollut Res Int 23:1732–1741

Mashock MJ, Kappell AD, Hallaj N, Hristova KR (2016) Copper oxide nanoparticles inhibit the metabolic activity of Saccharomyces cerevisiae. Environ Toxicol Chem 35:134–143

Massalha H, Korenblum E, Tholl D, Aharoni A (2017) Small molecules below-ground: the role of specialized metabolites in the rhizosphere. Plant J 90:788–807

Mastronardi E, Tsae P, Zhang X, Monreal C, DeRosa MC (2015) Strategic role of nanotechnology in fertilizers: potential and limitations. In: Rai M, Ribeiro C, Mattoso L, Duran N (eds) Nanotechnologies in food and agriculture. Springer International Publishing, Cham, pp 25–67

Mazur GA, Medvid GK, Gvigora IT (1986) Use of natural zeolite to increase the fertilizer of coarse soils. Soviet Soil Sci 16:105–111

Ming DW, Mumpton FA (1989) Zeolites in soils. In: Dixon LB, Weed SB (eds) Minerals in soil environments. Soil Science Society of America, Madison, pp 873–911

Mirzajani F, Askari H, Hamzelou S, Farzaneh M, Ghassempour A (2013) Effect of silver nanoparticles on *Oryza sativa* L. and its rhizosphere bacteria. Ecotoxicol Environ Saf 88:48–54

Moghaddasi S, Fotovat A, Khoshgoftarmanesh AH, Karimzadeh F, Khazaei HR, Khorassani R (2017) Bioavailability of coated and uncoated ZnO nanoparticles to cucumber in soil with or without organic matter. Ecotoxicol Environ Saf 144:543–551

Monreal CM, DeRosa M, Mallubhotla SC, Bindraban PS, Dimkpa C (2016) Nanotechnologies for increasing the crop use efficiency of fertilizer-micronutrients. Biol Fertil Soils 52:423–437

Morales-Díaz AB, Ortega-Ortíz H, Juárez-Maldonado A, Cadenas-Pliego G, González-Morales S, Benavides-Mendoza A (2017) Application of nanoelements in plant nutrition and its impact in ecosystems. Adv Nat Sci Nanosci Nanotechnol 8:013001

Mrakovcic M, Absenger M, Riedl R, Smole C, Roblegg E, Fröhlich LF, Fröhlich E (2013) Assessment of long-term effects of nanoparticles in a microcarrier cell culture system. PLoS One 8:e56791

Mukherjee A, Majumdar S, Servin AD, Pagano L, Dhankher OP, White JC (2016) Carbon nanomaterials in agriculture: a critical review. Front Plant Sci 7:172

Munir T, Rizwan M, Kashif M, Shahzad A, Ali S, Amin N, Zahid R, Alam MFE, Imran M (2018) Effect of zinc oxide nanoparticles on the growth and Zn uptake in wheat (*Triticum aestivum* L.) by seed priming method. Digest J Nanomater Biostr 13:315–323

Nair PM, Chung IM (2014) Impact of copper oxide nanoparticles exposure on *Arabidopsis thaliana* growth, root system development, root lignification, and molecular level changes. Environ Sci Pollut Res Int 21:12709–12022

Nair R, Varghese SH, Nair BG, Maekawa T, Yoshida Y, Kumar DS (2010) Nanoparticulate material delivery to plants. Plant Sci 179:154–163

Nair R, Mohamed MS, Gao W, Maekawa T, Yoshida Y, Ajayan PM, Kumar DS (2012) Effect of carbon nanomaterials on the germination and growth of rice plants. J Nanosci Nanotechnol 12:2212–2220

Narendhran S, Rajiv P, Sivaraj R (2016) Influence of zinc oxide nanoparticles on growth of *Sesamum indicum* L. in zinc-deficient soil. Int J Pharm Pharm Sci 8:365–371

Nowack B, Bucheli TD (2007) Occurrence, behavior and effects of nanoparticles in the environment. Environ Pollut 150:5–22

Oliveira JLD, Campos EVR, Pereira AES, Pasquoto T, Lima R, Grillo R, De Andrade DJ, Dos Santos FA, Fraceto LF (2018) Zein nanoparticles as eco-friendly carrier systems for botanical repellents aiming sustainable agriculture. J Agric Food Chem 6:1330–1340

Omanović-Mikličanin E, Maksimović M (2016) Nanosensors applications in agriculture and food industry. Bull Chem Technol Bosnia Herz 47:59–70

Palmqvist NG, Bejai S, Meijer J, Seisenbaeva GA, Kessler VG (2015) Nano titania aided clustering and adhesion of beneficial bacteria to plant roots to enhance crop growth and stress management. Nature 5:10146

Pan B, Xing B (2012) Applications and implications of manufactured nanoparticles in soils: a review. Euro J Soil Sci 63:437–456

Panpatte DG, Jhala YK, Shelat HN, Vyas RV (2016) Nanoparticles: the next generation technology for sustainable agriculture. In: Singh DP, Singh HB, Prabha R (eds) Microbial inoculants in sustainable agricultural productivity. Springer, India, pp 289–300

Parveen A, Rao S (2015) Effect of nanosilver on seed germination and seedling growth in *Pennisetum glaucum*. J Clust Sci 26:693–701

Patolsky F, Zheng G, Lieber C (2006) Nanowire-based biosensors. Anal Chem 78:4260–4269

Perreault F, Samadani M, Dewez D (2014) Effect of soluble copper released from copper oxide nanoparticles solubilisation on growth and photosynthetic processes of *Lemna gibba* L. Nanotoxicology 8:374–382

Prasad R, Kamal S, Sharma PK, Oelmuller R, Varma A (2013) Root endophyte *Piriformospora indica* DSM 11827 alters plant morphology, enhances biomass and antioxidant activity of medicinal plant *Bacopa monnieri*. J Basic Microbiol 53:1016–1024

Prasad R, Kumar V, Prasad KS (2014) Nanotechnology in sustainable agriculture: present concerns and future aspects. African J Biotech 13:705–713

Prasad R, Bhattacharyya A, Nguyen QD (2017) Nanotechnology in sustainable agriculture: recent developments, challenges and perspectives. Front Microbiol 8:1014

Preetha PS, Balakrishnan N (2017) A review of nanofertilizers and their use and functions in soil. Int J Curr Microbiol App Sci 6:3117–3133

Qureshi A, Singh DK, Dwivedi S (2018) Nano-fertilizers: a novel way for enhancing nutrient use efficiency and crop productivity. Int J Curr Microbiol App Sci 7:3325–3335

Rahale S (2011) Nutrient release pattern of nanofertilizer formulation. PhD (Agri.) Thesis. Tamilnadu Agricultural University, Coimbatore

Raliya R, Saharan V, Dimpka C, Biswas P (2018) Nanofertilizer for precision and sustainable agriculture: current state and future perspectives. J Agric Food Chem 66:6487–6503

Ramesh K, Reddy DD (2011) Zeolites and their potential uses in agriculture. Adv Agron 113:219–241

Rane M, Bawskar M, Rathod D, Nagaonkar D, Rai M (2015) Influence of calcium phosphate nanoparticles, *Piriformospora indica* and *Glomus mosseae* on growth of *Zea mays*. Adv Nat Sci Nanosci Nanotechnol 6:045014

Raskar SV, Laware SL (2014) Effect of zinc oxide nanoparticles on cytology and seed germination in onion. Int J Curr Microbiol App Sci 3:467–473

Rastogi A, Zivcak M, Sytar O, Kalaji HM, He X, Mbarki S, Brestic M (2017) Impact of metal and metal oxide nanoparticles on plant: a critical review. Front Chem 5:78

Read DS, Matzke M, Gweon HS, Newbold LK, Heggelund L, Ortiz MD, Lahive E, Spurgeon D, Svendsen C (2016) Soil pH effects on the interactions between dissolved zinc, non-nano- and nano-ZnO with soil bacterial communities. Environ Sci Pollut Res Int 23:4120–4128

Rico CM, Majumdar S, Duarte-Gardea M, Peralta-Videa JR, Gardea-Torresdey JL (2011) Interaction of nanoparticles with edible plants and their possible implications in the food chain. J Agric Food Chem 59:3485–3498

Rispail N, De Matteis L, Santos R, Miguel AS, Custardoy L, Testillano PS, Risueño MC, Pérez-de-Luque A, Maycock C, Fevereiro P, Oliva A, Fernández-Pacheco R, Ibarra MR, De la Fuente JM, Marquina C, Rubiales D, Prats E (2014) Quantum dot and superparamagnetic nanoparticle interaction with pathogenic fungi: internalization and toxicity profile. ACS Appl Mater Interfaces 6:9100–9110

Rodrigues DF, Jaisi DP, Elimelech M (2013) Toxicity of functionalized single-walled carbon nanotubes on soil microbial communities: implications for nutrient cycling in soil. Environ Sci Technol 47:625–633

Rodrigues SM, Trindade T, Duarte AC, Pereira R, Koopmans GF, Römkens PFAM (2016) A framework to measure the availability of engineered nanoparticles in soil: trends in soil tests and analytical tools. Trends Anal Chem 75:129

Romih T, Drašler B, Jemec A, Drobne D, Novak S, Golobič M, Makovec D, Susič R, Kogej K (2015) Bioavailability of cobalt and iron from citric-acid-adsorbed $CoFe_2O_4$ nanoparticles in the terrestrial isopod Porcellio scaber. Sci Total Environ 508:76–84

Roshanravan B, Soltani SM, Mahdavi F, Rashid SA, Yusop MK (2014) Preparation of encapsulated urea-kaolinite controlled release fertiliser and their effect on rice productivity. Chem Speciat Bioavailab 26:249–256

Roshanravan B, Soltani SM, Rashid SA, Mahdavi F, Yusop MK (2015) Enhancement of nitrogen release properties of urea–kaolinite fertilizer with Chitosan binder. Chem Speciat Bioavailab 27:44–51

Ruhil K, Sheeba, Ahmad A, Iqbal M, Tripathy BC (2015) Photosynthesis and growth responses of mustard *(Brassica juncea* L. cv. Pusa Bold) plants to free air carbon-dioxide enrichment (FACE). Protoplasma 252:935–946

Ruttkay-Nedecky B, Krystofova O, Nejdl L, Adam V (2017) Nanoparticles based on essential metals and their phytotoxicity. J Nanobiotechnol 15:33

Sabir A, Yazar K, Sabir F, Kara Z, Yazici MA, Goksu (2014a) Vine growth, yield, berry quality attributes and leaf nutrient content of grapevines as influenced by seaweed extract (*Ascophyllum nodosum*) and nanosize fertilizer pulverizations. Sci Hort 175:1–8

Sabir S, Arshad M, Chaudhari SK (2014b) Zinc oxide nanoparticles for revolutionizing agriculture: synthesis and applications. Sci World J 2014:925494

Sadeghzadeh B (2013) A review of zinc nutrition and plant breeding. J Soil Sci Plant Nutr 13:905–927

Savoy H (1999) Fertilizers and their use. PB-1637, agricultural extension service, The University of Tennessee, Knoxville

Schultz C, Powell K, Crossley A, Jurkschat K, Kille P, Morgan AJ, Read D, Tyne W, Lahive E, Svendsen C, Spurgeon DJ (2015) Analytical approaches to support current understanding of exposure, uptake and distributions of engineered nanoparticles by aquatic and terrestrial organisms. Ecotoxicology 24:239–261

Sharifi R, Mohammadi K, Rokhzadi A (2016) Effect of seed priming and foliar application with micronutrients on quality of forage corn (*Zea mays*). Environ Exp Biol 14:151–156

Sharma P, Jha AB, Dubey RS, Pessarakli M (2012) Reactive oxygen species, oxidative damage, and antioxidative defense mechanism in plants under stressful conditions. J Bot 2012:217037

Sharma R, Ragavan KV, Thakur MS, Raghavaro KSMS (2015) Recent advances in nanoparticle-based aptasensors for food contaminants. Biosens Bioelectron 74:612–627

Shrestha B, Acosta-Martinez V, Cox SB, Green MJ, Li S, Canas-Carrell JE (2013) An evaluation of the impact of multiwalled carbon nanotubes on soil microbial community structure and functioning. J Hazard Mater 261:188–197

Siddiqi KS, Husen A (2017) Plant response to engineered metal oxide nanoparticles. Nano Res Lett 12:92

Simarmata T, Hersanti, Turmuktini T, Fitriatin BN, Setiawati MR, Purwanto (2017) Application of bioameliorant and biofertilizers to increase the soil health and rice productivity. HAYATI J Biosci 23:181–184. https://doi.org/10.1016/j.hjb.2017.01.001

Singh MD, Chirag G, Prakash PO, Mohan MH, Prakasha G, Vishwajith (2017) Nano fertilizers is a new way to increase nutrients use efficiency in crop production. Int J Agri Sci 9:3831–3833

Sohair EED, Abdall AA, Amany AM, Houda RA (2018) Effect of nitrogen, phosphorus and potassium nano fertilizers with different application times, methods and rates on some growth parameters of Egyptian cotton (*Gossypium barbadense* L.). Biosci Res 15:549–564

Solanki P, Bhargava A, Chhipa H, Jain N, Panwar J (2015) Nano-fertilizers and their smart delivery system. In: Rai M, Ribeiro C, Mattoso L, Duran N (eds) Nanotechnologies in food and agriculture. Springer, Cham, pp 81–101

Soliman AS, Hassan M, Abou-Elella F, Ahmed AHH, El-Feky SA (2016) Effect of nano and molecular phosphorus fertilizers on growth and chemical composition of baobab (*Adansonia digitata* L.). J Plant Sci 11:52–60

Songkhum P, Wuttikhun T, Chanlek N, Khemthong P, Laohhasurayotin K (2018) Controlled release studies of boron and zinc from layered double hydroxides as the micronutrient hosts for agricultural application. Appl Clay Sci 152:311–322

Sturikova H, Krystofova O, Hedbavny J, Adam V (2017) The comparison of effect of zinc sulphate and zinc oxide nanoparticles on plants. Mendel Net 24:932–936

Subbarao CV, Kartheek G, Sirisha D (2013) Slow release of potash fertilizer through polymer coating. Int J Appl Sci Eng 1:25–30

Subramanian KS, Manikandan A, Thirunavukkarasu M, Rahale CS (2015) Nano-fertilizers for balanced crop nutrition. In: Rai M, Ribeiro C, Mattoso L, Duran N (eds) Nanotechnologies in food and agriculture. Springer International Publishing, Cham, pp 69–80

Sun D, Hussain H, Yi Z, Siegele R, Cresswell T, Kong L, Cahill D (2014) Uptake and cellular distribution, in four plant species, of fluorescently labeled mesoporous silica nanoparticles. Plant Cell Rep 33:1389–1402

Syu YY, Hung J-H, Chen J-C, Chuang HW (2014) Impacts of size and shape of silver nanoparticles on *Arabidopsis* plant growth and gene expression. Plant Physiol Biochem 83:57–64

Tarafdar JC, Raliya R, Mahawar H, Rathore I (2014) Development of zinc nanofertilizer to enhance crop production in pearl millet (*Pennisetum americanum*). Agribiol Res 3:257–262

Tepe N, Bau M (2014) Importance of nanoparticles and colloids from volcanic ash for riverine transport of trace elements to the ocean: evidence from glacial-fed rivers after the 2010 eruption of Eyjafjallajökull Volcano, Iceland. Sci Total Environ 488–489:243–251

Tripathi DK, Chauhan DK (2017) Plants and carbon nanotubes (CNTs) interface: present status and future prospects. In: Prasad R, Kumar V, Kumar M (eds) Nanotechnology: food and environmental paradigm. Springer Nature, Singapore, pp 317–340

Tripathi DK, Singh S, Singh S, Srivastava PK, Singh VP, Singh S, Prasad SM, Singh PK, Dubey NK, Pandey AC, Chauhan DK (2017) Nitric oxide alleviates silver nanoparticles (Ag NPs)-induced phytotoxicity in *Pisum sativum* seedlings. Plant Physiol Biochem 110:167–177

Umar S, Anjana, Iqbal M (2003) Potassium fertilization: a substitute to pesticide application in field crops. In: Prakash S (ed) Proceedings of national symposium on biochemical sciences, health and environmental aspects. Allied Publishers, New Delhi, pp 494–496

Umar S, Anjana, Iqbal M (2011) Interactive effects of potassium and nitrogen nutrition on physiological use efficiency of nitrogen and crop yield. In: Jain V, Kumar PA (eds) Nitrogen use efficiency in plants. New India Publishing Agency, New Delhi, pp 125–155

Van Aken B (2015) Gene expression changes in plants and microorganisms exposed to nanomaterials. Curr Opin Biotechnol 33:206–219

Van Breusegem F, Dat JF (2006) Reactive oxygen species in plant cell death. Plant Physiol 141:384–390

Vittori AL, Carbone S, Gatti A, Vianello G, Nannipieri P (2015) Uptake and translocation of metals and nutrients in tomato grown in soil polluted with metal oxide (CeO_2, Fe_3O_4, SnO_2, TiO_2) or metallic (Ag, Co, Ni) engineered nanoparticles. Environ Sci Pollut Res Int 22:1841–1853

Wang X, Han H, Liu X, Gu X, Chen K, Lu D (2012a) Multi-walled carbon nanotubes can enhance root elongation of wheat (*Triticum aestivum*) plants. J Nanopart Res 14:841

Wang Z, Xie X, Zhao J, Liu X, Feng W, White JC, Xing B (2012b) Xylem-and phloem-based transport of CuO nanoparticles in maize (*Zea mays* L.). Environ Sci Technol 46:4434–4441

Wang WN, Tarafdar JC, Biswas P (2013) Nanoparticle synthesis and delivery by an aerosol route for watermelon plant foliar uptake. J Nanopart Res 15:1–13

Wang D, Lin Z, Wang T, Yao Z, Qin M, Zheng S, Lu W (2016a) Where does the toxicity of metal oxide nanoparticles come from: the nanoparticles, the ions, or a combination of both? J Hazard Mater 308:328–334

Wang X, Yang X, Chen S, Li Q, Wang W, Hou C, Gao X, Wang L, Wang S (2016b) Zinc oxide nanoparticles affect biomass accumulation and photosynthesis in *Arabidopsis*. Front Plant Sci 6:1243

Wanyika H, Gatebe E, Kioni P, Tang Z, Gao Y (2012) Mesoporous silica nanoparticles carrier for urea: potential applications in agrochemical delivery systems. J Nanosci Nanotechnol 12:2221–2228

Werlin R, Priester JH, Mielke RE, Krämer S, Jackson S, Stoimenov PK, Stucky GD, Cherr GN, Orias E, Holden PA (2011) Biomagnification of cadmium selenide quantum dots in a simple experimental microbial food chain. Nat Nanotechnol 6:65–71

White PJ, Brown PH (2010) Plant nutrition for sustainable development and global health. Ann Bot 105:1073–1080

Wong MH, Misra RP, Giraldo JP, Kwak S-Y, Son Y, Landry MP, Swan JW, Blankschtein D, Strano MS (2016) Lipid exchange envelope penetration (LEEP) of nanoparticles for plant engineering: a universal localization mechanism. Nano Lett 16:1161–1172

Xu H, Jiang Q, Reddy N, Yang Y (2011) Hollow nanoparticles from zein for potential medical applications. J Mater Chem 21:18227

Yasmin J, Ahmed MR, Cho B-K (2016) Biosensors and their applications in food safety: a review. J Biosyst Eng 41:240–254

Yuvaraj M, Subramanian KS (2014) Controlled-release fertilizer of zinc encapsulated by a manganese hollow core shell. Soil Sci Plant Nutr 61:319–326

Yuvaraj M, Subramanian KS (2017) Development of slow release Zn fertilizer using nano-zeolite as carrier. J Plant Nutr 41:311–320

Zaytseva O, Neumann G (2016) Carbon nanomaterials: production, impact on plant development, agricultural and environmental applications. Chem Biol Tech Agri 3:17

Zhou JM, Huang PM (2007) Kinetics of potassium release from illite as influenced by different phosphates. Geoderma 138:221–228

Zhu H, Han J, Xiao JQ, Jin Y (2008) Uptake, translocation, and accumulation of manufactured iron oxide nanoparticles by pumpkin plants. J Environ Monit 10:713–717

Zhu X, Wang J, Zhang X, Chang Y, Chen Y (2010) Trophic transfer of TiO_2 nanoparticles from daphnia to zebrafish in a simplified freshwater food chain. Chemosphere 79:928–933

Chapter 20
Weed Control Through Herbicide-Loaded Nanoparticles

Amna, Hesham F. Alharby, Khalid Rehman Hakeem, and Mohammad Irfan Qureshi

20.1 Introduction

The term 'weed' applies to any plant that grows or reproduces aggressively or is invasive outside its native habitat (Zimdahl 2007). The term is not based on any taxonomical considerations and relates purely to the social status of the plant in a given scenario of phytosociology. Therefore, one can see extremely important crops and devastating weeds belonging to the same genus or family (Lenser and Theissen 2013). Economically, weeds are always unwanted because they can cause great damages to crop productivity (Fried et al. 2017; Stewart Jr 2017) and ecosystems (Pyšek et al. 2017). Weeds can be classified on the basis of their life habits as annual or perennial. The annual weeds regrow out of the seeds dropped in the soil or environment during the previous seasons, whereas the perennial weeds regrow from existing plants, dormant buds, roots, stolons, rhizomes, tubers, etc.

20.2 Impact of Weeds on Agriculture

Many weed species have moved out of their natural geographic ranges and spread around the world in tandem with human migrations and commerce. Weed seeds are often mixed inadvertently with crop seeds while harvesting. Consecutively, such lots of seeds are collected and transported with crop grain to various new geographical locations, so humans serve as vector of weed seeds transport (Hassan et al. 2005;

Amna · M. I. Qureshi (✉)
Department of Biotechnology, Jamia Millia Islamia, New Delhi, India

H. F. Alharby · K. R. Hakeem
Department of Biological Sciences, Faculty of Science, King Abdulaziz University, Jeddah, Saudi Arabia

A. Husen, M. Iqbal (eds.), *Nanomaterials and Plant Potential*,
https://doi.org/10.1007/978-3-030-05569-1_20

Fig. 20.1 Impact of weeds on various attributes of a social system

National Geographic 2011). Ultimately, nations suffer from multiple adverse impacts caused by weeds (Fig. 20.1).

Some weed species have been classified as noxious weeds by the government authorities, because if left unchecked, they often compete with the native or crop plants and/or cause harm to livestock. They are often foreign species accidentally or imprudently imported into a region where there are few natural controls to limit their population and spread. Some plants that are often considered to be weeds include amaranth (pigweed),, bermuda grass, bindweed, broadleaf plantain, burdock, common lambsquarters, creeping charlie, dandelion, goldenrod, Japanese knotweed, kudzu, leafy spurge, milk thistle, *Parthenium*, poison ivy, ragweed, sorrel, striga, St John's wort, sumac, tree of heaven, wild carrot, wood sorrel and yellow nutsedge, among others. Many invasive weeds were introduced deliberately in the first place and may have not been considered nuisances at that point of time (Smith 1995; Randall 2017).

20.3 Destructions Caused by Weeds

Weeds are naturally adapted to survive to a higher degree. They are called 'pioneer' plants, as they are the first to establish themselves even in very harsh conditions. Their seeds are especially adapted to survive and spread through unique mechanisms. For instance, many weed seeds have small 'hooks' which may penetrate animals' skin and cannot be removed easily. As the animal moves along, the seed penetrates deeper and deeper into the skin and often injures the flesh also. This also proves damaging to wool business, as the presence of seeds in the wool decreases its value (Sinden et al. 2004). Weeds have a negative impact on bottom line of the farming business; but growers often don't have enough knowledge about how much the weeds affect yields and quality of crops.

20.3.1 Competition with Crop Plants

Plants are always in competition with one another. The competition can be for light, water, nutrients and growing space. Competition for light is the first factor that influences the growth of young plants. In most cases weeds grow faster than commercial crops (Patterson 1995; Swanton et al. 2015; Gharde et al. 2018). As a result, weeds overshadow the commercial crop reducing the amount of photosynthetic active radiation (PAR) required by crop plants to grow and develop (Kropff and Spitter 1990). In dry land conditions, where means of irrigation are not available and the farmers rely on rain, plants compete for available water. Since most weeds are well adapted, they grow faster and develop fast-spreading root systems that are efficient in absorbing water and nutrients. The input of fertilizers, a source of nutrition, is thus consumed by unwanted plants. It results in high production cost; unintentional adulteration of grains, vegetables and other economically important plants; transport of allergens to consumers; and so on (Patterson 1995; Baysinger and Sims 1991).

Allelopathic compounds are chemicals that are excreted by plant roots or leaves in order to protect that plant from other neighbouring plant species that might reduce its chance of survival (Kunz et al. 2016). Most plants that are highly competitive have allelopathic compounds that influence other plants' growth negatively (Ahmed et al. 2018). For this reason also, weeds can have a significant effect on field crops. Contamination of weeds into agricultural lands leads to enormous enhancement of production costs, as the farmers have to manage additional activities, such as manual removal of weeds during growth season, application of chemicals, supply of additional fertilizers to meet out doses lost in competition to weeds and mulching, etc., in order to control weeds in the field.

20.3.2 Toxicity of Weeds

Depending on the sensitivity of humans and animals, a weed could be a dermal and/or respiratory allergen, neurotoxic, digestion-disruptive and sometimes even life-threatening. Some plants can cause serious problems throughout the year, while others exert poisonous effects only during certain specific periods in the year. Many a times, pollens have been reported to cause health hazards to both human and animals (Bajwa et al. 2016; Ahmed et al. 2018; Pablos et al. 2017). Seeds of *Senecio* can prove fatal if not removed from wheat before milling. The weeds known to cause harm to human and animal health include *Lantana camara*, *Nerium oleander*, *Hypericum perforatum*, *Tribulus terrestris*, *Fagopyrum esculentum*, *Trifolium* spp., *Medicago denticulata*, *Panicum schinzii*, *Crotalaria burkeana*, *Cotyledon orbiculata*, *Homeria* spp., *Moraea* spp., *Pachystigma pygmaeum*, *P. thamnus*, *Tribulus terrestris*, *Lolium temulentum* (carrying the fungus *Endoconidium temulentum*), *Parthenium hysterophorus*, etc. A number of medical cases are reported throughout the world due to exposure to weed and weed-associated products (Mack and Smith 2011; Mazza et al. 2014).

Grain crops are harvested when moisture of the plant material is at its lowest. The low water content ensures that grains can be stored for a long period. If weeds are mixed with dry grains, the high water content of weeds may cause fermentation in the grain storage facilities, causing high storage losses. Certain weeds, e.g. *Cleome viscosa*, when eaten by milking cattle, impart an undesirable flavour to milk. Similarly, *Tribulus terrestris* seeds get attached to the body of the sheep and impair the quality of wool. If seeds of weeds get mixed with the grains of the main crops, like cereals and pulses, it lowers the quality of the produce.

Water reservoirs such as dams, water tanks and rivers are susceptible to many weeds. Such places receive seeds of weeds through birds, bats or winds. Growth of weeds at such places has numerous disadvantages, such as increased water evaporation through leaves of weeds, water contamination causing undesirable taste and/or smell and oxygen scarcity for fish if present in the water body, among others (Pípalová 2006).

20.4 Weed Management and Controlling Methods

Weed management strategies (Fig. 20.2) encompass preventive as well as control methods. The preventive methods include all such measures through which the introduction of weeds into crop fields could be avoided, whereas the control methods include ways of destruction and eradication of weeds after they have come up in the crop field.

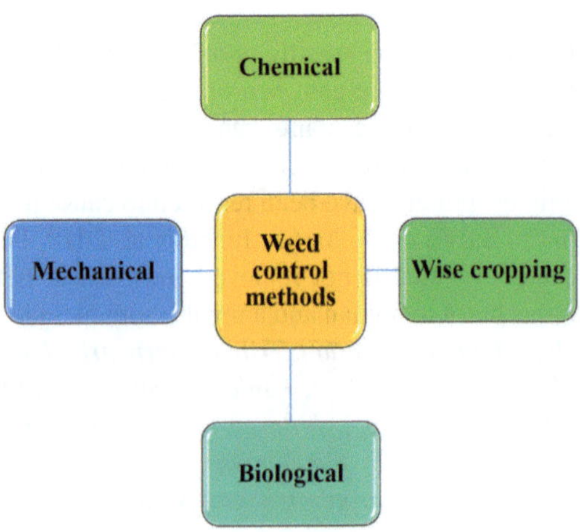

Fig. 20.2 Various conventional methods used to control weeds in a cropping system

20.5 Preventive Measures

Since weeds multiply at a much faster rate and are hardier than crop plants, they always win over crops, if allowed to establish in the field. Furthermore, it is difficult and costly to eradicate them. The potential preventive measures are (i) using clean seeds which are free from weed seeds; (ii) using well-decomposed dung or compost; and (iii) sanitizing fields of weed seeds before seeding the crops. By adopting various methods of weed prevention (Smith 1995; Smith et al. 2005; McDougall et al. 2011; Rao et al. 2017), the losses could be minimized.

20.6 Weed Control Methods

These methods, classified as mechanical methods, cropping or cultural methods, biological methods and chemicals methods, have different merits as shown in Fig. 20.3.

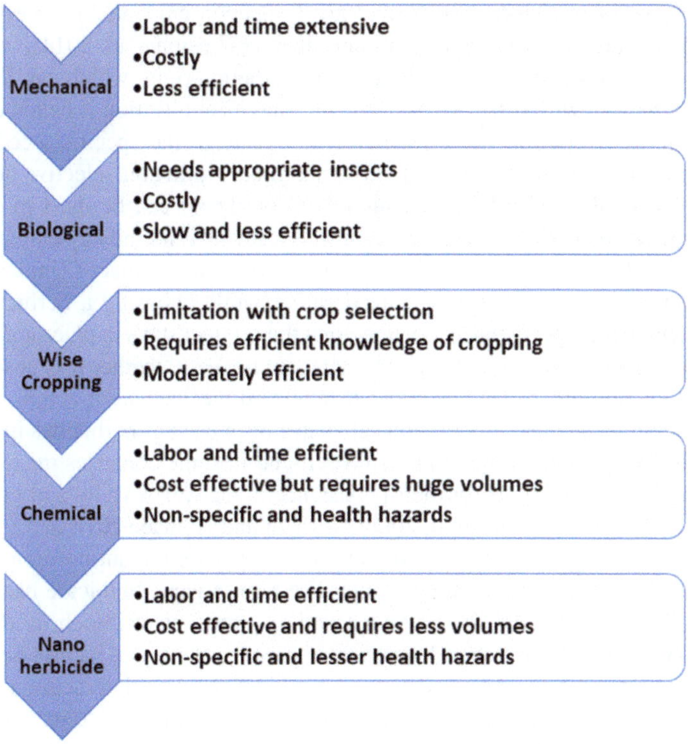

Fig. 20.3 A comparative account of technologies used to control weeds

The most common mechanical methods are hand pulling, hand weeding, burning, flooding, ploughing, harrowing, etc. Pulling weeds by hand or with the help of sickle is the oldest and most efficient method. Weeds can be easily uprooted after a good soaking irrigation or rain. However, this method is labour-intensive besides being costly and time-consuming (Walsh et al. 2013; Walsh and Powels 2014; Broster et al. 2013).

Wise cropping/culture methods include proper crop rotations so that neither annual nor perennial weeds have a free growth. Inter-tilled crops like cotton and crops like potato and groundnut, which necessarily require digging of soil, help in reducing the infestation of weeds. Growing crops, like sunn hemp, that have a very vigorous and leafy growth helps in smothering the weeds. A suitable combination of mechanical methods and crop rotation is very effective in controlling weeds (Darmency et al. 2017; Lowry and Smith 2018).

Under biological approach, the plant or animal enemies of weeds may be used for their control or eradication. The most notable example is the control of prickly pear by using the cochineal insects. Likewise, Kans grass is controlled by growing basket grass, the roots of which are supposed to excrete substances inhibitory to Kans. Biological control of weeds also includes the use of plant pathogens, nematodes, parasitic plants and other organisms. *Parthenium* can be controlled using beetles. Biological methods prove to be highly efficient and economical, provided the right type of predators, which may not feed upon cultivated crop even under starvation conditions, are selected and introduced (Winston et al. 2014).

Any chemical substance of biological or synthetic origin, which can kill plants or inhibit their growth, is known as herbicide, and its application is called chemical method of weed control. Selective herbicides damage the specific weed species while leaving the desired crop relatively unharmed, while non-selective herbicides (also called total weed killers in commercial products) can be used to clear the waste ground, industrial and construction sites, railways and railway embankments, as they kill all plant material with which they come into contact (Zimdahl 2007). Other factors that are taken into consideration while choosing a herbicide compound include their persistence, i.e. how long the product stays in place and remains active (also known as residual action), means of uptake (whether it is absorbed by foliage only, through roots or by other means) and mechanism of action. Products such as common salt and other metal salts have been used as herbicides in the past, but these have gradually fallen out of favour, and in some countries many of these are banned due to concerns for their persistence in the soil, toxicity and groundwater contamination (Guzzella et al. 2006; Noshadi and Homaee 2018).

Modernherbicides are often synthetic mimics of natural plant hormones, which interfere with growth of the target plants. The term organic herbicide has come to mean herbicides intended for organic farming. Some plants, such as the genus *Juglans* (walnuts) or the tree of heaven, also produce their own natural herbicides. Such action of natural herbicides, and other related chemical interactions, is called allelopathy. Due to herbicide resistance, a major concern in agriculture, a number of products combine herbicides with different modes of action. Integrated pest management may use herbicides along with pest control methods. In the United States in 2007, about 83% of all herbicide usage, determined by weight applied, was in agriculture. The world pesticide expenditures in that year totalled about $39.4 billion; herbicides constituted about 40% of those sales, followed by insecticides, fungicides and other types. Smaller quantities are used in forestry and pasture systems and for the management of areas earmarked as the wildlife habitat.

Herbicides are applied to crops in various ways, e.g. (a) pre-sowing treatment, wherein the soil is treated with herbicide before sowing or planting of the crop; (b) pre-emergence treatment, which involves herbicide application after sowing or planting of the crop but before the emergence of the crop; (c) post-emergence treatment in which herbicide is applied after full emergence of crop plants; (d) directed application, which comprises of spray of non-selective herbicides over the space left between the rows in wide-spaced-row crops, without treating the crop foliage or crop shoots; and (e) band application, wherein herbicides are sprayed only over the crop rows, and the interrow spaces are left unsprayed. In the case of expensive herbicides, normally soil-treating mode of application is preferred.

Herbicides are an integral component of most crop management systems. However, although weeds are targeted by herbicides application, sometimes herbicide sprays hugely affect the yields of food crops, which may transfer toxicity to consumers. Further, no single management system effectively controls all the weed species (Smith 1995; Owen 2016). Any shift in the selection of suitable herbicide might result into the growth of other weeds, which are less sensitive to the herbicide used. Certain herbicides produce long residual effect and hence limit the choice of next crop in the rotation. Herbicide drifts may harm the neighbouring crops, e.g. ester form of 2,4-D may harm the neighbouring crop of cotton, soybean, okra, etc. Besides, the use of herbicides may cause environmental pollution. All these limitations of herbicides in use compelled scientists to look for better alternatives for effectively eradicating weeds with no or negligible harm to cultivated crops, and this has culminated into the development of nanoherbicides (Yadav et al. 2015).

20.7 Nanoherbicidal Approach for Weed Control

Nanotechnology has shown potential for efficient delivery of chemical and biological pesticides using nanosized preparations or nanomaterial-based agrochemical formulations (Kothari and Wani 2019). Herbicides are loaded on nanomaterials (NMs) to facilitate the higher bioavailability and thus ensure better eradication of weeds. For producing the NM, which could load and deliver an impressive amount of herbicide to seeds of herbs in the soil during the non-crop stage or into the live plants (Abgail and Chidambaram 2017), a perfect match between nanostructure and herbicide molecules has to be worked out.

The active ingredient is adsorbed, attached, encapsulated or entrapped unto or into the nano-matrix. Controlled release of the active ingredient is achieved due to slow-release characteristics of NMs, bonding of the ingredients to the material and the environmental conditions. The benefits of NM-based formulations are the improvement of efficacy due to higher surface area, higher solubility, higher mobility and lower toxicity due to elimination of organic solvents.

A nanoherbicide consists of either minute particles of active ingredients of herbicide or other minutely engineered structures possessing herbicidal properties. Nanoherbicides can be a cause to enhance wettability and dispersion of agricultural formulations, hence the restricted movement of herbicide to other areas. Bio-composites and nanomaterials possess useful properties (thermal stability, stiffness, permeability, solubility, crystallity and biodegradability) needed for formulating nanoherbicides. Nanoherbicides also offer large specific surface area and hence increased affinity to the target. Nanoemulsions, nanoencapsules, nanocontainers and nanocages are some of the nanoherbicide delivery techniques capable of controlling or delaying the delivery (Roy et al. 2013; Khatem et al. 2016). The current constraints due to droplet size may be overcome by using NM-encapsulated or nanosized herbicides that will contribute to efficient spraying and reduction of spray drift and splash losses.

Basically, nanoformulations should possess the ability to degrade faster within the soil and at a slow rate within the plants that have the residue level below the criteria necessary for regulation in food stuff. Sodium dodecyl sulphate (SDS) is utilized to enhance the photodegradation of nanoparticles (NPs) in the soil. The SDS-modified Ag/TiO_2 imidacloprid nanoformulation has been developed by a microencapsulation technique that used chitosan and alginate (Baker et al. 2017). Many other nanomaterials could also be tried for safer development of nanoherbicides as, for instance, polymer stabilizers such as polyvinylpyrrolidone (PVP), polyvinyl alcohol (PVOH) and poly(acrylic acid)-b-poly(butyacrylate (PAA-b-PBA) have been tried for synthesizing bifenthrin, a stable nanopesticide (Maarouf et al. 2016).

Nanoherbicide in 1–100 nm range will try to mingle with the soil particles and destroy weed seeds and weeds via their roots. Herbicides like atrazine and triazine could be encapsulated to get efficient release to the plants. NPs of poly(epsilon-caprolactone) containing the herbicide atrazine were prepared, characterized and evaluated in terms of their herbicidal activity and genotoxicity (Pereira et al. 2014). Most of the herbicides available in the market are designed in such a way that either they kill or control the above-ground part of the weed plants; none of them inhibits the activity of viable underground plant parts like rhizomes or tubers, which act as a source for new weeds in the following season (Ali et al. 2014; Dashora and Kanika 2018).

Developing a target-specific molecule of herbicide encapsulated along with nanoparticle is aimed for specific receptor in the roots of target weeds, which penetrates into the roots system of the weeds and reaches the parts that inhibit glycolysis of food reserves in the root system, thus causing the specific weed plant to starve due to lack of food and ultimately get killed (Hess 2018). Detoxification of weed residues is necessary, because excessive use of herbicides for long periods of time leaves the residues in the soil and causes damage to succeeding crops (Chinnamuthu and Boopathi 2009). Also, a continuous use of the same herbicide for long durations renders the weed resistant to that particular herbicide. Up to 88% detoxification of atrazine herbicide by the carboxy methyl cellulose (CMC) NPs is on record (Satapanajaru et al. 2008).

The easiest way to eliminate weeds is to destroy their seed banks in the soil and prevent them from germinating when weather and soil conditions become favourable for their growth (Buhler 1997). Being very small, nanoherbicides will be able to blend with the soil, eradicate weeds in an eco-friendly way without leaving any toxic residues and prevent the growth of weed species that have become resistant to conventional herbicides. Weeds survive and spread through underground structures such as tubers and deep roots. Ploughing infects fields while removing weeds by hand can make these unwanted plants spread to uninfected areas (Prasad et al. 2014). Each weed plant produces thousands of seeds every growing season. In some cases, the buried seeds can sprout even after 20 years. Frequent tilling of the soil can multiply weeds that spread through root fragments.

20.8 Synthesis of Nanoparticles

The major components required for synthesis of nanoherbicides are the appropriate nanomaterial and effective herbicide. Crop protection companies have started manufacturing products composed of particles of active ingredients in the nanoscale range (1–100 nm). This size is 2000 to 50,000 times smaller than the size of particles used in the conventional crop protection products. For example, a nanoherbicide contains many trillions of particles of active ingredient per litre. The extra surface area that is formed by reducing the size of the particle boosts potency, accelerates uptake by the plant, increases solubility in the spray tank and reduces or even eliminates the risk of settling and separation (Manceau et al. 2008; Zaseybida and Wilkes 2016). In addition, nanotechnology is spawning a variety of breakthrough products such as herbicide sensors that can dramatically reduce the amount and toxicity of the products used.

Nanoparticles are synthesized using well-designed systematic methods, which start from grinding the crop protection products down to a couple of microns in a single pass through a large disk mill (300 or 500 litre machine). The method used by most manufacturers of such products to develop NPs is through the comminution of larger, coarse particles, which is referred to as the dispersion or top-down method (Mukhopadhyay 2014). Wet comminution with a media mill with high energy density is used to bring bigger particles down to submicron particle size, preferably using the popular NP-manufacturing method called media milling or grinding. For example, stirred media mills are used by ceramic manufacturers, lithium ion battery manufacturers and electronic component manufacturers to bring particle size down. Nanogrinding with a stirred media mill is suitable in particle size reduction, equipment scalability and cost-effectiveness (Hill 2010).

With rapid advancements of stirred media mills, the agitator design becomes a key factor. Rapidly agitating the beads with the pins or pegs on the agitator shaft creates grinding and dispersion. For effective results, it is important to maintain a uniform compression throughout the chamber's length. This process is used to achieve higher efficiency in particle size reduction by utilizing complete media charge and mill capacity. It is important to keep particles free from external contamination during the milling process. The equipment used in metal grinding can slough off metal particles, thus causing accidental contamination, leading to a poor-quality agrochemical. To exclude this probability, ceramic mills can be used which are free of metal-grinding shafts and chambers and ensure that the coatings are not compromised.

20.9 Nanoscale Carriers in Plant Protection

The potential of nanotechnology to create crop protection products with revolutionary properties is prompting many of the world's largest chemical companies such as Syngenta Crop Protection, BASF and Bayer CropScience to speed up their commercial development. Also, DuPont Crop Protection has provided expertise and support for the US National Nanotechnology Initiative, a 10-year $12 billion research program to date (Clunan and Rodine-Hardy 2014). These companies have been testing the performance of crop protection products containing particles of conventional size that can be boosted simply by adding a catalyst-containing nanoscale particles. BASF and Embrapa, the Brazilian Agricultural Research Cooperation, jointly developed herbicide-tolerant soybeans that were green-lighted by the Brazilian Biosafety Technical Commission also known as CTNBio. BASF and Embrapa launched the new production system with the brand name Cultivance to Brazilian farmers. They are also seeking necessary approvals to introduce this technology in key export markets, such as the United States and China (Hill 2010; Agostinetto et al. 2018).

The Cultivance Production System's herbicide-tolerant soybean varieties along with BASF's broad-spectrum's imidazolinone class of herbicides have been tailored, using nano techniques, to suit regional conditions (www.basf.com). Cultivance herbicides are designed for post-emergence applications that help cultivators control weeds during the first few weeks of crop growth. Cultivators get a season-long control of both broadleaf and grass weeds, including those difficult to control with just a one-time application. Fewer herbicide applications per hectare will reduce the use of machinery and labour, reducing costs for cultivators as well as diminishing the release of CO_2 into the surroundings (Janak and Grichar 2016).

For the last several years, crop protection research has been focusing more on ways to use nanotechnology to attack a weed's seed coating and prevent them from germinating. This approach destroys the weed even when it is buried in the soil and prevents it from growing in even the most favourable conditions (Melander et al. 2017). Due to the small proportions of the nanoscale herbicides, they can easily blend with the soil and attack seeds that are buried below the reach of tillers and conventional herbicides. This also allows growers to prevent the spread of weeds that will multiply through stem cuttings from tilling.

20.10 Mode and Mechanism of Nanoherbicide Action

Some of the common herbicides have been discussed ahead for their mode of action (Song 2014). Modern herbicides generally inhibit the activity of enzymes/proteins in the cell and, consequently, trigger a series of events that kill or inhibit the development of the cell or the organism (Vats 2015). The phytotoxic action of any herbicide can be separated into two phases: the mechanism of action and the mode of action. Mechanism of action is understood as the biochemical or biophysical process in the cellular interior to be inhibited by the herbicidal activity, including the one at the molecular levels. This initial process may be sufficient to kill sensitive plants, but normally several other chemical reactions or processes are also needed to kill the plant. Mode of herbicide action is targeting the weeds at functional, morphological or anatomical aspects. In fact, it is the outcome of mechanism of herbicide at molecular level.

Some examples can make this mode of action clear to understand. The herbicide bispyribac-sodium, belonging to the chemical group pyrimidil carboxy, is used for the post-emergence control of grasses, Cyperaceae and some dicotyledons (Rodrigues and de Almeida 1998). It acts like the ALS inhibitors, which inhibit the enzyme acetolactate synthase (ALS), also called acetohydroxyacid synthase (AHAS). ALS is the first enzyme in the route of synthesis of branched chain amino acids called valine, leucine and isoleucine. The ALS and three other enzymes are common to two metabolic routes that occur in plastids of young plant tissues and produce three amino acids. In the route that produces valine and leucine, ALS catalyses the reaction of two molecules of pyruvate to form acetolactate, the precursor of these amino acids. It also catalyses the reaction between pyruvate and ketobutyrate to produce acetohydroxybutyrate, the isoleucine precursor. ALS requires flavin adenine dinucleotide (FAD), thiamine pyrophosphate (TPP) and Mg^{+2} or Mn^{+2} to be activated.

Herbicides can start to inhibit the growth of weeds within a few hours of their application, long before there are effects on other processes, such as photosynthetic reactions, aerobic respiration or synthesis of RNA or proteins. Studies have shown that the herbicide rapidly blocks the cell cycle from G_2 to mitosis and from G_1 to DNA synthesis. Although there is no direct effect on the mitotic apparatus, there may be a regulatory function over the control of cell division (Hess 2018). A possible effect of inhibition of ALS by herbicides is the accumulation of the substrate of this enzyme, ketobutyrate (Brown 1990).

Another herbicide clomazone belonging to the chemical group isoxazolidinones is indicated for the pre-emergence control of mono- and dicotyledons and can be applied in the rice crop soon after the onset of emergence (Silva et al. 2017). This herbicide acts by inhibiting the synthesis of carotenoids. Initially it was thought that the primary site of action of clomazone was early enzymatic synthesis of carotenoids, such as isopentenyl pyrophosphate isomerase and prenyltransferase, the enzymes responsible for the elongation of the carotenoid chain to 10, 15 or 20 carbons, with subsequent generation of oxidative stress, which destroys the membranes of the cells, thus causing plants to die. After several works, it has been demonstrated that these enzymes are not inhibited by clomazone, leaving the possibility that a clomazone metabolite produced in vivo is the true inhibitor or that the site of inhibition is outside the carotenoid biosynthesis route (Mueller et al. 2000). It is possible that some of these metabolites are converted to toxic form in plants (Devine et al. 1993; Kaňa et al. 2004).

Quinclorac herbicide, belonging to the chemical group quinolines, indicated for the post-emergence control of mono- and dicotyledons, acts as an inhibitor of cell wall synthesis. In fact, there are three hypotheses regarding the mechanism of action of this compound: (1) mechanism similar to auxin mimics (growth regulators); (2) stimulation of ethylene synthesis with co-production of cyanide, which blocks the transport of electrons to reduce oxygen; and (3) inhibition of cell wall synthesis. Auxins are plant growth regulators that interfere with cell elongation growth inhibition (Fig. 20.4, Grossmann 2000).

Auxins control the activity of genes through a sequence of events. Initially, auxins activate receptor proteins present in the cell membrane. These receptors send secondary messengers that will cause two types of effect in the plant cell: rapid response effects and long-term effects (Grossmann 2000). Soon after the application of auxins, there is accumulation of calcium in the cytoplasm, stimulation of ethylene production and acidification of the cell wall. Ethylene promotes the formation of cellulase that degrades cellulose in the cell wall, leading to cell elongation.

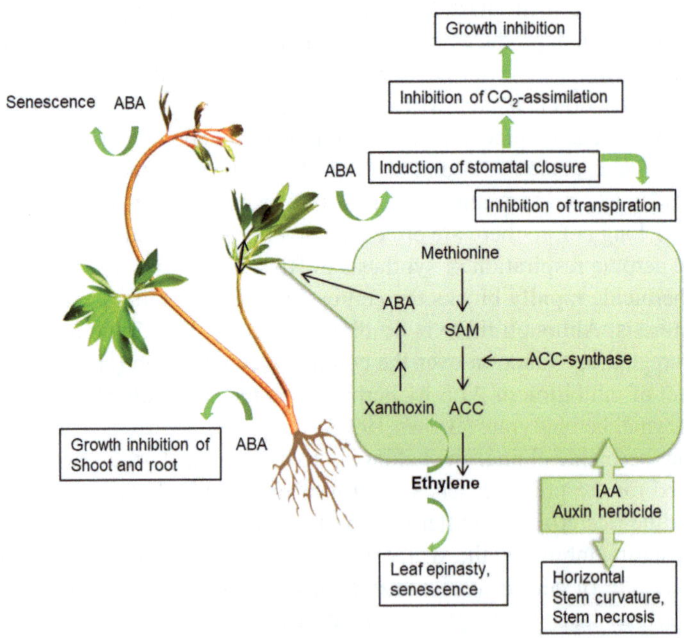

Fig. 20.4 Proposed model of the mode of action of auxin herbicides and the phytohormone indole-3-acetic acid (IAA) at high concentrations, which induce growth inhibition and senescence in dicot plant species. (Modified and adopted with permission, from Grossmann 2000)

In the long run, calcium activates the calmodulin protein, which in turn activates key regulating enzymes responsible for the metabolism of various cell components (Voxeur and Höfte 2016). Much of the symptoms observed after the application of auxins are caused by the action of ethylene. Twisting of stem and petioles, wilting, chlorosis and leaf fall are secondary effects, leading ultimately to death of sensitive plants. Propanil, belonging to the chemical group of chloroaniline derivatives, used to control monocot and dicot weeds, works by acting as photosynthetic inhibitor. The photosynthesis is inhibited by inhibition of ATP-synthase, flow of electrons and phosphate transport into the cell and chloroplast. Propanil also inhibits RNA and protein synthesis (Gealy et al. 2003). Symptoms of chlorosis occur a few days after treatment, followed by desiccation, foliar necrosis and death of plants.

20.11 Nanoherbicides Metabolism: Transport of the Metabolites

Herbicides affect a wide range of metabolism in plants involved in multiple processes (Fig. 20.5). Most techniques used for the study of absorption and translocation are based on using radiolabelled herbicide. This helps in tracing the movement of the herbicide through the plant as long as there is a radioactivity detector-type phosphorimager, although the latter does not give any information about the metabolite transport (Rojano-Delgado et al. 2012; Fernandez et al. 2015). Glufosinate hardly translocates, and, if it does, it takes a small proportion only, whereas imazamox is a systemic herbicide with a high translocation speed (Bukun et al. 2012; Busi et al. 2017). It was observed in two cultivars of *Triticum aestivum* with Clearfield technology (Rojano-Delgado et al. 2014) that the susceptible cultivar, which is unable to metabolize the herbicide, translocates imazamox from leaves to roots throughout the experiment, showing that translocation of imazamox occurs throughout the plant. On the contrary, in the resistant cultivar, the metabolites are the compounds translocated to roots, proportionally to dose and time. Some closely related cultivars exhibit differential response to a herbicide, which might be due to differential metabolism at target sites (Maroli et al. 2016). Nanoherbicides can be taken up both by foliar and root absorption, depending on the application method, and then can exert their phytotoxic effect near the point of entry or be translocated throughout the plant. Foliar uptake is typical for foliar herbicide (post-emergence) applications, whereas root absorption is mostly observed in soil-applied herbicides (Fig. 20.5). A mixture of these mechanisms is also observed in multiple herbicides. The achievement of foliar herbicide absorption is not an easy task. This is because, in contrast to root absorption, foliar uptake involves absorption of the herbicide by the plant parts not specifically designed to do so. Once deposited on the leaf surface, the active ingredients have to cross multiple barriers, such as the epicuticular waxes and leaf cuticle at the leaf surface before they reach the apoplast and finally penetrate the plant cells (Devine et al. 1993). Alternatively, the herbicide would penetrate through the stomata and reach the mesophyll cells (Eichert and Burkhardt 2001).

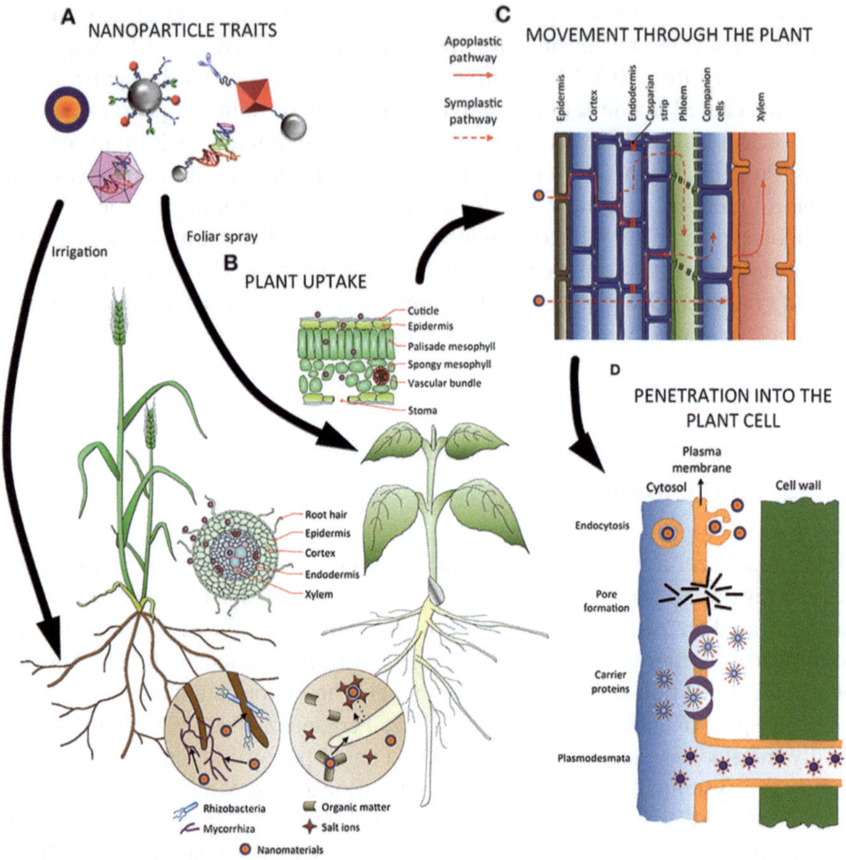

Fig. 20.5 Movement and impact of nanoherbicides in the plant system (Reproduced with permission; from Pérez-de-Luque 2017)

These events are governed by external parameters, such as the chemical and physical properties of the herbicide or the external environment, as well as by internal parameters such as the physiological nature of the leaf surface. While external parameters are not a source of (heritable) herbicide resistance, internal parameters are. The leaf cuticle is a thin biological layer covering the aerial parts of most plants, and its function is to prevent uncontrolled water loss through transpiration (Nawrath 2002). Some cuticles contain suberin, a shikimate-derived combination of cutin-like aliphatic polymers and aromatic moieties. Interestingly, while cutin and suberin are partly hydrophilic, the epicuticular waxes that form the outer surface of the cuticle are hydrophobic (Reiderer 2006). This means that the plant cuticle does not act as a homogeneous layer in terms of hydrophobicity because it becomes more 'water-friendly' deeper into the layer (Devine et al. 1993). Despite this lipophilic gradient, cuticles are considered to be lipophilic membranes (Nawrath 2006), so the resistance of the cuticle against the penetration of polar solutes is high. Thus, non-polar herbicide penetration is usually described following a solution-diffusion model (Riederer and Friedmann 2006).

Therefore, the ability of an herbicide to penetrate plant cuticles is proportional to the solubility and mobility of the herbicide through the cuticle (Schreiber 2005, 2006). This means that changes in cuticle composition may change herbicide susceptibility. This additional pathway, generally called 'aqueous pores' (Schönherr 2006), is supposed to be formed by adsorption of molecules of water via polar elements within the cuticular membrane and pectic cell wall (Schreiber 2005). The 'aqueous pores' pathway may involve lower size selectivity, compared with the lipophilic path. Aqueous pores should provide an additional penetration pathway exclusively available for water soluble, polar herbicides, whereas lipophilic herbicides should exclusively penetrate via the lipophilic pathway, i.e., by solution and diffusion into the cuticular matrix (Schönherr 2006). Open stomata also provide herbicides a fast way to reach leaf mesophyll cells, provided the spraying solution surface tension is below 30 mN m^{-2} of surface energy (Devine et al. 1993). The activated stomata are distinct from the inactive stomata based on wettability of the cuticle surrounding the guard cells (Eichert and Burkhardt 2001). This wettability depends on the presence or absence and the nature of the hygroscopic elements found on the surface of stomatal pores (Eichert et al. 2008).

It is often surprising to non-plant physiologists that roots are quite impermeable to water. In fact, roots account for two watertight barriers for the avoidance of water and solutes running down and out the plant once absorbed. One is an external suberized layer (the exodermis) that covers all but the active growing meristematic tissues. The other is the Casparian strip, another suberized layer located at the root endodermis. Herbicide absorption in roots mostly occurs through the root tips and root hairs, the same locations where solute absorption occurs and both the outer exodermis and the inner Casparian strip are not fully developed. However, the role of the later layer is not clear, as it appears to be a watertight but not 'solute-tight' barrier. Herbicide uptake into the roots is a result of the combination of a dual-step process: a rapid and perhaps apoplastic initial entry, mostly due to non-metabolic processes such as bulk water flow and herbicide diffusion along concentration gradients, is followed by a slower phase of entry and accumulation associated with metabolic activity. Whether water transpiration (together with water mass flow) is related to herbicide absorption is another controversy, with several studies either supporting (Araldi et al. 2011) or opposing the hypothesis.

20.12 Nanoherbicide Translocation

While nanoherbicide absorption can be considered as a short-range transport, herbicide translocation is a long-range delivery that requires two transport systems to be present in plants: the phloem and xylem vascular systems. Translocation is a desirable attribute because it allows the herbicide to reach both treated and untreated parts of the plant. The movement of water, sugars and other compounds such as amino acids and inorganic ions through the phloem is due to the active loading of sugars into sieve elements of the phloem by companion cells. The high concentration of sugars in the phloem causes the water potential in the tissue to drop, so water comes from the xylem and creates a pressure gradient that pushes the water from the source leaves to the sink organs in a two-osmometers-connected model (Henton et al. 2002; Kizilova 2008). Solutes enter the sieve elements via both the apoplast and symplast. The symplastic system implies the passage of solutes from one cell cytoplasm to another through cell channels, the plasmodesmata (Bel 2018), while the apoplastic system is constituted by the continuum of cell walls of adjacent cells, as well as by the extracellular spaces out of the plasmalemma. Thus, the apoplastic system involves the active pumping of sugars first out of the mesophyll photosynthetic cells and then into the companion cells through a H^+-sucrose co-transport mechanism that uses highly specific channels and is driven by the proton-motive force, generated by an H^+-ATPase pump.

Homeostasis within the cell is maintained by cell membranes, so much of the internal regulation can be attributed to both the permeability of the phospholipidic layers and these membrane-bound proteins acting as channels, carriers and pumps. In these terms, herbicides can move through the plasmalemma either by simple diffusion due to a concentration gradient or by a more specific transport mechanism, using suitable membrane channels and carriers. Regarding the putative existence of carrier-mediated, energy-requiring herbicide transport through the cell membrane, facts such as the positive effect of metabolic inhibitors on transport inhibition or the existence of saturable uptake kinetics in some herbicides support this hypothesis. However, herbicide diffusion appears to be the most important mechanism of transport, at least in the lipophilic fast-penetrant herbicides (Balke and Price 1988; Hess 2018).

20.13 Market of Nanoherbicides and Future Prospects

There is no accurate data available on the market share of nanoherbicides, as it is a latest technology which is still evolving. However, some reports on crop protection chemicals have been published in the recent past. According to recent report published in 2016, the total market for crop protection chemicals is estimated at $54.88 billion in the year 2016 (www.marketandmarket.com). It is projected to grow at a compound annual growth rate (CAGR) of 5.15% from 2016 and reach $70.57 billion by 2021, with herbicides occupying the largest market share (44.2%), followed by insecticides and fungicides.

The biopesticide is expected to become the fastest-growing category of products. According to the report 'Herbicides Market by Type (Glyphosate, 2, 4-D, Diquat), Crop Type (Cereals & Grains, Oilseeds & Pulses, Fruits & Vegetables), Mode of Action (Non-selective, Selective), and Region - Global Forecast to 2022', the global herbicides market has reached around USD 27.21 billion in 2016 and is expected to reach around USD 39.15 billion by 2022, at a compound annual growth rate (CAGR) of 6.25% during the period of forecasting. The market is driven by factors such as adoption of better farming practices and rise in production of cereals and grains, especially in the Asia-Pacific region. During this period, a significant trend involves a gradual transformation from using selective herbicide to non-selective herbicide. With a CAGR of 5.69%, glyphosate segment is assumed to grow during this period. Among the various types of herbicides, bio-herbicide is thought to grow with the fastest growth rate of around 23.5% CAGR.

In 2016, Asia-Pacific and Latin America were the top consumers of herbicides which accounted for more than half of the market share. North American market has reached a saturated phase due to which the growth rate is slow, while Asia-Pacific is expected to experience the fastest growth rate in the market. Increasing demand from end applications will drive demand for glyphosate-based herbicides. Countries such as the USA, Japan, China and Taiwan are leading consumers in terms of per capita usage of pesticide market as compared to low-consumption countries such as Greece, India and others.

South America dominated the herbicides market in 2015 due to the emergence of the South American countries (Brazil and Argentina in particular) as the agricultural hubs, growing above the global growth average. Additionally, the regulatory framework in South America is less stringent as compared to North America and Europe. The Pesticide Action Network (South America) handles regulatory control in the region. The herbicide products can be registered with minimum resistance from regulatory agencies in the South American region. The key companies profiled are BASF SE (Germany), E.I. du Pont de Nemours & Company (USA), The Dow Chemical Company (USA), Monsanto Company (USA), Syngenta AG (Switzerland), Platform Specialty Products Corporation (USA), FMC Corporation (USA), Nufarm Ltd. (Australia), Nissan Chemical Industries Ltd. (Japan) and Drexel Chemical Co. (USA).

In conclusion, nanoherbicides are an 'efficient form' of herbicides, which are required in low amounts with high bioavailability to the weeds. The worldwide campaigns emphasize on greener, safer and eco-friendly approaches even to eradicate weeds. In addition, development of nanosensors might play an effective role in controlling weeds and making food products safer for consumption. The nanosensors will indicate whether the food products obtained from nanoherbicide-applied fields are free from herbicidal residue and safe for consumption. New methods for analysis need to be developed for detection and validation and to access the effects of nanomaterials/nanofoods in the entire ecosystem. Improvement of databanks and international collaborations for idea, policy and regulation for manipulation of such knowledge are the need of the hour. The authorities concerned should provide transparent roadmaps and guidelines for reducing the risks associated with nanotechnological products.

References

Abgail EA, Chidambaram R (2017) Nanotechnology in herbicide resistance. Intech Open Sci https://doi.org/10.5772/intechopen.68355

Agostinetto D, Fraga DS, Oliveira ACB, Andres A, Villela FA (2018) Response of soybean cultivars in rotation with irrigated rice crops cultivated in Clearfield® system. Planta Daninha. https://doi.org/10.1590/S0100-83582018360100048

Ahmed J, Bagheri R, Bashir H, Baig MA, Al-Huqail A, Ibrahim MM, Qureshi MI (2018) Organ-specific phytochemical profiling and antioxidant analysis of *Parthenium hysterophorus* L. Biomed Res Int 2018:ID9535232

Ali MA, Rehman I, Iqbal A, Din S, Rao AQ, Latif A, Samiullah TR, Azam S, Husnain T (2014) Nanotechnology: a new frontier in agriculture. Intl J Adv Life Sci 1:129–138

Araldi R, Velini ED, Carbonari GM, Sampaio TF, Trindade MLB (2011) Relationship between water consumption and herbicide absorption in weeds and sugarcane. Planta Daninha 29:1045–1051

Bajwa AA, Chauhan BS, Farooq M, Shabbir A, Adkins SW (2016) What do we really know about alien plant invasion? A review of the invasion mechanism of one of the world's worst weeds. Planta 244:39–57

Baker S, Volova T, Prudnikova SV, Satish S, Prasad MNN (2017) Nanoagroparticles emerging trends and future prospect in modern agriculture system. Environ Toxicol Pharmacol 53:10–17

Balke NE, Price TP (1988) Relationship of lipophilicity to influx and efflux of triazine herbicides in oat roots. Pestic Biochem Physiol 30:228–237

Baysinger JA, Sims BD (1991) Giant ragweed (*Ambrosia trifida)* interference in soybeans. Weed Sci 39:358–362

Bel AJE (2018) Plasmodesmata: a history of conceptual surprises. In: Sahi VP, Baluška F (eds) Concepts in Cell Biology – History and Evolution, Plant Cell Monographs. Springer, Heidelberg, pp 221–270

Broster JC, Koetz EA, Wu H (2013) Herbicide resistance levels in annual ryegrass (*Lolium rigidum* Gaud.) and wild oat (*Avena spp.*) in southwestern New South Wales. Plant Protec Quart 28:126–132

Brown HM (1990) Mode of action, crop selectivity, and soil relations of the sulfonylurea herbicides. Pest Sci 29:263–281

Buhler DD (1997) Implications of weed seedbank dynamics to weed management. Weed Sci 45:329–336

Bukun B, Nissen SJ, Shaner DL, Vassios JD (2012) Imazamox absorption, translocation, and metabolism in red lentil and dry bean. Weed Sci 60(03):350–354

Busi R, Goggin DE, Heap I, Horak MJ, Jugulam M, Masters RA, Napier RM, Riar DS, Satchivi NM, Torra J, Westra P, Wright TR (2017) Weed resistance to synthetic auxin herbicides. Pest Manag Sci 74:2265–2276

Chinnamuthu CR, Boopathi PM (2009) Nanotechnology and agroecosystem. Madras Agril J 96:17–31

Clunan A, Rodine-Hardy K (2014) Nanotechnology in a globalized world: Strategic assessments of an emerging technology. The NPS Institutional Archive, Calhoun http://hdl.handle.net/10945/43101

Darmency H, Colbach N, Le Corre V (2017) Relationship between weed dormancy and herbicide rotations: implications in resistance evolution. Pest Manag Sci 73:1994–1999

Dashora A, Kanika S (2018) Green synthesis of nanoparticles and their applications. Adv Sci Engin Med 10:523–541

Devine MD, Duke SO, Fedtke C (1993) Physiology of herbicide action. Prentice Hall, Englewood Cliffs, p 441

Eichert T, Burkhardt J (2001) Quantification of stomatal uptake of ionic solutes using new model system. J Exp Bot 52:771–781

Eichert T, Kurtz A, Steiner U, Goldbach HE (2008) Size exclusion limits and lateral heterogeneity of the stomatal foliar uptake pathway for aqueous solutes and water-suspended nanoparticles. Physiol Plant 134:151–160

Fernandez P, Gauvrit C, Barro F, Menendez J, De Prado R (2015) First case of glyphosate resistance in France. Agron Sustain Dev 35:1469–1476

Fried G, Chauvel B, Reynaud P, Sache I (2017) Decreases in crop production by non-native weeds, pests and pathogens. In: Vilà M, Hulme PE (eds) Impact of biological invasions on ecosystem services, invading nature, Springer Series in Invasion Ecology, vol 12, pp 83–101

Gealy DR, Wailes EJ, Estorninos LE Jr, Chavez RSC (2003) Rice cultivar differences in suppression of barnyardgrass (*Echinochloa crus-galli*) and economics of reduced propanil rates. Weed Sci 51:601–609

Gharde Y, Singh PK, Dubey RP, Gupta PK (2018) Assessment of yield and economic losses in agriculture due to weeds in India. Crop Protec 107:12–18

Grossmann K (2000) Mode of action of auxin herbicides: a new ending to a long, drawn out story. Trends Plant Sci 5:506–508

Guzzella L, Pozzoni F, Giuliano G (2006) Herbicide contamination of surficial groundwater in Northern Italy. Environ Pollut 142:344–353

Hassan RM, Scholes R, Ash N (2005) Ecosystems and human well-being: current state and trends: findings of the condition and trends working group. Island Press, Washington DC, USA, p 570

Henton SM, Greaves AJ, Piller GJ, Minchin PEH (2002) Revisiting the munch pressure-flow hypothesis of long-distance transport of carbohydrates: modelling the dynamics of solute transport inside a semipermeable tube. J Exp Bot 53:1411–1419

Hess FD (2018) Herbicide effects on plant structure, physiology and biochemistry. In: Altman J (ed) Pesticide interactions in crop plants – beneficial and deleterious effects. CRC Press Taylor & Francis, Boca Raton

Hill J (2010) How to uniformly disperse nanoparticles in battery cathode coatings. Adv Mat Process 2010:34–36

Janak TW, Grichar WJ (2016) Weed control in corn (*Zea mays* L.) as influenced by preemergence herbicides. Intl J Agron 2016:2607671

Kaňa R, Špundová N, Ilík P, Lazár D, Klem K, Tomek P, Nauš J, Prášil O (2004) Effect of herbicide clomazone on photosynthetic processes in primary barley (*Hordeum vulgare* L.) leaves. Pestic Biochem Physiol 78:161–170

Khatem R, Bakthi A, Hermosín MC (2016) Comparison of the systemic nanoherbicide Imazamox-LDH obtained by direct synthesis and reconstruction: preliminary results. (IRNAS) Comunicaciones Congresos SII:P25

Kizilova NN (2008) Long-distance liquid transport in plants. Proc Estonian Acad Sci 57:179–203

Kothari R, Wani KA (2019) Environmentally friendly slow release nano-chemicals in agriculture: a synoptic review. In: Poonia R, Gao X, Raja L, Sharma S, Vyas S (eds) Smart farming technologies for sustainable agricultural development. IGI Global, Hershey, pp 220–240

Kropff MJ, Spitter CJT (1990) A simple model of crop loss by weed competition from early observation on relative leaf area of the weeds. Weed Res 31:97–105

Kunz C, Sturm DJ, Varnholt D, Walker F, Gerhards R (2016) Allelopathic effects and weed suppressive ability of cover crops. Plant Soil Environ 62:60–66

Lenser T, Theißen G (2013) Molecular mechanisms involved in convergent crop domestication. Trends Plant Sci 18:704–714

Lowry CJ, Smith RG (2018) Weed control through crop plant manipulations. Non-Chem Weed Control. https://doi.org/10.1016/B978-0-12-809881-3.00005-X

Maarouf S, Tazi B, Guenoun F (2016) Synthesis and characterization of new composite membranes based on polyvinylpyrrolidone, polyvinyl alcohol, sulfosuccinic acid, phosphomolybdic acid and silica. J Chem Pharmaceut Res 8:387–395

Mack RN, Smith MC (2011) Invasive plants as catalysts for the spread of human parasites. NeoBiota 9:13–29

Manceau A, Nagy KL, Marcus MA, Lanson M, Geoffroy N, Jacquet T, Kirpicht chikova T (2008) Formation of metallic copper nanoparticles at the soil-root interface. Environ Sci Technol 42:1766–1772

Maroli AS, Nandula VK, Duke SO, Tharayil N (2016) Stable isotope resolved metabolomics reveals the role of anabolic and catabolic processes in glyphosate-induced amino acid accumulation in *Amaranthus palmeri* biotypes. J Agril Food Chem 64:7040–7048

Mazza G, Tricarico E, Genovesi P, Gherardi F (2014) Biological invaders are threats to human health: an overview. Ethol Ecol Evol 26:112–129
McDougall KL, Alexander JM, Haider S, Pauchard A, Walsh NG, Kueffer C (2011) Alien flora of mountains: global comparisons for the development of local preventive measures against plant invasions. Divers Distrib 17:103–111
Melander B, Liebman M, Davis AS, Gallandt ER, Bàrberi P, Moonen A-C, Rasmussen J, van der Weide R, Vidotto F (2017) Non-chemical weed management. In: Hatcher PE, Froud-Williams RJ (eds) Weed research: expanding horizons. Willey, pp 245–270
Mueller C, Schwender J, Zeidler J, Lichtenthaler HK (2000) Properties and inhibition of the first two enzymes of the non-mevalonate pathway of isoprenoid biosynthesis. Biochem Soc Trans 28:792–793
Mukhopadhyay SS (2014) Nanotechnology in agriculture: prospects and constraints. Nanotechnol Sci Appl 7:63–71
National Geographic (2011) National geographic answer book: 10,001 fast facts about our world. National Geographic Society, Washington, DC, pp 175
Nawrath C (2002) The biopolymers cutin and suberin. The Arabidopsis Book. https://doi.org/10.1199/tab.0021
Nawrath C (2006) Unraveling the complex network of cuticular structure and function. Curr Opin Plant Biol 9:281–287
Noshadi E, Homaee M (2018) Herbicides degradation kinetics in soil under different herbigation systems at field scale. Soil Tillage Res 184:37–44
Owen MDK (2016) Diverse approaches to herbicide-resistant weed management. Weed Sci 64:57–584
Pablos I, Eichhorn S, Briza P, Asam C, Gartner U, Wolf M, Ebner C, Bohle B, Arora N, Vieths S, Ferreira F, Gadermaier G (2017) Proteomic profiling of the weed feverfew, a neglected pollen allergen source. Sci Rep 7:6049
Patterson DT (1995) Weeds in a changing climate. Weed Sci 43:685–701
Pereira AES, Grillo R, Mello NFS, Rosa AH, Fraceto LF (2014) Application of poly(epsilon-caprolactone) nanoparticles containing atrazine herbicide as an alternative technique to control weeds and reduce damage to the environment. J Hazard Mater 268:207–215
Pérez-de-Luque A (2017) Interaction of Nanomaterials with Plants: What Do We Need for Real Applications in Agriculture? Front Environ Sci 5, https://doi.org/10.3389/fenvs.2017.00012
Pípalová I (2006) A review of grass carp use for aquatic weed control and its impact on water bodies. J Aquat Plant Manag 44:1–12
Prasad R, Kumar V, Suranjit K, Prasad S (2014) Nanotechnology in sustainable agriculture: present concerns and future aspects. Afr J Biotechnol 13:705–713
Pyšek P, Blackburn TM, García-Berthou E, Perglová I, Rabitsch W (2017) Displacement and local extinction of native and endemic species. In: Vilà M, Hulme PE (eds) Impact of biological invasions on ecosystem services, invading nature, Springer Series in Invasion Ecology, vol 12, pp 157–175
Randall RP (2017) A global compendium of weeds. 3rd edn, ISBN 9780646967493, Perth, Australia
Rao AN, Brainard DC, Kumar V, Ladha JK, Johnson DE (2017) Preventive weed management in direct-seeded rice: targeting the weed seedbank. Adv Agron 144:45–142
Reiderer M (2006) Introduction: biology of the plant cuticle. In: Reiderer M, Müller C (eds) Biology of the plant cuticle, Annual Plant Review Series, vol 23. Blackwell Publishing, Oxford, pp 1–10
Riederer M, Friedmann A (2006) Transport of lipophilic non-electrolytes across the cuticle. In: Riederer M, Müller C (eds) Biology of the plant cuticle, Annual Plant Reviews, vol 23. Blackwell Publishing, Oxford, pp 250–279
Rodrigues BN, de Almeida FS (1998) Guide to herbicides, 4th edn. Ministry of Agriculture, Livestock and Food Supply, Brasília, Brazil

Rojano-Delgado AM, Cruz-Hipólito H, De Prado R, de Castro MDL, Rodríguez Franco A (2012) Limited uptake translocation and enhanced metabolic degradation contribute to glyphosate tolerance in *Mucuna pruriens* var. utilis plants. Phytochemistry 73:34–41

Rojano-Delgado A, Priego-Capote F, de Castro MDL, De Prado R (2014) Ultrasound-assisted extraction with LC–TOF/MS identification and LC–UV determination of imazamox and its metabolites in leaves of wheat plants. Agron Sustain Dev 25:357–363

Roy A, Singh SK, Bajpai J, Bajpai AK (2013) Controlled pesticide release from biodegradable polymers. Cent Eur J Chem 12:453–469

Satapanajaru T, Anurakpongsatorn P, Pengthamkeerati P, Boparai H (2008) Remediation of atrazine-contaminated soil and water by nano zerovalent iron. Water Air Soil Pollut 192:349–359

Schönherr J (2006) Characterization of aqueous pores in plant cuticles and permeation of ionic solutes. J Exp Bot 57:2471–2491

Schreiber L (2005) Polar paths of diffusion across plant cuticles: new evidence for an old hypothesis. Ann Bot 95:1069–1070

Schreiber L (2006) Review of sorption and diffusion of lipophilic molecules in cuticular waxes and the effects of accelerators on solute mobilities. J Exp Bot 57:2515–2523

Silva MM, Santos JB, Ferreira EA, Brito OG, Donato LMS, Santos MV (2017) Forage plants and weeds that are sensitive to atmospheric clomazone residuals. Planta Daninha 35:e017165078

Sinden J, Jones R, Hester S, Odom D, Kalisch C, James R, Cacho O (2004) The economic impact of weeds in Australia – report to the CRC for Australian weed management. CRC for Australian Weed Management, Glen Osmond

Smith AE (1995) Handbook of weed management system. Taylor & Francis, Oxford

Smith RG, Gross KL, Januchowski S (2005) Earthworms and weed seed distribution in annual crops. Agri Ecosys Environ 108:363–367

Song Y (2014) Insight into the mode of action of 2,4-dichlorophenoxyacetic acid (2,4-D) as an herbicide. J Integr Plant Biol 56:106–113

Stewart CN Jr (2017) Becoming weeds. Nat Genet 49:654–655

Swanton CJ, Nkoa R, Blackshaw RE (2015) Experimental methods for crop–weed competition studies. Weed Sci 63(Spl):2–11

Vats S (2015) Herbicides: history, classification and genetic manipulation of plants for herbicide resistance. Sust Agri Rev 15:153–192

Voxeur A, Höfte H (2016) Cell wall integrity signaling in plants: "To grow or not to grow that's the question". Glycobiology 26:950–960

Walsh MJ, Powels SB (2014) High seed retention at maturity of annual weeds infesting crop fields highlights the potential for harvest weed seed control. Weed Technol 28:486–493

Walsh M, Newman P, Powels S (2013) Targeting weed seeds in crop: a new weed control paradigm for global agriculture. Weed Technol 27:431–436

Winston RL, Schwarzländer M, Hinz HL, Day MD, Cock MJW, Julien MH (eds) (2014) Biological control of weeds: a world catalogue of agents and their target weeds, 5th edn. USDA Forest Service, Forest Health Technology Enterprise Team, Morgantown FHTET-2014-04. pp 838

Yadav SK, Lal SS, Srivastava AK, Bag TK, Singh BP (2015) Efficacy of chemical and non-chemical methods of weed management in rainfed potato (*Solanum tuberosum*). Ind J Agril Sci 85:382–386

Zaseybida LL, Wilkes BA (2016) Bio-available mineral fertilizer and derivative applications, including product processes. US Patent No. US 2016/0200634 A1, 10pp

Zimdahl RL (2007) Fundamentals of weed science, 3rd edn. Academic Press. Elsevier Inc, San Diago

Chapter 21
Impact of Fabricated Nanoparticles on the Rhizospheric Microorganisms and Soil Environment

Mokula Mohammed Raffi and Azamal Husen

21.1 Introduction

Nanoparticles (NPs) can be defined as objects ranging in size from 1 to 100 nm that due to their size may differ from the bulk material. NPs are broadly classified into three classes: (i) one-dimensional NPs, i.e., thin films or monolayers that have a common use in the field of solar cells; (ii) two-dimensional NPs, i.e., carbon nanotubes; and (iii) three-dimensional NPs, i.e., dendrimers, quantum dots, and fullerenes (Carbon 60) (Bhatia 2016). The available literature indicates that the most studied NPs are Ag, Au, TiO_2, and ZnO, whereas the least studied NPs include Al_2O_3, CeO_2, quantum dots, SiO_2, and SnO_2. Currently, a variety of metallic nanomaterials are being produced using copper, zinc, titanium, magnesium, gold, alginate, and silver. NPs are being used for diverse purposes such as medical treatments, production of solar and oxide fuel batteries for energy storage, and materials of daily use such as cosmetics or clothes (Biswas and Wu 2005; Mauter and Elimelech 2008; Husen and Siddiqi 2014a; Gowramma et al. 2015; Hussein 2016; Siddiqi and Husen 2017a; Siddiqi et al. 2018a, b, c, d). Presently, there is a growing need to use environment-friendly NPs that do not produce toxic wastes in their synthesis protocol (Husen and Siddiqi 2014b, c; Siddiqi and Husen 2016a, b, c; Siddiqi et al. 2016, 2018a; Husen 2017).

For the successful use of nanotechnology in agriculture, it is important to understand the phytotoxic and genotoxic effects of nanoparticle interactions with plants. A number of studies have been focused on such effects which reported positive as well as negative impacts of NPs on different or similar plant species. Tables 21.1

M. M. Raffi
Department of Natural Resource Management, College of Agriculture and Rural Transformation, University of Gondar, Gondar, Ethiopia

A. Husen (✉)
Department of Biology, College of Natural and Computational Sciences, University of Gondar, Gondar, Ethiopia

A. Husen, M. Iqbal (eds.), *Nanomaterials and Plant Potential*,
https://doi.org/10.1007/978-3-030-05569-1_21

Table 21.1 Impact of some NPs on different plant species

Key reference	Nanoparticles	Plant species	Impact
Yang and Watts (2005)	Al_2O_3	*Zea mays*, *Cucumis sativus*, *Glycine max*, *Brassica oleracea*, *Daucus carota*	Phytotoxic
Lin and Xing (2007)	MWNT, Al, ZnO, Zn, Al_2O_3	*Brassica napus*, *Raphanussativus*, *Lolium perenne*, *Lactuca sativa*, *Zea mays*, *Cucumis sativus*	Phytotoxic
Yang et al. (2007)	TiO_2	*Spinacia oleracea*	N_2 fixation
Canas et al. (2008)	SWCNTs	*Brassica oleracea*, *Daucus carota*, *Cucumis sativus*, *Allium cepa*, *Lycopersicon esculentum*, *Lactuca sativa*	Phytotoxic
Gao et al. (2008)	TiO_2	*Spinacia oleracea*	Fresh weight and dry weight increased by 60%; *Rubisco activase* increased by 42%, and its activity increased 2.5 times over the control
Zhu et al. (2008)	Fe_2O_3	*Cucurbita maxima*,*Phaseolus limensis*	Uptake, translocation, andaccumulation of Fe_2O_3
González-Melendi et al. (2008)	Carbon-coated Fe	*Cucurbita pepo*	NP transformation in plant
Battke et al. (2008)	Pb	*Hordeum vulgare* cv. *Barke*	Bioaccumulation and growth
Lin and Xing (2008)	ZnO	*Lolium perenne*	Uptake and toxicity of ZnO
Lee et al. (2008a, b)	Cu	*Phaseolus radiatus*,*Triticum aestivum*	Phytotoxic
Doshi et al. 2008	Al	*Phaseolus vulgaris*, *Lolium perenne*	No effect
Lee et al. (2008a, b)	Cu	Mung bean	Reduced root and seedling growth
Kumari et al. 2009	Ag	*Allium cepa*	Cytotoxic and genotoxic
Stampoulis et al. (2009)	Ag, MWCNT, Cu, ZnO, Si	*Cucurbita pepo*	Influenced seed germination, root elongation, and biomass
Stampoulis et al. (2009)	Ag-Nps	Zucchini	Reduced biomass
Stampoulis et al. 2009	CuO	Zucchini	Reduced biomass and root growth
Barrena et al. (2009)	Au, Ag, Fe_3O_4	*Lactuca sativa*, *Cucumis sativus*	Toxic effect on plant and microbes

(continued)

Table 21.1 (continued)

Key reference	Nanoparticles	Plant species	Impact
Shah and Belozerova (2009)	Si, Pd, Au, Cu	*Lactuca sativa*	Toxic effect on seed germination
Lin et al. (2009)	SWNT	*Nicotiana tobacum* cv.	Bright Yellow (BY-2) cellsSWNT transports in plant cells
Liu et al. (2009)	MWNT, C_{70}	Rice	Uptake and bioaccumulation of CNT in plant
Su et al. (2009)	TiO_2	Spinach	Improved energy utilization leading to growth improvement
Tan et al. (2009)	MWNT	Rice	Effect on rice cell
Khodakovskaya et al. (2009)	MWCT	Tomato	Increased seed germination and growth rate
Wild and Jones (2009)	MWNT, TiO_2, CeO_2	*Triticum aestivum*	No effect; proposed investigation method of NP in plants cell
Asli and Neumann (2009)	TiO_2	*Zea mays*	Inhibited hydraulic conductivity, leaf growth, and transpiration
Rostami and Shahsavar (2009)	Ag	Mission olive explant	Severe injuries and browning of the explants
Seeger et al. (2009)	TiO_2	Willow tree	No significant effects on transpiration rate, growth, and water use efficiency
Kurepa et al. (2010)	TiO_2	*Arabidopsis thaliana*	Uptake and distribution of TiO_2 in plants
Lee et al. (2010)	Al_2O_3	*Arabidopsis thaliana*	No significant effect on seed germination, root elongation, and number of seeds
Ma et al. (2010)	La_2O_3, Gd_2O_3, CeO_2,Yb_2O_3	*Brassica napus, Raphanus sativus, Triticum aestivum, Lactuca sativa, Brassica oleracea, Lycopersiconesculentum, Cucumis sativus*	The rare earth oxide nanoparticlesaffected the plant
Speranza et al. (2010)	Pb	Kiwi	Pb toxicity to kiwifruit pollen
Wang et al. (2010)	Fe_3O_4	*Lolium perenne L.,Cucurbita mixta*	Physiological effect of magnetic NP on plants

(continued)

Table 21.1 (continued)

Key reference	Nanoparticles	Plant species	Impact
Mazumdar and Ahmed (2011)	Ag-NPs	*Oryza sativa*	Broke the cell wall and damaged the vacuoles of root cells
Kumari et al. (2011)	ZnO NPs	*Allium cepa*	Lipid peroxidation, decreasing of mitotic index; increasing of the micronuclei and chromosomal aberration index
Larue et al. (2011)	TiO_2 NPs	*Triticum aestivum*, *Brassica napus*, *Arabidopsis thaliana*	No effect on seed germination and root elongation
Savithramma et al. (2012)	Ag	*Boswellia ovalifoliolata*	Increased germination and seedling growth
El-Temsah and Joner (2012)	Ag	Barley	Reduced germination and shoot length
Prasad et al. (2012)	ZnO	*Arachis hypogaea*	Up to 1000 mg L^{-1} the NP promoted seed germination and growth vigor, but 2000 mg L^{-1} was observed to be toxic for the plant
Priester et al. (2012)	CeO_2 ZnO_2	Soybean	Nano-ZnO accumulated in plant tissue; growth and yield of plant reduced, nitrogen fixation affected
Khodakovskaya et al. (2013)	MWCNTs	Tomato plants	Tomato plants grown in the soil supplemented with CNTs produced more flowers and fruits than the plants grown in the control soil
Feng et al. (2013)	Ag	Clover	Reduced aboveground biomass
Rico et al. (2013)	CeO_2	Rice	Enhanced membrane damage and photosynthetic stress
Shaw et al. (2014)	CuO	Barley	Reduction in growth; significantly low GSH/GSSG ratio; increase in H_2O_2 and lipid peroxidation with increased concentration of NP

(continued)

Table 21.1 (continued)

Key reference	Nanoparticles	Plant species	Impact
Judy et al. (2015)	Ag_2S NPs, polyvinylpyrrolidone-coated Ag-NPs (PVP-Ag)	*Solanum lycopersicum*	100 mg kg^{-1} PVP-Ag-NPs and Ag^+ reduced the plant biomass and mycorrhizal colonization
Judy et al. (2015)	Biosolids (WWTP) containing a mixture of NMs	*Medicago truncatula*	Reductions in nodulation frequency and plant growth
Sweet and Singleton (2015)	Ag	Bishop pine	Inhibition of root and shoot growth
Majumdar et al. (2016)	CeO_2	*Phaseolus vulgaris*	NP increased antioxidant enzyme activities in the aerial tissues
Tomacheski et al. (2017)	Ag-NPs	*Avena byzantine*, *Lactuca sativa*, and *Raphanus sativus*	Low root growth of oat in the soil loaded with AgNp_silica; enhancement of dry mass of radish plants; root growth and dry mass of lettuce unaltered
Jiang et al. (2017)	TiO_2	*Triticum aestivum*	The abscisic acid and jasmonic acid contents increased with increasing concentration of TiO_2 NPs
Tripathi et al. (2017)	ZnO	*Triticum aestivum*	Reduced photosynthesis; increased H_2O_2 and lipid peroxidation

and 21.2 and Fig. 21.1 summarize the possible role and/or the impact of NPs on plant growth and rhizospheric microorganisms. The unprecedented extensive use and the uncontrolled release of nanomaterials (NMs) into the environment and soil through various pathways, such as agricultural amendments of sewage sludge, atmospheric deposition, landfills, or accidental spills during industrial production, pose enormous threat to living organisms. The current models (material-flow model, probabilistic modeling) of environmental emissions of engineered nanomaterials estimate that in the sewage-sludge-treated soil, TiO_2-NP concentrations increase between 0.94 and 3.6 mg kg^{-1} per year, whereas for silver NPs (Ag-NPs) and fullerenes, the yearly predicted increases are more than 1000-fold lower, as they vary between 0.09 and 0.65 μg kg^{-1} and between 0.38 and 1.5 μg kg^{-1}, respectively (Sun et al. 2014). These NPs may be extremely resistant to degradation and have the potential to accumulate in water bodies of water or the soil. Ag-NPs are used in

Table 21.2 Impact of NPs on different microorganisms

Key references	Nanoparticles	Microorganisms	Impact
Dubchak et al. (2010)	Ag-NPs	Mycorrhiza	Inhibited mycorrhizal colonization of sunflower
Cherchi and Gu (2010)	TiO_2 NPs	*Cyanobacterium Anabaena variabilis*	Growth, N_2 fixation, and storage by the cyanobacterium were affected
He et al. (2011)	Fe_3O_4 NPs	*Actinobacteria*, *Duganella*, *Streptomycetaceae*, or *Nocardioides*	Stimulation of specific groups of bacteria
Bandyopadhyay and Peralta-Videa (2012)	CeO_2 and ZnO NPs	*Sinorhizobium meliloti*, *S. melba*	CeO_2 and ZnO NPs were toxic to *Sinorhizobium meliloti*; CeO_2 NPs were bacteriostatic, and ZnO NPs were bactericidal to *S. melba*
Feng et al. (2013)	Ag-NPs, pristine iron oxide (FeO)	AMF	Both affected AMF colonization of clover; FeO NPs lowered the glomalin content and acquisition of P by AMF causing significant biomass reduction
Yuan et al. (2013)	Ag-NPs	*Nitrosomonas europaea*	Alterations in expression of genes involved in energy production and the nitrification process, inhibiting bacteria
Karunakaran et al. (2013)	TiO_2	PGPRs, *Bacillus subtilis*, and *Pseudomonas fluorescens*	NPs inhibited the growth of PGPRs; bacteria were susceptible to toxicity
Allard et al. (2013)	WO_3 NPs (tungsten)	*Azotobacter vinelandii*	Detrimental to the growth of the N_2-fixing bacterium under Mo-limiting conditions
Karunakaran et al. (2014)	Nanosilica and bulk silica(SiO_2) and alumina (Al_2O_3)	*B. megaterium*, *B. brevis*, *P. fluorescens*, and *Azotobacter vinelandii*	Nano and bulk SiO_2 particles were nontoxic to PGPRs up to 1000 mg L^{-1}; Al_2O_3 NPs were highly toxic to bacteria
Fan et al. (2014)	TiO_2	*Rhizobium leguminosarum bv. viciae 3841*	TiO_2 NPs caused morphological changes in bacterial cells; nodule development slow, onset of N_2 fixation in symbiotic interaction with garden pea
Gurunathan (2015)	Graphene oxide (GO) NPs	*B. megaterium*, *B. cereus*, *B. subtilis*, *B. mycoides*, and *B. marisflavi*	Negative effect of GO NPs on isolated rhizobacteria

(continued)

Table 21.2 (continued)

Key references	Nanoparticles	Microorganisms	Impact
Wang et al. (2017)	Ag-NPs	*Acidobacteria*, *Actinobacteria*, *Cyanobacteria*, and *Nitrospirae*	Microbial population decreased with increasing Ag-NPs concentration
Wang et al. (2017)	Ag-NPs	*Proteobacteria* and *Planctomycetes*	Increase in microbe population

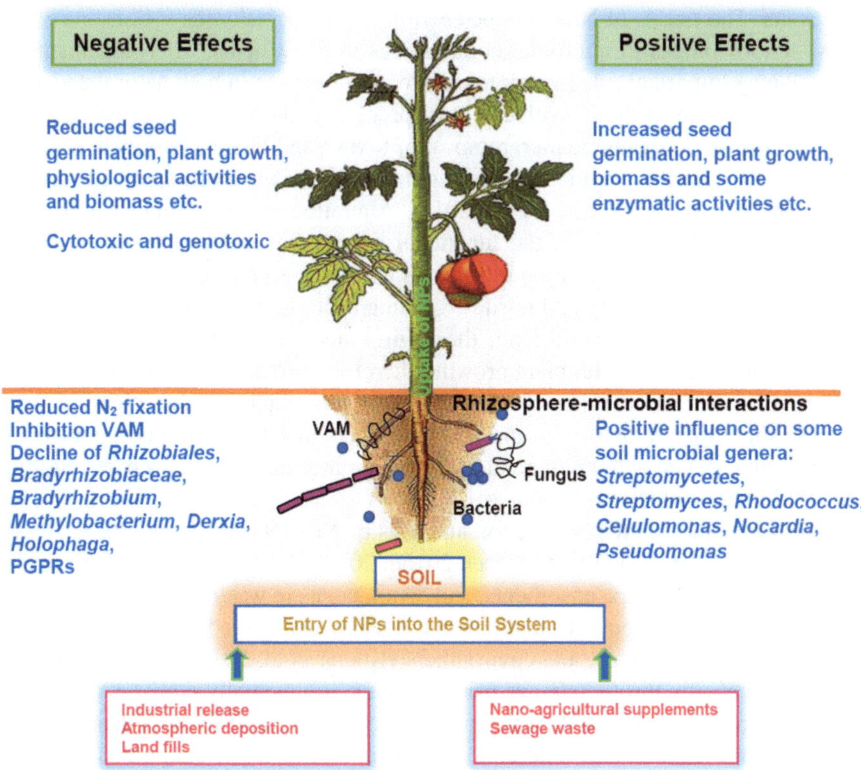

Fig. 21.1 Impact of NPs on plant growth and rhizospheric microorganisms

several industries, and their continuous release may hamper many physiological and biochemical processes in the living organisms including both autotrophs and heterotrophs (Tripathi et al. 2017). Application of NPs has shown promising results in the areas of delivery of pesticides, biopesticides, fertilizers, and genetic material for plant transformation. The use of nanomaterials for delivery of pesticides and fertilizers is expected to reduce the dosage and ensure a controlled slow delivery. Application of NMs to stabilize the biocontrol preparations is likely to go a long way in reducing the environmental hazard (Manjunatha et al. 2016). The range of nanotechnology products is now extensive and can be divided into several compound classes, such as carbonaceous nanomaterials, metal oxides, semiconductor materials (including quantum dots), zerovalent metals (iron, silver, and gold), and nanopolymers (dendrimers). A variety of products are now being generated, including NPs as well as nanofibers, nanowires, and nanosheets, and the range and types of NMs are consistently expanding. An estimated inventory of 1814 consumer products of the engineered NPs was identified in October 2013, with 435 products of Ag-NPs (Vance et al. 2015; Thul and Sarangi 2015). However, we remain ignorant of the effects of NPs that might be accumulating in our natural ecosystems like soil and water. The escape of NPs into the environment, especially the soil, is one of the most serious threats to microbial communities in ecosystems. Microorganisms in the soil play important role in maintenance of soil function in both natural and agricultural soils due to their involvement in processes like soil structure formation, decomposition of organic matter, removal of toxins, and biogeochemical cycles. In addition, they are involved in suppressing the soilborne plant diseases and promoting plant growth (Karimi and Fard 2017). Alteration in the composition of soil microflora can be critical for the functional integrity of soil. For instance, in the nitrification process, ammonium nitrogen is converted to nitrite and then to nitrate by the ammonia-oxidizing and nitrite-oxidizing bacteria, respectively. The deletion of any one of these bacteria from the environment leads to decreased nitrogen removal and interferes with plant growth. Therefore, protection of the environment and beneficial microorganisms from NPs is very important, and the scientific community should pay attention to the adverse effects of NPs on microorganisms, in spite of their beneficial commercial uses (Hajipour et al. 2012; Husen and Siddiqi 2014b; Siddiqi and Husen 2016d, 2017b).

Various studies on the production and uses of NPs (Navarro et al. 2008; Husen and Siddiqi 2014b; Siddiqi and Husen 2016d, 2017b) have highlighted their application as ion exchangers, adsorbents, and disinfectants in water and air for eliminating ions, organic compounds, and pathogens and also assessed the associated risks to human health, ecology, and environment (Kumar et al. 2014). More research is needed, especially through assays using more environmentally realistic concentrations of nanomaterials based on the predicted concentrations in modelization studies and more realistic exposure conditions (Simonin and Richaume 2015). It has been demonstrated that the toxic effects on soil microbial community are highly dependent on both the NMs considered and the soil properties. Soil properties, especially the clay and organic matter content, seem to play an important role for the bioavailability of NPs. Identification of soil parameters controlling the bioavailability of NPs is fundamental for a better environmental risk assessment (Karimi and Fard 2017). Information available so far on the effect of NPs on the rhizosphere and soil microorganisms is quite meagre. Hence, this chapter is focused on the current scientific evidences regarding the positive and negative (toxic) influences of different NPs on plants and the rhizospheric microorganisms.

21.2 Rhizosphere and Engineered NPs

In 1904, Lorenz Hiltner, a German agronomist and plant physiologist, coined the term "rhizosphere" derived in part from the Greek word "rhiza" that means "root," to describe the plant-root interface (Hiltner 1904; Hartmann et al. 2008). Hiltner described the rhizosphere as the area around a plant root that is inhabited by a unique population of microorganisms influenced by the chemicals released from plant roots, the root exudates. It involves interactions between plant roots, soil microbes, and plant pathogens. Soil microorganisms play a very significant role in maintaining the soil ecosystem, soil health, and crop productivity. NPs have both positive and negative effects on plant roots and rhizospheric microorganisms. Since the engineered NPs released to the environment go down the soil, the effects of engineered NPs on soil processes and the organisms that carry them out should be grasped. At present, inadequate information is available on how these engineered NPs affect the soil microorganisms. However, they influence the soil microorganisms via (i) a direct toxicity effect, (ii) changes in the bioavailability of toxins or nutrients, (iii) indirect effects resulting from their interaction with natural organic compounds, and (iv) interaction with toxic organic compounds causing synergism or antagonism (Haris and Ahmad 2017).

21.2.1 Effect of NPs on Rhizosphere Soil Population

Soil microbial communities have a direct impact on soil quality through processes such as nutrient cycling, decomposition of organic matter, and symbiotic relationships with terrestrial plant species (Kennedy and Smith 1995). Consequently, protection of soil microbial biomass and diversity is a major challenge in agriculture. We have limited information on the interaction between NPs and soil microbial community (Simonet and Valcarcel 2009; Dinesh et al. 2012; Rodrigues et al. 2013). The fabricated and/or engineered NPs enter the soil environment through sewage water, agriculture management practices, and other anthropogenic sources. The metallic oxide NPs constitute a group of engineered NPs that may be potentially hazardous for the environment, as reported by Frenk et al. (2013). The authors assessed the effects of engineered NPs on soil bacteria by measuring the activity, composition, and size of the bacterial community following an exposure to the copper oxide (CuO) and magnetite (Fe_3O_4) nanosized (<50 nm) particles in two different soil types, a sandy loam and a clay loam, under two ENP concentrations (1 and 0.1%). The bacterial community in the sandy loam soil was more susceptible to change due to exposure to these ENPs, relative to the clay loam soil. Particularly, CuO had a strong effect on the microbial community composition and size in the sandy loam soil, whereas few effects were noticed in the clay loam soil. Fe_3O_4 changed the bacterial community composition in the sandy loam soil only. Further, the abundance of bacteria of class Bacilli decreased on addition of 0.1% CuO but increased with 1% CuO in the sandy loam soil; in the clay loam soil, the abundance was reduced with 1% CuO. These authors also reported that the other important soil bacterial groups, *Rhizobiales* and *Sphingobacteriaceae*, were negatively affected by CuO addition to the soil. The bacterial populations reacting to a metal oxide-engineered NP, especially CuO, are highly dependent on the concentration of the NPs applied to the soil (Frenk et al. 2013).

Mirzajani et al. (2013) studied the phytotoxic effect of silver NPs (Ag-NPs) on *Oryza sativa* and some of its rhizosphere bacteria. Nine isolates of the genus *Bacillus* were isolated, identified, and used in these studies. The Ag-NPs treatment altered the populations of bacteria as *Bacillus thuringiensis* SBURR1 was totally eliminated, and *Bacillus amyloliquefaciens* SBURR5 became the most populated one and revealed that Ag-NPs may damage the cell wall of the bacteria and transform them to protoplasts resulting in the leakage of sugars and proteins through the bacterial membrane. Xu et al. (2015) reported the disturbing action of TiO_2 NPs and CuO NPs on soil community in a flooded paddy soil, emphasizing that high bioavailability of CuO NPs was the main source of microbial toxicity. The metabolic changes caused due to exposure of rhizosphere bacteria to zinc oxide (ZnO) NPs were investigated on cell signaling molecules, which regulate the intra- as well as intercellular activities. ZnO NPs effectively disrupt cell signaling in *Pseudomonas chlororaphis* O6, a rhizosphere-competent bacterium. Reduction in phenazine production with increasing doses of ZnO NPs correlated with the lowered levels of the quorum-sensing molecules, acyl-homoserine lactones (AHSLs). The ZnO NPs had a higher efficacy than Zn ions in these responses. Linked with changes to phenazine production was an effect on Fe metabolism manifested by an enhanced siderophore production, as the Zn concentration from NPs or ions increased. The ZnO NPs-induced phenotypes of lowered phenazines, AHSLs, and increased siderophores were observed with a mutant lacking the global regulator GacS (Goodman et al. 2016).

The effect of combined pre-seed treatment with microbial inoculants and molybdenum (Mo) NPs on the composition of root exudates and the taxonomic diversity of microorganisms in the chickpea plants rhizosphere were studied by Shcherbakova et al. (2016). The introduced bacteria were most abundant in variants with Mo NPs. The symbiotic effectiveness of *Mesorhizobium ciceri* strain ST282 was improved by co-inoculation with *Bacillus subtilis* helper bacteria and Mo NPs. A self-sufficient legume-rhizobium symbiosis improved the physiological status of the plant, increasing the structural diversity of the microbial community of the rhizosphere through changes in the activity of root exudates, and paved the way for the development of the most effective associative bacteria (Shcherbakova et al. 2016). Moll et al. (2017) tested two different TiO_2 NPs (P25, E171) and a bulk TiO_2 (particle size > 100 nm) for their effects on the diversity and community composition of soil microorganisms. To analyze the microbial diversity, Illumina MiSeq paired-end sequencing of ribosomal markers was used. The authors reported that application of TiO_2NPs altered the structure of the prokaryotic community but not that of the fungal community. Specific microbial taxa responded positively or negatively to the particular TiO_2 NP treatments and hence may be used as bio-indicators for TiO_2 NPs. They also examined the effect of increasing concentrations of TiO_2 NPs on wheat growth and yield. No negative effects on wheat growth and on the arbuscular mycorrhizal root colonization were detected in their study. These results reveal that prokaryotes are more sensitive to the TiO_2 NP than fungi. The influence of five different metal NPs present in the biosolids on soil microbial community as a function of time was investigated and found that ZnO and Zero Valent Cu NPs were not toxic to the soil bacterial community. Biosolids mixed with Ag-NPs and TiO_2 in contrast changed the bacterial richness and composition in wavering pattern as a function of time (Shah et al. 2014).

Akhilesh et al. (2013) examined the effect of chemically synthesized Ag-NPs on *Vigna radiata*, *Brassica juncea*, and their rhizobacteria and reported a huge reduction in germination potential of seeds so that only 11.07% and 27.74% of seeds of the two species, respectively, could germinate. The length of roots and shoots and the leaf number were also reduced in treated plants. The authors also found a reduction in the population of rhizosphere bacteria. Wang et al. (2017) investigated the effect of different concentrations of Ag-NPs (10, 50, and 100 mg kg^{-1}) on the soil microbial community exposed for 7 days. The number of *Acidobacteria*, *Actinobacteria*, *Cyanobacteria*, and *Nitrospirae* significantly decreased with increasing Ag-NPs concentration; at the same time, several other phyla, *Proteobacteria* and *Planctomycetes*, increased in number. After 12 h of exposure, the survival rate of the ammonia-oxidizing bacterium *Nitrosomonas europaea* treated with 1 mg L^{-1} Ag + ions was significantly lower than those treated with 1 mg L^{-1} and 10 mg L^{-1} of Ag-NPs, but a much lower survival rate was observed in the treatment with a high (20 mg L^{-1}) concentration of Ag-NPs.

Fullerene is a pure carbon molecule composed of at least 60 atoms of carbon. Because a fullerene takes a shape similar to a soccer ball or a geodesic dome, it is also referred to as a *buckyball* after Buckminster Fuller, the inventor of the geodesic dome, for whom the molecule is more formally named as fullerene. Carbon nanomaterials (CNMs) may be directly toxic to soil microorganisms or may alter the bioavailability of nutrients and the toxicity of organic compounds and/or toxins (Dinesh et al. 2012). In addition, toxicity to plants may have indirect impact, negative or positive, on the microbial communities.

A limited published literature is available on the effects of CNMs, which are generally negative, on rhizosphere microorganisms. However, a few studies have also revealed some neutral or positive biological effects. For instance, the impact of fullerenes (C_{60}) on the soil microbial community populations was evaluated using the total phospholipid-derived phosphate. The soil was treated with 1 and 1000 μg C_{60} g^{-1} soil for 180 days; the results showed that fullerenes had no impact on the structure or function of the soil microbial community or the soil enzymatic activities (Tong et al. 2007). Shrestha et al. (2013) reported diverse effects on soil (sandy loam) microbial communities after a 90-day exposure of multi-walled carbon nanotubes (MWCNT) to a wide range of CNT concentrations (10, 100, and 10,000 mg/kg). No observable effects were evidenced on the soil microbial composition and enzymatic activities at lower concentrations. However, at 10,000 mg kg^{-1}, a mixed response was observed; a decrease in abundance was detected in some select bacterial species (for instance, *Waddlia*, *Holophaga*, *Derxia*, and *Opitutus*), whereas the number of the polycyclic aromatic hydrocarbon (PAH)-degrading organisms (for instance, *Cellulomonas*, *Rhodococcus*, *Pseudomonas*, and *Nocardioides*) was markedly increased. These results suggest a potential shift toward more stress-tolerant organisms with increase in the soil-MWCNT concentration, although the findings are too limited to be convincing. The first exposure of CNTs lowered the microbial biomass immediately, but the values recovered to the level of the control by the end of the experiment despite the repeated addition of CNTs. The abundance and diversity of the ammonium-oxidizing archaea (AOA) were higher than those of the ammonium-oxidizing bacteria (AOB) under the exposure of CNTs (Chen et al. 2015). The mechanisms responsible for these interactions are unknown; however, the limited literature does suggest that under certain exposure scenarios, CNM may have neutral or perhaps modestly beneficial effects on microbial communities (Mukherjee et al. 2016).

Several authors have reported negative/toxic effects of the carbon-based materials on soil bacteria. Fang et al. (2007) showed that C_{60} aggregates in water at 0.01 mg L^{-1} significantly increased the levels of the iso- and anteiso-branched fatty acids from 5.8% to 31.5% and 12.9% to 32.3%, respectively, in *Bacillus subtilis* (gram-positive), suggesting an increase in membrane fluidity as an adaptation response to C60. Alternatively, the aq-C60 at 0.5 mg L^{-1} caused alteration in the *Pseudomonas putida* (gram-negative) phase-transition temperatures and levels of unsaturated fatty acids from bacterial membrane. Johansen et al. (2008) applied agglomerates of pristine C_{60} fullerenes (50 nm to micron-size) to the soil at 0, 5, 25, and 50 mg kg^{-1} dry soil to assess their effect on the soil microbiota by measuring total respiration, the biomass, number, and diversity of bacteria, as well as the total number and diversity of protozoans during 14 d. These authors reported that respiration and microbial biomass were unaffected by the fullerenes at any time, whereas the number of fast-growing bacteria was decreased by three- to fourfold just after incorporation of the nanomaterial. Tong et al. (2012) investigated the effects of surface coating of SWCNTs on soil microbial community under low and high organic matter concentration. Upon 6000 μg/g of functionalized SWNTs (fSWCNT, coated with polyethylene glycol or m-polyamino benzene sulfonic acid) exposure for 6 weeks, they observed some microbial community shift, keeping the total biomass unaffected. Another important factor that may influence the toxicity of CNTs to bacteria is the presence of residual impurities. Commercial CNTs are synthesized with strong acids and contain up to 4.5–15% of metals such as cobalt (Co), iron (Fe), nickel (Ni), yttrium (Y), and other impurities, which may exert toxic effects on microbes (Kang et al. 2007; Petersen et al. 2014). Another possible mechanism of CNT toxicity is the induction of ROS, which may then directly interact with organelles to induce DNA damage or protein inactivation that results in apoptosis and cell death (Jackson et al. 2013).

Rodrigues et al. (2013) investigated the effect of carboxyl-functionalized single-walled carbon nanotubes (SWNTs) on soil bacterial and fungal communities employing the culture-dependent and culture-independent methods. The soil was added with 0, 250, and 500 μg of SWNTs per gram of soil SWCNT for 14 days, and the populations were observed. The soil bacterial community was transiently affected by the presence of SWNTs. After 3 days of exposure, the number of colony-forming units (CFUs) was significantly decreased, but the bacterial community completely recovered after 14 days. Alternatively, higher doses of SWNTs had a similar biomass loss at 3 days, but the fungal community was unable to recover even after 14 days. Similarly, Jin et al. (2013) reported that at relatively low concentrations (300–1000 μg g^{-1} soil), SWCNTs significantly lowered the activities of most enzymes and microbial biomass, which showed a negative relationship with surface area of SWCNT. Additionally, in a 3-week study, the authors reported that SWCNTs showed similar toxicological responses to MWCNTs but at five-times lower concentrations due to the higher surface area of single-walled tubes than that of MWCNTs. Kerfahi et al. (2015) compared the effect of raw and acid-treated or fSWCNT on soil bacterial communities, at different concentrations of MWCNTs (0, 50, 500, and 5000 μg g^{-1}) to the soil. The soil DNA was extracted at 0, 2, and 8 weeks, and the V3 region of the 16S rRNA gene was PCR-amplified and sequenced to find out that the bacterial diversity was not affected by either type of MWCNT. On the other hand, the overall soil bacterial community composition was affected only by fMWCNT at high concentrations, and a detectable effect was noticed at 2 weeks, but the bacterial diversity remained unaffected. The authors ascribed this early effect to the acidic nature of fMWCNTs, which caused a decrease in soil pH at higher exposure concentrations and subsequently changed (temporarily) the soil bacterial communities (Kerfahi et al. 2015). The short-term effect of MWCNTs on the activity and biomass of microorganisms inhabiting two different soil types was investigated by Chung et al. (2011). Up to 5000 μg MWCNT g^{-1} soil was applied, and the activities of 1,4-β-glucosidase, cellobiohydrolase, xylosidase, 1,4-β-N-acetylglucosaminidase, and phosphatase as well as the microbial biomass were measured; most of the enzyme activities showed a tendency to be repressed under 500 μg MWCNT g^{-1} soil, and all enzymatic activities as well as microbial biomass *C* and *N* were significantly lowered under 5000 μg MWCNT g^{-1} soil, which suggested that high concentrations of MWCNTs could lower the microbial activity and biomass in soils. Chung et al. (2015) also studied the impact of GO at 0.5–1 mg kg^{-1} and noted a decrease up to 50% in the activity of key soil enzymes, including xylosidase, 1,4-ß-N-acetyl glucosaminidase, and phosphatase, after 21 days of exposure. All these reports indicate that CNMs may have a considerable negative effect on soil microbial communities.

Shan et al. (2015) investigated the effects of biochar, activated carbon (AC), WCNTs, and MWCNTs in various concentrations (0, 0.2, 20, and 2000 mg kg^{-1} dry soil) on the fate of ^{14}C-catechol and microbial community in the soil and reported that SWCNTs at 2000 mg kg^{-1} significantly reduced mineralization. The inhibitory effects of AC and SWCNTs on the mineralization were attributed to the inhibited soil microbial activities and the shifts in the microbial communities, as suggested by the reduced microbial biomass *C* and the separated phylogenetic distance. However, MWCNTs at 0.2 mg kg^{-1} significantly stimulated mineralization, compared with the control soil. Tong et al. (2016) evaluated the effects of nC_{60} aggregates of different particle sizes via organic solvents on soils with different organic matter contents by measuring the total microbial biomass, metabolic activity, and bacterial community structure and observed that nC_{60} aggregates, introduced as an aqueous suspension, had size-dependent effects on soil bacterial community composition in the low organic matter system but induced a minimal change in the microbial biomass and metabolic activity in soils with both high and low organic matter contents. The authors suggested that nC_{60} aggregates of smaller size may have negative impact on soil biota and soil organic matter and play a key role in modulating the environmental effect of nanomaterials. Karunakaran et al. (2013) evaluated the effect of nanosilica and silicon on soil properties, total bacterial population, and maize seed germination. Sodium silicate-treated soil inhibited plant-growth-promoting rhizobacteria in contrast to nanosilica. Bacterial population doubled in the presence of nanosilica from 4×10^5 CFU to 8×10^5 CFU per gram of soil. Nanosilica also promoted seed germination percentage in maize than the conventional Si sources. Rangaraj et al. (2014) reported enhanced microbial populations and total biomass content ($C = 1508$ µg g^{-1} and $N = 178$ µg g^{-1}) in maize rhizosphere due to application of nanosilica, which plays a vital role in influencing soil nutrient content and microbial biota and, hence, may promote the growth of maize crop.

21.2.2 *Effect of NPs on Soil Enzymes*

Soil enzyme activity, often treated as an indicator of soil health and soil biota, is also useful in determining the sustainability of agricultural ecosystems, particularly the physicochemical and microbiological processes in the soil (Caldwell 2005; Cao et al. 2016; Li et al. 2017). Fabricated NPs also show their impact on different soil enzymes in the rhizosphere. Du et al. (2011) established that TiO_2 and ZnO NPs affect protease, catalase, and peroxidase, but not urease. Dehydrogenase, urease, and phosphatases are among the most frequently evaluated soil enzymes (Burns et al. 2013). The influence of TiO_2 NPs on mung bean was investigated where activities of acid phosphatase, alkaline phosphatase, and dehydrogenase were noticed to rise by 67.3%, 72%, and 108.7%, respectively (Raliya et al. 2015). Ag-NPs (0, 0.024, 0.24, 4.80, and 9.60 µg g^{-1} dry soil) were found to inhibit the activity of soil exoenzymes dehydrogenase, urease, acid phosphatase, neutral phosphatase, and alkaline phosphatase in the rhizosphere of wetland plants (*Arundo donax*, *Iris wilsonii*, and *Typha orientalis*), effect being more pronounced with higher (4.80 and 9.60 µg g^{-1} dry soil) Ag-NPs levels. The lowest Ag-NPs concentration (0.024 µg g^{-1} dry soil) could cause adverse effects on the activity of exoenzymes in *T. orientalis* rhizosphere; the impact was limited in the rhizosphere of *I. wilsonii* and *A. donax*, suggesting that high concentrations of Ag-NPs could affect negatively all soil exoenzyme activities, while the impact of low Ag-NPs level was related mainly to plant species (Cao et al. 2017). High concentrations of Ag-NPs were toxic on rhizosphere microbes (Navarro et al. 2008, Choi et al. 2008, Das et al. 2012, Garcia et al. 2012), leading to a decline in the bulk of soil microorganisms and the capacity of enzyme production (Hemida et al. 1997; Cao et al. 2017).

Toxicity of NPs on microorganisms was given much emphasis, but studies on the effect of NPs on exoenzymes in the soil are relatively few. Experiments by Zheng et al. (2011) revealed the inhibitory effect of TiO_2 NPs on ammonia monooxygenase and nitrite oxidoreductase, the enzymes involved in the N_2 cycle, but no significant impact on the activities of exopolyphosphatase and polyphosphate kinase. The authors related these findings to the biological N and P removal as well as the depletion of NH_3-oxidizing bacterial population caused by the NPs. Du et al. (2011) noted inhibition of the activity of soil protease, catalase, peroxidase, but not of urease, in the presence of TiO_2 and ZnO NPs. Nonlethal doses of ZnO NP applied to *B. subtilis* and *Pseudomonas aeruginosa* completely inhibited or lowered the activities of enzymes involved in starch degradation, denitrification, and urea degradation (Santimano and Kowshik 2013). Increased dehydrogenase activity and a minimal effect of fluorescein diacetate hydrolase in the soil treated with nano-zerovalent iron (nZVI, 10,000 μg/g soil) were reported by Cullen et al. (2011). Studies on the effect of Ag-NPs (1, 10, 100, and 1000 μg g^{-1}) and silver ions on soil exoenzymes, urease, acid phosphatase, arylsulfatase, and β-glucosidase involved in nutrient cycles and on dehydrogenase and fluorescein diacetate hydrolase involved in microbial overall activity established that Ag-NPs could control all enzyme activities and particularly inhibit those of dehydrogenase and urease (Shin et al. 2012). Urease was most sensitive to Ag-NPs, which was inhibited at a concentration as low as 1 μg Ag-NP g^{-1}. It was reduced by approximately 90% at 1000 μg g^{-1} and could not recover after 7 days which indicates that Ag-NPs clearly inhibit the nitrogen cycling in the soil ecosystem. Dehydrogenase was inhibited by more than 50%, whereas the inhibitory effect of Ag-NPs was less than 50% on the rest of the enzymes.

Li et al. (2017) examined the impact of concentration and exposure time-dependent toxicity of CeO_2NPs (0, 100, 500, and 1000 mg kg^{-1} soil mixtures) on soil microorganisms by testing the activity of urease, phosphatase, and β-glucosidase in a soil-grass microcosm system. Higher NP concentration (100 mg kg^{-1} and above) inhibited the activity of urease and β-glucosidase but stimulated that of phosphatase. A longer contact time between NPs and the soil alleviated the NPs impact on soil enzymes and potentially reduced the NPs ecotoxicity in the soil environment. Simonin et al. (2016) exposed an agricultural soil to TiO_2 NPs at concentrations 1 and 500 mg kg^{-1} dry soil to know the response of the microbial community and to assess their impact on soil function, the nitrogen cycle. Simonin et al. (2016) found strong negative impacts on nitrification enzyme activities and the abundance of ammonia-oxidizing microorganism, TiO_2-NPs-triggered cascading negative effects on denitrification enzyme activity, and a deep modification of the bacterial community structure after just 90 days of exposure to even the lowest NPs concentration, realistic concentration of NPs. Results obtained from the various experiments plea for further research to assess how these emerging pollutants modify the soil health and broader ecosystem function.

proliferation, several new materials are expected to become a source of engineered NPs to the ecosystem. The rhizosphere has shown that NPs influence soil ecosystem, soil health, and plant productivity. The poorly understood interaction mechanisms of nanomaterials or engineered NPs with soil ecosystem and/or rhizospheric microorganisms have been discussed in this chapter. The regular accumulation and deposition of nanomaterials may influence the soil health both directly and indirectly. Thus, a framework allowing for the extrapolation of various results concerning the impact of nanomaterials or engineered NPs on soil ecosystem and rhizospheric microorganisms, and for assessment of threats of the prospective nanomaterials, is highly desirable.

References

Akhilesh BG, Shrivastava P, Privastava P, Pahu P (2013) Effect of recklessly disposed silver nanoparticles from laboratory on *Vigna radiata, Brassica juncea,* rhizosphere flora and drosophila melanogaster. Int J Pharm Bio Sci 4:590–601

Allard P, Darnajoux R, Phalyvong K, Bellenger JP (2013) Effects of tungsten and titanium oxide nanoparticles on the diazotrophic growth and metals acquisition by *Azotobacter vinelandii* under molybdenum limiting condition. Environ Sci Technol 47:2061–2068

Asli S, Neumann PM (2009) Colloidal suspensions of clay or titanium dioxide nanoparticles can inhibit leaf growth and transpiration via physical effects on root water transport. Plant Cell Environ 32:577–584

Bandyopadhyay S, Peralta-Videa JR (2012) Comparative toxicity assessment of CeO_2 and ZnO nanoparticles towards *Sinorhizobium meliloti*, a symbiotic alfalfa associated bacterium: use of advanced microscopic and spectroscopic techniques. J Hazard Mater 241:379–386

Barrena R, Casals E, Colón J, Font X, Sánchez A, Puntes V (2009) Evaluation of the ecotoxicity of model nanoparticles. Chemosphere 75:850–857

Battke F, Leopold K, Maier M, Schmidhalter U, Schuster M (2008) Palladium exposure of barley: uptake and effects. Plant Biol 10:272–276

Bhatia S (2016) Nanoparticles types, classification, characterization, fabrication methods and drug delivery applications. In: Natural polymer drug delivery systems: nanoparticles, plants, and algae. Springer International Publishing, Switzerland, pp 33–93

Biswas P, Wu CY (2005) Critical review, nanoparticles and the environment. J Air Waste Manag Assoc 55:708–746

Burns RG, Deforest JL, Marxsen J, Sinsabaugh RL, Stromberger ME, Wallenstein MD, Weintraub MN, Zoppini A (2013) Soil enzymes in a changing environment: current knowledge and future directions. Soil Biol Biochem 58:216–234

Caldwell BA (2005) Enzyme activities as a component of soil biodiversity: a review. Pedobiologia 49:637–644

Canas JE, Long M, Nations S, Vadan R, Dai L, Luo M, Ambikapathi R, Lee EH, Olszyk D (2008) Effects of functionalized and non functionalized single-walled carbon nanotubes on root elongation of select crop species. Environ Toxicol Chem 27:1922–1931

Cao C, Huang J, Cai W, Yan C, Liu J, Jiang Y (2016) Effects of silver nanoparticles on soil enzyme activity of different wetland plant soil systems, soil and sediment contamination. An In J 26:558–556

Cao C, Huang J, Cai WS, Yan CN, Liu JL, Jiang YD (2017) Effects of silver nanoparticles on soil enzyme activity of different wetland plant soil systems. Soil Sediment Contam Int J. https://doi.org/10.1080/15320383.2017.1363158

Chen Q, Wang H, Yang B, He F, Han X, Song Z (2015) Responses of soil ammonia- oxidizing microorganisms to repeated exposure of single-walled and multi-walled carbon nanotubes. Sci Total Environ 505:649–657

Cherchi C, Gu A (2010) Impact of titanium dioxide nanomaterials on nitrogen fixation rate and intracellular nitrogen storage in *Anabaena variabilis*. Environ Sci Technol 44:8302–8307

Choi O, Deng KK, Kim NJ, Ross L, Surampalli RY, Hu ZQ (2008) The inhibitory effects of silver nanoparticles, silver ions, and silver chloride colloids on microbial growth. Water Res 42:3066–3074

Chung H, Son Y, Yoon TK, Kim S, Kim W (2011) The effect of multi-walled carbon nanotubes on soil microbial activity. Ecotoxicol Environ Saf 74:569–575

Chung H, Kim MJ, Ko K, Kim JH, Kwon HA, Hong I (2015) Effects of graphene oxides on soil enzyme activity and microbial biomass. Sci Total Environ 514:307–313

Cullen LG, Tilston EL, Mitchell GR, Collins CD, Shaw LJ (2011) Assessing the impact of nano- and micro-scale zerovalent iron particles on soil microbial activities: particle reactivity interferes with assay conditions and interpretation of genuine microbial effects. Chemosphere 82:1675–1682

Das P, Williams CJ, Fulthorpe RR, Hoque ME, Metcalfe CD, Xenopoulos MA (2012) Changes in bacterial community structure after exposure to silver nanoparticles in natural waters. Environ Sci Technol 46:9120–9128

Dinesh R, Anandaraj M, Srinivasan V, Hamza S (2012) Engineered nanoparticles in the soil and their potential implications to microbial activity. Geoderma 173–174:19–27

Doshi R, Braida W, Christodoulatos C, Wazne M, O'Connor G (2008) Nano-aluminum: transport through sand columns and environmental effects on plants and soil communities. Environ Res 106:296–303

Du WC, Sun YY, Ji R, Zhu JG, Wu JC, Guo HY (2011) TiO_2 and ZnO nanoparticles negatively affect wheat growth and soil enzyme activities in agricultural soil. J Environ Monit 13:822–828

Dubchak S, Ogar A, Mietelski JW, Turnau K (2010) Influence of silver and titanium nanoparticles on arbuscular mycorrhiza colonization and accumulation of radiocaesium in *Helianthus annuus*. Span J Agric Res 8:S103–S108

El-Temsah YS, Joner EJ (2012) Impact of Fe and Ag nanoparticles on seed germination and differences in bioavailability during exposure in aqueous suspension and soil. Environ Toxicol 27:42–49

Fan R, Huang YC, Grusak MA, Huang CP, Sherrier DJ (2014) Effects of nano-TiO_2 on the agronomically-relevant Rhizobium-legume symbiosis. Sci Total Environ 466–467:503–512

Fang J, Lyon DY, Wiesner MR, Dong J, Alvarez PJJ (2007) Effect of a fullerene water suspension on bacterial phospholipids and membrane phase behavior. Environ Sci Technol 41:2636–2642

Feng Y, Cui X, He S, Dong G, Chen M, Wang J, Lin X (2013) The role of metal nanoparticles in influencing arbuscular mycorrhizal fungi effects on plant growth. Environ Sci Technol 47:9496–9504

Frenk S, Ben-Moshe T, Dror I, Berkowitz B, Minz D (2013) Effect of metal oxide nanoparticles on microbial community structure and function in two different soil types. PLoS One 8:e84441

Gao F, Liu C, Qu C, Zheng L, Yang F, Su M, Hong F (2008) Was improvement of spinach growth by nano-TiO_2 treatment related to the changes of *Rubisco activase*. Biometals 21:211–217

Garcia A, Delgado L, Tora JA, Casals E, Gonzalez E, Puntes V, Font X, Carrera J, Sanchez A (2012) Effect of cerium dioxide, titanium dioxide, silver, and gold nanoparticles on the activity of microbial communities intended in wastewater treatment. J Hazard Mater 199:64–72

Ge Y, Schimel JP, Holden PA (2012) Identification of soil bacteria susceptible to TiO_2 and ZnO nanoparticles. Appl Environ Microbiol 8:6749–6758

González-Melendi P, Fernández-Pacheco R, Coronado MJ, Corredor E, Testillano PS, Risueño MC, Marquina C, Ibarra MR, Rubiales D, Pérez-de-Luque A (2008) Nanoparticles as smart treatment-delivery systems in plants: assessment of different techniques of microscopy for their visualization in plant tissues. Ann Bot 101:187–195

Goodman J, Mclean JE, Britt DW, Anderson AJ (2016) Sublethal doses of ZnO nanoparticles remodel production of cell signaling metabolites in the root colonizer *Pseudomonas chlororaphis* O6. Environ Sci: Nano 3:1103–1113

Gowramma B, Keerthi U, Rafi M, Rao DM (2015) Biogenic silver nanoparticles production and characterization from native strain of *Corynebacterium* species and its antimicrobial activity. 3 Biotech 5:195–201

Gurunathan S (2015) Cytotoxicity of graphene oxide nanoparticles on plant growth promoting rhizobacteria. J Ind Eng Chem 32:282–291

Hajipour MJ, Fromm KM, Ashkarran AA, De Aberasturi D, De Larramendi IR, Rojo T, Serpooshan V, Parak WJ, Mahmoudi M (2012) Antibacterial properties of nanoparticles. Trends Biotechnol 30:499–511

Haris Z, Ahmad I (2017) Impact of metal oxide nanoparticles on beneficial soil microorganisms and their secondary metabolites. Int J Life Sci Scienti Res 3:1020–1030

Hartmann A, Rothballer M, Schmid M (2008) Lorenz Hiltner, a pioneer in rhizosphere microbial ecology and soil bacteriology research. Plant Soil 321:7–14

He S, Feng Y, Ren H, Zhang Y, Gu N, Lin X (2011) The impact of iron oxide magnetic nanoparticles on the soil bacterial community. J Soil Sediment 11:1408–1417

He S, Feng Y, Ni J, Sun Y, Xue L, Feng Y, Yu Y, Lin X, Yang L (2016) Different responses of soil microbial metabolic activity to silver and iron oxide nanoparticles. Chemosphere 147:195–202

Hemida SK, Omar SA, Abdel-Mallek AY (1997) Microbial populations and enzyme activity in soil treated with heavy metals. Water Air Soil Pollut 95:13–22

Hiltner L (1904) Under neuere Erfahrungen und probleme auf dem Gebiet der Bodenbaketriologie und unter besonderer Berucksichtingung der Grundungung und Brache. Arb Dtsch Landw Ges 98:59–78

Husen A (2017) Gold Nanoparticles from plant system: synthesis, characterization and their application. In: Ghorbanpour M, Manika K, Varma A (eds) Nanoscience and plant–soil systems, vol 48. Springer, Cham, pp 455–479

Husen A, Siddiqi KS (2014a) Carbon and fullerene nanomaterials in plant system. J Nanobiotechnology 12:16

Husen A, Siddiqi KS (2014b) Phytosynthesis of nanoparticles: concept, controversy and application. Nanoscale Res Lett 9:229

Husen A, Siddiqi KS (2014c) Plants and microbes assisted selenium nanoparticles: characterization and application. J Nanobiotechnology 12:28

Hussein AK (2016) Applications of nanotechnology to improve the performance of solar collectors – recent advances and overview. Renew Sust Energ Rev 62:767–792

Jackson P, Jacobsen NR, Baun A, Birkedal R, Kuhnel D, Jensen KA, Vogel U, Wallin H (2013) Bioaccumulation and ecotoxicity of carbon nanotubes. Chem Cent J 7:154

Jiang F, Shen Y, Ma C, Zhang X, Cao W, Rui Y (2017) Effects of TiO_2nanoparticles on wheat (*Triticum aestivum L.*) seedlings cultivated under super-elevated and normal CO_2 conditions. PLoS One 12(5):e0178088

Jin L, Son Y, Yoon TK, Kang YJ, Kim W, Chung H (2013) High concentrations of single- walled carbon nanotubes lower soil enzyme activity and microbial biomass. Ecotoxicol Environ Saf 88:9–15

Johansen A, Pedersen AL, Jensen KA, Karlson U, Hansen BM, Scott-Fordsmand JJ, Winding A (2008) Effects of C_{60} fullerene nanoparticles on soil bacteria and protozoans. Environ Toxicol Chem 27:1895–1903

Judy JD, McNear DH, Chen C, Lewis RW, Tsyusko OV, Bertsch PM, Rao W, Stegemeier J, Lowry GV, McGrath SP, Durenkamp M, Unrine JM (2015) Nanomaterials in biosolids inhibit nodulation, shift microbial community composition, and result in increased metal uptake relative to bulk/dissolved metals. Environ Sci Technol 49:8751–8758

Kang S, Pinault M, Pfefferle LD, Elimelech M (2007) Single-walled carbon nanotubes exhibit strong antimicrobial activity. Langmuir 23:8670–8673

Karimi E, Fard EM (2017) Nanomaterial effects on soil microorganisms. In: Ghorbanpour M, Khanuja M, Varma A (eds.) Nanoscience and plant–soil systems, vol 48, Springer, Cham, pp 137–200

Karunakaran G, Suriyaprabha R, Manivasakan P, Yuvakkumar R, Rajendran V, Kannan N (2013) Impact of nano and bulk ZrO_2, TiO_2 particles on soil nutrient contents and PGPR. J Nanosci Nanotechnol 13:678–685

Karunakaran G, Suriyaprabha R, Manivasakan P, Rajendran V, Kannan N (2014) Influence of nano and bulk SiO_2 and Al_2O_3 particles on PGPR and soil nutrient contents. Curr Nanosci 10:604–612

Kennedy AC, Smith KL (1995) Soil microbial diversity and the sustainability of agricultural soils. Plant Soil 170:75–86

Kerfahi D, Tripathi BM, Singh D, Kim H, Lee S, Lee J, Adams JM (2015) Effects of functionalized and raw multi-walled carbon nanotubes on soil bacterial community composition. PLoS One 10:e0123042

Khodakovskaya M, Dervishi E, Mahmood M, Xu Y, Li Z, Watanabe F Biris AS (2009) Carbon nanotubes are able to penetrate plant seed coat and dramatically affect seed germination and plant growth. ACS Nano 3:3221–3227

Khodakovskaya MV, Kim BS, Kim JN, Alimohammadi M, Dervishi E, Mustafa T, Cernigla CE (2013) Carbon nanotubes as plant growth regulators: effects on tomato growth, reproductive system, and soil microbial community. Small 9:115–123

Kumar P, Kumar A, Fernandes T, Ayoko GA (2014) Nanomaterials and the environment. J Nanomater 2014:528606

Kumari M, Mukherjee A, Chandrasekaran N (2009) Genotoxicity of silver nanoparticles in *Allium cepa*. Sci Total Environ 407:5243–5246

Kumari M, Khan SS, Pakrashi S, Mukherjee A, Chandrasekaran N (2011) Cytogenetic and genotoxic effects of zinc oxide nanoparticles on root cells of Allium cepa. J Hazard Mater 190:613–621

Kurepa J, Paunesku T, Vogt S, Arora H, Rabatic BM, Lu J, Wanzer MB, Woloschak GE, Smalle JA (2010) Uptake and distribution of ultrasmall anatase TiO_2 alizarin red S nanoconjugates in *Arabidopsis thaliana*. Nano Lett 10:2296–2302

Larue C, Khodja H, Herlin-Boime N, Brisset F, Flank AM, Fayard B, Chaillou S, Carrière M (2011) Investigation of titanium dioxide nanoparticles toxicity and uptake by plants. J Phys Conf Ser 304:012057

Lee W, An Y, Yoon H, Kweon H (2008a) Toxicity and bioavailability of copper nanoparticles to the terrestrial plants mung bean (*Phaseolus radiatus*) and wheat (*Triticum aestivum*): plant uptake for water insoluble nanoparticles. Environ Toxicol Chem 27:1915–1921

Lee WM, An YJ, Yoon H, Kweon HS (2008b) Toxicity and bioavailability of copper nanoparticles to the terrestrial plants mung bean (*Phaseolus radiatus*) and wheat (*Triticum aestivum*): plant agar test for water-insoluble nanoparticles. Environ Toxicol Chem 27:1915–1921

Lee CW, Mahendra S, Zodrow K, Li D, Tsai YC, Braam J, Alvarez PJ (2010) Developmental phytotoxicity of metal oxide nanoparticles to *Arabidopsis thaliana*. Environ Toxicol Chem 29:669–675

Li B, Chen Y, Liang W, Mu L, Bridges WC, Jacobson AR, Darnault CJG (2017) Influence of cerium oxide nanoparticles on the soil enzyme activities in a soil-grass microcosm system. Geoderma 299:54–62

Lin D, Xing B (2007) Phytotoxicity of nanoparticles: inhibition of seed germination and root growth. Environ Pollut 150:243–250

Lin D, Xing B (2008) Root Uptake and Phytotoxicity of ZnO Nanoparticles. Environ Sci Technol 42:5580–5585

Lin S, Reppert J, Hu Q, Hudson JS, Reid ML, Ratnikova TA, Rao AM, Luo H, Ke PC (2009) Uptake, translocation, and transmission of carbon nanomaterials in rice plants. Small 5:1128–1132

Liu Q, Chen B, Wang Q, Shi X, Xiao Z, Lin J, Fang X (2009) Carbon nanotubes as molecular transporters for walled plant cells. Nano Lett 9:1007–1010

Ma Y, Kuang L, He X, Bai W, Ding Y, Zhang Z, Zhao Y, Chai Z (2010) Effects of rare earth oxide nanoparticles on root elongation of plants. Chemosphere 78:273–279

Majumdar S, Peralta-Videa JR, Trujillo-Reyes J, Sun Y, Barrios AC, Niu G, Flores-Margez JP, Gardea-Torresdey JL (2016) Soil organic matter influences cerium translocation and physiological processes in kidney bean plants exposed to cerium oxide nanoparticles. Sci Total Environ 569–570:201–211

Manjunatha SB, Biradar DP, Aladakatti YR (2016) Nanotechnology and its applications in agriculture: a review. J Farm Sci 29:1–13

Mauter MS, Elimelech M (2008) Environmental applications of carbon-based nanomaterials. Environ Sci Technol 42:5843–5859

Mazumdar H, Ahmed GU (2011) Phytotoxicity effect of silver nanoparticles on *Oryza sativa*. Int J Chem Tech Res 3:1494–1500

Mirzajani F, Askari H, Hamzelou S, Farzaneh M, Ghassempour A (2013) Effect of silver nanoparticles on Oryza sativa L. and its rhizosphere bacteria. Ecotoxicol Environ Saf 88:48–54

Moll J, Klingenfuss F, Widmer F, Gogos A, Bucheli TD, Hartmann M, van der Heijden MGA (2017) Effects of titanium dioxide nanoparticles on soil microbial communities and wheat biomass. Soil Biol Biochem 111:85–93

Mukherjee A, Majumdar S, Servin AD, Pagano L, Dhankher OP, White JC (2016) Carbon nanomaterials in agriculture: a critical review. Front Plant Sci 7:172

Navarro E, Baun A, Behra R, Hartmann NB, Filser J, Miao AJ, Quigg A, Santschi PH, Sigg L (2008) Environmental behavior and ecotoxicity of engineered nanoparticles to algae, plants, and fungi. Ecotoxicology 17:372–386

Petersen EJ, Henry TB, Zhao J, Maccuspie RI, Kirschling TL, Dobrovolskaia MA, Hackley V, Xing B, White JC (2014) Identification and avoidance of potential artifacts and misinterpretations in nanomaterial ecotoxicity measurements. Environ Sci Technol 48:4226–4246

Prasad TNVKV, Sudhakar P, Sreenivasulu Y, Latha P, Munaswamy V, Raja Reddy K, Sreeprasad TS, Sajanlal PR, Pradeep T (2012) Effect of nanoscale zinc oxide particles on the germination, growth and yield of peanut. J Plant Nutr 35:906–927

Priester JH, Ge Y, Mielke RE, Horst AM, Moritz SC, Espinosa K, Gelb J, Walker SL, Nisbet RM, An YJ, Schimel JP, Palmer RG, Hernandez-Viezcas JA, Zhao L, Gardea-Torresdey JL, Holden PA (2012) Soybean susceptibility to manufactured nanomaterials with evidence for food quality and soil fertility interruption. Proc Natl Acad Sci U S A 109:E2451–E2456

Raliya R, Biswas P, Tarafdar JC (2015) TiO_2 nanoparticle biosynthesis and its physiological effect on mung bean (*Vigna radiata* L.). Biotechnol Rep 5:22–26

Rangaraj S, Gopalu K, Muthusamy P, Rathinam Y, Venkatachalam R, Narayanasamy K (2014) Augmented biocontrol action of silica nanoparticles and *Pseudomonas fluorescens* bioformulant in maize (*Zea mays* L.). RSC Adv 4:8461–8465

Rico CM, Morales MI, Barrios AC, McCreary R, Hong J, Lee WY, Nunez J, Peralta-Videa JR, Gardea-Torresdey JL (2013) Effect of cerium oxide nanoparticles on the quality of rice (*Oryza sativa* L.) grains. J Agric Food Chem 61:11278–11285

Rodrigues DF, Jaisi DP, Elimelech M (2013) Toxicity of functionalized single-walled carbon nanotubes on soil microbial communities: implications for nutrient cycling in soil. Environ Sci Technol 47:625–633

Rostami AA, Shahsavar A (2009) Nano-silver particles eliminate the in vitro contaminations of olive mission explants. Asian J Plant Sci 8:505–509

Santimano MC, Kowshik M (2013) Altered growth and enzyme expression profile of ZnO nanoparticles exposed non-target environmentally beneficial bacteria. Environ Monit Assess 185:7205–7214

Savithramma N, Ankanna S, Bhumi G (2012) Effect of nanoparticles on seed germination and seedling growth of *Boswellia ovalifoliolata* an endemic and endangered medicinal tree taxon. Nano Vis 2:61–68

Seeger EM, Baun A, Kästner M, Trapp S (2009) Insignificant acute toxicity of TiO_2 nanoparticles to willow trees. J Soil Sediment 9:46–53

Shah V, Belozerova I (2009) Influence of metal nanoparticles on the soil microbial community and germination of lettuce seeds. Water Air Soil Pollut 197:143–148

Shah V, Jones J, Dickman J, Greenman S (2014) Response of soil bacterial community to metal nanoparticles in biosolids. J Hazard Mater 274:399–403

Shan J, Ji R, Yu Y, Xie Z, Yan X (2015) Biochar, activated carbon, and carbon nanotubes have different effects on fate of 14C-catechol and microbial community in soil. Sci Rep 5:16000

Shaw AK, Ghosh S, Kalaji HM, Bosa K, Brestic M, Zivcak M, Hossain Z (2014) Nano-CuO stress induced modulation of antioxidative defense and photosynthetic performance of Syrian barley (Hordeum vulgare L.). Environ Exp Bot 102:37–47

Shcherbakova EN, Shcherbakov A, Andronov EE, Gonchar LN, Kalenskaya SM, Chebotar VK (2016) Combined pre-seed treatment with microbial inoculants and Mo nanoparticles changes composition of root exudates and rhizosphere microbiome structure of chickpea (*Cicer arietinum* L) plants. Symbiosis 73:57–69

Shen Z, Chen Z, Hou Z, Li T, Lu X (2015) Ecotoxicological effect of zinc oxide nanoparticles on soil microorganisms. Front Environ Sci Eng 9:912–918

Shin YJ, Kwak JI, An YJ (2012) Evidence for the inhibitory effects of silver nanoparticles on the activities of soil exoenzymes. Chemosphere 88:524–529

Shrestha B, Acosta-Martinez V, Cox SB, Green MJ, Li S, Canas-Carrell JE (2013) An evaluation of the impact of multiwalled carbon nanotubes on soil microbial community structure and functioning. J Hazard Mater 261:188–197

Siddiqi KS, Husen A (2016a) Fabrication of metal nanoparticles from fungi and metal salts: scope and application. Nanoscale Res Lett 11:98

Siddiqi KS, Husen A (2016b) Fabrication of metal and metal oxide nanoparticles by algae and their toxic effects. Nanoscale Res Lett 11:363

Siddiqi KS, Husen A (2016c) Green synthesis, characterization and uses of palladium/platinum nanoparticles. Nanoscale Res Lett 11:482

Siddiqi KS, Husen A (2016d) Engineered gold nanoparticles and plant adaptation potential. Nanoscale Res Lett 11:400

Siddiqi KS, Husen A (2017a) Recent advances in plant-mediated engineered gold nanoparticles and their application in biological system. J Trace Elem Med Biol 40:10–23

Siddiqi KS, Husen A (2017b) Plant response to engineered metal oxide nanoparticles. Nanoscale Res Lett 12:92

Siddiqi KS, Rahman A, Tajuddin, Husen A (2016) Biogenic fabrication of iron/iron oxide nanoparticles and their application. Nanoscale Res Lett 11:498

Siddiqi KS, Husen A, Rao RAK (2018a) A review on biosynthesis of silver nanoparticles and their biocidal properties. J Nanobiotechnology 16:14

Siddiqi KS, Rahman A, Tajuddin, Husen A (2018b) Properties of zinc oxide nanoparticles and their activity against microbes. Nanoscale Res Lett 13:141

Siddiqi KS, Husen A, Sohrab SS, Osman M (2018c) Recent status of nanomaterials fabrication and their potential applications in neurological disease management. Nanoscale Res Lett 13:231

Siddiqi KS, Rashid M, Rahman A, Tajuddin, Husen A, Rehman S (2018d) Biogenic fabrication and characterization of silver nanoparticles using aqueous-ethanolic extract of lichen (*Usnea longissima*) and their antimicrobial activity. Biomat Res 22:23

Simonet BM, Valcarcel M (2009) Monitoring nanoparticles in the environment. Anal Bioanal Chem 393:17–21

Simonin M, Richaume A (2015) Impact of engineered nanoparticles on the activity, abundance, and diversity of soil microbial communities: a review. Environ Sci Pollut Res 22:13710–13723

Simonin M, Richaume A, Guyonnet JP, Dubost A, Martins JMF, Pommier T (2016) Titanium dioxide nanoparticles strongly impact soil microbial function by affecting archaeal nitrifiers. Sci Rep 2016:33643

Speranza A, Leopold K, Maier M, Taddei AR, Scoccianti V (2010) Pd-nanoparticles cause increased toxicity to kiwifruit pollen compared to soluble Pd(II). Environ Pollut 158:873–882

Stampoulis D, Sinha SK, White JC (2009) Assay-dependent phytotoxicity of nanoparticles to plants. Environ Sci Technol 143:9473–9479

Su M, Liu H, Liu C, Qu C, Zheng L, Hong F (2009) Promotion of nano-anatase TiO2 on the spectral responses and photochemical activities of D1/D2/Cyt b559 complex of spinach. Spectrochim Acta Part A Mol Biomol Spectrosc 72:1112–1116

Sun TY, Gottschalk F, Hungerbühler K, Nowack B (2014) Comprehensive probabilistic modelling of environmental emissions of engineered nanomaterials. Environ Pollut 185:69–76

Sweet MJ, Singleton I (2015) Soil contamination with silver nanoparticles reduces Bishop pine growth and ectomycorrhizal diversity on pine roots. J Nanopart Res 17:448

Tan XM, Lin C, Fugetsu B (2009) Studies on toxicity of multi-walled carbon nanotubes on suspension rice cells. Carbon 47:3479–3487

Thul ST, Sarangi BK (2015) Implications of nanotechnology on plant productivity and its rhizospheric environment. In: Siddiqui MH, Al-Whaibi MH, Mohammad F (eds) Nanotech- nology and plant sciences. Springer, Cham, pp 37–54

Tomacheski DM, Pittol DN, Simões VF, Ribeiro RMC, Santana (2017) Impact of silver ions and silver nanoparticles on the plant growth and soil microorganisms. Global J Environ Sci Manage 3:341–350

Tong Z, Bischoff M, Nies L, Applegate B, Turco RF (2007) Impact of fullerene (C_{60}) on a soil microbial community. Environ Sci Technol 41:2985–2991

Tong Z, Bischoff M, Nies LF, Myer P, Applegate B, Turco RF (2012) Response of soil microorganisms to as-produced and functionalized single-wall carbon nanotubes (SWNTs). Environ Sci Technol 46:13471–13479

Tong ZH, Bischoff M, Nies LF, Carroll NJ, Applegate B, Turco RF (2016) Influence of fullerene (C60) on soil bacterial communities: aqueous aggregate size and solvent co-introduction effects. Sci Rep 6:28069

Tripathi DK, Tripathi A, Shweta SS, Singh Y, Vishwakarma K, Yadav G, Sharma S, Singh VK, Mishra RK, Upadhyay RG, Dubey NK, Lee Y, Chauhan DK (2017) Uptake, accumulation and toxicity of silver nanoparticle in autotrophic plants, and heterotrophic microbes: a concentric review. Front Microbiol 8:07

Vance ME, Kuiken T, Vejerano EP, McGinnis SP, Hochella MF, Rejeski D, Hull MS (2015) Nanotechnology in the real world: redeveloping the nanomaterial consumer products inventory. Beilstein J Nanotechnol 6:1769–1780

Wang H, Law N, Pearson G, van Dongen BE, Jarvis RM, Goodacre R, Lloyd JR (2010) Impact of silver (I) on the metabolism of *Shewanella oneidensis*. J Bacteriol 192:1143–1150

Wang J, Shu K, Zhang l SY (2017) Effects of silver nanoparticles on soil microbial communities and bacterial nitrification in suburban vegetable soils. Pedosphere 27:482–490

Wild E, Jones KC (2009) Novel method for the direct visualization of in vivo nanomaterials and chemical interactions in plants. Environ Sci Technol 43:5290–5294

Xu C, Peng C, Sun L, Zhang S, Huang H, Chen Y, Shi J (2015) Distinctive effects of TiO_2 and CuO nanoparticles on soil microbes and their community structures in flooded paddy soil. Soil Biol Biochem 86:24–33

Yang L, Watts DJ (2005) Particle surface characteristics may play an important role in phytotoxicity of alumina nanoparticles. Toxicol Lett 158:122–132

Yang F, Liu C, Gao F, Su M, Wu X, Zheng L, Hong F, Yang P (2007) The improvement of spinach growth by nanoanatase TiO_2 treatment is related to nitrogen photoreduction. Biol Trace Elem Res 119:77–88

Yuan Z, Li J, Cui L, Xu B, Zhang H, Yu CP (2013) Interaction of silver nanoparticles with pure nitrifying bacteria. Chemosphere 90:1404–1411

Zheng X, Chen YG, Wu R (2011) Long-term effects of titanium dioxide nanoparticles on nitrogen and phosphorus removal from wastewater and bacterial community shift in activated sludge. Environ Sci Technol 45:7284–7290

Zhu H, Han J, Xiao JQ, Jin Y (2008) Uptake, translocation, and accumulation of manufactured iron oxide nanoparticles by pumpkin plants. J Environ Monit 10:713–717

Chapter 22
Effect of Carbon-Based Nanomaterials on Rhizosphere and Plant Functioning

Javed Ahmad Wagay, Sanjay Singh, Mohammed Raffi, Qazi Inamur Rahman, and Azamal Husen

22.1 Introduction

Carbon-based nanomaterials (CBNMs), often called majic carbons, have shown potential application in various disciplines of science due to their exceptional physical, chemical, and mechanical properties (Hurt et al. 2006; Bennett et al. 2013; Srivastava et al. 2015). Their synthesis was started around the year 1985 (Baughman et al. 2002). Carbon is considered to be the most stable atom with balanced thermodynamic configurations (sp^3 to sp^2) and has given rise to different allotropic structures including carbon nanotubes (CNTs), fullerenes, nano-diamonds, graphene sheets, nano-onions, nano-cones, nano-horns, carbon dots, nano-beads, and nano-fibers with change in the reaction conditions (Mauter and Elimelech 2008; Sharon and Sharon 2010; Cha et al. 2013; Baptista et al. 2015). These entities have a vast heterogeneity in their morphology and size, including the outer range of dimensions within the nanoscale measurement. The exclusive nanoscale dimensions of CBNMs can be illustrated by taking the

J. A. Wagay
Department of Plant Sciences, College of Agriculture and Rural Transformation, University of Gondar, Gondar, Ethiopia

S. Singh
Department of Plant Science, College of Agriculture and Natural Resources, Mizan-Tepi University, Mizan, Ethiopia

M. Raffi
Department of Microbiology, College of Agriculture and Rural Transformation, University of Gondar, Gondar, Ethiopia

Q. I. Rahman
Department of Chemistry, College of Natural and Computational Sciences, University of Gondar, Gondar, Ethiopia

A. Husen (✉)
Department of Biology, College of Natural and Computational Sciences, University of Gondar, Gondar, Ethiopia

A. Husen, M. Iqbal (eds.), *Nanomaterials and Plant Potential*,
https://doi.org/10.1007/978-3-030-05569-1_22

example of graphene that has only length and breadth with sp^2-C-atoms in a checkered six crystal lattice with sp^3-C-atoms at the defect sites. Because of this dimensional character, graphene is considered as an important fundamental compound for its other nanoallotropes (Georgakilas et al. 2015). Fullerene is any allotrope (molecular form) of carbon, in the form of a hollow sphere, ellipsoid, or tube. A 20-polyhedron facet symmetry (nC_{60}) is the most common structure of fullerene, including C_{28} and C_{36} (Ugarte 1992; Wang et al. 2001). Nano-onions are coaxial poly-layered fullerenes with shining prospects in bioindustry (Chichiriccò and Poma 2015) because of having a promising thermal and electrical conductivity with prominent mechanical properties. Similarly, CNTs are cylindrical in morphology with several nanometers in diameter and grapheme-rolled sheets showing varying chiralities. They are of two main types: single-walled nanotubes (SWCNTs) and multi-walled nanotubes (MWCNTs).

In plant system, these materials cause increased yield; influence plant functioning, nutrient uptake, and protection from disease; and may also lead to cytotoxicity and genetic alteration (Husen and Siddiqi 2014a, b; Mukherjee et al. 2016). They affect the bioavailability and toxicity of organic contaminants in the soil due to their adsorption properties and also have the ability to influence soil rhizosphere microbial community (Khodakovskaya et al. 2013; Shrestha et al. 2013; Mukherjee et al. 2016). These nanomaterials enter the plant cells mostly through stomata and cell wall as the size of CBNMs is less than its pore diameter of stomata (Torney et al. 2007; Zhang et al. 2007). The CBNMs have shown positive as well as negative impacts on plant growth and functioning (Husen and Siddiqi 2014a, b; Mukherjee et al. 2016). They may accumulate in some plant species and may not in some others (Husen and Siddiqi 2014a, b; Mukherjee et al. 2016).

MWCNT-encapsulated fungicides have also been used against *Alternaria alternata* (Sarlak et al. 2014). Similarly, graphene oxide (GO)-encapsulated fertilizers have caused slow, steady, and efficient discharge for longer periods (Zhang et al. 2014). Potassium nitrate encapsulated in GO has resulted in extended discharge and its longer accessibility to plant roots (Zhang et al. 2014). CBNMs have synergistic impact on most of the mycocides because of their antifungal character (Wang et al. 2014). Burlaka et al. (2015) found the CBNMs as good delivery agents of gene to cells during plant transgenic techniques, because of their specific chemical structure (Liu et al. 2009). This chapter aims at elucidating CBNMs and their impact on plant functioning and on the rhizosphere and its associated soil microflora.

22.2 Plant Growth and Physiological Response to CBNMs

CBNMs have shown potential to increase seed germination and plant growth (Husen and Siddiqi 2014a, b), the extent of the effect being dependent on plant species, nature of the growth medium, CNM type/concentration, and growth conditions (Table 22.1). The roots of wheat showed fast growth (10×) on application of 150 mg L^{-1} of CBNMs (carbon nano-dots) up to 10 days as compared to control group (Tripathi and Sarkar 2015). Sonkar et al. (2012) synthesized water-soluble carbon nano-onions (wsCNOs) from wood wool (waste product of wood) and examined their interaction with *Cicer arietinum*, which appeared positive

Table 22.1 Effect of CBNMs on various plant growth parameters in different plant species at different concentrations, incubation periods, and culture media

Key references	Plant	CBNMs	Concentration	Culture medium	Incubation period	Effects on plants
Joshi et al. (2018a)	Oat	MWCNT	70, 80, 90 μg mL^{-1}	Distilled water/ soil-based	10 days/6 h, and later raised in earthen pots	Stimulated seed germination and seedling growth; increased number of roots and root hair, amount of vascular tissues, chlorophyll content, stomatal conductance, relative water status, and photosynthetic efficiency; uptake of MWCNT inside the primed seeds and vascular bundles confirmed by TEM
Fathi et al. (2017)	Castor plant	MWCNTs	50 and 100 μg mL^{-1}	NA	7 days	Increased seed germination by 96.7%, seedling vigor index was 400, increased fresh and dry weights
Hasaneen et al. (2016)	French bean	CNTs	20,50 μg L^{-1}	NA	30 days	Significant increase in plant growth and biomass; TEM images confirmed CNTs in leaves, in vascular bundles specifically in sieve tubes of the phloem, improved water absorption and nutrient uptake
Lahiani et al. (2015)	Barley Corn Rice Soybean Switch grass Tomato	CNHs	25, 50 and 100 mg mL^{-1}	Tobacco cell culture	10–20 days	Growth increased by 78%, uptake confirmed by TEM
Liu et al. (2015)	Rice	Graphene, GO, graphene ribbon	5, 50, 100, 200 mg L^{-1}	NA	16 days	Increased germination at lower concentrations but delayed at higher concentrations of graphene
Tripathi and Sarkar (2015)	Wheat	Carbon nano-dots	150 mg L^{-1}	NA	10 days	Enhanced root growth (10×)
Zhang et al. (2015)	Tomato	CNTs	40 mg L^{-1}	Distilled water	11 days	Accelerated germination

(continued)

Table 22.1 (continued)

Key references	Plant	CBNMs	Concentration	Culture medium	Incubation period	Effects on plants
Haghighi and Teixeira da Silva (2014)	Tomato Onion Turnip Radish	CNTs	10, 20 and 40 mgL^{-1}	Ultra-pure water	12 days	Improved germination in tomato and onion
Hu and Zhou (2014)	Wheat	Graphene, GO, graphene ribbon	200 mg L^{-1}	–	5 days	Graphene ribbon-enhanced germination but graphene and GO in combination delayed it
Saxena et al. (2014)	Wheat	wsCNPs	10–150 mg L^{-1}	NA	20 days	Optimum growth observed at 50 mg L^{-1} treatment where root and shoot lengths increased up to 3 times
Srivastava and Rao (2014)	Wheat Maize Peanut Garlic bulb	oMWCNTs	20, 50 mg L^{-1}	Distilled water	5 days	Enhanced germination
Tiwari et al. (2014)	Maize	MWCNTs	5, 10, 20, 40, 60 mg L^{-1}	Nutrient agar medium	7 days	At 20 mg L^{-1} increased dry weight along with Ca and Fe contents but depressed growth at higher dose
Khodakovskaya et al. (2013)	Wheat Maize Peanut Garlic	wsMWCNTs	20, 50 mg L^{-1}	MS medium, later soil	72 days	Increased flowering/fruit ratio and growth
Kole et al. (2013)	Bitter melon	Fullerols $C_{60}(OH)_{20}$	0.943, 4.72, 9.43, 10.88 and 47.2 mg L^{-1}	NA	NA	Increased plant biomass and phytomedicine content
Lahiani et al. (2013)	Barley Soybean	MWCNT	50, 100, and 200 mg L^{-1}	NA	10–11 days	Germination increased by 50% (barley and soybean) and 90% (corn). Root length increased up to 26% (soybean); shoot and leaf length increased by 40% (corn). MWCNT internalization was visualized by Raman spectroscopy and TEM images

Liang et al. (2013)	Tobacco	CNPs	25,75, 125 mg pot^{-1}	NA	NA	Enhanced primary growth, leaf area, increased dry matter and chlorophyll, soluble protein, nitrogen and potassium contents
Jiang et al. (2012)	Rice	MWCNTs	50, 100, 150 mg L^{-1}	MS medium	6 days	Enhanced germination
Nair et al. (2012)	Rice	CNTs	NA	Basal growth medium	NA	Increased seed water content and germination rate
Sonkar et al. (2012)	Gram	Water-soluble carbon nano-onions	10, 20, and 30 mg L^{-1}	NA	5 and 10 days	Enhancement in growth
Wang et al. (2012)	Wheat	o-MWCNT	0, 80, and 160 mg L^{-1}	NA	3 and 7 days	Increase in root length
Khodakovskaya et al. (2011)	Tomato	SWCNT and MWCNT	50 mg L^{-1}	NA	10 days	Enhanced the total fresh biomass
Mondal et al. (2011)	Mustard	Pristine, oxidized-MWCNT	2.3–46.0 mg L^{-1}	NA	NA	Enhanced germination, increased root and shoot growth
Tripathi et al. (2011)	Gram	Citrate-coated water-soluble CNTs	6.0 µg mL^{-1}	NA	10 days	Visualized internalization of coated wsCNTs by SEM and TEM images
Shen et al. (2010)	*Arabidopsis* Rice	SWCNTs	5–250 mg L^{-1}	Cell culture	6–72 h	Increased cell aggregation, chromatin condensation, plasma membrane deposition, H_2O_2 accumulation
Khodakovskaya et al. (2009)	Tomato	MWCNTs	10, 20, 40 mg L^{-1}	MS medium	20 days	Increased seed moisture content and accelerated germination rate
Wild and Jones (2009)	Wheat	MWCNT	0, 160, 900, and 5000 mg L^{-1}	NA	24 h and 48 h	CNTs were adsorbed onto the root surface, pierced the root epidermal cells, and accumulated within the tissue

(continued)

Table 22.1 (continued)

Key references	Plant	CBNMs	Concentration	Culture medium	Incubation period	Effects on plants
Cañas et al. (2008a)	Cucumber Onion	Uncoated and PABS coated SWCNTs	0, 104, 315, and 1750 mg L^{-1}	NA	24 h and 48 h	Uncoated CNTs increased root length in onion and cucumber in comparison to coated CNTs
Lin and Xing (2007)	Rye grass	MWCNT	2000 mg L^{-1}	NA	NA	Increased root length by 17%

throughout the life cycle of the plant. The authors used four treatments (0, 20, 30 mg L^{-1}) of wsCNOs along with control (without wsCNOs) and noted an increase in plant growth (shoot length, branching, number of roots and length, water uptake, etc.) with increase in concentration of wsCNOs. The phenotypes of *C. arietinum* plants, as found after 5 days (A) and 10 days (B), are shown in Fig. 22.1. They claimed that improvement in *C. arietinum* growth and yield was connected with the uptake of wsCNOs (embedded in the xylem vessels) as shown by scanning electron microscopy. Cañas et al. (2008a) have reported that SWCNTs enhanced the root growth of *Cucumis sativus* and *Allium cepa* but had adverse effects on *Lactuca sativa*, *Daucus carota*, and *Brassica oleracea*. Zhang et al. (2015) treated *Lycopersicon esculentum* seeds cotton-cushioned glass bottles with 40 mg L^{-1} graphene. They have reported that germination rate increased from 26% on the second day to 43% on the fourth day in comparison to control. CBNMs could cause not only root emergence but also shoot emergence in various plant species (Table 22.1).

Khodakovskaya et al. (2009) have also investigated *L. esculentum* from germination to flowering and found that MWCNTs caused a fast increase of shoot height and enhanced flowering by two times than in the control. The treatment increased the yield of tomato fruit by 200% at a minimum dose of 50 µg mL^{-1}. Some other studies also mentioned that SWCNTs (10–40 mg L^{-1}) increased the rate of seed germination and caused perforation in seed testa in many plants such as *Capsicum annuum*, *Festuca arundinacea*, and *Salvia macrosiphon* (Pourkhaloee et al. 2011).

Fig. 22.1 Phenotypic images of gram plants treated with different concentrations of water-soluble carbon nano-onions: (**a**) plants after 5 days; (**b**) plants after 10 days; (*a*) control plant; (*b–d*) plants treated with 10, 20, and 30 mg mL^{-1} wsCNOs, respectively. (Adapted from Sonkar et al. 2012)

The best germination rates were shown by *Capsicum annuum* at 10 mg L^{-1} and *Salvia macrosiphon* at 30 mg L^{-1} concentration of SWCNTs (Yan et al. 2013). MWCNTs also exhibited reflective stimulatory germination property in some plants such as *Oryza sativa* and *L. esculentum* (Khodakovskaya et al. 2009; Biris and Khodakovskaya 2012). SWCNTs caused a fast germination response in *Hordeum vulgare*, *Glycine max*, *Zea mays*, and *Brassica juncea* (Lahiani et al. 2013; Mondal et al. 2011). The cause of germination in these plants by SWCNTs was linked with the process of imbibition caused by the rupturing of seed testa which in turn could switch on *SLR1 and RTCS* aquaporin genes (Lahiani et al. 2013; Yan et al. 2013). Tripathi et al. (2011) showed the effect of citrate-coated wsCNTs in *Cicer arietinum* after a 10-day exposure to 6 mg L^{-1} concentration and observed internalization of CNTs by electron microscopy. They further reported that once these materials enter into the vascular tissue, they make an aligned network that results in a high water uptake which has a positive correlation with plant growth. In a recent study Kumar et al. (2018) have also reported that CNP treatment increased seed germination and caused elongated hypocotyl, larger cotyledon area, and enhanced chlorophyll content in *Arabidopsis thaliana* seedlings.

Villagarcia et al. (2012) determined the physiological response of tomato by exposing seeds to five types of CNTs. The results indicated correlation between the physiological responses of seeds and surface chemistry of CNTs, which causes a significant change in seed germination rates. In an experiment it was found that fullerol accelerates biomass accumulation and amounts of medicinal compounds in *Momordica charantia* (Kole et al. 2013). Recently, Joshi et al. (2018b) have reported an augmented growth of MWCNT-primed *Triticum aestivum*, increased stomatal density, xylem-phloem size, epidermal cells, and water uptake, whereas no DNA damage was recorded. In another investigation, Fan et al. (2018) have shown that the 50 mg L^{-1} MWCNT was beneficial for photosynthesis and lateral root emergence, but toxic for root growth in *A. thaliana*. The authors also claimed that the relative electron transport rate and photochemical quantum yield of PSII were increased ~12% while the lateral root production up to nearly fourfold in comparison to control plants. Zhao et al. (2017) recorded the MWCNT content in *A. thaliana*, rice, maize, and soybean by using ^{14}C-labeled MWCNTs. The MWCNT content varied in different plant parts possibly because of different plant species used and the different sizes of MWCNTs. They also noticed that changes in biochemical features were more sensitive than physiological attributes. It has been reported that CNTs induced positive effects on flowering (Khodakovskaya et al. 2013), fruiting (Khodakovskaya et al. 2013), and water channel protein production in tomato (Villagarcia et al. 2012) and photosynthetic activity in spinach (Giraldo et al. 2014) along with root growth in maize and wheat (Miralles et al. 2012; Begum et al. 2012; Yan et al. 2013). On the other hand, reduced growth of shoot and root (Cañas et al. 2008b; Begum et al. 2012; Yan et al. 2013), enhanced production of reactive oxygen species (Lin et al. 2009; Tan et al. 2009), and chromosomal aberrations (Ghosh et al. 2011) due to CNT exposure have also been reported. Environmental abiotic stresses such as salinity and drought negatively affect the overall plant growth and development (Husen et al. 2014, 2016, 2017, 2018, 2019;

Getnet et al. 2015; Embiale et al. 2016), whereas CBNMs move simultaneously with other nutrients in soil (Nair et al. 2010; Rico et al. 2011; Husen and Siddiqi 2014a, b; Martínez-Ballesta et al. 2016) and may have an effect on plant stress response mechanisms under salinity and/or drought stress. For instance, Martínez-Ballesta et al. (2016) used MWCNTs in control and 100 mM NaCl-treated broccoli plants. It was shown by TEM images that MWCNTs could enter cells in adult plants, with a higher accumulation under salt stress. MWCNTs increased water uptake under stressed condition, promoted growth, and increased the net assimilation of CO_2. MWCNTs also induced changes in the lipid composition, rigidity, and permeability of the root plasma membranes, relative to the salt-stressed plants. Moreover, the increased aquaporin transduction was observed, which improved water uptake and transport and mitigated the negative impacts of salt stress. The role of graphene-based nanomaterials has also been investigated. In a recent experiment, Ghorbanpour et al. (2018) used various doses (100–800 μg mL^{-1}) of nano-graphene oxide and observed their response influence on morphological, physiological, and biochemical features of *Plantago major* calli cultures under normal and polyethylene glycol-induced drought stress conditions. They found that lower doses (100 and 200 μg mL^{-1}) of nano-graphene oxide mitigated the adverse effects of drought stress through enhanced proline content and decreased H_2O_2 level. However, higher doses caused oxidative stress and showed toxic effects on the physiological regulations of calli. Several factors related to nanotube, plant species, and experimental conditions were found to be responsible for modulating the impact of MWCNTs on plant growth. In an experiment, He et al. (2017) analyzed the effect of GO on spinach and chive plants and found that the lower concentration of GO promoted seed germination and plant growth. They suggested that the oxygen-containing functional groups of GO collected water, and the hydrophobic sp^2 domains transported water to seeds to accelerate germination. Additionally, GO was not detected either on the surface or inside the plant cells; and they claimed that it was not phytotoxic. Recently, Joshi et al. (2018a) used functionalized MWCNT treatment to *Avena sativa* plants at 70, 80, and 90 μg mL^{-1}. MWCNTs were absorbed by seed and taken up by roots into shoots (Fig. 22.2). The authors have suggested that application of MWCNTs stimulated seed germination and plant growth and facilitated more water uptake by increasing the number of roots, root hair, vascular tissues (enhanced the growth of xylem cells by about 1.85-fold in the shoot vasculature), stomatal conductance, chlorophyll content (by 57%), and photosynthetic efficiency (by 15%). Moreover, no toxic effects of MWCNT were observed on DNA of the primed plants and in the human cell lines treated with grains harvested from the MWCNT-primed plants (Joshi et al. 2018a).

22.3 Biomass Production and Yield in Response to CBNMs

CBNMs, such as SWCNT and MWCNT (50 μg mL^{-1}), added to Murashige and Skoog growth medium increased the biomass of *L. esculentum* seeds by 75% and 110%, respectively, in comparison to activated carbon and graphene (Khodakovskaya

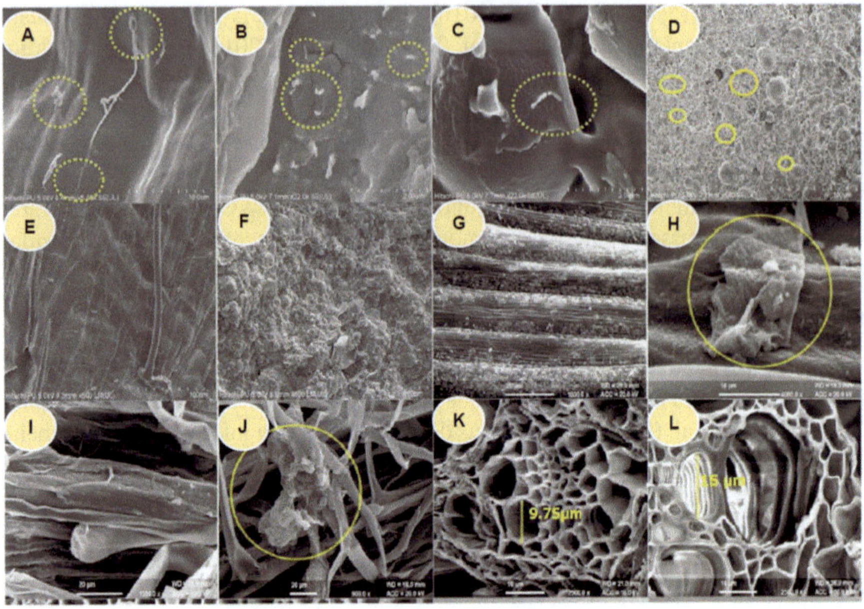

Fig. 22.2 Field emission scanning electron microscope images of the surface of oat seeds primed with MWCNT showing their entry into the seed (yellow circles; **a–d**); no similar activity was seen in seeds of control samples (**e**, **f**); SEM images of the surface of main shoot of oat; control (**g**), FW90 primed (**h**), inside the main roots of control (**i**) functionalized multi-walled carbon nanotubes 90 μg mL^{-1} primed (**j**; yellow circle), xylem vessels in cross sections of shoots of control (**k**) and functionalized multi-walled carbon nanotubes 90 μg mL^{-1} primed (**l**, enlarged vessels). (Adapted from Joshi et al. 2018a)

et al. 2011). In another study, Khodakovskaya et al. (2012) have compared the influence of MWCNT and activated carbon on the tobacco cells and recorded the growth enhancement of tobacco cells 55–64% higher at 5–500 μg mL^{-1} MWCNT in comparison to untreated controls. However, activated carbon could increase the cell growth by 16% at low concentration (5 μg mL^{-1}), while their growth was suppressed by ~25% at higher concentration (100–500 μg mL^{-1}). The authors claimed that the upregulation of several genes due to CNT treatments, including an aquaporin (*NtPIP1*) and two other genes (*CycB* and *NtLRX*), facilitates water transport, cell wall formation, and cell division.

Another CBNMs, fullerol, has shown its positive role not only in relation to biomass and yield of *Momordica charantia* but also to the production of metabolites of therapeutic value, as reported by Kole et al. (2013). They found that different concentrations of fullerol increased the biomass by more than 54%, water content by 24%, fruit length by 20%, fruit number by 59%, and fruit weight by 70%, besides the total fruit yield improvement up to 128%. Similarly, fullerol caused an increase of 74% in cucurbitacin-B7, 82% in lycopene, and 20% in charantin, used for the treatment of AIDS, diabetes, and cancer, respectively (Ng et al. 1992; Raman and Lau 1996; Basch et al. 2003). Kole et al. (2013) further confirmed the uptake, translocation, and accumulation of fullerol in different parts of the tested plant by bright-field imaging and FTRI spectroscopy. Saxena et al. (2014) analyzed the impact of wsCNPs present in biochar on wheat and found an optimum growth of plants at 50 mg L^{-1} among all concentrations (10–150 mg L^{-1}) applied up to 3 weeks, with a prominent increase of root and shoot length (up to three times with respect to untreated controls).

Lahiani et al. (2015) reported that carbon nano-horns (CNHs) were actively involved in increasing growth and biomass of some terrestrial plants. Sonkar et al. (2012) applied wsCNOs (30 μg mL^{-1}) isolated from wood wool (pyrolysis waste product of wood) on *Cicer arietinum* plants for 10 days and observed higher growth rate of the wsCNOs pre-treated plants in soil, in terms of biomass accumulation. The carbon, hydrogen, and nitrogen (CHN) content of shoot and fruits in plants with and without wsCNO treatment showed a significant difference. The significant increase in percentage of carbon and hydrogen in shoots reflects an enhanced synthesis of organic biomass in wsCNO-treated plants. The results also indicated that wsCNOs are nontoxic to plant cells and play a good role as active growth stimulants and could be used as benign growth promoters. Recently, Joshi et al. (2018b) have reported an augmented growth of MWCNT-primed *Triticum aestivum*, showing improved grain number and biomass.

22.4 Photosynthetic Response to CBNMs

CBNMs have been reported to promote the functioning of photosynthetic apparatus in leaves and enhance the rate of photosynthesis (Schmitz-Linneweber 2001). SWCNTs are able to activate the function of chloroplast (Heller 2009). They accumulate in the chloroplast and increase the rate of reduction of dichlorophenol-indophenol (DCPIP), an oxidizing agent, as observed in the spinach chloroplast (Schmitz-Linneweber 2001). DCPIP catches electrons released from the reaction center of photosystem II to photosystem I (Kim 2009). The chloroplast DCPIP was three times reduced than in the control after 6 h exposure, thus elevating the photosynthetic efficiency significantly. The reason for high photosynthetic activity by SWCNTs is the fast electron transfer between SWCNTs and chloroplasts, as the SWCNTs reportedly generate a photoelectrochemical current on in vitro extracted reaction centers and nano-disc lipids (Ham 2010), as confirmed by the study of spinach thylakoids (Boghossian 2013; Calkins et al. 2013).

In a recent study, Joshi et al. (2018a) have assessed the impact of MWCNTs on the photosynthetic ability and chlorophyll concentration in the leaves of *A. sativa*. They found that chlorophyll concentration was higher in CNT-primed plants and suggested that this might be related to an improvement in the overall plant growth by CNT treatments. In this experiment, *Fv/Fm* was also increased by the CNT treatments, which obviously increases chlorophyll concentration and photosynthesis. Earlier studies have also shown that SWCNTs promote the photosynthetic activity and enhance the electron transport rate, as compared to control (Ham 2010;

Boghossian 2013; Calkins et al. 2013). It has also been suggested that the major encouraging quality of CNTs is to absorb a wide range of wavelengths that allow two bandgaps to change the solar light into excitons, which facilitates the transfer of more electron flow in photosynthesis and increases chlorophyll concentrations (Wong et al. 2017).

22.5 CBNMs as Plant Fertilizers and Disease Protectors

A pretty number of reports are available justifying the importance of different CBNMs as fertilizers called as nanofertilizers. Most of these are altered mineral and organic fertilizers with nano-carbons that ensure fast nutrient availability and flow, prevent nutrient loss, and improve plant growth. CBNMs are being used to wrap fertilizers for their high stability and efficiency in different crops (González-Melendi et al. 2008). Films of GO are used to encapsulate for slow release of fertilizers (Zhang et al. 2014) and soil-mobile nutrients such as potassium nitrate that causes delayed process of fertilizer release, which results in a maintained nutrient uptake of plants (Zhang et al. 2014). CBNMs are being remarcably used in dapper delivery mechanism of fertilizers and organic molecules such as oligonucleotides or DNA into plant cells, because of their fast penetrating properties into the plant cells through cellulosic cell wall and plasma lemma (Liu et al. 2009).

The impact of CBNMs on pesticide residue in zucchini, corn, tomato, and soybean was reported to reduce in the presence of carbon nanotubes (Torre-Roche et al. 2013). The CBNMs were reported to have a slow-releasing effect on different pesticides on crops (Perez-de-Luque and Rubiales 2009). Sarlak et al. (2014) demonstrated that fungicides, encapsulated in MWCNTs functionalized with citric acid, had a higher toxicity against *Alternaria alternata* fungi.

CBNMs are substantial compounds for synthesis of broad-spectrum fungicides to control different plant diseases caused by fungi (Wang et al. 2014). CNTs, fullerenes, and GO were reported to have strong antifungal properties and used as fungicides against *Fusarium graminearum* and *F. poae* (Wang et al. 2014). Having an intimate association with fungal spores, CBNMs cause exosmosis and plasmolysis, leading to the death of fungal spores. In some other reports, the underlying antifungal property of GO has been ascribed to the initiation of membrane catalysis which creates imbalance of membrane potential (Chen et al. 2014) and electron transport (Liu et al. 2017) and generation of reactive oxygen species (Hui et al. 2014, Mangadlao et al. 2015). The antibacterial properties of CBNMs also depend on their size, and basal plane larger sheets of GO were found to have higher antibacterial properties than its smaller sheets (Akhavan & Ghaderi 2010; Liu et al. 2011). SWCNTs and MWCNTs were also reported to have an antibacterial activity against some bacterial strains (Liu et al. 2009; Kang et al. 2008). Nano sheets of GO and its reduced form have the most inhibitive effect against *Escherichia coli* bacterium but no cytotoxicity to A_{549} cells (Hu et al. 2010). SWCNTs and GO also have a prominent wound healing property to prevent infection, allow oxygen to the wound site,

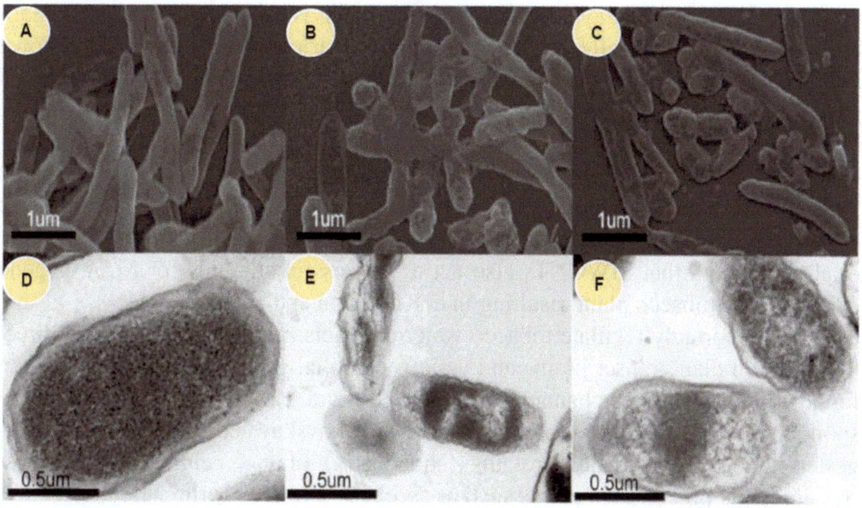

Fig. 22.3 SEM images of *Ralstonia solanacearum* cells treated (**a**) without CNMs, (**b**) with SWCNTs, and (**c**) with GO suspensions, respectively. TEM images of *R. solanacearum* cells treated (**d**) without CNMs, (**e**) with SWCNTs, and (**f**) with GO suspensions, respectively. 100 μL of SWCNTs and GO dispersion (250 μg mL^{-1}) were incubated with 1 mL of different bacterial suspensions (108 cfu mL^{-1}) for 2 h at a 120 rpm shaking speed and 30 °C. (Adapted from Wang et al. 2012)

and stimulate the tissue growth (Simmons et al. 2009). CBNMs cause bacterial resistance and reveal poly site antibacterial mechanisms, due to breaking of bacterial cell wall and making oxidative stress, which in turn results in the arrest of cell growth (Kang et al. 2007; Lyon et al. 2008).

Wang et al. (2012) investigated the antibacterial activity of CBNMs against the copper-resistant plant pathogenic bacterium *Ralstonia solanacearum* and showed that SWCNT dispersion had the strongest antibacterial activity, followed by GO, MWCNTs, reduced GO, and fullerene. The antibacterial activity of SWCNTs and GO caused rupture to cell membrane, which in turn resulted in release of cytoplasmic content from bacterial cell and hence inactivated *R. solanacearum* (Fig. 22.3). Therefore, SWCNTs and GO are suggested as appropriate pesticides for the control of plant bacterial diseases.

22.6 CBNMs and Their Molecular Mechanism of Action

Khodakovskaya et al. (2012) reported that CNTs revealed an upregulation of several genes, including an aquaporin (*NtPIP1*) and two genes, namely, *CycB* and *NtLRX*, involved in water transport, cell wall formation, and cell division. Cell cycle regulation of CBNMs was also reported by Schnittger et al. (2002), where he indicated that ectopic *CYCLIN B1;2* gene expression is a basic requirement for mitosis in plant cells and showed the mechanism of nuclear division induction between endoreduplication and fast cell divisions in *Arabidopsis* trichomes. This was also shown that *NtLRX1* gene (extensin1) played a prominent role in synthesis of cell wall when the cells were boosted with MWCNTs (Tire et al. 1994; Merouropoulos et al. 1999). Transcript abundance of the *NtLRX1* gene between the control and MWCNT-supplemented plants was shown, with the highest rate of gene expression at 100 μg mg L^{-1} concentration in 25 days of incubation. It is suggested that extensins are produced by plant cells as a defense during stress (Salva and Jamet 2001; Bucher et al. 2002) and that MWCNTs also act as a stressful stimulus for *NtLRX1* gene expression in tobacco plant resulting in cell division and growth.

CNTs reportedly regulate tobacco water channels (aquaporins) and have direct influence on plant transcriptome and various essential genes. MWCNTs have been found to express water channel gene (*NtPIP1*) and the production of the corresponding *NtPIP1* protein in tobacco cells (Khodakovskaya et al. 2012). It was suggested that water channel proteins (aquaporins) play a central role in most physiological processes in plants such as ascent of sap, seed germination, cell elongation, flowering, and photosynthesis (Kaldenholff and Fischer 2006; Maurel 2007). The expression of water channel gene (*PIPlb*) in tobacco plants resulted in fast cell division and respiration (Aharon et al. 2003). The report has shown *SLTIP2;2* gene expression that caused a direct positive effect on water potential and permeability of cells, which also increased the rate of transpiration in genetically modified plant species under stressful conditions (Sade et al. 2009). *SLTIP2;2* and *OsPIP1;3* (aquaporin) gene expression was also reported to cause high yield of some fruit biomass and seed germination under water stress (Liu et al. 2007). In a study with *A. thaliana*, CNP treatments have induced early flowering in both long-day and short-day growth conditions which indicates a photoperiod-dependent effect (Kumar et al. 2018). The authors found that despite the upregulation of repressor of gibberellic acid 1 *(RGA1)*, the early flowering is most likely gibberellic acid-independent. They suggest that the early flowering in CNP-treated *A. thaliana* seedlings was phytochrome *B* and photoperiod-dependent.

22.7 CBNMs and Rhizosphere

CBNMs may be toxic to soil microorganisms, may alter the bioavailability of nutrients, or may affect the toxicity level of organic compounds and/or toxins (Dinesh et al. 2012). There are reports on both progressive and regressive effects of CBNMs on rhizosphere microorganisms. CBNMs generally appear to show negative effects on soil/rhizosphere microbes, but studies have also revealed neutral or positive biological effects. For instance, the impact of fullerenes on soil microbial community populations was evaluated using total phospholipid-derived phosphate. The soil was treated with 1 and 1000 μg C_{60} g^{-1} soil for 180 days; fullerenes showed no impact on the structure or function of the soil microbial community or on soil enzymatic activities (Tong et al. 2007). Shrestha et al. (2013) reported diverse changes in soil (sandy loam) microbial communities after 90 days of exposure to MWCNT. CNT concentrations (10, 100, and 10,000 mg kg^{-1}) caused no observable effects on soil microbial composition and enzymatic activities at lower concentrations, but a decreased abundance was detected in select bacterial species such as *Waddlia*, *Holophaga*, *Derxia*, and *Opitutus* at 10,000 mg kg^{-1}. Notably, the amount of polycyclic aromatic hydrocarbon (PAH)-degrading organisms, such as *Cellulomonas*, *Rhodococcus*, *Pseudomonas*, and *Nocardioides*, was markedly increased. Chen et al. (2015) reported that the abundance and diversity of ammonium-oxidizing archaea (AOA) were higher than that of ammonium-oxidizing bacteria (AOB) under the influence of CNTs. The mechanisms responsible for these interactions are unknown; however, the limited literature does suggest that under certain exposure scenarios, CBNMs may have neutral or perhaps modestly beneficial effects on microbial communities (Mukherjee et al. 2016). Fang et al. (2007) showed that C_{60} aggregates in water at 0.01 mg L^{-1} significantly increased the levels of iso and anteiso branched fatty acids from 5.8% to 31.5% and 12.9% to 32.3%, respectively, in *Bacillus subtilis* (Gram positive), suggesting an increase in membrane fluidity as an adaptation response to C_{60}. Alternatively, the aq-C_{60} at 0.5 mg L^{-1} resulted in an alteration on *Pseudomonas putida* (Gram negative) phase transition temperatures and levels of unsaturated fatty acids from bacterial membrane. Johansen et al. (2008) reported that respiration and microbial biomass were unaffected by the fullerenes at any time, whereas the number of fast-growing bacteria was decreased by three- to fourfold just after incorporation of the nanomaterial, when they applied agglomerates of pristine C_{60} fullerenes (50 nm to micron-size) to soil at 0, 5, 25, and 50 mg kg^{-1} dry soil to assess their effect on soil microbiota by measuring total respiration, biomass, and the number and diversity of bacteria and protozoans during the 14-day incubation. Tong et al. (2012) quantitatively investigated the effects of surface coating of SWCNTs on soil microbial community under low and high organic matter concentration. Upon 6000 μg g^{-1} of functionalized SWCNT exposure for 6 weeks, they observed some microbial community shift keeping the total biomass unaffected. Another important factor that may influence the toxicity of CNTs to bacteria is the presence of residual impurities.

The induction of ROS, which may then directly interact with organelles to induce DNA damage or protein inactivation that results in apoptosis and cell death, might be another possible mechanism of CNT toxicity (Jackson et al. 2013). Rodrigues et al. (2013) investigated the effect of carboxyl-functionalized SWCNTs on soil bacterial and fungal communities employing the culture-dependent and culture-independent methods. The soil was added with 0, 250, and 500 μg of SWNTs per gram of soil for 14 days, and the microbial populations were observed. The bacterial soil community was transiently affected by the presence of SWCNTs, but a major impact was observed after 3 days of exposure when the number of colony-forming units (CFUs) was significantly decreased but the bacterial community completely recovered after 14 days. Alternatively, higher doses of SWCNTs had a similar biomass loss at 3-day exposure, but the fungal community was unable to recover even after 14 days. Similarly, Jin et al. (2013) reported that at relatively low concentrations (300–1000 μg g^{-1} soil), SWCNTs significantly lowered the activities of most enzymes and microbial biomass which showed a negative relationship with surface area of SWCNTs. Kerfahi et al. (2015) compared the effect of raw and acid-treated or fSWCNTs on soil bacterial communities, at different concentrations (0 μg g^{-1}, 50 μg g^{-1}, 500 μgg^{-1}, and 5000 μgg^{-1}). The soil DNA was extracted at 0, 2, and 8 weeks, and the V3 region of the 16S rRNA gene was PCR-amplified and sequenced; the bacterial diversity was not affected by either type of MWCNTs. On the other hand, the overall soil bacterial community composition was affected only by fMWCNTs at high concentrations, and a detectable effect was noticed after 2 weeks, and the bacterial diversity remained unaffected.

Chung et al. (2011) reported short-term effects of MWCNTs on the activity and biomass of microorganisms inhabiting two different soil types. Upon application of MWCNT (up to 5000 μg g^{-1} soil), activities of 1,4-β-glucosidase, cellobiohydrolase, xylosidase, 1,4-β-N-acetylglucosaminidase, and phosphatase and microbial biomass were measured. Most of the enzyme activities showed a tendency to be repressed under 500 μg MWCNT g^{-1}soil, and all enzymatic activities as well as microbial biomass C and N were significantly lowered under 5000 μg MWCNT g^{-1} soil, suggesting that high concentrations of MWCNTs could lower the microbial activity and biomass in soils. They also studied the impact of GO at 0.5–1 mg kg^{-1} and noted a decrease up to 50% in the activity of select key soil enzymes, including xylosidase, 1,4-ß-N-acetylglucosaminidase, and phosphatase after 21 days of exposure. Shan et al. (2015) investigated the effects of biochar, activated carbon (AC), and SWCNTs and MWCNTs in various concentrations (0, 0.2, 20, and 2000 mg kg^{-1} dry soil) on the fate of ^{14}C-catechol and microbial community in soil and reported that SWCNTs at 2000 mg kg^{-1} significantly reduced mineralization. The inhibitory effects of AC and SWCNTs on the mineralization were attributed to the inhibited soil microbial activities and the shifts in microbial communities, as suggested by the reduced microbial biomass C and the separated phylogenetic distance. However, MWCNTs at 0.2 mg kg^{-1} significantly stimulated mineralization, compared with the control soil. Tong et al. (2016) evaluated the effects of nC_{60} aggregates of different particle sizes via organic solvents on soils with different organic matter contents by measuring the total microbial biomass, metabolic activity, and bacterial community structure and concluded that nC_{60} aggregates, introduced as an aqueous suspension, exhibited size-dependent effects on soil bacterial community composition in the low organic matter system, but induced a minimal change in the microbial biomass and metabolic activity in soils with both high and low organic matter contents. These authors also suggested that nC_{60} aggregates of smaller size may have negative impact on soil biota and that soil organic matter may play a key role in modulating the environmental effect of nanomaterials. Karunakaran et al. (2013) evaluated the effect of nanosilica and silicon on soil properties, total bacterial population, and maize seed germination. The sodium silicate-treated soil inhibited plant growth-promoting rhizobacteria in contrast to nanosilica. Bacterial population doubled in the presence of nanosilica from 4×10^5 CFU to 8×10^5 CFU per gram of soil. Rangaraj et al. (2014) also reported enhanced microbial populations and total biomass content ($C = 1508$ μg g^{-1} and $N = 178$ μg g^{-1}) in maize rhizosphere caused by nanosilica, which plays a vital role in influencing the soil nutrient content and microbial biota and, hence, may promote the growth of maize plant.

22.8 Conclusion

Most of the investigations concerning CBNMs vis-à-vis plants have been focused on seed germination seedling growth responses. CBNMs show beneficial effects on optimum concentration by hormesis and stress defense systems, whereas no effects or adverse effects stem from higher concentrations. Variability in CBNM impact on different plant species may be due not only to CBNM concentration but also to their heterogeneity in size, structure, chemical composition, and surface area. Studies have demonstrated that CBNMs increase the root and shoot growth at different growth stages, improve biomass, and also promote cell elongation and enhance crop yield. The stimulatory role of CBNMs in plants on water uptake, germination, photosynthesis, biomass, crop yield, and disease protection is of significant economic importance for agriculture, horticulture, as well as bioenergy sectors. Besides, CBNMs have an overall healthier impact on rhizosphere and its associated microfloa, increasing the activity of soil anaerobic ammonium-oxidizing bacteria as well as their involvement in protein and carbohydrate synthesis. However, their increased application over the last few years has raised concerns about the probable biological and environmental risks, warranting optimization and safety assessment considerations. Compared to application-oriented investigation of CBNMs, studies on their ecotoxicity are inadequate, and data on their interaction with plant species and materials, growth media, and analytical techniques are insufficient. Since CBNMs exhibit some unique and reactive properties, they are likely to affect the individual plant or microbial species but also to upset the ecological system. Various reports presented in this chapter indicate that the CBNM interactions with biota and the co-existing impurities are amazingly complicated and may have noteworthy consequences. Risk assessment studies involving the CBNM sources, pathways, and sinks are highly desirable.

References

Aharon R, Shahak Y, Wininger S, Bendov R, Kapulnik Y, Galili G (2003) Over expression of a plasma membrane aquaporin in transgenic tobacco improves plant vigor under favorable growth conditions but not under drought or salt stress. Plant Cell 15:439–447

Akhavan O, Ghaderi E (2010) Toxicity of graphene and graphene oxide nanowalls against bacteria. ACS Nano 4:5731–5736

Baptista FR, Belhout SA, Giordani S, Quinn SJ (2015) Recent developments in carbon nano material sensors. Chem Soc Rev 44:4433–4453

Basch E, Gabardi S, Ulbricht C (2003) Bittermelon (*Momordica charantia*): are view of efficacy and safety. Am J Health Syst Pharm 60:356–359

Baughman RH, Zakhidov AA, de Heer WA (2002) Carbon nanotubes—the route toward applications. Science 97:787–792

Begum P, Ikhtiari R, Fugetsu B, Matsuoka M, Akasaka T, Watari F (2012) Phytotoxicity of multi-walled carbon nanotubes assessed by selected plant species in the seedling stage. Appl Surf Sci 262:120–124

Bennett SW, Adeleye A, Ji Z, Keller AA (2013) Stability, metal leaching, photoactivity and toxicity in fresh water systems of commercial single wall carbon nano tubes. Water Res 47:4074–4085

Biris AS, Khodakovskaya MV (2012) Method of using carbon nanotubes to affect seed germination and plant growth. Patent US0233725

Boghossian AA (2013) Application of nanoparticle antioxidants to enable hyperstable chloroplasts for solar energy harvesting. Adv Energy Mater 3:881–893

Bucher M, Brunner S, Zimmermann P, Zardi GI, Amrhein N, Willmitzer L, Riesmeier JW (2002) The expression of an extensin-like protein correlates with cellular tip growth in tomato. Plant Physiol 128:911–923

Burlaka OM, Pirko YV, Yemets AI, Blume YB (2015) Plant genetic transformation using carbon nanotubes for DNA delivery. Cytol Genet 49:349–357

Calkins JO, Umasankar Y, O'Neill H, Ramasamy RP (2013) High photo-electrochemical activity of thylakoid–carbon nanotube composites for photosynthetic energy conversion. Energy Environ Sci 6:1891–1900

Cañas JE, Long MQ, Nations S, Vadan R, Dai L, Luo MX (2008a) Effects of functionalized and nonfunctionalized single-walled carbon nanotubes on root elongation of select crop species. Environ Toxicol Chem 27:1922–1931

Cañas JE, Long M, Nations S, Vadan R, Dai L, Luo M, Ambikapathi R, Lee EH, Olszyk D (2008b) Effects of functionalized and non functionalized single-walled carbon nanotubes on root elongation of select crop species. Environ Toxicol Chem 27:1922–1931

Cha C, Shin SR, Annabi N, Dokmeci MR, Khademhosseini A (2013) Carbon-based nano materials multifunctional materials for biomedical engineering. ACS Nano 7:2891–2897

Chen J, Peng H, Wang X, Shao F, Yuan Z, Han H (2014) Graphene oxide exhibits broad-spectrum antimicrobial activity against bacterial phytopathogens and fungal conidia by intertwining and membrane perturbation. Nanoscale 6:1879–1889

Chen Q, Wang H, Yang B, He F, Han X, Song Z (2015) Responses of soil ammonia-oxidizing microorganisms to repeated exposure of single-walled and multi-walled carbon nanotubes. Sci Total Environ 505:649–657

Chichiriccò G, Poma A (2015) Penetration and toxicity of nanomaterials in higher plants. Nanomaterials (Basel) 5:851–873

Chung H, Son Y, Yoon TK, Kim S, Kim W (2011) The effect of multi-walled carbon nanotubes on soil microbial activity. Ecotoxicol Environ Saf 74:569–575

Dinesh R, Anandaraj M, Srinivasan V, Hamza S (2012) Engineered nanoparticles in the soil and their potential implications to microbial activity. Geoderma 173-174:19–27

Embiale A, Hussein M, Husen A, Sahile S, Mohammed K (2016) Differential sensitivity of *Pisum sativum* L. cultivars to water-deficit stress: changes in growth, water status, chlorophyll fluorescence and gas exchange attributes. J Agron 15:45–57

Fan X, Xu J, Lavoie M, Peijnenburg WJGM, Zhu Y, Lu T, Fu Z, Zhu T, Qian H (2018) Multiwall carbon nanotubes modulate paraquat toxicity in *Arabidopsis thaliana*. Environ Pollut 233:633–641

Fang J, Lyon DY, Wiesner MR, Dong J, Alvarez PJJ (2007) Effect of a fullerene water suspension on bacterial phospholipids and membrane phase behavior. Environ Sci Technol 41:2636–2642

Fathi Z, Nejad RAK, Zadeh HM, Satari TN (2017) Investigating of a wide range of concentrations of multi-walled carbon nanotubes on germination and growth of castor seeds (*Ricinus communis* L.). J Plant Pro Res 57:228–236

Georgakilas V, Perman JA, Tucek J, Zboril R (2015) Broad family of carbon nanoallotropes: classification, chemistry, and applications of fullerenes, carbon dots, nanotubes, graphene, nanodiamonds, and combined superstructures. Chem Rev 115:4744–4822

Getnet Z, Husen A, Fetene M, Yemata G (2015) Growth, water status, physiological, biochemical and yield response of stay green sorghum {*Sorghum bicolor* (L.) Moench} varieties-a field trial under drought-prone area in Amhara regional state, Ethiopia. J Agron 14:188–202

Ghorbanpour M, Farahani AHK, Hadian J (2018) Potential toxicity of nano-graphene oxide on callus cell of *Plantago major* L. under polyethylene glycol-induced dehydration. Ecotoxicol Environ Saf 148:910–922

Ghosh M, Chakraborty A, Bandyopadhyay M, Mukherjee A (2011) Multi-walled carbon nanotubes (MWCNT): induction of DNA damage in plant and mammalian cells. J Hazard Mater 197:327–336

Giraldo JP, Landry MP, Faltermeier SM, McNicholas TP, Iverson NM, Boghossian AA (2014) Plant nanobionics approach to augment photosynthesis and biochemical sensing. Nat Mater 13:400–408

González-Melendi P, Fernández-Pacheco R, Coronado MJ, Corredor E, Testillano PS, Risueño MC, Marquina C, Ibarra MR, Rubiales D, Pérez-de- Luque A (2008) Nanoparticles as smart treatment-delivery systems in plants: assessment of different techniques of microscopy for their visualization in plant tissues. Ann Bot 101:187–195

Haghighi M, Teixeira da Silva JA (2014) The effect of carbon nanotubes on the seed germination and seedling growth of four vegetable species. J Crop Sci Biotechnol 17:201–208

Ham MH (2010) Photo electrochemical complexes for solar energy conversion that chemically and autonomously regenerate. Nat Chem 2:929–936

Hasaneen MNAG, Abdel-aziz HMM, Omer AM (2016) Effect of foliar application of engineered nanomaterials: carbon nanotubes NPK and chitosan nanoparticles NPK fertilizer on the growth of French bean plant. Bioch Biot Res 4:68–76

He Y, Hu R, Zhong Y, Zhao X, Chen Q, Zhu H (2017) Graphene oxide as a water transporter promoting germination of plants in soil. Nano Res 11:1928. https://doi.org/10.1007/s12274-017-1810-1

Heller DA (2009) Multimodal optical sensing and analyte specificity using single-walled carbon nanotubes. Nat Nanotech 4:114–120

Hu X, Zhou Q (2014) Novel hydrated graphene ribbon unexpectedly promotes aged seed germination and root differentiation. Sci Rep 4:3782

Hu W, Peng C, Luo W, Lv M, Li X, Li D, Huang Q, Fan C (2010) Graphene-based antibacterial paper. ACS Nano 4:4317–4323

Hui L, Piao J, Auletta J, Hu K, Zhu Y, Meyer T, Liu H, Yang L (2014) Availability of the basal planes of graphene oxide determines whether it is antibacterial. ACS Appl Mater Interfaces 6:13183–13190

Hurt RH, Monthioux M, Kane A (2006) Toxicology of carbon nanomaterials: status, trends and perspectives on the special issue. Carbon 44:1028–1033

Husen A, Siddiqi KS (2014a) Carbon and fullerene nanomaterials in plant system. J Nanobiotechnol 12:16

Husen A, Siddiqi KS (2014b) Phytosynthesis of nanoparticles: concept, controversy and application. Nano Res Lett 9:229

Husen A, Iqbal M, Aref MI (2014) Growth, water status and leaf characteristics of *Brassica carinata* under drought stress and rehydration conditions. Braz J Bot 37:217–227

Husen A, Iqbal M, Aref IM (2016) IAA-induced alteration in growth and photosynthesis of pea (*Pisum sativum* L.) plants grown under salt stress. J Environ Biol 37:421–429

Husen A, Iqbal M, Aref IM (2017) Plant growth and foliar characteristics of faba bean (*Vicia faba* L.) as affected by indole-acetic acid under water-sufficient and water-deficient conditions. J Environ Biol 38:179–186

Husen A, Iqbal M, Sohrab SS, Ansari MKA (2018) Salicylic acid alleviates salinity-caused damage to foliar functions, plant growth and antioxidant system in Ethiopian mustard (*Brassica carinata* A. Br.). Agric Food Secur 7:44

Husen A, Iqbal M, Khanum N, Aref IM, Sohrab SS, Meshresa G (2019) Modulation of salt-stress tolerance of niger (*Guizotia abyssinica*), an oilseed plant, by application of salicylic acid. J Environ Biol 40:40:94–104

Jackson P, Jacobsen NR, Baun A, Birkedal R, Kuhnel D, Jensen KA, Vogel U, Wallin H (2013) Bioaccumulation and ecotoxicity of carbon nanotubes. Chem Cent J 7:154

Jiang Y, Hua Z, Zhao Y, Liu Q, Wang F, Zhang Q (2012) The effect of carbon nanotubes on rice seed germination and root growth. In: Proceedings of the 2012 international conference on applied biotechnology. Springer, Berlin, pp 1207–1212

Jin L, Son Y, Yoon TK, Kang YJ, Kim W, Chung H (2013) High concentrations of single-walled carbon nanotubes lower soil enzyme activity and microbial biomass. Ecotoxicol Environ Saf 88:9–15

Johansen A, Pedersen AL, Jensen KA, Karlson U, Hansen BM, Scott-Fordsmand JJ, Winding A (2008) Effects of C_{60} fullerene nanoparticles on soil bacteria and protozoans. Environ Toxicol Chem 27:1895–1903

Joshi A, Kaur S, Singh P, Dharamvir K, Nayyar H, Verma G (2018a) Tracking multi-walled carbon nanotubes inside oat (*Avena sativa* L.) plants and assessing their effect on growth, yield, and mammalian (human) cell viability. Appl Nanosci 8:1399–1414

Joshi A, Kaur S, Dharamvir K, Nayyar H, Verma G (2018b) Multi-walled carbon nanotubes applied through seed-priming influence early germination, root hair, growth and yield of bread wheat (*Triticum aestivum L.*). J Sci Food Agri 98:8

Kaldenholff R, Fischer M (2006) Aquaporins in plants. Acta Physiol 187:169–176

Kang S, Pinault M, Pfefferle LD, Elimelech M (2007) Single-walled carbon nanotubes exhibit strong antimicrobial activity. Langmuir 23:8670–8673

Kang S, Mauter MS, Elimelech M (2008) Physicochemical determinants of multiwalled carbon nanotube bacterial cytotoxicity. Environ Sci Technol 42:7528–7534

Karunakaran G, Suriyaprabha R, Manivasakan P, Yuvakkumar R, Rajendran V, Kannan N (2013) Impact of nano and bulk ZrO_2, TiO_2 particles on soil nutrient contents and PGPR. J Nanosci Nanotechnol 13:678–685

Kerfahi D, Tripathi BM, Singh D, Kim H, Lee S, Lee J, Adams JM (2015) Effects of functionalized and raw multi-walled carbon nanotubes on soil bacterial community composition. PLoS One 10:1230–1242

Khodakovskaya M, Dervishi E, Mahmood M, Xu Y, Li Z, Watanabe F, Biris AS (2009) Carbon nanotubes are able to penetrate plant seed coat and dramatically affect seed germination and plant growth. ACS Nano 3:3221–3227

Khodakovskaya M, de Silva K, Nedosekin D, Dervishi E, Biris AS, Shashkov EV, Galanzha EI, Zharov VP (2011) Complex genetic, photothermal, and photoacoustic analysis of nanoparticle-plant interactions. Proc Natl Acad Sci U S A 108:1028–1033

Khodakovskaya M, de Silva K, Biris AS, Dervishi E, Villagarcia H (2012) Carbon nanotubes induce growth enhancement of tobacco cells. ACS Nano 6:2128–2135

Khodakovskaya MV, Kim BS, Kim JN, Alimohammadi M, Dervishi E, Mustafa T, Cernigla CE (2013) Carbon nanotubes as plant growth regulators: effects on tomato growth, reproductive system, and soil microbial community. Small 9:115–123

Kim JH (2009) The rational design of nitric oxide selectivity in single-walled carbon nanotube near-infrared fluorescence sensors for biological detection. Nat Chem 1:473–481

Kole C, Kole P, Randunu KM, ChoudharyP PR, Ke PC, Rao AM, Marcus RK (2013) Nanobiotechnology can boost crop production and quality: first evidence from increased plant biomass, fruit yield and phytomedicine content in bitter melon (*Momordica charantia*). BMC Biotechnol 13:37

Kumar A, Singh A, Panigrahy A, Sahoo PK, Panigrahi KCS (2018) Carbon nanoparticles influence photomorphogenesis and flowering time in *Arabidopsis thaliana*. Plant Cell Rep 37:901–912

Lahiani MH, Dervishi E, Chen J, Nima Z, Gaume A, Biris AS, Khodakovskaya MV (2013) Impact of carbon nanotube exposure to seeds of valuable crops. ACS Appl Mater Interfaces 5:7965–7973

Lahiani MH, Chen J, Irin F, Puretzky AA, Green MJ, Khodakovskaya MV (2015) Interaction of carbon nanohorns with plants: up take and biological effects. Carbon 81:607–619

Liang T, Yin Q, Zhang Y, Wang B, Guo W, Wang J, Xie J (2013) Effects of carbon nanoparticles application on the growth, physiological characteristics and nutrient accumulation in tobacco plants. J Food Agric Environ 11:954–958

Lin D, Xing B (2007) Phytotoxicity of nanoparticles, inhibition of seed germination and root growth. Environ Pollut 150:243–250

Lin C, Fugetsu B, Su YB, Watari F (2009) Studies on toxicity of multi-walled carbon nanotubes on *Arabidopsis* T87 suspension cells. J Hazard Mater 170:578–583

Liu HY, Yu X, Cui DY, Sun MH, Sun WN, Tang ZC, Kwak SS, Su WA (2007) The role of water channel proteins and nitric oxide signaling in rice seed germination. Cell Res 17:638–649

Liu SB, Wei L, Hao L, Fang N, Chang MW, Xu R, Yang Y, Chen Y (2009) Sharper and faster "nano darts" kill more bacteria: a study of antibacterial activity of individually dispersed pristine single-walled carbon nanotube. ACS Nano 3:3891–3902

Liu S, Zeng TH, Hofmann M, Burcombe E, Wei J, Jiang R, Kong J, Chen Y (2011) Antibacterial activity of graphite, graphite oxide, graphene oxide, and reduced graphene oxide: membrane and oxidative stress. ACS Nano 5:6971–6980

Liu S, Wei H, Li Z, Li S, Yan H, He Y, Tian Z (2015) Effects of graphene on germination and seedling morphology in rice. J Nanosci Nanotechnol 15:2695–2701

Liu R, Zhu X, Chen B (2017) A new insight of graphene oxide-Fe(III) complex photochemical behaviors under visible light irritation. Sci Rep 7:40711

Lyon DY, Brunet L, Hinkal GW, Wiesner MR, Alvarez PJJ (2008) Antibacterial activity of fullerene water suspensions (nC60) is not due to ROS-mediated damage. Nano Lett 8:1539–1543

Mangadlao JD, Santos CM, Felipe MJL, De Leon ACC, Rodrigues DF, Advincula RC (2015) On the antibacterial mechanism of graphene oxide (GO) Langmuir-Blodgett films. Chem Commun 51:2886–2889

Martínez-Ballesta MC, Zapata L, Chalbi N, Carvajal M (2016) Multiwalled carbon nanotubes enter broccoli cells enhancing growth and water uptake of plants exposed to salinity. J Nanobiotechnology 14:42

Maurel C (2007) Plant Aquaporins: novel functions and regulation properties. FEBS Lett 581:2227–2236

Mauter MS, Elimelech M (2008) Environmental applications of carbon-based nanomaterials. Environ Sci Technol 42:5843–5859

Merouropoulos G, Bernett DC, Shitsat AH (1999) The Arabidopsis Extensin gene is developmentally regulated, is induced by wounding, methyl jasmonate, abscisic and salicylic acid and codes for a protein with unusual motifs. Planta 208:212–219

Miralles P, Johnson E, Church TL, Harris AT (2012) Multiwalled carbon nanotubes in alfalfa and wheat: toxicology and uptake. J R Soc Interface 9:3514–3527

Mondal A, Basu R, Das S, Nandy P (2011) Beneficial role of carbon nanotubes on mustard plant growth: an agricultural prospect. J Nanopart Res 13:4519–4528

Mukherjee A, Majumdar S, Servin AD, Pagano L, Dhankher OP, White JC (2016) Carbon nanomaterials in agriculture: a critical review. Front Plant Sci 7:172

Nair R, Varghese SH, Nair BG, Maekawa T, Yoshida Y, Kumar DS (2010) Nanoparticulate material delivery to plants. Plant Sci 179:154–163

Nair R, Mohamed MS, Gao W, Maekawa T, Yoshida Y, Ajayan PM, Kumar DS (2012) Effect of carbon nanomaterials on the germination and growth of rice plants. J Nanosci Nanotechnol 12:2212–2220

Ng TB, Chan WY, Yeung HW (1992) Proteins with abortifacient, ribosome inactivating, immune modulatory, antitumor and anti-AIDS activities from Cucurbitaceae plants. Gen Pharmacol 23:575–590

Perez-de-Luque A, Rubiales D (2009) Nanotechnology for parasitic plant control. Pest Manag Sci 65:540–545

Pourkhaloee A, Haghighi M, Saharkhiz MJ, Jouzi H, Doroodmand MM (2011) Carbon nanotubes can promote seed germination via seed coat penetration. Seed Technol 33:155–169

Raman A, Lau C (1996) Anti-diabetic properties and phytochemistry of *Momordica charantia* L. (*Cucurbitaceae*). Phytomedicine 2:349–362

Rangaraj S, Gopalu K, Muthusamy P, Rathinam Y, Venkatachalam R, Narayanasamy K (2014) Augmented biocontrol action of silica nanoparticles and *Pseudomonas fluorescens* bioformulant in maize (*Zea mays* L.). RSC Adv 4:8461–8465

Rico CM, Majumdar S, Duarte-Gardea M, Peralta-Videa JR, Gardea-Torresdey JL (2011) Interaction of nanoparticles with edible plants and their possible implications in the food chain. J Agric Food Chem 59:3485–3498

Rodrigues DF, Jaisi DP, Elimelech M (2013) Toxicity of functionalized single-walled carbon nanotubes on soil microbial communities: implications for nutrient cycling in soil. Environ Sci Technol 47:625–633

Sade N, Vinocur BJ, Diber A, Shatil A, Ronen G, Nissan H, Wallach R, Karchi H, Moshelion M (2009) Improving plant stress tolerance and yield production: is the tonoplast aquaporin SlTIP2;2 a key to isohydric to anisohydric conversion. New Phytol 181:651–661

Salva I, Jamet E (2001) Expression of the tobacco Ext 1.4 extensin gene upon mechanical constraint and localization of regulatory regions. Plant Biol 3:32–41

Sarlak N, Taherifar A, Salehi F (2014) Synthesis of nanopesticides by encapsulating pesticide nanoparticles using functionalized carbon nanotubes and application of new nano composite for plant disease treatment. J Agric Food Chem 62:4833–4838

Saxena M, Maity S, Sarkar S (2014) Carbon nano particles in 'biochar' boost wheat (*Triticum aestivum*) plant growth. RSC Adv 4:3994–3998

Schmitz-Linneweber C (2001) The plastid chromosome of spinach (*Spinacia oleracea*): complete nucleotide sequence and gene organization. Plant Mol Biol 45:307–315

Schnittger A, Schobinger U, Stierhof YD, Hulskamp M (2002) Ectopic B-type cyclin expression induces mitotic cycles in endo reduplicating *Arabidopsis Trichomes*. Curr Biol 12:415–420

Sharon M, Sharon M (2010) Carbon Nano forms and applications. McGraw Hill Professional, New York

Shen C, Zhang Q, Li J, Bi F, Yao N (2010) Induction of programmed cell death in *Arabidopsis* and rice by single-wall carbon nanotubes. Am J Bot 97:1602–1609

Shrestha B, Acosta-Martinez V, Cox SB, Green MJ, Li S, Cañas-Carrell JE (2013) An evaluation of the impact of multiwalled carbon nanotubes on soil microbial community structure and functioning. J Hazard Mater 261:188–197

Simmons TJ, Lee SH, Park TJ, Hashim DP, Ajayan PM, Linhardt RJ (2009) Antiseptic single wall carbon nanotube bandages. Carbon 47:1561–1564

Sonkar SK, Roy M, Babara DG, Sarkar S (2012) Water soluble carbon nano-onions from wood wool as growth promoters for gram plants. Nanoscale 4:7670–7675

Srivastava A, Rao DP (2014) Enhancement of seed germination and plant growth of wheat, maize, penut and garlic using multiwalled carbon nanotubes. Eur Chem Bull 3:502–504

Srivastava V, Gusain D, Sharma YC (2015) Critical review on the toxicity of some widely used engineered nanoparticles. Ind Eng Chem Res 54:6209–6233

Shan J, Ji R, Yu Y, Xie Z, Yan X (2015) Biochar, activated carbon, and carbon nanotubes have different effects on fate of 14C-catechol and microbial community in soil. Sci Rep 5:16000

Tan X, Lin C, Fugetsu B (2009) Studies on toxicity of multi-walled carbon nanotubes on suspension rice cells. Carbon 47:3479–3487

Tire C, de Rycke R, de Loose M, Inze D, Van M, Englr G (1994) Extensin gene expression is induced by mechanical stimuli leading to local cell wall strengthening in *Nicotiana plumbaginifolia*. Planta 195:175–181

Tiwari DK, Dasgupta-Schubert N, Cendejas LM, Villegas J, Carreto Montoya L, García SE (2014) Interfacing carbon nanotubes (CNT) with plants: enhancement of growth, water and ionic nutrient uptake in maize (*Zea mays*) and implications for nanoagriculture. Appl Nanosci 4:577–591

Tong Z, Bischoff M, Nies L, Applegate B, Turco RF (2007) Impact of fullerene (C_{60}) on a soil microbial community. Environ Sci Technol 41:2985–2991

Tong Z, Bischoff M, Nies LF, Myer P, Applegate B, Turco RF (2012) Response of soil microorganisms to as-produced and functionalized single-wall carbon nanotubes (SWNTs). Environ Sci Technol 46:13471–13479

Tong Z, Bischoff M, Nies LF, Carroll NJ, Applegate B, Turco RF (2016) Influence of fullerene (C_{60}) on soil bacterial communities: aqueous aggregate size and solvent cointroduction effects. Sci Rep 6:28069

Torney F, Trewyn GB, Lin VSY, Wang K (2007) Mesoporous silica nanoparticles deliver DNA and chemicals into plants. Nat Nanotechnol 2:295

Torre-Roche RDL, Hawthorne J, Deng Y, Xing B, Cai W, Newman LA, Wang Q, Ma X, Hamdi H, White JC (2013) Multiwalled carbon nanotubes and C_{60} fullerenes differentially impact

the accumulation of weathered pesticides in four agricultural plants. Environ Sci Technol 47:12539–12547
Tripathi S, Sarkar S (2015) Influence of water soluble carbon dots on the growth of wheat plant. Appl Nanosci 5:609–616
Tripathi S, Sonkar SK, Sarkar S (2011) Growth stimulation of gram (*Cicer arietinum*) plant by water soluble carbon nanotubes. Nanoscale 3:1176–1181
Ugarte D (1992) Curling and closure of graphitic networks under electron-beam irradiation. Nature 359:707–709
Villagarcia H, Dervishi E, de Silva K, Biris AS, Khodakovskaya MV (2012) Surface chemistry of carbon nanotubes impacts the growth and expression of water channel protein in tomato plants. Small 8:2328–2334
Wang BC, Wang HW, Chang JC, Tso HC, Chou YM (2001) More spherical large fullerenes and multi-layer fullerene cages. J Mol Struct Theochem 540:171–176
Wang XP, Han HY, Liu XQ, Gu XX, Chen K, Lu DL (2012) Multi-walled carbon nanotubes can enhance root elongation of wheat (*Triticum aestivum*) plants. J Nanopart Res 14:10
Wang X, Liu X, Chen J, Han H, Yuan Z (2014) Evaluation and mechanism of antifungal effects of carbon nano materials in controlling plant fungal pathogen. Carbon 68:798–806
Wild E, Jones KC (2009) Novel method for the direct visualization of in vivo nanomaterials and chemical interactions in plants. Environ Sci Technol 43:5290–5294
Wong MH, Giraldo JP, Kwak SY, Koman VB, Sinclair R, Lew TT, Bisker G, Liu P, Strano MS (2017) Nitroaromatic detection and infrared communication from wild-type plants using plant nanobionics. Nat Mater 16:264–272
Yan SH, Zhao L, Li H, Zhang Q, Tan JJ, Huang M (2013) Single-walled carbon nanotubes selectively influence maize root tissue development accompanied by the change in the related gene expression. J Hazard Mater 246:110–118
Zhang L, Su M, Liu C, Chen L, Huang H, Wu X, Liu X, Yang F, Gao F, Hong F (2007) Nanoparticles in medicine: therapeutic applications and developments. Trace Elem Res 109:68–73
Zhang M, Gao B, Chen J, Li Y, Creamer AE, Chen H (2014) Slow-release fertilizer encapsulated by graphene oxide films. J Chem Eng 255:107–113
Zhang M, Gao B, Chen J, Li Y, Zhang M, Gao B, Chen J, Li Y (2015) Effects of graphene on seed germination and seedling growth. J Nanopart Res 17:78–84
Zhao Q, Ma C, White JC, Dhankher OP, Zhang X, Zhang S, Xing B (2017) Quantitative evaluation of multi-wall carbon nanotube uptake by terrestrial plants. Carbon 114:661–670

Chapter 23
Progress in Research on Nanomaterial-Plant Interaction

Mohammad Babar Ali

23.1 Introduction

Nanotechnology is one of the key technologies of the present century. Application of nanomaterials (NMs) in agriculture/horticulture sector is likely to ensure a tremendous improvement in farming techniques, crop development, and environmental protection and curtail the use of chemical fertilizers, pesticides, herbicides, etc. (Sekhon 2014; Liu and Lal 2015; Kah 2015; Sertova 2015). Human population in the world is expected to exceed nine billion by 2050. Nanotechnology is poised to play a vital role in providing more and better food for this rapidly growing world population by improving the quality and quantity of plant produce and by enhancing their nutrient contents and reducing the uptake of metals (Kole et al. 2013; Sagadevan and Periasamy 2014; Husen and Siddiqi 2014a; Siddiqi and Husen 2017a, b; Karny et al. 2018; Chen et al. 2018; Ogunkunle et al. 2018; Sheykhbaglou et al. 2018). Therefore, the development of new NPs is important for exploitation of the strengths of nanotechnology for improving plant performance (Fraceto et al. 2016). The chemically synthesized NPs are toxic and limit plant growth performance (Mattiello et al. 2015; Navarro et al. 2015; Cox et al. 2016; Wang et al. 2016; Du et al. 2017; Tiwari et al. 2017; Spengler et al. 2017). Therefore, plant-mediated biosynthesis of NPs is rapidly gaining the ground. This green synthesis of NPs is safe, quick, energy-efficient, cost-effective, and environment-friendly, with limited waste production and greenhouse gas emissions (Husen and Siddiqi 2014b, c; Fraceto et al. 2016; Husen 2017; Siddiqi et al. 2018).

There are indications that investment in the global agribusiness market will soon increase to 3.4 trillion dollars (Hooley et al. 2012). According to the National Nanotechnology Initiative (NNI), the United States of America (USA) is spending 3.7 billion dollars for a 4-year term for developing useful nanotechnology (Hirsh

M. B. Ali (✉)
Rhizosphere Science Laboratory, Department of Plant and Soil Sciences, University of Kentucky, Lexington, KY, USA

A. Husen, M. Iqbal (eds.), *Nanomaterials and Plant Potential*,
https://doi.org/10.1007/978-3-030-05569-1_23

et al. 2014; Banterle et al. 2014), followed by Japan and the European Union (Sodano and Verneau 2014). The Nanotechnology Research Review (2014) proclaimed the consumption of nanoscale ceramic powders (oxides, carbides, nitrides, and borides used for starting materials for solid commercial articles) costing 12.1 billion dollars over the years 2013–2018. The consumption of nanocomposites was estimated to grow from 225,060 metric tons in 2014 to nearly 584,984 metric tons over a period of 5 years.

23.2 Future Prospects of Nanoparticle-Plant Interaction

The main aim of NP application on plants is to increase plant growth and productivity and improve plant protection without polluting the environment. Recent researches, based almost entirely on laboratory experiments, have shown both positive and negative effects of NPs on plant physiology. Nanoparticles not only enhance plant productivity but also induce accumulation of secondary metabolites in plants and can possibly be used as elicitors of secondary metabolites. However, certain groups of NPs have shown toxicity at the cellular and molecular level, normally in a concentration- or duration-dependent way. Their toxicity often relates to their size, shape, chemistry, and surface area to volume ratio, the factors that determine the biological characteristics of NPs. Redesigning of NPs using new methods and standard protocols is thus imperative to ensure a healthy plant performance and improve the plant productivity; it is also to be seen that the cost and time involved do not become the limiting factors.

Various NP-based products have already been developed, e.g., nano-sized nutrients made from ZnONPs and TiO_2 NPs promoted growth and antioxidant potential in tomatoes (Raliya et al. 2015), macronutrient fertilizers coated with ZnO NPs caused enhanced nutrient absorption by plants and their quick transfer to specific sites (Milani et al. 2015), and acetamiprid-loaded alginate-chitosan nanocapsules improved the transfer of agrochemicals and control of dosage (Kumar et al. 2015). However, it is yet to be confirmed whether the positive effects of these products are consistent in different plant species and varied soil conditions, particularly during the large-scale cultivation. It may be noted that carbon nanotubes and nanosilver-based products did not show potential for large-scale agricultural applications (Aschberger et al. 2015). Further, these products are yet to be evaluated repeatedly for their impact on plant performance, and hardly few NP-based products have been officially registered as yet (EFSA 2009; USDA 2015; Gewin 2015; JRC scientific and policy reports 2014). However, according to the European Patent Office (EPO) and Patent Statistical Database (PATSTAT), about 200,000 patents were identified related to nanotechnology from 76 patent organizations; majority (47%) of these patents were assigned to the USA, followed by Japan (25%), EU (20%), and Korea (4%) (Parisi et al. 2014). Manufactured nanoparticles, nano-emulsions, and nanocapsules are no longer new to agricultural chemicals (Gewin 2015).

23.2.1 Plant-Based Biosynthesis of Nanoparticles

NP biosynthesis using plant extracts has received tremendous attention; the NPs so produced have less toxicity and are preferably used to improve plant productivity and plant protection. The phyto-mediated biosynthesis of NPs and the mechanisms involved have been examined by many researchers (Husen and Siddiqi 2014b; Siddiqi and Husen 2016a, b; Siddiqi et al. 2016, 2018). Ag NPs synthesized through biological method, using *Bryophyllum pinnatum* leaf extract, have been able to improve crop production by preventing the bacterial pathogen attack (Tareq et al. 2017). Kumari et al. (2017a) studied in vitro and in vivo the effect of biosynthesized Ag NPs on tomato plant infected with *Alternaria solani*, the causative agent of early blight disease of tomato. They noted increase in fresh weight (32.58%) and total chlorophyll content (23.52%) in tomato plants treated with 5 μg mL^{-1} biosynthesized Ag NPs. Also, application of these Ag NPs on infected plants decreased the fungal spore count (48.57%), lipid peroxidation (30%), proline content (39.59%), and superoxide dismutase activity (28.57%), indicating that the Ag NPs were able to prevent further fungal growth, decrease the stress parameters, and increase the chlorophyll content. Another study by this group (Kumari et al. 2017b) revealed that foliar application of 5 μg mL^{-1} of biosynthesized Ag NPs reduced the number of fungal spores and number of lesions per leaf in *Alternaria brassicicola*-infected *Arabidopsis thaliana* plants, without causing any perceptible change in soil properties (pH, electric conductivity, soil organic carbon, soil microbial biomass carbon, soil enzymes activities, microbial diversity). The authors concluded that biosynthesized Ag NPs have the potential to act as strong antimicrobial agent and protect plants from damage caused by pathogens, without altering the native soil microflora.

In contrast, foliar application of 5 μg mL^{-1} Ag NPs reduced the number of fungal spores, indicating that the biosynthesized Ag NPs control the fungal spore formation, and hence the number of lesions got reduced from 2.9/leaf in the pathogen-infected plants to 0.9/leaf in Ag NP-treated plants. These studies suggest that application of biosynthesized Ag NPs may be able to control the fungal infection of crop plants with proper management. Therefore, application of biosynthesized Ag NPs may improve the floral diversity and increase plant defense against the pathogen attack. Thus, controlling the spread of plant diseases by applying biosynthesized Ag NPs may increase plant productivity and undo the 10–30% loss in crop yield that occurs per year due to incidence of plant diseases (Savary et al. 2012). Raliya et al. (2014) reported the beneficial effect of biosynthesized MgO NPs prepared from *Aspergillus flavus* on cluster bean (*Cyamopsis tetragonoloba*). MgO NP (15 mg L^{-1})-treated plants increased the root and shoot growth and chlorophyll content in cluster bean. Similarly, TiO_2 NPs biosynthesized from *Aspergillus flavus* induced the plant growth of *Vigna radiata* and rhizospheric microbial population (Raliya et al. 2015). Prasad et al. (2018) demonstrated that application of phytogenic Ag NPs (synthesized from *Stevia rebaudiana* leaf extract) caused improvement in seed germination, plant growth, contents of chlorophyll, carotenoids and protein, and antifungal activity of *Aspergillus niger* in rice (*Oryza sativa*), maize (*Zea mays*), and peanut (*Arachis hypogaea*). Balashanmugam et al. (2016) biosynthesized Ag NPs using ten different species of *Cassia* and obtained the most efficient and stable ones with *Cassia roxburghii* aqueous leaf extract, which were markedly effective in controlling various plant diseases caused by fungi. Ponmurugan (2017) biosynthesized gold and silver NPs with *Trichoderma atroviride* and tested them against a tea pathogenic fungus *Phomopsis theae*. Application of these NPs in soil significantly reduced canker size and improved leaf yield in tea plants, indicating that productivity of tea leaves can be increased and the crop be protected from the disease caused by *Phomopsis theae.*

Ahmad et al. (2017) reported that fabrication of Fe NPs from aqueous extract of neem leaf (*Azadirachta indica*) induced biocidal activity against phytopathogens of apple such as *Alternaria mali*, *Botryosphaeria dothidea*, and *Diplodia seriata* indicating that these NPs may be useful to prevent fungal attacks, which harm apple productivity. Sathiyabama and Charles (2015) isolated cell-wall polymer (chitosan) from *Fusarium oxysporum* f. sp. *lycopersici* and cross-linked it with sodium tripolyphosphate (TPP) to synthesize nanoparticles (CWP-NPs). Application of these biosynthesized NPs that caused delay in the expression of wilt disease symptom in tomato plants minimized the wilt disease severity and increased the ultimate yield. Phytochemical-capped Au NPs synthesized using the extracts of galangal rhizome and considered as being nontoxic and biocompatible markedly increased seed germination, seedling vigor index, and physiological and biochemical properties of maize seedlings (Mahakham et al. 2016). Phyto-mediated Fe NPs, fabricated from coconut husk, increased the biomass, quantum yield of photochemistry, and chlorophyll content in *Oryza sativa* (Sebastian et al. 2018).

Green synthesis of CuONPs (20–40 nm), using *Morus alba* leaf extract, was done by Singh et al. (2017). Higher concentrations (100 and 500 mg L^{-1}) of these NPs reduced the chlorophyll and sugar content in cauliflower (*Brassica oleracea* var. *botrytis*) and tomato (*Solanum lycopersicum*), *whereas* 10 mg L^{-1} of NPs could increase the chlorophyll and sugar contents slightly in tomato plants only. Increased concentration of NPs enhanced lipid peroxidation, electrolyte leakage, and superoxide and hydrogen peroxide accumulation, thus showing toxicity at higher concentrations. Deposition of lignin was found in both plants, but NP accumulation was higher in tomato than in cauliflower plants, which might be due to the difference in root morphology of the two species. These studies suggest that all phyto-mediated NPs are not safe for plant physiology. Fe NPs biosynthesized with the help of *Eucalyptus* leaf extracts were able to remove Cr (VI) under optimal experimental conditions (Jin et al. 2017). The removal efficiency was 55.7% after four repeats. Such NPs also increased the oxidative degradation of herbicide ametryn (Sangami and Manu 2017). Fe NPs synthesized by the green method from the green tea and eucalyptus leaf extract were able to remove nitrate from swine wastewater, indicating that the plant-based NPs may also be useful for cleaning the environmental pollution and the damaged plant-growing fields (Wang et al. 2014).

Application of phyto-mediated NPs to enhance plant growth is rapidly growing in field conditions. Several studies have suggested that the phyto-mediated NPs alter plant metabolism, improve plant protection against pathogen, and increase tolerance. However, since most of these studies are based on short-term experiments in controlled laboratory conditions, generalizations should not be drawn unless these results have been endorsed by long-term studies in open fields under varied environmental factors such as variation in soil pH, diverse temperatures, soil nutrient variation, improper irrigation management, and many more. Thus, limited knowledge about the mechanism of NP action and lack of long-term studies of NP impact on plants cause a major concern and certainly deserve more attention.

23.2.2 Nutrient Carriers

The main task in developing sustainable crops is to minimize the inputs of NPs and maximize the crop productivity. This is feasible if the engineered molecules are able to make the soil more fertile and improve the nutrient utilization by plants for greater productivity and better environmental security. Nanotechnology has the potential to improve the profile of nutrients and their efficiency. There are various nutrient carriers available in the medical field for drug delivery, like micelles, liposomes, nano-emulsions, protein-carbohydrate nanoscale complexes, solid nanolipid particles, and others; these can possibly be utilized to carry nutrients or specific NPs for improving the plant productivity. These materials may interfere with NPs in the soil to facilitate ion exchange (e.g., NH_4^+ $H_2PO_4^-$, HPO_4^{2-}, PO_4^{3-}, Zn^{2+}) and promote nutrient translocation to plants. Thus, the presence of nutrient carriers and newly developed agromolecules may help improve the overall soil profile (absorption, bioavailability, solubility, and dispersion) and finally the plant productivity. Such products may have the potential to improve the nutritional uptake, increase the soil integrity of agricultural lands, and minimize the toxicity of NPs.

23.2.3 Company-Based Products

Various companies in agricultural sector have developed a vast range of NP-based products. Syngenta, a leading agrochemical corporation, developed a nano-encapsulated product, which is available under the name "Karate Zeon®" (Syngenta 2018). It contains lambda-cyhalothrin that is released quickly on contact with leaves, for the control of biting, chewing, as well as sucking insect pests in a variety of crops. The product is currently available for use in most countries including India. Another product of Syngenta, called "Gusbuster," is a nano-insecticide that releases its content when exposed to alkaline environment (e.g., insect stomach). Nano Green Sciences (USA) has developed an organic NPs-based plant tonic called "Nano Green" that promotes nutrient uptake and plant growth. Agro Nanotechnology Corp. (USA) has also produced a similar product called "Nano-Gro" which increases plant growth, yield, and resistance.

23.3 Conclusion

Future research extended to open-field experiments may help to understand better the mechanism of NP effect on plants. Production of phyto-mediated NPs looks highly promising as they are environment-friendly, cost-effective, and less time-consuming and ensure long-term safety of products. However, for use in agricultural sector, these NPs need to be designed carefully keeping in view their mode of treatment (soil or foliar) so that high impact with easy routes may be made possible to ensure improved quality of crops.

There is no doubt about the beneficial effects of NPs and their contributions to various fields including medical and agricultural sectors. However, it is not advisable to use NPs in developing food stuffs unless their toxicity has been tested on a large scale through repeated short-term as well as long-term studies under varied sets of conditions, and their mechanism of action is fully understood. Rigorous testing of nanoproducts should be done through governmental agencies also before bringing them to the market.

References

Ahmad H, Rajagopal K, Shah AH, Bhat AH, Venugopal K (2017) Study of bio-fabrication of iron nanoparticles and their fungicidal property against phytopathogens of apple orchards. IET Nanobiotechnol 11:230–235

Aschberger K, Gottardo S, Amenta V, Arena M, Botelho MF, Bouwmeester H, Brandhoff P, Mech A, Quiros PL, Rauscher H, Schoonjans R, Vittoria VM, Peters R (2015) Nanomaterials in food current and future applications and regulatory aspects. J Phys Conf Ser 617:12–32, conference 1.

Balashanmugam P, Balakumaran MD, Murugan R, Dhanapal K, Kalaichelvan PT (2016) Phytogenic synthesis of silver nanoparticles, optimization and evaluation of *in vitro* antifungal activity against human and plant pathogens. Microbiol Res 192:52–64

Banterle A, Cavaliere A, Carraresi L, Stranieri S (2014) Food SMEs face increasing competition in the EU market: marketing management capability is a tool for becoming a price maker. Agribusiness 30:113–131

Chen R, Zhang C, Zhao Y, Huang Y, Liu Z (2018) Foliar application with nano-silicon reduced cadmium accumulation in grains by inhibiting cadmium translocation in rice plants. Environ Sci Pollut Res Int 25:2361–2368

Cox A, Venkatachalam P, Sahi S, Sharma N (2016) Silver and titanium dioxide nanoparticle toxicity in plants: a review of current research. Plant Physiol Biochem 107:147–163

Du W, Tan W, Peralta VJR, Gardea TJL, Ji R, Yin Y, Guo H (2017) Interaction of metal oxide nanoparticles with higher terrestrial plants: physiological and biochemical aspects. Plant Physiol Biochem 110:210–225

European Food Safety Authority (2009) Updating the opinion related to the revision of Annexes II and III to Council Directive 91/414/EEC concerning the placing of plant protection products on the market-Toxicological and metabolism studies. EFSA J 1166:1–6

Fraceto LF, Grillo R, de Medeiros GA, Scognamiglio V, Rea G, Bartolucci C (2016) Nanotechnology in agriculture: which innovation potential does it have? Front Environ Sci 4:20

Gewin W (2015) Everything you need to know about nanopesticides. Modern Farmer Article. http://modernfarmer.com/2015/01/everything-need-know-nanopesticides/ [Accessed, 2015]

Hirsh S, Schiefer J, Gschwandtner A, Hartmann M (2014) The determinants of firm profitability differences in EU food processing. J Agric Econ 65:703–721
Hooley G, Piercy NF, Nicoulaud B (2012) Marketing strategy and competitive positioning. Prentice Hall/Financial Times, London
Husen A (2017) Gold nanoparticles from plant system: synthesis, characterization and their application. In: Ghorbanpour M, Manika K, Varma A (eds) Nanoscience and plant–soil systems, vol 48. Springer, Cham, pp 455–479
Husen A, Siddiqi KS (2014a) Carbon and fullerene nanomaterials in plant system. J Nanobiotechnol 12:16
Husen A, Siddiqi KS (2014b) Phytosynthesis of nanoparticles: concept, controversy and application. Nano Res Lett 9:229
Husen A, Siddiqi KS (2014c) Plants and microbes assisted selenium nanoparticles: characterization and application. J Nanobiotechnol 12:28
Jin X, Liu Y, Tan J, Owens G, Chen Z (2017) Removal of Cr (VI) from aqueous solutions via reduction and absorption by green synthesized iron nanoparticles. J Clean Prod 176:929–936
JRC scientific and policy reports (2014). Proceedings of a workshop on Nanotechnology for the agricultural sector: from research to the field. European Commission Joint Research Centre, Institute for Prospective Technological Studies. ISBN 978-92-79-37917-8, http://doi.org/10.2791/80497
Kah M (2015) Nanopesticides and nanofertilizers: emerging contaminants or opportunities for risk mitigation? Front Chem 3:64
Karny A, Zinger A, Kajal A, Shainsky-Roitman J, Schroeder A (2018) Therapeutic nanoparticles penetrate leaves and deliver nutrients to agricultural crops. Sci Rep 8:7589
Kole C, Kole P, Randunu KM, Choudhary P, Podila R, Ke PC, Rao AM, Marcus RK (2013) Nanobiotechnology can boost crop production and quality: first evidence from increased plant biomass, fruit yield and phytomedicine content in bitter melon (*Momordica charantia*). BMC Biotechnol 13:37
Kumar S, Chauhan N, Gopal M, Kumar R, Dilbaghi N (2015) Development and evaluation of alginate-chitosan nanocapsules for controlled release of acetamiprid. Int J Biol Macromol 81:631–637
Kumari M, Pandey S, Bhattacharya A, Mishra A, Nautiyal CS (2017a) Protective role of biosynthesized silver nanoparticles against early blight disease in *Solanum lycopersicum*. Plant Physiol Biochem 121:216–225
Kumari M, Pandey S, Mishra SK, Nautiyal CS, Mishra A (2017b) Effect of biosynthesized silver nanoparticles on native soil microflora via plant transport during plant pathogen nanoparticles interaction. 3 Biotech 7:345
Liu R, Lal R (2015) Potentials of engineered nanoparticles as fertilizers for increasing agronomic productions. Sci Total Environ 514:131–139
Mahakham W, Theerakulpisut P, Maensiri S, Phumying S, Sarmah AK (2016) Environmentally benign synthesis of phytochemicals-capped gold nanoparticles as nanopriming agent for promoting maize seed germination. Sci Total Environ 573:1089–1102
Mattiello A, Filippi A, Poscic F, Musetti R, Salvatici MC, Giordano C, Vischi M, Bertolini A, Marchiol L (2015) Evidence of phytotoxicity and genotoxicity in *Hordeum vulgare* L. exposed to CeO_2 and TiO_2 nanoparticles. Front Plant Sci 6:1043
Milani N, Hettiarachchi GM, Kirby JK, Beak DG, Stacey SP, McLaughlin MJ (2015) Fate of zinc oxide nanoparticles coated onto macronutrient fertilizers in an alkaline calcareous soil. PLoS One 10:e0126275
Nanotechnology Research Review (2014) https://www.bccresearch.com/market-research/nanotechnology/2014-nanotechnology-research-review-report-nan047f.html
Navarro E, Wagner B, Odzak N, Sigg L, Behra R (2015) Effects of differently coated silver nanoparticles on the photosynthesis of *Chlamydomonas reinhardtii*. Environ Sci Technol 49:8041–8047

Ogunkunle CO, Jimoh MA, Asogwa NT, Viswanathan K, Vishwakarma V, Fatoba PO (2018) Effects of manufactured nano-copper on copper uptake, bioaccumulation and enzyme activities in cowpea grown on soil substrate. Ecotoxicol Environ Saf 155:86–93

Parisi C, Vigani M, Rodríguez-Cerezo E (2014) Proceedings of a workshop on "Nanotechnology for the agricultural sector: from research to the field". JRC Sci Policy Rep 1:40

Ponmurugan P (2017) Biosynthesis of silver and gold nanoparticles using *Trichoderma atroviride* for the biological control of *Phomopsis* canker disease in tea plants. IET Nanobiotechnol 11:261–267

Prasad A, Astete CE, Bodoki AE, Windham M, Bodoki E, Sabliov CM (2018) Zein nanoparticles uptake and translocation in hydroponically grown sugar cane plants. J Agric Food Chem 66:6544–6551

Raliya R, Biswas P, Tarafdar JC (2015) TiO_2 nanoparticle biosynthesis and its physiological effect on mung bean (*Vigna radiata* L.). Biotechnol Rep 5:22–26

Raliya R, Tarafdar JC, Singh SK, Gautam R, Choudhary K, Maurino Veronica G, Saharan V (2014) MgO nanoparticles biosynthesis and its effect on chlorophyll contents in the leaves of Cluster bean (*Cyamopsis tetragonoloba* L.). Adv Sci Eng Med 6:538–545

Sagadevan S, Periasamy M (2014) Recent trends in nanobiosensors and their applications – a review. Rev Adv Mater Sci 36:62–69

Sangami S, Manu B (2017) Synthesis of green iron nanoparticles using laterite and their application as a fenton-like catalyst for the degradation of herbicide ametryn in water. Environ Technol Innov 8:150–163

Sathiyabama M, Charles RE (2015) Fungal cell wall polymer based nanoparticles in protection of tomato plants from wilt disease caused by *Fusarium oxysporum* f. sp. *lycopersici*. Carbohydr Polym 133:400–407

Savary S, Ficke A, Aubertot JN, Hollier C (2012) Crop losses due to diseases and their implications for global food production losses and food security. Food Sec 4:519–537

Sebastian A, Nangia A, Prasad MNV (2018) A green synthetic route to phenolics fabricated magnetite nanoparticles from coconut husk extract: Implications to treat metal contaminated water and heavy metal stress in *Oryza sativa* L. J Clean Prod 174:355–366

Sekhon BS (2014) Nanotechnology in agri-food production: an overview. Nanotechnol Sci Appl 7:31–53

Sertova NM (2015) Application of nanotechnology in detection of mycotoxins and in agricultural sector. J Cent Eur Agric 16:117–130

Sheykhbaglou R, Sedghi M, Fathi-Achachlouie B (2018) The effect of ferrous nano-oxide particles on physiological traits and nutritional compounds of soybean (*Glycine max* L.) seed. An Acad Bras Cienc 90:485–494

Siddiqi KS, Husen A (2016a) Engineered gold nanoparticles and plant adaptation potential. Nano Res Lett 11:400

Siddiqi KS, Husen A (2016b) Green synthesis, characterization and uses of palladium/platinum nanoparticles. Nano Res Lett 11:482

Siddiqi KS, Husen A (2017a) Recent advances in plant-mediated engineered gold nanoparticles and their application in biological system. J Trace Elem Med Biol 40:10–23

Siddiqi KS, Husen A (2017b) Plant response to engineered metal oxide nanoparticles. Nano Res Lett 12:92

Siddiqi KS, Rahman A, Tajuddin, Husen A (2016) Biogenic fabrication of iron/iron oxide nanoparticles and their application. Nano Res Lett 11:498

Siddiqi KS, Husen A, Rao RAK (2018) A review on biosynthesis of silver nanoparticles and their biocidal properties. J Nanobiotechnol 16:14

Singh A, Singh NB, Hussain I, Singh H (2017) Effect of biologically synthesized copper oxide nanoparticles on metabolism and antioxidant activity to the crop plants *Solanum lycopersicum* and *Brassica oleracea* var. botrytis. J Biotechnol 262:11–27

Sodano V, Verneau F (2014) Competition policy and food sector in the European Union. J Int Food Agribusiness Mark 26:155–172

Spengler A, Wanninger L, Pflugmacher S (2017) Oxidative stress mediated toxicity of TiO_2 nanoparticles after a concentration and time dependent exposure of the aquatic macrophyte *Hydrilla verticillata*. AqToxico 190:32–39
Syngenta (2018) Karate with Zeon Technology insecticide. http://www.syngenta-us.com/insecticides/karate-with-zeon-technology
Tareq FK, Fayzunnesa M, Kabir MS (2017) Antimicrobial activity of plant-median synthesized silver nanoparticles against food and agricultural pathogens. Microb Pathog 109:228–232
Tiwari M, Sharma NC, Fleischmann P, Burbage J, Venkatachalam P, Sahi SV (2017) Nano titania exposure causes alterations in physiological, nutritional and stress responses in tomato (*Solanum lycopersicum*). Front Plant Sci 8:633
USDA (2015) U.S. Department of agriculture awards $3.8 million in grants for nanotechnology research. Available online at: http://nifa.usda.gov/press-release/usda-awards-38-million-grants-nanotechnology-research
Wang T, Lin J, Chen Z, Megharaj M, Naidu R (2014) Green synthesized iron nanoparticles by green tea and eucalyptus leaves extracts used for removal of nitrate in aqueous solution. J Clean Prod 83:413–419
Wang X, Yang X, Chen S, Li Q, Wang W, Hou C, Gao X, Wang L, Wang S (2016) Zinc oxide nanoparticles affect biomass accumulation and photosynthesis in *Arabidopsis*. Front Plant Sci 6:1243

Index

A. Husen, M. Iqbal (eds.), *Nanomaterials and Plant Potential*,
https://doi.org/10.1007/978-3-030-05569-1

F

G

H

T

MIX
Papier aus verantwortungsvollen Quellen
Paper from responsible sources
FSC® C105338

FSC
www.fsc.org

If you have any concerns about our products,
you can contact us on
ProductSafety@springernature.com

In case Publisher is established outside the EU,
the EU authorized representative is:
Springer Nature Customer Service Center GmbH
Europaplatz 3, 69115 Heidelberg, Germany

Printed by Libri Plureos GmbH
in Hamburg, Germany